Civil Engineering Conference (CEC 2022)

IOP Conference Series: Materials Science and Engineering
Volume 1252

Kosice, Slovakia
6 – 8 June 2022

ISBN: 978-1-7138-6418-9
ISSN: 1757-8981

Printed from e-media with permission by:

Curran Associates, Inc.
57 Morehouse Lane
Red Hook, NY 12571

Some format issues inherent in the e-media version may also appear in this print version.

Institute of Physics
Dirac House, Temple Back
Bristol BS1 6BE UK

Phone: 44 1 17 929 7481
Fax: 44 1 17 920 0979

techtracking@iop.org

Additional copies of this publication are available from:

Curran Associates, Inc.
57 Morehouse Lane
Red Hook, NY 12571 USA
Phone: 845-758-0400
Fax: 845-758-2633
Email: curran@proceedings.com
Web: www.proceedings.com

TABLE OF CONTENTS

Preface

Peer Review Statement

Potential Use of Transparent Wood in Spatial Folded Structures 1
V Karľa, V Bajzecerová, J Kanócz

Experimental Form-Finding: A Review 5
T Baroš, L Kabošová, M Baroš, D Katunský

Experimental Verification of a Simulation Model of High-Speed Train Aerodynamic Impact on Railway Infrastructure 13
N A Labutin, A V Benin, L K Diachenko

New Test Equipment for Testing the Watertightness of Facade Structures In-Situ 20
B Bielek, D Szabo, J Klem

Aerial Work with Helicopters in the Construction of Cableways and Their Impact on the Concrete Mixture Design 26
A Bisták, Z Hulínová, I Vavrík, P Šťastný

Life Cycle Assessment of Residential Buildings and Scenarios for Prolonged Life Span 31
Z Blichova, S Vilcekova, E Kridlova Burdova, D Katunský

Resistance of a Sealed Joint Subjected to Tensile Stress at Different Temperatures 37
A Boháček, J Šlanhof, A Průcha

Comparison Between Flood Damages Evaluation Methods and Selection of the Appropriate Method for Application in Slovakia 42
M Hlinková, P Bujanský, H Abd-Elhamid

Assessment and Adaptation to Climate Change and Sea Level Rise Impacts on Egypt's Northern Coasts 49
M Zelenakova, H Abd-Elhamid, K Barańczuk, J Barańczuk

Recycling Cement Bypass Dust and Granulated Blast Furnace Slag into Hydraulic Road Binder 58
N Stevulova, J Junak, J Strigac, N Junakova

The Carbon Dioxide Generation Rate from Burning of Candle and Its Effect on Room Ventilation 64
P Kapalo, A Eštoková, O Voznyak

Size Effect of Mortar Specimens Under Uniaxial Compression 69
M Kovac

Investigation of the Thermal Properties of Candle Wax Material 76
A Eštoková, P Kapalo

The Seismic Response of Cylindrical Steel Tank at Al Hoceïma City in Morocco 81
K Kotrasova, E Kormanikova, M Loukili, S Mansour, D Foti

Long Term Accumulation of Heat Energy from the Sun 88
M Gorás, J Domanický, F Vranay

Transit Smart Solution for Bratislava – Example of Tramway Network Development 97
T Schlosser, P Schlosser, M Korfant, G Bálint, J Takács

Climate Change Adaptation in Pavement Design 115
D Hodakova, A Zuzulova, S Capayova

Experimental Determination of Dynamic Modulus of Elasticity of Concrete with Ultrasonic Pulse Velocity Method and Ultrasonic Pulse Echo Method 129
I Ivanchev

Economic Efficiency: Yield on Cost as Core Tool of Benchmarking 140
S Pagac

Dosimetric Measurement of Lighting Conditions in Office Environment 144
P Hanuliak, P Hartman

Plastic Zones in the Pin Connection of the Laminated Lattice Girder 149
M Tomko, I Demjan

Survey of Consumer Preferences in the Field of Wooden Buildings 157
J Švajlenka, M Kozlovská

Quantification of Acoustic Risk from the Tram Ride in Track Curves 162
E Panulinova, S Harabinova

Applications of a Phase-Field Fracture Model to Materials with Inclusions 167
R Vodička

The Potential of Lightweight Panel Systems in Building Industry 175
M Kozlovská, D Mokrenko, Z Struková, M Špak

The Effect of Manufacturing Imperfections in the Slip Joints 184
M Mojto, J Brodnianský, J Brodnianský

Elastic Exploring of Human Lung Parenchyma 192
B Bošanská, M Minárová, K Tvrdá

Early-Age Strains and Stresses in Bi-Axially Restrained Expansive Concrete Elements 201
V V Tur, V H Sannikava

Design and Evaluation of Building Structures Planned for Harsh Conditions 211
D Katunský, E Dolníková, M Lopušniak, M Vertaľ, T Baroš, K Hutkai, P Jaroš, L Kabošová, J Katunská, V Karla, M Zozulák

On the Load-Carrying Capacity Produced by Different Load Models for Road Bridges 217
J Gocal, J Odrobinak, J Vican

Evaluation of Daylight Comfort in Industrial Building 227
E Dolnikova, D Katunsky, M Lopusniak

Analysis of Alternative Designs for the Reconstruction of an Administrative Building – Case Study 233
A Sedlakova, V Merjava, S Vilcekova, M Gmitter, D Burak

Stationary Dissipative Systems of Classical Mechanics in the Basis of Unscientific Principles of the Theory of Creep of Reinforced Concrete 242
A Beglov, R Sanjarovsky, T Ter-Emmanuilyan

Quantification of Temporal and Spatial Changes in Soil Suspension Density, Specific Surface Area and Parameters of the Distribution of the Texture of the Dispersed Soil Phase 249
M Gomboš, D Pavelkova, B Kandra, A Tall

Numerical Analysis of Shear Resistance on Composite Beams. 257
M Weissova, V Kvocak, D Dubecky, P Vaňová

Influence of the Longitudinal Steel Ration in RC Beams and Steel Reinforcement Ration in RC Slabs on the Seismic Energy Dissipation Mechanisms for a MR RC Frame Structure 266
I Sococol, T C Petrescu, P Mihai, D T Babor

Sewage and Coronavirus; Maribor Case Study 284
M Nekrep Perc, J Kramer Stajnko

Water Vapor Transmission Parameters of the Kežmarok Sandstone 289
P Jaroš, M Vertaľ

Mechanical Fracture Parameters of Hemp Fibre Reinforced Cement-Based Composites with Recycled Aggregate 295
H Simonova, B Kucharczykova, B Poletanovic, I Merta, Z Kersner

Digital Preservation of the Jesuitas Church in Valencia (Spain) Using 3D Laser Scanning 301
J Moreno-Puchalt, A Almerich-Chulia, X Laumain

Renewable Energy in a Dense Urban Fabric: Solar Gain Estimation in the Case of a Turn of the Century District 307
V Sugár, Zs Fáczányi

Corrosion Acceleration Test on Weathering Steel and Paint-Coated Steel for Bridge Structural Members 318
J Liu, K Ojima, F Jiang, M Hirohata, T Nakayama, T Sakamoto

An Investigation on Combined Section by Steels with Different Strengths for Bridge Girder 327
M Hirohata, K Jármai

Influence of Takt Time Planning on Construction Efficiency 333
M Kozlovská, D Klosova

Renovation of Dormitory Campus Using the Principles of Sustainable Development 340
E Luzay, D Shafieva, M Zeleňáková, Š Petkanič

Production of Polymer-Modified-Asphalt Mixture as an Alternative Method to the Conventional Polymer-Modified-Bitumen Mixtures 348
R Kovács, Ľ Polakovič, A Czímerová, J Mandula

Studying the Potential Microbial Risk of Construction Cementitious Materials in the Urban Environment 354
M Hološová, A Eštoková, A Luptáková

The Influence of Inclusive Deformation Mapping on the Kinematic Equations of Nonlinear Elasticity 364
R Flajs

Effects of Hydro Power Plant Releases on Navigation Conditions in the Lower Lock Approach 370
P Fošumpaur

Analytical Determination of Longitudinal Forces on the Vessels During Filling of the Navigation Lock 378
T Kašpar

Mechanical Behavior of Strengthened Reinforced Concrete Beams an Experimental Study 384
G Bou Abdallah, I Ivanova, J Assih, C Diagana, D Dontchev

Thermodynamic Aspects of Viscoelastic Materials' Linear and Nonlinear Behavior 391
M Minárová, K Tvrdá

Use of Building Information Modelling (BIM) in Planning and Managing the Life Cycle of Buildings 399
J Smetanková, P Mésároš, K Krajníková, V Biolek

Indoor Environmental Quality in Green Certified Office Buildings 409
K Harčárová, S Vilčeková

Moving Load Effect on a Bridge 414
J Melcer, V Valašková

Natural Vibration of a Membrane Hypar in a Steel Frame 420
L Stulerova, L Kapolka, S Kmet

Methods of Strengthening the Grains of Recycled Aggregates 427
A Sicakova

VR-Based Safety Training Research in Construction 440
V Chellappa, P Mésároš, M Špak, M Spišáková, P Kaleja

The Effect of Physicochemical Properties of Bottom Sediments on Nitrogen and Phosphorus Sorption 449
N Junakova, J Junak, M Balintova

Renewal of the Roof Structure of the Winter Stadium in Prešov 455
J Vojtuš

Indoor Air Quality Assessment in the Apartment – Case Study 465
P Kapalo, M Kováč, R Nagy, Z Baka

Overview of Meteorological and Hydrological Conditions in the Hornád River Basin 471
P Nagy, M Garaj, M Hlinková

Nonvisual Aspects of Daylight in the Built Environment 476
Jozef Hraška

Gradual Transformation of an Administrative Building into a Green Structure 484
M Vertal, Z Vranayova, K Cakyova, A Vargova, F Vranay

The Possibilities of Using Digital Twin Technology in the Life Cycle of Buildings 495
Z Struková, M Kozlovská, A Tažiková, J Čabala

Construction Segments Applying Modern Methods of Construction in Slovakia 505
M Spišáková, J Galla, T Mandičák

Analysis of Stress Concentration at the Edge of Hole in Plates with Different Widths by Using FEM 511
B Safaei, Z Pezeshki, K Kotrasova, E Kormanikova

Modelling of Stormwater Runoff and Wastewater Flow in the Sewer Network 522
R Wittmanová, I Makro, J Hrudka, M Šutús, Š Stanko

Internal Insulation Systems and Their Assessment for Historic Buildings by Hygrothermal Simulation 528
K Hutkai, D Katunský, M Zozulák

Analysis of Vibrations in a Dynamic Damper 532
A Jaskot

Practical Deflection Analysis of Tubular Telecommunication Towers Subjected to a Temperature Gradient 540
P A Król

Incorporating New Technologies in Road Industry: Test Track in Antwerp, Belgium 548
S R Omranian, W Van den bergh, B Golchin, K Couscheir, C Vuye, J Braspenninckx

Time-Space Modelling of Construction 559
A Tazikova, Z Strukova, J Cabala

Analysis of Building Materials Used for the Construction of Family Houses in the Boundaries from Cradle to Gate with Options 563
E Kridlova Burdova, S Vilcekova

Quality of First Flush of Rainwater from Urban Area: Short Review 573
I Marko, R Wittmanová, J Hrudka, A Raczková, I Škultétyová

Local Loading of Sandwich Panels - Assessment According to Standards and Experimental Verification 578
M Magura, J Brodniansky

Exploitation of Document Management Systems in Construction Industry 590
T Mandicak, P Mesaros, M Spisakova, A Kanalikova

Tools of 4D Modeling of Construction Projects 597
Yaroslav Nechyporchuk, Renáta Bašková

Benefits of BIM Models and Mixed Reality in the Implementation Phase of Construction Projects 603
J Čabala, M Kozlovská, Z Struková, A Tažíková

Green Roofs on Buildings – Different View on Its Construction Solution 609
M Kováč

Stability of Travertine Barrier on Milino Lake 615
I Štimac Grandić, J Drobac, N Bede, I Ružić

Different Ways of Mechanical Wave Propagation Damping in Ackerman Ceilings Using Recycled Rubber 622
K Kuliński

Numerical Simulations for Time-Dependent Corrosion Surface of Unpainted Structural Steel Plates Using Spatio-Temporal Data 630
F Jiang, M Hirohata

Numerical Analysis of Slope Stability 640
S Harabinova, E Panulinova

The Impact of the Exterior Green Wall on the Summer Regime in the Building – Energy Simulation 644
M Kováč

Author Index

IOP Conf. Series: Materials Science and Engineering 1252 (2022) 011001 doi:10.1088/1757-899X/1252/1/011001

PREFACE

The International Scientific Conference CIVIL ENGINEERING CONFERENCE 2022 (CEC 2022) on the occasion of the 45th anniversary of the establishment of the faculty under auspices of
assoc. prof. Ing. Peter Mésároš, PhD., the Dean of the Faculty of Civil Engineering, Technical University of Kosice,
and
Ing. Andrej Doležal, the Minister of Transport and Construction of the Slovak Republic,
was held in Košice, Slovakia on 7th - 8th June 2022.
The conference was organised by the Faculty of Civil Engineering, the Technical University of Košice, Slovakia. It issued topics of civil and environmental engineering for colleagues, graduates, teachers, researchers, students, and the professionally cooperating public.

The Conference assembled more than 118 delegates, the representatives from the universities and institutions from:

- Technical University of Košice, Slovakia,
- Slovak University of Technology in Bratislava, Slovakia,
- University of Žilina, Slovakia,
- Czech Technical University in Prague, Czech Republic,
- Brno University of Technology, Czech Republic,
- Charles University, Praha, Czech Republic,
- University of Ljubljana, Slovenia,
- Czestochowa University of Technology, Poland,
- University of Rijeka, Croatia,
- University of Chemical Technology and Metallurgy, Bulgaria,
- Óbuda University, Hungary,
- Osaka University, Japan,
- Universitat Politècnica de València, Spain,
- Brest State Technical University, Belarus,
- University of Antwerp, Belgium,
- Russian University of Transport, Russia,
- Indian Institute of Technology Guwahati, India,
- Moscow State University of Civil Engineering, Russia,
- University of Avignon, France,
- Politechnic University of Bari, Italy,
- Warsaw University of Technology, Poland,
- TU Wien, Austria,
- Lviv Polytechnic National University, Ukraine,
- University of Reims Champagne-Ardenne, France,
- Slovak Academic of Science, Košice,
- University of Architecture, Civil Engineering and Geodesy, Bulgaria,

- University of Miskolc, Hungary,
- Eastern Mediterranean University, Turkey,
- University of Johannesburg, South Africa,
- Zagazig University, Egypt,
- Railway Technical Research Institute, Tokyo, Japan,
- University of Mohaghegh Ardabili, Iran,
- University of Maribor, Slovenia,
- L.N. Gumilyov Eurasian National University, Kazakhstan,
- Technical University of Iasi, Romania,
- Emperor Alexander I St. Petersburg State Transport University, Russian Federation,
- Povazska Cement Factory Ltd., Ladce, Slovakia,
- Department of Traffic Engineering, City of Bratislava, Slovakia,
- VUIS - CESTY, Ltd. Bratislava, Slovakia,
- Port of Antwerp, Belgium,
- DOTIS Consult, Ltd., Slovakia, Slovakia,
- Centre for Energy Research, Hungary,
- CTP Invest SK, Bratislava, Slovakia.

The scientific contributions involved a wide range of research fields:

- Building Physics and Services
- Sustainable Architecture and Energy Efficiency
- Statics, Dynamics and Modelling
- Structural Engineering and Bridges
- Transport and Geotechnical Engineering
- Construction Technology, Organization and Management
- Construction Economics, Marketing and Management
- Building Information Modelling
- Innovations in Construction Design and Technology
- Environmental Engineering
- Healthy Indoor Environment
- Hydrotechnical Engineering
- Material Engineering and Recycling
- Sustainable Water Management
- Sustainable Civil and Environmental Engineering
- Urbanism and Architecture

IOP Conf. Series: Materials Science and Engineering 1252 (2022) 011001 doi:10.1088/1757-899X/1252/1/011001

The CEC 2022 offered an international platform for the dissemination of the original research results. It provides a pleasant environment to present a new, great ideas of researchers and discovered advances in the field of civil and environmental engineering and related interdisciplinary topics.
We strongly believe that CEC 2022 will have a significant impact on the development of contemporary analytical, numerical and experimental methods in structural and physical problems, and will also create an opportunity for opening a forum for discussion and collaboration amongst the participants.
The conference provided an opportunity for discussion of scientific topics and research results, to share their knowledge and experiences, to collaborate in innovative designs and proposals.
Ninety-four contributions have been selected to be published in the volume of IOP Conference Series: Materials Science and Engineering. The organizing committee appreciates the enormous work of the both the editorial board and reviewers in the preparation of this conference proceedings. Many thanks also to all authors for their flexibility and meeting deadlines.
We wish to express our gratitude to the reviewers, contributors and participants for their contribution to the success of CEC 2022.

Issue editors
Peter Mésároš
Vinayakam Jothiprakash
Kamila Kotrasová
Abdelazim Negm,
Martina Zeleňáková

EDITORS

Peter MÉSÁROŠ, Technical University of Košice, Slovakia
Vinayakam JOTHIPRAKASH, Indian Institute of Technology in Mumbai, India
Kamila KOTRASOVÁ, Technical University of Košice, Slovakia
Abdelazim NEGM, Zagazig University, Egypt
Martina ZELEŇÁKOVÁ, Technical University of Košice, Slovakia

IOP Conf. Series: Materials Science and Engineering 1252 (2022) 011001 doi:10.1088/1757-899X/1252/1/011001

COMMITTEES

SCIENTIFIC COMMITTEE

ORGANISING GUARANTEE

ORGANISING COMMITTEE

REVIEW STATEMENT

All papers published in this volume of IOP Conference Series: Materials Science and Engineering have been peer reviewed per two reviewers through processes administered by the Editors. Reviews were conducted by expert referees to the professional and scientific standards expected of a proceedings journal published by IOP Publishing.

IOP Conf. Series: Materials Science and Engineering 1252 (2022) 011002 doi:10.1088/1757-899X/1252/1/011002

Peer Review Statement

All papers published in this volume have been reviewed through processes administered by the Editors. Reviews were conducted by expert referees to the professional and scientific standards expected of a proceedings journal published by IOP Publishing Publishing.

- **Type of peer review:** Double Anonymous
- **Conference submission management system:** Morressier
- **Number of submissions received:** 97
- **Number of submissions sent for review:** 97
- **Number of submissions accepted:** 94
- **Acceptance Rate (Submissions Accepted / Submissions Received × 100):** 96.9
- **Average number of reviews per paper:** 2
- **Total number of reviewers involved:** 100
- **Contact person for queries:**
  Name: Kamila Kotrasová
  Email: kamila.kotrasova@tuke.sk
  Affiliation: Faculty of Civil Engineering, Technical University of Kosice, Slovak Republic

IOP Conf. Series: Materials Science and Engineering 1252 (2022) 012001 doi:10.1088/1757-899X/1252/1/012001

Potential Use of Transparent Wood in Spatial Folded Structures

V Karľa[1], V Bajzecerová[2], J Kanócz[3]

[1] Center for Research and Innovation in Construction, Faculty of Civil Engineering, Technical University of Košice, Park Komenského 10A, 042 00 Košice
[2] Department of Steel and Timber Structures, Faculty of Civil Engineering, Technical University of Košice, Vysokoškolská 4, 042 00 Košice
[3] Department of Architecture, Faculty of Arts, Technical University of Košice, Letná 9, 042 00 Košice

E-mail: viktor.karla@tuke.sk; viktoria.bajzecerova@tuke.sk; jan.kanocz@tuke.sk

Abstract. Transparent wood is a relatively new material, the manufacturing starts with removing the lignin from wood veneer and infiltrating the delignified wood by a polymer thereafter. Transparent wood usually proves better mechanical properties than original wood and polymer. The article presents the potential use of transparent wood in bearing structures. The transparent wood could be used as I-beams structures, box beams, trusses, or folded structures. A spatial folded structure was numerically analyzed. Various types of transparent wood or transparent plywood were used. A comparison with commonly used materials, such as glass and other wood-based materials were made. The comparison shows that transparent wood has great potential in the use as a structural material. The plywood made from transparent wood veneer with the alternating direction of wood fibers acts as a laminate and allows a biaxial load transfer, which can be successfully used in a thin plate of the spatial folded structure while maintaining relatively good transparency.

1. Introduction
Recently, new types of transparent wood composites have emerged, having still better and better properties, considering their potential use in civil engineering and architecture. In a previous article, we were comparing I-beam structures with web from transparent wood, transparent plywood, OSB board and glass [1]. The aim of this article was to compare new transparent wood composites with the materials mentioned before at a different type of structure, which might produce some striking new forms while using transparent wood composites as thinly as possible. For the analysis, a spatial folded structure was designed. The structure could serve as a pavilion for exhibition or other purposes. Basic dimensions are shown in figure 1. The structure was built of various TW materials, OSB board and glass. The results of structural analysis of all models were compared.

2. Transparent wood materials
The first considered material was transparent wood veneer from densified wood infiltrated by polymethyl methacrylate (PMMA) [2]. Some properties were listed in [2], however, Poisson's ratio and shear modulus values were missing. Poisson's ratio μ in longitudinal direction was approximated

by content percentage of PMMA (88%) and balsa (12%) in the final transparent wood sample. In transverse direction, it was calculated by equation:

$$\mu_{transverse} = \mu_{longitudinal} \times E_{transverse} \div E_{longitudinal} \quad (1)$$

Shear modulus G for transverse and longitudinal directions were then calculated by equations:

$$G_{longitudinal} = \frac{E_{longitudinal}}{2(1+\mu_{longitudinal})} \quad (2)$$

$$G_{transverse} = \frac{E_{transverse}}{2(1+\mu_{transverse})} \quad (3)$$

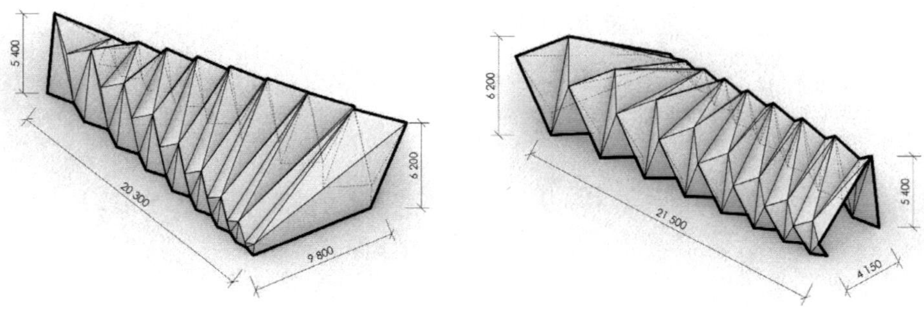

Figure 1. Shape and external dimensions of the investigated origami structure.

Next material was transparent wood infiltrated by polyvinyl alcohol (PVA) [3]. Poisson's ratio and shear modulus were calculated based on percentage of contents of cellulose and polymer in the final sample. Young's modulus and tensile strength were obtained from [3]. Another two materials were all bio-based transparent wood [4]. One was infiltrated by chitosan (CTS) and other by cellulose nanofibers (CNF). For both materials, Poisson's ratio and shear modulus were estimated similarly to the previous type. The final transparent fibre wood material (PMMA T fibre) was taken from [5]. Poisson's ratio was approximated, and shear modulus was calculated using equation (2). All the materials and their properties are summarised in table 1.

Table 1. Mechanical properties of 2 mm thick veneers.

Material	Direction	Young's modulus E (GPa)	Shear modulus G (GPa)	Poisson's ratio μ (-)	Tensile strenght f (MPa)
PMMA TW veneer [2]	longitudinal	4.3	1.57	0.37	62.5
	transverse	2.4	0.99	0.21	14.6
PVA TW veneer [3]	longitudinal	3.85	1.33	0.45	143
	transverse	3.1	1.14	0.36	67
CTS TW veneer [4]	longitudinal	10.81	4.09	0.32	171.85
	transverse	3.48	1.58	0.1	26.5
CNF TW veneer [4]	longitudinal	15.33	5.51	0.39	258.84
	transverse	4.07	1.85	0.1	31.27
PMMA T fibre W [5]	-	2.2	0.9	0.23	46.8
OSB board	-	3.8	1.08	0.3	9.9
Glass	-	70	28.7	0.22	20

3. Numerical model

Software DLUBAL RFEM 5.25.02 [6] was used for the structural analysis. The panels of the folded structure were modelled as a laminate, with the orientation of the longitudinal wood fibres in the direction of the local x-axis. The local x-axis and the wood fibre orientation of the roof panels were in the span direction, in the case of the wall panels it was in the vertical direction. The gable panel was designed to be twice as thick as other walls for every scenario. The density of transparent wood was 700 kg/m^3. The structure was loaded with a climatic load corresponding to the location of Košice. Load combinations were considered according to the Eurocode.

The layers of the panels were modelled in an RF laminate additional module of the DLUBAL software. Each layer of transparent plywood was 2 mm thick from the respective veneers. Various fibre orientation of layers was used. An odd number of layers was considered for each case. A total of 15 models with different materials were created. The linear calculation was chosen only to evaluate the materials used and the effect of their orthotropic properties. Analyses such as stability analysis, shear stress assessment, design and consideration of joint stiffness were not included.

4. Results

The results are summarized in table 2. It is clear from the obtained data that transparent wood composites exceed the standard materials in their mechanical properties. While the glass would have to be 52 mm thick and the OSB board 58 mm thick with the given construction geometry, the most advantageous types of transparent wood composites would suffice with a thickness of only 18 mm.

In the case of the investigated origami structure, it is observable that the lamination of transparent composites greatly aids their properties when used in such types of load-bearing structures. This is mainly because in this type of construction, not all of the supporting elements are oriented in the same way, they are at different angles and therefore the reduction of the non-isotropic properties of transparent wood by lamination is very desirable.

Table 2. Results of analysis of 3D structure from different materials. The values are related to the smallest possible suitable panel thicknesses.

Panel parameters			Decisive results for panel design				Informative results of structural analysis				
Material	No. of layers	Thickness (mm)	Normal tensile stress in longitudinal (x) direction (MPa)	Tensile strength f_x (MPa)	Normal tensile stress in transverse (y) direction (MPa)	Tensile strength f_y (MPa)	Normal compressive stress in longitudinal (x) direction (MPa)	Normal compressive stress in transverse (y) direction (MPa)	Shear stress (MPa)	Vertical deformation (mm)	Relative deformation (mm) *
PMMA TW veneer [2]	27	54	6.1	62.5	14.0	14.6	-9.3	-9.7	2.4	9.5	1/956
PMMA TW_CP [2]	13	26	33.9	50.1	32.8	44.9	-31.8	-44.3	4.8	23.5	1/387
PMMA TW_Qi [2]	11	22	39.9	45.4	48.9	50.1	-50.7	-44.6	4.2	28.5	1/319
PVA TW veneer [3]	11	22	41.5	143	64.2	67	-43.7	-53.1	5.8	25	1/363
PVA TW_CP	9	18	67.4	112.6	79.8	97.4	-82.1	-78.5	6.8	32.3	1/281
PVA TW_Qi	9	18	60.2	112.6	85.6	97.4	-70.6	-81.4	5.5	32.1	1/283
CTS TW veneer [4]	21	42	13.9	171.9	25.3	26.5	-16.3	-18.9	4.0	7.2	1/1262
CTS TW_CP	9	18	97.9	113.7	50.6	84.64	-122.3	-50.6	7.1	20.1	1/452
CTS TW_Qi	11	22	45.2	113.7	52.6	84.64	-54.2	-44.1	4.8	14.7	1/618
CNF TW veneer [4]	21	42	12.4	258.8	25.5	31.27	-16.9	-19.4	4.2	5.7	1/1594
CNF TW_CP	9	18	103.2	167.8	43.4	122.3	-129.3	-43.6	6.9	15.3	1/594
CNF TW_Qi	9	18	78.5	167.8	110.1	122.3	-72.9	-97.9	6.0	16.9	1/538
PMMA TFW [5]	-	26	28.1	46.8	44.7	46.8	-36.0	-34.1	4.9	33.4	1/272
OSB board	-	58	3.9	9.9	9.7	9.9	-7.5	-5.5	1.8	9.4	1/966
Glass	-	52	9.3	20	19.3	20	-17.4	-12.3	3.9	0.9	1/10094

CP – plywood with 9 veneer layers (0/90/90/0/-90/0/90)

Qi – plywood with 9 veneer layers (0/0/0/45/90/-45/0/0/0); – plywood with 11 veneer layers (0/0/45/90/-45/0/45/90/-45/0/0)
* Relative deformation is related to the 9.085 m most deformable panel

If we compare the normal stresses in the x and y directions and the resulting minimum construction thicknesses for individual materials, we find that CP-type plywood with orthogonal fibres orientation behaves best. Composites infiltrated with cellulose nanofibers (CNF TW_CP) and polyvinyl alcohol (PVA TW_CP) have the best properties overall.

In addition, it is worth mentioning that even better results could be achieved for all materials by further optimizing the 3D structure, especially with regard to deformation from snow loads, where it is the largest on the longest boards. Even while maintaining the length of these boards, it is possible to adjust the geometry so that they are rotated more vertically, which would have resulted in less deformation in this area of the structure.

5. Conclusion

Based on the performed analyses of mechanical properties of transparent wood structures, we evaluate these as very promising materials for future use in the field of load-bearing structures of buildings. Despite their thinness, these materials can carry the necessary load with a sufficient reserve. In the case of applications of more complex 3D structures, the plywood types of these composites behave better.

The best mechanical properties were found for transparent wood composites that are infiltrated with cellulose nanofibers (CNF) and polyvinyl alcohol (PVA). In the case of polyvinyl alcohol, the very good optical properties of these composites are also an advantage. The main advantage of cellulose nanofibers is their low impact on the environment and human health. The disadvantage so far are their weaker optical properties compared to other types of transparent wood composites.

6. References

[1] Kanócz J, Bajzecerová V, Karľa V 2020 *Analysis of mechanical properties of I-beam with web from transparent wood* (12th International Scientific Conference of Civil and Environmental Engineering for PhD. Students and Young Scientists. IOP Publishing s. [1-6] [online])

[2] Fu Q, Yan M, Jungstedt E, Yang X, Li Y, Berglund L A 2018 *Transparent plywood as a load-bearing and luminescent biocomposite* (Composites Science and Technology. 164. 296-303)

[3] Mi R, Li T, Dalgo D, Chaoji C, Kuang Y, He S, Zhao X, Xie W, Gan W, Zhu J Y, Srebric J, Yang R, Hu L 2019 *A Clear, Strong, and Thermally Insulated Transparent Wood for Energy Efficient Windows* (Advanced Functional Materials. 30. 1907511. 10.1002/adfm.201907511)

[4] Le Van H, Mwongeli R, Panicker P, Agumba D, Pham H, Kim J 2021 *All-biobased transparent-wood: A new approach and its environmental-friendly packaging application* (Carbohydrate Polymers. 264. 118012. 10.1016/j.carbpol.2021.118012)

[5] Wang X, Zhan T, Liu Y, Shi J, Pan B, Zhang Y, Cai L, Shi S 2018 *Large-Size Transparent Wood for Energy-Saving Building Applications* (*ChemSusChem*. 5. 3855-3862)

[6] DLUBAL RFEM 5.25.02. Structural Analysis Software, Version 5.07.

Acknowledgments

This contribution was supported by the Scientific Grant Agency and projects KEGA 046TUKE-4/2019 and KEGA 030TUKE-4/2022 supported by the Cultural Grant Agency of the Ministry of Education, Science, Research and Sport of the Slovak Republic and the Slovak Academy of Sciences. This paper was elaborated with financial support project with ITMS code: 26220220182 (TECHNICOM).

Experimental Form-finding: A review

T Baroš[1], L Kabošová[2], M Baroš[3] and D Katunský[1]

[1] Institute of Architectural Engineering, Faculty of Civil Engineering, Technical University of Košice, Vysokoškolská 4, 040 23 Košice, Slovakia
[2] Center for Research and Innovation in Construction, Technical University of Košice, Park Komenského 10A, 042 00 Košice, Slovakia
[3] Bachelor of Computer Science student, Faculty of Mathematics and Physics, Charles University, Ke Karlovu 3, 121 16 Praha 2, Czech Republic
[4] Institute of Architectural Engineering, Faculty of Civil Engineering, Technical University of Košice, Vysokoškolská 4, 040 23 Košice, Slovakia

E-mail: tomas.baros@tuke.sk; lenka.kabosova@tuke.sk; martin.baros@mff.cuni.cz; dusan.katunsky@tuke.sk

Abstract. This paper deals with the review and categorization of form-finding techniques. Their utilization in architectural design, consisting of realized as well as theoretical concepts, is studied. Employing experiments, parametric and simulation-based designing, or artificial intelligence in the design process provides a platform for novel, architecturally unique, unexpected structures.

1. Introduction

The search for pioneering ideas in architecture has driven such architects as Antoni Gaudí, Frei Otto, or Richard Buckminster Fuller to experiment with yet uncharted methods to create their amazing realizations. The first examples of form-finding (parametric) design were the upside-down models of churches by Antoni Gaudi. String models weighted down with birdshot to create complex vaulted ceilings and arches. Through playing with the weights position or strings length, the arch shape changed and the change affected the neighbouring arches. A mirror under the model help to see how it would look inverted. German architect Frei Otto in collaboration with biologist J.G. Helmke conducted studies on Radiolaria microorganisms to explore their natural equilibrium and to recreate the principle on lightweight building structures. As a result of these studies, he developed the so-called Form-Finding, a process in which a predefined material system in predefined conditions becomes naturally self-organized. Otto used analogical models, like hanging chains or nets of cables to investigate the structural behaviour of grid shell structures. The form-finding technique has evolved since [1].

Today, form-finding represents form-improvement or even form-exploration. The morphogenesis of free-form shapes (bio-constructivism) involves architectural generative models derived from nature [1]. The technique of parametric form-finding means approaching the problem based on numerous operations applied to a representational element or geometry entity. These operations are transformations whose influence can be limited based on the user's limitations and restrictions [2].

2. Categorization of form-finding

Form-finding can be divided into two fundamental problems:
a) the choice of a set of succeeding shapes and b) the connection of the shapes using any operation [2].

To design parametric architecture, the logic of parameters should be specified. Within this logic, several factors act as driving forces to stimulate architecture's formation. Each would then have its specific set of parameters controlling the variations. When multiple layers of parameters and factors are intertwined, a more complex logic may result.

Four main types of form-finding factors (parameters) are identified: a) fabrication, b) structure, c) environment, and d) spatial organization. Through the variations of the specific parameters, new designs are generated. In architecture, parametric variations are employed at different levels, with various spatial organization implications and logical approaches [1]. The history of architecture abounds with examples where architectural elements and features with technical functions become the object of articulatory or "ornamental" endeavours [3].

2.1. Form-finding through realization (process/production/fabrication)

The process of designing, exploring, and finding optimal structural shapes using experimental strategies and tools is identified in architecture as Form-finding. Robert Hooke, in 1675 first mentioned the principle when the funicular net was inverted and subjected to gravitational force to obtain a structural state of pure tension as identification of compression-only situations. This reverse hanging method is thus the oldest and the most diffused form-finding technique [4].

For finding structural form and generating geometry of cable nets or tensile structures, a procedure that applied physical models and simulated the pre-stress state typical for these structures was developed and then applied. The small-scale model's closed frame was dipping into a membrane-forming liquid to find the shape of tensile structures. The formed soap films represented mathematically defined minimal surfaces. The concept of the Pneumatic or inflated hill method is an alternative to the use of hanging models, whereas the structural form is obtained through inflation of membranes and can be applied to design reinforced-concrete shells and air-supported membrane halls [4].

Peter Zumthor completed his Bruder Klaus Field Chapel in 2007. The work has evident respect for the construction materials which create this sensuous space. The construction principle is intriguing as the Chapel is made with a wigwam consisting of 112 tree trunks as a framework. Upon completion of the frame, layers of concrete were poured and rammed atop the existing surface, each around 50 cm thick. Afterward, the wooden frame was set on fire, leaving behind a hollowed blackened cavity and charred walls [5].

'Waterflux', a museum by R&Sie(n), designed with nature-inspired organic forms, is a transcription of the research into an architectural experience through a transformation of contextual parameters. The art museum/alpine ice research station was designed gradually: from the digitization of the envelope of a traditional habitat to burrowing hollows within its wooden volume by a five-axis drill machine. The water states; the cycle of the water turning into ice and then melting again inspired the dynamic form-finding process [6].

The Spanish Ensamble Studio approaches the creation of their works through experimental processes. To accurately control the creation process, their idea is to experience and seek while finding the logic in the development. The structure is the architecture itself. It not only works to beat gravity but also frames the space around and defines architecture. In the Tower of Landscape project, the magnetic relationship between people, nature, and the space that incarnates nature, is explored. In this work, stacked polystyrene boards form the inner mould, surrounded by wooden formwork. A concrete mixture is subsequently poured to reinforce the layers. That creates observable space, where the natural and the artificial blur. The architecture resembles the creations of nature [7].

The Truffle is another project made through experimentation. To create an atypical living space, the soil was excavated and subsequently filled with hay bales, leaving air cavities for the concrete to be poured. The soil was removed after the concrete had matured to reveal an unexpected, amorphous mass with a texture, color, form, and essence of concrete. The hay bales from the inside were removed in a very unconventional way; the calf ate 50 m3 of this food. The mass that has emerged has truly unique contours [8].

The relationship between biological and digital fabrication and the involvement of animals in the design process could be explored in MIT's The Silk Pavilion. Silkworm's swarm was positioned at the bottom rim of the primary structure scaffold (by silk threads CNC machine fabricated polygonal panels).

The pavilion's geometry was created using an algorithm inspired by the silkworm's ability to generate a cocoon via a single multi-property silk thread. This algorithm assigns a single continuous thread across patches providing various degrees of density. Density variation was created, informed, ordered by the silkworm (a biological printer) to create a secondary structure by spinning flat non-woven silk patches, as they locally reinforced the gaps across CNC-deposited silk fibres.

The fibre structures created by silkworms triggered an idea for creating a computational schema for determining the shape and material optimization of fibre-based surface structures. Variations in material organization across the surface formed apertures within the structure. Their location, size, and density were derived from mapping the season-specific sun path [9].

Another strong example of form-finding techniques is undoubtedly the fabrication of ice shells. Kokawa, for instance, has proposed a strategy for creating architecture through spraying layers of snow and water onto a pneumatic formwork and thus creating ice shells. The principle lies in the absorbing properties of snow, acting as a water sponge and turning the water into a solid ice layer almost instantly. Inflatable formworks for ice domes are yet another intriguing construction method. An ice shell is created utilizing the principle of minimal surface or equilibrium surface. To enhance the structural properties, a cable net was integrated into the inflatable membrane. Because the membrane has to be smoothly released to leave only the ice dome, the bottom foundations are simple, with only the cable net attached to it [10].

2.2. From Form-finding to Form-improvement and Form-exploration

Ancient Roman architects' knowledge was a significant wellspring of knowledge was the perception and observation of construction defects and failures and applied remedies and employed solutions. The theory was pursuance of observation of strain, buckling, distortion, shift and breakdown, as in current practice limit examination. The Pantheon complexity with the underlying geometrical simplicity is shared a static idea: the balance of the central vault is set up by an arrangement of abutments and barrel vaults [11].

Form-finding techniques for this category can be assigned to three categories: physical hanging models, equilibrium methods, and optimization schemes [12].

Physical hanging models, like those used by Antoni Gaudi (Church of Colonia Güell.), Heinz Isler (Deitingen service station, Switzerland, 1968; Roof of the open-air theatre in Grötzingen, 1977) and Frei Otto (Munich Olympic Stadium, 1972; The Multihalle in Mannheim, 1973-1974), typically rely on inextensible cable networks to create purely axial systems under a gravitational load [12]. At the beginning of the 19th century, F.J. Gerstner and H. Moseley introduced a theoretical basis for the catenary forms calculation. Before that, the catenary concept was used by Italian scientist Giovanni Poleni to study the structural stability of St. Peter's dome and to propose the works to its rehabilitation. Funicular models are a great representation of catenary structures [13].

The shape of a funicular model interacts with the forces of each component. Using this technique, Gaudí designed the geometry of his extraordinary works which were later built in masonry and stone. Through a trial and error process, this technique allowed him to correct the structure's shape following the aesthetic or functional principles while concerning the equilibrium state. We could consider this as a from-exploration of physical model shape transformation [13].

Thin-shell concrete structures inspired the invention of the inflatable-formwork construction method. This technique was developed in 1940 by Californian architect Wallace. Inflated sailcloth cushions provide flexible reinforcement for the concrete shell. Shotcrete is sprayed on the inflated formwork resulting in a solid and stable structure [10]. Swiss engineer Heinz Isler in the 1950s worked with extending these concepts through physical models of free-form shell roofs. Isler developed experimental forms from manipulating a sheet or membrane subjected to various external forces and handling techniques. The method of the hanging membrane, the pneumatic/ inflated hill method, or the flow method, were used to create the freeform shapes [14]. In the 1960s, fabric formworks were utilized to design and subsequently construct shell structures made of ice. Reinforced concrete (RC) shells - a continuous structure - were simulated through suspension of wet pieces of fabric or membrane. The derived forms were then frozen and finally inverted. Every winter, he conducted experiments by freezing a fabric or spraying water over objects like plants, inflated balloons, hanging ropes [10] [4].

During the 60-ties Italian architect, Dante Bini developed and patented a curious construction technique that leveraged the form-finding process to erect RC domes rapidly. Bini explored this technology with his patented 'Binishell' system for RC shells in 1969. Concrete was poured over a flat pneumatic preformed formwork, and then inflation allowed the rising and completing of the roof within a couple of weeks [10] [4].

The 20th century and a wave of new practices and developments in computational models and form-finding methods brought the need for the true theory of form, entirely based on the potential of information processing, enabled by electronic computers [15].

Italian engineer Sergio Musmeci believed that, when designing structures, sophisticated design tools can be incorporated for making crucial design decisions, which influence the outcome. He states that the engineer must not start from the given form and thus limit himself, but on the contrary, he should make the acting forces and one or more performance parameters fixed and consider the final shape of the structure as unknown. The minimal form, in which every part is equally important for the structure, is indeed a set of possible forms, not only one particular. Within this limited yet rich group of architectural possibilities, the intuition, and creativity of the engineer are demonstrated [16]. His form-finding approach manifests the best in The bridge over the Basento River project. For this complexity project, Musmeci used a new design method based on experiments evoked by the search for essentiality and the desire to concretize invisible lines of force with a continuous structure in space. The study required particular calculations and model tests of structures. He sought to move structural logic and reliance on the study of static equilibria to the point where, through them, they created a form that combined natural adhesion to its function with an expressive charge. It is best described by "Form is unknown, not tension!". It is a concept of searching for form or minimal forms, i.e. those that respond to their structural function using a minimum amount of material. The continuous form is the organic form precisely because it represents the solution of forces and equilibria that come from a much larger complex of events in a single structure. Based on a dual process, a specific tension regime is assigned, and a shape suitable for its production is obtained. The shape of the vault arose from a lengthy design process, following different approaches (empirical, with the study of the behaviour of physical models; analytical, based on continuous mathematical models; geometric, based on graphical statics; numerical with finite elements). He used these methods in a non-sequential manner and presented the experimental evidence to validate the calculations. Specifically, several models were made: first, a study was performed with a film of soap solution mixed with glycerin formed between cotton fibres to sketch the shape, then a neoprene membrane was stretched vertically to create a second model that reproduces the bridge support and two adjacent half-arcs. The subsequent methacrylate model at a scale of 1:100 allowed the study of elasticity to improve the shape and was exposed to various load conditions, while deformations were measured using electric strain gauges. Prior to implementation, a model of two fields in micro concrete was made on a scale of 1:10, length 14 meters for load tests by a load experiment using hydraulic jacks with several tons. The performed tests confirmed the results of the calculations and made it possible to determine the exact shape of the vault and the geometry of the edges [15].

2.3. Form-finding through algorithms

As the form-finding leads the design towards the structural optimum, one might imagine that this process cannot result in a free-form shape, unbounded by structural and construction principles. Structural optimization is a fusion of mathematical programming and the finite element technique. Laws of nature and, in particular, Heinz Isler's design methods were a driving force behind architectural optimization. Gaudi's philosophy and the funicular technique inspired generating the first physical models of free-form shells. Nowadays, the progress in hardware and software enables radical improvement of architectural form-finding. Novel fast numerical calculation techniques are dominant over purely experimental form-finding. Mathematical optimization is employed to leverage the computational power for interactively searching for optimal solutions to a problem from among a line of possible candidates. The original form-finding concept, focused on finding the optimal form, can be transformed into the pursuit of form improvement (performance improvement). This process aims at improving the performances of an already-existing spatial configuration, which does not necessarily mean achieving the structural optimum. Consequently, optimization is an intriguing form-exploration method, enriching conceptual design [4] [14]. This concept is most often represented by digital simulation usually using particle-spring systems. By simulating the physical behaviour of deformable bodies we can find and investigate results in real-time and parametrically edit input data influenced the creating form.

It is worth mentioning other techniques as well. Dynamic relaxation, force density, stress distribution, thrust-network analysis, and particle-spring method use iteration algorithms to control nodal geometry through finding an equilibrium of internal forces with applied external loads. One- or multiple-criteria optimization schemes optimize parameters such as nodal coordinates in the structural system for given boundary conditions [12]. A digital upgrade to Gaudi's funicular tool is the Homeostatic Model Technique (HMT), which can continuously generate a surface shape [13]. Combinatorial Equilibrium Modelling (CEM) is a form-finding approach for air-supported structures, controlling the connectivity and the equilibrium state of structural networks via an operative topological diagram [17]. The Force Density Method (FDM) is utilized for form-finding of prestressed cable systems.

Creating parametric architecture does not necessarily need to involve generative computation techniques as it is primarily about form-finding. Complex, unpredictable forms can originate through algorithmic design. Generative algorithms are still an unexplored field of parametric designing [18]. The process is an analogy to natural selection.

Funes and Pollack [19] tested the behavior of LEGO structures created through the evolutionary algorithm employing mutation. Genotypes (genetic representation) are required, representing the variable parameters. Genotypes mate and are recombined and mutated in the following generations to create all possible options.

During the optimization process, all possibilities are tested with a physics simulator to find the fittest option for the specified conditions. The actual physical properties of LEGO bricks were programmed, and the evolutionary process combined bricks of various shapes into buildable 3D forms. The real-life behavior of the resulting forms was evaluated through physical models. The algorithms used did not constrain the pool of potential solutions, therefore the evolved 3D forms were surprisingly alien-looking, as the computer found unique ways through the evolutionary search process [20].

Another case study by Lamb, Voon and Healy provide a generative technique for accessing and manipulating the genetic structure of any form in order to maximize the likelihood of an emergent architectural result. The information encoded as state changes between cells in a 3D grid and the execution of a 3D Cellular Automaton (CA) constitutes the foundation of such a mechanism. The CA cells are similar to those of a normal organism in two regards. In first, instructions encoded equally in all cells of the growing creature determine the material form and organism's behaviour. Second, using a discrete generative mechanism to convert 1D information to 3D form due to control local transitions between units made up of a series of low-level instructions as of transition rules set and initial cell configuration conjunction formulated in the Cellular Automata [21].

The use of mathematical concepts of fractal geometry random and self-similar fractals can be applied in architectural structures. In the case study, Rian and Asayama focused on applying the principle of random fractals as a design tool when designing a canopy branching structure. They thus arrived at a different form of the whole structure compared to commonly designed, existing shelter structures, where most canopy structures achieve only an abstract impression nature applied to the branched structure of the columns with a flat roof. After hanging model experiments, Falk et al. proposed a non-flat roof for such a tree-like column supported canopy structure. Their roof is an origami-style folded plate roof supported by designed branching columns. However, their proposal's folded roof does not match any natural item, suggesting that their design sense may not be mimicking nature [22].

Form-finding with swarm intelligence is also a fascinating approach for design. In Asli Agirbas' case study, the principle of swarm intelligence in nature has been assimilated, and the formation of complex geometry has been made using swarm intelligence simulation to design façades. Through these form-finding studies, we can demonstrate that we can create complex emergent forms beyond Euclidean geometry by implementing an understanding of the functioning of natural principles. The multi-agent systems (MAS), an open, variable, and flexible computer system that lives in a virtual environment, can show unpredictable, autonomous behaviours for its designed target, where passive data objects turn into active agents and act according to available data. MAS-based swarm intelligence, a self-organising system (as a morphogenetic perspective of design) and the solid-void proportion and desired light intake determination of the designer (as a morphodynamic perspective of design), take concepts of biological development and the mechanisms of nature as a model in the design process [23].

2.4. Time for Self-Form-finding?

As per formal computer science definition, the algorithm is a finite sequence of elementary commands, implementation of which allows getting relevant output data, by mechanical means, for each admissible input after a finite number of steps [24]; however, this term is relatively abstract. The primary purpose of any algorithm is to solve a given problem. Generally, even a simple procedure such as making coffee can be perceived as an algorithm. The vast majority would not consider designing shape and form as a set of precise steps for it being a highly creative process, but why could not the algorithm be creative, and why would it not be able to create yet unseen things?

Creativity is the process of bringing something new into being [25], though it is based mainly on previous knowledge (such as what was seen working or not and what was liked) and empirical research. This provides a perfect parallel to the basic notion of machine learning.

Webpage ThisPersonDoesNotExist.com demonstrates the power of this concept to come up with something new by creating photo-real images of people that do not exist. The algorithm is based on the research published by NVIDIA using a generative adversarial network (GAN). There is also the first theatre play written by artificial intelligence that has shown the ability of machine learning to deal with such a complicated topic as the natural language is. The play's script was generated automatically by an AI system based on the GPT-2 neural language model [26].

So far, the algorithms in the field of form-finding were used only as a tool and not as independent creators. This leaves their full potential yet to be discovered.

3. Conclusion

People have always tried to find a form for architectural work using the technology or knowledge offered by the time. Thus, we can consider such experimental approaches in design as always innovative. By mixing the old principles of form-finding techniques with the current ones, there is a clash and the creation of new mutations of such methods. By knowing the applied ways of achieving the expected formable processes and discovering new possibilities, we can push the boundaries of what is possible. This gradually evolving process from primitive imitation, through experiences as traditional knowledge, to expertise practice is constantly evolving. In every type/method/technique with experimental aspects, there is some kind of ambiguity - a little bit of improvisation - and such phenomena we could characterize as inevitable, desirable and accompanying. Scientific understanding enables extreme freedom to operate outside the preconceived processes. We should not ignore history but study it and reread it, use existing technology, and face the problems of our present times with knowledge of the past. Nowadays, the mixture of old - sometimes appeared as primitive - methods with advanced technology we are capable of bringing an utterly new angle of view on the design process. Each time offers possibilities and hides opportunities to look at problems differently. It is up to us to meet this challenge. Every precedent can start changes in approach and set the direction for future generations. Maybe it is time not to be afraid to experiment more and beyond the conventional.

IOP Conf. Series: Materials Science and Engineering 1252 (2022) 012002 doi:10.1088/1757-899X/1252/1/012002

4. Graphic appendix A

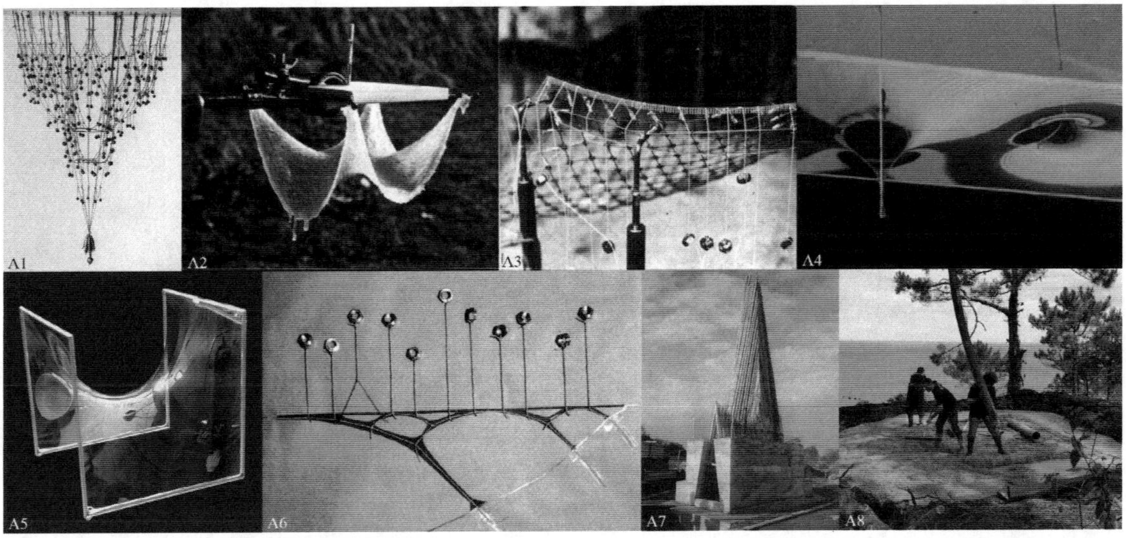

Figure A1. Poly-funicular model of the Church of la Colonia Güell (© E.Casanelles, Barcelona, La Poligrafia, 1965); Figure A2: Heinz Isler hanging model (© Heinz Isler); Figure A3, A4: Frei Otto hanging model and soap film model (© ETH Zürich archives); Figure A5: Basento river soap blow bubble check it is the minimal surface, set from this peculiar edge frame; Figure A6: Study-models for the bridge over the Astico river(1956); Figure A7: Realization of Bruder Klaus Field Chapel (© Peter Zumthor, Mechernich, Germany, 2007); Figure A8: Realization of The Truffle (© Ensamble studio, Costa da Morte, 2010).

5. References

[1] Siddhaarth S 2016 *Marg Institute of Design and Architecture SWARNABHOOMI (Doctoral dissertation, ANNA UNIVERSITY).*

[2] Kourkoutas V 2007 *Parametric form finding in contemporary architecture.*

[3] Schumacher P and Lei Z 2017 *From Typology to Topology: Social, Spatial, and Structural. Architectural Journal* **11**(590), 1-18

[4] Tedeschi A 2014 *AAD, Algorithms-aided design: parametric strategies using Grasshopper. Le penseur publisher*

[5] Archdaily website [online] https://www.archdaily.com/106352/bruder-klaus-field-chapel-peter-zumthor

[6] Archdaily website [online] https://www.archdaily.com/10157/waterflux-rsien

[7] Ensamble website [online] https://www.ensamble.info/towers-of-landscape

[8] Ensamble website [online] https://www.ensamble.info/thetruffle

[9] MIT website [online] https://www.media.mit.edu/projects/silk-pavilion/overview/

[10] Wu Y, Liu X, Chen B, Li Q, Luo P and Pronk A 2019 *Design, construction and monitoring of an ice composite shell structure. Automation in Construction* **106**, 102862

[11] Martines G 2009 *The Conception and Construction of Drum and Dome. Marder and Wilson Jones (eds.)*

[12] Bagrianski S and Halpern A B 2014 *Form-finding of compressive structures using prescriptive dynamic relaxation. Computers & Structures* **132**, 65-74

[13] Bellés P, Ortega N, Rosales M and Andrés O 2009 *Shell form-finding: Physical and numerical design tools. Engineering structures* **31**(11), 2656-2666

[14] Vizotto I (2010). *Computational generation of free-form shells in architectural design and civil engineering. Automation in construction* **19**(8), 1087-1105

[15] LinkFang website [online] https://it.linkfang.org/wiki/Ponte_sul_Basento
[16] Raffaella M 2020 *Pier Luigi Nervi and the School of Italian Engineering: Science or the Art of Building?. In IOP Conference Series: Materials Science and Engineering. IOP Publishing* (Vol. **809**, No. 1, p. 012001)
[17] Wan Z, Ohlbrock P O, D'Acunto P, Cao Z, Fan F and Schwartz J 2021 *A form-finding approach for the conceptual design of air-supported structures using 3D graphic statics. Computers & Structures* **243**, 106401
[18] Çalışkan O 2017 *Parametric design in urbanism: A critical reflection. Planning Practice & Research* **32**(4), 417-443
[19] Funes P and Pollack J 1998 *Evolutionary Body Building: Adaptive Physical Designs for Robots. Artif Life* **4** (4), 337–357 doi: https://doi.org/10.1162/106454698568639
[20] Funes P and Pollack J. B. 1999 *Computer evolution of buildable objects for evolutionary design by computers. Evolutionary design by computers* 387-403
[21] Coates P, Healy N, Lamb C and Voon, W L 1996 *The use of Cellular Automata to explore bottom up architectonic rules*
[22] Rian I M and Asayama S 2016 *Computational Design of a nature-inspired architectural structure using the concepts of self-similar and random fractals. Automation in Construction* **66**, 43-58
[23] Agirbas A 2019 *Façade form-finding with swarm intelligence. Automation in Construction* **99**, 140-151
[24] Drózd J and Kryl R 1992 *Začínáme s programováním. Grada Publishing as.*
[25] Cunningham R J 2007 *The courage to create Rollo May: The courage to be Paul Tillich. Journal of humanistic psychology,* **47**(1), 73-90
[26] 1.0 THEaiTRobot, Košťák D, Hrbek D, Rosa R and Dušek O 2021 *AI: When a Robot Writes a Play (technical report)*

Acknowledgments
The paper presents partial research results of project VEGA 1/0674/18 "Theoretical and experimental analysis of architectural-structural shapes and fragments of building envelope structures designed for harsh weather conditions".

Experimental verification of a simulation model of high-speed train aerodynamic impact on railway infrastructure

N A Labutin[1], A V Benin[2] and L K Diachenko[1]

[1] Bridges Department, Faculty of Transport Construction, Emperor Alexander I St. Petersburg State Transport University,
[2] Mechanics and Strength of Materials and Structures Department, Faculty of Transport Construction, Emperor Alexander I St. Petersburg State Transport University, 9, Moskovsky pr., Saint Petersburg, 190031, Russian Federation

E-mail: n_labutin@outlook.ru

Abstract. The article shows the results of an experimental measurement of the air pressure change occurring due to the passing of an electric high-speed train. These data are then used to verify the elaborated simulation model of a high-speed train aerodynamic impact on the infrastructure facilities. According to the outcomes of research, the proposed model is quite reliable and suitable for further application.

1. Introduction

High-speed railway infrastructure should meet especially strict requirements. One of them is capability of railway facilities to withstand the aerodynamic impact generated by passing trains. In Russia, the issues of aerodynamic impact are in focus due to the design and construction of Moscow—St. Petersburg high-speed railway line, which has to serve trains at speed up to 400 km/h. As the existing methods of structural calculations for bearing aerodynamic stress are limited with the speed of 300 km/h, it is necessary to carry out extra research in the aerodynamic impacts in order to specify these methods.

This task mainly involves determining the intensity of aerodynamic impact on the surface at speed of more than 300 km/h. It is the experimental measurement that can provide the most reliable results and identify the magnitude of the impact. However, as the maximum train speed in Russia is limited to 250 km/h, there is no opportunity to carry out the measurements needed. So, one has to apply methods of computational simulation and verify simulation models within the speed range available for experimental research.

2. Description of an aerodynamic impact

When a high-speed train moves along the track a reverse airwave appears against the first carriage, which is due to compaction of the front air mass and its further decompaction. A comparable airwave, though of less pressure amplitude, is formed against the last carriage [1-9].

The main parameters of the airwave are values of pressure peaks and distance between them. According to the measurements conducted in The Railway Research Institute (VNIIZhT) during the Sapsan electric high-speed train certification tests, a zone of excessive pressure is formed at a distance of 15–20 m in front of the head aerodynamic panel and the peak pressure values alternate violently in the period between 0.02 and 0.035 seconds (which corresponds to the distance of 1.4–2.5 m given the speed of 250 km/h) [10].

While an infrastructure facility is being flowed around by this airwave, the pressure gets various values in different points on its external surfaces, which in its turn forms an aerodynamic force acting upon the facility.

3. Experimental measurements

Experimental measurements of the aerodynamic impact involve measuring the change of air pressure in a particular point when a train passes in close vicinity to it. The Sapsan electric high-speed train was taken as a reference train. The measurements were carried out at several sections of St. Petersburg–Moscow Oktyabr'skaya railway in Tver' and Novgorod regions, the train speed being in the range of 140–250 km/h.

The measuring equipment included high-frequency diaphragm pressure sensors that were mounted on tripods or attached with G-clamps to various infrastructure facilities (catenary supports, signals, railings, etc.). The process of measuring is showed in figures 1a and 1b.

a) b)

Figure 1. The process of measuring
(a – sensor attached to a catenary support; b – sensor mounted on a tripod).

Once the measurements had been accomplished, their results were presented in diagrams showing the air pressure change as a function of train position against the measuring point. This was done for speed values of 200, 220 and 250 km/h (figure 2). Based on this, the distribution of the peak pressure values was visualized against the distance from track center line and the height above the level of head rail (figures 3, 4).

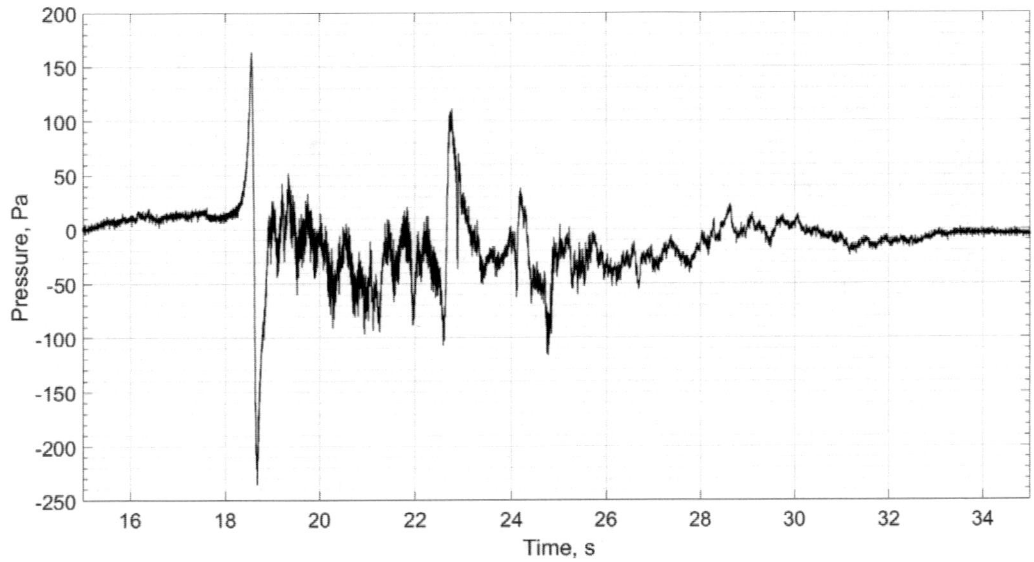

Figure 2. Diagram showing the pressure change at train speed of 220 km/h.

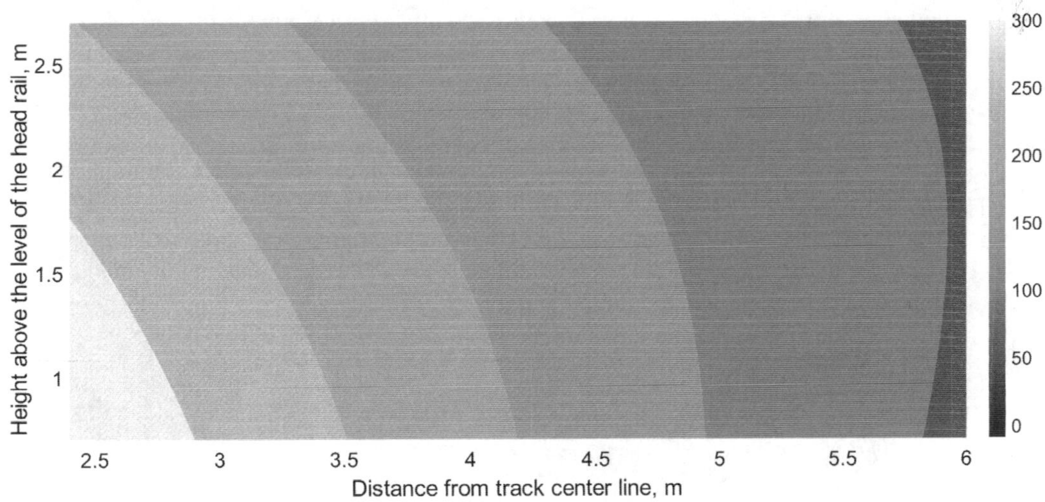

Figure 3. Distribution of maximum pressure at train speed of 250 km/h.

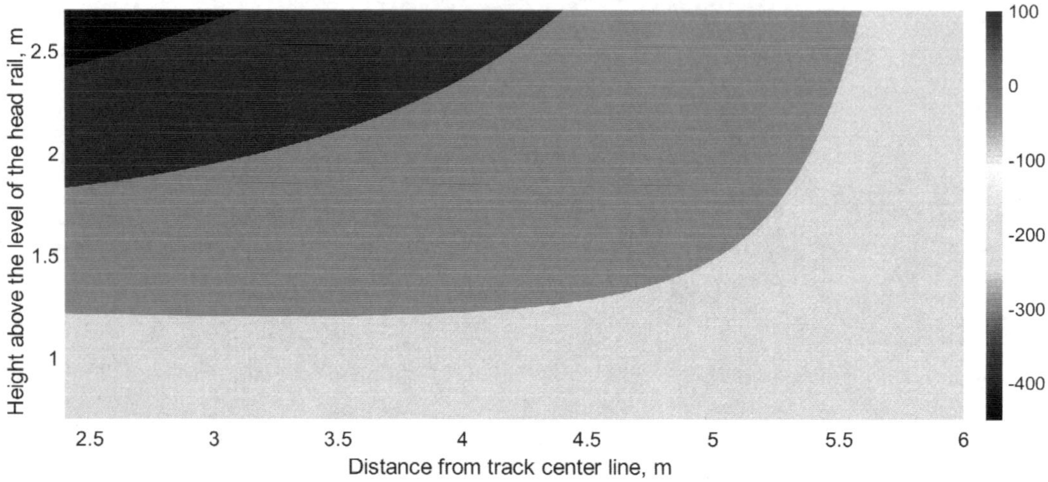

Figure 4. Distribution of minimum pressure at train speed of 250 km/h.

4. Elaboration of a simulation model

The simulation model was elaborated by means of ANSYS CFX computational flow dynamics software. Calculation of the aerodynamic impact involves solving three-dimensional unsteady-state nonlinear Navier-Stokes equations with account of viscosity of the medium. The software resolves gas flow problems using the finite-volume method, which aims at dividing the elements into reference volumes with central nodal points.

The obtained model consists of two elements: a stationary domain simulating the air environment and a mobile domain presenting a high-speed train within this environment. The conditions at the upper and lateral boundaries of the air domain were assigned as "opening", the lower boundary was assigned as "wall". The SST turbulence model, which had proved optimal for structural aerodynamics tasks [11-13], was applied to demonstrate high reliability of the method compared to simulations in aerodynamic tubes. The simulation model is showed in figure 5.

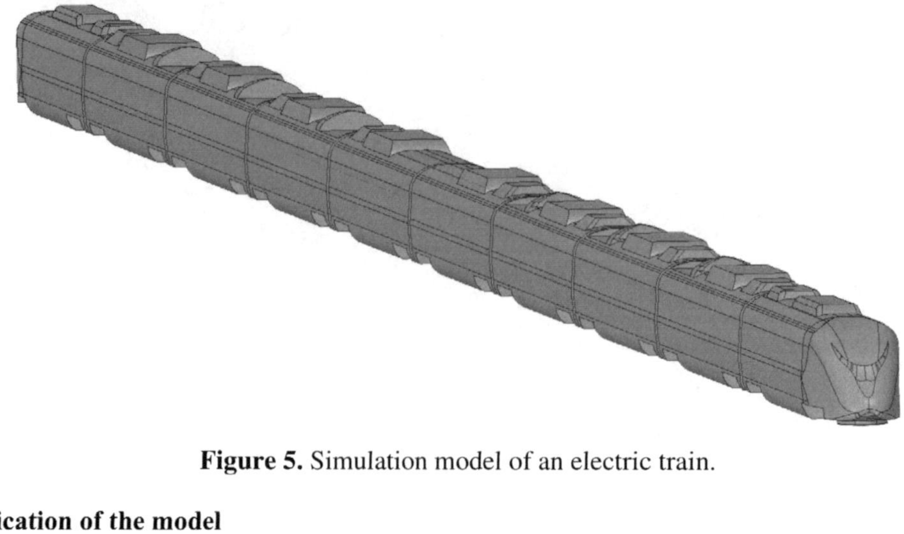

Figure 5. Simulation model of an electric train.

5. Verification of the model

In order to verify the simulation model some fixed points of monitoring were set up within the air domain, whose task was to record changes of air pressure occurring due to the passage of the train domain. Co-ordinates of these points were set according to the positions of sensors used in experimental measurements at each value of the train speed.

Apart from that, for the purpose of rigorous verification the results of applying the simulation model to the train speed of 300 km/h were then compared with the experimental data (figure 6) obtained for analogous ICE3 electric high-speed double trainset (see [14]). Results of the simulation modelling are given in figure 7.

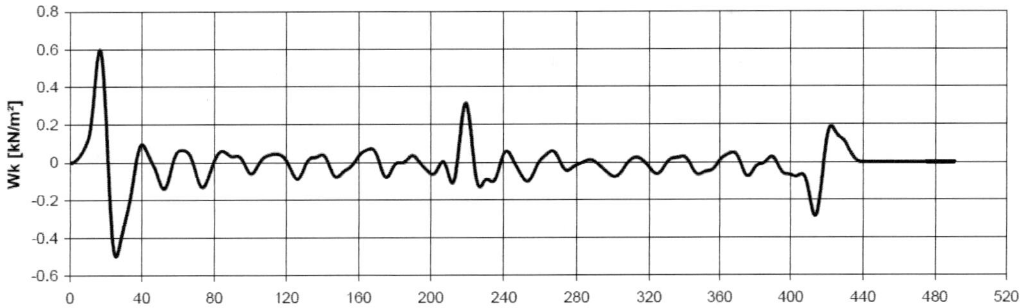

Figure 6. Diagram showing pressure distributed along the length of an electric high-speed train at the speed of 300 km/h [14] (x-axis – length of an electric train, y-axis – pressure).

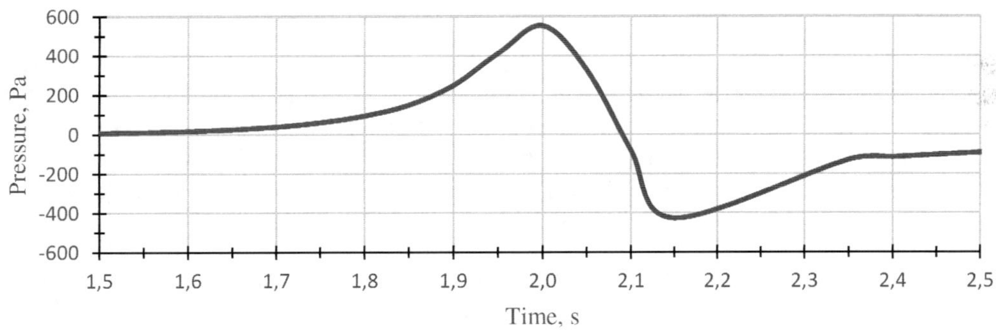

Figure 7. Diagram showing pressure at the passing of the electric train head obtained by means of software.

6. Conclusion

Based on experimental data compared to the results of simulation modelling, we have drawn the relative values of deviations for certain measuring points (see Table 1).

Table 1. Relative differences between the experimental and simulation.

Speed, km/h	Distance from track center line, m	Height above the level of the head rail, m	Difference	
			maximum pressure	minimum pressure
200	3.4	2.15	13%	12%
	7.6	1.35	3%	1%
	7.1	4.2	9%	12%
220	7.6	3.2	14%	1%
	3.5	3.15	5%	3%
	7.6	3.15	10%	11%
	7.5	2.4	4%	2%
250	6.5	1.8	14%	15%
	6.5	2.5	8%	15%
	3.5	1.7	5%	11%
	9.2	1.5	14%	15%

The range of relative differences is determined by local conditions and facilities in close vicinity to the track center line at certain measuring points (e.g., service buildings, power supply infrastructure, station equipment, etc.). The gap between the peak values in the simulation model and in [14] counts to 8%, which is due to an insignificant difference between ICE3 and *Sapsan* head car shapes.

7. Summary

Within the scope of this research, we have conducted an experimental measurement of peak pressure values of a head airwave generated by the *Sapsan* electric high-speed train in the speed range of 200–250 km/h. The results of the measurement have been used to verify the simulation model of a high-speed train aerodynamic impact. The gap between the experimental and simulation data is 1 to 15%. Then, another verification was undertaken against analogous experimental data [14] characterizing the train speed of 300 km/h; the relative difference here counts to 8%. Thus, we have proved that the elaborated simulation model is quite reliable and can be applied for calculating aerodynamic impact induced on infrastructure facilities by a high-speed train.

8. References

[1] Baker CJ 2014 A review of train aerodynamics. Part 1 – Fundamentals. *The Aeronautical Journal **117***(1201):201-228.

[2] Baker CJ 2010 The flow around high-speed trains. *Journal of Wind Engineering and Industrial Aerodynamics* **98(6):**277-298.

[3] Oranello A, Schober M 2006 Aerodynamic Performance of a Typical High-Speed Train. *4th WSEAS International Conference on Fluid Mechanics and Aerodynamics. Elunda, 21-23 august 2006. p.18-25.*

[4] Baker C, Johnson T, Flynn D et al 2019 Aerodynamic loads on trackside structures, pass-ing t rains and people. Train Aerodynamics*:151–179.*

[5] Schetz JA 2001 Aerodynamic of high-speed trains. *Annual review of fluid Mechanics* **33***:* 371-422.

[6] Dyachenko L., Benin A. 2017 An assessment of the dynamic interaction of the rolling stock and the long-span bridges on high-speed railways. *MATEC Web of Conferences: p. 00014.*

[7] Diachenko L., Benin A., Smirnov V., Diachenko A. 2018 Rating of dynamic coefficient for simple beam bridge design on high-speed railways. *Civil and Environmental Engineering.* **14.** N 1. P. 37-43.

[8] Soper D, Gillmier S, Baker C, Morgan T, Vojnovic L 2019 Aerodynamic forces on railway acoustic barrier. *Journal of Wind Engineering & Industrial Aerodynamic* **191**:266-278.

[9] Soper D, Gillmeier S, Baker C, Morgan T, Vojnovic L 2019 Aerodynamic forces on rail-way acoustic barriers. *Journal of Wind Engineering and Industrial Aerodynamics* **191**:266-278.

[10] Lazarenko Yu M, Kapuskin A N 2012 Aerodinamicheskoye vozdeistviye vysokoskorostnogo elektropoezda "Sapsan" na passazhirov na platformakh i na vstrechnyye poezda pri skreshchenii [Aerodynamic impact of the *Sapsan* electric high-speed train on passengers standing on platforms and on oncoming trains at meeting points], *Vestnik VNIIZhT* [VNIIZhT Scientific Journal]. **№4**. *p.11-14.*

[11] Zampieri A, Rocchi D, Schito P, Somaschini C 2019. Numerical-experimental analysis of the slipstream produced by a high speed train. *Journal of Wind Engineering and Industrial Aerodynamics 104022.*

[12] Xiong XH, Yang B, Wang KW et al 2020. Full-scale experiment of transient aerodynamic ressures acting on a bridge noise barrier induced by the passage of high-speed trains operating at 380–420 km/h. *Journal of Wind Engineering and Industrial Aerodynamics* **204:**104298.

[13] Maleki S, Burton D, Thompson MC 2017. Assessment of various turbulence models (ELES, SAS, URANS and RANS) for predicting the aerodynamics of freight train container wagons. *Journal of Wind Engineering and Industrial Aerodynamics* **170**:68–80.

[14] Hoffmeister B 2007 Lärmschutzwände an Hochgeschwindigkeitsstrecken der Bahn – eine Herausforderung für den Leichtbau. *-A-CH Tagung der Österreichischen Gesellschaft für Erdbebeningenieurwesen und Baudynamik. Vienna, 27-28 September 2007. p. 1-11.*

Acknowledgments
This research was supported by the Slovak Scientific Grant Agency (VEGA) in collaboration with Slovak Ministry of Education, Science, Research and Sports and Slovak Academy of Sciences (SAS), under Grant number 2/0017/20.

IOP Conf. Series: Materials Science and Engineering 1252 (2022) 012004 doi:10.1088/1757-899X/1252/1/012004

New test equipment for testing the watertightness of facade structures in-situ

B Bielek[1], D Szabo[1] and J Klem[1]

[1] Slovak University of Technology in Bratislava, Faculty of Civil Engineering, Radlinského 11, 810 05 Bratislava, Slovakia

E-mail: boris.bielek@stuba.sk; daniel.szabo@stuba.sk; josip.klem@stuba.sk

Abstract. With the current trend of construction of high-rise buildings in urban agglomerations, ensuring watertightness of their facade structures exposed to the effects of high-intensity wind-driven rain is coming to the fore. Laboratory experimental testing of water impermeability of elements of building envelope structures and their joints is performed in the process of preparation of the facade implementation in accordance with STN EN 1027: 2002 in rain chamber equipment. Due to the fact that the quality and functional properties of the finished facade are significantly influenced by the assembly technology and work discipline in its implementation, successful testing of the facade and its components in the rain chamber may not guarantee the watertightness of the finished facade. For this reason, the investor has the opportunity to check the finished facade in terms of its hydrodynamic safety and reliability by a functional test of its watertightness according to the standard regulation STN EN 13051 Curtain walling - Watertightness - Site test. The article describes in detail the new technological device for the implementation of the in-situ watertightness test of the facade, the measurement methodology, including its functional verification on the facade of the high-rise administrative building Nivy Tower in Bratislava.

1. Introduction

At present, especially in large urban agglomerations, we are witnessing intensive construction of high-rise buildings with exterior walls in the form of light suspended exterior cladding or transparent facades. As the height of the building increases, so does the wind speed and thus the load on the exterior walls by the complex effects of wind-driven rain.

If the exterior wall is loaded on the outer surface by water (eg from the effect of wind-driven rain) and if it is simultaneously loaded by a force causing the water to move towards the dry-interior surface, conditions are created for water penetration to the exterior wall. The forces causing water penetration through building envelope structures include capillary suction, air pressure difference, gravity, kinetic energy of raindrops and wind air currents – Figure 1. If we neglect the penetration of rainwater through disturbed or defective fragments of building envelope structures (cracks in concrete or reinforced concrete exterior walls, cracks in plaster layers of aerated concrete walls, etc.), we can limit rainwater penetration to joints (between prefabricated parts of prefabricated lightweight buildings, between parts of facades of buildings, between opaque and transparent parts of exterior walls, etc.) or functional joints (between the window frame and the window sash) in the exterior structures of buildings.

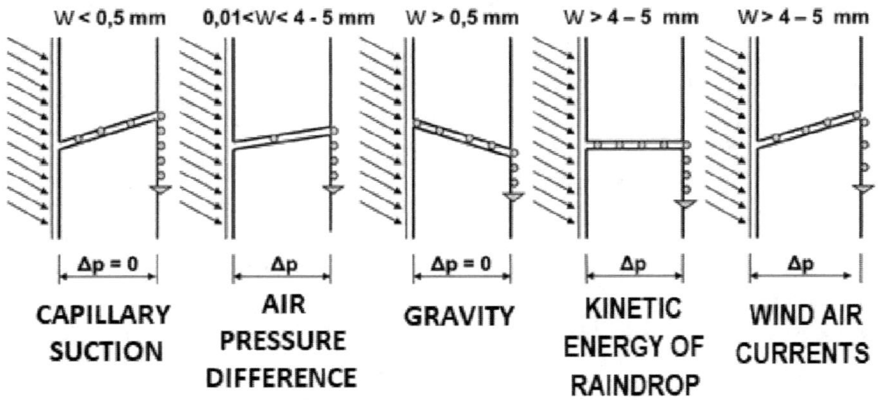

Figure 1. Forces causing water penetration through the exterior wall structure.

2. Reasoned design of joints in the concept of two stages of sealing

In the correct, theoretically justified structural creation of the contact detail in the building envelope structures, the effect of rain and the effect of wind must be eliminated separately. This principle leads to the exterior walls and their contact with the two stages of sealing. In contact with two stages of sealing, their outer zone eliminates the effect of wind-driven rain and creates a rain obstacle of contact due to its reasoned shaping. Only behind a suitably oversized rain obstacle there is a dimensioned and suitably compressed wind barrier sealing profile in the middle contact zone. Theoretically-experimentally justified structural creation of joints of a light exterior cladding conceived in accordance with the modern theory of hydrodynamics of buildings in the concept of two stages of sealing is documented in Figure 2.

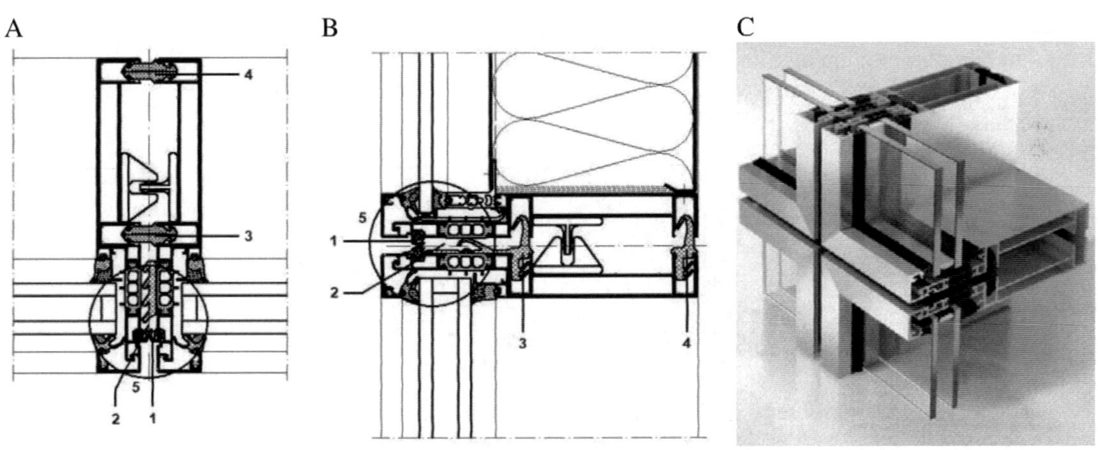

Figure 2. Joints in the construction of a light exterior cladding - element transparent suspended facade of the Schüco system USC 65 series conceived in accordance with the modern theory of building hydrodynamics in the concept of two stages of sealing A – vertical contact of elements, B – horizontal contact of elements, C – axonometry of contact crossing detail, 1 – contact entry overlap, 2 – decompression cavity, 3 – decompression cavity closure, 4 – dimensioned wind obstacle, 5 – rain obstacle system.

We consider as a rain obstacle of vertical contact not only the element overlapping the contact, but also the internal space behind it, the function of which is:

- in conjunction with the air of the outdoor climate to eliminate the pressure difference between the air of the outdoor climate and the interior of the rain barrier,

- to ensure the transport of a small amount of water, which by other effects of water movement got behind the covering of the contact into the inner space of the rain barrier of contact and its drainage to the outdoor environment.

This internal space of the rain barrier of contact in accordance with its function is called the decompression space or decompression cavity of contact.

The rain barrier system is a basic dimensioned structural element of the contact. Its dimensioning is based on the modern theory of building hydrodynamics, which eliminates the theory of the capillary suction force, the force of air pressure difference, the force of gravity, the kinetic energy of raindrops and the critical force of wind air currents by the principles of theoretical-experimentally justified structural dimensioning of joints.

The second basic dimensioned structural element of the joints between the parts of the exterior walls is the wind barrier (air seal) of the joint, which significantly affects the air filtration by contact (from the effect of wind and air pressure difference) and thus indirectly affects its watertightness.

The aim of the application of the modern theory of hydrodynamics of buildings to the structural creation of joints or gaps in the exterior structures of buildings is to ensure the criterion of water impermeability of the joint. This criterion tells us that rainwater from the complex effects of wind-driven rain must not hit the wind barrier (air seal) or the wind barrier must be situated behind a properly dimensioned rain barrier.

3. Laboratory testing of water impermeability of exterior structures

Laboratory experimental testing of water impermeability of elements of building envelope structures and their joints is carried out in accordance with STN EN 1027: 2002 in rain chamber equipment. A new integrated pressure and rain chamber is located at the Faculty of Civil Engineering STU in Bratislava, which serves for experimental research of the general phenomenon of air filtration and water penetration through details, parts and systems of building envelope structures.

Watertightness is the ability of a closed test specimen (eg a window) to withstand the penetration of water under load up to a certain test pressure p_{max} (Pa), which is called the watertightness limit. Water penetration is the wetting of the inner surface of the test specimen. The watertightness of the test specimen is evaluated and classified by watertightness classes in accordance with STN EN 12208: 2001. The view of the experimentally examined window in terms of watertightness in the integrated large pressure and rain chamber is documented in Figure 3.

Figure 3. View of the measured window installed in the mask panel of the large pressure and rain chamber during the watertightness test.

4. In-situ testing of water impermeability of exterior structures

Due to the fact that the quality and functional properties of the finished facade are significantly influenced by the assembly technology and work discipline in its implementation, successful testing of the facade and its components in the rain chamber may not guarantee the watertightness of the finished facade. For this reason, investors can check the finished facade in terms of its hydrodynamic safety and reliability with a functional test. The standard regulation STN EN 13051 - Curtain Walling -

Watertightness - Site test makes it possible to test the watertightness of light transparent facades even on an already finished facade by the in-situ method.

The essence of the test is to load the facade with a constant and determined amount of water, which will form a water film on its outer surface in order to determine the time when the penetration of water on its inner surface is detected. The test device for water spraying a facade consists of a set of nozzles placed in one line at a distance of less than 400 mm from each other, which is placed at a distance of 250 mm from the outer surface of the facade at the upper level of the test area. The facade is sprayed with a test device with a constant water flow of 5 l per meter of conventional spray device under a pressure of 2 to 3 bar for 30 minutes and it is observed whether water has penetrated its inner surface. The test can also be supplemented by creating an air pressure difference between the exterior and interior of the tested facade by creating an airtight chamber on the inner surface of the facade using an airtight foil, from which air is extracted by a fan, creating the required test air pressure differences according to EN 12155.

5. New test equipment developed at the Faculty of Civil Engineering STU

The Department of Building Construction of the Faculty of Civil Engineering STU in Bratislava has developed a functional technological device for the implementation of the subject waterproofing test of the facade by the in-situ method. The test equipment consists of the following parts:

- spraying device with nozzles at a spacing of 400 mm with a flow rate of 2 l / min at a water pressure of 2.3 bar – Figure 4,
- water pressure regulator with filter and pressure gauge type Caleffi 5351 - 3/4”- Figure 5,
- pressure fan Stäfa Ventilator AG type MKV 12.22-008 with pressure range 0 to 1700 Pa – Figure 6,
- case with Starvert iC5 frequency converter for pressure regulation by means of precise setting of fan speed, with differential pressure gauge DPT2500-R8-AZ-D with ranges from 0-100 Pa to 0-2500 Pa and with circuit breaker and current protector – Figure 7,
- 50 m length hose diameter 3/4“ on drums for connection to a local water source, two pieces – Figure 8.

Figure 4. Forces causing water penetration through the exterior wall structure.

Figure 5. Caleffi 5351 Water Pressure Regulator Assembly with Gardena 8188-29 Flowmeter.

6. Watertightness test of the facade of the Nivy Tower office building in Bratislava

The functionality of the technological equipment for the in-situ watertightness test was verified in cooperation with the company Ingsteel, spol. s.r.o. Bratislava on the facade structures of the Nivy Tower administrative building in Bratislava. Its cladding consists of a suspension wall in the form of a modular transparent element facade in a basic floor plan module of 1350 mm and a vertical module of 3750 mm (floor height). The construction of the element facade is designed from the Schüco aluminium modular system with a thermal bridge break. The elements are designed as two-part construction with width of 2700 mm, with three levels of functional seals, with controlled drainage of rainwater and condensate in each segment of the facade blocks. The element is glazed with insulating triple glazing and anchored to the upper surface of the exterior reinforced concrete girders.

Figure 6. Stäfa AG pressure blower.

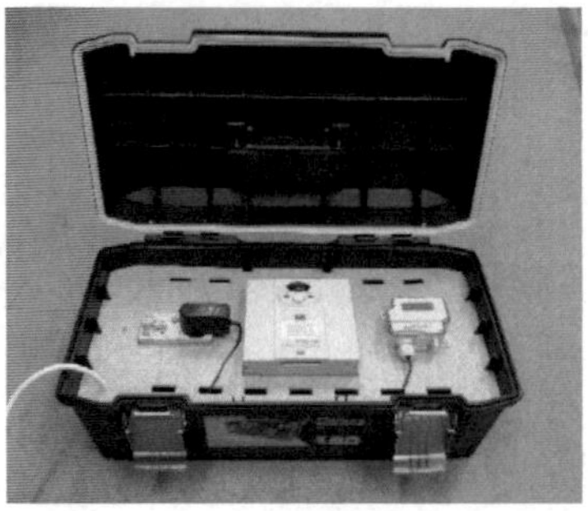

Figure 7. Measuring and control part.

Figure 8. Drums with hoses for connection to a water source.

To test the watertightness of the transparent element facade of the Nivy Tower high-rise building in Bratislava, facades of two world orientations, north and west, were selected from the 15th floor to the 17th floor. The order of watertightness tests of individual facade surfaces must proceed from the lower-lying surfaces to the higher-lying surfaces. In this case, we proceeded from the facade elements on the 15th floor gradually to the facade elements on the 17th floor.

The spraying device was placed at a distance of 250 mm from the outer surface of the light cladding to be examined in the upper level of the test area so that the axes of the nozzles are situated in a horizontal direction. In our case, the spraying device was controlled from the footbridge (nacelle) of the system for maintenance of facades, which was suspended on the rocker arm of the crane and attached to the facade at the required distance through two suction cups along its edges – Figure 9.

The test was carried out by opening the water supply and adjusting it so as to ensure a constant flow rate corresponding to 5 l/min per meter of spraying device with an accuracy of ± 10%. Each surface (in our case a facade element) was tested in a total time of 30 min. During the testing, it was checked from the interior of the facade to see if water had penetrated its inner surface. At the end of the test, the water penetration was checked for another 30 minutes. If water penetration into the inner surface of the facade is detected, all places with this defect shall be recorded.

On one test surface, it was decided to perform a watertightness test of the facade at air pressure difference in accordance with Annex B of STN EN 13 051. For this test, it was necessary to create a "pressure chamber" from the interior of the facade in the form of a wooden frame covered with transparent polyethylene foil, which is attached from the inside to the exterior frame profiles of the facade element and sealed with sealing tape or sealant – Figure 10. Then, by means of a fan controlled by a frequency converter, air is pumped out of the pressure chamber thus formed by a sequence of test pressures according to EN 12 155, thus creating a pressure difference between the interior of the facade element and the exterior corresponding to the watertightness classification. This must not be less than 0.25 times the design wind load determined according to Eurocode 1: Part 2.7.

Figure 9. View of the attachment of the spraying device to the tested transparent façade.

Figure 10. Creating a pressure difference on both sides of the test specimen by pumping air out of the created "pressure chamber".

In our case, the procedure was to perform 3 pressure shocks up to 1155 Pa with an increase from 0 to 1155 Pa in more than 1 second, while the duration of the vacuum lasts at least 3 seconds. Then the spraying device is started with a constant flow rate of 5 l of water/min per 1 m of normal joint length (in our case at an element length of 2.7 m it was 13.5 l/min) and it works for 15 minutes without vacuum. Subsequently, the vacuum is gradually increased by 150 Pa every 5 minutes according to the load diagram in Figure 11 up to a maximum value of 1050 Pa, while the possible penetration of water into the interior is monitored at each pressure stage. After completing the watertightness test of the facade, the possible penetration of water on the inner surface of the facade is monitored for another 30 minutes.

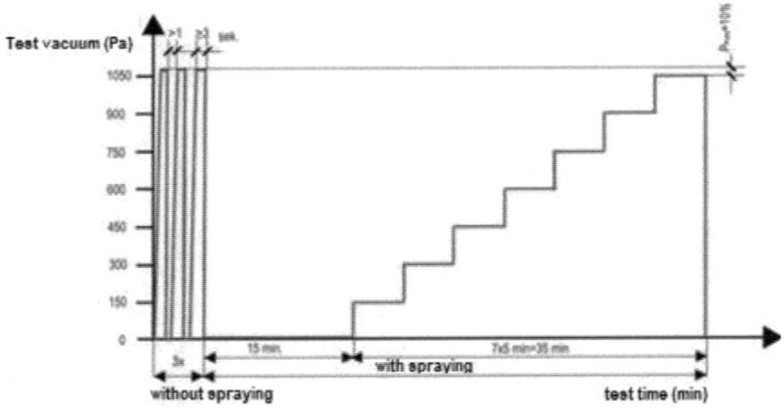

Figure 11. Load diagram for facade watertightness test at air pressure difference in accordance with Annex B STN EN 13 051.

A total of 18 facade modules were tested for watertightness. Based on the performed watertightness tests of the facade in accordance with STN EN 13 051, we can state that all tested areas of the modular transparent facade passed the test without evidence of water penetration on its inner surface.

7. Conclusion

Violation of the watertightness of the facade structures of the building and the penetration of rainwater on their inner surface represents a serious hygienic defect of the construction, which is also reflected in the life of the facade and its elements. Checking the functional condition of the facade after its implementation before handing over the premises to the tenant or new owner by in-situ watertightness test gives the investor a guarantee of quality work and elimination of unpleasant complaints during the period of operation of the building.

8. References

[1] Bielek B Szabo D and Čurpek J 2020 *Development of test equipment and in-situ experimental verification of the watertightness of the facade of the high-rise administrative building Nivy Tower in Bratislava in accordance with STN EN 13 051*

[2] STN EN 13051 *Curtain Walling - Watertightness - Site test* (2003)

[3] STN EN 12155 *Curtain walling - Watertightness - Laboratory test under static pressure* (2002)

[4] STN EN 12154 *Curtain walling – Watertightness - Performance requirements and classification* (2001)

[5] STN EN 12152 *Curtain walling - Air permeability - Performance requirements and classification* (2003)

[6] STN EN 1027 *Windows and doors - Water tightness - Test method.* (2016)

[7] STN EN 12208 *Windows and doors – Watertightness - Classification.* (2001)

Acknowledgments

This research was supported by Scientific Grant Agency MŠVVŠ SR and SAV under VEGA 1/0113/19 and by the Slovak Research and Development Agency under the contract No. APVV-16-0126.

IOP Conf. Series: Materials Science and Engineering 1252 (2022) 012005 doi:10.1088/1757-899X/1252/1/012005

Aerial work with helicopters in the construction of cableways and their impact on the concrete mixture design

A Bisták[1], Z Hulínová[1], I Vavrík[1] and P Šťastný[1]

[1] Department of Building Technology, Faculty of Civil Engineering, Slovak University of Technology in Bratislava, Radlinského 11, 810 05 Bratislava, Slovakia

E-mail: andrej.bistak@stuba.sk; zdenka.hulinova@stuba.sk; ivan.vavrik@stuba.sk; patrik.stastny@stuba.sk

Abstract. Slovakia is a rugged country, a significant part of which consists of mountain areas. Construction activities in these localities should consider specific natural conditions, while taking into account economic issues. In these cases, the efficient way of construction is the use of aerial work, so the execution of constructions using helicopters. However, compared to conventional technologies, helicopters are limited in their work by several factors, which are mainly related to their movement in a three-dimensional environment (atmosphere). The aim of this article is to show how the choice of fresh concrete mixture is related to the work of helicopters, to design a suitable mixture for the currently valid Slovak requirements and to test it in laboratory conditions. The selection of the mixture is narrowly specialized for the construction of reinforced concrete footings of cableways, which are typically realized in mountain areas. The results showed that by using a suitable superplasticizer based on polycarboxylates, it is possible to ensure the required workability of fresh concrete within 180 minutes. However, the condition must always be met that the onset of setting of the used cement does not occur earlier than 180 minutes. Longer workability time can be achieved by adding a retarding additive.

1. Introduction

Slovakia is a rugged country, a significant part of which consists of mountain areas with varying degrees of nature protection. Construction activities in these localities should take into account specific natural conditions, in particular in relation to environmental protection, while taking into account significant economic issues. Construction work in such exposed areas can be effectively carried out using helicopters. The activity of helicopters in construction is commonly referred to as the broader term *aerial work* [1]. Helicopters have an unusually wide range of uses in the construction industry, which can be divided into two main areas - the execution of assembly processes and the execution of concreting processes. Compared to concreting with conventional technologies, concreting using helicopters is characterized by a fundamental specificity, which represents a long time of emptying the mixer truck on the construction site.

The aim of this article is to show how the choice of fresh concrete mixture is related to the work of helicopters, and thus to the quality of the construction. The steps will focus on the description of the requirements for the concrete mixture during its transport by helicopter, the design of a suitable mixture for the current conditions in Slovakia and its verification in laboratory conditions. Design of the concrete mixture is for the execution of reinforced concrete footings of single-rope circulating ropeway towers (figure 1).

IOP Conf. Series: Materials Science and Engineering 1252 (2022) 012005 doi:10.1088/1757-899X/1252/1/012005

2. Methods

During the execution of concrete processes using helicopters, the concrete in the construction site is transported by means of a concrete bucket, which is suspended in the helicopter's undercarriage. Due to the fact that the load capacity of the helicopter is strictly limited and in addition to the design characteristics of the helicopter is also determined by the current weather conditions [1], a concrete bucket with a volume of max. 1 m^3 of fresh concrete (or less, often 0.75 m^3) is used. The emptying of a common truck mixer with a drum volume of 8 m^3 thus takes several hours on the construction site. If we implement the construction using this technology, concrete mixtures intended for common constructions are unsuitable and there is a need to use special concrete with a significantly extended workability time. In addition, constructions concreted by helicopters are naturally exposed to extreme weather conditions due to their location. Such an exposure also places additional requirements on the design of fresh concrete, especially in terms of the frost resistance of concrete. Due to the transport of concrete by bucket, it is also necessary to choose the right consistency, as workability affects other properties of the concrete throughout the life of the structure. However, designers often do not specify the workability of concrete, which leads to inappropriate adjustment of workability by customers - by adding water directly on the construction site [2].

Figure 1. Cableway footing with pre-assembled anchor bolts (authors).

However, despite the correct choice of concrete components and their relative proportions, the finished structure may not yet meet our expectations, especially in terms of strength and durability. Every single phase of concreting must be consistently ensured: mixing and transport of concrete, its storage, compaction, stripping and, above all, treatment [2]. These activities are not the subject of our research, but this requirement applies to every concrete structure and we take it for granted even in the case of helicopter construction technology.

The use of helicopters in construction is diverse and in the conditions of our region they are often used in the construction of cableways [3]. For example, in the years 2011 - 2012, the construction of a cableway connection on the northern and southern sides of Chopok in the Low Tatras took place. On a hard-to-reach construction site, situated on both sides of the Low Tatras ridge and just below the top of Chopok at an altitude of 2003 m above sea level, helicopters were used in concreting and assembly processes. In the years 2011 - 2014, the construction of a new cable car to the highest Czech mountain Sněžka (1603 m above sea level) was carried out [4]. The construction site was for the most part located in the 1st zone of the Krkonoše National Park. It is an area of arctic-alpine tundra, which is very sensitive to any human activities. Construction work in this area cannot be carried out with heavy construction equipment, so the only acceptable solution was to use helicopters. The concrete mixers transported fresh concrete for the second section of the new cableway route to the Lví důl locality, where it was transferred to concrete buckets with a volume of 0.7 m^3 and further transported by helicopter (figure 2). More than 500 m^3 of fresh concrete was gradually processed. The longest unloading of the mixer truck took almost 5 hours. A Mi-8T helicopter with a maximum load capacity of 2500 kg was used on the construction site, which was limited to only 2000 kg at the top of Sněžka [4].

Figure 2. Helicopter departure with a filled concrete bucket [4].

3. Theory / Calculation

When designing a suitable concrete mixture, we relied on current documentation from Leitech s.r.o., Bratislava, representative of the supplier Leitner ropeways, Sterzing (Italy), to ensure that the proposed fresh concrete will correspond to the actual concrete used in Slovakia in the execution of this kind of project. The actual design of the mixture took place in cooperation with the technology department of the Danucem Slovensko a.s., Rohožník Technical Competence Center in Bratislava. In the conditions of our region, concrete of strength class C25/30 with degrees of environmental influence XC3, XF1 or XC4, XF1 in accordance with the standard STN EN 206+A2 [5] is typically designed for the footing of the cableway tower. In some cases, there is also concrete of higher strength class C30/37. Therefore, in accordance with the standard in question, we specify the proposed concretes as follows:

- concrete STN EN 206 - C25/30 - XC3, XF1 (SK) - Cl 0.2 - Dmax16 - S3 (respectively S4), workability time 360 min,
- concrete STN EN 206 - C30/37 - XC4, XF1 (SK) - Cl 0.2 - Dmax16 - S3 (respectively S4), workability time 360 min.

In order to design the mixture as closely as possible to that used in the actual construction, we provided components for the mixing of test mixes, which are used in concrete plants in northern Slovakia. Due to the geographical distribution of sites with the potential for such construction, it is highly probable that concrete would be supplied from these concrete plants (region of Low Tatras, Fatra, Choč Mountains, etc.) After evaluating the practical possibilities of the testing laboratory, especially the availability of suitable materials and also time constraints, we proceeded to the construction of only one test facility with modified requirements, which we specify as follows:

- concrete STN EN 206 - C25/30 - XC3, XA1 (SK) - Cl 0.2 - Dmax16 - S3, max. penetration 50 mm according to STN EN 12390-8, workability time at least 180 min.

We assume that this mixture design will be a full-fledged replacement for the mixture for strength class C25/30, mentioned above, with the exception of a shorter workability time. However, this time can be easily extended. According to Danucem's experience, the recommended consistency for transporting concrete through a concrete bucket is S4 or at least S3. Consistency S5 is not recommended,

IOP Conf. Series: Materials Science and Engineering 1252 (2022) 012005 doi:10.1088/1757-899X/1252/1/012005

as concrete could segregate during transport by helicopter in the bucket, neither consistency S2, as concrete could clog the drain hole of the bucket.

3.1. Production process

The test mix of concrete with the above mixture was made in cooperation with the testing laboratory of the Danucem Technical Competence Center in Bratislava and its volume was 60 l of fresh concrete. CEM II/A-LL 42.5 N cement, mined aggregates of fractions 0/4, 4/8 and 8/16 mm from Veľké Úľany gravel pit, drinking water and Stachement 2305.1 superplasticizing additive were used. The mixture had a water content of 0.41. Subsequently, 10 standard cube-shaped test specimens with a 150 mm edge were prepared and compacted on a laboratory vibrating table. After mixing the test mix, a change in the consistency of the fresh concrete over time was also observed and recorded. Consistency was measured by a slump test according to STN EN 12350-2 [6]. All samples were stored in an air-conditioned humid environment with an air temperature of (20 ± 2) °C and a relative humidity of (95 ± 5) %. Tests of hardened concrete were also performed in the Danucem laboratory - compressive strength determined after 7 days and after 28 days from the date of production.

4. Results

All test specimens (concrete cubes) were measured, weighed and crushed perpendicular to the direction of compaction in a laboratory hydraulic press. The test results are shown in table 1 and table 2.

Table 1. Properties of fresh concrete - test mix.

Bulk density (kg/m^3)	Air content after 5 min (%)	Temperature of the concrete sample after mixing (°C)	Concrete sample temperature during production (°C)	Consistency over time slump test (mm)				
				5 min	45 min	90 min	135 min	180 min
2360	2.90	23.3	24.0	250	240	220	180	110

Table 2. Properties of hardened concrete (compressive strength) - test mix.

Sample number	Age of test sample	Compressive strength (MPa)	Average compressive strength f_{cc} (MPa)
1	7 days	45.0	44.9
2	7 days	44.7	
3	7 days	45.5	45.8
4	7 days	46.0	
5	28 days	46.2	48.6
6	28 days	50.6	
7	28 days	49.1	
8	28 days	52.2	50.9
9	28 days	50.4	
10	28 days	50.3	

The test results show that the designed concrete met, respectively exceeded the required value of compressive strength already after 7 days from production; the test specimens met the requirements for a higher strength class (C30/37) than required. The same conclusion was shown when testing samples at the age of 28 days from production. The consistency of the fresh concrete also met the requirement for consistency grade S3 within 180 minutes of mixing. It was not possible to perform watertightness and water absorption tests due to time constraints, but it is possible to supplement them additionally.

5. Discussion

Modelling of construction processes may point to the need to develop new elements, systems and technologies that will speed up and simplify construction work and reduce its impact on changing conditions during construction. This case also occurred in the case of modelling of construction processes realized by helicopters [3]. Innovative elements or systems can have a material character (for example, the use of assembly jigs in assembly processes), or they can exist at the conceptual level, which is often the case in concrete practice in the case of concrete mixtures. These mixtures can be "finalized" for the conditions of a specific construction in a specific place and time, so we proceeded to make a test mix of fresh concrete with the required parameters, especially with extended workability time, which meets the demanding conditions of construction in mountain environments.

The concreting process is one of the key factors that affect compliance with the execution deadlines, affect the final quality of the construction work and, last but not least, affect the cost side of the investment process. For this reason, its modelling has been studied by several authors, such as Fedorko and Ficzeri [7], who used computer simulation methods to analyse the transport of fresh concrete from the central concrete plant to the construction site using truck mixers. The authors proposed several possible solutions, while meeting the defined conditions would shorten the concreting process by 86 minutes in terms of simulation results. This time represents a saving of 20 % of the time compared to the actual situation.

6. Conclusions

The aim of the research was to design a fresh concrete mixture that would meet the demanding conditions of construction in a mountain environment using the work of helicopters. The obtained results show that by using a suitable superplasticizer based on polycarboxylates, it is possible to ensure the required workability of fresh concrete within 3 hours (180 minutes). However, the condition must always be met that the onset of setting of the used cement does not occur earlier than 180 minutes. Longer workability time can be achieved by modifying the mixture - by adding a retarding additive. The specific type of additive depends on several factors, especially the temperature of the concrete and the environment. It is therefore necessary to carry out verification tests for specific conditions and to consult the supplier on the type and dosage of retarder. The properties of the concrete mixtures designed in this way fully meet the requirements of primary transport by helicopter.

7. References

[1] Bisták A, Hulínová Z and Neštiak M 2021 *Slovak J. Civ. Eng.* **29** 20-26
[2] Collepardi M 2009 *Moderní beton* (Prague: IC ČKAIT)
[3] Bisták A, Hulínová Z, Neštiak M and Chamulová B 2021 *Sustainability* **13** 13671
[4] Veselý J 2013 *Beton Technol. Konstr. Sanace* **13** 26-29
[5] STN EN 206+A2 *Concrete. Specification, performance, production and conformity.* (2021)
[6] STN EN 12350-2 *Testing fresh concrete. Part 2: Slump test.* (2020)
[7] Fedorko G and Ficzeri Š 2013 *Perner´s Contacts* **8** 46-52

Acknowledgments

The authors of the paper thank Tomáš Kamas from Danucem Slovensko a.s., Rohožník and Leitech s.r.o., Bratislava for their cooperation. This paper presents a partial output of the solution of the project VEGA 1/0101/22.

Life cycle assessment of residential buildings and scenarios for prolonged life span

Z Blichova[1], S Vilcekova[2], E Kridlova Burdova[2] and D Katunský[1]

[1] Institute of Architectural Engineering, Faculty of Civil Engineering, Technical University of Košice, Vysokoškolská 4, 042 00 Košice, Slovak Republic
[2] Department of Environmental Engineering, Faculty of Civil Engineering, Technical University of Košice, Vysokoškolská 4, 042 00 Košice, Slovak Republic

E-mail: zelmira.blichova@tuke.sk; silvia.vilcekova@tuke.sk; eva.kridlova.burdova@tuke.sk; dusan.katunsky@tuke.sk

Abstract. This study deals with determination of the share of building materials impacts in whole building life cycle considering longer life span. Our aim is to investigate differences between widely used 50-year life span compared to100 and 150 years. Purpose of this study is thinking about aging residential building fond in Slovakia and their future. It is needs to be said that highest peak in building development occurred between 1955 till 1983 when prefabricated residential buildings were mostly made. These buildings take almost 38% share of all buildings finished till 2020 and youngest of them are almost 40 years old in 2022. It is needed to re-evaluate 50-year life span in all cases. Based on this situation this work focuses on residential buildings located in Slovak republic. For the assessment One Click LCA software was used, boundaries were set for cradle to grave. The functional unit was chosen 1 m^2 of the total floor area in duration of determined life span. Results are presented for modelled scenarios which investigates changes in global warming potential impact category.

1. Introduction

The building industry is a very important sector to human beings and also at the same time is responsible for significant environmental degradations [1]. According to the International Energy Agency (IEA) the construction and operation of buildings belong to major consumers of energy and material resources and also major polluters of the environment during all stages of life cycle. Share of building construction is 35% of final energy consumption in 2020 and contribute up to 38% of global annual greenhouse gas emissions. The slightly lower proportion of buildings emissions compared with the 39% seen in 2018 was due to the increases in transport and other industry emissions relative to buildings (GSR_ES, 2020) [2]. The basic vision of Environmental Strategy 2030 mentioned also in Integrated National Energy and Climate Plan for 2021 to 2030 (INECP, 2021) is to achieve a better quality of environment. A sustainable circular economy based on the consistent protection of environmental components using as few non-renewable natural resources and hazardous substances as possible, leading to improvement in population health is the significant vision too [3]. Buildings consume a significant share of global energy, mostly from non-renewable sources, and emit significant amounts of CO_2, which is not in line with the goals of sustainable development, therefore one of the main changes needed is to make buildings more energy efficient by isolating them from the environment [4]. In the past decades, life cycle assessment (LCA) has been increasingly applied to the building sector [1]. LCA is together with life cycle cost (LCC) the primary method used to assess the environmental and economic in case of LCC feasibility of building construction. An estimation of the building's life span is essential to carrying out these methods. However, given the diverse factors that affect the building's life span, it was estimated typically based on its main structural type, but different buildings have different life spans [5]. However, the validity of the LCA results is directly related to how representative is the inventory data to regional characteristics [6]. Several published studies consider the lifespan of residential buildings to be 50 years [4][6][7][8]. However, highest peak in building development in Slovak Republic occurred between 1955 till 1983 when prefabricated residential buildings were mostly made. These buildings take almost 38% share of all buildings finished till 2020 and youngest of them are almost 40 years old in 2022. It is needed to re-evaluate 50-year life span which has been considered in previous study. Based on this situation this paper is focused on residential buildings located in Slovak republic. The aim of this study is to investigate differences between widely used 50-year life span compared to 150 years because of current state of art in age of Slovak building fond by OneClick LCA software with boundaries for cradle to grave.

2. Materials and method

This work deals with a residential building consisting of 50 apartments in Slovakia, with focus on different life span scenarios. This building was previously analysed with three other residential building samples from Slovakia using eTool LCD tool. It represents construction approach in around 2020. Aim is to analyse possible benefits of building preservation for longer period.

2.1. Life cycle assessment

Life cycle assessment, or LCA, is tool which take systematic approach to quantify the potential environmental impact caused by analysed product, service or building. It considers whole life cycle from gathering the raw material for manufacturing the analysed product up to its disposal, including recycling potential. Buildings have different position among product in terms of LCA. They are very complex as a product, as it was mentioned their life span differ and they constantly require energy for the use, which can also differ in amount required.

2.2. Goal and scope

In this study are presented three scenarios for residential building in eastern Slovakia. Goal of the work is to calculate a percentual differences between total impacts of different life span scenarios and evaluate possible benefits of building preservation. For this purpose we picked residential building in eastern Slovakia and modelled four scenarios. One with 50 year life span, second with 100 year life span, third with 150 year life span. Work of Sukwon at al. [5] recognize among other studies, that building's life span strongly influences both the LCA and LCC outcomes, it is unrealistic to consider all the factors that affect it. Due to this, given life spans are simple estimations. Inspected impact is global warming potential (GWP) which is measured in CO_2 equivalent. Information was gathered from previously analysed project in eTool LCD. Data was obtained from material summary report, construction budget included in technical documentation of building. In time of conducting the analysis, building was not fully finished and energy demand for heating and hot water, is taken from calculations in technical documentation. Electricity demand is calculated to be 2400 kWh/y per apartment. This number is referential demand used for two occupants living in 3 room apartment with no electric heating or electric preheating of water. Data was obtained from electricity provider. Construction scenario is average scenario for northern Europe climate together with demolition scenario. All energy inputs are calculated for emission factors and manufacturing processes used in 2021. With this approach we evaluate worst case scenario with no change towards environmental goals. Used program is OneClick LCA and the study has been conducted in accordance with approach level(s).

2.3. Inventory analysis

Building material information and masses are obtained from construction budget. Specific supplier was not listed and was estimated as nearest possible choice. Basic characteristics are listed in **Table 1**.

Table 1. Building information.

Basic information	**Unit**
Build up area	1286 m^2
Gross floor area	5770.2 m^2
Gross internal floor area	4608.1 m^2
Terrace area/balconies	1162.1 m^2
Number of apartments	50
Number of polyfunctional spaces	2
Number of floors	7

Main construction materials are reinforced concrete foundation with piles, reinforced concrete skeleton and stairs, POROTHERM infill masonry, YTONG blocks and monolithic reinforced concrete roof. Thermal insulation is provided by contact mineral insulation, EPS100S, EPS150S. Waterproofing is secured by PVC foils. In shafts are used gypsum plaster boards. Windows are insulated triple glazed with aluminium frame and interior doors are wooden with solid core. Apartment entrance doors are security class 2 doors. Surface finishes consists of ceramic tyles in areas with wet operation, gypsum plaster, gypsum-cement plaster, and water based interior paint. Exterior finishes are concrete frost resistant tiles, silicone plaster. Also among used materials are OSB boards, asphalt strips, cement screeds, PE foils, geotextiles, steel railings, gravel and other materials in smaller quantities which were also added. Description of services used in building is described in **Table 2**.

Table 2. Services used in building.

Services used	**Type**
Water supply	Public water supply
Wastewater	Public sewerage
Tap hot water	District heating system
Heating source	District heating system and heat exchanger station
Heating device	floor heating, radiant heating panels
Electric supply	Commercial grid

3. Results and discussion

Results of the analysis are presented in Table 3. Since prediction about emission factor was not included, it represents the worst case scenario with no change made in manufacturing of goods, nor energy demand for building, nor energy production method. Module D is also included in Table 3. but it is not counted to the total impact value. Amount of impact is graphically shown in Figure 1. Scenario with 100 years life span, is 81.96% larger than scenario with 50 years life span. Energy use is the biggest contributor to GWP, as it was also shown in Table 3. In case of 100 year it doubled since no improvement is done also around one third of listed materials have calculated life span of 50 years which lowers the impact in replacement and refurbishment category in case of building with 50 years life span to which we compare to. Impact of building with 150 years life span is 164.72% grater compared to 50 years life span. Impact in replacement and refurbishment category grew more than six and a half time and emissions linked to energy use tripled. This shows that in such long period more material needed to be replaced or refurbished and this affected results. During this time replaced material can have better properties and reduce energy demand but in this scenarios, worst options are calculated with no change at all. Work of Marsh [9] states that on average, a building lifespan of 80 years reduces environmental impact by 29%, 100 years by 38%, and 120 years by 44%, in comparison to a lifespan of 50 years. Presented findings show 9.02% smaller impact per year for lifespan of 100 years and 12.92% smaller impact per year for 150 years life span compared to impact per year of 50 years life span. If we also consider D module into total, which is in Figure 2., impact decreases significantly and also difference between scenarios gets smaller. Impact per year for building with 100 year life span is 1.85% lower and impact per year for building with 150 year life span is 2.32% lower than impact per year of building with 50 year life span. Again it needs to be pointed out that these calculations don't consider any improvement done during whole life span. This difference between compared results with and without D module shows possibilities in mitigation of building impact by recycle and reuse of materials after demolition, especially in cases designed for short time period. In other words if building needs to have short life span, more attention should be given to reusable or renewable components. In case of 50 year life span D module represents 16%. CO2 emissions which could be reduced.

Table 3. Global warming potential impacts for individual phases of building life. Whole building during selected life span.

Modules		50 years kg CO_2e	100 years kg CO_2e	150 years kg CO_2e
A1-A3	Construction Materials	1370900.18	1370900.18	1370900.18
A4	Transportation to site	34596.44	34596.44	34596.44
A5	Construction/installation process	228840.81	228840.81	228840.81
B1	Use Phase			
B3	Repair	0	0	0
B4-B5	Material replacement and refurbishment	145718.08	515502.13	952005.3
B6	Energy use	6137052.55	12274105.1	18411157.64
B7	Water use	338895.78	677791.55	1016687.33
C1	Deconstruction/demolition	19618.68	19618.68	19618.68
	Waste transportation	42571.78	42571.78	42571.78
C3	Waste processing	33863.81	33863.81	33863.81
C4	Waste disposal	280.48	280.48	280.48
D	External impacts (not included in totals)	-1336036.63	-1424539.51	-1550575.31
Total		8352338.59	15198070.96	22110522.45

Impact keys: over 40% (red); 31%-40% (blue); 21%-30% (orange); 10%-20% (yellow)

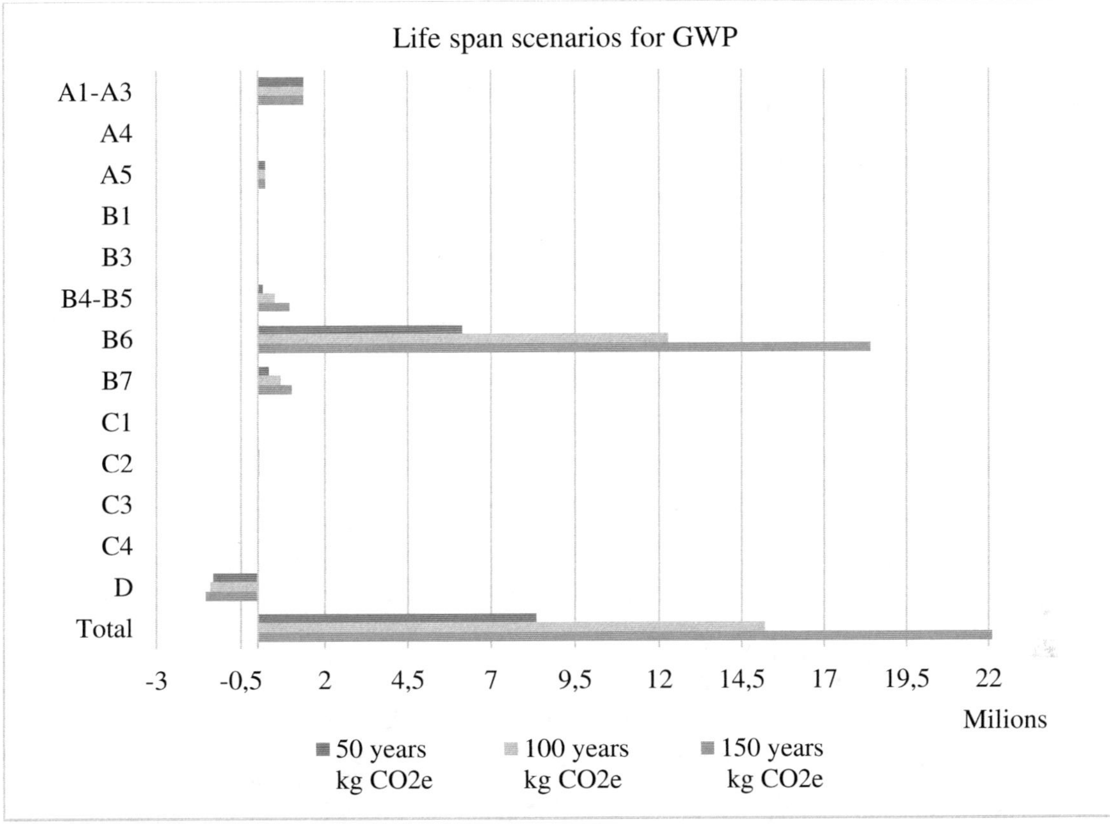

Figure 1. CO_{2e} emissions of building scenarios for each module. Total amount without D module.

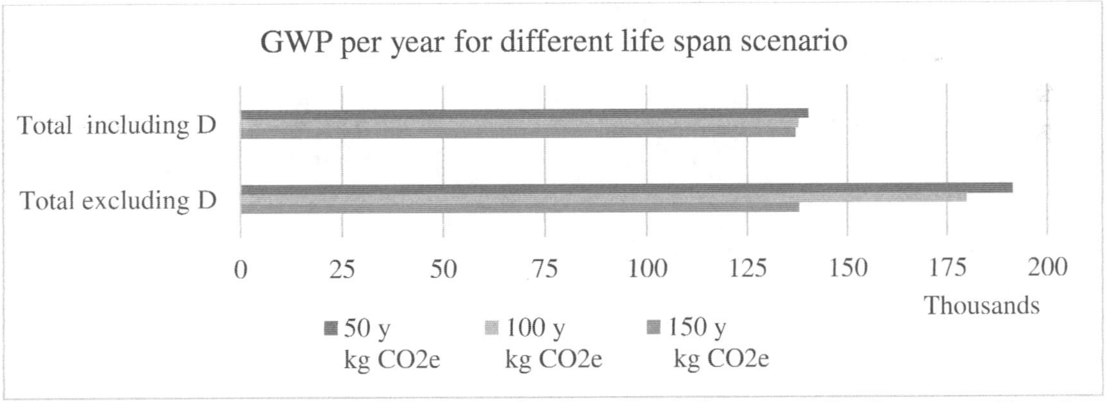

Figure 2. CO_{2e} emissions of building scenarios for each result category. Total amount with and without D module.

Chosen lengths of building life spans were simplified in these results to show possibilities linked to longer life span, together with addressing the concern, what should be done if great part of our building stock reaches estimated life span. Marteinssons study show how in real practice, the main cause of building demolition is not physical degradation itself. 44% was caused by subjective perception, then 26% by change in use and 17% caused by deterioration [10]. According to this numbers buildings are demolished before the physical end of life. Review studies also points out differences between assessed life spans. This raise the question if we should calculate life span linked to economic and social changes or give the longest estimation as possible with given effort of careful preservation. Variety of life spans should be presented as necessity to better inform about choices which can be taken.

4. Conclusion

Question of life span arose during the LCA assessment of different buildings. Specially in case of prefabricated reinforced concrete panel residential buildings, which form large part of the Slovak housing stock. First taken measures were insulations and change of openings, which protected building envelope and prevented damage caused by weather conditions and also were designed mainly to reduce energy demand. Historic city centres are showing us how long can preserved building last and it is more than 50 years. On top of that it is unimaginable to evocate so many people and let them lost a lot of value linked to the apartments. Investigation of life cycle assessment of different life span scenarios favour approach of preservation, as in case of 150 years life span, which should produce 12.92% less emissions compared to building with life span of 50 years counted per year. This brings up solid argument for developing construction solutions to improve these buildings, and prolong their life and also to improve recycling and reuse which will reduce impact of buildings built with shorter life span.

5. References

[1] Dong Y Ng S T Liu P 2021 *A comprehensive analysis towards benchmarking of life cycle assessment of buildings based on systematic review, Build Environ* **204** 108162
[2] United Nations Environment Programme, *Global Status Report for Buildings and Construction: Towards a Zero emission, Efficient and Resilient Buildings and Construction Sector. Nairobi* (2020)
[3] Slovak Ministry of Economy. *Integrated National Energy and Climate Plan for 2021 to 2030, prepared pursuant to Regulation (EU) 2018/1999 of the European Parliament and of the Council on the Governance of the Energy Union and Climate Action, Bratislava* (2019)
[4] Moradibistouni M Vale B Isaacs N 2021 *A life cycle study of insulation in a case study building with a focus on the effect of the national energy profile, J. Build. Eng.* **43** 103178
[5] Sukwon J Bumbo L Mun Y Y 2021 *Building life-span prediction for life cycle assessment and life cycle cost using machine learning: A big data approach, Build Environ* **205** 108267
[6] Morales M Moraga G Kirchhein A P Passuello A 2019 *Regionalized inventory data in LCA of public housing: A comparison between two conventional typologies in southern Brazil, J. Clean. Prod.* **238** 117869
[7] Balasbaneh A T Marsono A K B Gohari A 2019 *Sustainable materials selection based on flood damage assessment for a building using LCA and LCC, J. Clean. Prod.* **222** 844-855
[8] Ansah M K, Chen A X, Yang H Lu L Lam P T I 2020 *An integrated life cycle assessment of different façade systems for a typical residential building in Ghana, Sustain. Cities Soc* **53** 101974
[9] Marsh R 2017 *Building lifespan: effect on the environmental impact of building com-ponents in a Danish perspective, Architect. Eng. Des. Manag.* **13** 80–100
[10] Marteinsson B 2005 *Service Life Estimation in the Design of Buildings, University ofGävle*

Acknowledgments

This research was financially supported by the Grant Agency of the Slovak Republic to support projects VEGA 1/0512/20 and VEGA 1/0674/18.

IOP Conf. Series: Materials Science and Engineering 1252 (2022) 012007 doi:10.1088/1757-899X/1252/1/012007

Resistance of a sealed joint subjected to tensile stress at different temperatures

A Boháček[1] and J Šlanhof[1] and A Průcha[1]

[1]Institute of Technology, Mechanization and Construction Management, Faculty of civil Engineering, Brno University of Technology, Veveří 331/95, Brno 602 00 Czech Republic

E-mail: bohacek.a@fce.vutbr.cz

Abstract. In this paper, research on the topic of Resistance of a sealed joint subjected to tensile stress at different temperatures is presented. Tensile stress is one of the physical effects acting on conventional structures. This research specifically deals with the sealing of problematic material. Aquapanel cement board is chosen as the problematic material. This board has a large amount of fine dust particles on its surface, which prevent the sealant from adhering properly to the base material. Next, two sealants and the appropriate primers are selected. The sealants used were selected based on the manufacturer's recommendations on suitability for use in sealing cementitious materials. From these three components, the Aquapanel cement board, the sealant and the primer, test specimens are made, which are tested according to the valid European standard. The results of this test are carefully recorded and further analysed. Based on the results of this test, it is possible to recommend both tested sealants for sealing cement materials that are exposed to tensile stress at different temperatures.

1. Introduction

The issue of sealing individual materials is still a current topic in the field of construction. The main functions of the sealed joint are to allow a certain range of movement of the substrates, to prevent the entry of water, air, and other substances into the structure. It also provides thermal and sound insulation, or fire resistance. [1] [2]

Sealed joints are exposed to a large number of external influences. It is therefore necessary to choose a sealant based on what influences it should be resistant to. This information is provided by the sealant manufacturer as a recommendation for the use of the sealant. However, this recommendation does not always correspond to reality and the sealed joint is able to retain its properties and not degrade even after exposure to these influences. [3]

In this article, the author specifically deals with the sealing of problematic material, adhesion, and cohesion of the sealed joint after its exposure to tensile stress at different temperatures.

2. Material

Basic materials are selected for the needs of the research published in this article. One of them is the material from which the base plates are made. Aquapanel cement board is chosen as this material. This material is characterized by the problem of sealing due to the fact that on its surface there are fine dust particles that disrupt the sealed joint.

Other important materials are sealants, which are selected based on the manufacturer's recommendations for sealing such a problematic base material, which Aquapanel cement board undoubtedly is. They are also selected according to availability on the market and, last but not least, according to the purchase price. Two sealants are selected for this research. One representative of polyurethane sealants and one MS polymer. Appropriate primers are also selected for them, which have the task of partially eliminating the problem of sealing the selected cementitious base material. These primers or primers are chosen exclusively according to the recommendations of the manufacturer of the respective sealant [4-12].

3. Methods

The test methodology used for the research published in this article is based on the valid European standard ČSN EN ISO 8339 entitled Building structures - Sealants - Sealants - Determination of tensile properties (elongation at break). This standard precisely defines both the production of test specimens and the test procedure itself.

The test specimen consists of two Aquapanel baseboards measuring 30 x 50 mm and 12.5 mm thick. It also consists of two wooden spacers measuring 9 x 12 x 50 mm, which together with the base plates delimit the space for the application of the selected sealant. The applied sealant therefore has floor plan dimensions of 12 x 12 mm and a height of 50 mm. The material of the spacers is chosen with emphasis on a sufficient supply of air to the applied sealant so that it can mature properly.

The test specimens are manufactured at a constant temperature of (23 ± 2) ° C. Five test specimens are prepared for each test. During their production, the manufacturer's recommendation to use a primer must be followed. In addition, the three basic conditions set out in the aforementioned European standard must be met, and these are the following conditions:

a) Eliminate the formation of air bubbles;
b) Press the sealant against the contact surfaces against the base material;
c) Smooth the surface of the sealant to the plane of the base bodies and spacers.

The test specimens thus prepared are stored for 28 days at a temperature of (23 ± 2) ° C and a relative humidity of (50 ± 5)% to ensure the correct maturation of the applied sealant. After this storage, the test specimens are subjected to three storage cycles according to the prescribed procedure before testing. This procedure is as follows:

- 3 days in an oven at (70 ± 2) ° C;
- 1 day in distilled water at (23 ± 2) ° C;
- 2 days in an oven at (70 ± 2) ° C;
- 1 day in distilled water at (23 ± 2) ° C.

Before starting the test, the test pieces are left for 24 hours at a temperature of (23 ± 2) ° C and a relative humidity of (50 ± 5)%. The test is performed at (23 ± 2) ° C and (-20 ± 2) ° C.

When tested at (23 ± 2) ° C, the wooden spacers are removed, and the test specimens are placed in a test rig where they are stretched at (23 ± 2) ° C at a rate of (5.5 ± 0.7) mm / minute to violation of the test sample. A plot of elongation with applied force is recorded.

Before the test, (-20 ± 2) ° C, the test pieces shall be stored for not less than 4 hours at (-20 ± 2) ° C. The wooden spacers are removed, and the test specimens are placed in a test rig where they are stretched at (23 ± 2) ° C at a rate of (5.5 ± 0.7) mm / minute until the test specimen fails. A plot of elongation with applied force is recorded.

For each test sample, the secant modulus is calculated according to the following formula:

$$\sigma = F / s \tag{1}$$

σ secant modulus N / mm2 [N / mm2];
F force at selected compression [N];
s initial cross-sectional area of the test specimen [mm2].

4. Results

For each selected sealant, five test specimens were tested at both a positive temperature value and a negative temperature value. Table 1 shows the test results at (23 ± 2) ° C and Table 2 shows the test results at (-20 ± 2) ° C. The result tables show the force values required to break the test specimen. The breakage in millimetres and the calculated secant modulus are also given here. During the testing itself, the elongation is recorded depending on the applied force.

Table 1. Test according to ČSN EN ISO 8339 at temperature (23±2) °C.

Sealant	Test sample nr.	Force at failure [N]	Average force at failure [N]	Sekant modulus of elasticity [N/mm^2]	Stretching in case of failure [mm]
Sealant I (MS polymer)	1	153.2	189.8	0.26	1.62
	2	141.2		0.24	1.70
	3	260.7		0.43	5.02
	4	205.2		0.34	3.22
	5	188.7		0.31	2.65
Sealant II (polyurethan sealant)	1	74.1	53.0	0.12	1.07
	2	18.6		0.03	0.31
	3	10.8		0.02	0.60
	4	68.4		0.11	1.22
	5	93.4		0.16	1.12

Table 2. Test according to ČSN EN ISO 8339 at temperature (-20±2) °C.

Sealant	Test sample nr.	Force at failure [N]	Average force at failure [N]	Sekant modulus of elasticity [N/mm^2]	Stretchin g in case of failure [mm]
Sealant I (MS polymer)	1	302.6	249.2	0.50	2.47
	2	254.8		0.42	4.21
	3	240.5		0.40	2.47
	4	197.0		0.33	2.04
	5	251.2		0.42	1.94
Sealant II (polyurethan sealant)	1	122.8	92.7	0.20	1.88
	2	58.0		0.10	0.96
	3	38.0		0.06	0.61
	4	90.4		0.15	2.06
	5	154.4		0.26	2.79

5. Analysis of results

For sealant I, it is necessary to apply a force of about 200 N to break the test specimen, both for testing at a positive temperature value and at a negative one. Sealant I is so resistant and strong to tensile stress that during testing it did not break the sealed joint, but mechanically damaged the base material. The failure of the base material occurred in all tested test specimens and is shown in Figure 1.

Sealant II needs to develop a force for breaking the sealed joint significantly lower than sealant I. The value of the force used to break the test specimens, also for testing at positive and negative temperatures, is on average 53 N for testing at (23 ± 2) ° C and for testing at a temperature of (-20 ± 2) ° C an average of 93 N. In the case of sealant II, the sealed joint was broken in all test specimens by unilateral tearing of the applied sealant in the entire contact area with the base material. This violation is shown in Figure 2.

The measured values of maximum elongation at failure of test specimens, which are also given in the table of results, confirm that both selected sealants can transfer tensile stress to the extent that occurs during temperature changes in real use of sealed joints such as building facades.

Figure 1. Sealant I, Sample nr. 3, t = (23±2) °C.

Figure 2. Sealant II, Sample nr. 5, t = (-20±2) °C

6. Conclusion

Based on the results of the test performed at a positive and negative temperature value, it can be said that both selected sealants fulfilled their purpose and passed the performed tensile test in which the test specimens are exposed to a positive and negative temperature value. In conclusion, it can be said that based on the performed tensile test, both selected sealants can be recommended for sealing the selected cementitious base material.

7. References

[1] Petrie E M *Handbook of adhesives and sealants*. New York: McGraw-Hill company, 2009, ISBN 0-07-049888-1.
[2] Klosowski J and Wolf A T *Sealants in construction*. 2nd edition. Boca Raton: Taylor and Francis Group, 2016, ISBN 978-1-4200-1785-4.
[3] Lal Mittal K and Pizzi A *Handbook of sealant technology*. Boca Raton: Taylor and Francis Group, 2009, ISBN 978-0-8493-9162-0.
[4] Chew M Z L and Zhou X and Tan S T, On-site weathering of sealants under tropical conditions, Construction and Building Materials, Vol. 18, Issue 4, 2004, pp. 287-293.
[5] Stögbauer H and Wolf A T The influence of heat ageing on one-part construction silicone sealants, Construction and Buildings Materials, Vol. 5, Issue 1, 1991, pp. 27-32.
[6] Wolf A T *Movement capability of sealants. Construction and Building Materials*, 1991, 5, pp.127-134. DOI: 10.1016/0950-0618(91)90063-Q. ISSN 09500618.
[7] ČSN EN ISO 8339 Building construction - Jointing products - Sealants - Determination of tensile properties (elongation at break) Prague: Czech Standards Institute, 2006
[8] Šlanhof J and Liška P and Nečasová B Evaluation of test methods for testing of sealants, Advances and Trends in Engineering Sciences and Technologies II - Proceedings of the 2nd International Conference on Engineering Sciences and Technologies, ESaT 2016, 2017, pp. 643-648, ISBN: 978-131539382-7.
[9] Nečasová B and Liška P and Šlanhof J "*Performance of selected polyurethane joint sealants in concrete structures*," MATEC Web of Conferences, vol. 146, 02015, 2018.
[10] Liška P and Nečasová B and Šlanhof J and Šimáčková M Determination of Tensile Properties of Selected Building Sealants in Combination with High-pressure Compact Laminate (HPL), In: Procedia Engineering.,2015, pp. 199-205, ISSN 18777058.
[11] Information on https://cze.sika.com
[12] Information on https://www.mapei.com

Acknowledgment
The contribution was made with the support of junior specific research of the Faculty of Civil Engineering of the Brno University of Technology conducted under the designation FAST-J-22-7872 and support of standard specific research of the Faculty of Civil Engineering of the Brno University of Technology conducted under the designation FAST-S-22-7865.

IOP Conf. Series: Materials Science and Engineering 1252 (2022) 012008 doi:10.1088/1757-899X/1252/1/012008

Comparison between flood damages evaluation methods and selection of the appropriate method for application in Slovakia

M Hlinková[1], P Bujanský[1] and H Abd-Elhamid[2,3]

[1] Department of Environmental Engineering, Faculty of Civil Engineering, Technical University of Kosice, 040 01 Košice, Slovakia
[2] Department of Water and Water Structures Engineering, Faculty of Engineering, Zagazig University, Zagazig, 44519, Egypt
[3] Center for Research and Innovation in Construction Faculty of Civil Engineering, Technical University of Kosice, 040 01 Košice, Slovakia

E-mail: maria.hlinkova@tuke.sk; peter.bujansky@tuke.sk; hany_farhat2003@yahoo.com

Abstract. Flood damages come immediately with flood events. There is a wide range of models and methods that can be used for its evaluation, which bring satisfactory but not exact results. The rate of flood damages estimation can differ significantly from the real damages measured after the real flood. The current paper presents three methodologies for assessing flood damages including INSYDE model, AGRIDE-c model and RASOR platform. These methodologies are compared and their suitability for application to available data is analysed. The available data includes the number of damaged elements due to flood as well as the extent of the flooded area. Results of comparison shows that the INSYDE model and RASOR platform could be possibly modified and applied to the available data, but there are significant changes and further analysis needed. The analysis of selected methods shows the data that are needed must be first obtained from other source, then modified, and then could be applied. But we do not have the same data, so the further studies need significant changes.

1. Introduction

Each country facing the risks of emergencies comes up with its own techniques for assessing the risks and damage caused by floods. The assessment process then enters data that are left to be important in the country and that contribute to the possibilities of assessing and predicting flood damages. The estimation of flood losses can be made immediately after the flood event, and later the regime situation can be compared with the real consequences. As one of the effective methods of estimating flood damage is DEM with FLEMOps+ model [1]. Specific views on the economic impacts that can be seen in the results of evaluation method in the city of Florence. The indicators of vulnerability include the works of art and the cultural heritage of the city. The results of the study offered a financial assessment of flood damages as well as a proposal for mitigation measures [2]. When proposing measures to mitigate the adverse results that are needed, they are not included in the implementation, but also with the information that is more appropriate for the affected areas. Differences in the application of measures are visible in areas that are more contractual than others in floods [3-7].

The topic of flood damages and flood risks is a recent and not yet exhausted topic. Each flood event brings with it a different degree of flooding, it always occurs under different hydrological and technical conditions. Due to the wide area of damages, there is still room for dissemination of knowledge in the field of floods. The aim of the submitted contribution is to compare proposed methodologies with required data to data we have available obtained from Reports on the course and consequences of floods in the Slovak Republic. According to the data from Reports, there will be a new flood damage evaluation intended. Data are collected post-event and describes mostly quantification of exposed and damaged elements.

2. Description of the compared methodologies

The current study provides a brief overview to selected case study with developed methods or models to determine economic values of flood damages. In the following sections, a comparison between three models is presented in addition to Slovak conditions to which the models can be modified.

2.1. INSYDE model

Proposed probabilistic methodology offers an opportunity to determine economic values of the residential physical building damages by analysing components of damages. The method considers the available knowledge on damage mechanism. According to the authors of the proposed methodology, the INSYDE model (In-depth Synthetic Model for Flood Damage Estimation) is clear and modified to actual knowledges of relevant hazard and vulnerability variables. The calculation of the total economic damages in this model is mathematically expressed as the sum of the elements of common costs for cleaning and removal of wastes, damages to non-structural and structural, cost of damages to finishing works and damaged windows and doors. The mathematical expression of damages is later converted into economic value through the determined market value of the damaged elements. The following formula can check the mathematical expression [8]:

$$D = \sum_{i=1}^{n} C_i \tag{1}$$

where:
D is the total economic damages for each building,
C_i includes the mentioned costs. The complete list of costs is presented in Table 1:

Table 1. List of costs [8].

Item	Description
Clean-up	pumping, waste disposal, cleaning, dehumidification
Removal	Screed, pavement, skirting, partition walls, plasterboard, external plaster, internal plaster, doors, windows, boiler
Non-structural	partition replacements, screed replacements, plasterboard replacement
Structural	soil consolidation, local repair, pillar repair
Finishing	external plaster replacement, internal plaster replacement, external painting, internal painting, pavement replacement, skirting replacement
Windows and doors	door replacements, window replacements
Building system	boiler replacement, radiator painting, heating replacement, electrical system replacement, plumbing system replacement

According to Table 1, the scale of cost is considering expressing total for economic damages for each residential building *(D)*. The geometric properties of the building are used to estimate the damages to buildings. Unit prices are at least per unit of measure, and the total cost estimate is determined as the replacement of damaged components per unit of measure. The cost of each component expresses an extension to the unit price. Hence, the costs *C* can be express as follows [8]:

$$C_{ij} = \text{unit price}_{ij} \times \text{extension}_{ij} \tag{2}$$

This approach can be applied to different elements in different countries or geographic areas. On the other hand, the extension can be replaced by any different relevant value according to the country of applicable. The method was developed in Italian conditions, and the results showed good estimates of post-event flood losses compared with other damage models [8].

2.2. AGRIDE-c model

The AGRIDE-c model is a conceptual model that assesses the damages itself to floods on agricultural crops. The type of crop is considered the crown is planted in the monitored area and its vegetative and generative development. The assessment should also consider the climatic conditions of the region where the assessment is carried out. The type of vulnerability is associated with the type of agricultural crop and in the extreme case, agricultural land use. The economic evaluation involves determining the physical security of the crops, which are converted into monetary value in the next step. An important factor in the economic evaluation is the phase in which the crop is at the time of the flood. The assessment does not consider damages to machinery and equipment, but only damages to planted crops [9]. In the model, there are only short-term effects on the soil that occur during floods. Based on the evidence, we can say that most of the land and crops damage occur. Damage to other components is not taken into account. Then, absolute damage D can be obtained as follows [9]:

$$D = (\Delta GO) - (\Delta PC) \tag{3}$$

where:
D is absolute damage,
ΔGO is gross output,
ΔPC is the increase or decrease in production.
Variables included in the calculation and the structure of the AGRIDE-c model is shown in Figure 1:

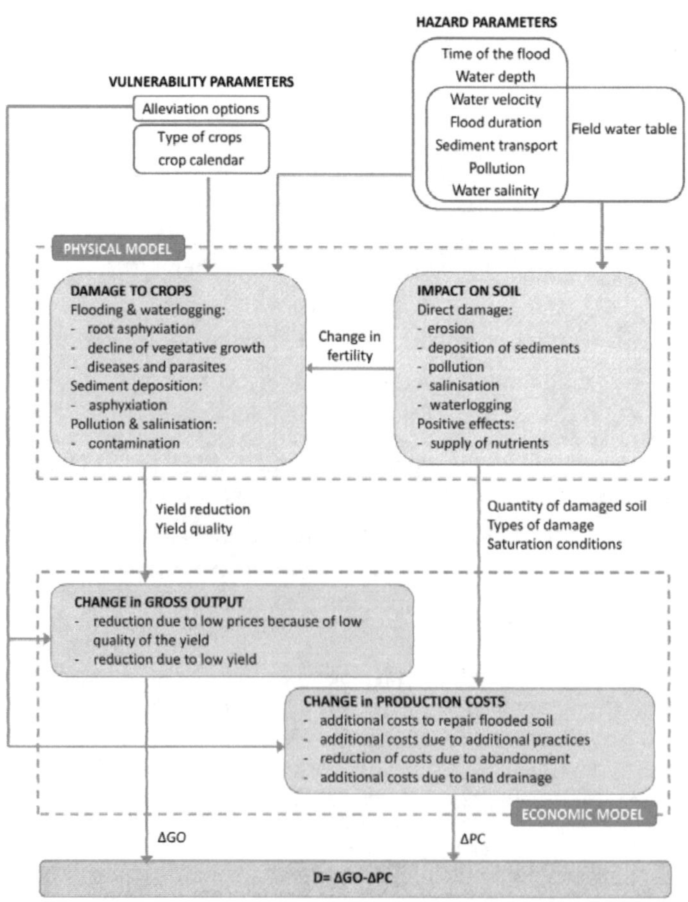

Figure 1. Conceptual model of AGRIDE-c [9].

AGRIDE-c combines a physical and an economic model to evaluate the absolute damage. The physical model is identified in yellow boxes, and it evaluates physical damages to crops and impact of soil. The model for the assessment of the physical damage to crops calculates the reduction in the amount and quality of the harvest due to the flood. The green boxes evaluate economic model, the first model evaluates reduction ΔGO (gross outputs) and the second model assess ΔPC of the increase or decrease in production costs in comparison with the no-flood scenario. The presented methodology is in the design stage, its extension to other agricultural aspects is being considered [9].

2.3. The RASOR platform

The RASOR platform enables extensive characterization of features and their vulnerabilities through security level curves. The first step is to identify the risk and natural hazard, vulnerability, and exposure. The hydrological and climatic characteristics of the studied area enter the assessment process. The exposure elements that belong to the curves are eligible and the costs of repairing the damage after the flood [10].

Exposure analysis aims to identifying at building scale the vertical distribution of the unit use. Exposure is intended as the ensemble of distinctive parameters which allows for properly assigning damage curves and recovery or replacement costs. Considering the parameters of building exposed to floods (number of floors, the surface area, the building use, the presence of cellar). Among variables is taken into account also use or type of the buildings. By the application of the Point in Polygon vector tool was in the study area obtained using characterization of buildings. For the risk analysis were identified 14 categories of buildings: residential, commercial, commercial/residential, hospital, school, industrial, place of worship, offices, sport, parking, transport, agriculture, theaters and leisure activities, camping and temporary lodging. Economic values were used to estimate the replacement and recovery costs for structures and contents [10].

The RASOR allows update exposure data and to make scenario-based predictions to support long- and short-term risk related decisions. In the case study of Florence, the mitigation measures were assessed, and four flood scenarios were considered. As the results represents, damages to structures were estimated according to flood scenarios inundated areas. The results of the case study shown estimation of theoretical investment related to flood mitigation measures and exposed damaged buildings [10]. According to the authors, the methodology should be adaptable and adaptable to any Meta context when geographic exposure data are available. The scale of aid on buildings can also track changes in the use of buildings, but also changes in market values [10].

2.4. Slovak conditions

An important element in the development of a new methodology for flood damages assessment is to simplify the approach for evaluator as much as possible. The proposed methodology for the evaluation of flood damages is based on simple access to data from the reports on the course and consequences of floods in the Slovak Republic for the period addressed. The unit measures of the available data are number of affected subjects, how many components were affected by flood. There are 5 categories of subjects, and each of them is divided into some sub-categories. The main categories including; flooded residential buildings, flooded non-residential buildings, damaged engineering networks, affected animals, other flooded damage. Each of the mentioned subjects contains sub-categories.

The data are available online, it can be found on the website of Ministry of the Environment of Slovak Republic. These data can be applied to any of the compared methods to develop new flood damages evaluation methodology with relevant results and practical use. All the data in the mentioned reports are related to self-districts in Slovakia. For the first draft of the methodology there will be used the division to self-districts, so the first results will be obtained on the large scale. Later, the scale can be modified and narrowed to evaluate and assess flood risk and flood damage on detailed scale with detailed results. We can consider the range of flooded area too. There are available three types of flooded area in the reports, urban area, agriculture area and forest area. All are three types are related to self-district, too.

IOP Conf. Series: Materials Science and Engineering 1252 (2022) 012008 doi:10.1088/1757-899X/1252/1/012008

3. Results and discussion

As mentioned above, there are available only range of flooded areas and exposure elements to flood obtained from reports on the course and consequences of floods in the Slovak Republic. So, a comparison of the methods that will be most suitable for our conditions can be done.

3.1. INSYDE model

INSYDE model allows determining the total economic damages for each building by costs of each component. The unit price of the work performed in the damaged range is considered. Table 2 summarize the available data from reports and requested data from INSYDE model.

Table 2. Available and requested data.

	Available data		Requested data
Quantity	Flooded residential buildings Flooded non-residential buildings Damaged engineering networks Affected animals Other flooded damage	Costs	Clean-up Removal Non-structural Structural Finishing Windows and doors Building system

INSYDE model is based on the quantification of the costs required to carry out the cleaning work of the elements affected by the flood. The model approach requires quantification of the damages of each element and the unit price for the damaged elements replacements. The model can be modified to our conditions provided that the input parameters change. Our available data (Table 2) contains sub-categories, which describes quantity of smaller elements, for example flooded residential buildings contains description about family houses, flat houses, and other houses. But it is only quantification, how many buildings were affected. The costs can be theoretically modified according to JKSO [11], which is document primarily designed to classification of constructions for statistical purposes, for the purposes of international comparison of statistical indicators and for the classification of types of constructions.

The JKSO classifies buildings and engineering networks according to the purpose of their main use. Subjects of the classification are buildings and engineering networks. Building types are more detailed into sections, groups and classes. Buildings are classified according to the purpose of their main use. The main use of the building is determined by the percentage of the purpose of use the usable area of the building [11]. According to the classification of buildings, the list price of the building type can be determined. Several indicators are included in the price calculation, there are several organizations on the market that offer price valuation of selected buildings. Finally, we can say, that the INSYDE model can be modified to determine costs for damaged buildings and engineering networks.

3.2. AGRIDE-c model

Reports on the course and consequences of floods in the Slovak Republic offers review of area exposed to flood. The flooded area is classification as an urban, forestry and agricultural area. The description offers extend of flooded in hectares and it is related to 8 self-regions in Slovak republic. Table 3 represents post-event flood data in the time period from January to the end of June 2020. The AGRIDE-C model is based on specify analysis of crops and their life and vegetation cycles. Given the availability of data, we can immediately state that the methodology cannot (yet) be applied to our input data. The authors of AGRIDE-C model consider the vegetation and life cycle of crops, which have different costs. In our available data, we can consider only extent of flooded agriculture area. Because of the lack of data, proposed AGRIDE-c model we cannot modified to our study.

Table 3. Extent of flooded area in Slovak Republic from January to the end of June 2020.

Self-region	Extent of flooded area [ha]			SUM [ha]
	Urban area	Agriculture area	Forrest area	
Bratislava	-	-	-	0.00
Trnava	19.00	15.00	-	34.00
Trenčín	20.50	204.50	-	225.00
Nitra	5.06	14.55	-	19.61
Žilina	2.00	0.50	-	2.50
Banská Bystrica	1.65	1.05	0.80	3.50
Prešov	5.20	0.40	-	5.60
Košice	60.30	-	-	60.30
SUM	**113.71**	**236.00**	**0.80**	**350.51**

3.3. The RASOR platform

To use RASTOR platform, the most important we need to know are the depth of the flood and types of damaged buildings. In our data documents, we have available data about monthly total precipitation. However, more detailed data can be obtained from Slovak hydro-meteorological institute (SHMÚ). The flood reports from SHMÚ are described in more detail and contain several necessary hydrological as well as meteorological data. The data in flood reports contain information on the culmination in individual river basins in each calendar month. In the proposed RASOR platform, we can replace the depth of the flood by the culmination of the water flow in the river basin. On the other hand, we can determine the type and number of damaged buildings. If we did a more detailed survey, we could also get data on the number of floors in damaged buildings. However, our available culmination data describe water levels that have occurred in specific river basins. This means that flood values would be the same for each type of building in each river basin, distinguishing only the numbers of water levels in specific river basins and building types. The offered methodology could be applied to our conditions under certain conditions, but it is questionable what detailed results it would bring.

4. Conclusion

This study presented a comparison of three selected methodologies to evaluate flood damages in different scales. The INSYDE model abets an opportunity to determine economic values of the residential physical building damages by analysing components of damages. The AGRIDE-c model is a conceptual model that assesses the flood damage on specific agricultural crops. The RASOR is used to determine damage to buildings. An important indicator is the depth of the flood, the number of floors of the exposed building, the use of the building, whether the damaged building has a basement. These three selected methods were compared with the data available for Slovak Republic. The analysis of the AGRIDE-c model data shown, we do not have available as such specific data about crops as it is needed. The further analysis shows, the INSYDE model and RASOR platform can be modified to Slovak conditions, but they are need significant changes. Because of the lack of data, or because of the variation of the data, we don´t dispose such as specify/exact data as needed. Further studies are needed to confirm this assumption.

5. References

[1] Apel H, Aronica G T, Kreibich H, Thieken A H 200 *Flood risk analysis-how detailed do we need to be?, Nat Hazards* **49** 79-98
[2] Arrighi C, Brugioni M, Castelli F, Franceschini S, Mazzanti B *Flood risk assessment in art cities: the exemplary case of Florence (Italy), Flood Risk Management*
[3] Poussin J K, Wouter Botzen W J, Aerts J C J H 2015 *Effectiveness of flood damage mitigation measures: Empirical evidence from French flood disasters, Global Environmental Change* **31** 74-84
[4] Zelenakova, M Preliminary flood risk assessment in the Hornad watershed. In: *5th International Conference on River Basin Management*, 2009, pp.15-24

[5] Nagy, P et al Identification of dry and wet years in eastern Slovakia using indices. In: *Advances in Environmental Engineering* (AEE 2019), 2020, 444: 012041
[6] Kubiak-Wojcicka, K et al Identification of Extreme Weather Events Using Meteorological and Hydrological Indicators in the Laborec River Catchment, Slovakia. *Water*, 2021, **10,** 1413
[7] Zelenakova, M et al Flood occurrence evaluation based on the heights of water level in the rivers. In: *3rd International Conference Water Resources and Wetlands*, 2016, pp 153-157
[8] Dottori F, Figueiredo R, Martina M L V, Molinari D and Scorzini A R 2016 *INSYDE: a synthetic, probabilistic flood damage model based on explicit cost analysis, Nat. Hazards Earth Syst. Sci* **16** 2577-2591
[9] Molinari D, Scorzini A R, Gallazzi A, Ballio F 2019 *AGRIDE-c, a conceptual model for the estimation of flood damage to crops: development and implementation, Nat. Hazards Earth Syst. Sci* **19** 2565-2582
[10] Ariighi C, Rossi L, Trasforini E, Rudari R, Ferraris L, Brugioni M, Franceschini S, Castelli F 2018 *Quantification of flood risk mitigation benefits: A building-scale damage assessment through the RASOR platform, Journal of Environmental Management* **207** 92-104
[11] JKSO, Decree no. 323/2010 Z. z. Decree of the Statistical Office of the Slovak Republic, which issues the Statistical Classification of Buildings

Acknowledgments

This work was supported by the Slovak Research and Development Agency under the Contract no. APVV-20-0281. This work was supported by project of the Ministry of Education of the Slovak Republic VEGA 1/0308/20 Mitigation of hydrological hazards, floods and droughts by exploring extreme hydroclimatic phenomena in river basins.
The authors are grateful to support of the projects HUSKROUA/1702/6.1/0072 and HUSKROUA/1901/8.1/0088.

Assessment and adaptation to climate change and sea level rise impacts on Egypt's northern coasts

M Zelenakova[1], H Abd-Elhamid[2,3], K Barańczuk[4] and J Barańczuk[4]

[1] Department of Environmental Engineering, Faculty of Civil Engineering, Technical University of Kosice, 040 01 Košice, Slovakia
[2] Department of Water and Water Structures Engineering, Faculty of Engineering, Zagazig University, Zagazig, 44519, Egypt
[3] Center for Research and Innovation in Construction, Faculty of Civil Engineering, Technical University of Kosice, 040 01 Košice, Slovakia
[4] Miejskie Laboratorium Badań Środowiskowych, Wydział Oceanografii i Geografii, Uniwersytet Gdański

E-mail: martina.zelenakova@tuke.sk; hany_farhat2003@yahoo.com; katarzyna.baranczuk@ug.edu.pl; jacek.baranczuk@ug.edu.pl

Abstract. Climate change and sea level rise (SLR) represent many challenges for coastal areas such as coastal erosion and saltwater intrusion (SWI). This study aims to highlight the effect of climate change and SLR on Egypt's northern coasts and present appropriate adaptations. A number of techniques for adapting with climate change and SLR have been presented. Two numerical models have been used for simulation PLAXIS and 2D-FEST. The results showed that SLR will affect a number of cities on Egypt's northern coasts and the Nile Delta aquifer will be affected by SWI. A number of scenarios have been applied to adapt with climate change and SLR. Precast Diaphragm Wall (PDW) is used to protect Egypt's northern coasts from SLR. The results showed that PDW is applicable and cost efficient. The cost of building PDW along the north coasts is about 1.0% of the likely losses due to SLR in 2100. However, to control SWI into the Nile Delta aquifer a number of measures have been presented and discussed. The results revealed that using treated wastewater to increase the recharge to the Nile Delta aquifer combined with decreasing the abstraction from the aquifer could help in protecting the aquifer from probable SLR.

1. Introduction

Climate change is accelerating as a result of human activities and natural processes. Climate change could have negative consequences if it continues. Weather patterns could shift, resulting in flooding, drought, erosion, wetlands loss and seawater intrusion into fresh groundwater aquifers. Global warming is regarded main cause of climate change, with the potential to melt enough polar ice to raise the sea levels. Global mean sea level rise was projected (10-20 cm) over the last century (IPCC, 1996) [1]. The rate of future sea level rise (SLR) is expected to outpace that of the past. SLR is expected to be (20 - 88 cm) by 2100 (IPCC, 2001) [2]. Long-term effects of SLR on coastal areas include; increase the coastal erosion, submergence of shore cities, saltwater intrusion (SWI), loss of agricultural land and increasing the coastal water table [3].

The Mediterranean region is more vulnerable to climate change. These areas are suffering from water scarcity, and its availability is decreasing in several countries as a result of population growth, which has increased abstraction and contamination of groundwater. Climate change could have direct impact on water resource, salinization of coastal aquifers, soil degradation, and agricultural production at the Mediterranean basin. Climate change impacts on coastal cities should be addressed with appropriate adaptation measures.

Several studies have been conducted to protect coastlines from SLR (Baric et al., 2008; Fenger et al., 2008; Taormina and Chau, 2015) [4,5,6]. In addition, some studies in Egypt carried out to protect the coasts from SLR (Fanos et al., 1995; El-Raey et al., 1999; Heikal et al., 2012) [7,8,9]. A number of researchers have also investigated SWI in the Nile Delta aquifer while taking into account the SLR (e.g. Abd-Elhamid et al., 2016, 2019, 2011, Javadi et al., 2012) [10,11,12,13]. A few studies have been conducted to protect Egypt's coasts from the anticipated SLR. The current study aims to assess the impact of climate change and SLR on Egypt's northern coasts and present appropriate adaptation measures to protect coastal areas from expected SLR. In the simulation of groundwater flow and SWI into coastal aquifers, the PLAXIS and 2D-FEST models are used. Various techniques are also applied to protect Egypt's northern coasts from climate change and SLR.

2. Methodology

This research includes two stages. The PLAXIS software is used in the first stage to test the efficacy of using Precast Diaphragm Walls (PDW) to protect coastal cities on Egypt's northern coasts from SLR. Controlling the coastal erosion and shoreline inundation could be accomplished through a variety of methods. Offshore breakwaters, perched beaches, groins, revetments, dikes, floodwalls, seawalls, bulkheads, and dams are examples of hard structures. Artificial beach nourishment, dune building, and marsh building are examples of soft structures. Miscellaneous include structure elevation and strengthening of the existing structure. PLXIS (2-D) is used in the analysis of using PDW to study seepage around diaphragm wall and carry out the necessary analysis for the diaphragm wall design; more information on PLAXIS software can be found in (Abd-Elhamid et al. (2016) [14].

2D-FEST model was used in the second stage to investigate SWI in the Nile Delta aquifer. Climate change impact and SLR on SWI was studied and various SWI control scenarios were used to mitigate SWI. The differential equations for transient fluid flow and solute transport in saturated/unsaturated soils are given in terms of two primary variables; pore-water pressure (u_w) and solute concentration (c). The detailed information on differential equations and numerical solution of coupled fluid flow and solute transport can be found in (Abd-Elhamid, 2010) [3]. Water flow and solute transport differential equations are written as [3]:

Water flow equation:

$$C_{ww}\frac{\partial u_w}{\partial t}+C_{wT}\frac{\partial T}{\partial t}+C_{wa}\frac{\partial u_a}{\partial t}=\nabla[K_{ww}\nabla u_w]+\nabla[K_{wT}\nabla T]+\nabla[K_{wa}\nabla u_a]+\rho_w\nabla(K_w\nabla z) \qquad (1)$$

Solute transport equation:

$$\begin{aligned}
\frac{\partial((\theta_w+H\theta_a)c)}{\partial t} &= -\left[\left(\frac{\partial}{\partial x}(v_{w_x}c)+\frac{\partial}{\partial y}(v_{w_y}c)\right)+\left(\frac{\partial}{\partial x}(v_{a_x}c)+\frac{\partial}{\partial y}(v_{a_y}c)\right)\right] \\
&+\left[\left\{\frac{\partial}{\partial x}\left(D_{w_{xx}}\frac{\partial}{\partial x}(\theta_w c)+D_{w_{xy}}\frac{\partial}{\partial y}(\theta_w c)\right)+\frac{\partial}{\partial y}\left(D_{w_{yy}}\frac{\partial}{\partial y}(\theta_w c)+D_{w_{yx}}\frac{\partial}{\partial x}(\theta_w c)\right)\right\}\right. \\
&+H\left.\left\{\frac{\partial}{\partial x}\left(D_{a_{xx}}\frac{\partial}{\partial x}(\theta_a c)+D_{a_{xy}}\frac{\partial c}{\partial y}(\theta_a c)\right)+\frac{\partial}{\partial y}\left(D_{a_{yy}}\frac{\partial}{\partial y}(\theta_a c)+D_{a_{yx}}\frac{\partial}{\partial x}(\theta_a c)\right)\right\}\right] \\
&-\tfrac{\partial}{\partial t}(\rho_b K_d c)-(\lambda_w\theta_w+H\lambda_a\theta_a)(1+\rho_b K_d/\theta_w)c=0
\end{aligned} \qquad (2)$$

3. Climate change and SLR impacts on the Egypt's northern coasts

This section of the study discusses the impact of climate change and SLR on coastal cities as well as coastal aquifers (Nile Delta aquifer).

3.1 Impact of climate change and SLR on Egypt's coastal cities

SLR will result in the loss of many coastal areas and increases SWI into coastal aquifers. It will also raise the water table and reduce the agricultural productivity. Because a large portion of Egypt's Mediterranean coastline is below sea level, it will be affected by sea level rise. Egypt's important cities, such as Alexandria, Marina, Marsa Matrouh, Rosetta, and Port Said, are located in or near these areas (Figure 1a). Also, the Mediterranean Sea water intruded into 100 Km in the Nile Delta aquifer (Figure 1b).

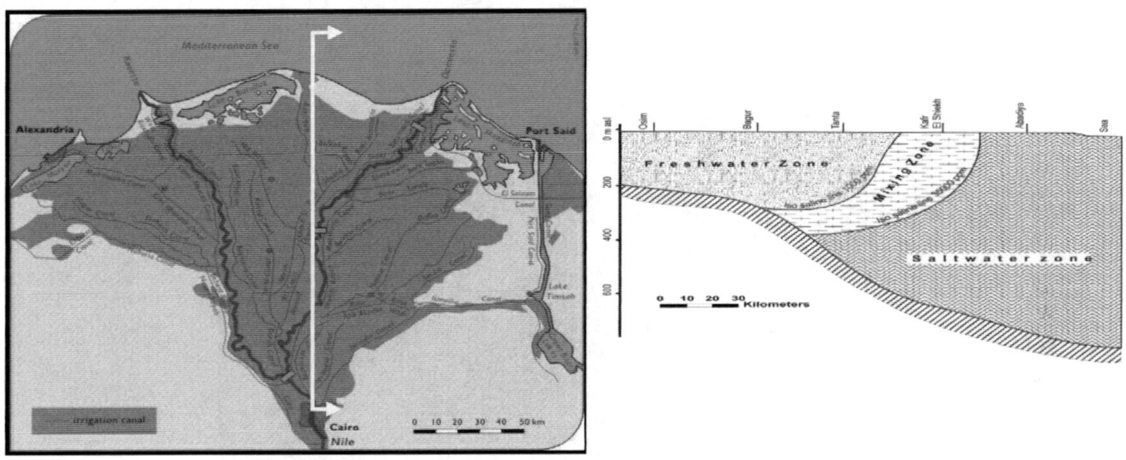

(a) Location map (b) Vertical cross section

Figure 1. The Nile Delta location map and vertical cross section.

A number of studies discussed SLR impacts on these areas of Egypt (e.g. El-Raey et al., 1995-1996-1999a,b; Frihy, 2003-2009) [15,8,16,17,18,19]. These studies predicted that if not suitable actions are taken, Egypt would face the following consequences: loss of beaches, tourism, agricultural and fishing land. Over the current century, up to 15% of the delta could be inundated that extend to 20 km inland. Also, SWI could lead to contamination of fresh water, soil salinity and loss of land productivity. Egypt is among the top countries predictable to be affected by a 1.0 m SLR (Dasgupta et al., 2007) [20]. Satellite maps of the Nile delta clarifying the impact of 0.5 and 1.0 m SLR is shown in Figure 2 (Simonett, 2002) [21]. Some studies were conducted to assess the impact of SLR on Egyptian coasts. The Nile delta is the areas of high vulnerability, as well as potential socioeconomic consequences, have been broadly defined. Alexandria, Beheira, Port Said, and Damietta are among the high-risk areas. El-Raey, (2010) [22] presented SLR impact on environmental and socioeconomic aspects of these areas. These cities are vulnerable to SLR and they will face significant challenges such as shoreline erosion, fisheries stress and SWI. Previous studies of these cities used various techniques and different scenarios of SLR 0.5 and 1.0 m to appraisal the probable losses (Elray, 1995-1999a,b,c) [15,8,16,17]. The losses at some of these cities in 2100 are expected as following: Alexandria ($3.5 billion) and Rosetta ($2.9 billion), [15,8,16,17].

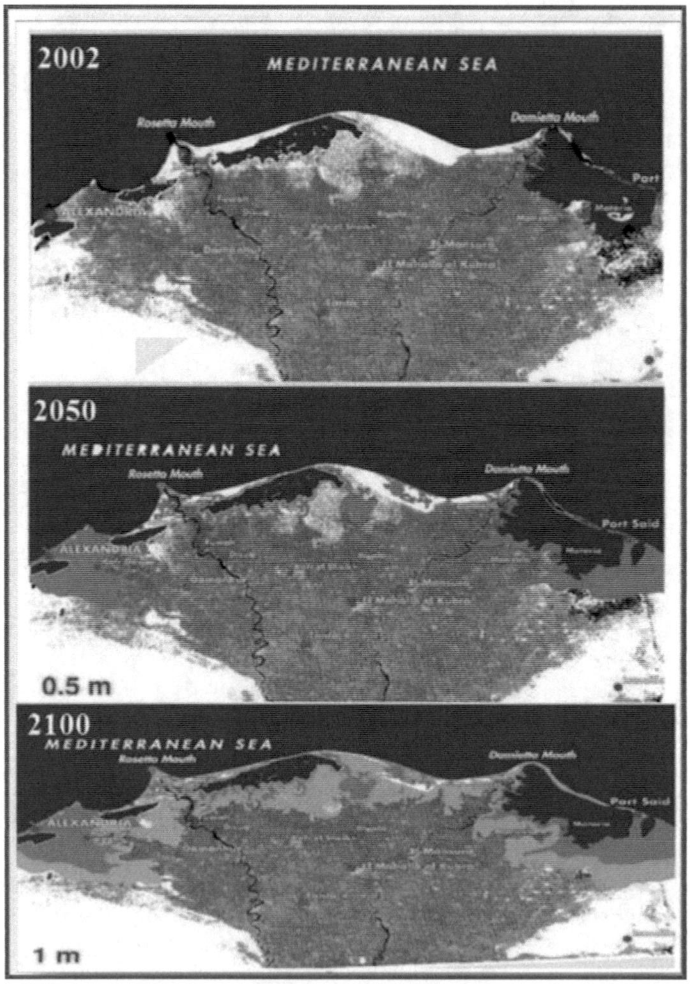

Figure 2. The impact of sea level rise (0.5 and 1m) on the Nile Delta (Simonett, 2002) [21].

3.2 Impact of climate change and SLR on SWI into the Nile Delta aquifer

Seawater intrusion threatens several coastal aquifers in the world. The Nile Delta aquifer is one of the largest freshwater reservoirs in the world. Previous research found that SWI intruded to more than 100 Km in the Nile Delta aquifer. 2D-FEST model is applied to examine SWI into the Nile Delta aquifer. Figure 1a shows a vertical section through the mid of Nile Delta. The length of the section is 150 km, depths are 250 m on landside and 700 m on seaside. The equiconcentration line 1.0 moved in the aquifer to 112 km from the seashore (Figure 3a). Two scenarios are applied to investigate SLR impact on SWI in the Nile Delta aquifer (50 and 100 cm). Figures 3b and c show the results of the two scenarios. In scenario one when sea level is raised by 50 cm, equiconcentration line 1.0 advanced 4.5 km and reached 116.5 km (Figure 3b). In the second scenario, when sea level raised by 100 cm, equiconcentration line 1.0 advanced 8.25 km and reached 120.25 km (Figure 3c).

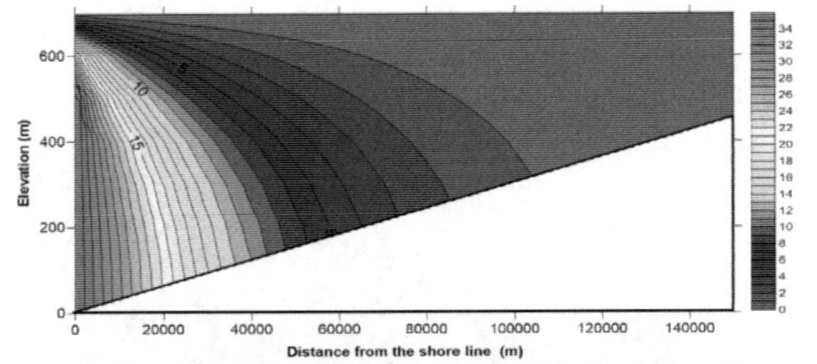

(a) current situation of SWI in the Nile Delta aquifer

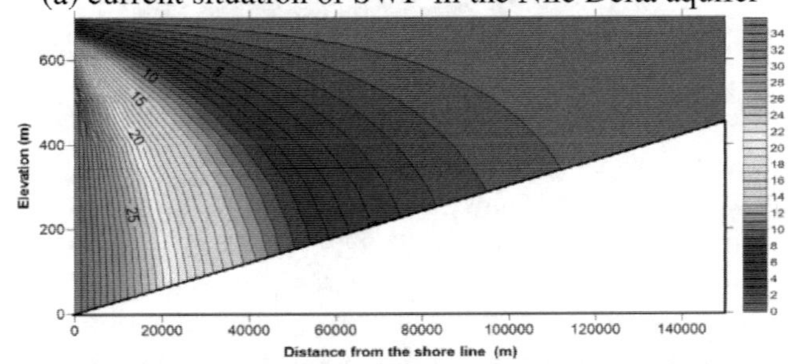

(b) SWI in the Nile Delta aquifer at SLR = 50 cm

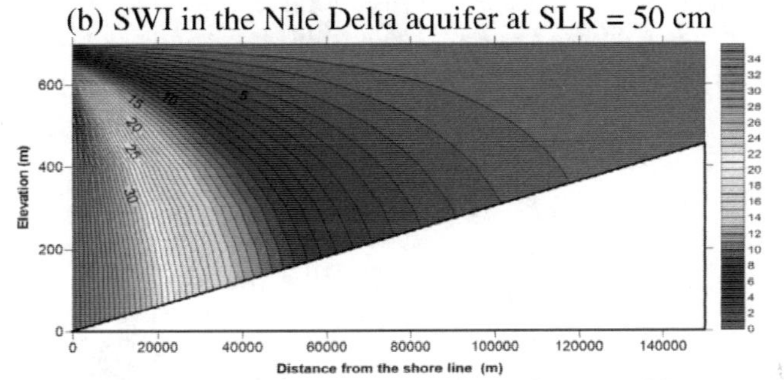

(c) SWI in the Nile Delta aquifer at SLR = 100 cm

Figure 3. SWI in Nile Delta aquifer due to SLR.

The results indicated that SLR tremendously increased SWI in the Nile Delta aquifer especially if the second scenario is considered (SLR=100 cm) that increased SWI into 120.25 km into the aquifer which will lead to damage a large quantity of fresh water. The aquifer should be managed to protect the available freshwater using appropriate measures.

4. Adaptation to climate change and SLR effects on Egypt's northern coasts

This part of the study presents the adaptation techniques that could be applied to protect both coastal shore and coastal aquifers in the Egypt's northern coasts from climate change and SLR.

4.1 Coastal shore protection from SLR

Adaptation begins with identifying climate change impacts followed by evaluating the available options for dealing with the effects through changes in policies and/or building protection structures. Several adaptation management choices, including beach reinforcement, nourishment, construction of seawalls and breakwaters, change in land use, and development of comprehensive monitoring and decision support systems, can be used to protect shoreline from SLR [14]. Egyptian has started building coastal protection structures such as jetties, groins, seawalls, and breakwaters to face beach erosion. Fanos et al. (1995) [7] and El-Raey et al. (1999b) [16] provided examples of protection methods used on Egypt's northern coasts. Figure 4 shows sea barriers for combating piracy at Damietta for combating SLR (Russmerritt, 2008) [23].

Figure 4. Sea barriers at Damietta for combating SLR. (Russmerritt, 2008).

PDW is used in this study to protect coastal shores from SLR. PDW will provide stability, permanence, and water retention. As shown in Figure 5, PDW height is 10 m, 2.0 m above sea level and 8.0 m fixed in the soil. The efficiency of using PDW in stopping saltwater seepage is evaluated using PLAXIS model. A numerical study was carried out to investigate the effect of using PDW on head losses in leaking water produced by 1.0 m SLR. PDW thickness is 30 cm for a length of 10 m, with adequate reinforcement. The results show that PDW is effective at preventing seawater seepage beneath the ground's surface and has increased percolation length to compensate to 1.0 m water level difference produced by SLR. More information about the PDW design can be found in (Abd-Elhamid et al., 2016) [14].

(a) PDW

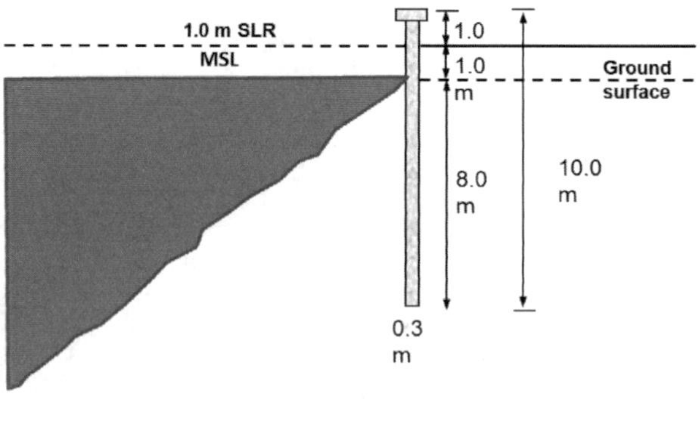

(b) Cross section for PDW

Figure 5. PDW and designed cross section.

The cost of making 1.0 m of PDW is estimated to be $1000. This cost include investigations, drilling, supports, walls, water stop, and finishing. This means each kilometre of PDW costs $1.0 million. The expected losses for Alexandria in 2100 is $3.5 billion (if no action is taken) [16]. According to this study, building PDW along Alexandria coasts $35.0 million that represent only 1.0% of the likely losses. In addition, the cost of constructing PDW along North Coasts (1000 Km) length is $1.0 billion. However, building a PDW with a length of 1000 km takes time, but it can be done on stages depend on priorities. The cost of constructing PDW with 1000 km length may be high, but it represents only 1% of the expected losses if nothing is done.

4.2 Nile Delta aquifer protection from SLR

2D-FEST model is applied to control the SWI into the aquifer using various scenarios such as decreasing abstraction, increasing recharge and combination of the two. To avoid the increase in SWI due to SLR, the pumping rate should be reduced (50%). When this is applied to the model, SWI has retarded by 4.5 km toward the sea (Figure 6a). Another scenario is used in which the aquifer's recharge has increased (50%). This resulted in a 5.0 km retardation in SWI (Figure 6b). However, combining the two scenarios resulted in a greater retardation of SWI to 8.5 km (Figure 6c). The results indicated that to adapt with the SLR 1.0 m that resulted in increasing SWI into 120.25 km, the abstraction should be reduced by 50% and recharge should be increased by 50%. This scenario could be able to mitigate the impact of climate change and SLR.

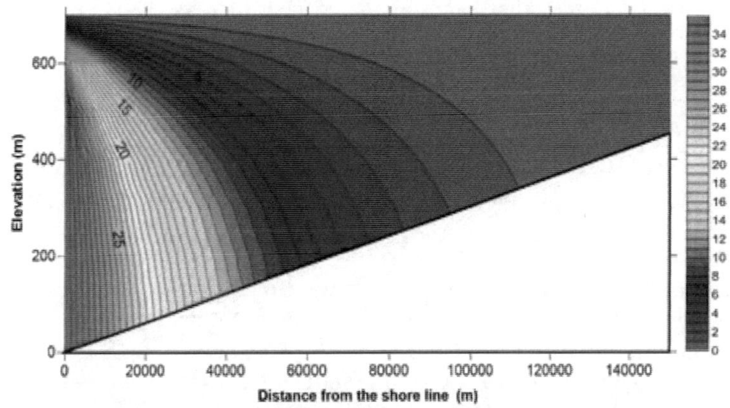

(a) Decreasing abstraction from the aquifer (50%)

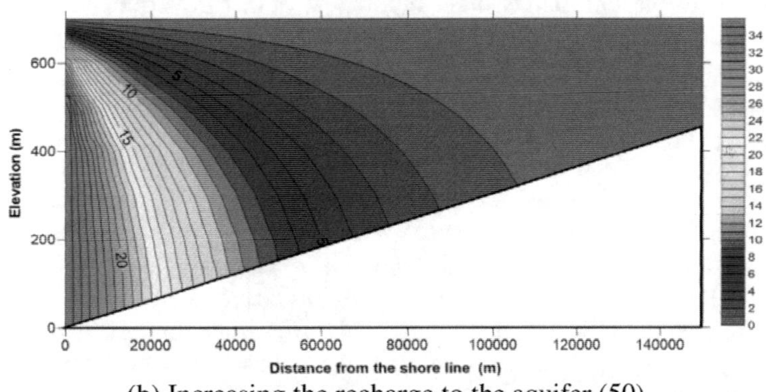

(b) Increasing the recharge to the aquifer (50)

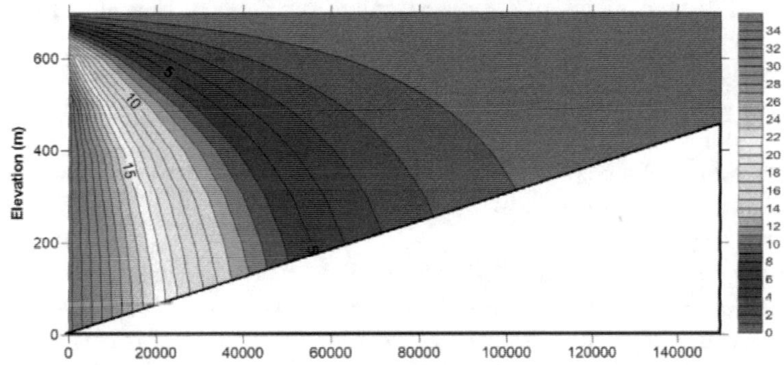

(c) Combination of decreasing abstraction and increasing the recharge to the aquifer

Figure 6. Mitigation of SWI in Nile Delta aquifer.

5. Conclusion

Climate change and SLR are expected to have significant effect on Egypt's coastal areas. According to previous research, sea levels will rise by (20 - 88 cm) at 2100. If nothing is done, this rise could have far-reaching consequences for Egypt's northern coasts in the long run. This paper examined impacts of climate change and SLR on coastal cities as well as the Nile Delta aquifer. Protection measures for Egypt's northern coasts have also been proposed. A detailed study of using PDW to protect Egypt's coasts from SLR is presented. The cost is evaluated against the economic losses that would result due to SLR if no act is taken. According to the results, the cost of constructing one kilometre of PDW is $1.0 million. Alexandria city could be protected with a 35 kilometre coast length at a cost $35.0 million. However, if nothing is done, the predictable losses at 2100 are around $3.5 billion. Using PDW along Egypt's northern coast of length 1000 km will cost $1.0 billion, which represent 1% of the economic losses caused by SLR. On the other hand, SLR of 50 cm increased SWI by 4.5 km and SLR of 100 cm increased SWI by 8.25 km. Different scenarios have been applied to protect the Nile Delta aquifer SWI increase due SLR. Three scenarios have been applied including; reducing abstraction from the aquifer (50%), decreasing recharge to the aquifer (50%) and combination of the two. Decreasing abstraction from the aquifer resulted in retarding SWI toward the sea by 4.5 km. Increasing the recharge to the aquifer resulted in retarding SWI 5.0 km. However, combination of the two scenarios could result in more retardation to SWI to (8.5 km), which could mitigate the likely impact of SLR.

6. References

[1] IPCC 1996 Climate Change: The science of climate change: contribution of working group I to the second assessment report of the Intergovernmental Panel on Climate Change, Houghton, J. T., Meira Fihlo, L. G., Callander, B. A., Harris, N., Kattenberg, A. and Maskell, K. Cambridge University Press, Cambridge, UK and New York, NY, USA.

[2] IPCC Climate Change 2001 Impacts, Adaptations, and Vulnerability: contribution of working group II to the third assessment report of the Intergovernmental Panel on Climate Change, McCarthy, J. J., Canziani, O. F., Leary, N. A., Dokken, D. J. and White, K. S. Cambridge University Press, Cambridge, UK and New York, NY, USA.

[3] Abd-Elhamid H 2010 A Simulation-optimization model to study the control of seawater intrusion in oastal aquifers. *PhD Thesis, University of Exeter, UK.*

[4[Baric A, Grbec B and Bogner D 2008 Potential Implications of Sea-Level Rise for Croatia. *Journal of Coastal Research* 24 299–305.

[5] Fenger J, Buch E, Jakobsen P R and Vestergaard P 2008 Danish Attitudes and Reactions to the Threat of Sea-Level Rise. *Journal of Coastal Research* 24 394–402.

[6] Taormina R and Chau K 2015 Neural Network River Forecasting with Multi-Objective Fully Informed Particle Swarm optimization. *Journal of Hydroinformatics* 17 99-113.

[7] Fanos A M, Khafagy A A and Dean R G 1995 Protective works on the Nile Delta. *Journal of Coastal Research* 11 516-528.

[8] El-Raey M 1999a Impact of Climate Change on Egypt. Environmental Software and Services. *http://www.ess.co.at/GAIA/CASES/EGY/impact.html.*

[9] Heikal E M, Koraim A S and Abozaid A A 2012 Protecting coasts from probable sea level rise by using porous vertical seawall. *Egyptian International Journal of Engineering Sciences and Technology* 15 1233-1243.

[10] Abd-Elhamid H F, Javadi A, Abdelaty I and Sherif M 2016 Simulation of seawater intrusion in the Nile Delta aquifer under the conditions of climate change, *Hydrology Research*, 47 (6): 1198-1210.

[11] Abd-Elhamid H F, Abd-Elaty I M and Sherif M M 2019 Evaluation of potential impact of Grand Ethiopian Renaissance Dam on Seawater Intrusion in the Nile Delta Aquifer", *International Journal of Environmental Science and Technology (IJEST),* 16(5): 2321-2332

[12] Abd-Elhamid H F and Javadi A A 2011 A Cost-effective Method to Control Seawater Intrusion in Coastal Aquifers, *Journal of Water Resources Management*, 25: 2755-2780.

[13] Javadi A A, Abd-Elhamid H F and Farmani R 2012 A Simulation-Optimization Model to Control Seawater Intrusion in Coastal Aquifers Using Abstraction/Recharge Wells, *International Journal for Numerical and Analytical Methods in Geomechanics*, 36 (16):1757-1779

[14] Abd-Elhamid H F, El-Kilany M E and Javadi A A (2016) A cost-effective method to protect the coastal regions from sea level rise A case study: Northern coasts of Egypt *Journal of Water and Climate Change* 7(1): 114-127.

[15] El Raey M, Nasr S, Frihy O, Desouk S and Dowidar Kh 1995 Potential impacts of accelerated sea-level rise on Alexandria Governorate, Egypt *J. Coastal Research* 51 190–204.

[16] El Raey M, Dewidar Kh and El Hattab M 1999b Adaptation to the impact of sea level rise in Egypt. *J. Climate Research* 12 117–128.

[17] El-Raey M, Frihy O, Nasr S M and Dewidar K H 1999c Vulnerability assessment of sea level rise over port said governorate, Egypt. *Environmental Monitoring and Assessment* 56 113–128.

[18] Frihy, O.E. 2003 The Nile Delta - Alexandria coast: vulnerability to sea-level rise, consequences and adaptation. Mitigation and Adaptation Strategies for Global Change 8 115–138.

[19] Frihy O F, Deabes M, Shereet S M and Abdalla F A 2010 Alexandria-Nile Delta coast, Egypt: update and future projection of relative sea-level rise. *Environ. Earth Sciences* 61 253–273.

[20] Dasgupta S, Laplante B, Meisner C and Yan J 2007 The impact of Sea Level Rise on Developing Countries: A Comparative Study. *World Bank Policy Research Working Paper 4136, February 2007, World Bank, New York.*

[21] Simonett Otto (2002) Nile Delta: Potential Impact of Sea Level Rise", UNEP/GRID-Arendal, *http://maps.grida.no/go/graphic/nile_delta_potential_impact_of_sea_level_rise*

[22] El-Raey M 2010 Impacts and Implications of Climate Change for the Coastal Zones of Egypt. *Coastal Zones and Climate Change* 31- 50.

[23] Russmerritt (2008) Egypt, Yahoo company, *http://www.flickr.com/ photos/29835954@N04/2788615751/ (accessed July 11, 2009).*

Acknowledgments

This work was supported by the Slovak Research and Development Agency under the Contract no. APVV-20-0281, and a project funded by the Ministry of Education of the Slovak Republic VEGA1/0308/20 "Mitigation of hydrological hazards, floods, and droughts by exploring extreme hydroclimatic phenomena in river basins".

This work has received support from the SCORE (Smart Control of the Climate Resilience in European Coastal Cities) project, funded by the European Union's Horizon 2020 research and innovation programme under grant agreement No 101003534.

IOP Conf. Series: Materials Science and Engineering 1252 (2022) 012010 doi:10.1088/1757-899X/1252/1/012010

Recycling cement bypass dust and granulated blast furnace slag into hydraulic road binder

N Stevulova[1], J Junak[1], J Strigac[2] and N Junakova[1]

[1] Institute of Environmental Engineering, Faculty of Civil Engineering, Technical University of Kosice, Vysokoskolska 4, 042 00 Kosice, Slovakia
[2] Povazska Cement Factory Ltd., Janka Krala 1, 018 63 Ladce, Slovakia

E-mail: nadezda.stevulova@tuke.sk; jozef.junak@tuke.sk; strigac.j@pcla.sk; natalia.junakova@tuke.sk

Abstract. This paper is aimed at the possibility of recycling the cement bypass dust (BPD) and ground granulated blast furnace slag (GGBS) into a normal hardening hydraulic road binder in accordance with standard EN 13282-2. 10 wt.% addition of BPD as potentially hazardous waste formed during the clinker production by burning alternative fuels was used into cement mixtures. Binder mixtures' composition was designed according to the recommended sheet of Slovak commercial product (DoroCem). The main constituents of the experimental hydraulic road binder mixture were Portland cement clinker, ground limestone and GGBS. Effect of additional component of BPD on behaviour of fresh mixtures and properties of mortar specimens after 56 days of hardening was studied. The results prove that only one of the tested mixtures (BM2) with the lowest SO_3 content and clinker but with the highest amounts of GGBS, ground limestone and alkalis (Na_2O + K_2O) met the standard requirement for initial setting time for tested fresh mixtures. The higher values of compressive strength found for all hardened specimens are most likely related to the formation of the hydrated and geopolymer products in the binder system. Based on these conclusions, BM2 could be used as hydraulic road binder.

1. Introduction

The cement-manufacturing sector is one of significant contributor to the global environmental impacts related to anthropogenic carbon dioxide emissions [1]. The present strategy in the cement clinker production is based on increasing replacement of traditional fossil fuels with clearly more cost-effective alternative fuels. Bypass technology with different configurations is installed to separate dust particles from the flue gases leaving the production of Portland clinker. However, fine dust particles are a complex mixture of inorganic substances, often containing alkaline compounds, sulphates, and chlorides originated from alternative fuels used in the cement kiln atmosphere [2]. Different bypass positions lead to two kinds of bypass dust. While cement kiln bypass dust (CBPD) is collected from the bypass only, BPD comes from the flue gas leaving the preheater tower at lower temperatures. Therefore, these two bypass dust products are characterized by different chemical composition. While CBPD is rich in calcium, BPD has high content of alkalis, chlorides, and sulphates and may also contain heavy metals [3]. CBPD waste can be recycled and replace the calcium component in raw mixture, and thus contribute to sustainable production of cementitious clinker phases [4]. The mostly decarbonized form of the Ca component in CBPD can also lead to CO_2 emissions reduction. Unlike CBPD, BPD with high alkali content cannot be utilized in clinker manufacturing as a result of exceeding the cement clinker standard limits. The total volume of BPD generated in the Slovak Cement Factories is about 6% of the total annual production of clinker. The existence of BPD and its disposal in hazardous waste landfills represent a significant environmental problem. Therefore, interest of researchers is in finding of appropriate recycling applications and high-value products. As literature sources indicate, considerable efforts for utilization of BPD waste are in construction sector [5, 6] for development of sustainable building materials and commercially available.

In this article, the potential of using BPD as additional component in making normal hardening hydraulic road binder is investigated. The key properties of fresh cement pastes and hardened specimens were tested according to the standard [7].

2. Materials and Methods

2.1. Constituents and composition of hydraulic road binder mixture

The main constituents of three experimental hydraulic road binder mixtures (BMs) were Portland cement clinker, ground limestone, and GGBS. BPD was used as an additional constituent ensuring alkaline activation of GGBS. Chemical composition of BPD is represented by high content of alkaline oxides $K_2O + Na_2O$ (48.27 %; of which 5.54 % belongs to Na_2O), CaO (7.05 %), chlorides (33.69 %), and sulfates expressed as SO_3 (11.13%). The presence of small amounts (below 1%) of other oxides such as SiO_2, MgO, Al_2O_3, P_2O_5, and Fe_2O_3 in BPD was also confirmed. Loss on ignition of BPD was 0.52 %. As mineral phases in BPD were determined: sylvite (KCl), arcantite (K_2SO_4), halite (NaCl), free lime (CaO), quartz (SiO_2), and anhydrite ($CaSO_4$). Mineral phases present in cement clinker were C_3S, C_2S, C_3A, C_4AF, lime, periclase, quartz, arcanite, and portlandite. Limestone contained calcite, dolomite, and quarz. Complex aluminosilicates such as melilite and merwinite were identified in GGBS.

BPD as key alkaline binder component was used in a constant amount (10 wt.%. of the total weight of the mixture) in all binder mixtures in accordance with the standard [7]. The contents of binder components in BMs were designed according to the range of hydraulic binder constituents recommended by the DoroCem product data sheet [8] (Table 1). The fineness of experimental binder mixtures and SO_3 content in them are in line with the standard requirements, as seen in Table 2. Contents of the important oxides in BMs are in Table 3.

Table 1. Designed binder mixtures with variation of constituents (10 wt.% BPD).

Mixture	Clinker (wt.%)	Limestone (wt.%)	GGBS (wt.%)
BM1	66	17	7
BM2	63	17.5	9.5
BM3	67.3	16.2	6.5
Reference	100	-	-
DoroCem	58.5-71.5	16.2-17.8	6.5-9.5

2.2. Preparing the fresh binder mixtures and hardened mortar specimens

The preparation of the reference mix and experimental mixes BM1-BM3 was carried out by mixing of the dry binder components with deionized water in a standard mixer (MATEST, Treviolo, Italy) at low speed for 120 s according to the standard [9]. A higher amount of water due to the high fineness of BPD was needed to prepare BMs in accordance with standard [10]. Therefore, the standard water coefficient (w/c = 0.5) had to be optimized to w/c = 0.23 to improve process ability of BMs. After stirring (90 s) of mixture, interruption and removal of any cement paste adhering to the walls of the mixer and stirrer, the mixing process was performed at a higher speed for 120 s.

Table 2. Proportion of +90 μm particles and SO_3 content in experimental binder mixtures with 10wt.% BPD.

Mixture	Proportion +90 μm particles (wt.%)	SO_3 (wt.%)
BM1	10.46	0.372
BM2	10.38	0.363
BM3	10.56	0.380
Standard	≤15	< 4

Table 3. CaO/SiO_2 ratio and contents of important oxides (wt.%) in designed binder mixtures BM1-BM3.

Mixture	CaO/ SiO_2	Al_2O_3	Na_2O	K_2O
BM1	3.07	4.17	0.75	5.79
BM2	2.99	3.09	0.87	5.90
BM3	3.12	4.14	0.76	5.81

Fresh mix was put in a steel mold for three bodies with dimensions of 40 x 40 x 160 mm, and a laboratory vibrating table (MATEST, Treviolo, Italy) was used for compaction of the mortar specimens under conditions of 3,000 vibrations per minute. Vibration of pastes lasted until no air bubbles formed on their surface and the specimen surface was relatively smooth. After vibration, storage of vibrated pastes in a humid environment (covered with polyethylene terephthalate (PET) foil) followed. The mortar bodies were demolded after 48 h and placed in an aqueous medium until compressive strength test.

Mixing and hardening processes were carried out in laboratory conditions (temperature of 23°C; relative humidity of 55–65%).

2.3. Methods

2.3.1. Testing of the fresh pastes

Initial setting time and consistency as the main parameters of fresh binder mixtures were determined. The initial setting time for fresh experimental pastes was determined by the penetration method using a Vicat apparatus (ELE International, Sheffield, UK) according to the standard [9]. Although the standard for a normal hardening hydraulic road binder does not require the consistency determination, this parameter was tested for all binder pastes using a flow table test in accordance with [11] for freshly mixed mortars including minerals binders.

2.3.2. Testing of the hardened specimens

The bulk density of the test specimens was evaluated according to the standard [12]. The compressive strength of experimental specimens based on alternative binder constituents after 56 days of hardening was determined according to the standard [13] (ADR ELE 2000, International Ltd., UK). The final value of this strength parameter represents the average of six tests with an error deviation of less than 10%.

3. Results and Discussion

3.1. Properties of fresh pastes

The initial setting time for the experimental fresh BMs was determined as a function of the depth of needle penetration. According to the standard [7], the distance between the cylindrical steel needle and the pad has to be 6 ± 3 mm in setting time ≥ 150 min. Value of 120 min for initial setting time was found for the reference paste. This value is related to the cement class 32.5 specified in [14]. The composition of experimental BM1 and BM3 led to a shortening of their initial setting times (130 min and 120 min, respectively) required by the standard for a normal hardening hydraulic road binder [7]. Only BM2 resulted in achieving the value required by the standard (150 min).

In terms of consistency of all experimental BMs, values of flow diameter ranged from 146 mm to 164.5 mm compared to the reference sample (150 mm). These values place them in the class of plastic mortars. The highest flow diameter was obtained for BM3 sample. It should be noted, differences in flow diameter between BM1 and BM2 did not exceed 2 mm.

3.2. Properties of hardened cement mortar
The values of bulk density and compressive strength of reference and experimental specimens of BM1-BM3 after 56 days of hardening are presented in Table 4. The results indicate that experimental binder mixtures exhibited bulk density values in the range of 2100–2160 kgm^{-3}.

The compressive strength after 56 days of hardening is a key factor in meeting standard requirements for the mechanical properties of a normal hardening hydraulic binder. It is mentioned, that all experimental specimens have higher value of compressive strength by 14 to 23% after 56 days of hardening than reference sample. The BM3 sample achieved the highest value of compressive strength, where two of the main binder constituents in the mixture (limestone, and GGBS) were at the minimum level of the recommended range for the DoroCem product. The compressive strength for class N 4 given by the standard [7] is in the range of 32.5 to 52.5 MPa. As shown in Table 4, the measured values of strength parameter for experimental specimens BM1-BM3 after 56 days of hardening (58.9 to 64.9 MPa) were 13–22 higher than the upper limit of the range for class N4. The highest increase in compressive strength (BM3 sample) compared with the reference specimen corresponds to the highest CaO/SiO_2 ratio (3.11 wt.%) due to the highest clinker representation in binder mixture (BM3).

Table 4. Bulk density and compressive strength of 56 days hardened specimens compared to reference sample.

Specimen	Bulk density (kgm^{-3})	Compressive strength (MPa)
BM1	2138 ± 15	58.9 ± 3
BM2	2120 ± 8	60.9 ± 5
BM3	2163 ± 10	64.9 ± 5
Reference	2090 ± 12	51.9 ± 4

The higher compressive strengths found can be attributed to the formed products during the hydration reactions taking place in BMs between their components with water molecules. Also formation of geopolymeric compounds through alkaline reaction of GGBS with BPD is expected. The hydration process involving a complex of chemical reactions leads to the changes in the properties of the binder system, and mainly in the setting and hardening conditions. This process is sensitive to many factors of a physical and chemical nature. One of the critical factors in achieving the final strength of the binder in the allowable time is its particle size distribution. The smaller particle size results in a higher surface to volume ratio, with a larger area per unit volume for the interaction of water molecules with the clinker particles and the alternative binder components. Therefore, finer materials react faster with water molecules, resulting in a higher rate of strength development.

On the other hand, the chemical composition of the binder mixture also plays an important role in the formation of various hydrated phases. Clinker composition is of particular importance, and the presence of free CaO in the binder mixtures is very important for the formation of the main hydration products C-S-H and C-S-A-H. We assume that the strength of the specimens was significantly affected by the relatively high content of free CaO capable of hydration on portlandite, and its subsequent pozzolanic reaction with the active components SiO_2 and Al_2O_3 present in GGBS. The CaO presence in GGBS tends to form a disordered Ca/Mg aluminosilicate phase which is responsible for the reactivity of GGBS during alkaline activation by sodium and potassium oxides present in BPD. Bypass dust is considered to be a potent alkaline activator with high pH value which promotes the dissolution of GGBS, accelerates the chemical reactions taking place in the binder system, and thus the strength development [15]. The content of sodium and potassium oxides content in BMs (5.79-5.81 wt.%) is responsible for the formation of silicon-based geopolymers [16]. Alkaline activation is based on the dissolution of ingredients containing silicon and aluminum [17]. It is believed that the Si-O-Si, Al-O-Al and Al-O-Si bonds present in the silicates, aluminates and/or aluminosilicates first break down and then the bonds polymerize, and a geopolymer network forms with participation of alkaline cations. As a result, formation of hydration products similar to natural zeolites (Na-A-S-H or K-A-S-H) occurs [18]. These arguments support the above discussion regarding the composition of experimental binder mixtures with respect to their mechanical properties. However, confirmation of this assumption requires microstructure-oriented research.

4. Conclusions

The evaluation of the results of the properties of the fresh mixtures and mortar specimens after 56 days of hardening according to the standard requirements for a normal hardening hydraulic road binder led to following conclusions:

- only the binder mixture BM2 with the lowest SO_3 content and clinker but with the highest amounts of GGBS, ground limestone and alkalis ($Na_2O + K_2O$) among the tested binder mixtures met the standard requirement for the initial setting time,
- 56 days' compressive strength of all experimental samples reached a higher strength than that specified in the standard STN EN 13282-2 for class N4 (≥32.5 MPa, ≤52.5 MPa), exceeding the upper limit by 13-22 %. Based on literature data, we assumed that the high values of compressive strength were due to the formation of the hydrated and geopolymer products.

5. References

[1] Andrew RM 2018 *Earth Syst. Sci. Data* **10** 195
[2] Strigac J 2015 *Selected Scientific Papers – Journal of Civil Engineering* **10** 81
[3] Siddique R, Kunala and Mehta A 2020 *Utilization of industrial by-products and natural ashes in mortar and concrete development of sustainable construction materials. In Nonconventional and Vernacular Construction Materials Characterisation, Properties and Applications* (Cambridge: Woodhead Publishing (Series in Civil and Structural Engineering) pp. 247–303
[4] Hanein T, Hayashi Y, Utton C, Nyberg M. Martinez J, Quintero-Mora N and Kinoshita H 2020 *Constr. Build. Mater.* **259** 120420
[5] Abdel-Gawwad HA, Heikal M, Mohammed MS, El-Allem SA, Hassan HS, Garcia V and Alomayri T 2019 *J. Clean. Prod.* **232** 1218
[6] Shubbar AA, Jafer H, Abdulredha M, Al-Khafaji ZS, Nasr MS, Al Masoodi Z and Sadique M 2020 *J. Build. Eng.* **30** 101327
[7] STN EN 13282-2 2015 *Hydraulic Road Binders—Part 2: Normal Hardening Hydraulic Road Binders—Composition, Specifications and Conformity Criteria* (Bratislava: Slovak Office of Standards, Metrology and Testing)
[8] DoroCem available online: https://crh-white.com/wp-content/uploads/Product_documentation/SpecialneSpojiva/dorocem-turňa.pdf (accessed on 27 November 2020).
[9] STN EN 196-3 2017 *Methods of Testing Cement—Part 3: Determination of Setting Times and Soundness* (Bratislava: Slovak Office of Standards, Metrology and Testing)
[10] Heikal M, Aiad I and Helmy IM 2002 *Cem. Concr. Res.* **32** 1805
[11] STN EN 1015-3 2007 *Methods of Test for Mortar for Masonry—Part 3: Determination of Consistence of Fresh Mortar (by Flow Table)* (Bratislava: Slovak Office of Standards, Metrology and Testing)
[12] STN EN 12390-7 2020 *Testing Hardened Concrete. Density of Hardened Concrete* (Bratislava: Slovak Office of Standards, Metrology and Testing)
[13] STN EN 12390-7 2019 *Methods of Testing Cement—Part 1: Determination of Strength* (Bratislava: Slovak Office of Standards, Metrology and Testing)
[14] STN EN 197-1 2012 *Cement. Part 1: Composition, Specifications and Conformity Criteria for Common Cements* (Bratislava: Slovak Office of Standards, Metrology and Testing)

[15] Criado M, Walkley B, Ke X, Provis JL and Bernal SA 2018 *Sustainability* **10** 4709
[16] Wang W and Noguchi T 2020 *Constr. Build. Mater.* **252** 119105
[17] Heikal M, Zaki MEA and Ibrahim SM 2020 *Constr. Build. Mater.* **260** 119947
[18] Provis JL and Van Deventer JSJ 2009 *Geopolymers: Structures, Processing, Properties and Industrial Applications* (Sawston: Woodhead Publishing)

Acknowledgements
This research was supported by the Agency of the Ministry of the Education, Science, Research and Sport of the Slovak Republic, Grant No. VEGA 1/0222/19 and Grant No. 1/0419/19.

IOP Conf. Series: Materials Science and Engineering 1252 (2022) 012011 doi:10.1088/1757-899X/1252/1/012011

The carbon dioxide generation rate from burning of candle and its effect on room ventilation

P Kapalo[1], A Eštoková[2] and O Voznyak[3]

[1] Institute of Architectural Engineering, Faculty of Civil Engineering, Technical university of Kosice, Vysokoskolska 4, 042 00 Kosice, Slovakia
[2] Institute of Environmental Engineering, Faculty of Civil Engineering, Technical university of Kosice, Vysokoskolska 4, 042 00 Kosice, Slovakia
[3] Department of Heat and Gas Supply and Ventilation, Lviv Polytechnic National University, St. Bandery 12, 79013 Lviv, Ukraine

E-mail: peter.kapalo@tuke.sk; adriana.estokova@tuke.sk; orest.voznyak@i.ua

Abstract. In construction, the emphasis is currently on high energy efficiency of buildings. A ventilation system ensuring the indoor air quality has a significant share of energy consumption in buildings. Open fire sources, such as gas stoves and candles result in the air pollution in the indoor environment that causes more intensively operation of ventilation systems. The aim of this research is to quantify the pollutants arising from the burning of a randomly selected wax candle that commonly is used in households. Experimental measurements were performed with a special regard to carbon dioxide (CO_2) production. The weight loss of the wax during the burning of the candle was has been measured and subsequently the production of CO_2 was estimated using a chemical formula. Based on the amount of CO_2 produced, the intensity of air exchange using ventilation system has been calculated to achieve the required quality of the indoor environment. The calculated air volume flow rate required to remove the pollutants generated during the burning of the candle is in this case approximately 54% of the air volume flow required per one person.

1. Introduction

At present, candles are used as decoration of the house or at celebrations. The products of candles´ burning, such as CO_2, belong to the harmful pollutants which need to be removed due to ventilation system. To design a ventilation system correctly, it is necessary to know the amount of air to remove pollutants from indoors. Nowadays, a great emphasis is put on the energy efficiency of buildings and thus there is an effort to minimize air exchange in buildings. When designing buildings, controlled mechanical or hybrid ventilation of buildings is increasingly being considered. To design a ventilation system correctly, it is necessary to know the amount of air to remove pollutants from indoors, which was the subject of research presented in papers [1, 2].

There are several studies around the world investigating the pollutants arising from the process of burning candles. Massoudi and Hamidi [3] have found out that the studied paraffin candles had emitted several dangerous compounds, including hazardous alkenes, and aromatics like toluene and benzene. Derudi et al. [4] characterized the emission factors of different pollutants originating from the candles with special regard to volatile organic compounds (VOCs), which the EU identifies as main indoor pollutants arising from the burning of scented paraffin candles. Another study [5] has reported that products of burning, originating from both cooking and candles burning had represented up to 65% of indoor air pollutants. The study has been performed in 56 housing units over a period of 45 hours. According to study [6], toxic substances such as formaldehyde were produced in large amounts from various scented candles, primarily but not only when they were lit. The linalool and limonene detection rates, representing the most abundant fragrant allergens, was higher than 50% [7]. As it is recommended in [8], it is necessary to ensure sufficient air exchange when burning candles.

Therefore, the objective of this paper is to estimate the sufficient volume flow rate of air needed to ensure the removal of the indoor pollutants originating from candle burning. Because co2 is the major pollutant in candle burning and human respiration, these two sources of pollutants are compared in this study.

2. Materials and methods

In order to determine the volume flow rate of air required to ventilate the room in which the candle was burning, it was necessary to quantify the pollutants arising during the burning of the candle. For this purpose, the carbon dioxide as the major burning product was chosen to be monitored. A simple, scented, uncoloured wax candle commonly used in households and restaurants was selected for the investigation. During the experimental measurement, the candle burned in the selected room and the weight loss of the candle, the CO_2 concentration and the interior air temperature were continuously measured over a period of 130 minutes. The type of wax has been determined based on the rate of the wax loss and the mass flow of CO_2 has been calculated using a chemical formula describing the wax decomposition. From the produced mass flow of carbon dioxide and the permissible concentration in the indoor air, it was possible to calculate the required air flow volume.

2.1. Characteristics of the experimental room

The experimental measurement proceeded in summer season in a five-storey building, in a room situated on the third floor. The floor plan of the room is shown in Figure 1. The dimensions in the figure are given in meters.

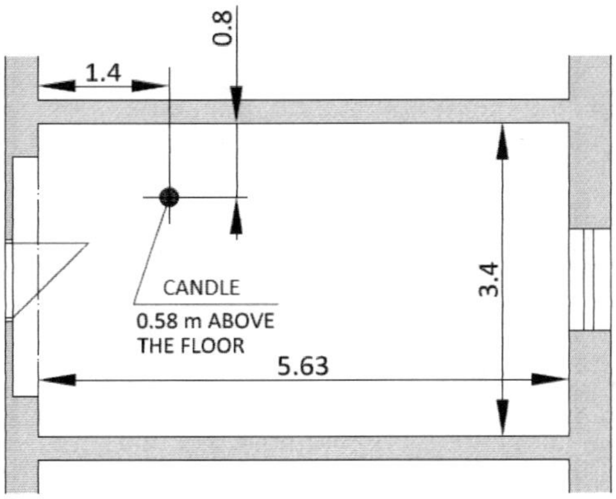

Figure 1. Experimental room.

2.2. Measuring devices

A laboratory balance with a separate platform KERN PCD 3000-2 [9], which was used to measure the weight loss of candle wax due to burning, was of these technical parameters: weighing capacity max 3500.00g, readability 0.02g, repeatability 0.02g, linearity 0.05g minimal piece weight 0.2g/piece. The ZG106 [10] instrument with following technical parameters was used to measure CO_2 concentration and air temperature around the burning candle: measurement range of CO_2: 0 to 3,000ppm, accuracy ± 5%, repeatability ± 20 ppm, temperature range: 0 to 50 °C, display resolution and accuracy ± 1 °C. Since the window was open during the entire measurement time, the measurement of indoor air temperature was for information only.

2.3. *Characteristics of candle*

A commonly available ordinary non-perfumed candle has been selected for the burning experiment. The original parameters of the candle (diameter, height, weight), measured by a caliper with a measuring range of 0-150 mm with a scale of 0.05 mm, before the experiment were as follows: candle diameter: 21.84 ± 0.01 mm, candle height: 166.62 ± 0.13 mm, candle weight: 55.25 ± 0.003 g, wax color: white. The candle volume has been calculated to be 6.24×10^{-5} m^3 with a relative error of ± 0.14 m^3. The calculated candle density was 885.93 ± 1.28 kg/m^3.

3. Results and discussion

3.1. *The measured wax loss*

The mass losses of the candle wax were determined every 5 minutes. Figure 2 presents the trend in mass losses during the 130-minute measuring.

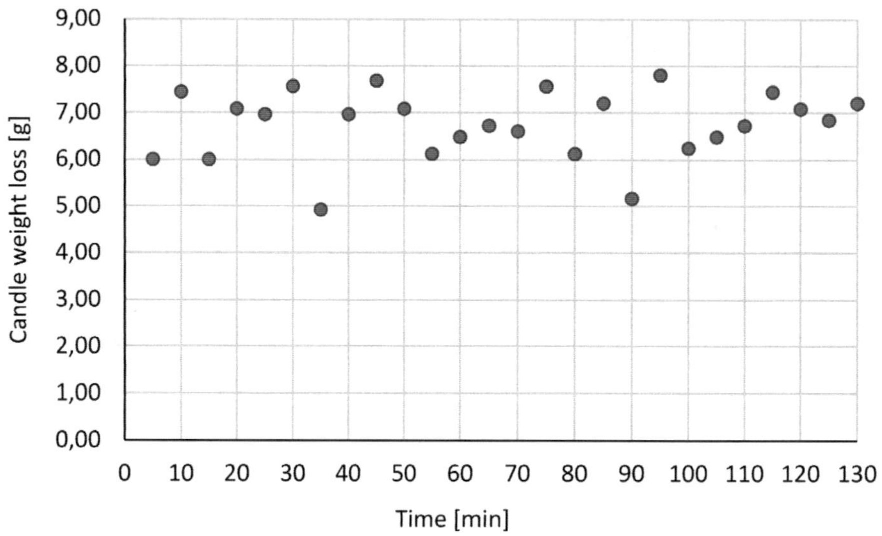

Figure 2. The measured losses in wax weight during burning of candle.

As it can be seen in Figure 2, the mass loss rate was uniform during whole experimental period. Based on the 26 data measured, a mean weight loss of wax was calculated of 6.75 g/h with a standard deviation of ± 0.73 g/h and a relative error of 2.12%.

The type of candle wax has been determined according to [11] considering the calculated average wax loss rate. Comparising of the burning rates of the different types of candles reported in the study by Hansen [11], a candle material was identified as palm stearin (vegetable fatty acids from palm oil). Palm stearin wax loss rate reported in Hansen study was 6.78 g/h. The deviation of the wax loss rate for the material from the study [11] and our measured material was 0.44%.

3.2. *Calculation of the mass flow rate of carbon dioxide*

The measured average air temperature was 28.97 °C with a standard deviation of ± 0.08 °C and a relative error of 0.05%. The measured values of CO_2 concentration oscillated around 608.67 ppm with a standard deviation of ± 18.56 ppm and a relative error of 0.59%. This concentration corresponds to the 1095.36 mg/m^3, which, when converted to the concentration in the whole model room of volume of 50m^3 (Figure 1), gives 54.49 g of CO_2 in the room air.

To calculate the theoretical amount of CO_2 produced, the chemical equation (1) was used, in which the process of burning a candle made of palm stearin is described [12].

$$2\ C_{57}H_{110}O_6 + 163\ O_2 \rightarrow 114\ CO_2 + 110\ H_2O \tag{1}$$

where $C_{57}H_{110}O_6$ is palm stearin, O_2 is gaseous oxygen, CO_2 is gaseous carbon dioxide and H_2O is gaseous water. Based on the identified mass loss of the candle (6.75 g/h), the rate of carbon dioxide formation was estimated as 5.28 mg/s. This value corresponds to the theoretical overall produced amount of carbon dioxide originating from burning of 41.2 g per the 130-minute experiment.

The produced CO_2 mass flux of a person weighing 80 kg has been calculated according to a study [13] from formula (2).

$$q_m = 0.120406 \cdot m_p \tag{2}$$

where: q_m is CO_2 mass flow rate [mg/s] and m_p is body weight of occupants [kg].

The mass flow rate of carbon dioxide produced by a person weighing 80 kg was found as 9.63 mg/s. It can be stated that the CO_2 production resulting from candle burning represents approximately 54.83% of the CO_2 amount produced by one student when writing a school assignment.

3.3. Calculation of the air volume flow rate

Based on the identified CO_2 mass flow rate linked to the occupant and the candle, the overall outdoor air volume flow rate can be calculated by applying the formula (3). The estimated air volume is required to ensure a sufficient quality of the indoor air in the room.

$$q_V = \frac{q_m}{C_{IDA} - C_{SUP}} \, . 3600 \tag{3}$$

where: q_v is required outdoor air volume flow rate [m^3/h]; q_m is CO_2 mass flow rate [mg/s]; C_{IDA} is indoor air CO_2 concentration [mg/m^3] and C_{SUP} is outdoor air CO_2 concentration [mg/m^3].

As q_{m} data input for equation (3), we used the CO_2 mass flow rates calculated for candle according to chemical equation (1) and for person according to equation (2). The average carbon dioxide concentration of 400 ppm was considered as the background regarding the outdoor air and the CO_2 concentration of 1,000 ppm was considered as the maximum in the indoor air.

In this experiment, the required volume flow of ventilation air, that is needed for the reduction of CO_2 concentration due to the candle burning was calculated of 17.61 m^3/h. Similarly, the required volume flow of air for the reduction of CO_2 arising from human respiration was estimated to be 32.13 m^3/h (Figure 3).

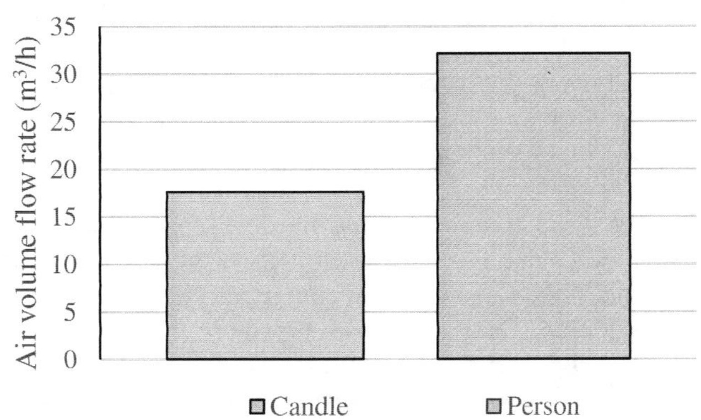

Figure 3. The comparison of air volume flow rate required for ventilation due to candle burning and human respiration.

4. Conclusion

The experimental findings revealed that burning of the candle has a significant influence on the carbon dioxide pollution in the indoor air. In spite of the fact that the weight of the burning candle represented only 0.07% of the human body weight when considering the average weight of 80 kg, the emissions of carbon dioxide linked to the candle burning represented more than 54% of the CO_2 emissions produced by human respiration. It was found that each indoor user of the average weight of 80 kg, performing a sedentary work, produced about 9.63 mg/s CO_2 while a burning candle produced 5.28 mg/s.

The needed air volume flow rate is directly proportional to the production of CO_2. So, the required air volume flow rate for ventilation of the room connected exclusively to burning of one candle is estimated to 17.61 m^3/h, whereas the need for air exchange linked to a person's stay in a room is 32.13 m^3/h. It can be stated that when one person is sitting in a room while a candle is burning, approximately 35% of the total ventilation air is required to remove carbon dioxide originating from the candle burning.

5. References

[1] Kapalo P, Vilceková S, Domnita F and Voznyak O 2014 *Determine a methodology for calculating the needed fresh air, The 9th International Conference "Environmental Engineering"* Vilnius Lithuania

[2] Kapalo P, Vilcekova S and Voznyak O 2014 *Using experimental measurements the concentrations of carbon dioxide for determining the intensity of ventilation in the rooms. Chemical Engineering Transactions* **39** 1789–1794

[3] Massoudi R and Hamidi A 2017 *Some Candles Emit Hazardous Materials for Human Health and are Indoor Air Pollutants, International Journal of TROPICAL DISEASE & Health* **24**(2) 1-10

[4] Derudi M, Gelosa S, Sliepcevich A, Cattaneo A, Rota R, Cavallo D and Nano G 2012 Emissions of air pollutants from scented candles burning in a test *chamber, Atmospheric Environment* **55** 257-262

[5] Bekö G, Weschler Ch J, Wierzbicka A, Karottki D G, Toftum J, Loft S and Clausen G 2013 *Environmental Science & Technology* **47**(18) 10240-10248

[6] Ahn J H, Kim K H, Kim Y H and Kim B W 2015 *Characterization of hazardous and odorous volatiles emitted from scented candles before lighting and when lit, Journal of Hazardous Materials* **286** 242-251

[7] Bartsch J, Uhde E and Salthammer T 2016 *Analysis of odour compounds from scented consumer products using gas chromatography-mass spectrometry and gas chromatography-olfactometry, Analytica Chimica Acta* **904** 98-106

[8] Heinze J E and Silverman A 2016 *Comment on "Characterization of hazardous and odorous volatiles emitted from scented candles before lighting and when lit", Journal of Hazardous Materials* **303** 171

[9] Kern official website *KERN PCD 3000-2 Products* [online] https://www.kern-sohn.com/en/PCD

[10] Zyaura official website *ZG106 General Introduction* [online] http://www.zyaura.com/products/zg106.asp#General_Introduction

[11] Hansen J A, Pedersen P B, Jensen T N, Poulsen M, Thomann S, Kilstrom O, Roque P and Ventura R 2018 *Ministry of environmental and food of Denmark - Environmental Protection Agency. Environmentally friendly candles with reduced particle emissions, Environmental project No. 2053.* The Danish Environmental Protection Agency

[12] Lide D R ed. 2009 *CRC Handbook of Chemistry and Physics* (90th ed.). Boca Raton, Florida: CRC Press.

[13] Kapalo P, Domnita F, Bacotiu C and Podolak M 2018 *The influence of occupants' body mass on carbon dioxide mass flow rate inside a university classroom – case study, Int J Environ Health Res* **28**(4) 432–447

Acknowledgments

This work was supported the Slovak Scientific Grant Agency under the contract No. VEGA 1/0230/21.

IOP Conf. Series: Materials Science and Engineering 1252 (2022) 012012 doi:10.1088/1757-899X/1252/1/012012

Size effect of mortar specimens under uniaxial compression

M Kovac[1]

[1] Institute of Structural Engineering, Faculty of Civil Engineering, Technical University of Kosice, Vysokoskolska 4, 042 00 Kosice, Slovakia

E-mail: maros.kovac@tuke.sk

Abstract. This paper investigates the phenomenon of the size effect of quasi-brittle lime-based material. This research aims to confirm the existence of a size effect on a homogeneous material - test specimens of mortar, which are subjected to uniaxial compression and provide additional knowledge to explain the causes of the size effect. This paper deals with the analysis of the results of experimental tests of mortar specimens without inclusion and with one centrally located steel inclusion. Two series of tests with different slenderness and size were prepared. The results of the experiment showed that the compressive strength of the test specimens decreased with increasing specimen size. The larger interface area between the inclusion and the mortar matrix negatively affected the compression strength.

1. Introduction

The phenomenon of the size effect of quasi-brittle materials such as normal concrete and high-strength concrete was extensively investigated under various stress methods. The heterogeneity of concrete of small structures and/or laboratory-sized specimens due to coarse aggregate grains complicates the use of the continuum mechanics. As is well known, concrete is a very complex heterogeneous material consisting of cement, gravel, sand, and water. After mixing, setting, and hardening of the concrete mixture, this material can be considered as two-component consisting of coarse aggregate grain and mortar matrix, which is the most important characteristic of concrete heterogeneity. Both aggregate grain and mortar matrix exhibit almost linear elastic brittle behaviour under uniaxial compression [1]. Heterogeneity, in which coarse aggregate grains prevail, is thought to control the crack propagation process. It is the initiation and development of these cracks in the heterogeneous material that influence the mechanisms at a higher level (macro-level). The mortar matrix is also heterogeneous in nature, but the size of the fine aggregate grain in its volume will not direct the course of the force flow under the action of uniaxial compression. A brittle failure of a test specimen of relatively homogeneous material is assumed [2].

The aim of this research was to investigate whether the use of a fine aggregate grain in the volume of the mortar test specimen confirms the assumption of the linear course of the σ - ε curve of the specimens without inclusion (relatively perfect homogeneous material) and the specimens with one inclusion (inhomogeneous material with coarse aggregate grain and mortar matrix). On the other hand, the aim of the experimental tests was to verify the existence of the size effect phenomenon and to analyse the effect of inclusion on the achieved compressive strength of the test specimens. In general, the principle of dimensional effect has been adopted from previous research [3], in which the strength of specimens decreased with increasing specimen size and therefore larger specimens showed lower strength and, conversely, smaller specimens showed higher strength.

2. Experimental research

2.1. Material and specimens

The article presents the results of research that included two series of experimental tests, each of which consisted of 5 test specimens. Both series formed specimens of lime mortar with natural hydraulic lime as a binder - NHL 2. The first series consisted of test specimens without inclusion and the second with one centrally placed inclusion with respect to the direction of loading. Class M 2.5 dry mortar mix was used for specimen preparation. The test specimens were cast into a plastic formwork, which was printed using a 3D printer. The test specimens were removed from the formwork on the 5th day after production. During this time, they were covered with foil in an enclosed space to limit the shrinkage of the specimen from its free surface. The position of the centric inclusion for the test specimen with one inclusion was secured in advance by gluing to the formwork base. The steel inclusions strength class was S235. The height of the steel inclusions was identical to the depth of the test specimens. The ratio of the side of the test specimen to the diameter of the inclusion (ratio h/ϕ) was constant with a value of 5. The diameter of the inclusions was 20, 15 and 10 mm. To refine the results of experimental tests, 3 replicates of both series were performed. Uniaxial compression tests were performed on day 10 after production on five test specimens with different sizes, and were divided according to their slenderness (ratio h/d) into two groups (figure 1):

- Variable slenderness of the test specimens (h/d = 4, 3, 2).
- Constant slenderness of the test specimens (h/d = 4).

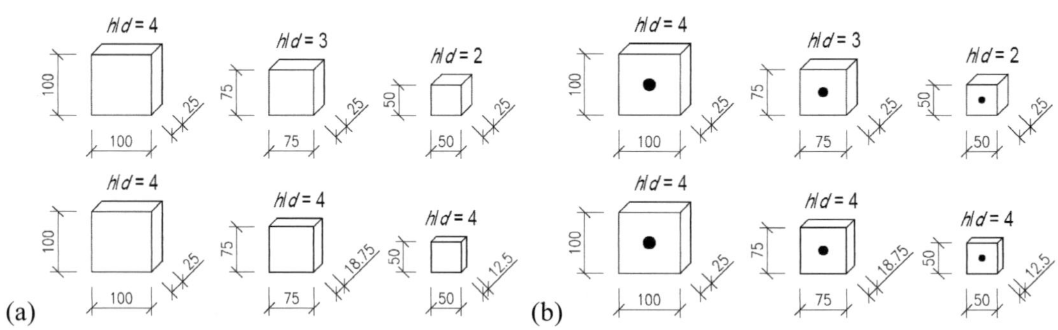

Figure 1. Experimental analysis of test specimens (a) without inclusion, (b) with inclusion.

When the size effect is analysed, the ratio of the aggregate grain size to the smallest dimension of the test specimen is usually the key parameter. Material averaging was found to be appropriate if the elastic material properties are considered and the experimental results are evaluated from a representative specimen volume [4]. Van Mier [5] stated that in the case of normal concrete with solid grains of natural aggregates, the representative volume must be greater than 8 times the maximum grain size of the aggregate. For the continuum assumption with considering the same properties in each point of the volume, the representative volume should be at least 3 to 5 times the maximum aggregate grain size [5]. These principles were also applied to the mortar test specimens used in the experiment, where the maximum aggregate grain size after sieving was 1 mm. The selected maximum aggregate grain in the mortar matrix met the criteria of a representative volume of elements for the smallest dimension from the whole series of test specimens d = 12.5 mm.

2.2. Arrangement of the test

Compression tests were performed using a test set-up with manual control (figure 2). The device consisted of a solid base, a steel frame, guide rails and a crosshead for holding the rigid steel plates into which the test specimens have been placed. The vertical deformation was measured using a gauge held on a vertical bar with an accuracy of 0.01 mm. The test specimens were subjected to uniaxial compression. The tests were controlled by deformation at an average speed of approximately 0.0025 mm/s. The value of compressive force during the test was measured using a load cell with a capacity of

1000 N. Figure 3 shows the test specimen SQ-10-75-15-2 after the test from the front and back sides. It was assumed that the test set-up will prevent transverse deformation on the bases of steel plates. This can lead to the formation of confined zones under the action of triaxial compressive stress, where the matrix tends to remain intact and can thus create an area without cracks [6].

Figure 2. Arrangement of the uniaxial compression test.

(a)

(b)

Figure 3. Specimen after the test (a) front side, (b) back side.

3. Results and discussion

Figure 4 shows the fracture of 3 selected test specimens without inclusion with dimensions of 100×100×25 mm ($h/d = 4$), 75×75×25 mm ($h/d = 3$) and 50×50×25 mm ($h/d = 2$). It can be seen from the figure that the cracks formed mostly parallel to the direction of the compressive load. Initiation and propagation of cracks in the direction of loading occurred in the later stages of loading, while their development continued until the failure of the specimen. All specimens without inclusion showed brittle failure.

Figure 4. Fracture of mortar specimens without inclusion (Scale 1:0.75:0.5).

The results of the experimental research were evaluated based on the stress-strain relationships ($\sigma - \varepsilon$) shown in the graphs in figures 5 to 9. The vertical axis of the graphs represents the contact compressive stress of the test specimen and the horizontal axis the vertical deformation, i.e., the compression value divided by the initial height of the test specimen. The graphs were prepared separately for specimens with and without inclusion and specimens with constant and variable slenderness.

Figures 5 and 6 show graphs $\sigma - \varepsilon$ for test specimens without inclusion. Figure 5 represents test specimens of variable slenderness and figure 6 specimens of identical slenderness. When analysing the whole set of specimens, a linear ascending branch can be observed for all test specimens, which confirmed the previous findings [1]. As can be seen from both figures, the different slenderness affected not only the achieved ultimate load of the test specimens, but also the slope of the ascending branch of the curve $\sigma - \varepsilon$. While the slope of the ascending branch was almost identical at the variable slenderness until the failure of the specimens, in the case of the constant slenderness the slope was steeper for smaller specimens.

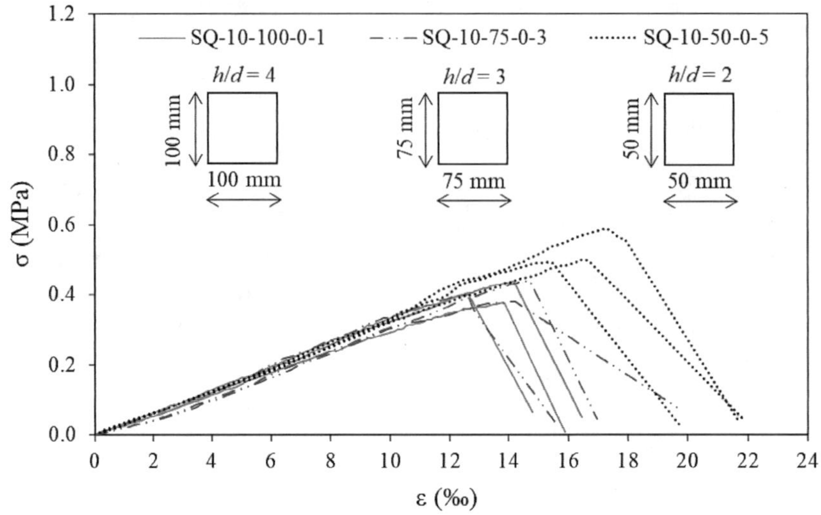

Figure 5. Stress-strain curves for mortar specimens without inclusion with variable slenderness.

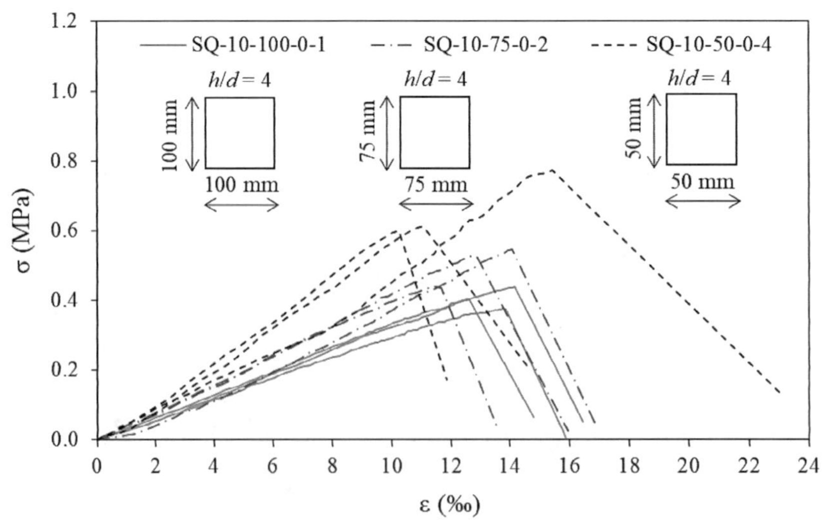

Figure 6. Stress-strain curves for mortar specimens without inclusion with constant slenderness.

Figure 7 shows the fracture of 3 selected test specimens with inclusion with dimensions of 100×100×25 mm, 75×75×25 mm and 50×50×25 mm. It can be seen from the figure that the initiation and development of cracks in the specimen during built-in inclusion was significantly different. The solid inclusion in the relatively soft mortar matrix acted as a stress concentrator, significantly controlling the crack propagation process in the specimen. Weak bonding at the interface of the steel inclusion and the mortar matrix led to the initiation of cracks at the interface. With a further increase in load, the cracks propagated into the matrix until the total failure of the test specimen.

Figure 7. Fracture of mortar specimens with inclusion (Scale 1:0.75:0.5).

Figures 8 and 9 show the $\sigma - \varepsilon$ curves of test specimens with inclusion. As with the test specimens without inclusion, a significant effect of slenderness on the ascending branch of the curve $\sigma - \varepsilon$ was observed for the specimens with one inclusion. The slope of the ascending branch increased as the size of the test specimen decreased in the case of constant slenderness. In the case of the variable slenderness, again, the almost identical inclination of the ascending branch applies. It was found that the inclusion in the mortar matrix caused a reduction in the strength of the test specimen under uniaxial compression but did not change the slope of the ascending branch of the curve. On the other hand, when reaching higher load values (before the ultimate load), the effect of inclusion was manifested by the curvature of the curve in front of its peak, partly also by the decreasing branch of the curve $\sigma - \varepsilon$.

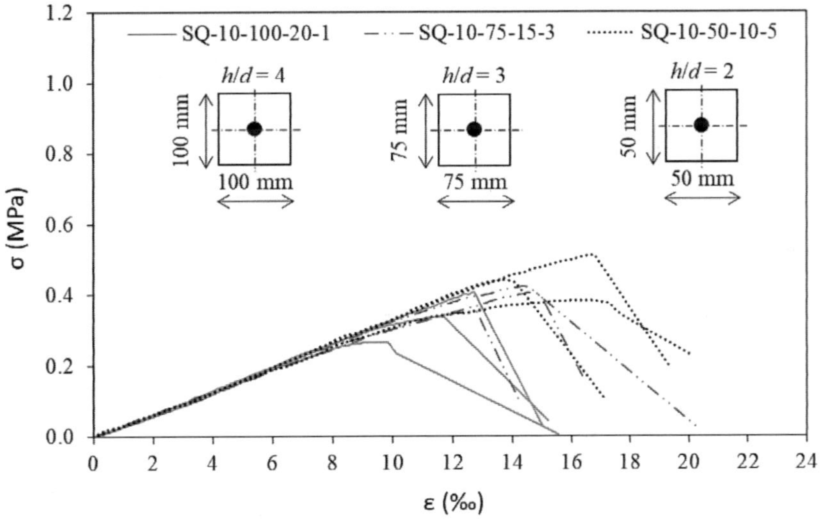

Figure 8. Test specimens with inclusion and variable slenderness.

When analysing the whole set of specimens, it can be observed that the phenomenon of size effect was manifested in both test series. Analysis of the test specimens of double slenderness and with and without inclusion showed, that the compressive strength of the test specimens decreased with increasing specimen size. Moreover, the results of the experiment showed a reduction of the compressive strength of the specimens with inclusion compared to the specimens without inclusion, which confirmed the test specimens of constant and variable slenderness. The conclusions of the analysis are presented in figures 10 and 11. The figures show graphs of the measured values of ultimate stresses as well as their dependence on the changing size of the test specimen.

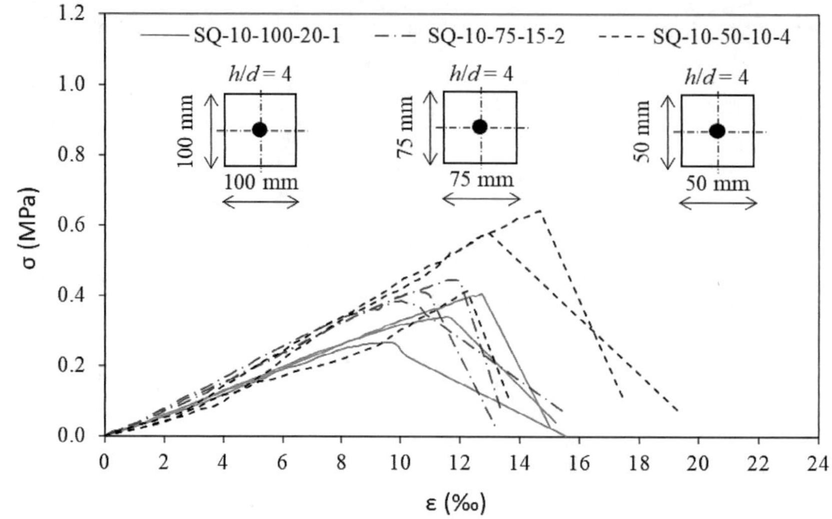

Figure 9. Test specimens with inclusion and constant slenderness.

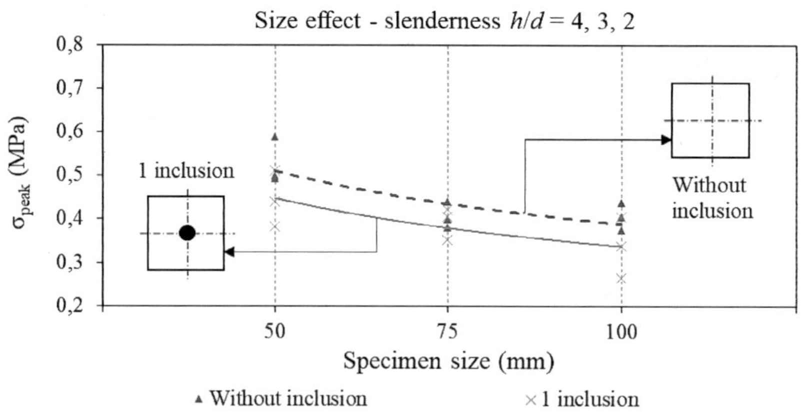

Figure 10. Comparison of the size effect of variable slenderness.

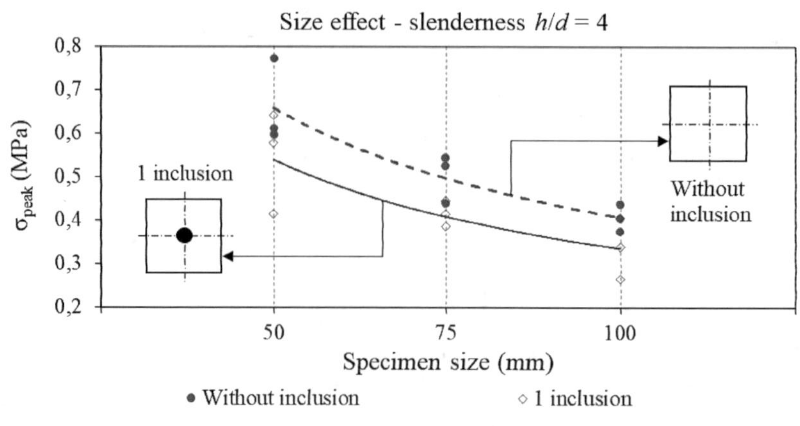

Figure 11. Comparison of the size effect of constant slenderness.

The mean values of the measured ultimate stress from the three series are given in table 1. A test specimen measuring 100×100×25 mm was taken as a reference value for comparing the experimental results.

Table 1. Mean values of failure stress of test specimens.

Specimen (mm)	*h*/*d*	Without inclusion σ_{peak} (MPa)	Without inclusion $\Delta\sigma_{peak}$ (%)	With inclusion σ_{peak} (MPa)	With inclusion $\Delta\sigma_{peak}$ (%)
100×100×25	4	0.406	-	0.337	-
75×75×25	3	0.407	0.16	0.393	16.68
50×50×25	2	0.527	29.82	0.445	32.15
100×100×25	4	0.406	-	0.337	-
75×75×18.75	4	0.503	24.06	0.415	23.37
50×50×12.5	4	0.660	62.67	0.545	61.87

4. Conclusion

Based on a series of experimental tests, it can be stated that the experiment confirmed the phenomenon of size effect. The results of the experiment showed that the compressive strength of the test specimens decreased with increasing specimen size. In the case of specimens without inclusion, the ultimate stress values with respect to the reference specimen measuring 100×100×25 mm decreased as the size of the test specimens decreased. The reduction in compressive strength of 0.16% and 29.82% was measured on the specimens of variable slenderness, and on the specimens of constant slenderness by 24.06% and 62.67%. Equivalent strength reductions were observed in the inclusive test specimens, except for the 75×75×25 mm test specimen, where a decrease in strength of 16.68% was observed. The experiment also showed that the interface between the inclusion and the mortar matrix negatively affected the compression strength. An average reduction in compressive strength of the test specimens for variable slenderness of 14.16%, resp. by 20.98% in the case of constant slenderness. This can be explained by the relative change in the volume of inclusions with respect to the changing size of the test specimen. On the other hand, changing the size of the inclusions will affect the interface area. Finally, an important factor may be the wall effect in the case of the constant slenderness of elements.

5. References

[1] Neville A M 1997 Aggregate bond and modulus of elasticity of concrete *ACI Materials Journal.* **94** 71-74
[2] Vonk R A 1992 *Softening of concrete loaded in compression* PhD Thesis (Netherlands: Eindhoven University of Technology)
[3] Van Mier J G M 1984 *Strain-softening of concrete under multiaxial loading conditions* PhD Thesis (Netherlands: Eindhoven University of Technology)
[4] Van Mier J G M 1998 A fundamental explanation of softening *7nd Int. Conf. on Fracture Mechanics of Concrete and Concrete Structures: Proc. FraMCoS* (Jelu, Korea) ed B H Oh et al (Seoul: Korea Concrete Institute) pp 299-305
[5] Van Mier J G M 1997 *Fracture processes of concrete* (Florida: CRC Press) p 464
[6] Van Mier J G M 1998 Failure of concrete under uniaxial compression: An overview *3nd Int. Conf. on Fracture Mechanics of Concrete and Concrete Structures: Proc. FraMCoS* (Gifu, Japan) ed H Mihashi and K Rokugo (Freiburg: Aedificatio) pp 1169-82

Acknowledgments

This work was supported by The Slovak Scientific Grant Agency under the contract No. VEGA 1/0363/21 and also by The Slovak Research and Development Agency under the contract No. APVV-15-0777.

Investigation of the thermal properties of candle wax material

A Eštoková[1] and P Kapalo[2]

[1] Institute of Environmental Engineering, Faculty of Civil Engineering, Technical university of Kosice, Vysokoskolska 4, 042 00 Kosice, Slovakia
[2] Institute of Architectural Engineering, Faculty of Civil Engineering, Technical university of Kosice, Vysokoskolska 4, 042 00 Kosice, Slovakia

E-mail: adriana.estokova@tuke.sk

Abstract. Indoor air quality is a key aspect of people staying inside buildings. Combustion processes have been identified as one of the most important sources of pollutants in the indoor environment. Proper HVAC design must account for all sources and variations in their intensity, such as also the burning of candles, which produce not only increased amounts of carbon dioxide as a result of the combustion of the organic matrix, but also often dangerous organic substances. On the other hand, the candle burning represents an energy source as well. This article deals with the study of thermal decomposition of selected candle material using methods of thermal analysis - thermal gravimetric analysis (TG) and differential scanning calorimetry (DSC). The wax material analyzed was identified as a mixture of palm stearin and paraffin. During the candle wax material heating, the energy up to 772.6 J per 1 g of material can be released in dependence on the heating rate.

1. Introduction

The use of candles in the home has a long tradition. Candles were originally used as a light and heat sources in a room. In today's over-technological world, candles are used more to create a pleasant and cozy environment to relax. The use of candles in the world is not negligible. According to the European candle association [1], 738,530 tons of candles were consumed in the European Union in 2018, which is 1.44 kg of candles per capita. Several scientific studies point to the harmfulness of candle products. According to various studies [2, 3], some types of paraffin candles produced various hazardous products, including benzene, toluene and alkenes. Toxic compounds such as formaldehyde can be emitted in large quantities from some scented candles, especially in significant rates when they burned [4]. In order to avoid pollution of the indoor environment of buildings during the burning of candles, candle manufacturers strive to produce products that produce as few harmful substances as possible. Great efforts are being made by manufacturers and candle associations to protect consumers. As reported in [5], reduction in dangerous emissions from candles can be achieved by using high quality candles, trimming long wicks and preventing candles from burning when very high air exchange is in the room, e.g. draughty in the room. However, according to [6], burning candles do not pose significant health risks if they are used in rooms with adequate ventilation and if only one candle burns in a long time period.

On the other hand, wax material could be potentially a source of energy or source of energy storing. Amount of heat production from renewable sources is often not in time line with its consumption. For this reason, the possibility of efficient storage of thermal energy is an important aspect. Candle wax is one of the potential materials for storing energy using wax phase change. Several authors have been investigated the thermal conductivity of wax for its perspective use in thermal energy storage [7-9]. The paper is aimed at investigation of the thermal parameters of wax material of the commonly used type of candle by using thermal analysis methods. These methods allow not only to study the thermal properties of material, however, the energy released during the decomposition of the material can be also determined.

2. Materials and methods

2.1. Candle material

A commonly available non-perfumed palm stearin candle with density of 885.93 ± 1.28 kg/m^3 was selected for the experiment. Stearin which occurs naturally in animal fats and vegetable oils is used in the manufacture of candles to improve the appearance and burning properties of candles. The main components are stearic acid and palmitic acid, the chemical formula is $C_{57}H_{110}O_6$. White stearin is characterized by a higher content of stearic acid (40 - 60%) and a nice white color. The flakes of the candle were heated to analyze the decomposition intervals and enthalpy during the decomposition.

2.2 Thermal analysis

Thermal analysis of the samples was performed on a STA 449 F3 Jupiter (Netzch, Germany) in the temperature range 25-1000 ° C in a N_2 atmosphere (rate 60 ml/min). A sample with weight of 26.5 ± 0.5 mg was heated in an Al_2O_3 crucible at three different heating rates: 5, 10 and 30 K/min. Thermal gravimetric analysis (TG) and differential scanning calorimetry (DSC) methods were applied to analyse the mass losses and decomposition peaks.

3. Results and discussion

The thermal changes of the wax material under three different heating rates are shown in DSC curves in Figure 1. DSC curves present the heat transferred to or from the material undergoing a physical or chemical change.

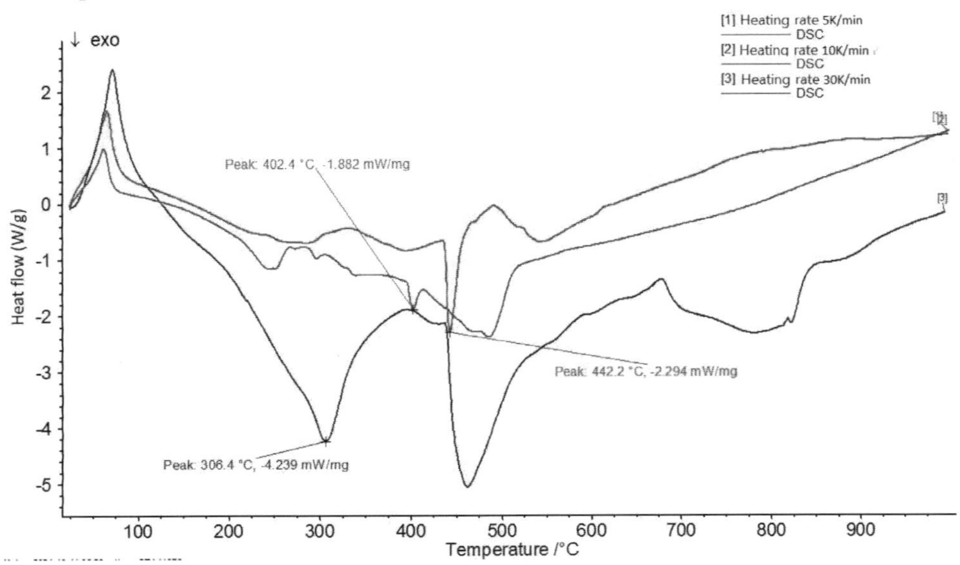

Figure 1. DSC curves of the wax material.

As seen in the Figure 1, four temperature intervals were identified. The DSC peaks and enthalpies measured in our study are reported in Table 1. The peak (T_i) demonstrated the temperature at which the transition or decomposition process reaches their maximum or where the process occurs the fastest. The area of the peak accounted for the consumed or released amount of heat ΔH (enthalpy).

Table 1. The DSC peaks and enthalpies during sample heating.

Heating rate (K/min)	T_1 (°C)	ΔH_1 (J/g)	T_2 (°C)	ΔH_2 (J/g)	T_3 (°C)	ΔH_3 (J/g)	T_4 (°C)	ΔH_4 (J/g)
5	62.3	192.3	283.7	-271.8	402.4	-43.7	486.2	-772.6
10	65.9	164.2	332.7	-106.9	442.2	-127.1	541.3	-259.9
30	72.0	136.4	306.4	-295.0	462.0	-478.6	786.3	-260.4

Solid-liquid transition of the material analyzed was observed at temperature above 60°C, which is slightly higher than reported in [10]. Teles dos Santos et al. presents a similar curve and indicate recrystallization of palm stearin at 50 °C [10]. Similarly, in [11] the melting temperature for palm stearin was reported 46.7 °C and melting heat 128.4 J/g applying the heating rate of 5K/min. Comparing our results to those of mentioned, we can assume, that the material measured in our study was not composed only of palm stearin. Pradhan and Ramaswamy [12] reported the temperature about 60 °C for paraffine wax. Therefore, we assume the material was mixed of paraffin and palm stearin. This finding can be in consistence with the higher melting heat of paraffin (151.62 J/g) given in [12]. Authors in [13] present even higher melting heat of various paraffin materials up to 196 J per gram.

The following peak can be linked to volatilization/decomposition of palm stearin/paraffin wax material at 283.7 °C, 332.7 °C and 306.4 °C, respectively. The remaining peaks are likely related to the decomposition of the oxidised waxes where any remaining carbonaceous residue is transformed to CO_2. As observed, the heating rate effects the temperatures of the individual peaks whereas in general, it moves them to higher values. As seen in Table 1, except from the melting process, all other processes are accompanied by heat releasing during exothermal changes. The energy released depends on the type of the chemical reaction running and the rate of heating. Anyway, during the candle wax material heating, the energy up to 772.6 J per 1 g of material can be released.

The mass losses during heating the wax material are presented in Figure. 2. As seen in Figure 2, the mass losses varied from 99.21 to 92.32 % depending on the rate of heating.

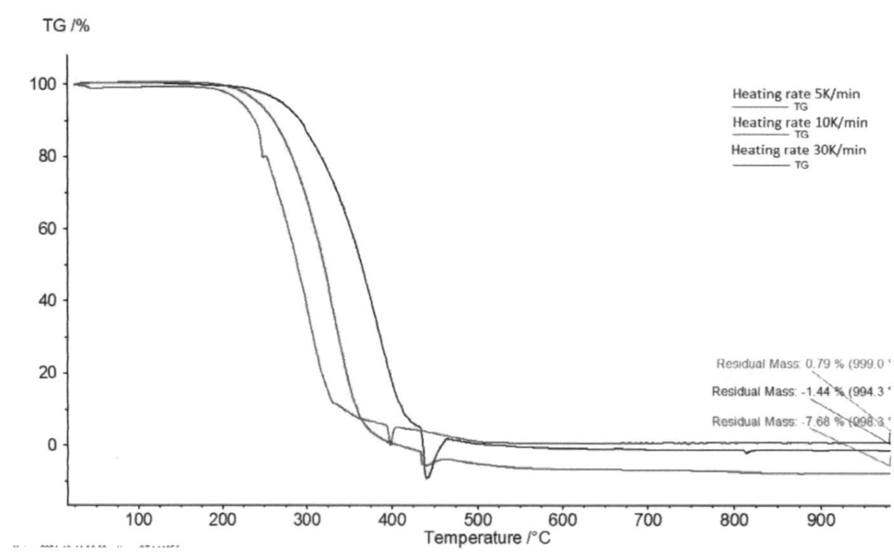

Figure 2. Mass losses of the wax material during heating.

The solid-liquid transition at about 60°C (Table 1) was not accompanied by mass loss mass. The organic material decomposition occurring in the interval 200-500 °C was responsible for whole mass loss during the heating period. According to the heating rates, the rate of decomposition of the organic matrix differed. The Figure 3 gives the DTG (Difference Thermogravimetry) curves representing decomposition rates of the wax material per individual heating rates.

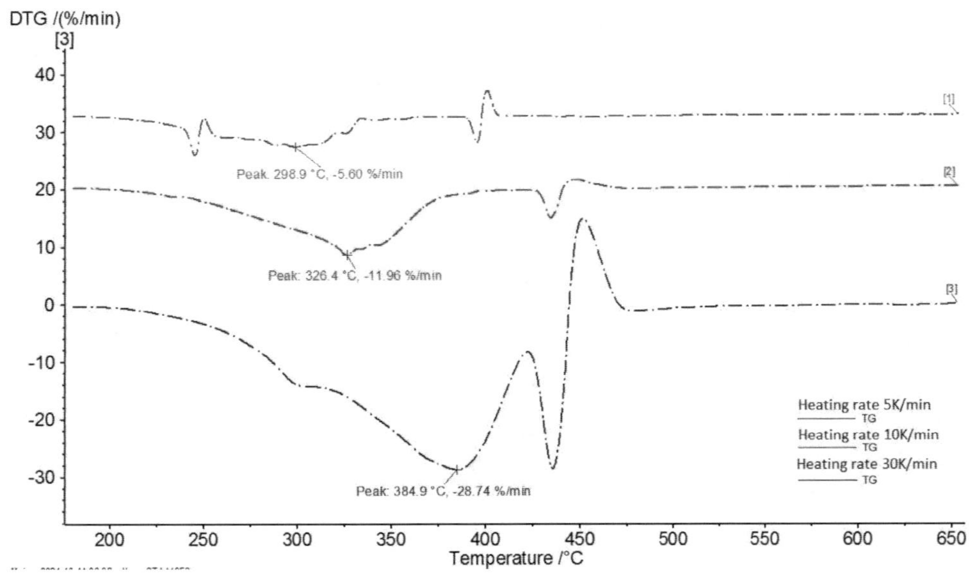

Figure 3. Decomposition rates of the main process of the wax material.

The DTG maxima detected in our study correspond to those of reported in [13]. Depending on the type of paraffin, the authors observed the DTG maxima in range 288 to 483 °C.

4. Conclusion

The methods of thermal analysis were used to evaluate thermal performance of the candle wax material. The findings revealed, that the material was composed not only of the palm stearin as assumed before the analyse, but was likely a mixture of palm stearin and paraffin. Another instrumental method, e.g. infrared analysis will be performed to find the actual composition. The exothermal peaks in the DSC curves confirmed the amount of the energy released by decomposition up to 772 J/9 depending on the rate of heating. Thermal analysis was proved to be useful tool in evaluation of thermal properties of candle wax materials to support the calculation of optimal ventilation rates in room and to predict a potential of candle wax material for energy storage.

5. References

[1] European Candle Association ASBL 2019 (Stuttgart, Germany) 169
[2] Massoudi R and Hamidi A 2017 *International Journal of Tropical Disease & Health* **24(2)** 1
[3] Derudi M, Gelosa S, Sliepcevich A, Cattaneo A, Cavallo D, Rota R and Nano G 2013 *Environmental Science and Pollution Research International* **21**
[4] Ahn JH, Kim KH, Kim YH and Kim BW 2015 *Journal of Hazardous Materials* **286** 242
[5] Pagels J, Wierzbicka A, Nilsson E, Isaxon C, Dahl A, Gudmundsson A, Swietlicki E and Bohgard M 2009 *Journal of Aerosol Science* **40** 193
[6] Kim KH, Szulejko JE, Ahn JH 2016 *Journal of Hazardous Materials* **303** 172
[7] Kumar PM, Saravanakumar PT, Mylsamy K, Kishore P, and Prakash KB 2019 *AIP Conference Proceedings* **2128** *020027*
[8] Hoffmann DJ, Burr MT, Kroll MJ and Logan LM 2014 *Fire Technology* **50(6)** 1379
[9] Harikrishnan S, Magesh S and Kalaiselvam S 2013 *Thermochimica Acta* **565** 137
[10] Teles dos Santos M, Viana IS, Ract JNR and Le Roux GAC 2016 *Journal of Food Engineering* **185** 17
[11] Nusantoro BP 2009 *Agritech* **29** 154
[12] Pradhan R and Ramaswamy A 2019 *SN Appl. Sci.* **1** 859

[13] Webber GV 2000 *Dissertation* (University of Stellenbosch, South Africa) 132

Acknowledgments
This work was supported the Slovak Scientific Grant Agency under the contract No. VEGA 1/0230/21.

IOP Conf. Series: Materials Science and Engineering 1252 (2022) 012014 doi:10.1088/1757-899X/1252/1/012014

The seismic response of cylindrical steel tank at Al Hoceïma city in Morocco

K Kotrasova[1], E Kormanikova[1], M Loukili[2], S Mansour[3] and D Foti[3]

[1] Technical university of Kosice, Faculty of Civil Engineering, Institute of Structural Engineering, Vysokoskolska 4, 042 00 Kosice, Slovakia
[2] University of Avignon, INRAE, UMR408 SQPOV, 84000 Avignon, France
[3] Politechnic University of Bari, Dept. of Civil Engineering Sciences and Architecture, 70125 Bari, Italy

E-mail: kamila.kotrasova@tuke.sk; eva.kormanikova@tuke.sk; mohammed.loukili@univ-avignon.fr; sulyman.mansour@poliba.it; dora.foti@poliba.it

Abstract. The aim of this article is determination of seismic response of cylindrical steel tank situated at Al Hoceïma city in Morocco. When a storage facility containing liquid starts moving due to earthquake loading, the fluid filling exerts additional hydrodynamic pressure on storage facility solid domain. The seismic effect of a liquid–filled storage facility converts the contained liquid into and equivalent spring-mass system representing impulsive mode together with convective mode of vibration of the storage facility - liquid system. The seismic response of cylindrical steel tank subjected to seismic loading was calculated by using the elastic response spectrum for locality Al Hoceïma city in Morocco. The rules of Eurocode 8-4 were respected for documented seismic analysis of storage facility containing liquid.

1. Introduction

The people have always had to store their supplies, both liquid and solid, including water and food. Over the centuries, however, social life and industry have evolved, thus the warehouses and the storage facilities have also developed. At the beginning of the 21st century, steel and reinforced concretes were later used for construction of storage facilities. The steel facilities have begun to use because they could be easily dismantled and reassembled [1]. The prestressed concrete materials were later added to the available construction technologies too for storage facilities. Recently, mixed systems have been used [2]. These facilities are mainly used, for example, for the storage of hazardous and flammable liquids.

The seismic performance of these reservoirs was of special importance, which exceeds the economic value of the storage facility in order to maintain functionality even after a serious earthquake [3]. The main reason is the potential danger associated with equipment failure among highly flammable products, which can lead to large-scale uncontrolled fires, and leakage and spillage of these liquids can cause serious injury to persons and livestock, extensive serious environmental damage, and life in affected populations [5]. In the past, a number of fluid reservoirs have failed [1, 6,7], see Figures 1 and 2. Such failures have aroused great interest and many scientists have studied the dynamic behaviour of fluid reservoirs analytically and experimentally to accurately predict the seismic responses of fluid reservoirs [8].

Early studies on the seismic response of cylindrical liquid filled storage facilities, such as Jacobsen and Housner [9-11], considered the rigid storage facilities [12-15]. Analytical approaches to analysis flexible storage facilities were determined by Veletsos [16-18]. Malhotra, Wenk, Wieland [19-21] developed simple procedure for seismic response of flexible liquid-storage facilities [22,23].

Figure 1. Damages of steel tanks a) 'Elephant's foot buckling [6], b) 'Diamond' buckling [7].

Figure 2. Total collapses and damages of steel tanks [1].

2. Simple procedure for seismic analysis

The knowledge of influence of liquid filling on solid domain of storage facilities during seismic even contributes to better design of these structures to be resistant to earthquakes and reliable in terms of safety and long-term use of these specific engineering structures – storage systems [24-27]. The seismic srudy of liquid storage facilities is essential in the seismic design of storage facilities, and it has been adopted the standards and codes as are IITK-GSDMA Guidelines for seismic design of liquid storage tanks [28,29], NZSEE guidelines, ACI 350.3, and Eurocode 8-4 too [30]. These codes use the different kind of models developed by G. W. Housner, A, S. Velestos, P. K. Malhotra and others. Only Eurocode is recommended to be used of absolute summation rule for the combination of impulsive and convective hydrodynamic effects. The mechanical model representing impulsive and convective modes of vibration of liquid filling in the tank – liquid system, see Figure 3.

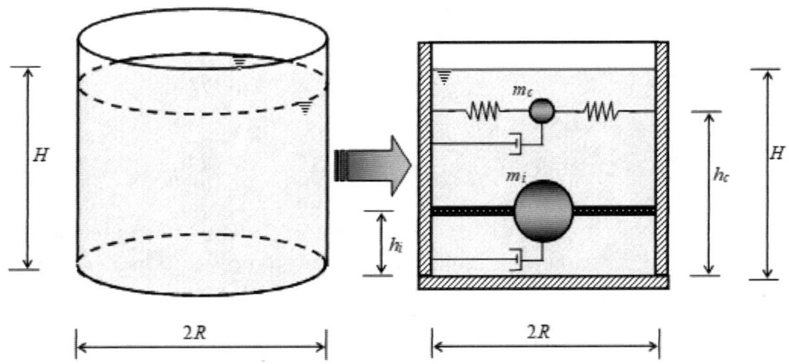

Figure 3. The ground supported cylindrical storage facilities and its single degree of freedom systems.

The time period of impulsive mode T_i [s] and the time period of convective mode T_c [s] of vibration for a ground supported cylindrical storage facility are given by

$$T_i = C_i \frac{H\sqrt{\rho}}{\sqrt{s/R}\sqrt{E}}, \tag{1}$$

$$T_c = C_c\sqrt{R} \tag{2}$$

where

- C_i [-] is coefficient for the natural period in impulsive mode,
- C_c [$ms^{-1/2}$] is coefficient for the natural period in convective mode,
- H [m] is height of the liquid filling,
- R [m] is inner radius of storage facility,
- s [m] is equivalent uniform wall thickness of the storage facility, it is weighted average for the wet height of the the storage facility wall,
- ρ [$m \cdot kg^{-3}$] is liquid mass density,
- E [Pa] is numerical constant described elastic material property of storage facilities called as Young modulus or modulus of elasticity.

The total base shear V [N] at the bottom of the storage facility wall is given by algebraically sum of the total base shear in impulsive mode and the total base shear in convective mode:

$$V = (m_i + m_w + m_r)S_e(T_i) + (m_c)S_e(T_c), \tag{3}$$

where

- m_l is total mass of liquid filling,
- m_i is impulsive mass of liquid filling,
- m_c is convective mass of liquid filling,
- m_w is mass of storage facility wall, and
- m_b is mass of storage facility base slab,
- m_r is mass of storage facility roof,
- h_i is the height from the base of the point of application of the wall impulsive component pressure resultant,
- h_c is the height from the base of the point of application of the wall convective component pressure resultant,
- h_w is the height of the centre of gravity of the the storage facility wall,
- h_b is the height of the centre of gravity of the the storage facility bottom,

- h_r is the height of the centre of gravity of the the storage facility roof,
- $S_e(T_i)$ is value horizontal elastic response spectrum in impulsive mode,
- $S_e(T_c)$ is value horizontal elastic response spectrum in convective mode.

The total bending moment M [Nm] at the bottom of the storage facility wall is given by algebraically sum of the bending moment in impulsive mode and the bending moment in convective mode

$$M = (m_i h_i + m_w h_w + m_r h_r) S_e(T_i) + (m_c h_c) S_e(T_c). \quad (4)$$

The total overturning moment M' [Nm], given by

$$M' = \left(m_i h_i' + m_w h_w' + m_r h_r' + m_b \left(\frac{t_b}{2} \right) \right) S_e(T_i) + (m_c h_c') S_e(T_c), \quad (5)$$

it is be obtained by algebraically sum of the total overturning moment in impulsive and the total overturning moment in convective mode. The total overturning moment M' is used to check the stability of the tank at the bottom of base slab.

Seismic coefficients $S_e(T)$ will be calculated separately for impulsive mode and convective mode, respectively. The impulsive spectral accelaration $S_e(T_i)$, obtained from an elastic response spectrum for damping value corresponding to the limit state. $S_e(T_c)$ is convective spectral acceleration obtained from a 0.5%-damped elastic response spectrum.

The values of C_i, C_c, the masses, m_i and m_c (as ratios m_i/m_l and m_c/m_l), as well as their corresponding heights h_i, h_c, h_i', h_c' (as ratios h_i/H, h_c/H, h_i'/H and h_c'/H), are documented as function of the tank slenderness parameter γ in Figure 4. The tank slenderness parameter γ is given by H/R.

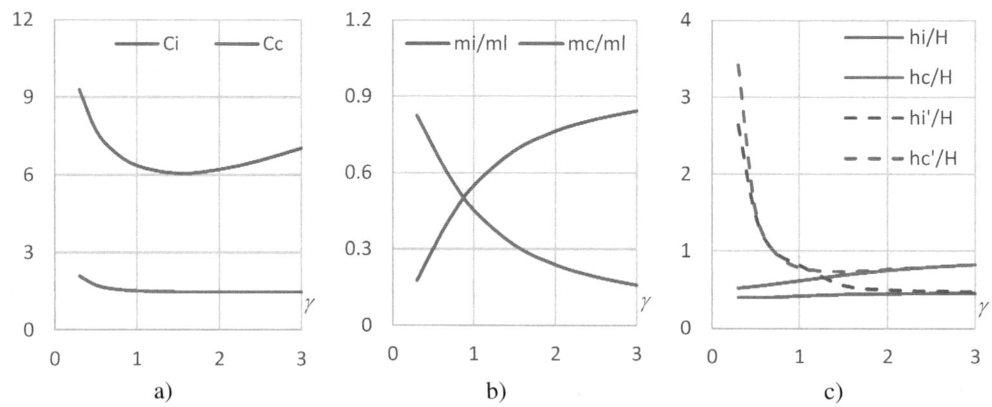

Figure 4. Seismic parameters as function of parameter γ (the slenderness) a) coefficients C_i and C_c; b) ratios m_i/m_l and m_c/m_l; c) ratios h_i/H, h_c/H, h_i'/H and h_c'/H.

3. Seismic effects of liquid filled storage facility

A roofted stainless steel tank with diameter 8.0 m and wall height of 8.8 m is anchored to a concrete foundation.

- *Liquid filling:* The tank is filled with Niger seed oil with the variable fluid filling heights. The fully filling height was considered 8m and the partially filling heights were considered: 6 m, 4 m, and 2 m. The mass density of oil is $\rho = 924$ kgm^{-3}.
- *Wall:* The tank walk is made from four parts by height, each part is 2.2 m high. The lower two layers are 1.0 cm thick and the upper two layers have 0.8 cm thickness. The tank stainless steel material properties are given: Young modulus (modulus of elasticity) of storage facility material is $E = 2.1 \cdot 10^{11}$ Nm^{-2}, and storage facility mass density is $\rho = 7930$ kgm^{-3}.
- *Roof:* The mass of the storage facility roof structure is 7.0 t.
- *Bottom:* Tank has the stainless steel base plate of 1.0 cm thickness supported on steel prestressed concrete base plate of 0.5 m thickness.

The seismic responses of the ground supported cylindrical stainless steel tank filled with Niger seed oil situated at Al Hoceïma city in Morocco for considered full filling of the tank 8 m, and the partially fillings 6 m, 4 m, and 2 m are presented in tables 5 and 6.

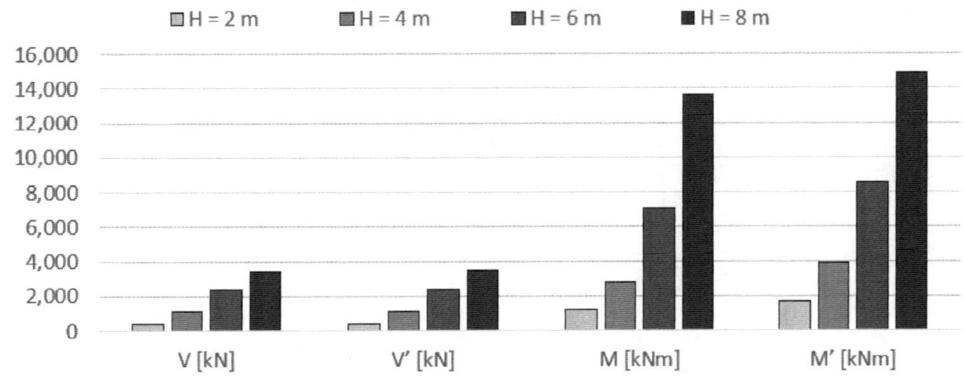

Figure 5. The total seismic response of fluid filled stainless steel tank for variable heights of fluid filling.

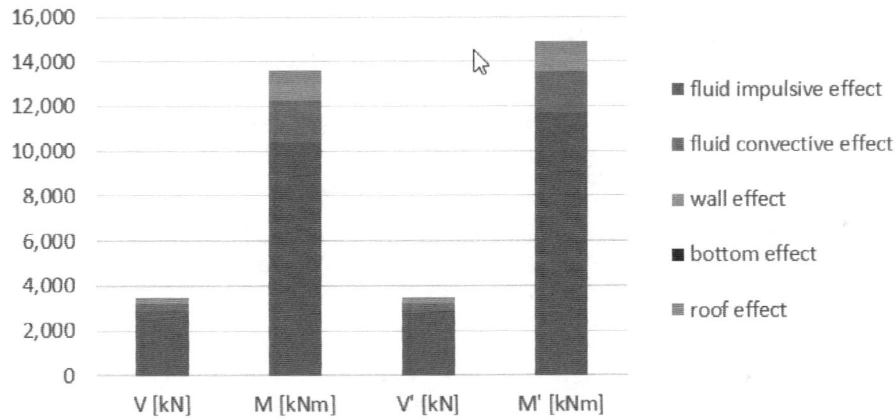

Figure 6. The components of the seismic response of stainless steel tank for 8 m heigh of fluid filling.

4. Conclusions

The ground supported cylindrical stainless steel tank filled with Niger seed oil situated at Al Hoceïma city in Morocco was excited by ground motion. The variable fluid filling heights are considered. The full filling of the tank is 8 m, and the partially fillings are 6 m, 4 m, and 2 m. The monitored responses of the interest were

- the total base shears,
- the total bending moments, and
- the total overturning moments.

It is seen from the components of the seismic response of stainless-steel tank filled to heigh 8 m that seismic effect of fluid is markedly dominant in comparing with seismic effect of tank solid. The peak values of seismic response fluid filled tank are given in case of full fluid filling of the stainless steel tank. The elastic response spectrum was considered as seismic loading for locality Al Hoceïma city in Morocco. The rules of Eurocode 8-4 were respected for this seismic analysis of liquid storage tank.

5. References

[1] Mermer AS, Kaya M, Arslan AS 2011 Using Seismic Isolation Elements to Protect Cylindrical Steel Liquid Storage Tanks from Destructive Forces of Earthquakes. *GU J Sci* **25** (1) 165-173

[2] Zdravkov L, Pantusheva M 2019 Typical damage in steel storage tanks in operation. *Procedia Structural Integrity* **22** 291-298

[3] Monteiro M, Svet V, Sandilands D, Tsysar S 2015 Experimental Investigations of Various Methods of sludge measurements in storage oil tank. *Advances in remote sensing* **4** 119-137

[4] Codex Alimantarius, International food standards 1987 Code of practice for the storage and transport of edible fats and oils in bulk CAC/RCP 36-1987

[5] Kotrasova K, Michalcova V, 2020 Numerical Modelling of Fluid Domain Flow in Open Top Channel. *Management of Water Quality and Quantity*, Springer Nature s. 287-306.

[6] Haroun MA 2004 Mitigation of elephant-foot bulge formation in seismically excited steel storage tanks. SMIRT18-K16-6, China

[7] NZSEE: Seismic design of storage Tanks 2009 197

[8] Loukili M, Dutykh D, Kotrasova K, Ning N 2021 Numerical Stability Investigations of the Method of Fundamental Solutions Applied to Wave-Current Interactions Using Generating-Absorbing Boundary Conditions. *Symmetry* **13**(7) 1153

[9] Priestley MJN, Wood JH, Davidson BJ 1986 Seismic design of storage tanks Bulletin of the New Zelands Nation Society for Earthquake Engineering **19** (4) 272-284

[10] Kotrasová K 2018 Vibration Analysis of Simply Supported Rectangular Tank Partially Filled with Water. *MATEC Web of Conferences* **210** 1-6 04003

[11] Silvestri S, Gasparini G, Trombetti T, Foti D. 2012 On the evaluation of the horizontal forces produced by grain-like inside silos during earthquakes. *Bulletin of Earthquake Engineering* **10** 1535-1560

[12] Komariah LN, Dewi TK, Ramayanti C 2018 *IOP Conf. Series: Material Science and Engineering* **543** 012033

[13] Trombetti T, Gasparini G, Silvestri S, Foti D, Ivorra S 2013 Analytical Developments and Shaking Table Tests on Dynamic Behavior of Squat Silos Containing Grain-Like Material. *Proceedings of the Conference "New Developments in Structural Engineering and Construction* Yazdani, S. and Singh, A. (Eds.), ISEC-7, Honululu, June 18-23, 1743-1748

[14] Kotrasová K, Loukili M 2020 The seismic response of fluid filling in rectangular reservoir at Nador city Morocco. *Civil and Environmental Engineering* **16** (2).267-275

[15] Seleemah AA, El-Sharkawy M 2011 Seismic response of base isolated liquid storage ground tanks. *Ain Shams Engineering Journal* **2** 33-42

[16] Michalcova V, Kotrasova K 2020 The Numerical Diffusion Effect on the CFD Simulation Accuracy of Velocity and Temperature Field for the Application of Sustainable Architecture Methodology. *Sustainability* **12**(23) 1-19

[17] Housner GW 1954 Earthquake pressures on fluid containers. California institute of technology, Pasadena California.

[18] Sho G 2012 Cost Optimal Selection of storage tanks in LPG vaporization station. *Natural Resources* **3** 164-169

[19] Foti D, Trombetti T, Silvestri S, Di Chiacchio L, Ivorra S, Tayolr C, Dietz M 2015 Analysis of the dynamic behaviour of squat silos containing grain-like material subjected to shaking table tests—ASESGRAM final report. *Geotechnical, Geological and Earthquake Engineering* **35** 437-457

[20] Kotrasova K 2017 Study of hydrodynamic pressure in wall of tank. *Proceedia Engineering* **190**, 2-6

[21] Totorean AF, Bernard SI, Susan-Resiga RF 2016 Fluid dynamic in helical geometries with applications for by-pass grafts. *Applied Mathematics and Computation* **272** 604-613

[22] Silvestri S, Ivorra S, Di Chiacchio L, Trombetti T, Foti D, Gasparini G, Pieraccini L, Dietz M, Taylor C 2016 Shaking-table tests of flat-bottom circular silos containing grain-like material. *Earthquake Engineering and Structural Dynamics* **45**(1) 69-89

[23] Jaiswal OR, Rai DC, Jain SK 2004 Review of code provision on design seismic forces for liquid storage tanks. IITK-GSDMA-EQ04-V1.0. Kanpur, Indian Institute of Technology Kanpur
[24] Livaoglu R, Dogangun A 2006 Simplified seismic analysis procedures for elevated tanks considering fluid-structure-soil interaction. *Journal of fluid and structures* **22** 421-439.
[25] Zulfiqar Y, Hyder MJ, Jehanzeb A, Ahmad HW, Zulfiqar A, Chaudry UM 2021 Structural Assessment of a Cylindrical Liquid Tank due to Shape Change of Roof during Sloshing Induced by Seismic Activity. *Arabian Journal for Science and Engineering* **46**, 8075–8085
[26] Kotrasová K, Kormaníková E 2017 The study of seismic response on accelerated contained fluid. *Advances in Mathematical Physics* **2017** 1-9
[27] Malhotra PK, Wenk P, Wieland M. 2000 Simple procedure for seismic analysis of liquid-storage tanks. *Structural Engineering International* **3** 197-201
[28] IS 1893: (Part 2), 2006: Criteria for earthquake resistant design of structures: Part 2 Liquid retaining tanks (revision of IS 1893 (part 2))
[29] IITK-GSDMA Guidelines for seismic design of liquid storage tanks – provisions with commentary and explanatory examples. Kanpur, Indian Institute of Technology Kanpur, 2005
[30] Eurocode 8 - Design of structure for earthquake resistance - Part. 4: Silos, tanks and pipelines. 2006

Acknowledgments
This research was supported by the Scientific Grant Agency of the Ministry of Education of Slovak Republic and the Slovak Academy of Sciences project VEGA 1/0374/19.

IOP Conf. Series: Materials Science and Engineering 1252 (2022) 012015 doi:10.1088/1757-899X/1252/1/012015

Long term accumulation of heat energy from the sun

M Gorás[1], J Domanický[1] and F Vranay[1]

[1] Institute of Architectural Engineering, Faculty of Civil Engineering, Technical University of Kosice, Vysokoskolska 4, 040 23 Kosice, Slovakia

E-mail: michal.goras@tuke.sk; jan.domanický@tuke.sk; frantisek.vranay@tuke.sk

Abstract. The source of hot water production for the investigated storage tanks are solar collectors. Solar collectors provide recharging of underground storage tanks all year round. Due to the large active area of solar collectors - 225 m2, there are also large tanks for long-term accumulation. Accumulation tanks are in 3 sizes with a total volume of 160 m3. With regard to the size of the tanks, the mode is switched between the tanks according to the season. The tanks are made of a reinforced concrete packaging structure and are stored under the terrain of the Center of Research anf Innovation in Construction (CVIS laboratory).

1. Introduction

This article deals with energy sources and their storage. It is based on a combination of the strongest source - solar energy with its capture in the form of storage tanks. The article discusses the topic of heat accumulation according to the size of the tank and the length of heat storage. Design and solution in CVIS laboratory.

2. Solar energy and its potential

By far the largest energy resource is the sun. Annually, 3.9×10^{24} J = 1.08×10^{18}kWh of solar energy reaches the surface of the Earth. This is about 10000 times more than the annual global primary energy demand and much more than all available energy reserves on earth. In other words, using one ten-thousandth part of the incoming sunlight would cover the whole energy demand of mankind. Figure 1 illustrates these values with energy cubes.[1]

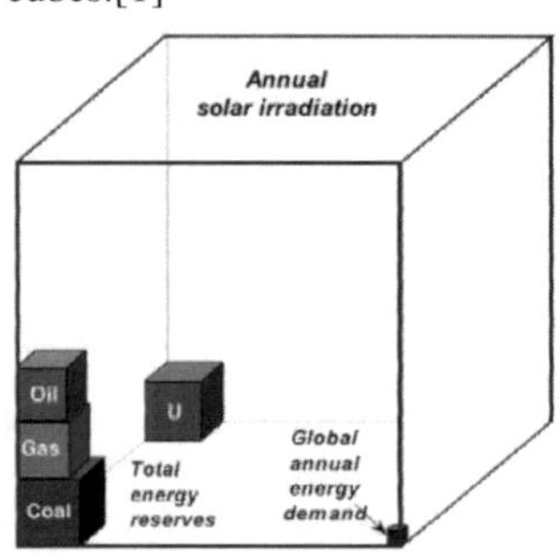

Figure 1. Energy Cubes: the Annual Solar Irradiation Exceeds Several Times the Total Global Energy Demand and All Fossil Energy Reserves [1].

3. Thermal storage tank

Immediately after obtaining energy, it is necessary to think about its transfer. It rarely happens that the energy gained is needed at some time. Therefore, it is necessary that the storage tank not only accumulates energy but also stores it, for a time when for example the sun's rays are very weak.

The ideal storage tank should look like it would receive all the energy (heat) and keep it for as long as possible, of course without heat loss. However, the closer we get to this requirement, the higher the price of the device. Technologies in today's world are very advanced, but a reservoir that meets these features has not yet been invented.

The choice of tray consists of 4 aspects. Capacity, size, heat loss and price. [2]

3.1 Thermal capacity of tank

The capacity of the tank depends on the storage medium and the method of storage. We know two basic types of accumulation. Chemical and thermal accumulation. The connection is shown in Figure 2. Aquifer - in this case it is an underground natural space as a reservoir of water for heat accumulation.[2]

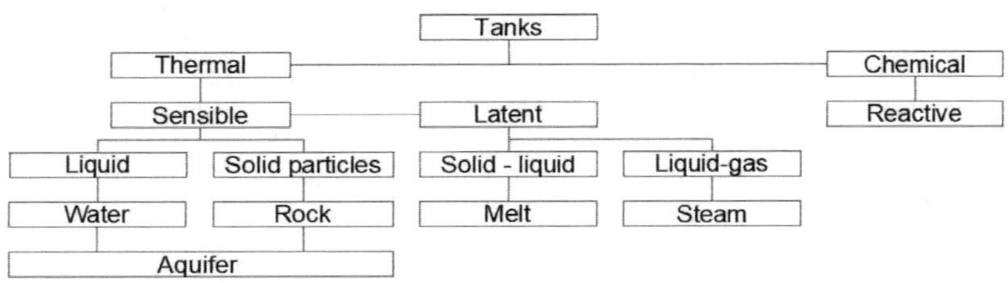

Figure 2. Types of tanks [2].

Chemical energy is in the form of oil, gas or gasoline. There is bound energy released by combustion. However, so far it has not been possible to convert this energy into pirate conditions and efficiencies.

Nowadays, only thermal storage tanks are used. They are divided into storage tanks for sensible and latent heat. Latent heat is generated when the state changes. E.g. when switching from ice to water, just as much energy is needed as heating the water to 80 ° C. Sensible heat represents the behavior of the temperature during the supply and consumption of energy. Sensible heat is derived according to the equation. [2]

$$Q_{Sp} = m.c.(\theta_2 - \theta_1) \tag{1}$$

The volume of the tank depends on the heat capacity of the materials. Several of them are described in Table 1.

Table 1. Specific and volumetric heat capacity of materials for heat accumulation at 20 ° C [1].

Medium	Temperature range (°C)	Specific heat capacity (Wh / kg.K)	Volume heat capacity (Wh / m3.K)	Density (kg/m3)
water	0-100	1.16	1160	998
air	-273->1000	0.28	0.31	1.1
oil	0-400	0.44-0.5	350-450	800-900
gravel, sand	0-800	0.2	360-390	1800-2000
granite	0-800	0.21	570	2750
concrete	0-500	0.24	460-560	1900-2300
brick	0-1000	0.23	330-440	1400-1900
iron	0-800	0.13	1000	7860
gravel backfill (37% water)	0-100	0.37	810	2200
mix of salts (53KNO3 + 4NaNO2 + 7NaNO3)	150-450	0.36	480-550	2561-2243

The heat in the Q_{Sp} storage tank depends on the weight of the medium (m), the heat capacity (c) and the temperature difference. If we have a 200 liter tank and needed to heat it from 10 ° C to 60 ° C, the energy needed would be:

$$Q_{Sp} = 200.1{,}16.(60 - 10) = 11600 \text{ Wh} = 11{,}6 \text{ kWh} \quad (2)$$

3.2 Tank size

The size of the tank depends on water consumption. It is necessary to take into account the morning and evening peaks. Table 2 shows the energy needs.

Table 2. Guideline values for dimensioning solar thermal storage tanks.

	The need for energy (kWh/d)	Storage medium	Accumulation time (days)	Range of usable temperatures
hot water preparation	2-3kWh/d per person	water	2-3	45-60°C
Low temperature heating	0.5-1kWh/d per m2 of living space	water, rock	2-20	30-60°C
Solar house heating	0.2-0.4kWh/d per m2 of living space	water, rock	60-120	25-40°C

Conventional solar home storage tanks range from 150-1000 liters (1.5 - 2 times the daily requirement). Therefore, they are much larger than hot water tanks, which have a maximum capacity of one day.

3.3 Heat loss

Heat loss is the main enemy of any accumulation. They occur whenever the medium temperature is higher than the ambient temperature (usually the temperature of the technical room, etc.)

Heat loss is the main enemy of any accumulation. They occur whenever the medium temperature is higher than the ambient temperature (usually the temperature of the technical room, etc.) To reduce heat loss, it is necessary to insulate the entire surface of the tank well and tightly so that there are no air gaps between it and the insulation. The problem occurs with outlets from the tank. Figure 3 shows the tank and possible heat leaks. [3]

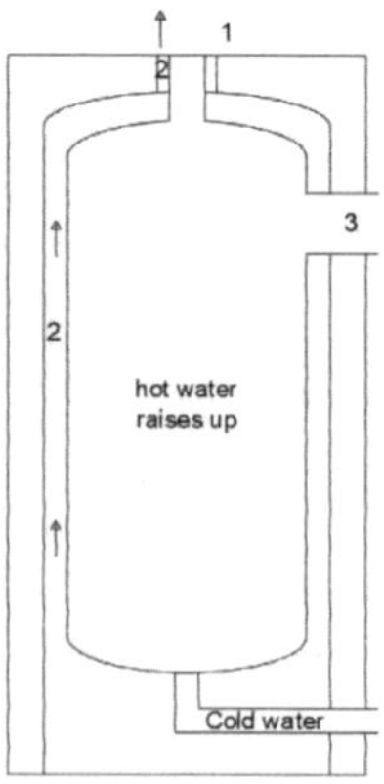

Figure 3. Heat loss in tank.

1. When hot water is drawn through a pipe led directly upwards, self-gravity circulation occurs, leading to heat loss.
2. The tank can be cooled by air flow in the joints between the tank and the thermal insulation.
3. Protruding flanges that are insufficiently insulated are often seen.

IOP Conf. Series: Materials Science and Engineering 1252 (2022) 012015 doi:10.1088/1757-899X/1252/1/012015

The heat losses on the storage tank are calculated using the equation:

$$W_{Sp} = k_{Sp}.A_{Sp}.(\theta_{Sp} - \theta_L) \tag{3}$$

In a given equation, k_{Sp} is the storage heat loss coefficient, analogous to the collector coefficient. Usually, the value of ***k.A*** is given in W / K - quantifies the heat flow that the tank transmits to its surroundings with each ° C of heat difference. For a standard 250-500l tank, this value reaches the size of 1.5 - 3.5 W / K.

In this case, the time constant is also interesting, which indicates the time during which the accumulated energy drops to 37% without consumption. It depends on the volume to surface area of the container. Table 3 shows the time constants for certain types of containers. Figure 4 shows the cooling with the tank switched off.

Table 3. Time constants for certain types of container

Storage	Capacity (m^3)	Surface area (m^2)	Capacity / Area (m)	Storage time constant
250 L steel cylinder tank r= 0.5m ; 1.3m high	0.25	2.4	0.104	241 h = 10 d
500 L steel cylinder tank r= 0.6m ; 1.6m high	0.50	4.4	0.113	262 h = 11 d
1000 L steel cylinder tank r= 0.8m ; 2.0m high	1.00	6.1	0.164	380 h = 16 d
5 m^3 cubic steel tank 1.7 x 1.7 x 1.7m	5.00	17.6	0.284	659 h = 27 d
20 m^3 cubic tank 2 x 2.5 x 4m	20.0	46	0.435	1009 h = 42 d
1000 m^3 cylinder tank r = 11m ; 10m high	1000	554	1.8	4176 h = 174 d

The time constant is the time during which the energy content of the storage tank without consumption drops to 37% of the flood value. The stated values are in the theoretical area, in practice they are usually less favorable. Time constant = capacity. Specific heat / surface. k-value
Specific heat (water) = 1.16 kWh / m^3K, k (insulation) = 0.5W / m^2 K

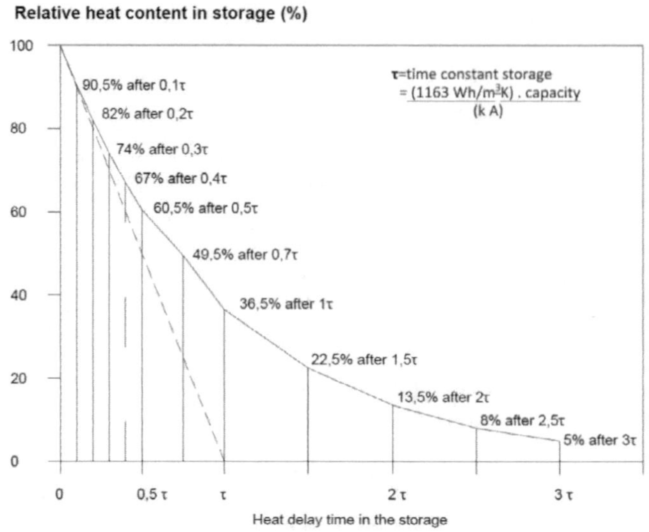

Figure 4. Relative heat content in storage.

4. Thermal accumulation in CVIS TUKE laboratory

The source of hot water production for the investigated storage tanks are solar collectors. Solar collectors provide recharging of underground storage tanks all year round. Due to the large active area of solar collectors - 225 m2, there are also large tanks for long-term accumulation. Accumulation tanks are in 3 sizes with a total volume of 160 m3. With regard to the size of the tanks, the mode is switched between the tanks according to the season. The tanks shown on figures 5-7 are made of a reinforced concrete packaging structure and are stored under the terrain of the CVIS laboratory. [4]

Figure 5. Thermal storage (CVIS).

Figure 6. Thermal storage (CVIS).

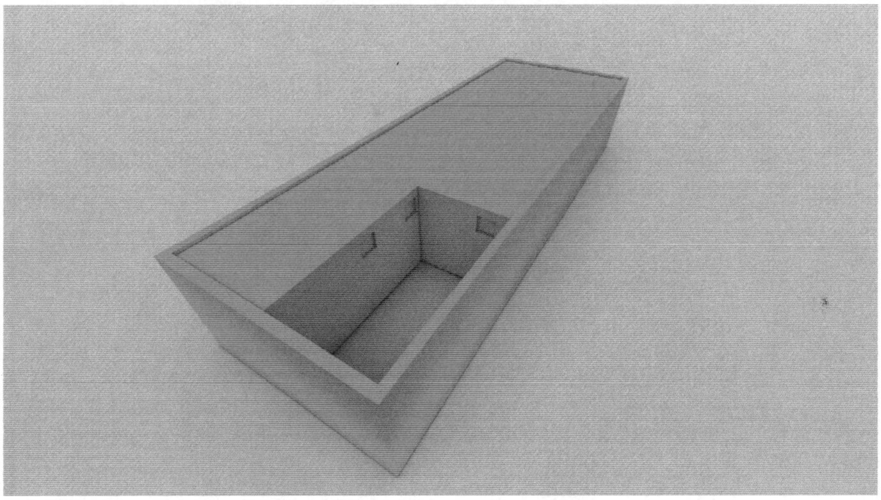

Figure 7. Thermal storage (CVIS).

The design of the tank consisted in dividing the capacity into 3. types of tank size. Ongoing measurements will indicate whether the data in Table 3. are correct and applicable to this type of tank as well. The use of the tank will be used for a water-to-water heat pump. The primary source in the primary tent is groundwater with a temperature of approx. 8°C-10°C. If the heat pump will have a primary water source warmer from the tank this will also adjust the COP coefficient. In this way, we will reduce energy consumption. [5]

4.1 Control scheme of CVIS TUKE laboratory

The storage tank control scheme allows (Figure 8) for thumb control and adjust charging or discharging as needed. The tanks need to be switched according to the season. An experimental house (scheme shown figure 9) on which we can simulate the season is built above the reservoirs for energy consumption. These conditions are perfectly prepared for the real situation.

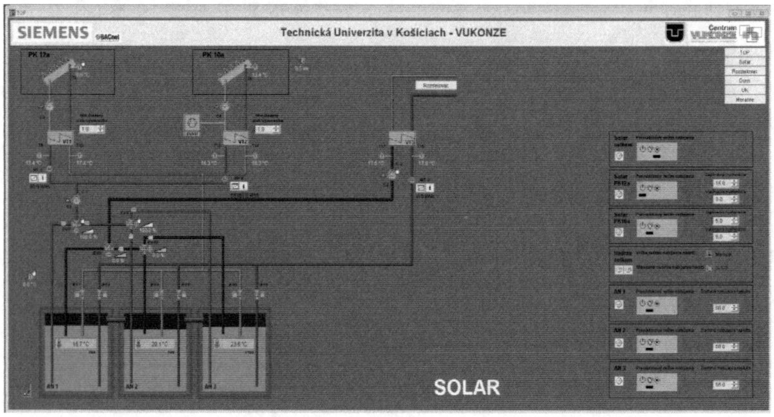

Figure 8. Scheme system of thermal energy storage.

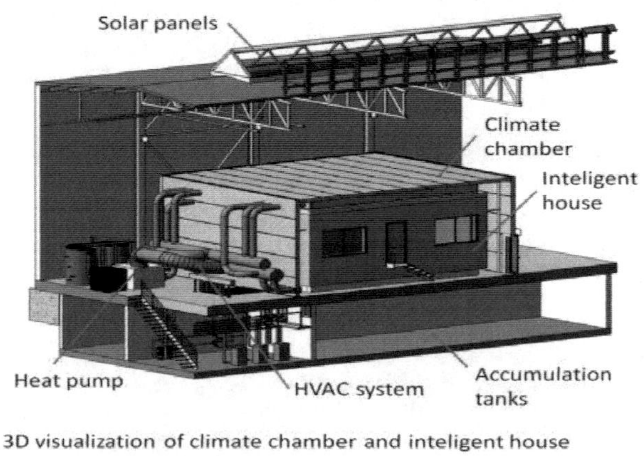

Figure 9. Experimental house for thermal energy storage.

4.2 Measurement CVIS TUKE laboratory

Measurement and measurement methodology are definitely. To evaluate the data, it is necessary to measure several different seasons and observe the charging, discharging of the tank. Charging and discharging of the tank is currently without consumption and the course of temperatures during discharging is monitored. The resulting data needs to be taken as a whole and not just a few days. The graphs (Figures 10 and 11) show the 50-day and 200-day curves. Figure 12 shows the system diagram and measuring devices.

IOP Conf. Series: Materials Science and Engineering 1252 (2022) 012015 doi:10.1088/1757-899X/1252/1/012015

The course of temperatures measured on the system of solar collectors and storage tanks (200 days).

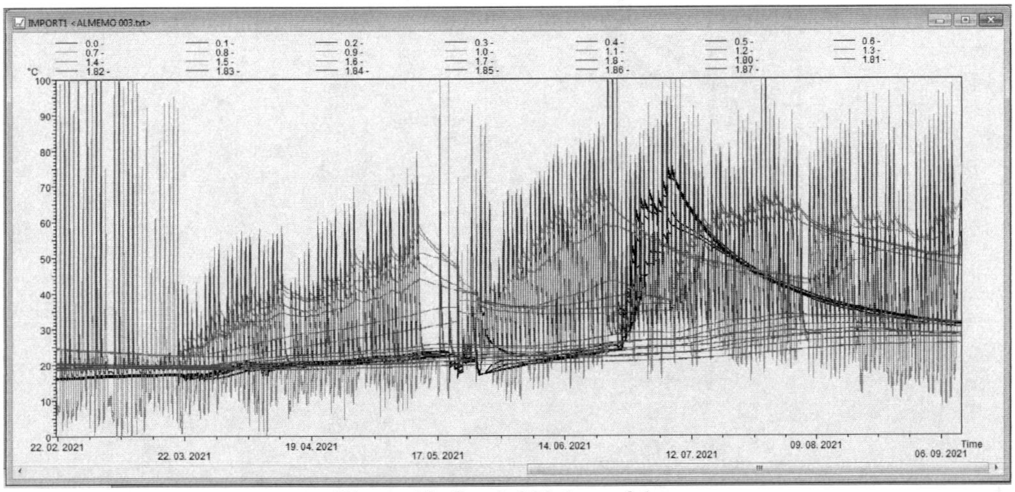

Figure 10. Graph 200 days of data.

The course of temperatures measured on the system of solar collectors and storage tanks (50 days).

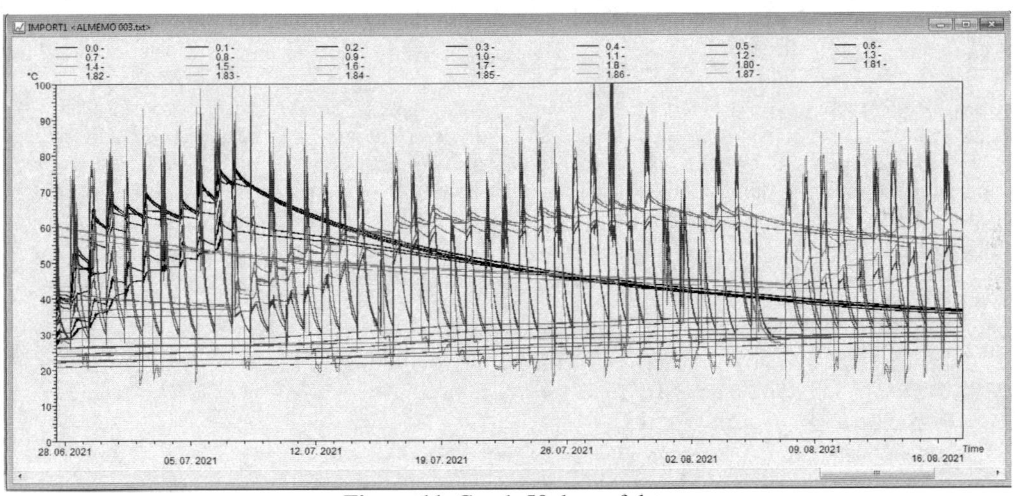

Figure 11. Graph 50 days of data.

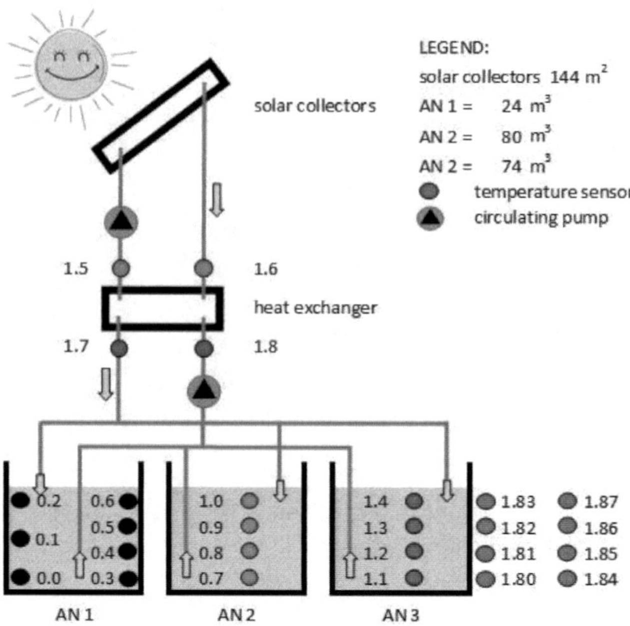

Figure 12. Location of sensors.

5. Future research

In future research, it is necessary to deal with real data of discharging and charging the tank with or without energy consumption. The research will point to the length of accumulation and the necessary improvements in water quality adjustment, water evaporation solutions. The sensors are located at different heights of the reservoir and also in places. These data will show the stratification of hot water and energy use.

6. References

[1] Hanser Verlg C Gmbh & Co KG 2005 *Understanding Renewable Energy Systems*

[2] Ladener H and Späte F 2001 *Solaranlagen* (Staufen bei Breisgau) ISBN: 80-247-0362-9

[3] Sarbu I and Sebarchievici C 2018 *A Comprehensive Review of Thermal Energy Storage. Sustainability* ***10***, 191.

[4] Kapalo P 2021 Nesica conferences presentation of *Thermal energy storage*

[5] Dahash A, Ochs F, Bianchi Janetti M and Streicher W 2019 *Advances in seasonal thermal energy storage for solar district heating applications: A critical review on large-scale hot-water tank and pit thermal energy storage systems, Applied Energy* **239** pp 296-315

Acknowledgement

The article was created with the support of the project Hungary-Slovakia-Romania-Ukraine ENI CBC Programme 2014-2020 HUSKROUA/1702 New Energy Solutions in Carpathian area.

IOP Conf. Series: Materials Science and Engineering 1252 (2022) 012016 doi:10.1088/1757-899X/1252/1/012016

Transit Smart Solution for Bratislava – Example of Tramway Network Development

T Schlosser[1], P Schlosser[2], M Korfant[3], G Bálint[1] and J Takács[1]

[1] Department of Transportation Engineering, Faculty of Civil Engineering, STU Bratislava
[2] DOTIS Consult, Ltd.
[3] Department of Traffic Engineering, City of Bratislava

E-mail: tibor.schlosser@stuba.sk; peter.schlosser@dotis.sk; matus.korfant@bratislava.sk; gabriel.balint@stuba.sk; jakub.takacs@stuba.sk

Abstract. The research presented was on the table more years and covers the cooperation of the University staff with City of Bratislava binding with the activity of private company. The base task was how to change the tremendous destruction of modal split in Bratislava, which growing on the side of private car usage. Results of couple of traffic studies show, that how could be this change re-define the current traffic behaviour and create for the new area new functional interest and certainly the new quality of relationships of land use. The system solution is in the amplification by a capacitive new tramway network. In the conditions of Bratislava, we need to change the behaviour of the people in the mobility over the area of the Capital city of Slovakia. The task was to solve the problem in Bratislava agglomeration as well. The solution is shown as a part of the integrated public transport services possibility to operate regular all days' mobility problems in the city and not only to cover the rush hours. This article is an overview of 5 years work of traffic engineers and planners to keep clear systematic approach and solutions with developers and municipality administration.

1. Introduction - Bratislava and tramway deployment

The fact that the size of the city of Bratislava as well as the average density of the population of the city does not allow us to economically solve the dense system of vertical segregation (f.e. for subway or elevated automatic PT system), it should be proposed tramway routes considered from the point of view of streets capacity through the traffic lights controlled junctions. Comprehensive system of vertical segregation in Bratislava is not expected due to the low values of the PT passenger volume (daily, peak period or hour's volumes). The objective value of the mobility by public transport in the top hours on the route should extend at least 15.000 passengers in one direction. Then we can reflect on the PT routes with the vertical segregation – on the subway or over ground track. In a gradual analysis and technical preparation of finding new tramway lines for the city of Bratislava is recommended to create a view on the options for changing the modal split in favour of public transport. The carrying urban public transport must be segregated, because the bus and trolleybus transport, which uses the same infrastructure of road infrastructure, is part of the saturated flows, with which are connected the creation of jams and congestions.

Bratislava has 440,000 inhabitants, but real lives in the city occurs up to 670,000 inhabitants and another 100,000 as daily commuters [1] and its average density of the settlement of the urbanized part is on the interface of the Old Town and Ružinov boroughs approx. 4000 inhabitants/km^2 (figure 1). More than 2 million m^2 of new features will be added in the new city centre with a prerequisite of up to 250,000 people present daily. The issue of PT operation is given in this area by bus and trolleybus lines. Capacity in one direction is about 4000 pass.h^{-1}. Tramway reaches the passengers capacity up to 9000 pass.h^{-1}, which is more than twice. Naturally, tramway creates unambiguous competition and in Bratislava conditions, is horizontally segregated from car transport. Bus and trolleybus transport, even when they have bus-lanes in certain sections, have been limited by saturated automotive flows in a dense network of traffic lights junctions.

In the *Land Use Master Plan (LMP)* [2] is defined the public transport such that track (rail) transport is the carrier system of the area of Bratislava and bus and trolleybus transport is an additional system to it. In *Transport Master Plan (TMP)* [3], this public transport scheme is also developed in the transport model. The part of the whole problem of the quality of public transport services was analysed according [4]. The city of Bratislava has resolved the basic strategic traffic service by radial-ring system, where the main urban public transport system – tramway is only some of the radials. The advantage is that in the obligatory part of LMP a rail track infrastructure is suitably defined. In this case, LMP includes the railway classic infrastructure and the tramway track as well as a special railway – in the distant future underground one. Only this rail infrastructure must be still to add on new sections, which was studied by technical as well as traffic-engineering tools.

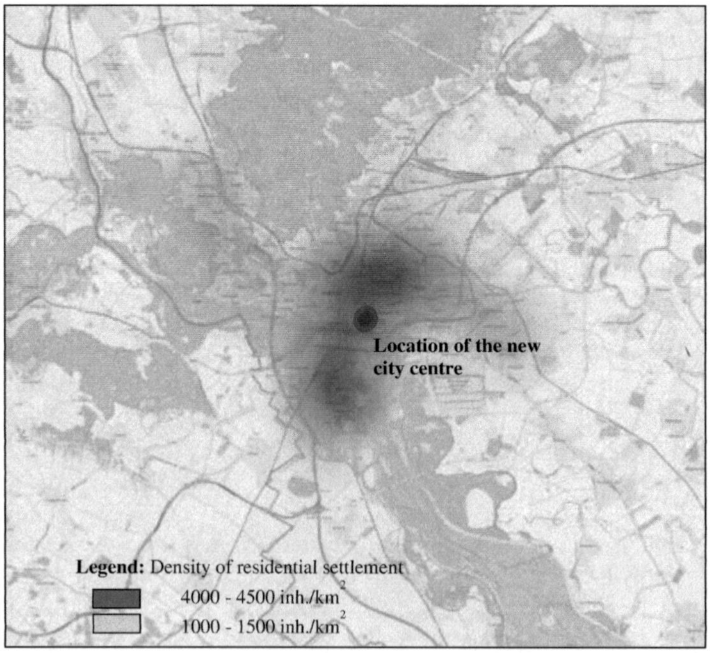

Figure 1. Density of settlement of Bratislava and location of the new city centre.

Tramway transport in Bratislava is almost everywhere horizontal segregated, most often in the middle of the median strip of the streets. Additional public transport is represented by the bus and trolleybus transit. Required street space, however, is not everywhere available. In our case, it means the tramways have to be defined in new routes in capital city of Slovakia. The added task is according to [5 and 6] how to determine the possibility to change the modal split in favour of the carrying system of public transport (CS PT). This is a main issue, which represents the new management of the tram routes close to the newly urbanized area of the new city centre of Bratislava. With this task is to reasonable demonstrate a reduction of traffic volume of individual transport by cars (ITC) because of the use of new routes of tram transit, and so change the modal split in favour of public transport. The basic method was used similarly as ultimately proposed in the project of Petrzalka trams – south radial of the city of Bratislava according to [7]. The similar procedure was used for developing areas of Bratislava in [8] for Nord-Western Area of Bratislava for creating a new PT system designing and with two new city bridges across the river Danube [9].

2. Public transport model for Bratislava – overview and methodology

The transport model is based on the transport model of city of Bratislava from TMP [3]. Subsequently, in the PT section of the model, it has been updated, which included the update of all PT line management and timetables. Weighting of the PT Transport Network was performed using "timetable assignment procedure". In case if the public passenger transport is modelled on the basis of exact timetables (ie Timetable Assignment Method) or time intervals of lines, edges are every section of the joint drive (lines) between two stops. Each PT link (line) represents a sequence of follow-up edges and nodes (public transport stops). For edges (links, lines) of public transport, these parameters need to be included in the modelled area defined in [10], as:

- length of PT line,
- capacity (means of transport),
- demand of passengers,
- driving time between stops,
- departure and arrival times in each stop.

For public transport stops, there is a need to include these parameters in the model area:

- capacity,
- accessibility (access time to and from the stop),
- the capacity of the passage in individual directions,
- delay in stop including delay of transfer (movement time within a stop and waiting time on transfer).

From the perspective of the potential of the development area of the new city centre, new sub-zones were created in the form of separate investment intentions, where data on the daily number of persons present in individual zones, which were subsequently inserted into a new origin-destination (OD) matrix. Estimation of the passenger's volume of new journeys in the public transport is based on the envisaged number of daily persons present in the new zone of the new city centre (detailed investors' data) and the current modal split of PT system in Bratislava. Subsequently, a new number of PT journeys among the individual sub-zones of the model (17 new zones in the OD matrix) on the basis of "Main Zone", which was presented by the original Zone. Size of the personal car trips was found based on extensive and detailed transport surveys. Thus, we know how much personal car journeys generate every intention in the new city centre because the city of Bratislava has an approved methodology *MTCA* [11] to determine the number of journeys from new intentions by different functions during the day.

To express changes in the number of passengers transported on the transport network after the establishment of new tram lines is a suitable tool the transport model for the right to display these changes to show whether in graphic form or numerical (counting, exiting stops, persons, vehicles, etc.). In Bratislava is the *PTV VISUM* [12] used for modelling. *PTV VISUM* software works in multiple ways when loading. For the calculation of the load, we chose the timetable assignment procedure, where every traveller is looking for a connection from the origin to the destination based on the leaving means of the exact timetable. Parameters in the process of selecting a means of transport are as follows:

- In-vehicle time,
- PuT-Aux ride time,
- Access time,
- Egress time,
- Walk time,
- Transfer wait time,
- Number of transfers,
- Transport system impedance,
- Perceived journey time.

Parameters for the selection of the means of transport remained unchanged, i.e. according to the original default software settings, as there are no exact studies that would know to define what changes to the results are chased by individual parameters.

2.1. The extent of the zone in terms of details and the whole city

The basic definition of the problem and its solutions with the new approach is derived from [7]. The development of the new city centre of Bratislava is in detail enough published in terms of the development of the land use and the issues of dynamic traffic in [5, 13, 14, 15], and [16]. The main problem of the city of Bratislava is, however, to establish permanently (persistent) system, continuous and well-defined solution of mobility relationships so that the people were to decide to use to a greater extent the public transport. The problem of public transport and ensuring the serviceability of the land use with public transport as the main – carrying system is proposed currently on the area of the new centre. Traffic-Urban design principles, which are part of the traffic planning work, are defined in [17]. Another solution to comply with major quality parameters on the operation of the new centre is not even possible. According to [18] "*the hierarchy and structure of public spaces in the territory is further compounded by the fact that this land use with the agglomeration civic amenities is the objective of not only the citizens who in this territory live and work and as a priority - they spend their free time but also visitors from other parts of the city, the surroundings of Bratislava and in connection with specific functions well as visitors from the whole of Slovakia*", or the eastern part of Austria or from the south of Hungary. The initial element from the point of view of the service of the area – the public spaces of the new city centre – is accessed from the point of view of the "mobility volume" of the population of Bratislava in the sequence of different modes of transport [19]:

1. walking,
2. public transport,
3. alternative transport (cycling and other),
4. car dynamic traffic.

One of the reasons of the technical solution of the servicing this area of the new centre is to link lines of tramway transport to address the transport from/to the centre and at the same time provide the possibility to perform transport tangential relations. In this way, it is about the change of modal split in favour of public transport at the expense of individual car transport. In this case, it can be mentioned that several studies [20, 21, 22 and 23] examined and created a space where we can reflect the technical solutions for the routing (tracking) of tramways around the area of the new centre. To put inside the problem of the quality of public transport service for changing the modal split could be used some criteria defined by [4]. The guiding principle must be the opinion, however, that supports expert decision-making, that the new city centre needs all of the proposed tramway lines. This decision has to confirm the present traffic-engineering analysis by creating a transport model, which was tested in the several of scenarios of the time-development of the area. The essence is the progressive attribution of the origin/destination relationships for individual investments, as well as tangential (through traffic) citywide transport flows in relation to the new centre of the city. This detailed solution has the objective to demonstrate the necessary implementation of carrying PT system, not only for the new centre of the city, but with the possibility of qualitatively higher level of servicing Bratislava by public transport.

A set of investments inside of the new city centre of Bratislava (more than 2 mil. m^2 of new functions) with their detailed entrances/exits (72) of buildings were tested. The accessibility was significantly assessed through the solution of servicing the centre area by the public transport system, which were oriented for tramway transit [19] and [24]. In particular, the detailed analysis of accessibility by pedestrians was performed in study by [18].

The strategy of redistribution of the division of modal split from personal cars to tramway public transport was actively addressed in the modelling. Isochrones and isodistances in detail by the mapping street accessibility (no ideal circle solution) were specifically monitored in various parameters in terms of time and distance. The basic parameters in terms of sustainable mobility for city centre are the mapping of pedestrian transport, public transport and parking policy. These three attributes are particularly important area for sustainable servicing of all kinds of traffic flows. Complex analysis of multimodality for the new city centre and its results can define if we have really got a „smart solution". The example of a basic analysis of pedestrian flows mentioned above according [25] was a good impulse as a basic step. However, it lacks measurable parameters and a link to other modes of transport. Other parameters of the quality of operability to be smart in city centre are the classification of the reduction of the number of trips to work and the system of public cycling (bike sharing), which increase the suitability of the total cycling transport in the city centre.

Among the basic steps of complex analytical activities meet the principles as defined from [18] and confirmed in [19], which can be determined as follows:

1. a complex design of the transport model of the public transport from the Bratislava *TMP* [3] according to the new principles for the design of urban roads, and the working versions of the various existing materials, e.g. [10], and [13, 24 and 26],
2. evaluate the modal split and the transition from individual car traffic to public transport according to [19],
3. calculate the perspective of the traffic flows in several variants and scenarios of tramway tracks/lines around the new city centre, which were coming out on detailed scenarios of the car traffic for testing of changing the modal split,
4. the demonstration of the collapse and saturated car flows in the street network in the new city centre without the solutions of the new tramway lines,
5. identify the sections with the highest traffic volumes or overloaded street network and the prospective network of urban public transport in the vicinity of the city centre,
6. to test the new tramway lines, which creates a new supply and demand for passengers and the inhabitants of the city according to the current conditions.

A comprehensive analysis was preceded by a very detailed capacity assessment of the dynamic traffic flows on each link and junctions on the concerned urban road network, which were based on the long-term and detailed traffic surveys published in [5 and 16].

The scope of the new city centre operation is characterized by the inside city through traffic – passage (of all transport modes) between the west and southern part of the city, which are homogeneous zones of housing and eastern part of the city that presents the job opportunities. The total volumes of passengers in all modes of transport reach values up to 220,000 per day. These are indirect passenger flows and these values will remain also in future. The new city centre is in the former brownfield and part of the river port and creating a new down town with administration, services and housing will become a new object of activities. In addition, in this area is the new bus station [27] with daily turnover is about 25,000 passengers. An additional 250,000 people are estimated in this space that will have its destination in the new city centre, what are direct passenger flows.

The tourist flows according the extent of the transport model, which is oriented to abroad towards Austria and Hungary, because of the geographical position of Bratislava, is classified as indirect attraction and do not have any influence to the analysis of the new city centre.

2.2. Methods of the tramway modelling for the new city centre in Bratislava

Changes in the number of passengers transported by other transport modes after the launch of new trams were not included in the study. However, these changes, respectively traffic volumes is possible to quantify, as is written above - whether in the form of cartograms, or subsequently numerically in the form personal km, vehicle km, average journey length etc. It is recommended for changes in car mode to use another type of transport model, i.e. for changes to the personal car usage a multimodal model that can express changes in transport behaviour and selection of means of transport (PT vs. IPC). For the needs of this study it was worked with a uni-modal model.

Bratislava suffers as every city with saturated traffic flows of individual transport with congestions at intersections exceeding the distance itself between the intersections. The goal is to reduce performance in personal car traffic. In the research used a transport model serves to express changes in the number of transported persons during the day in the tram after the construction of new tram lines. The PT transport model is built on data from transport surveys in public transport and hereby supplements the original 4-stage model from *TMP* [3]. There is no method of calculation of change in the traffic volume of other public transportation after the operation of trams, the increase in the volume of public transportation, and the decrease in the volume of cars.

In Bratislava conditions, transport modelling is used for more than 40 years to determine the transport prognosis. The older model was conducted classically as a multimodal demand transport model, just its results not convinced the traffic engineering experts and the city administration. This one was „flipped“

into PTV environments. Macro model that had through 250 zones without solving the external territory of agglomeration had untrustworthy outputs. Current transport surveys and calibrating them to the original (old) model did not show almost any responses and for individual urban parts had unbalanced values.

In case of the city of Bratislava is used by the transport model, macro model, because of that it also includes the territory of the Bratislava and Trnava self-governing region, as well as parts of Austria and Hungary. For example, they also report foreign resources [28], that mesoscopic model, resp. mesoscopic models (NAP Aimsun) tools should be used for smaller territories, e.g. Municipalities, cities. Macro models and macro model formation tools (e.g., VISUM) are suitable in assessing larger territories, e.g. regions.

The main effect of research on this project is to demonstrate the possibilities of changing the modal split for the new city centre because otherwise without solving a new light rail track system - represented by trams will bring the city of Bratislava to immense collapse. In addition, the will of the city or his management or City Parliament is to use a reasonable solution to the CPTS for the new city centre. The task was clear. The model was based on empirical experiences, from the lack of suitable outputs from the original – old model, towards to create a new uni-modal model for car transport and public transport as was defined in *TMP* [3].

2.3. Polemic multi- or uni- modal transport model

The analysis for the transport model has been carried out from different types of transport surveys, including a classic household questionnaire survey. The transport model is also made of supporting transport surveys for MTP and was created separately for car transport and especially for public transport and has supplemented over the years other PT or car special surveys in many details for calibration and validation of each transport mode. Already under the *TMP* [3] conditions decided to address the uni-modal transport model on the city's transport network. One of the objectives of research was to point out the fundamental problem that it cannot be served only by simultaneous operation of public transport and the car transport only to collapse throughout the area of the eastern part of the city at the internal traffic ring. The reason why it was used for the city of Bratislava uni-modal model for PT and car is too many of marginal conditions for each transport mode that are inserted as resistance features into the model itself. The result of a comprehensive model historically in Bratislava did not provide relevant answers to compare the current scenarios to the values of the results of the transport survey, in time. This condition is defined in Slovakia in the Technical requirements [29] (equivalent to *HBS* [30] in Germany). Within the project, a research method has been implemented, which deepens the details of a model specifically for public transport to achieve scenarios outputs that meet the actual operating conditions at the Bratislava Transit Network.

The expert team strategically decided just at the level of the creation of the new model whether to use multi-modal transport model or respect a detailed quality for each type of transport mode separately and to address the partial issue of dynamic car traffic and particularly the quality of PT. The issue of "orderliness and dynamics" of transport model in the conditions of Bratislava adopted already to the reason, that an expert team of traffic engineers and planners decided to hold the quality and the details separately for each mode of transport. The base rules were covered according of [31]. The transport model in detail is used for the needs of the transport-capacity assessment for each new investment. New investment has defined the limits of the area of the investment and its features (urban functions), the number of parking spaces and access to public transport. The simple reason is just complexity and functionality of the transport model. If your model has a few tens of boundary conditions (too detailed) for each kind of transport, they cannot be reflected in the comprehensive quality of service of the land use in each specific scenario. Each output scenario, which is studied in the detailed conditions, has to create an expert output for the decision-making process. This is according to capacity and serviceability calculation separately for each mode of transport in Slovak Technical requirements [29]. In practise a successful outcome cannot be reached from the multimodal model. The model with each detailed requirement inside for all transport modes is too rugged, slow and forfeiting the real conditions.

Therefore, for the conditions of Bratislava it was decided in the next steps to use the uni-modal access, which is further for the needs of public transport presented in this contribution. As has been said above of the dynamic transport is published in sufficient details in [5 and 32]. A rich discussion is underway around the world, and the decision-making process for this project was made of all the relevant documents that are presented and referred to the various studies and researches published in [24] and [27]. From them, for the Slovak conditions we rather looking for European solutions, which are also in certain specifics adopted for the transport model in Bratislava. The North American recommendations just because of the different and quality other approaches for public transport and the low value of modal split in advance of public transport cannot be used for Central European conditions. The development and operation of public transport in Central Europe has a high standard, which was used in the historical times of the 60-80-ies of the 20. century. The modal split between individual and public transport was in those times in Central and Eastern Europe for more than 80% in favour of public transport. Currently, the loss of state support and external regulations, from the 90-ties in the cities has dramatically changed to the disadvantage of public transport and fall on the level of approximately 50 – 50% between the two main modes of transport car and public transport. Nevertheless, the volumes of passengers in public transport of the Central European cities are still high compared to most of Western European cities as well as North American ones. For the conditions of the capital city of Slovakia – Bratislava is this problem sufficiently analysed in [5 and 6].

3. Public transport model for the city of Bratislava

In general, the macroscopic transport model is used to solve the optimal scenario of the transport system of a city or region development. These are mostly complex transportation-engineering solutions with transport planning tools (needs analysis and design of conceptual solutions) and assessment of costs and benefits of new infrastructure, transport-organizational and technical solutions and measures (e.g. new road or bicycle path, public transport line, park and ride, shopping centre, residential complexes, restrictions or changes in the organization and traffic management, introduction of parking policy, etc.). Specialities for modelling of public transport which were guided the authors for uni-modal PT transport model more oriented on time as a main parameter for passengers are motivated from [33].

The main purpose of the transport model is to analyse the current problems in transport and to predict the future state of service in the area. Macroscopic transport models are often used as a tool to qualify various projects and find the most appropriate solution to solve transport problems. In terms of creating first-rate scenarios for comprehensive service of urbanized areas, the offer of public transport includes the design of lines that connect stops at a certain time defined by the timetable. The transport model used in the process of analysing and planning the public transport system must take into account the characteristics in order to determine the impact of the transport system on passengers and operators. As a result, it must meet the following requirements:

- The attributes of public transport lines must be included in the integrated network of the transport model, which contains relevant data for all private and public transport systems. This condition for the public transport model of Bratislava is executed, because in the same detail it has all the attributes of urban public transport (tram, bus, trolleybus), suburban bus transport as well as complete state railway passenger transport
- The traffic model is intended to meet provide interfaces for importing digitized network data from various sources. The interface can accelerate the development of an integrated network model using basic interconnection data from GIS systems and public transport supply data from passenger information systems. This condition is also met
- Traffic - planning software has to help the planner in the planning by providing options to adjust the offer of public transport and thus the service of area. Ideally, the designer can "draw" the line on a computer screen and at the same time obtain information about the basic indicators of the public transport line. In this way, individual new routes and tram lines were inserted and tested
- To determine the impact on passengers, it is important to calculate indicators describing service quality and travel costs. This requires a transport impact model from *PTV* [12], which derives public transport connections from precise timetables and considers fares and fares systems

- The model must identify indicators, such as vehicle-kilometres, number of vehicles or operating costs to determine the impact on public transport operators. In combination with demand data, revenue from individual lines should be estimated as it is indicated from [3 and 13].

The traffic model of Bratislava city and its region is processed in the transport planning software *PTV Visum* [12]. The design of zonal structure of the transport model is based on definition of territory for the transport model and consists of internal and external zones (so-called external). These zones serve as origins and destinations for journeys made within one day (or part of a day). The territory of the capital city of Bratislava has 263 internal zones in the transport model, which correspond to the basic settlement units. The external zones include the municipalities of the catchment area of the capital and further aggregated municipalities of more distant areas. In Slovakia, the model covers the entire Bratislava and Trnava regions - which represent 347 zones. The model also includes 16 zones from Austria (part of Lower Austria and Burgenland) and 5 from Hungary (Moson). At the edges of the region, the model enters / exits traffic using the so-called cordon zones. Figure 2 shows the model area of the transport model of the city of Bratislava according to *TMP* [3], which was modified and changed according to the principles from [19] for the solved new city centre.

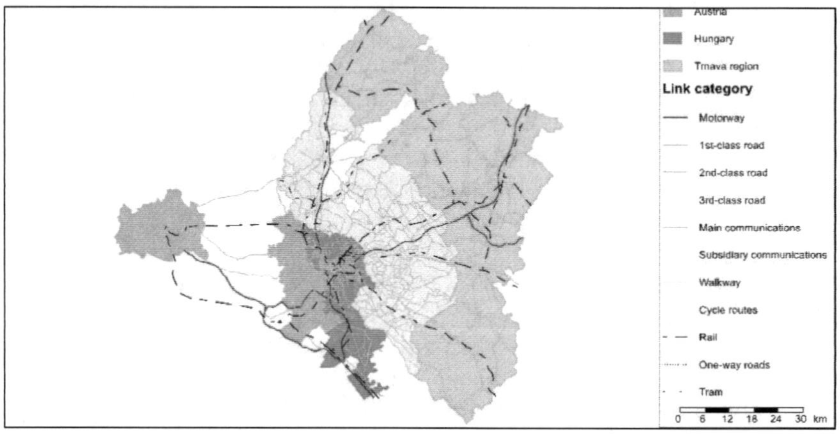

Figure 2. Model area of the transport model of the city of Bratislava in the territory of the agglomerations Vienna – Bratislava.

3.1. Modelling urban public transport

Public transport in Bratislava forms a network of bus, tram and trolleybus lines. Within the transport model of urban public transport, all tramway, trolleybus and bus lines are updated according to [10]. All public transport stops were also updated due to the fact that the original transport model contained duplicate stops at one point. The names of all public transport stops have also been updated. Figure 3 shows public transport line in the transport model.

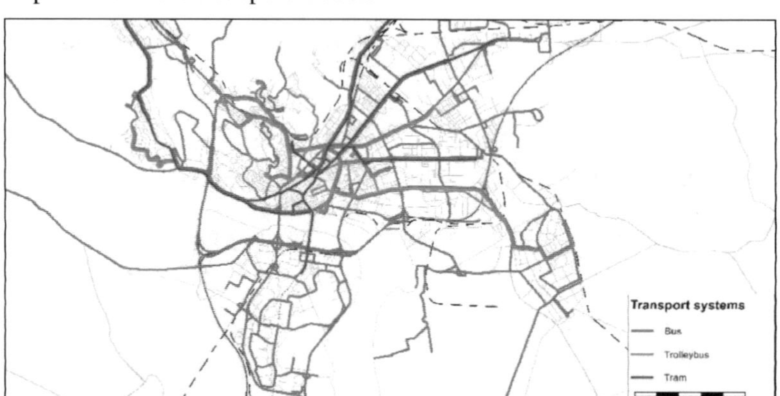

Figure 3. Scheme of public transport line in Bratislava – cut-out.

For the needs of updating the transport model, travel times among tram stops on selected lines were obtained, as average values from monitoring the movement of public transport vehicles using GPS in second intervals. These travel times have been imported into the "Line routes" section (figure 4), so after performing the calculation in the "Procedure sequence" section, the model will show more accurate results of the public transport load on the transport network and the procedure can be used for the calculation (TBA - "Timetable based assignment"). The TBA procedure, which is based on timetables, considers precise timetables. This method can only be used if a valid network of public transport lines and detailed timetables is used. Thanks to the coordination with timetables, this method shows very accurate results of the calculations. This method uses precise timetables and assigns each origin-destination relationship on the network separately. During the calculation, the relationships between the individual zones are allocated to the public transport network based on the route impedance, which is given by the so-called "perceived journey time" (PJT). Perceived journey time is the sum of time spent in public transport vehicles, data of the first and last stop, and times of all pedestrian journeys, waiting time for the first connection and waiting at transfers. These data are available from specific values of public transport movement from GPS in real time and passenger values from extensive surveys used for area general of transport.

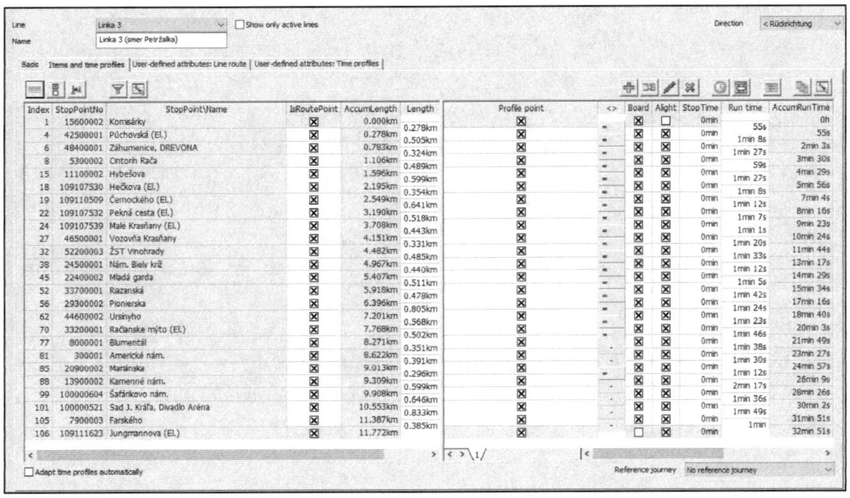

Figure 4. Import of travel times in the Line route section for inter-stop sections in tram transport.

Figure 5 shows the resulting timetable for a particular tram connection in a transport model.

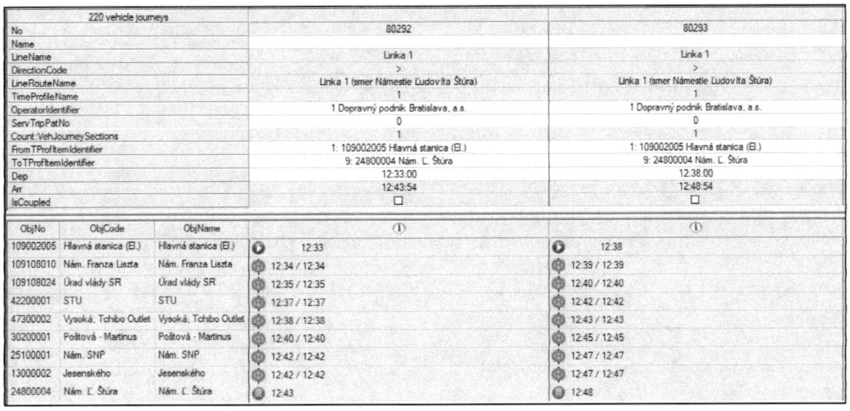

Figure 5. Display of travel times for a specific connection of the transport model.

3.2. *Load distribution methods for the transport network and for multiple routes*

This method simulates number of impacts under which the passenger decides choosing a transport connection. These impacts are the quality of transportation, psychological aspects and others. The calculation searches for more routes, although the transportation simulation is still according to the time criterion. The transport relationship is then assigned to found routes in a certain ratio, analogous to the ratio of travel times. Various restrictions are applied into the calculation, which prevent finding inappropriate routes according to [18], which are:

- number of transfers,
- accuracy of transport,
- regularity of transport, etc.

Regarding the use of the allocation of relationships (passengers) to the transport network, timetable based procedure (TBP) was used in the urban transport model. This is a procedure that is based on the implementation of timetables and is used when the connection intervals in the system are long and the coordination of timetables is important for transfers among lines. It considers precise timetables and is therefore suitable for modelling rural areas or railway networks. However, we used this principle for the entire rail transport in the city as well as in the agglomeration of Bratislava. These are tram and rail transport. This method can be used only for a network of public transport lines and a detailed timetable. Thanks to coordination with timetables, this method shows very accurate final results of the calculations. The method uses precise timetables and assigns each origin-destination relationship to the network separately. During the calculation, the relationships between the individual zones are allocated to the public transport network on the basis of impedance (resistance function) of a route, which is defined by the so-called perceived journey time (PJT). Perceived travel time is formed by the sum of time spent in public transport vehicles, time of availability to the first and last stop, times of all pedestrian journeys, time of waiting for a first connection and waiting for transfers.

Search for the shortest route in the *PTV VISUM* software is performed in individual criteria, which search for the optimal route at a given time from the traffic zone "i" to the traffic zone "j". Search procedure identifies the path with the lowest impedance as the optimal path. Impedance - the resistance function of the route is measured in minutes and is a linear combination of travel time and number of passes. It consists of the following time components according to *PTV* [12]:

- access time (min),
- time in vehicle (min),
- walking time between two stops (min),
- transfer waiting time (min),
- output time (min),
- number of transfers or transfer time (min) (adjustable).

3.3. *Scenario of tram lines in the New Centre of Bratislava*

Plan of new tram lines was analysed in the following sections and variants on streets (figure 6):

- Pribinova – Kosicka,
- Landererova – Kosicka,
- Dunajska - Mlynske nivy,
- Petrzalka - Karadzicova - Racianska.

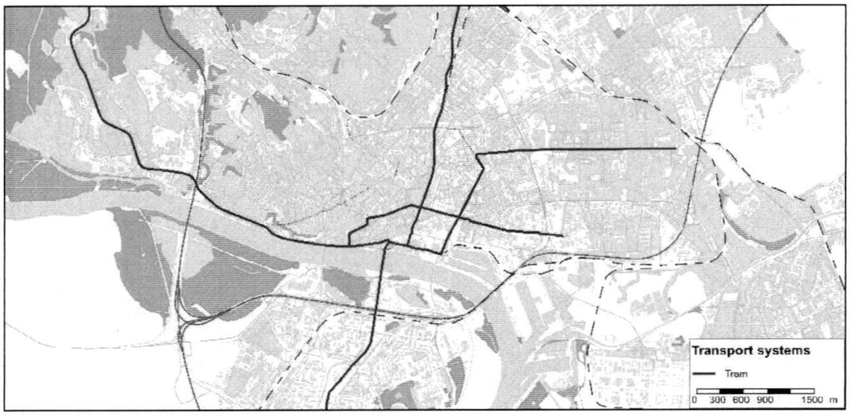

Figure 6. Plan of new tram lines for the new city centre in the PTV VISUM environment.

Linking on new tram lines was used to determine the capacity in the transport model, always for 2 tram lines with an average interval of 8 min. between the two routings. As a part of creation the public transport model for the new city centre in Bratislava, the individual elements of the transport model were modified. These were the basic elements:

- determination of time losses when travelling on a tram line (determination of the value of resistance function among stops on each section of the new line),
- update of public transport stops,
- elimination of duplicate stopping by public transport vehicles at stops,
- edit of public transport linking,
- insertion of current timetables for individual public transport lines into the transport model (exclusions are not modelled),
- creation of 17 new zones (new investments) in the new centre of Bratislava,
- connection of new zones to the transport network using connectors according to the standards of creating transport models,
- matrix of transport relations based on the assumed data on the daily number of persons arriving and departing to individual zones,
- loading transport network with the procedure of allocating public transport passengers.

The plan of individual lines and their sections determines the loss of time caused by traffic facilities on the street network of the area. Loss of time is entered into the transport model as a resistance function in the value of individual losses sum among stops. One stop is always included in the section. In general, the resistance function can be stated as the sum of individual time losses, which arise from the tram ride. This can be defined as:

$$T_{OF} = \sum T_i\,(s) \tag{1}$$

where:

T_{OF} – sum of all time losses from technical facilities on one section of the route,
T_i – specific parameter of delay from technical / technological facilities tram way,
T_Z – delay time caused by standing at the stop = 20 s,
T_K – delay time caused by standing at the intersection with traffic lights = 50 s,
T_V – delay time caused by passing the switch = 20 s,
T_O – delay time caused by passing through an arc up to a radius of 50 m = 20 s,
T_X – dwell time caused by crossing the track = 10 s.

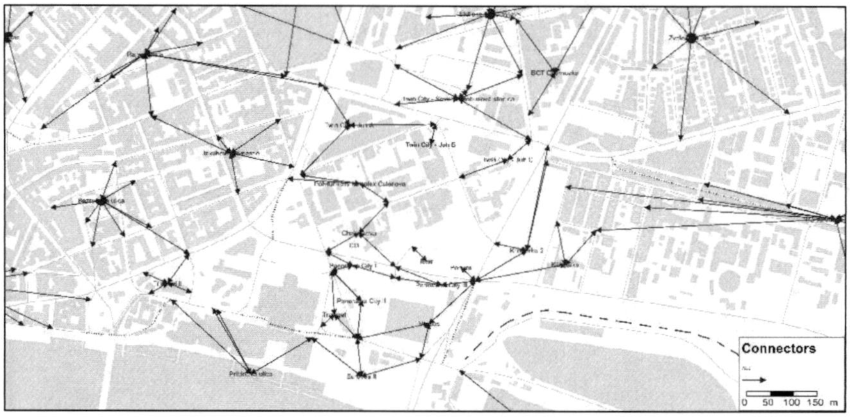

Figure 7. Scheme of new zones with connectors in the model within the new city centre.

Zones are connected using the so-called connectors to the tram stops, which are shown in figure 7. At the same time, bus and trolley bus stops have been remodelled around the city centre. Figure 8 shows the line for the planned tram line along Pribinova Street from the southern part and on Košická Street, from the eastern part of the New Centre. From the west it connects the city boroughs of Dúbravka - 35,000 inhabitants and Karlova Ves - 25,000 inhabitants) to the east of the city (the city borough of Ružinov - 60,000 inhabitants) in the background of PTV Visum. The line is 12.47 km long.

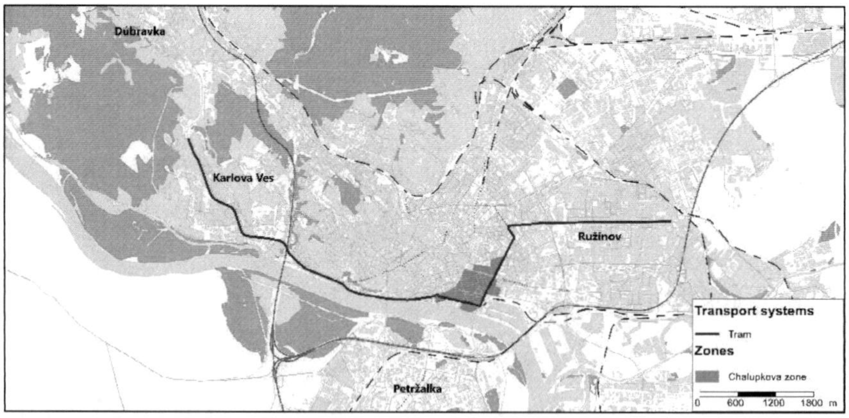

Figure 8. Scheme of the line route of the new tram line in the PTV VISUM environment.

Figure 9 shows the line for proposed tram line along the western part of the new centre and through the Filialka railway station in the PTV Visum. It is the route connecting the south of the city - Petrzalka (115,000 inhabitants) with the north-western part of the city - Nove mesto and Raca (a total of 48,000 inhabitants). The route is 15.35 km long.

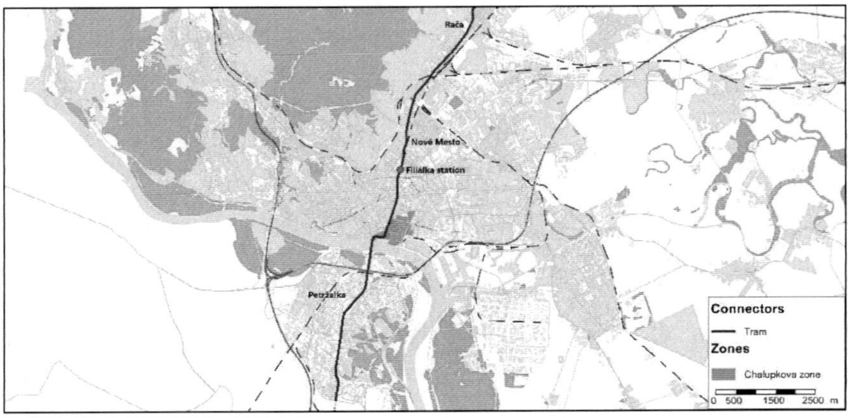

Figure 9. Scheme of the line route of the new tram line across the railway station Filialka.

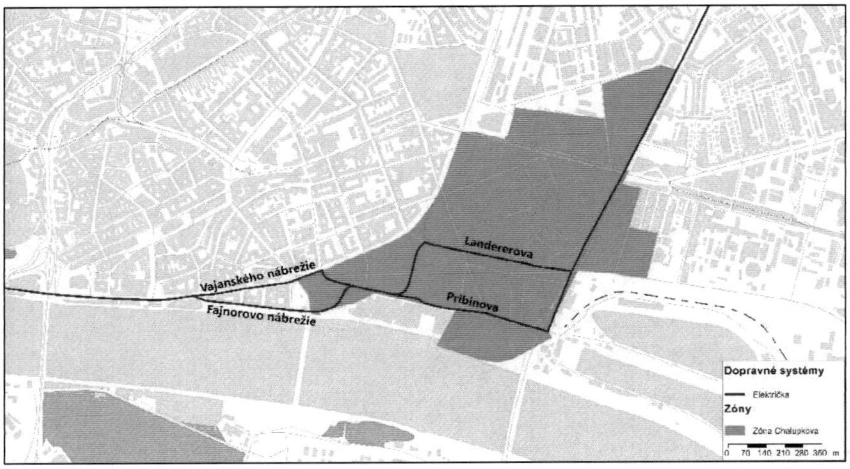

Figure 10. Scheme tested and planned tram lines along Vajanskeho Emb., Fajnorovo Emb. and after Landererova, resp. Pribinova Street in the new city centre.

Scenarios for new tram lines were performed in all-day passenger traffic intensities. After the introduction of resistance functions into the transport model, the basic values of transported passengers were significantly reduced by more than 50%, which were previously published to the public by private investors and the city of Bratislava.

Route C in the new city centre was analysed in detail - Pribinova versus Landererova section (figure 10).

Scenario via Vajanskeho Emb. and Fajnorovo Emb. reported according to the results of the transport model between the stops SND - Eurovea 2 20 756 resp. 20 354 passengers / 24 hours. In this case, however, the total losses of time have not been compared yet by narrowing the cross-section of the inner city ring road along the entire length of Vajanskeho Emb. The length of the column in rush hour showed more than 6 km column of vehicles. Figure 11 shows a flow map for the tram line in front of the SND building.

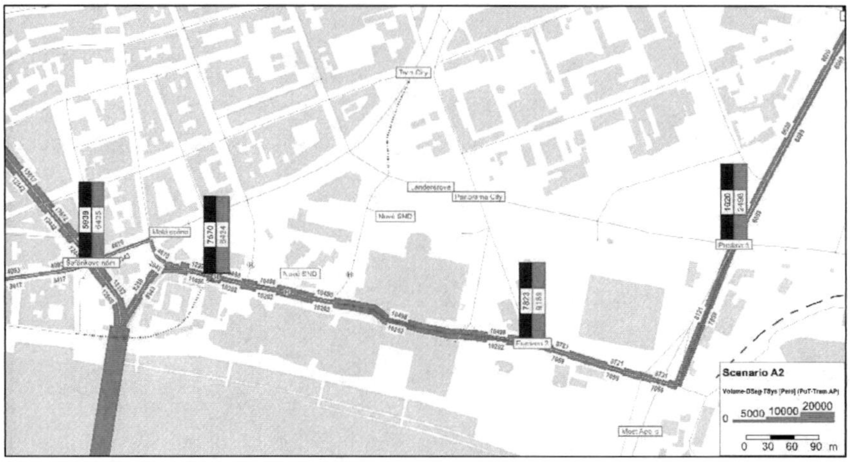

Figure 11. Flow map of the scenario tram line Pribinova - Kosicka via Pribinova.

Figure 12 shows assessed scenario of comprehensive service for the new centre by the public transport system with all three new tram routes serving the new territory (more than 2.2 million m^2 of new duties). In this case, a sub scenario of scenario A after Pribinova St., Landererova St. (scenario B) was not used after several objective causes were identified.

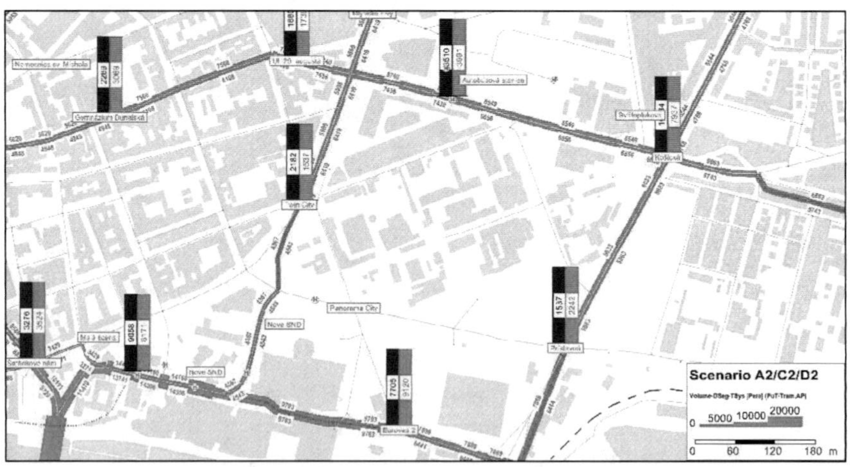

Figure 12. Flow map of the tram line scenario for the new city centre (detailed view by tram transport with the conversion of passengers at the stops).

Very interesting are the scenarios of a comprehensive network of tram lines for 2034 around the entire new city centre. The basic outputs point to an important conclusion, which clearly demonstrates the need to deal comprehensively with all routes at once. In addition, the new Filialka railway station is not included in this scenario of all tram lines, which will radically change the load values on the route Petrzalka - Racianska St. An example of flow map operating the whole area is shown in figure 13.

The maximum intensity of the traffic flow according to the results of the transport model between the stops Krizna - Mlynske nivy is 15,826 passengers / 24 hours. These data about transported passengers are still without the involvement of the railway station Filialka on Sancova St. It is recommended to continue the analyses of this scenario, where the data from the feasibility study of the Bratislava railway junction according to [26] supported by the Government of the Slovak Republic and the EU will be taken over. It is possible to assume a significant increase in passengers on this tram line, because it is estimated in the model railway station Filialka with a full-day transport value of about 65,000 passengers, which solves the interconnection of regional lines from the three main railway axes of Slovakia.

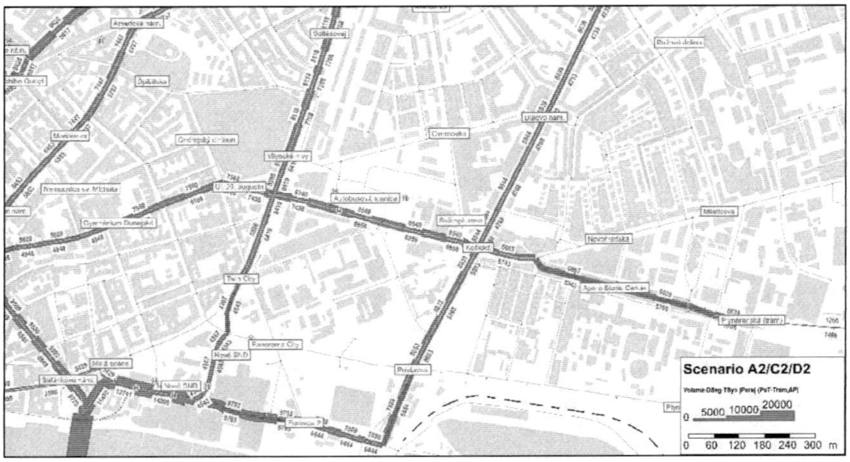

Figure 13. Flow map of three tram lines for the new city centre of Bratislava.

4. Conclusion

Historically in Bratislava was used a demand transport model on household surveys. In 2014 according to MTP, more than 10% of all households in Bratislava (in Bratislava is more than 192,000 households) addressed. The basic model that has been created as a multimodal model did not show a good feedback for large-scale additional transport surveys from both public and car traffic oriented. Therefore, it has decided to create uni-modal models that unambiguously exhibit high accuracy conformity for purpose-oriented surveys by transport mode. Originally, the base OD-matrix was created on the principles of a 4-step model whose result of the redistribution policy has taken over to the uni-modal model for each transport mode. This was then calibrated according to detailed transport surveys from for IAD and PT. The network assignment was preserved in the OD matrix only the load values were clarified according to long-term and precise surveys.

The selection of traffic allocation parameters was based on car traffic scenarios, where they assessed driving lanes for 2000 veh.h^{-1} and 2400 veh.h^{-1}. The assessment was selected according to the adjustment of the premises on each intersection of the number of input lanes and the queue length was determined before the traffic lights junction. The key factor of uni-modal model is a change in lane capacity to determine the size of the change - reduction of traffic volume whose difference is a saving that is used to change the modal split to the benefit of new trams.

The whole research was oriented to demonstrate a change in the modal split not in terms of competition between the transit modes bus/trolleybus versus tramway but especially competition between cars oriented land use, where it is clearly pointing the benefits of the PT rail-track system to operate a newly developing city centre.

Expert activities of the transport modelling of public transport by assessing the drawn up studies of technical solutions of new tramway tracks for the new city centre of Bratislava clearly demonstrated the need to implement these tramway lines as main part of public transport service. In addition, for the first time in great detail the transport model of the city of Bratislava was created in the uni-modal version. The PT model was managed to implement the real operating parameters of tramway transport. In second part there were involved and implemented the details for bus and trolley bus lines of *Public Transport Company of Bratislava* (www.dpb.sk), next the operating details of suburb bus PT of *Slovak Lines* private operator and the last was the *Slovak Railway Company* (https://www.zssk.sk/en/) where on its infrastructure are operated not only the state operator but RegioJet a private operator, as well.

Output are the dozens of detailed scenarios of the model, that "they play" with other details of the tram tracks. These are for example the accurate position of the tram stops and scrolling them to the right position, due to existing traffic lights junctions, intermediate new walking and cycling routes which starting/ending on tram stops. Operating zones of the new city centre were connected with all links to other urban areas with demonstration of exiting and entering passengers at the stops. A separate issue that helps the modelling is the creation of detailed stops isochrones/isodistances of public transport, which is in detail, however, published for this project in another article.

After many years of detailed analysis of the dynamic car transport (individual one), experts could already publish richly on the Central European International Conferences cited above and proceed under the same conditions the analysis and modelling of public transport in Bratislava. These models can actually compare the modal split for all developing projects. It can be observed that in the professional exchange of views with colleagues and experts with the neighbouring cities of Vienna, Brno, Budapest, Prague, we have in Bratislava sufficiently precise and professionally amply supported the uni-modal transport model for dynamic car traffic as well as for an integrated public transport system. This model can be applied for the city itself as well as its agglomeration. Moreover, this model can create high-quality cross-border activities and can be applied, when sufficient passenger flows among Slovakia, Austria, Hungary and the Czechia, because Bratislava lies on the border of 3 states.

These results help to open new approaches in the traffic-engineering and traffic-planning design works in terms of the development of new public transport projects, whether the actual rail, which is studying the possibilities to link the infrastructure of the tram and railways in the tram-train solutions, or improving the quality of suburban bus transport, which, in addition to trifling with the question of the implementation of the express bus lines on the territory of the city. These are the basic "blocks" to Bratislava to stop depressing decrease of modal split of public transport, which dropped for one human generation (25 years) by more than 30 %. However, Bratislava has 50% of the performance of passenger transport by public transport. However, it must be doing something more ...

5. References

[1] Statistical Office of the Slovak Republic 2020 [online] http://datacube.statistics.sk/#!/lang/en

[2] City of Bratislava 2007-2021 "Územný plán mesta Bratislavy" [Land Use Master Plan of Bratislava] [In Slovak] [online] https://bratislava.sk/sk/uzemny-plan. Územný generel dopravy hl. m. SR Bratislavy" [Transport Master Plan of Capital of Slovakia – Bratislava] 2015 [In Slovak], Transport Research Centre, Brno

[3] Transport Research Centre Brno 2015 "Územný generel dopravy hl. m. SR Bratislavy" [Transport Master Plan of Capital of Slovakia – Bratislava] [In Slovak]

[4] Davidich N, Halina F, Galkin A, Melenchuk T, Chumachenko I and Davidich Y 2018 "Rationale of Complex Indicator of Quality of Public Transport" *Romanian Journal of Transport Infrastructure* **7**(1) 18–33.

[5] Schlosser T, Schlosser P, Cápayová S and Hodáková D 2017 "Complex traffic analysis of new down town of Bratislava for environmental assessment" *17th International Multidisciplinary Scientific GeoConference* Volume 17 Energy and Clean Technologies Albena, Sofia.

[6] Schlosser T 2015 "Deľba prepravnej práce v Bratislave" [Modal split in Bratislava] [In Slovak] In *XV. Traffic Engineering Days*, Czech Road Association and Slovak Road Association Mikulov

[7] Schlosser P, Mališová D and Schlosser T 2010 „Nosný systém MHD 1. etapa, dopravno-inžinierske prieskumy dopravy a dopravný model v mestskej časti Petržalka" [Carrying Urban Public Transport System - phase 1., Traffic-engineering Surveys and Transport Model in the City Borough Petrzalka]. [In Slovak] Dotis Consult, Ltd. Bratislava

[8] Koštiaľ M, Schlosser T and Schlosser P 2017 "Development of Public Rail Track Transport in Nord-Western Area of Bratislava" In *Procedia Engineering: World Multidisciplinary Civil Engineering-Architecture-Urban Planning Symposium, WMCAUS. Prague.*

[9] Schlosser T and Schlosser P 2019 "New accessibility in Bratislava with study of two city bridges". In *10th International Conference on Bridges in Danube Basin.* Austrian Society for Construction Technology Vienna, Technische Universität Wien, Austria, p. 232-238

[10] M Korfant 2019 "Aktualizácia dopravného modelu mesta Bratislava – Rozbory a návrh nových podmienok" [Updating of transport model of Bratislava - Analysis and proposal of new conditions]. [In Slovak]. Magistrate of Bratislava

[11] City of Bratislava 2014 "Metodika dopravno-kapacitného posudzovania vplyvov investičných projektov". [Proposal for Amendment of the Methodology of Traffic Capacitive Impact

Assessment of Investment Projects]. Technical requirements for traffic planning and traffic modelling. [In Slovak]
[12] PTV VISUM manual 2015 PTV AG Karlsruh Germany https://www.ptvgroup.com/en /solutions/products/ptv-visum/knowledge-base/faq/visfaq/show/ VIS19902
[13] Schlosser T, Schlosser P, Hodáková D, Zuzulová A and Cápayová S 2019 „Modelling of Capacity and Public Transport Modal Split for New City Centre in Bratislava". In *4th World Multidisciplinary Civil Engineering-Architecture-Urban Planning Symposium, WMCAUS Prague*, **603** 5
[14] Schlosser T, Schlosser P and Schlosser Mi 2018 „Traffic model as a tool for changing of modal split in Bratislava new city centre". In *Conference of Transport Science - Technology in Service of Sustainable Transport*, Publisher Universitas-Győr Nonprofit Ltd., Hungary, Szécsényi University Győr.
[15] Schlosser T, Schlosser P and Schlosser Mi 2019 „Experiences of development of public transport service in the new city centre of Bratislava". In *Connecting Cultures, Enabling Economies, 26th World Road Congress*. Abu Dhabi - Paris: World Road Association (PIARC), article N° 0111
[16] Schlosser T and Schlosser P 2016 "New Tramway Infrastructure in Bratislava - How Can Influence the Well Prepared Traffic Engineering Analysis and Planning Tools the Decision Making". In *Procedia Engineering: World Multidisciplinary Civil Engineering-Architecture-Urban Planning Symposium*. Prague. Vol. 161.
[17] Schlosser T and Schlosser P, Korfant M 2019 „Dopravno-urbanistické zásady električkových tratí v novom centre Bratislavy. [Traffic-urban design principles of the tram lines in the new centre of Bratislava]. [In Slovak]. In *Verejná osobná doprava (Public Transport) - VOD 2019 Proceedings of 22nd International Conference of Public Transport,* Kongres STUDIO Bratislava, p. 35-44.
[18] Šujan J et al 2019 "Zásady rozvoja verejného priestoru - Rozvojová zóna Chalupkova" [The principle of the development of the public space - Development zone Chalupkova], Magistrate of Bratislava
[19] Schlosser T, Schlosser P, Korfant M et al 2019 „Električkové trate v novom centre mesta Bratislavy – Dopravné modelovanie a zmena deľby prepravnej práce". [Tramway lines in the new city centre of Bratislava – Transport modeling and a change of modal split]. [In Slovak]. 96 pp. Faculty of Civil Engineering - STU Bratislava and DOTIS Consult, Ltd. Bratislava.
[20] Májek V 2017 „Nové električkové trate v Bratislave, úsek Košická – Pribinova, Investičný zámer" [New tramway lines in Bratislava, Kosicka - Pribinova section, Investments study]. [In Slovak]. Prokos, Ltd. Bratislava
[21] Podolec O et al 2018 „Štúdia alternatívnych vedení električkových tratí: Pribinova, Landererova, Mlynské Nivy". [Study of alternative railtracks of tramway lines: Pribinova, Landererova, Mlynské Nivy streets]. [In Slovak]. Reming Consult, Bratislava
[22] Schlosser T, Borš M and Nemec N 2017 „Tram track layout options in the area of Danube riverbank – Pribinova – Karadžičova in Bratislava". *International Journal of Advanced Information Science and Technology (IJAIST).* Vol. 6, No.5, p. 11-16
[23] Schlosser T, Pomichal E and Bálint G 2017 „The new tram line design in the urban area of Bratislava". *International Journal of Advanced Information Science and Technology (IJAIST)*, Vol. 6, No. 5, p. 17-22.
[24] Schlosser T and Schlosser P 2018 "Traffic engineering analysis in the preparation and reconstruction of urban roads". *Slovak Journal of Civil Engineering*. Vol. 26, No. 2, p. 35-39.
[25] Rose A, Versluis L, Vlachou A and Alfrihat D 2016 "Bratislava Twin City South". London: Space Syntax Limited for HB Reavis
[26] Plišková J et al 2019 „Dopravný uzol Bratislava – štúdia realizovateľnosti". [Transport Node Bratislava – Feasibility study]. [In Slovak]. Consortia Bratislava Railway Node (Združenie Uzol Bratislava)

[27] Schlosser T and Schlosser P 2017 Suburban public transport network optimization and operation strategy of the new bus station in Bratislava, *Journal of Advanced Engineering, Volume: Inaugural Issue*, IGPRS – Gudapuris group CA USA

[28] Friedrich M 1999 "Modelling Public Transport - a European Approach". 78th Annual TRB Meeting, Transport Research Board, Washington D.CPTV AG Karlsruhe, https://www.researchgate.net/publication/238270544

[29] Ministry of Transport and Constructions of Slovak Republic 2006, 2010, 2016 "Výpočet kapacity pozemných komunikácií". [Roads Capacity Calculation]. [In Slovak]. TP 102 – Technical Requirements.

[30] HBS (Handbuch für die Bemessung von Straßenverkehrsanlagen) 2015 [Manual for the design of road traffic systems]. [In German]. Forschungsgesellschaft für Straßen- und Verkehrswesen e V Köln

[31] Ondráčková J et al 2017 "Metodika pro tvorbu a hodnocení makroskopických dopravních modelu" [The methodology for the creation and evaluation of the macroscopic transport model]. [In Czech], Transport Research Centre Brno.

[32] Schlosser T, Schlosser P and Braniš M 2018 „Strategy of traffic planning for new developments together with rail track public transport - Experiences from Bratislava". In *International Journal of Trend in Research and Development. Vol. 5, special issue ICTIMESH-18*, p. 137-142. Dubai.

[33] Borndörfer R, Grötschel M and Pfetsch M E 2008 „Models for Line Planning in Public Transport" In: Hickman M., Mirchandani P., Voß S. (eds) Computer-aided Systems in Public Transport. Lecture Notes in Economics and Mathematical Systems, vol 600. Springer, Berlin, Heidelberg.

Climate change adaptation in pavement design

D Hodakova[1], A Zuzulova[1] and S Capayova[1]

[1] Department of transportation engineering, Faculty of civil engineering, Slovak university of technology in Bratislava, Radlinskeho street 11, 81005, Slovakia

E-mail: silvia.capayova@stuba.sk

Abstract. Since 2011, the change in climatic characteristics has emerged as one of the three most common environmental risks today. Global warming forecasts warn that by 2100, our planet could warm by an average of 3.5 to 7.5 °C. In Europe, over the last 140 years, the average annual air temperature has risen by about 1.5 °C. In Slovakia, it was an increase of 1.7 - 1.8 °C on average. The consequences of these changes in road construction are immediate and intense and require adequate adaptation measures.

Long-term measurements of climatic conditions in Slovakia were the basis for assessing changes in average daily air temperatures in individual seasons. With the design of road pavements and calculations of pavement structure models, we have also tested road construction materials - especially asphalt mixtures. The results were used to correct the values of input data, design criteria, as well as in measures to reduce the impact of changes in climatic conditions.

The paper focuses on the characterization of the climatic conditions in Slovakia and the changes reflected in the design of the road constructions; also provides some insights into the solution of partial problems in the field of asphalt pavement and cement concrete pavement.

1. Introduction

The design of pavement structure is as a set of several activities related to the design of road construction, dimension and model calculations. This includes calculations of load effects, taking into account the properties of the materials, the subgrade conditions, and the climatic conditions.

The development of design methods of pavement structure can easily be described as a development from empirical methods to analytic-theoretical methods. This construction was replaced with a two-layer, later three- and multilayer system, and with a mathematical solution to calculations of stresses and strains in the layer system was applied. In analytical design methods, individual pavements layer of different materials is characterized by deformation properties, the layer system is laid on the linear-elastic half-space. It is assumed that the layers are of homogeneous and isotropic material.

For the calculation of stresses and strains in the individual layers of the physical model - which are subject to assessment, we apply the theory of the layered half-space. The calculation program LAYMED is part of the used design method. The analytical design method used is a complex system. Input data includes traffic load (calculate an equivalent number of design axles), classification of the water and temperature regime of the subgrade, characteristics of climatic conditions, and properties of the material layers - their strength and fatigue. The system procedure used to design and evaluation of the asphalt pavement is in the scheme in Figure 1.

Three criteria are used to assess the pavement design (block No. 3):

- ratio strength of bounded materials and radial stresses at the bottom of the layer,
- subgrade stability as the ration of vertical stresses and compressive strength,
- protection against frost penetration.

In the moderate climate zone, the ratio of the stress and strength of bounded material in the critical layer will be decisive. Pavements with cement concrete (CC) cover have properties other than asphalt pavements, behave differently and other procedures and criteria in the calculation of stresses and strains are used.

However, in this case, there is also a system design solution for CC pavements, in the scheme in Figure 2.

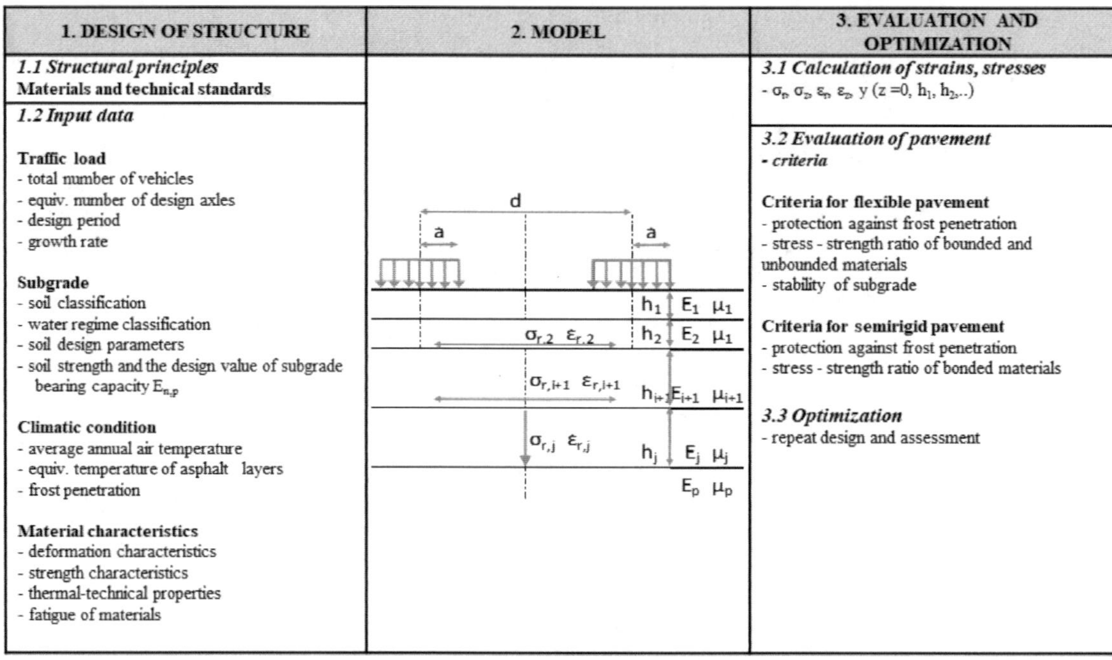

Figure 1. Design and evaluation of asphalt pavement.

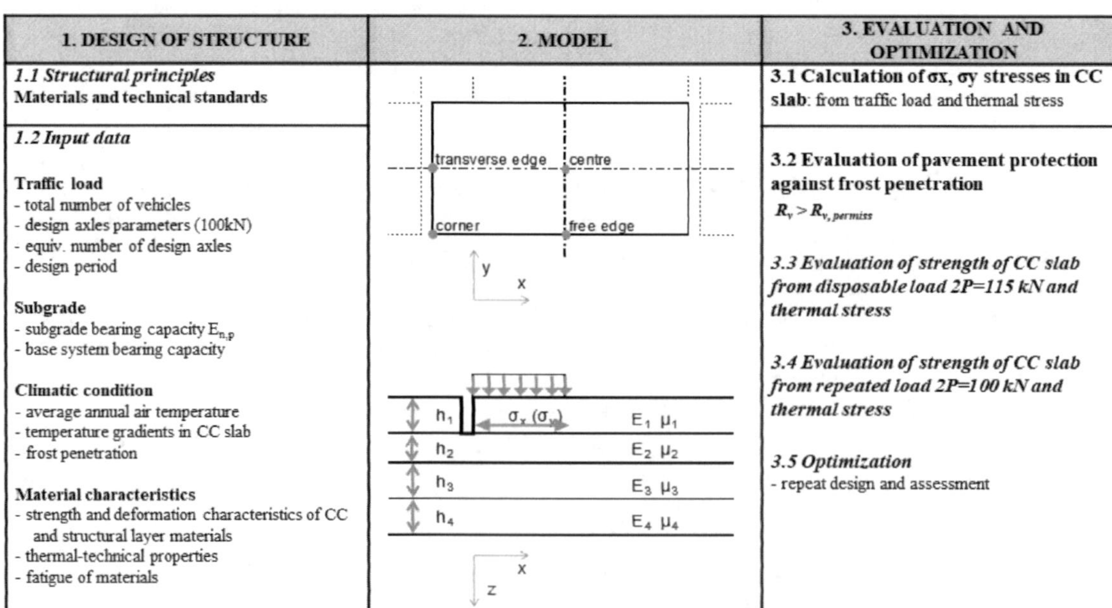

Figure 2. Design and evaluation of cement concrete pavement.

2. Characterization of climatic conditions in the pavement design

The territory of Slovakia is located approximately in the center of Europe and in an area affected by the influence of the continental climate, ocean influence, too. Climatic conditions often change in the area. One of the manifestations is a relatively large change in air temperature during the year (in the seasons), as well as the intensity of rainfall. For some parts of Slovakia, were recorded up to 200 - annual series of air temperature measurements. Data are available from the expanded number of meteorological stations of 100 – annual observation period. In earlier data, it is reported that winter (average air temperature 0 ° C) lasts about 2 months and summer 132 days, with hot summer with average daily air temperatures of more than 20 ° C (only) 49 days. Very valuable are data on the daily air temperature between 1901 and 1950 at various locations in the area. [1], [2]

In road construction, many building materials are used in pavements structures whose strength and deformation properties are temperature-dependent. These properties are very important as input data in analytical design methods and therefore have been the subject of measurement and evaluation. [3]

The temperature regime of the waveforms under different conditions of observation (its characteristics were measured) with different altitudes. Asphalt pavements and pavements with a CC cover were built at the measuring stations.

2.1. Asphalt pavement temperature regime

Long-term measurements of surface temperatures and pavements structure have been carried out continuously for 24 hours and all year. [4] Asphalt pavement with layers of asphalt mixtures with 20 - 25 cm thickness was built in four locations (locations) with different altitudes. From measurements at intervals of 30 min. to plot the course of daily temperatures at five points as well as the annual course of temperature, the course of the maximum pavement surface temperature, and the minimum temperature of the bottom surface of the asphalt layers. An example of the graphical representation of the annual course of asphalt pavement temperatures in one locality is in Figure 3.

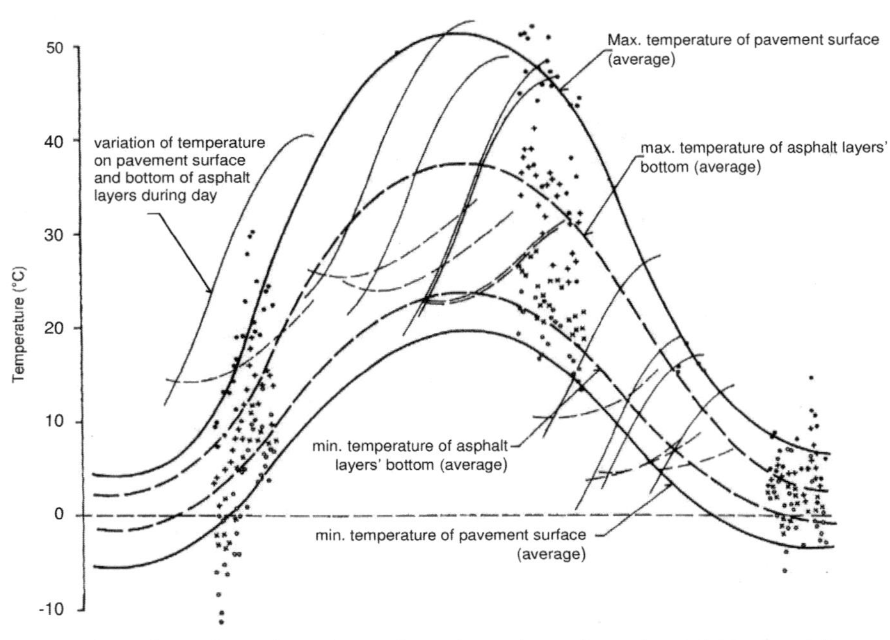

Figure 3. Temperature changes in asphalt pavement during the year.

Surface temperatures and individual asphalt pavements layers are dependent on air temperature. For the average and maximum values, the general Equation 1 applies:

$$T_{m,\,asf} = k.\,T_{m,r} + \mathrm{C} \quad (°\mathrm{C}) \tag{1}$$

where $T_{m,r}$ average annual air temperature,
C a constant, altitude function.

Average temperatures of asphalt pavements layers from measurements in locations with different altitudes are in Table 1.

Table 1. Average asphalt pavement temperatures 22 cm in thickness in different locations.

Location	Height above sea level (m a. s. l.) Average annual air temperature (°C)	Average asphalt pavement temperatures in 24 hours	6.00am to 6.00pm
A	130 m a. s. l. $T_{m.r} = 9.8$ °C	26.6	29.1
B	370 m a. s. l. $T_{m.r} = 7.4$ °C	23.8	26.1
C	700 m a. s. l. $T_{m.r} = 5.8$ °C	25.0	27.4
D	1050 m a. s. l. $T_{m.r} = 4.0$ °C	21.8	24.0

The results of the measurements and the data on the annual course of air temperature, surface, and asphalt pavement layers were derived and designed important characteristics for asphalt pavement calculations and design:

- division of the year into three periods,
- average temperatures of the entire thickness of the asphalt layers,
- equivalent (calculation) temperatures of the entire thickness of asphalt layers.

As an important characteristic of the asphalt pavement temperature regime, we considered the equivalent daily temperature of the layers (not average temperature for the season of the year), when five-sixths of the traffic load is carried by the road. By a special study, we estimated that the damage to the pavement asphalt layers over 24 hours is the same as at a temperature equivalent during day time between 06 and 18 hours. The designed values of the equivalent temperature of asphalt layers and the annual dimensional periods are in Table 2. [5]

Table 2. Dimensional periods of asphalt pavement.

Season of the year	Number of days	The proportion of the year	Equivalent temperature
spring, autumn	186	0.5	11 °C
summer	104	0.3	27 °C
winter	75	0.2	0 °C

2.2. Cement concrete pavement temperature regime

The temperature regime characteristics of CC pavements are:

- the average annual temperature of CC slabs and amplitude of this temperature in the annual cycle,
- the average and highest values of positive and negative temperature differences between the top and bottom surface of the CC slab.

An example of the graphical representation of the annual temperature changes of the CC pavement in one location is in Figure 4.

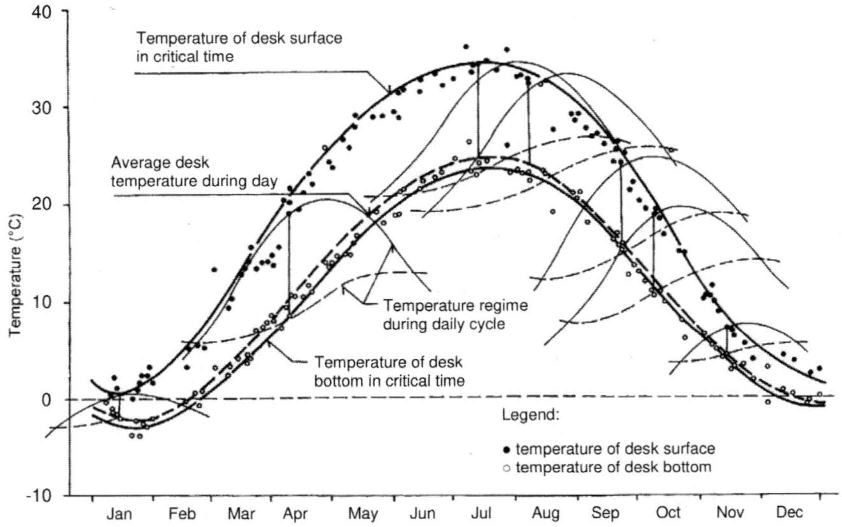

Figure 4. Temperature changes in cement concrete pavement during the year.

These characteristics of the temperature regime are important for stress and strain calculations of CC slabs. They were derived from measurements on constructions in terrain, from the sum of temperature ranges for 97 % of cases.

The average annual temperature of slabs is considered as the annual average of daily air temperatures. The amplitude of the annual air temperature is calculated according to the Equation 2 [6]:

$$A_{r,h} = 17.82 + 1.2\ T_{m,r} - 0.29\ h_B\ (°C) \tag{2}$$

where $T_{m,r}$ average annual air temperature (°C),
h_B thickness of the CC slab (mm).

The design (calculation) values of the temperature differences of the top and bottom surfaces of the CC slabs are calculated from the relationship to the average annual air temperature, considering the thickness of the slab.

Empirically derived Equations 3 a 4 have the form:

- for positive temperature difference:

$$+\Delta\ T_n = 12.440 - 0.6\ T_{m,r} + 0.028\ h_B\ (°C) \tag{3}$$

- for negative temperature difference:

$$-\Delta\ T_n = 6.214 - 0.3\ T_{m,r} + 0.0113\ h_B\ (°C) \tag{4}$$

3. Current and expected climate changes conditions

The basic data sources for the assessment of climatic conditions in Slovakia are the measurements of the Slovak Hydrometeorological Institute. Major experts use them and analyse the conditions for the needs of different sectors (industry, agriculture, energy, etc.). The evidence is publications and various scientific papers, including assessments of past changes and development scenarios.

From published analyses, we can point out that since the 19th century we have seen in Slovakia the average air temperature increase by 1.5 °C, regime changes, and total atmospheric precipitation. Since 1988, climate change has begun to emerge more rapidly, has been globally registered since 1985. After 1987, only one year had the average air temperature below the long-term average, otherwise, the average temperature was higher. [1], [7]

In addition to the frequency of incidence of intensive rain showers, other cases which may adversely affect the various sectors of the economy have also occurred. However, no measures were taken in road construction.

Currently, there are climate change scenarios in Slovakia until 2100, which have been prepared using global and regional models of the general circulation of the atmosphere (GCMs). It is assumed in them, that by 2100, our planet could warm by an average of 3.5 to 7.5 °C. In Slovakia, climate change manifests itself throughout the country. Extremely dry and humid years alternate in a short time. In Europe, over the last 140 years, the average annual air temperature has risen by about 1.5 °C. In Slovakia, it was an increase of 1.7 - 1.8 °C on average, in the summer months even more than 2.0 °C. Scenarios of precipitation sums are expected to increase in rainfall in the cold half, but their decline in the summer. An increase in the intensity of torrential rain is expected. [1], [2]

The expected changes to the conditions are that road construction has to respond to these changes, not only with adaptation measures. The need for highway and motorway design measures, designing pavement structure, and selection of road-building materials and technological processes highlights the increasing number of asphalt pavements failures and defects (track deviations), defects of CC pavements (slab shifts and faults on joints). [8]

With designing and dimensioning of pavement constructions, it is necessary to address the partial problems of improving the properties of road-building materials, in particular, asphalt mixtures, the correction of the values of the input data into the calculations as well as the criteria for assessing the design of flexible pavements and rigid pavements.

Demonstration of the significance of some results and measures to reduce the impact of climate changes in climatic conditions includes examples of the use the following are examples of the use of road mechanics and the results of research on the properties of asphalt mixtures.

4. Findings of partial problem solving and its applications

Several problems are related to the design of pavement structures, taking into account changes in climatic conditions. It is very important to define the improvement of the properties of road-building materials and the composition of pavement structures. In the case of asphalt pavement - pavement having multiple layers of the asphalt, it was necessary to expand the knowledge of the deformation properties of standard mixtures and to characterize the properties of new mixtures. In the case of CC pavements, the analysis of the influence of the temperature difference on the top and bottom surface of the slabs on the stresses in the slabs contributed to new findings.

4.1. Properties of asphalt mixtures

Design of the composition, production, and processing of standard asphalt mixtures, testing, and verification of their properties are defined in technical standards (STN, STN EN) as well as in technical regulations. In 2017, technical-quality conditions no. TKP 41 [9] for Warm Mix Asphalt (WMA), with specifications for compacted asphalt mixture following EN 13 108-1 [10] including high modulus asphalt concrete (HMAC), came into force.

The great influence on the physical-mechanical properties of asphalt mixtures has the temperature required for their production and processing. Before adopting European standards, the optimum composition of the mixture was determined by the Marshall test and the tensile strength test. At present, the mix is proposed in two possible ways - empirically or based on functional tests. In our country, the empirical proposal to determine the volumetric properties of the mixture, its resistance to permanent deformations, and water sensitivity is still preferred. Even though functional tests can more realistically simulate the actual behaviour of the mixture in the road construction, there are only a few laboratories with the necessary equipment. In recent years, when designing the asphalt mixture composition, we focus primarily on their temperature sensitivity, resistance to permanent deformations, water, and fatigue. Also, the requirement for the workability of the mixture and the temperature required for the production and processing of the mixture is included in the design.

When considering the properties of asphalt mixtures, the following scheme in Table 3 can be applied [11].

Table 3. Possibilities for modifying the properties of asphalt mixtures.

Intention/Objective	Basic measure	Supplementary measure
Increase the softening point of the binder	Changing the ratio of filler/asphalt binder	Special fillers and additives
Greater friction of mineral components	Crushed aggregates and max. grain size	Crushed sand (fraction 0/2)
Increase the stiffness of the mixture	Use of a higher stiffness binder	Additives to binder
The higher stability of asphalt mixture	The smaller fill rate of the aggregate mixture gap	Optimization of air voids of asphalt mixture

For the required quality of the asphalt mixture, it is crucial:

- using a suitable asphalt binder,
- the suitable ratio of filler and asphalt binder(F/A),
- use of crushed aggregate,
- optimum air voids of the mixture.

4.2. Recycling issues

In 2019 the total production of Hot (HMA) and warm mix asphalt (WMA)in Slovakia was 2.4 mil. tonnes: 35,000 tons of WMA, 2,000 tons of HWMA (Half-Warm Mix Asphalt), and 3,000 of CMA (Cold Mix Asphalt). Compared with other European countries, it's not a big production. Most of the produced mixture is applied mainly to the surface course of the pavement structure (Figure 5). In particular, Asphalt Concrete mixes (AC) predominate, followed by Stone Mastic Asphalt (SMA). The use of different types of asphalt mixtures in surface layers (%) in 2018 (no available data in 2019) was as follows [12]:

- Asphalt concrete (EN 13108-1) 95.5 %;
- Stone mastic Asphalt (EN 13108-5) 2.7 %;
- Mastic Asphalt (EN 13108-6) 0.8 %;
- Porous Asphalt (EN 13108-7) 0.6 %;
- Asphalt concrete for very thin layers (EN 13108-2) 0.4 %.

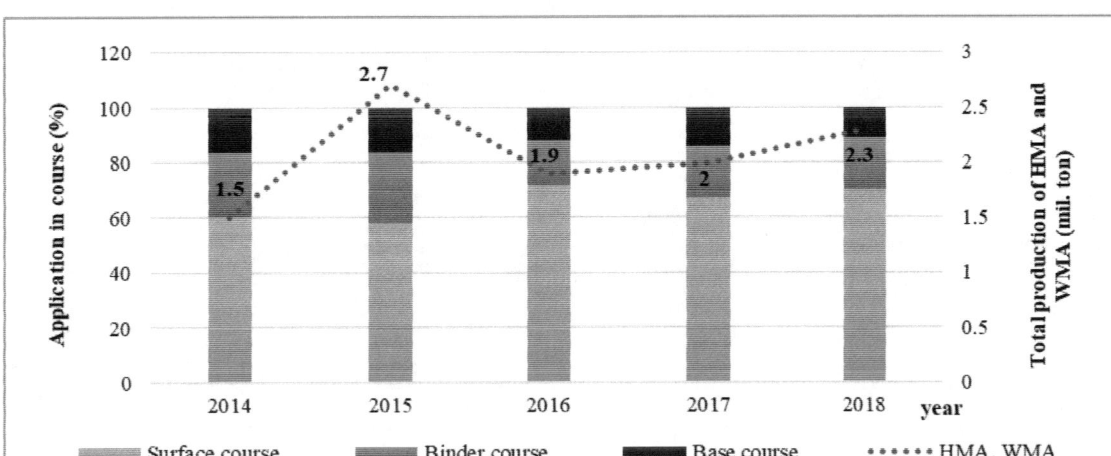

Figure 5. Application of HMA and WMA to the road structure.

By recycling and reuse of road construction materials, material and natural resources savings, environmental protection, and energy savings were followed. The mixture with reclaimed material is of sufficient quality for the new road construction layer and can also affect the resistance to permanent deformation of the cover.

In technical regulations, recycling technologies are defined by:

- the processing temperature of the material (hot, cold);
- type of material (single layer, multiple layers-maintaining protection layers);
- place of processing (in-situ, on a plant).

The application of individual technologies is also adapted to the area of legislation. In Slovakia, as well as in many other European countries, hot recycling is the most commonly used technology, namely Remix plus. Despite the availability of efficient machinery and the ability to work up to a thickness of 500 mm, cold recycling technology in Slovakia is minimal. Specific requirements are governed by these technical regulations:

- TP 043 Recyklácia asfaltových zmesí za horúca v obaľovacích súpravách (Hot recycling of asphalt at a central plant) (2020) [13],
- TP 044 Recyklácia asfaltových zmesí na mieste za horúca pre vozovky s dopravným zaťažením triedy II. až VI. (Hot in-place recycling of asphalt for pavement with a traffic load class II. to VI.) (2017) [14],
- TP 046 Opätovné spracovanie vrstiev netuhých vozoviek za studena na mieste (Cold in-place reuse of layers of flexible pavements) (2017) [15].

Figure 6 is an overview of the available amount of recycled material in Slovakia and its maximum application in the production of standard HMA and WMA (T = 100-150 °C). In low percentage, it was used in cold recycling or unbound pavement layers. [12]

From the point of view of the use of the recycled material and its recommended quantity to a particular road construction layer, there is a tendency to return to the higher recommended dosage. Following EN 13 108-8 [16], TP 043 [13], and TKP 6 [17], it is not necessary to prove the properties of aggregate and binder extracted from recycled material if:

- < 10 % of this material is used in the surface course, respectively
- < 20 % of this material is used in binder-, base- and leveling course of the pavement structure.

However, many research studies confirm that even in mixtures SMA can be used properly treated recycled material (pre-crash, softening with regenerative substance – rejuvenator) to the amount of up to 50 %. When used in asphalt concrete in an amount of up to 50 % or 60 % (depending on the particular road construction layer).

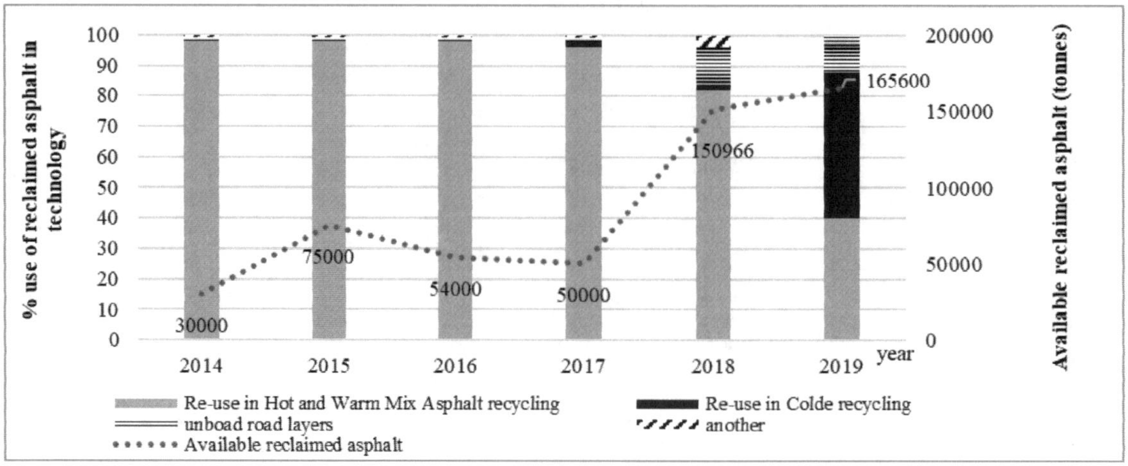

Figure 6. Available amount of reclaimed material.

Maximum utilization of the potential of the recycled material is achieved during the hot production of the new asphalt mixture on the plant. Parallel drum technology was also introduced, in which the recycled material was heated to 120 ° C. In such a process, up to 20 % of the recycled material can be added to the new mixture, up to 30 % in certain mixtures and conditions. Research and laboratory tests of mixtures have recently been focused on their properties with a larger proportion of recycled material. Some findings are interesting from the test results. Asphalt mixture AC 11 surf produced with 30 % of recycled material, adding 4.4 % of PMB 45/80-65 binder together with rejuvenator (mixture A) or modified PMB 45/80-65 binder (total amount of binder in the mixture was 6.1 %). Some important properties of mixtures A and B are shown in Table 4 [18].

Table 4. Properties of asphalt mixtures with reclaimed material.

Property	**Mix**	
	A	**B**
Air voids (%)	3.8	3.9
Stiffness modulus, IT – CY, 15°C (MPa)	9,753	9,819
Tensile bending stress, 0°C (MPa)	5	6.1
Resistance to permanent deformation:		
Max. proportional rut depth (%)	2.4	2.3
Max. value of rut depth increment- wheel tracking slope (mm/10^3)	0.041	0.021

The properties of the A and B mixtures are comparable, with very good resistance to permanent deformations. These mixtures are suitable for the surface course under varying climatic conditions.

4.3. Temperature gradient and stresses of CC slabs

An important characteristic of the temperature regime of the CC covers is the temperature difference between the top and bottom surfaces of CC slabs. It is most often the temperature gradient, this is - by changing the temperature to the unit thickness of the slab °C/mm, or simply a difference in the temperature of the top and bottom surfaces. The temperature difference is usually positive during the day, negative during the night. The temperature gradient in the CC slab causes deformation, at a positive temperature gradient, the slab has a convex shape. Depending on the thickness of the slab, its dimensions (width and length), and friction on the contact of the slab and the subbase, this can lead to stresses as large as those caused by the vehicle loading. [19]

Average temperatures of CC slabs and the value of temperature gradients were determined by experimental structures dependent on the altitude of the experiment locality.

The value of compressive and tensile stresses caused by the temperature difference, we have made calculations with a very simple application of Westergaard relationships.

To calculate the stresses in the geometric center of the slab we used Equation 5:

$$\sigma_{T,s} = \frac{E_{B,T}.\alpha_T.\Delta T}{2(1-\mu^2)}\left(C_x + \mu C_y\right) \text{(MPa)} \tag{5}$$

where $E_{B,T}$ the modulus of elasticity of the concrete at temperature, MPa,
α_T the coefficient of thermal expansion of concrete = 0.000012,
ΔT temperature difference at the surface and the bottom of the slab,
μ Poisson's ratio of the concrete = 0,20,
C_x, C_y coefficients determined for the dimensions slab and its relative slab stiffness, which is a function of the thickness of the slab hB (m) and the base and subgrade modulus of reaction k (MN.m^{-3}).

Calculations of stresses for different ΔT for:

- thickness of CC slabs (220, 250 a 280) mm,
- temperature difference (+10, +15, +20, +25) °C,
- base and subgrade modulus of reaction (80, 100, 150, 200) MN.m^{-3},

- and square slabs 3.75 x 3.75, resp. rectangular slabs 3.75 x 5.5 m.

The calculated radial stresses in the center of the slab from the positive temperature gradient + ΔT dependence on the thickness of square slab with 3.75 x 3.75 m dimensions are in Table 5.

Table 5. Radial stresses from the positive temperature gradient.

h_B (m)	k (MN.m^{-3})	+ΔT (°C) 10	15	20	25
	80	1.377	2.419	2.756	3.445
	100	1.515	2.661	3.032	3.789
0.22	150	1.732	3.041	3.465	4.331
	200	1.928	3.387	3.858	4.823
	80	1.222	1.969	2.244	2.805
	100	1.239	2.177	2.480	3.100
0.25	150	1.515	2.661	3.032	3.789
	200	1.633	2.868	3.268	4.085
	80	0.905	1.589	1.811	2.264
	100	0.984	1.728	1.968	2.461
0.28	150	1.279	2.246	2.559	3.199
	200	1.476	2.592	2.953	3.691

From the calculated values of stresses, a diagram was compiled (for k = 100 MN.m^{-3}). There is in Figure 7.

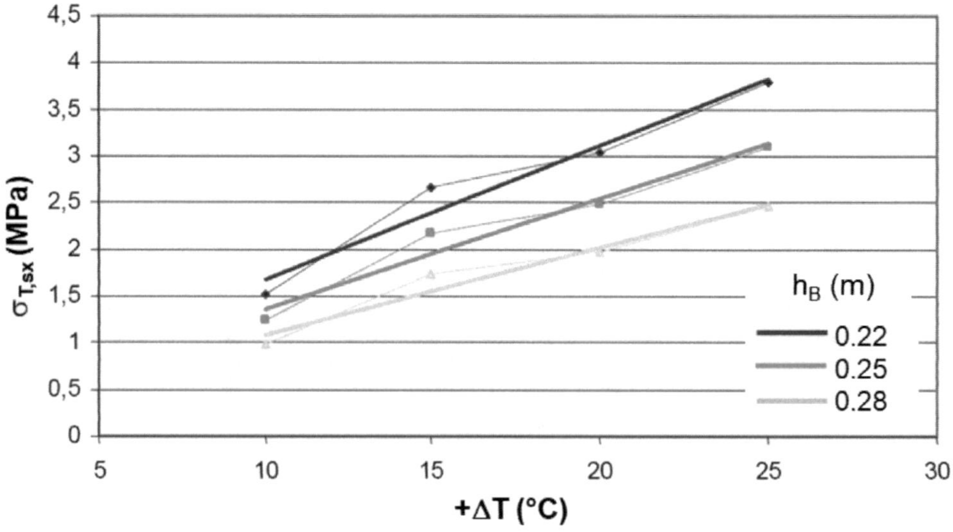

Figure 7. Radial stresses in a CC slab 3.75 x 3.75 m induced by temperature gradient ΔT.

Calculation models of CC slabs showed that the temperature-induced stresses are smaller in square panels as compared to rectangular slabs (with an aspect ratio up 1:2).

The analysis calculations also showed that the increased bearing capacity of the subbase also increased stresses in the slabs (Table 5).

5. Recommended pavement structures

In the recent period, stages of road constructions development can be characterized by expanding demands and then research solutions with the mitigation of the effects of climate change. In the first phase, the energy and material savings, as well as the extension of the lifecycle of the structures, were resolved. In the second phase, in particular, the improvement of the properties of asphalt mixtures and the resistance of pavement to permanent deformation were solved. We have also registered experiments to reduce the temperature of the cover using the color and reflectivity of the radiation from the asphalt pavement surface (Solar–heat blocking pavements). The optimistic result was when the maximum temperature reached 58 °C on the standard surface and of approximately 42 °C on the treated surface.

In experiments with road construction, especially, asphalt mixtures with a modified binder and various mixtures of asphalt concrete were used:

- HMAC – High Modulus Asphalt Concrete (high stiffness, standard binder),
- HMAC PMB (high stiffness, modified binder),
- RBL (Rich Bottom Layers) (higher amount of binder), and also
- PA (Porous Asphalt) and BBTM (Very thin asphalt concrete). [20]

The impact of the pavement layer composition with different quality and thicknesses of asphalt can show the results of stress and strain calculations in the K1 and K2 structure models (Figure 8).

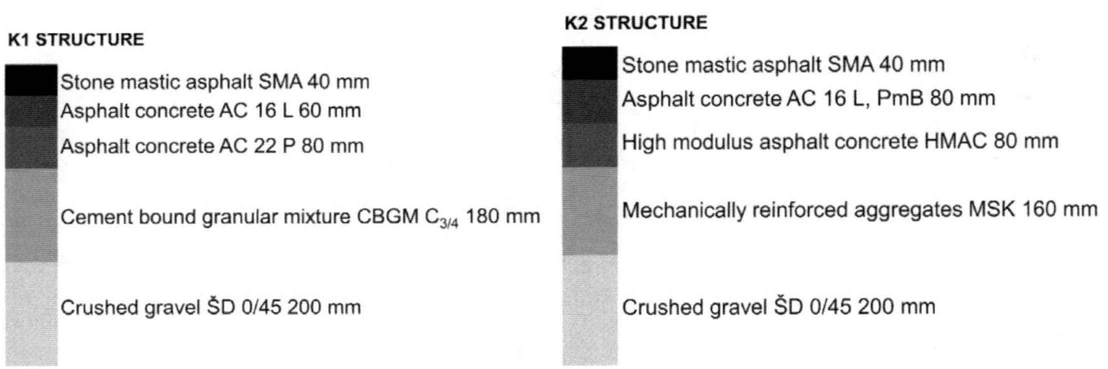

Figure 8. Pavement layer composition of K1 and K2 structures.

The deflections of the structures are similar but the influence of the temperature on the K2 structure deflection was smaller. In the K1 structure is the critical cement bound layer CBGM $C_{3/4}$, in the K2 construction is the critical layer HMAC. The radial stresses in these layers are high, so the strength of materials must be appropriately high. Stress values were:

K1 structure	0°C.........0.110 MPa,	27°C.........0.133 MPa
K2 structure	0°C.........0.754 MPa	27°C.........0.792 MPa

and the strength of asphalt should be 2.40 MPa.

The results of stress and strain calculations of various structures have been used in pavements design for medium and high traffic loads. An example may be the design and construction of pavement structure with mechanically reinforced aggregates (MSK) (Figure 9) or an alternative in which MSK has been replaced with cement bond granular mixture CBGM $C_{3/4}$, layer with a lower thickness. In each particular case, the dimensions are dependent on the traffic load, the design period, and the bearing capacity of the subgrade.

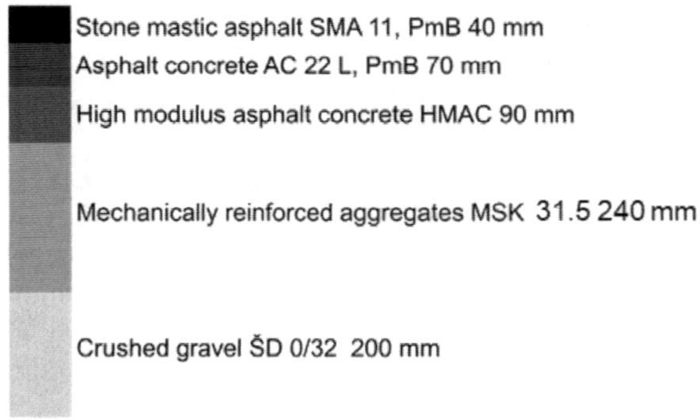

Figure 9. Pavement structure with mechanically reinforced aggregates.

6. Conclusion

Changes in climatic conditions in the territory of Slovakia, which is at the interface between the oceanic and continental climate, have been reflected in the operational ability of asphalt pavements and also CC pavements.

After 1970, climate change began to manifest strongly. The growth of the mean annual air temperature was greater than 1.5 ° C. According to climate change scenarios in the next period (until 2100), the increase in average monthly air temperatures may be higher, e.g. in the cold half of the year about 1.8 °C, in the warm half of the year up to 3.8 ° C.

In the system solution procedure for design, calculations, and assessment of flexible asphalt pavements and rigid cement concrete pavements, climatic conditions are considered. But expected climatic changes have raised demands for solutions and related issues.

In the field of designing asphalt pavements, it is proceeding systemically. Important input data includes characteristics of climatic conditions. The calculations take into account the temperatures of the asphalt layers (0 ° C, + 11 ° C, + 27 ° C) and the respective deformation properties of the mixtures.

The partial problems that were the subject of the solution concerned the properties of asphalt mixtures depending on temperature. The solution to their modification was followed by a scheme in which the basic and supplementary measures are. We also consider the recycling of materials and the use of recycled material as a partial problem.

For CC pavements, slab stresses from temperature (temperature changes) for different thicknesses and dimensions of the slabs were calculated. The results showed that lower stresses from temperature gradients were in thicker and square slabs. Very valuable results have been given by experiments with the measurement of slabs deformation at positive and negative temperature gradients when the slabs were laid on unbounded and bounded subbase. From the deformation of the slabs, we deduced that the stresses from + ΔT (-ΔT) gradient at the center of the CC slab were smaller in the slabs on the less rigid subbase.

In the recommended composition of pavement structures, several asphalt mixtures with a modified binder, new mixtures with higher binder contents, and also high modulus mixtures are used.

7. References

[1] Lapin M 2008 Scenáre zmien klímy do roku 2100 a možné dôsledky na Slovensku (Climate change scenarios by 2100 and possible consequences in Slovakia) *XIIIth Seminar of Ivan Poliaček with international participation* (Bratislava, Slovakia: Kongres STUDIO LLC) pp 8 – 19.

[2] Pecho J 2021 Changes in Diurnal Cycle of Air Temperature in Projections of the Future Climate in Slovakia until 2050 – 2100 Presentation at the 5th international conference *Airport pavement* (Bratislava, Slovakia: Kongres STUDIO LLC).

[3] Zuzulova A, Hodakova D, Capayova S, Schlosser T and Grosek J 2020 Climatic influences considered in pavement design methodology *International Multidisciplinary Scientific*

GeoConference Surveying Geology and Mining Ecology Management SGEM 2020 (Sofia, Bulgaria: STEF92 Technology) **20** 4.1 pp 349-356.

[4] Staňo R and Benko J 2008 Teplotný režim cementobetónových vozoviek (Temperature regime of cement concrete pavements) *XIIIth Seminar of Ivan Poliaček with international participation* (Bratislava, Slovakia: Kongres STUDIO LLC) pp 119 – 125.

[5] Gschwendt I and Zuzulova A 2013 Climatic conditions and their impact on road pavement design *9th International Conference on the Bearing Capacity of Roads, Railways and Airfields BCRRA* (Trondheim, Norway: Academika publishing) Volume 1 pp 523-536.

[6] Gschwendt I, Bačová K and Staňo R 1998 *Mechanika vozoviek, Návody na cvičenia* (*Pavement mechanics, Exercise instructions*) Scripts (Bratislava, Slovakia: Slovak technical university in Bratislava) p 108.

[7] Remišová E, Decký M, Podolka Ľ, Kováč M, Vondráčková T and Bartuška L 2015 Frost Index from Aspect of Design of Pavement Construction in Slovakia *World Multidisciplinary Earth Sciences Symposium* WMESS *Earth and Planetary Science* **5** pp 3–10.

[8] Grosek J, Zavrel T and Stryk J 2021 Mitigation possibilities of concrete pavement degradation IOP Conference Series: *Materials Science and Engineering* 1039 art. no. 012018

[9] Ministry of transport and construction of the Slovak Republic 2017 Technical regulation *TKP 41 Nízkoteplotné asfaltové zmesi* (Low temperature asphalt mixtures, in Slovak) document available at https://www.ssc.sk/files/documents/technicke-predpisy/tkp/tkp_41_2017.pdf.

[10] Technical standard 2019 *EN 13 108-1 Bituminous mixtures - Material specifications - Part 1: Asphalt Concrete*.

[11] Sekera M 2008 Zásady návrhu asfaltovej zmesi s ohľadom na klimatické podmienky Slovenska (Principles of asphalt mix design with respect to the climatic conditions of Slovakia) *XIIIth Seminar of Ivan Poliaček with international participation* (Bratislava, Slovakia: Kongres STUDIO LLC) pp 176 – 177.

[12] Europaen asphalt pavement association 2015-2019 *Asphalt in figures* pdf. documents available at www.EAPA.org.

[13] Ministry of Transport, construction and regional development of the Slovak Republic 2020 Technical regulation *TP 043 Recyklácia asfaltových zmesí za horúca v obaľovacích súpravách* (Hot recycling of asphalt at a central plant, in Slovak) document available at https://www.ssc.sk/files/documents/technicke-predpisy/tp/tp_043_2020.pdf.

[14] Ministry of transport and construction of the Slovak Republic 2017 Technical regulation *TP 044 Recyklácia asfaltových zmesí na mieste za horúca pre vozovky s dopravným zaťažením triedy II. až VI.* (Hot in-place recycling of asphalt for pavement with a traffic load class II. to VI., in Slovak) document available at https://www.ssc.sk/files/documents/technicke-predpisy/tp/tp_044_2017.pdf.

[15] Ministry of transport and construction of the Slovak Republic 2017 Technical regulation *TP 046 Opätovné spracovanie vrstiev netuhých vozoviek za studena na mieste* (Cold in-place reuse of layers of flexible pavements, in Slovak) document available at https://www.ssc.sk/files/documents/technicke-predpisy/tp/tp_046_2017.pdf.

[16] Technical standard 2020 *EN 13 108-8 Bituminous mixtures - Material specifications - Part 8: Reclaimed asphalt*.

[17] Ministry of transport and construction of the Slovak Republic 2019 Technical regulation *TKP 6 Hutnené asfaltové zmesi* (Compacted asphalt mixtures, in Slovak) document available at https://www.ssc.sk/files/documents/technicke-predpisy/tkp/tkp_06_2019.pdf.

[18] Valentin J and Soukupová L 2015 Vlastnosti asfaltového spojiva s využitím jemne mleté pryže. *XXth Seminar of Ivan Poliaček with international participation* (Bratislava, Slovakia: Kongres STUDIO LLC) pp 63 – 70.

[19] Hodáková D, Zuzulová A, Cápayová S and Schlosser T 2019 Temperature changes as decisive loads on cement concrete pavements, *International Multidisciplinary Scientific GeoConference Surveying Geology and Mining Ecology Management SGEM* 2019 (Sofia, Bulgaria: STEF92 Technology) **19** 4.1 pp 1093-1100.

[20] Mondschein P and Valentin J 2015 Konstrukce netuhých vozovek se směsmi VMT a RBL (Design of flexible pavements with HMAC and RBL mixtures) *XXth Seminar of Ivan Poliaček with international participation* (Bratislava, Slovakia: Kongres STUDIO LLC) pp 59 – 62.

Acknowledgments
We thank Professor Ing. Ivan Gschwendt, DrSc. for valuable consultation on the article. The publication of the paper was supported by funds of the Department of Transportation Engineering, Faculty of Civil Engineering, Slovak University of Technology in Bratislava, Slovakia.

IOP Conf. Series: Materials Science and Engineering 1252 (2022) 012018 doi:10.1088/1757-899X/1252/1/012018

Experimental determination of dynamic modulus of elasticity of concrete with ultrasonic pulse velocity method and ultrasonic pulse echo method

I Ivanchev[1]

[1] University of Architecture, Civil Engineering and Geodesy (UACEG), Faculty of Structural Engineering, Reinforced Concrete Structures, 1000 Sofia, Bulgaria

E-mail: ivanchev_fce@uacg.bg

Abstract. In this article are presented the author's experimental studies for determining the dynamic modulus of elasticity of concrete in tests of reinforced concrete beams at the age of concrete 1926 days. For two years the reinforced concrete elements were indoors, and for the next more than three years they were left outdoors, exposed to external atmospheric influences. The results of the measured velocities of P-longitudinal and S-transverse waves generated by piezoelectric transducers are presented. The research used a combination of two modern non-destructive methods in the study of concrete - ultrasonic pulse velocity and ultrasonic pulse echo. The dynamic Poisson's ratio and the dynamic modulus of elasticity of concrete are determined on the basis of the measured velocities of the longitudinal and transverse waves. The measurements were made respectively by means of portable instruments Proceq Tico and Pundit Live Array Pro of the company Proceq. The obtained results are compared with the theoretically calculated modulus of elasticity according to BDS EN 1992-1-1:2005/NA:2015.

1. Introduction

Concrete is one of the most commonly used building materials. The control of concrete quality and the determination of its characteristics and parameters are important both in the construction of new and in the assessment of existing reinforced concrete buildings and facilities. In design of reinforced concrete structures, the elastic modulus of concrete EC is main parameter for estimating the deformations of the structural elements in serviceability state [1,2,3]. The theoretical modulus of elasticity for a given age can be calculated according to BDS EN 1992-1-1:2005/NA:2015, and the experimental one can be determined using destructive or non-destructive methods. The destructive tests of specimens, prepared on the day of laying the concrete mix or of cores, taken from existing structures provide accurate information about the elastic modulus of the concrete [1]. Drilling cores is expensive, time consuming, causes additional damage and the results obtained cannot be used as information for the entire volume of the reinforced concrete structure. More and more non-destructive techniques are used for assessing the elastic properties of concrete in reinforced concrete buildings in serviceability state. These tests are fast, efficient, easy to perform and low cost. Many of them are standardized in many countries around the world [2]. The wide use of concrete as a basic building material requires the development of new equipment and techniques for its reliable and effective research [4].

In the present study, two different non-destructive methods were used simultaneously - Ultrasonic Pulse Velocity (UPVM) and Ultrasonic Pulse Echo (UPEM) for determining the dynamic and then calculating the static modulus of elasticity of concrete by measuring the propagation velocities of P-waves and S-waves of the ultrasonic signal [5]. Ultrasonic waves are mechanical vibrations and have a frequency greater than 20 kHz. It is assumed that in these methods the dynamic modus is the initial tangential modus at zero stress, because during the measurement of the ultrasonic velocity a negligible stress is applied [1]. P-waves (compression or longitudinal) propagate in a manner analogous to the propagation of sound waves in the air, i.e. they move longitudinally and cause the particles to vibrate in the direction of motion. In solid bodies, they propagate with highest velocity. For S-waves (shear or transverse) in concrete, the particles vibrate perpendicular to the direction of wave motion. The velocity is lower and is 60% of the velocity of the longitudinal waves. These waves also provide images of the inside of the test specimen and are used in instruments of Ultrasonic Pulse Echo Technology (shear wave tomography). Experimental researches for evaluating the dynamic and static properties of concrete by using measured S-waves velocities are still scarce [2]. These experimental results show a better relationship between the velocity of S-waves and the mechanical properties of concrete (compressive strength and modulus of elasticity) and less dependence on factors influencing the use of UPVM (air inclusions, water-cement ratio, age of concrete, conditions of hardening of the concrete mix, type of aggregates, presence of reinforcement, etc. factors) [1,2,5]. According to [1], measuring the velocity of S-waves propagation has the potential as an in-situ non-destructive method for estimating the elastic modulus of concrete.

1.1. Non-destructive ultrasonic method (UPVM)

UPVM uses the propagation of longitudinal ultrasonic waves introduced into concrete [3,5,6,7,8,9,10]. In this method it is measured the time or velocity of propagation of the ultrasonic signal. The devices consist of an ultrasonic oscillation generator, a transmitter, a receiver, an amplifier and a reading device [5,11]. The generator generates high frequency signals with a frequency of 25-150 kHz. The transmitter is a piezoelectric crystal that converts these signals into ultrasonic mechanical pulses that propagate through the concrete, undergoing multiple reflections at the boundaries of the various concrete components within the reinforced concrete element. The velocity of mechanical waves is very high and depends on the elastic properties and density of the medium in which they propagate. The receiving piezoelectric transducer registers the start of the longitudinal ultrasonic waves and converts them into electrical signals [6,12]. These signals are amplified by an amplifier and converted into a suitable analog or digital signal to visualize the results. There are three ways for placing the two piezoelectric transducers (figure 1): direct (opposite), semi-direct (at an angle) and indirect (one-sided) [3,5,7,8,10,11].

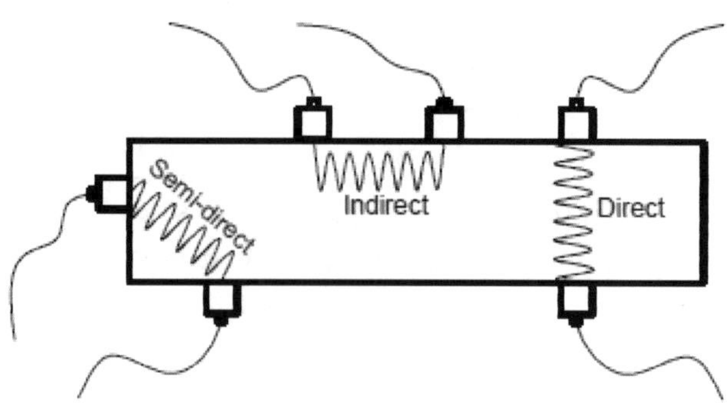

Figure 1. Location of the transducers on the reinforced concrete element.

The dynamic modulus of elasticity $E_{c,din}$ can be determined by the following formula [8,13]:

$$E_{c,din} = k\rho_c V_{UP}^2 , \qquad (1)$$

where:

V_{UP} is velocity of propagation of ultrasonic P-waves in concrete in [m/s];

k is factor that is assumed to be equal to $(1+\nu)(1-2\nu)/(1-\nu)$ in case of longitudinal waves transition [5,8];

ρ_c is the acoustic density of the material in $[\text{kNs}^2/\text{m}^4]$:

$$\rho_c = \frac{\gamma_c}{g}, \tag{2}$$

where:

γ_c is the weight per unit volume of the concrete;

g is the gravity acceleration.

For the Poisson's ratios ν the following values can be accepted [13]:

$\nu = 0{,}3$ at age of concrete from 2 days to 14 days;

$\nu = 0{,}2$ at age of concrete 28 days;

$\nu = 0{,}15$ at age of concrete equal to or greater than 90 days.

According to [5] the modulus of elasticity of concrete E_c could be determined by the following formula:

$$\text{E}_\text{C} = k_l E_{c,din}, \tag{3}$$

where k_l is a factor, which is accepted 0.87 for low strength concrete and for a high-strength concrete is 0.954 [13].

1.2. Ultrasonic pulse echo method

Ultrasonic pulse echo method [3,4] is a non-destructive testing technique that uses ultrasonic signals to detect defects in materials. With devices based on this method (Pundit Live Array Pro) when testing concrete objects, their internal structure can be displayed in different areas and depths. This is done through a variety of 2D images taken from different angles and the creation of an array of 3D data. This method is suitable when we do not have access to opposite or adjacent sides of the reinforced concrete element [14]. The pulses are transmitted and received by means of single- or multi-channel piezoelectric transducers.

The principle of this method is shown in figure. 2. A pulse generator sends an electrical pulse to the probe of the transmitter, which produces an ultrasonic pulse wave. This wave propagates in the sample and is reflected by the back wall of the sample or by a defect (inhomogeneity) in the volume (figure 2-a) and is received by the receiver (figure 2-b), which converts the wave into an electrical signal.

Ultrasonic pulse echo method uses two types of waves - longitudinal and transverse. For longitudinal waves, the motion of the particles is in the same direction as the motion of the wave. If the motion of the particles is perpendicular to the direction of motion of the wave, then it is called transverse wave, which is possible only in solid materials. The velocity of the two types of waves is different for different materials. The percentages of transmitted and reflected energy depend on the acoustic characteristics of the studied medium.

The probes can be of two types: vertical, which transmit longitudinal waves in a direction perpendicular to the sample, and angular, which have bevels and transmit longitudinal waves at an angle to the sample. At the boundary with the sample, the waves are refracted into longitudinal and transverse waves. The angle of the probes is chosen so that only the transverse waves enter the sample. Standard angular probes are made for different angles of transverse waves, for example 35°, 45°, 60°, 70° and 80° (Figure 2-b).

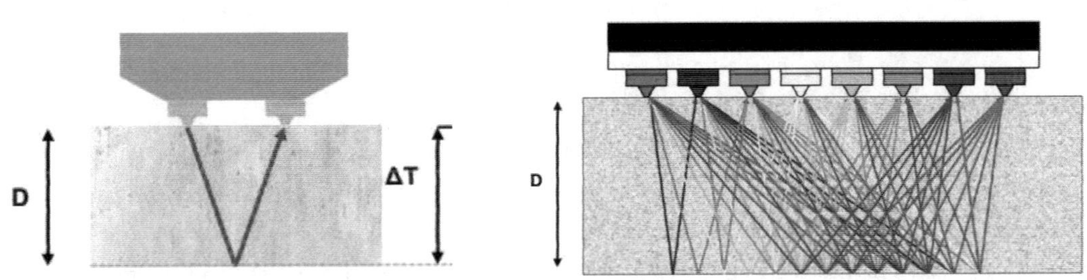

Figure 2. Wave motion in the ultrasonic pulse echo method [15].

With the equipment based on this method, the velocities of the transverse waves can be measured and thus it has the potential as an in-situ non-destructive method for determining the elastic modulus. It is still difficult to be made a conclusion on the appropriateness of this application due to the lack of a sufficient number of published experimental studies. The aim of this study is to be determined the dynamic and static modulus of elasticity by measuring the velocity of S-waves and to be demonstrated the practical potential for in-situ evaluation of concrete in reinforced concrete structures. The results obtained for the modulus of elasticity determined by the velocity of propagation of longitudinal and transverse waves will be compared with the results obtained from the velocity of P-waves with UPVM and the theoretically determined modulus of elasticity according to BDS EN 1992-1-1:2005/NA:2015.

The calculation of the dynamic modulus of elasticity of concrete in existing reinforced concrete structures using the velocities of P and S waves (V_P and V_S) is done by the following formulas [1,2,3,5,9,14,16,17]:

$$V_P = \sqrt{\frac{E_{c,din}(1-\nu)}{\rho_c(1+\nu)(1-2\nu)}} \tag{4}$$

$$V_S = \sqrt{\frac{E_{c,din}}{2\rho(1+\nu)}}\,, \tag{5}$$

where:

$E_{c,din}$ - dynamic modulus of elasticity of concrete;

ρ - acoustic density of concrete;

ν - Poisson's ratio.

From equations (4) and (5) we obtain:

$$E_{c,din} = 2\,\rho V_S^2(1+\nu)\,, \tag{6}$$

where:

$$\nu = \frac{V_P^2-2V_S^2}{2(V_P^2-V_S^2)} \tag{7}$$

From the dynamic modulus of elasticity, the static modulus of elasticity of concrete can be determined according to formula (3).

1.3. Theoretical calculation of the modulus of elasticity according to EC2

The modulus of elasticity can be calculated theoretically according EC2 [18] at different age of concrete by the following formula:

$$E_{cm}(t) = (f_{cm}(t)/f_{cm})^{0,3}E_{cm}\ [\text{MPa}]\,, \tag{8}$$

where $E_{cm}(t)$ and $f_{cm}(t)$ are the values at the age *t* days, E_{cm} and f_{cm} are the values at the 28th day of the modulus of elasticity and mean cylindrical compressive strength of concrete. The relation between $f_{cm}(t)$ and f_{cm} is determined by:

$$f_{cm}(t) = \beta_{cc}(t)f_{cm}\ [\text{MPa}] \tag{9}$$

The factor $\beta_{cc}(t)$ is calculated by:

$$\beta_{cc}(t) = e^{s\left(1-\sqrt{\frac{28}{t}}\right)}\,, \tag{10}$$

where *s* is factor, depending on the type of cement.

2. Experimental samples

The tests for the modulus of elasticity of concrete were made on reinforced concrete beams, which were prepared more than 5 years ago (1926 days). The different elements differ in the provided longitudinal reinforcement, concrete cover and reinforcement ratio (figure 3). Their structural parameters are selected so as to correspond to the characteristic parameters of the beams designed in practice in industrial and civil construction. They are prepared of concrete class C25/30, fine fraction of coarse aggregate (d_{max} = 12 mm) and consistency S3. For longitudinal tensile reinforcement of the beams of the different specimens, ribbed reinforcing bars are used at the usual reinforcement ratios. The distance between the stirrups is chosen so as to ensure the failure of the elements due to bending and not by shear. For previous studies, some of the beams were loaded to the yielding of longitudinal tensile reinforcement, and others to the stage of loading before yielding, reaching a maximum crack width of 0.3 mm. For two years the reinforced concrete elements were indoors, and for the next more than three years they were left outdoors, exposed to external atmospheric influence.

Visually, the visible concrete was assessed to identify and determine the visible defects (pores, cracks, etc.). These regions are excluded from further studies of the concrete. For each of the beams, the surfaces where the non-destructive tests are performed for determining the dynamic modulus of elasticity have been chosen.

Figure 3. Reinforced concrete beams on which the measurements were made [Source: Author I.Ivanchev].

3. Experimental determination of the dynamic modulus of elasticity of concrete with non-destructive methods

The experimental studies were conducted in an accredited mobile construction and testing laboratory at Novostroy Control Ltd., Sofia.

This article illustrates the results of experimental tests to determine one of the most important properties of concrete - modulus of elasticity. Measurements of longitudinal (P-waves) with UPVM were performed according to EN 12504-4:2004 [19] on the reinforced concrete beams with a portable ultrasonic control device Proceq TICO (figure 4) [20]. The operating frequency of the transmitter and receiver is 54 kHz, the resolution is 0.1 μs, and the distance between the transducers is 150 mm. For optimal contact between the piezoelectric transducers and the surface of the reinforced concrete element, the surfaces of the transmitter, receiver and concrete are covered with a special coupling paste.

The transmitter and the receiver of the ultrasonic device are placed symmetrically against each other (figure 4). Before each measurement, the equipment is calibrated with a reference cylindrical body with a known signal velocity.

Figure 4. Measurement of P-waves velocity with device Tico [Source: Author I.Ivanchev].

The longitudinal and transverse ultrasonic wave velocity measurements given in Table 2 and Table 3 are made in the 11 test regions of reinforced concrete beams. In the tables for each of the test regions and for each measurement are given mean velocity values of ten results. After making these measurements from the 11 regions are drilled concrete cores (figure 5-a) with a diameter (*D*) of 99 mm and a depth ranging from 151 mm to 156 mm from places without visible on the surface of the beams cracks, pores, caverns, defects and reinforcement according to EN 12504-1:2019 [21].

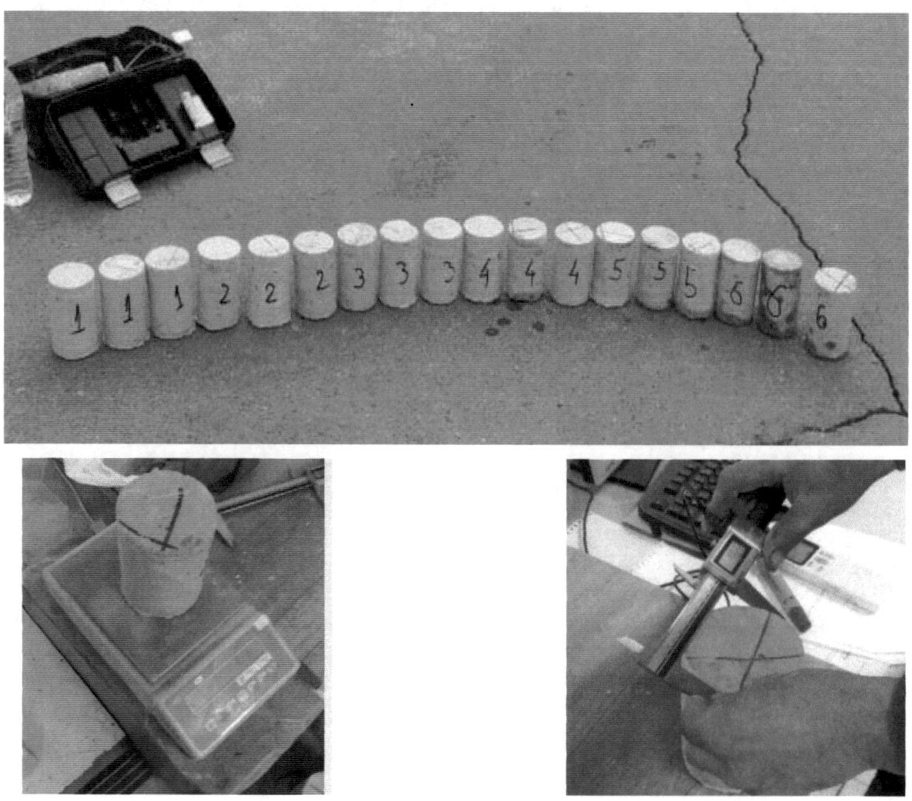

Figure 5. Drilled cores (a), weight measurement (b) and the measurement of geometric dimensions of the cores (c) [Source: Author I.Ivanchev].

Their geometric dimensions and weight are determined (figure5-b,c), and then the weight per unit volume of concrete is calculated (Table 1). The choice of core diameter is regulated by the *D/A* ratio, where *A* is the maximum size of the coarse aggregate. According to [21] this ratio must be equal to or greater than 3.0.

Table 1. Geometric dimensions, weight of cores and weight per unit volume of concrete for the 11 specimens.

№ of specimen	Diameter *D* [mm]	Length [mm]	Weight [kg]	Weight per unit volume [kN/m^3]
1	99	154	2.728	22.57
2	99	151	2.707	22.84
3	99	152	2.805	23.51
4	99	152	2.795	23.43
5	99	156	2.876	23.49
6	99	155	2.858	23.49
7	99	152	2.777	23.28
8	99	151	2.728	23.02
9	99	151	2.786	23.51
10	99	154	2.816	23.30
11	99	153	2.795	23.27

In Table 2 for the 11 test regions are given the values of weight per unit volume γ_c, volume density ρ_c, Poisson's ratio ν of concrete, the measured velocity of the longitudinal waves V_P, the probabilistic dynamic modulus of elasticity $E_{c,din}$ and static modulus of elasticity E_c of concrete.

Table 2. Probabilistic modulus of elasticity, determined by UPVM at age of concrete 1926th day.

№ of test region	γ_c [kN/m^3]	ρ_c [kNs^2/m^4]	υ [-]	V_P [km/s]	$E_{c.din}$ [N/mm^2]	E_c [N/mm^2]
1	22.57	2.300	0.15	4.605	46206.0	40199.3
2	22.84	2.328	0.15	4.608	46819.7	40733.2
3	23.51	2.397	0.15	4.739	50972.3	44345.9
4	23.43	2.388	0.15	4.876	53778.4	46787.2
5	23.49	2.394	0.15	4.692	49923.7	43433.6
6	23.49	2.395	0.15	4.660	49245.1	42843.2
7	23.28	2.373	0.15	4.578	47102.3	40979.0
8	23.02	2.346	0.15	4.648	48011.5	41770.0
9	23.51	2.396	0.15	4.772	51684.6	44965.6
10	23.30	2.375	0.15	4.762	51008.5	44377.4
11	23.27	2.372	0.15	4.853	52908.4	46030.3

The minimum, maximum and mean values of the experimentally determined static modulus of elasticity with P-waves with UPVM are respectively 40199.3 N/mm^2, 46787.2 N/mm^2 and 43315.0 N/mm^2.

Measurements of transverse (shear or S-waves) by ultrasonic pulse echo method are made in the same regions of the reinforced concrete beams with a portable device Pundit Live Array Pro of the company Proceq (figure 6) [15].

With the obtained P-wave and S-wave velocities with the two methods - ultrasonic pulse velocity method and ultrasonic pulse echo method, the dynamic Poisson's ratio and the dynamic modulus of elasticity according to formulas (6) and (7) are calculated. The results for the transverse wave velocities obtained with ultrasonic pulse echo method, the dynamic Poisson's ratio, volume density and dynamic modulus of elasticity are given in Table 3.

Figure 6. Measurement with Pundit Live Array Pro [Source: Author I.Ivanchev].

Table 3. Probabilistic modulus of elasticity, determined by ultrasonic pulse echo method at age of concrete 1926th day.

№ of test region	V_P [m/s]	V_S [m/s]	υ [-]	ρ_c [kNs^2/m^4]	$E_{c.din}$ [N/mm^2]	E_c [N/mm^2]
1	4605	2229	0.35	2.300	30785.6	26783.5
2	4608	2231	0.35	2.328	31214.1	27156.3
3	4739	2354	0.34	2.397	35496.7	30882.2
4	4876	2440	0.33	2.388	37902.1	32974.9
5	4692	2354	0.33	2.394	35335.4	30741.8
6	4660	2276	0.34	2.395	33332.9	28999.6
7	4578	2229	0.34	2.373	31706.8	27584.9
8	4648	2231	0.35	2.346	31535.0	27435.5
9	4772	2395	0.33	2.396	36603.1	31844.7
10	4762	2395	0.33	2.375	36256.5	31543.1
11	4853	2440	0.33	2.372	37588.2	32701.7

The dynamic Poisson's ratios are in the range of 0.33 - 0.35. These values of the dynamic Poisson's ratio are higher than the typical values and according to [2] one of the possible reasons is the difficulty of recording the exact arriving time of the first shear wave and hence the lower values of the measured shear wave velocity.

The minimum, maximum and mean values of the experimentally determined static modulus of elasticity with P-waves and S-waves with UPEM are respectively 26783.5 N/mm², 32974.9 N/mm² and 29877.1 N/mm².

The theoretically calculated static modulus of elasticity $E_{cm}(t)$ at age of concrete 1926th day according to EC2 is 33113.6 N/mm².

From the results of the two non-destructive methods, the relative error was determined according to formula (11). The result calculated according to EC2 is taken as the reference value of the concrete static modulus of elasticity.

$$\varepsilon_{\%} = \frac{E_c - E_{cm}(t)}{E_{cm}(t)} 100\%, \tag{11}$$

where E_c is the static modulus determined by one of the two non-destructive methods.

The minimum, maximum and mean relative error of the measurements by the two methods for each test region is determined (Table 4).

Table 4. Relative error, minimum, maximum and mean error for the static modulus of elasticity determined by both methods on age of concrete 1926 days.

№ of test region	E_c UPEM [N/mm²]	E_c UPVM [N/mm²]	E_c EC2 [N/mm²]	$\varepsilon_{\%}$ UPEM [%]	$\varepsilon_{\%}$ UPVM [%]
1	26783.5	40199.3	33113.6	19.12	21.40
2	27156.3	40733.2	33113.6	17.99	23.01
3	30882.2	44345.9	33113.6	6.74	33.92
4	32974.9	46787.2	33113.6	0.42	41.29
5	30741.8	43433.6	33113.6	7.16	31.17
6	28999.6	42843.2	33113.6	12.42	29.38
7	27584.9	40979.0	33113.6	16.70	23.75
8	27435.5	41770.0	33113.6	17.15	26.14
9	31844.7	44965.6	33113.6	3.83	35.79
10	31543.1	44377.4	33113.6	4.74	34.02
11	32701.7	46030.3	33113.6	1.24	39.01
	min $\varepsilon_{\%}$			0.42	21.40
	max $\varepsilon_{\%}$			19.12	41.29
	mean $\varepsilon_{\%}$			9.77	30.81

4. Conclusion

An experimental study is performed for determining the probabilistic dynamic and static modulus of elasticity, dynamic Poisson's ratio of concrete in reinforced concrete beams at age of concrete 1926 days. Ultrasonic pulse velocity method and ultrasonic pulse echo method are used for determining the propagation velocities of P-waves and S-waves in concrete. The values of the dynamic modulus of elasticity vary depending on the measurement method.

UPVM uses only P-wave velocity data, and UPEM uses P-wave and S-wave velocities. From the measurements conducted for the age of the concrete 1926th day the following conclusions can be made:

- The determined dynamic and static modulus of elasticity of concrete from the measurements of P-waves with UPVM is greater than that obtained with UPEM;
- The relative errors of the experimentally determined static modulus of elasticity with UPVM compared to the theoretically calculated according to EC2 vary from 21.40% to 41.29%. The

relative errors of the experimentally determined static modulus of elasticity with UPEM compared to the theoretically calculated according to EC2 vary from 0.42% to 19.12% .

- The mean relative error with UPVM is 30.81%, and with UPEM is 9.77%.
- According to [5], the greater error in using UPVM is due to the inaccurate estimate of the Poisson's ratio and that equation (1) is only suitable for homogeneous materials, leaving the validity in doubt for inhomogeneous materials, such as concrete;
- Using only one non-destructive assessment method can lead to errors and the combination of two or more non-destructive methods is more accurate and effective for assessing the specific condition of the concrete;
- Experimental studies confirm the findings of other researchers that there are many factors that affect ultrasonic pulse velocity of P-waves when is used UPVM (type and size of the aggregates, class and type of cement, water-cement ratio, presence of reinforcement, etc.), unlike S-waves. More research is needed in this area.

Nowadays, the UPVM continues to be reliable and promising non-destructive method based on rapid measurements using available equipment.

5. References

[1] Lee B, Kee S, Oh T and Kim Y 2017 *Evaluating the Dynamic Elastic Modulus of Concrete Using Shear-Wave Velocity Measurements, Hindawi, Advances in Materials Science and Engineering,* Article ID 1651753

[2] Candelaria D, Kee S , Yee E and Lee J 2021 *Effects of Saturation Levels on the Ultrasonic Pulse Velocities and Mechanical Properties of Concrete, Materials* **14,** 152

[3] Hadianfard M, Marzouk H and Shaoceyanb C 2017 *Strength and elastic moduli of a concrete bridge using advanced nondestructive techniques, Scientia Iranica, Transactions A: Civil Engineering* **24**, 3, 942-952

[4] Bishko A, Samokrutov A and ShevaldykinV 2008 *Ultrasonic echo-pulse tomography of concrete using Shear waves low-frequency phased antenna arrays, 17th World Conference on Nondestructive Testing,* Shanghai, China

[5] Naik T, Malhotra V and Popovics J 2004 *The Ultrasonic Pulse Velocity Method, Handbook on Nondestructive testing of Concrete*

[6] Lorenzi A, Tisbierek F and Pinto L 2007 *Ultrasonic Pulse Velocity Analysis in Concrete Specimens, IV Conferencia Panamericana de END,* Buenos Aires

[7] Karaiskos G, Deraemaeker A, Aggelis D and Hemelrijck D 2015 *Monitoring of concrete structures using the ultrasonic pulse velocity method, Smart Materials and Structures, vol.* **24**, 11

[8] Bungey J, Millard S and Grantham M 2006 *Testing of Concrete in Structures, 4th ed.; Publisher Taylor & Francis e-Library,* London & New York

[9] Chakraborty J, Wang X and Stolinski M 2021 *Damage Detection in Multiple RC Structures Based on Embedded Ultrasonic Sensors and Wavelet Transform, Buildings* **11**

[10] Savaliya K, Thaker K and Dave U 2014 *Comparison between Different Methods of Ultrasonic Pulse Velocity Tests on Concrete, International Journal of Engineering Research and Applications (IJERA) ISSN: 2248-9622 National Conference on Advances in Engineering and Technology*

[11] *Guidebook on Non-Destructive Testing of Concrete Structures IAEA* 2002 Vienna, Austria, IAEA–TCS–17 ISSN 1018–5518 IAEA

[12] Kumar S and Santhanam M 2006 *Detection of Concrete Damage Using Ultrasonic Pulse Velocity Method, Proc. National Seminar on Non-Destructive Evaluation,* Hyderabad, *Indian Society for Non-Destructive Testing Hyderabad Chapter,* 301-308

[13] Dimov D 2011 *Non-destructive testing of Building Structures, Direct Services,* Sofia, Bulgaria

[14] Carino N and Sansalone M 1984 *Pulse-echo method for flaw detection in concrete, National Bureau of Standards,* Washington

[15] Proceq official website [online]

https://www.proceq.com/uploads/tx_proceqproductcms/import_data/files/Pundit%20Live%20Array%20Cue%20Cards.pdf

[16] Popovics J 2008 *A Study of Static and Dynamic Modulus of Elasticity of Concrete, ACI-CRC Final Report*

[17] Kee S, Kang J, Choi B, Kwon J and Candelaria M 2019 *Evaluation of Static and Dynamic Residual Mechanical Properties of Heat-Damaged Concrete for Nuclear Reactor Auxiliary Buildings in Korea Using Elastic Wave Velocity Measurements, Materials,* **12**, 2695

[18] BDS EN 1992-1-1 *Eurocode 2: Design of concrete structures - Part 1-1: General rules and rules for buildings - National annex to BDS EN 1992-1-1.* (2005/NA:2015)

[19] EN 12504-4 *Testing concrete. Determination of ultrasonic pulse velocity.* (2004)

[20] Abmbv Manual website [online] http://www.abmbv.nl/files/proceq_tico_user_manual_en.pdf

[21] EN 12504-1 *Testing concrete in structures - Part 1: Cored specimens - Taking, examining and testing in compression.* (2019)

Acknowledgment

This research was supported by the research grant No KP-06-OPR01/3-2018 for the project "Creation of a flexible methodology for development and management of innovative project in scientific organizations" (2018/2021) funded by National Science Fund, "Competition for Financial Support for Basic Research Projects on Societal Challenges-2018", Ministry of Education and Science, Bulgaria.

Economic efficiency: yield on cost as core tool of benchmarking

S Pagac[1,2]

[1] Institute of Technology, Economics and Management in Construction, Faculty of Civil Engineering, Technical university of Kosice, Vysokoskolska 4, 042 00 Kosice, Slovakia
[2] CTP Invest SK, Laurinska 18, 811 01 Bratislava, Slovakia

E-mail: stanislav.pagac@ctp.eu

Abstract. The aim of the paper is to introduce and evaluate the yield of cost as tool of measurement and comparison in construction economy. Yield on cost evaluates the ratio of investment or development costs to project income over given period. Evaluation of effect time indexation and discount rates on comparative metrics is required. As result the base equation is proposed to use in comparing economic performance of peer project across development industry

1. Introduction

In development environment of industrial real estate, is much emphasis put on one of the most important numerical values used in financial evaluations of projects and business cases: yield on cost [1]. The number by its own, is unique in the volume of information it represents [2,3]. When presented alone, it offers immediate, although not elaborate, evidence on developers´ performance on both given market and reported segment [4].

Economic efficiency is a state that represents the optimum allocation of resource, the minimum cost of production goods and services, and maximum outcome [5]. Economic efficiency refers the absence of overproduction or underproduction [6-9].

An efficient situation means that it is not possible to improve the situation of one entity without worsening another [10]. In contrast, an inefficient scenario suggests that one entity can be better off without worsening another. Thus, if the reallocation of resources can produce increased outcomes, there is inefficiency [11].

2. Yield on Cost

The formula for calculation of yield on cost is:

$$YoC = \frac{PC}{PI}, \tag{1}$$

where

- *YoC* is yiel on cost,
- *PC* is project costs,
- *PI* is project income.

The formula is simple, however, the discrepancies of meaning and content of denominator and divider are of high importance as these two values comprise most often vast amount of data on projects. Examples of these data on costs side are:

- hard costs (construction),

- soft costs (design, management etc),
- finance costs,
- overheads,
- developers margin,
- risk premium,`
- land cost,
- and potentially many more.

Examples on income side:

- headline rent (month or year depending on form of calculation),
- rent on other areas,
- one off payments on increased standards,
- and potentially many more.

The chosen approach for paper calculation is the following definition:

Yield on Cost is calculated as weighted average of estimated or actual rent, without considering any rental time discounts, per year divided by hard construction costs including land costs of the project completed in the year of evaluation.

From the above definition it is obvious that the simpler path was selected for economical reporting on development project, not for detailed economic analysis. For introduction to the topic, the presented paper elaborates on the simplistic approach to Yield on Cost.

Operational use of the yield on cost economical model is for purpose of speed and compatibility simplified to large extend. It omits with full acknowledgment discount rates and indexations on both sides of the formula. This is justified with generalisation of assumption that the indexation of rents, also referred to as like for like growth, will compensate for amortisation or discount rate of the currency and depreciation of the properties.

The reason simple model as presented is functional in development practise can be explained in economic theory of modelling. Economic modelling was designed to build and explain economic reporting rather than provide forecasting.

The subject of further study and modelling is forecasting of the yield on cost, i.e. economic efficiency of a proposed rental development project during its life cycle. Main focus is on prediction of *YoC* for project considered for future execution while considering the location and market specifics of tenant and asset.

3. Discussion on formula

The formula for calculation of yield on cost is

$$YoC = \frac{12 \cdot MTHR}{ACC+PVLC+ASP+BF+ALC}, \tag{2}$$

where

- *YoC* is yiel on cost,
- *MTHR* is monthly total headline rent generated on the project,
- *PVLC* is present valuation land costs,
- *ASPC* is attributed soft project costs,
- *BC* is brookers fees,
- *ALC* is attributed legal costs.

The source of headline data are market data from real estate agencies and own historical developers project data. Source of these data are publicly accessible markets reports.

Data from Market outlook report of CBRE for greater Bratislava are set to maximum of 4.9 Eur/m^2 with outlook of 7% growth until end of 2022. This increase in headline rent from 4.9 to 5.2 Eur/m^2.

Looking at the present steep increase in prices of construction materials according to Slovak Statistical office nearing 28% on material side, more than 20% on labour costs and above 75% in energy costs. The cost side of the equation is considerably challenged. The rising costs on input side cause nervosity and delays in decision processes.

Table 1. Price indexes in Slovakia in first quarter of 2022.

Index	January	February	March
Delivery of electricity, gas, steam, cold air	141.8	182.1	175.4
Cost of construction work	115.0	120.4	121.7
Cost of construction materials	123.4	125.8	127.7

The shock increase of construction materials costs is markable across CEE region and it has direct impact on actual yield on cost data for first quarter of 2022 [3], see Table 1. The data presented publicly available from peers in CEE in industrial development are illustrating spread of yield on cost among peers, but more importantly the decreasing trend of this value [4], Figure 1.

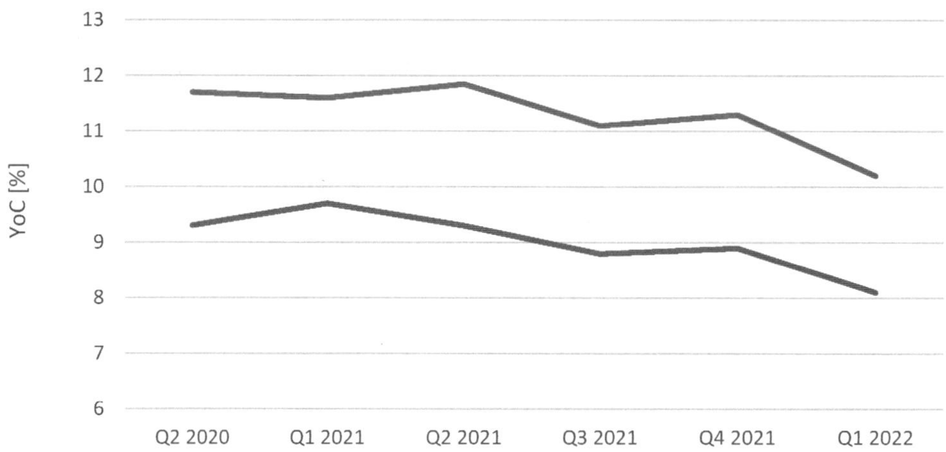

Figure 1. Yield on cost in CEE Region.

Over five quarters since Q2 2020 has the yield on cost managed to be resistant to inflation and increase in construction costs mainly due to parallel indexation of rental levels. The year 2021 has impacted slightly the construction costs, however it also peaked in demand due to pandemic reasons, same as caused brief period of shortage in construction materials, Figure 1.

The characteristic of yield on costs are universal across the development business and various branches. Attractivity of industrial development, not considering residential market while its product is not developed for operation and long term rents, has proven across the region.

Volume of investments to industrial property has maintained strong position of 36% in Q1 2022 in comparison to offices that dropped year to year from 37.6% to 14% share on total yearly investments in Slovakia [2], Figure 2.

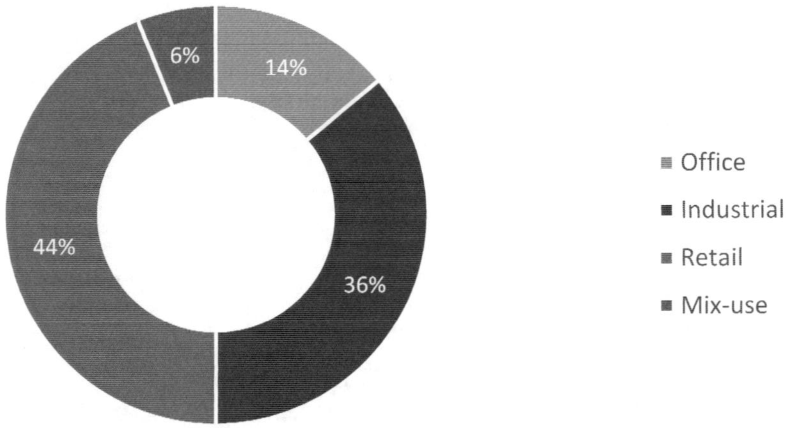

Figure 2. Share of investment volume per use.

4. Conclusion

The use of yield on cost as economic tool for decision on development is not new concept. However, to be used further as reporting value for communication towards investors, it is not clearly defined and on market is lack of a definition widely accepted. Nor is there any definition presented by an international organization to be recognized as Europe or world standard. It is the plan to further analyse the yield on cost formula and to create a model not only for reporting purposes but to provide a precise forecasting tool for projects that span across regions and one year project cycle.

5. References

[1] Boland LA 2014 *Model building in economics: its purpose and limitation* New York Cambridge university press 137

[2] CBRE Slovakia 2021 *Market outlook 2022 report Slovakia real estate* Bratislava, CBRE research Slovakia 8 [online] https://www.cbre.sk/-/media/cbre/countryslovakia/documents/slovakia- market-outlook-2022.pdf

[3] STATdat 2022 *Indexy spotrebiteľských cien oproti bázickému obdobiu* [online] http://statdat.statistics.sk/cognosext/

[4] Various, 2021, 2022, *Quarterly and annual announcements*, CEE industrial developers web pages

[5] Wallstreetmojo Editorial Team 2022 *Economic Efficiency* [online]https://www.wallstreetmojo.com/economic-efficiency/

[6] May RG 1971 The influence of quarterly Earnings Announcements on Investor Decisions as Reflected in Common Stock Price Changes. *Journal of Accounting Research* **9**, 119-163

[7] Gonchruk AG, Lazareva NO, Alsharf IAM 2015 Benchmarking as a performance management method. *Polish journal of managements studies* **11**(2) 26-36

[8] Makalesi A 2021 Quarterly earnings announcement and its impact on corporate financial performance. *Muhasebe Bilim Dünyası Dergisi* **23**(4) 797-815

[9] Deilman C, Lehman I, Reissman D, Hennerdorf J 2016 Data envelopment analysis of cities – Investigation of the ecological and economic efficiency of cities using a benchmarking concept from production management. *Ecological Indicators* **67** 798-806

[10] Subhash CR, Subal CK, Pami D 2015 *Benchmarking for Performance Evaluation A Production Frontier Approach* 292

[11] Mesosi A, Szabo L, Szabo S 2018 Cost-efficiency benchmarking of European renewable electricity support schemes. *Renewable and Sustainable Energy Reviews* **98** 217-226

IOP Conf. Series: Materials Science and Engineering 1252 (2022) 012020 doi:10.1088/1757-899X/1252/1/012020

Dosimetric measurement of lighting conditions in office environment

P Hanuliak[1] and P Hartman[1]

[1] Department of Building Structures, Faculty of Civil Engineering STU in Bratislava, Radlinského 2766/11, 810 05 Bratislava, Slovakia

E-mail: peter.hanuliak@stuba.sk; peter.hartman@stuba.sk

Abstract. The aim of the paper is to present partial outputs from continual measurement of light conditions in office environment during common work activity. The volunteers worn head mounted dosimetric devices – the Lightwatchers, to collect individual and realistic data in contrary to fixed photometric measurement devices. The outputs confirm that the natural activity of office workers has a great influence on recorded illuminance levels. It shows an overall decrease of light levels recorded on vertical plane in facial area when compared to standalone fixed sensors. Since the total amount of the light received through the eyes is important for a proper entrainment of our biological system, an adjustments for office´s lighting design needs to be taken into consideration.

1. Introduction

Lighting conditions in the environment in which a person is located not only affect his ability to perform visual tasks, but also directly affect health and well-being, for which there is already a strong scientific evidence. This field of research is developing very rapidly and with growing interest. A new term was introduced into the International Lighting Vocabulary: Integrative Lighting (CIE 17-29-028) as the lighting integrating both visual and non-visual effects and producing physiological and/or psychological benefits upon humans [1]. This term is applicable on the humans only, since for example there is a different action of the light being studied in the field of agriculture, named Horticulture Lighting. Integrative lighting is intended to be used in real life conditions, to be distinguished from the lighting used for therapeutic purposes - light therapy, used in medical applications. While for visual tasks it is important to provide illumination at the work plane, usually horizontally levelled, when evaluating light for non-visual action, it is important to quantify the illumination at eye level in the vertical plane. Both human´s light processing systems, visual and non-visual, have different characteristics in terms of spectral sensitivity as well as the required overall levels of illumination [2]. Spatial distribution and timing is also very important in assessing the non-visual effects of light. Due to these differences, it is also necessary to change the methods used for measuring lighting in internal spaces of buildings. Standard photometric measurement devices do not provide the necessary technical properties in this respect. Given the need to evaluate the spectral composition of illumination incident on the human eye over time, a portable spectrophotometer with the ability of continuous data recording would be an ideal instrument. Full spectrum spectrophotometric data gives a better opportunity to evaluate the light conditions for non-visual effects. Various models have been proposed to such evaluation [3, 4]. In 2018, a model that uses five spectral weighting functions as action spectra of all effective retinal photoreceptors was introduced (CIE S 026:2018) [5] requiring spectrophotometrical data as basic inputs.

IOP Conf. Series: Materials Science and Engineering 1252 (2022) 012020 doi:10.1088/1757-899X/1252/1/012020

However, such a spectrophotometer, which would be compact in size to be worn head-mounted, is not available on the market, since its production would be too expensive with a limited range of use.

2. Measurements

Measurements were performed in the office room of the administrative building of Slovak University of Technology in Bratislava. The testing room was located on 21st floor to minimize the effect of surrounding shading objects. The orientation of the windows was northeast no minimize the direct sunlight incidence. The room was occupied with five adult persons during the whole day from 7:30 to 16:30 in course of two weeks. Subjects were asked to perform their habitual office work tasks. Persons that required an use of computer were asked to set the screens on minimal levels of output screen power. The office room was equipped with static sensors: two photometric illuminance probes FLA 623 VL with datalogger ALMEMO, to measure horizontal illuminance levels every minute on the work desks in the height of 850 mm and four spectrophotometers mounted on tripods to measure vertical illuminance levels in the height of sitting person´s eyes, 1200 mm. Two spectrometers, those oriented to the left wall, were Konica Minolta CL500-A with recording step about 5 min and remaining two were Gigahertz-Optik MSC15 with recording step 1 min. A layout of the installed sensors is pictured on figure 1. A head mount Lightwatchers, in figure 2, were used by each volunteer, position 1-5, to record illuminance levels in the facial area throughout the day in step of 10 seconds. These measurements were part of a larger experiment, where lighting conditions in internal environment are being studied to find a correlation to evaluated biological markers. This study is still in the phase of evaluation and results will be published.

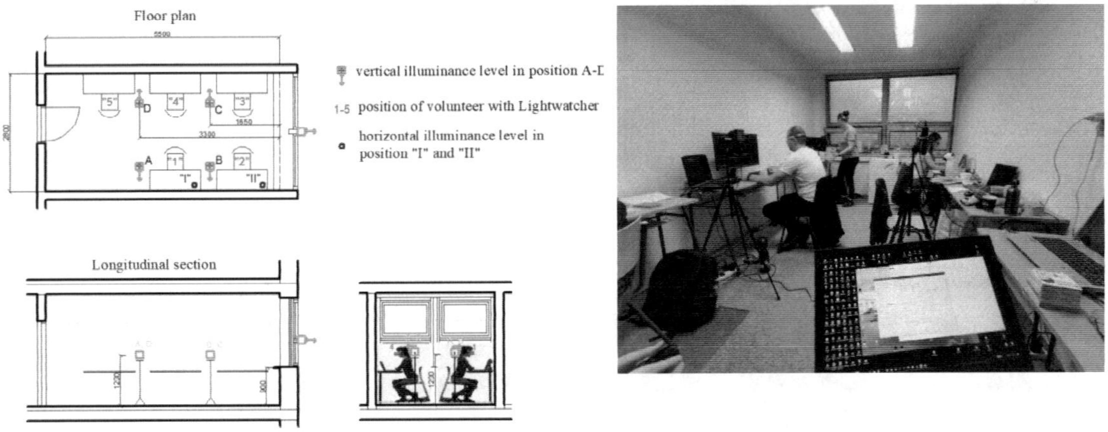

Figure 1. Layout of office equipped with sensors.

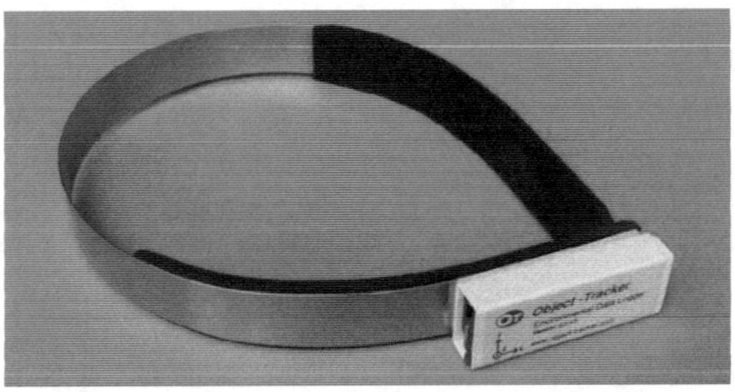

Figure 2. Lightwatcher device with headband.

3. Outputs

A typical output from sensoric system installed in the room and sensors worn by volunteers is shown on figure 3. A large set of numerical data was obtained for further processing. In this paper we aimed at evaluation of light levels in three main domains: horizontal illuminance levels and vertical illuminance levels as commonly measured variables in contrary to vertical illuminance levels measured in facial area of every volunteer. As previous studies suggested that the levels measured in facial area, using personal dozimetric devices will be lower in comparison to the horizontal levels [6]. Also the study [7] showed the big difference of obtained numerical values of illumination in simulated conditions of offices, where models that comprised the model of human body brought overall lower values in illumination level when compared to models without a human body. There was even a large variation in data as a result of person´s facial typology.

A typical data set of acquired data was examined. A day when a fully overcast sky occurred with very low external illuminance level was selected in figure 4. The volunteers used artificial light during the whole day as can be seen also in course of internal illuminance level from the CL500-A in "C". The levels of illumination in the fixed and head mounted photometric device, Lightwatcher in position "4" is shown in figure 5. Even the variations in data from personal device is large and caused by individual movement and activity, the overall level is lower than illuminances recorded by fixed device in corresponding position. This trend was observed in all positions.

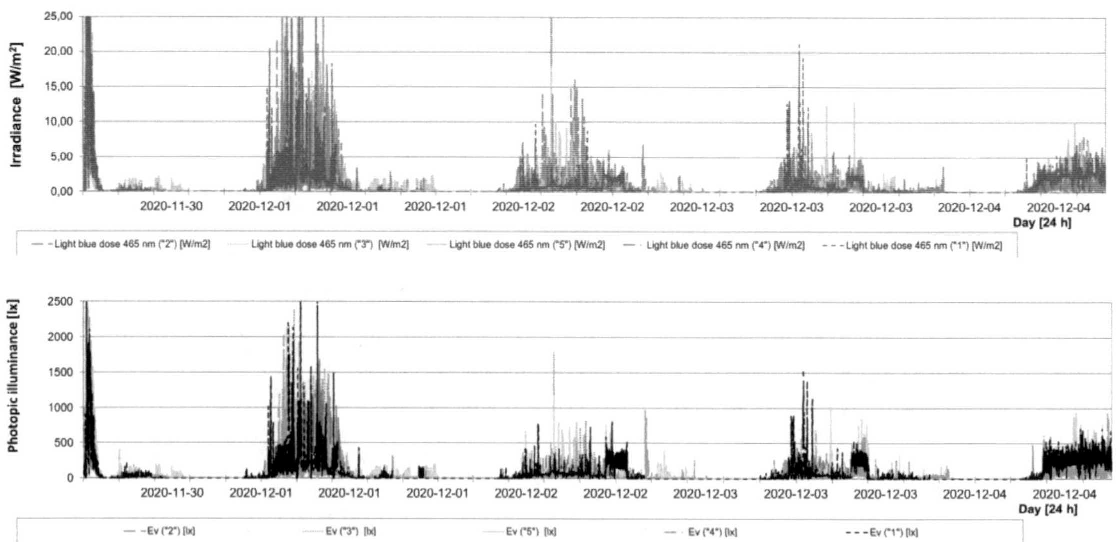

Figure 3. A typical output from a photopic and blue sensor in personal dosimeter in duration of one week with recording step 10 seconds to include random head movement and different eye focusing.

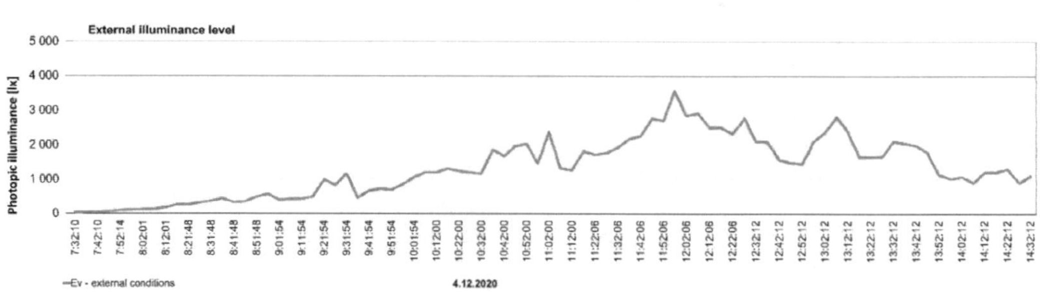

Figure 4. External illuminance levels measured on external glazing surface.

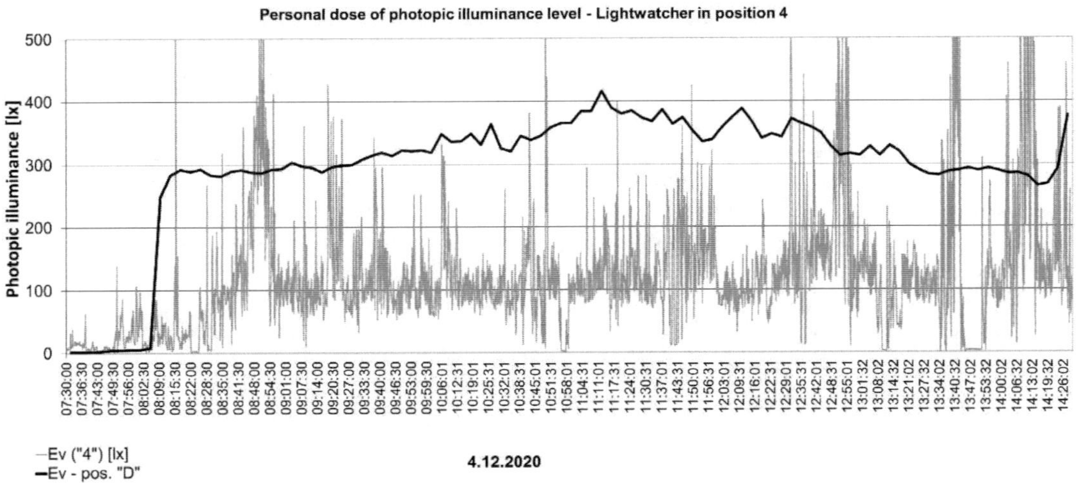

Figure 5. A light levels recorded by Lightwatcher in comparison to fixed device on vertical plane in position 4.

The graph on figure 6 shows an averaged values for all the three evaluated domains in all monitored positions. The levels of the horizontal illuminances recorded by fixed sensors located on work desks, position "I" and "II", were averaged on 469 lux. The levels of vertical illuminances recorded by fixed sensors located on tripods were averaged on 309 lux and the levels of vertical illuminances recorded by personal dosimeters in facial area were averaged on 186 lux. This is a large decrease in the total amount of the light that enters the eyes in reality and it differs significantly to the situation when we would rely solely on the fixed sensors without the presence of the human in assessed office environment.

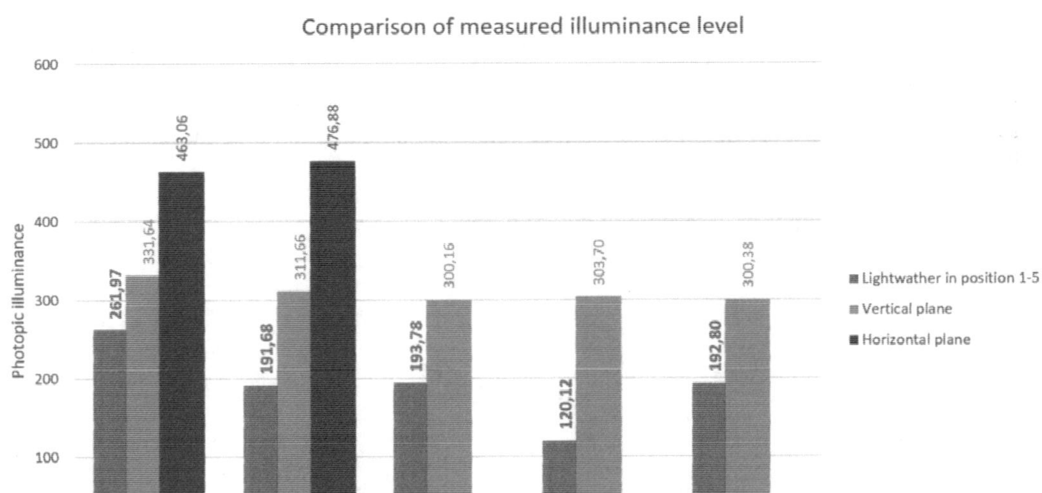

Figure 6. Averaged illuminance levels for different positions.

4. Conclusion

As we can see the real life measurements in the office environment differs significantly to the laboratory like setups. The presence of office workers with combination of their activity in space influence the total dose of light received in facial area. The light levels of the light recorded by personal dosimeters were lower as the levels recorded by the fixed devices, ranging from 20 % to 60 % depending on the position of the worker. Also the type of the workload has a crucial impact on measured values of illuminance in facial plane. The presence of artificial light also provides relatively uniform lighting levels, as can be seen in figure 6 from vertical and horizontal illuminance levels. Nevertheless, the illuminance levels recorded from Lightwatchers, which comprises the effect of random head movements, showed significantly different levels of facial illuminance levels and confirms previous studies described in the introductory part.

5. References

[1] Commision Internationale de L'Eclairage (CIE) *CIE S 017/E:2020* (ILV: International Lighting Vocabulary, 2nd Edition) CIE Central Bureau Vienna p. 2020

[2] Vetter C, Pattison P M, Houser K, Herf M, Phillips A J K, Wright K P, Skene D J, Brainard G C, Boivin D B and Glickman G 2021 *A Review of Human Physiological Responses to Light: Implications for the Development of Integrative Lighting Solutions* (LEUKOS) p 1-28

[3] Rea M S, Figueiro M G, Bierman A and Bullough J D 2010 *Circadian Light* (Journal of Circadian Rhythms) vol 8 p 1-10

[4] Gall D and Bieske K 2004 *Definition and measurement of circadian radiometric quantities* (Light and health - non-visual effects: proceedings of the CIE symposium '04) p 129-32

[5] Schlangen L 2018 *CIE S 026/E:2018 "System for Metrology of Optical Radiation for ipRGC-Influenced Responses to Light* (CIE international Standard)

[6] Vandahl C, Bieske K, Wolf S and Schierz CH 2011 *Light and Health in Factory Work Places* (CIE 27th Session) p 846-52

[7] Kucserova E, Janák M and Hanuliak P 2011 *Analysis of the Human Eye's Illumination conditions in Internal Environment* (Society for Light Treatment and Biological Rhythms, Program and Abstracts) vol 23 p 43

Acknowledgments

This research was supported by the Slovak Scientific Grant Agency (VEGA) in collaboration with Slovak Ministry of Education, Science, Research and Sports and Slovak Academy of Sciences (SAS), under Grant number 1/0042/21 and Slovak Research and Development Agency (APVV) under Grant number 18-0174.

Plastic zones in the pin connection of the laminated lattice girder

M Tomko[1] and I Demjan[1]

[1] Department of structural mechanics, Institute of Structural Engineering, Faculty of Civil Engineering, Technical university of Kosice, Vysokoskolska 4, 042 00 Kosice, Slovakia

E-mail: michal.tomko@tuke.sk; ivo.demjan@tuke.sk

Abstract. The paper deals with a static reanalysis of the existing ice-rink bearing steel structure which proves all the bearing members of the object after additional loading by new roof coat. In addition to experimental measurements carried out to diagnose the bearing roof steel structure and its elements needed for static assess the structure's current state, pressure coefficients in individual parts of the structure were defined by means of ANSYS software and consequently used in static analysis of wind effects. According to the static analysis results, the repair of initial state was designed and realized, with designing of details needed to be modified. By means of ANSYS software, the detail of pin joint support of laminated lattice-girders was 3D-modeled with volume elements. The deformation-stress analysis of the mentioned detail was also performed by considering of elastic-plastic material with linear hardening.

1. Introduction

When it is necessary to realize a reconstruction of buildings, their adequate resistance against new loading should be proved. Nowadays, the large-scale computational apparatus for static calculation of a structure is available, which can practically help with design of new structures. With certain safety margin it is possible to design the structures on the basis of standard regulations and procedures. Reverse process, i.e. calculation approximating to the stress of already existing structure and giving a true picture of its reliability rate is more difficult.

The mentioned ice-rink was built and given into usage 34 years ago, figure 1. At present, the ice-rink is no more in a condition to fulfil the standards required by a user and so, it was decided to realize a reconstruction of the ice-rink focused on the replacement of unsatisfactory, damaged and leaked roof coat. This fact requires static re-calculation of the structure at a present state with new designed roof coat, [1].

Figure 1. Ice-rink view.

The ice-rink roof structure can be defined as a steel latticed arch shell, which was unusual and very courageous lightened bearing structure of the roof covering, at the time of realization. The ice-rink covering is designed as laminated vault with ribs held obliquely with reciprocal intersections. Mono-layer vaults act as short cylindrical shells with distances less than the span and stiffened by transversal stiffeners (diaphragms, lattice-girders), figure 2. Structure of vault is made from tubes CHS B95/4, interconnected by verticals from tubes CHS B70/2.9. The ribs are built-up from uniform, always repeated and reciprocally interchanged segment up to 8.0 m long, [2], [3].

Figure 2. General view on the ice-rink's roof steel bearing structure (original state).

2. Static calculation

Basic assumption of good static assessment of a structure is knowledge of physical and mechanical properties of particular structural elements and their reciprocal static influences in a structure, i.e. parameters, which are also included in theoretical calculation model, if equivalency should be reached between the actual behaviour of buildings and the theoretical response of the static calculation model.

Provided that, input characteristics describing the mentioned steel structure are not known, then in technical practice, they are needed to be determined on the basis of diagnostic experimental measurements, [4].

In the spots of local damage, experimental measurement of webs thickness of steel elements was realized, i.e. statically important elements for the assessment of building structure (M1 – roof laminated lattice-girder, M2 – external support of roof laminated lattice-girder, M3 – internal support of roof laminated lattice-girder, M4 – beam, supported the internal shore of roof laminated lattice-girder, M5 – gable column), figure 3. The experimental measurement results are filed in [5].

The results from the geodetic survey demonstrated no relevant deviation of the mentioned laminated roof cupola, in comparison with designed state, which in the end would influence the results of the static calculation, and so they were not considered by creating the calculation model, figure 3 (A, B, C, D, E – points of geodetic survey).

Layout dimensions of the object were taken from design documentation [2] at the development of calculation model and the actual thicknesses of cross sections found out by experimental measurement were assigned to each element.

By means of IDA NEXIS software, the static calculation model of ice-rink was done. Three dimensional model contains 14452 nodes and 22357 bars members, figure 3. Mutual connection between several bar elements was hinged or rigid. Calculation model was analysed using geometrical non-linear calculation and 20 loading increments were used.

Taking into account the measured corrosion decreases of bearing structural members in the model, the static analysis of wind effects on the ice-rink structure was done. To specify the aerodynamic coefficients, air flow module in ANSYS was used, with defined pressure coefficients in certain parts of structure and subsequently used in static analysis of wind effects realized by means of IDA NEXIS software. Air flow module CFD (Computational Fluid Dynamics) permits to define accurately aerodynamic coefficients, speed and wind pressure on the structures, which's shape is not clearly defined in the standards.

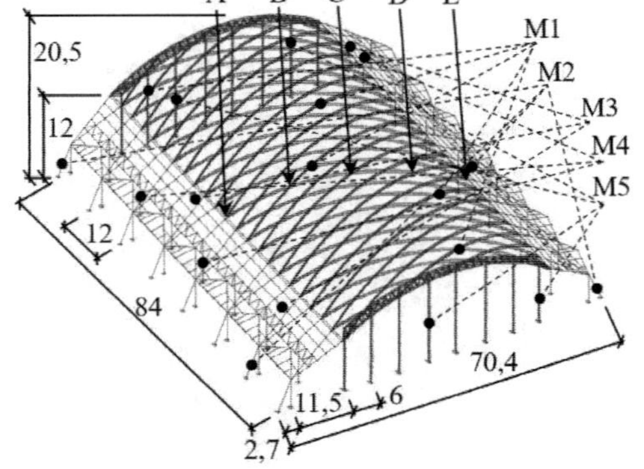

Figure 3. Calculation model of ice-rink (dimensions in [m]).

By means of ANSYS software, FEM model was created, with observation of wind speed and aerodynamic coefficients on the structure surface, figure 4, 5, 6. Calculation model of air flow contains 12 064 elements using finite element „Fluid 142“. Calculation was done in CFD module with turbulent flow RNG (Renormalisation Group Theory). It was considered with wind speed 29.66 m/s what corresponds with wind pressure 0.55 kN/m^2.

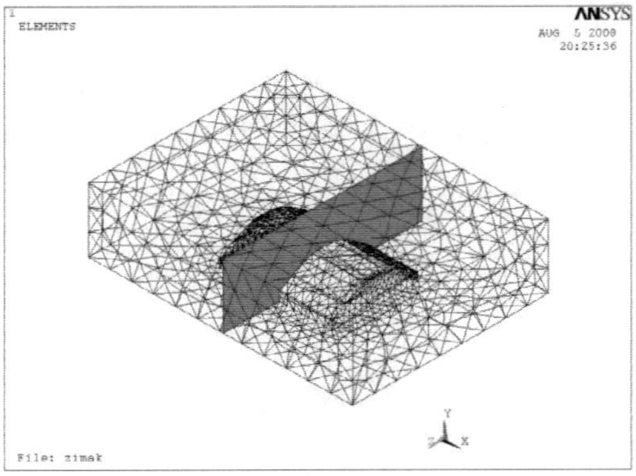

Figure 4. Calculation model.

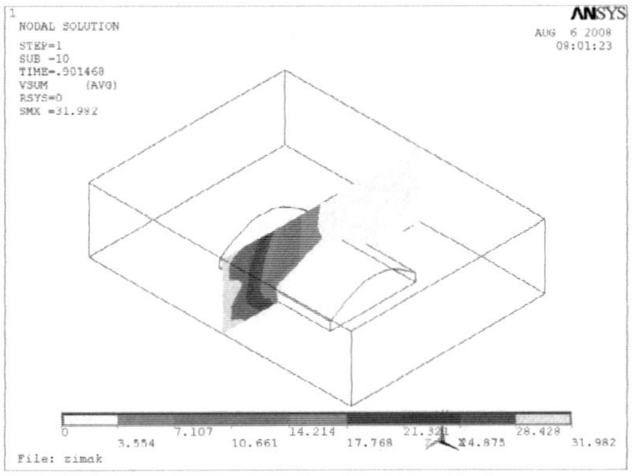

Figure 5. Air flow – wind speed.

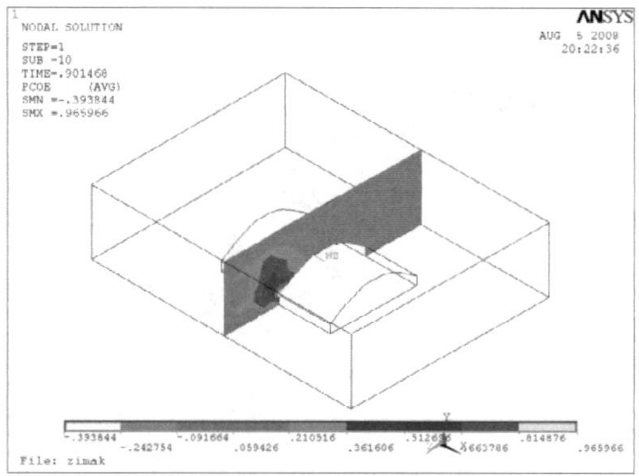

Figure 6. Aerodynamic coefficients.

The load was intended according to [6]:
– permanent (roof beam, sheet and isolating layers, suspend cube, footbridge over the tribunes),
– snow (III. snow area according to STN 73 0035, it was considered with non-symmetric snow, i.e. full value on first half of the roof and half value on second half of the roof),
– wind.

It was considered with 6 load cases, which create 4 non-linear load combinations, i.e. by application of non-linear calculation; it is not possible to apply principle of superposition.

3. Static calculation results

The assessment was carried out according to [7]. Building structure, after designed and realized structure works is satisfactory for 1st and 2nd limit state according to actual valid standard [7]. By assessment, values of ultimate stresses in the edges of upper and lower chord of laminated arches were exceeded from 20 % to 30 %, figure 7. Calculated stresses of some verticals of roof bearing structure arch were exceeded by 40 % in comparison with ultimate stresses in elastic area, figure 7. Other steel structure elements were satisfactory in 75 % up to 95 % of limit stress. Maximal deformation in the middle of laminated cupola is 142 mm.

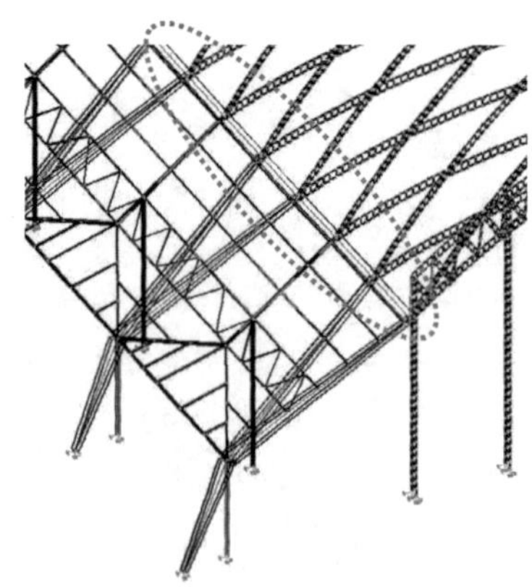

Figure 7. Part of render calculation model of ice-rink, schematic figuration of unsatisfactory verticals and a part of first field of lower and upper chord of bearing roof structure arch.

According to the static analysis results of ice-rink, the repair of initial state was designed and realized, with designing of details, which need to be modified (8 mm thick sheets need to be welded to first part of all lattice-girders, that create the roof structure, figure 8), so that the mentioned ice-rink, after designed and realized modifications will be satisfactory for 1st and 2nd limit state.

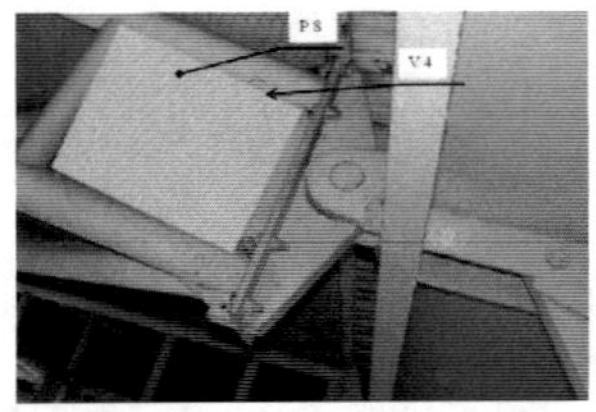

Figure 8. Connection of laminated lattice-girder to hinged pin, schematic figuration after strengthening of laminated lattice-girder by steel plate th. 8 mm.

4. Static analysis of lattice - girders pin connection

The process of transformation during which plastic deformations appear and develop is a plastic transformation, i.e. plastic deformations of the entity are permanents. Real dependences between stress and deformation in the plastic transformation process are quite difficult, so in the technical applications of plasticity theory, the properties of materials are characterized by idealized stress-strain diagrams. Non-linear relation of stress and transformation beyond the plasticity limit is considered as linear and the elastic-plastic material with linear strengthening is considered.

The criterion which defines the conditions for the entity point to get from elastic into plastic state is called plasticity condition and is formulated on the basis of experimental measurements, [8].

The aim of strengthening the first field of laminated lattice girders with steel plate, 8 mm thick is to redistribute the stresses in individual unsatisfactory parts of bearing steel roof structure, figure 8.

The FEM 3D-model of laminated lattice-girder's first part with pin connection was created by means of software ANSYS:

1) without strengthening – 42156 nodes, 20514 elements, figure 9a),
2) with strengthening by steel plate, 8mm thick – 42692 nodes, 20577 elements, figure 9b).

During the modelling of above mentioned detail, the volume element SOLID 186 (3-D space, 20-Node Tetrahedral Structural Solid or Layered Solid) and SOLID 187 (3-D space, 10-Node Tetrahedral Structural Solid) was applied.

Contact of pin plates with pin are intended only in compression. The mortised plates are welded to bearing steel frame. The contact elements (type "Frictionless") are modelled in the place, where pin is connected with mortised plate. At plastic zone method was considered the elastic-plastic material with hardening in tension characterised by bilinear approximation of stress-strain diagram of steel S 235, i.e. E = 209821 MPa, H = 8944 MPa, figure 14, [9]. Model with physically non-linear material was considered in the software ANSYS. This model was defined as "rate independent – isotropic hardening plasticity – Mises plasticity – Bilinear diagram.

The calculation was done by Newton-Raphson incremental method.

The axial forces were taken from the static calculation [1], i.e. the force in the upper chord = 152.11 kN, the force in the lower chord = 41.67 kN and subsequently the forces were calculated for the circumferential area of the tube modified by corrosive decreases $A_r = 1041.9\ mm^2$ (acting load on the tube A_r area of the upper chord is $14.6x10^4$ kPa and of the lower chord is $4x10^4$ kPa).

The Von Mises stresses and plastic strains were watched, figure 10, 11, 12, 13, 14 and table 1.

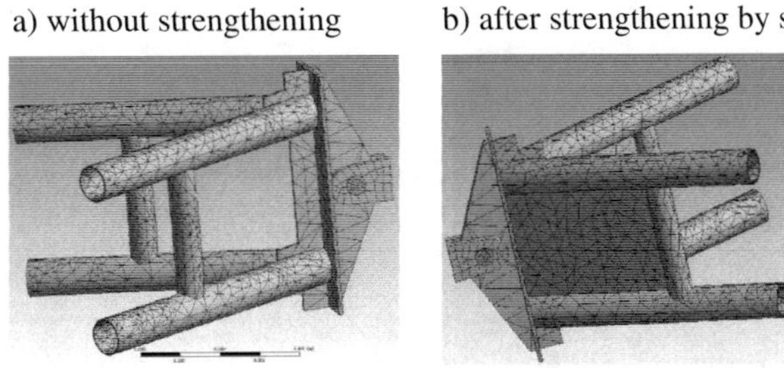

Figure 9. Calculation model of laminated lattice-girder's first field with hinged pin connection.

a) without strengthening b) after strengthening by steel plate, th. 8 mm

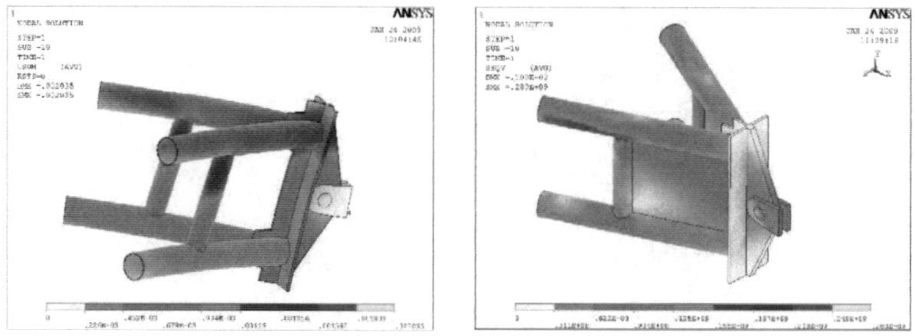

Figure 10. Von Mises stresses.

a) without strengthening b) after strengthening by steel plate, th. 8 mm

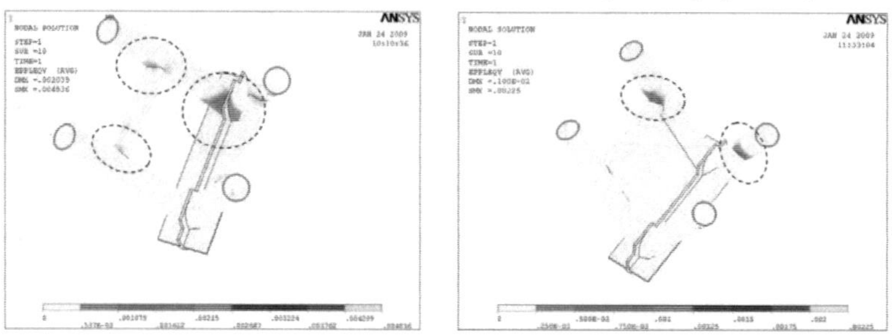

Figure 11. Plastic strains.

a) without strengthening b) after strengthening by steel plate, th. 8 mm

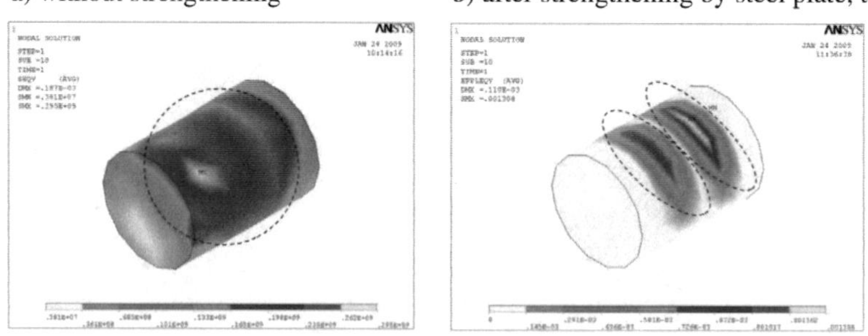

Figure 12. Plastic strains – pin.

a) without strengthening b) after strengthening by steel plate, th. 8 mm

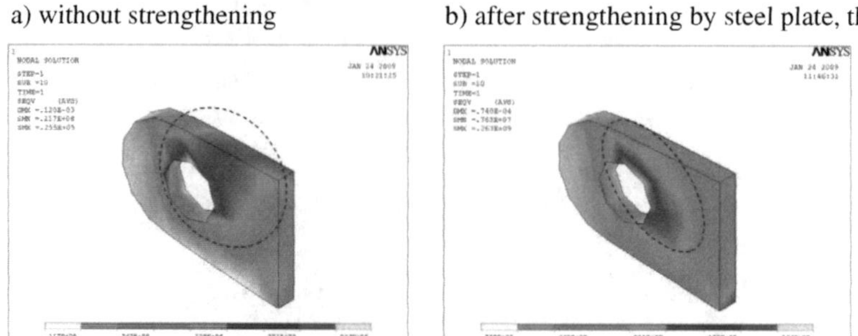

Figure 13. Von Mises stresses – pin plate.

Table 1. Values of maximum deformations, plastic strains and stresses.

Maximum values of:	without strengthening	with strengthening
- deformations detected for modelled detail	2.035mm	1.0mm
- stresses detected for modelled detail	318MPa	280MPa
- plastic strains detected for modelled detail	0.0048	0.0025
- stresses in the pin	295MPa	245MPa
- plastic strains in the pin	0.0048	0.0013
- stresses in pin plates	295MPa	262MPa
- plastic strains in the pin plate	0.00015	non-arise
- plastic strains in the upper chord area, at the contact of two tubes and the joining sheet	0.0026	0.0015

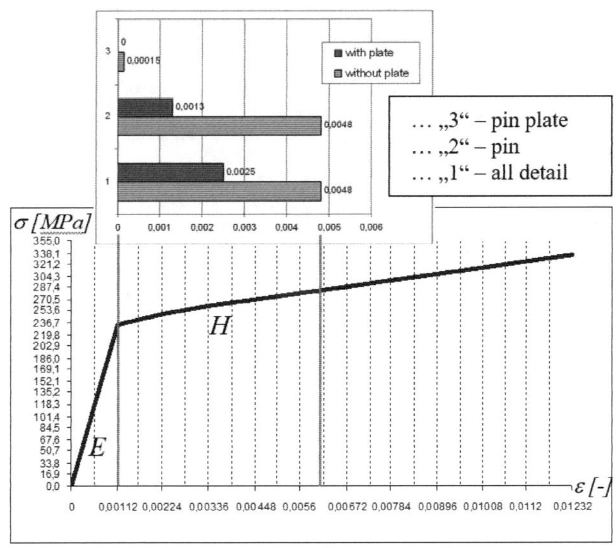

Figure 14. Stress-strain diagram of S 235 steel and plastic.

5. Conclusion

The load history was not considered. Maximum loading derating was taken into account, i.e. only dead load of bearing steel roof structure without snow and roof covering was considered in the calculation model.

In the case, when the snow fell during the 34 years of object's usage in the maximum load amount considered [6], then a dominant loading is the snow and it is possible to say that local plastic zones are created in the upper, respectively lower chord of the first field, in the pin plates and in the pin. This presumption can not exactly say, if the plastic zones appearance can only be supposed, then the static analysis results presented for the detail of strengthend laminated lattice-girder's first field by steel plate, 8 mm thick are affected by not considering the plastic zones created in the state before strengthening.

Then, the calculation presented for the first field of laminated lattice-girder strengthened by steel plate is only the information how the stress would be redistributed in case that steel plate is applied in the laminated lattice-girder:

- suggested structural modification „helped“ the laminated lattice-girder with its static impact within the structure, i.e. the tube created the upper chord was suitable strengthened as it does not comply with the standards without being strengthened by a steel plate;
- pin and pin plate were „helped“, as the forces and stresses were redistributed, in spite of existing permanent deformation.

In the case when snow fall would not be in the amount considered as maximum by the standard [9] during 34 years, there would be no plastic deformations and the presented calculation of the laminated lattice-girder strengthened by steel plate can be considered in term of the loading conditions as relevant.

It would be appropriate to recalculate the finite element mesh with denser division to verify the numerical solutions.

6. References

[1] Tomko M and Marko L and Brda J 2008 *Static calculation of steel structure roof ice hockey stadium in Poprad* (Kosice)
[2] Zeman J 1971 Design and production documentation - Part statics (incomplete)
[3] Marek P et al. 1985 *Steel structures of buildings*, SNTL/ALFA (Prague)
[4] Juhas P 2007 Quality and geometric tolerances rods and sections of steel structures, Faculty of Civil Engineering, University of Kosice (Kosice)
[5] Demjan I 2008 Test report no. 01/08/8, Laboratory of Civil Engineering, Department of Civil Engineering, Faculty of Civil Engineering, University of Kosice (Kosice)
[6] STN 73 0035 1988 *Load of building structures*
[7] STN 73 1401 1998 *Design of steel structures*
[8] Sejnoha J and Kufner V 1990 Elasticity, strength, plasticity III., CVUT (Prague)
[9] Kmet S, Tomko M and Brda J 1998. Probabilistic assessment of the reliability of systems based on of suspension bendig-rigid elements (Prague)

Acknowledgments
This work was supported by The Slovak Research and Development Agency under the contract No. APVV-15-0777 and also by The Slovak Scientific Grant Agency under the contract No. VEGA 1/0129/20.

IOP Conf. Series: Materials Science and Engineering 1252 (2022) 012022 doi:10.1088/1757-899X/1252/1/012022

Survey of consumer preferences in the field of wooden buildings

J Švajlenka[1], M Kozlovská[1]

[1] Department of Technology, Economy and Management of Construction, Faculty of Civil Engineering, Technical University of Košice, Slovakia.

E-mail: jozef.svajlenka@tuke.sk

Abstract. The future of future generations of humanity is conditioned by the decisions, actions and behavior of contemporary society towards the surrounding nature. The current critical state of the environment caused by the human factor forces one to think about a change in thinking in relation to ecology. Progress in this direction and the pursuit of a sustainable way of life is also felt in the construction sector. Modern wooden buildings are an environmental alternative to buildings made of energy-intensive materials. This type of construction is currently experiencing great success worldwide and is already part of the standard of living in many countries. Progress in this ecological way of construction has been recorded in recent years by the construction industry in Slovakia, but in its development it is progressing much more slowly than other countries of the European Union. The potential behind the development of wood-based structures is currently not fully exploited and it is therefore necessary to pay increased attention to research and innovation in the timber construction sector. The work deals with a survey of consumer preferences in the field of wooden buildings.

1. Introduction

The development of living standards in the past two centuries, in connection with the unprecedented development of technology, has been associated with a considerable use of non-renewable resources. But the current trend in the use of renewable resources also has its limits. Therefore, it is necessary to look for such ways of development in individual areas of life, including construction, which will optimize on the one hand the requirements of building users and on the other hand the requirements of sustainable development. According to Hájek [1], the sustainable development of construction will soon be affected by the following general requirements:

- reducing the overall complexity of buildings,
- reducing the material intensity of buildings,
- environmental Protection,
- recycling of building materials,
- a higher degree of construction industry.

2. Wooden buildings as modern methods of construction

In response to the above requirements as well as the overall trend of sustainable and efficient construction are modern wood-based construction methods. Own production and construction of wooden buildings in terms of materials used and production technology has a clearly lower impact on the environment compared to other traditionally preferred technologies in Slovakia (ceramic or silicate masonry, monolithic reinforced concrete ...). Modern Methods of Construction (MMC) according to Smith and Timberlak [2] are technologies that provide efficient procedures for the preparation and implementation of buildings, which result in a larger volume of products (production), higher quality and shorter time to obtain them. According to Burwood and Jess [3], their ambition is to increase the efficiency of construction by shortening the construction time, improving the quality, sustainability and impact of the construction and the construction itself on the environment.

Modern construction methods are usually technologies that use structures or their parts manufactured in factories. The preparation of more or less completed parts of building structures in production facilities has a high potential for increasing the efficiency of construction both in the phase of production of building elements and in the phase of their installation. In connection with the increasingly significant lack of construction capacity, but especially qualified human resources, there is already a real pressure in Slovakia to increase the efficiency of construction work. As a result, even in areas where conventional but not always effective practices have been preferred so far, modern, innovative and sustainable building solutions are being sought and used.

The advantages of MMC are shorter construction time, fewer implementation errors, reduced energy requirements, or reduced waste generation [4,5]. Their ambition is to increase the efficiency of construction, by improving quality, customer satisfaction, shortening construction time, sustainability and the impact of construction and the construction itself on the environment [6-8]. Within the use of modern construction methods in Slovakia, they appear to be the most widespread building systems for prefabricated wood-based buildings. In line with the trend of sustainability and implementation of wood-based constructions, the work deals with a survey of consumer preferences in the field of wooden buildings.

3. Materials and Methods

A total of 64 respondents took part in the survey, while a total of approximately 140 companies from Slovakia were contacted. The questionnaire was aimed not only at companies as a whole, but also at specific experts as individuals from the departments of design, budgeting, production, as well as suppliers of materials and users of structural systems of wooden buildings. Only 9.8% of the survey participants were female, despite the growing use of women in the construction sector. A representative sample consisted of different age categories, with the largest number of respondents aged 46-49 46.7%, the second relatively large age category was 25-34 years 36.9%, aged 50-64 12.0 %, at the age of 18-24 years 2.2% and over 65 years 2.2% of respondents also participated. In terms of belonging to the region, 33.7% were the most interviewed from central Slovakia, the professional public from western Slovakia had a similar representation 32.6%, significantly fewer were from eastern Slovakia 14.1% and the sample of respondents from the Czech Republic was 19.6%. %.

4. Results and Discussion

4.1. The importance of the parameters of wooden buildings from the perspective of respondents

An integral part of deciding what to focus on in the product or service innovation process is to know the importance of their individual criteria or parameters. In this question, based on a comparison of several publications, 8 basic properties characterizing wooden construction were selected and the task of respondents was to rank these parameters from the most important to the least important on a scale from 1 (least important) to 8 (most important).

Using the weighted average method, it was found from the obtained data that the most important parameter is considered to be the construction time, in second place, according to their importance, are thermal-technical properties and are followed by the price of wooden buildings. Equally important properties are the ecological footprint and energy balance during the use of buildings with a wood-based construction system, and behind them, in this ranking, there is a service life of a wooden building. According to the respondents, the acoustic properties are a slightly less important parameter and the fire resistance of wood-based buildings is considered to be the least important criterion (figure 1).

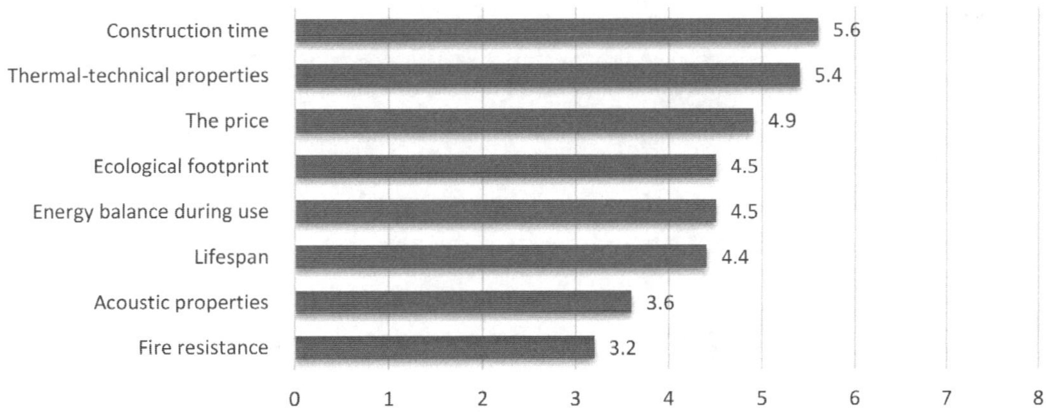

Figure 1. A survey of the importance of the basic parameters of wooden buildings from the perspective of respondents.

4.2. Survey of shortcomings of wooden buildings in need of improvement

This issue examined the shortcomings of the timber sector, which from the professional perspective requires more attention in relation to innovation and improvement. Respondents had a choice of several possible answers or could express their own views and experiences they encountered during their practice.

By summarizing the answers, it was found that the biggest shortcoming of a wooden building is considered fire resistance, with up to 48.9% of respondents thinking and acoustic properties with a similar percentage agreement in 47.8% of respondents. A relatively lower percentage of survey participants rate 15.2% as a weakness of wood-based structures and 7.6% as thermal-technical properties. Although these characteristics are generally considered to be a shortcoming in the timber construction sector, this is not entirely the case. Several experts took the view on this issue that the problem as such is not in the wooden construction system itself, but in unprofessional work or saving money at the expense of quality and price on the part of the contractor. For this reason, the insufficient professionalism of some companies in Slovakia creates a bad name for the entire wooden building sector. Up to 13.0% of respondents perceive unprofessionalism not only in construction, but also in design, and 3.3% of respondents consider the use of low-quality materials to be a shortcoming. In the case of professional and professional work in the implementation of wooden constructions and with the use of quality materials, all requirements are met in the same way as in masonry or steel constructions. Other shortcomings perceived by survey participants are the energy balance during use 4.3% and construction time 3.3%. 3.3% of respondents also agree that the basic shortcoming in wooden constructions is insufficient education among builders. 4.3% of the professional public do not see any shortcomings in the wood-based construction system (figure 2).

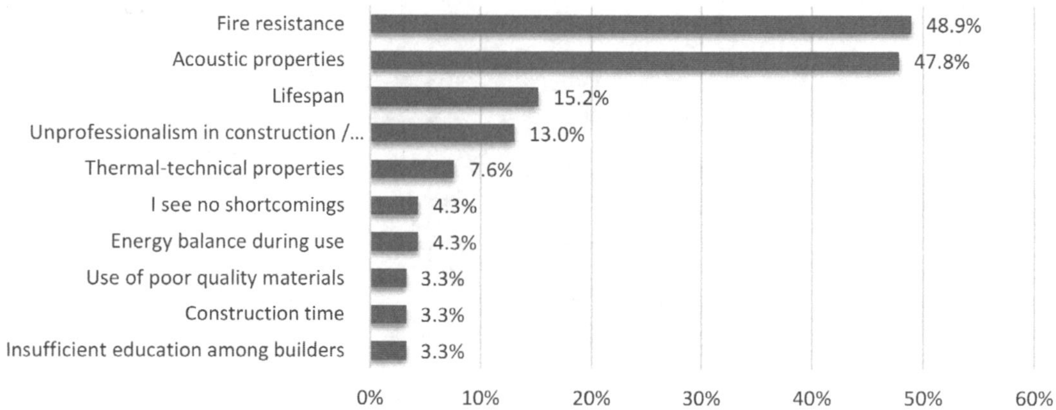

Figure 2. Survey of shortcomings of wooden buildings in need of improvement.

5. Conclusion

Wood as a renewable building material is currently experiencing great success and interest in wooden constructions around the world is growing significantly every year. This ecological and fast way of construction is starting to compete with constructions made of the most used materials such as steel and reinforced concrete. The work was carried out using the methodology of a questionnaire survey, through which the opinions of the professional public on the innovative potential in the field of wooden buildings in Slovakia were examined. The first part of the questionnaire was carried out in the form of closed questions aimed at surveying the application of wooden buildings and their innovative potential based on the experience of respondents from practice. The biggest shortcomings of wood-based buildings in need of improvement were identified as acoustic properties and fire resistance, the most common causes of which are poor design and use of poor quality materials.

6. References

[1] Hájek P a kol 2011 *Konstrukce pozemních staveb – komplexní přehled* Praha [online] [cit 2021-05-16] Available on: <http://www.ib.cvut.cz/1242KPKP>

[2] Smith R E ad Timberlake J 2011 *Prefab Architecture: A Guide to Modular Design and Construction*; John Wiley & Sons: Hoboken NJ USA ISBN 978-0-470-27561-0

[3] Burwood S and Jess P 2005 *Modern Methods of Construction Evolution or Revolution? A BURA Steering and Development Forum Report*; American Research Institute for Policy Development: New York NY USA Available online: http://www.buildicf.co.uk/pdfs (accessed on 1 October 2017)

[4] Kempton J 2009 *Modern methods of construction: Maintenance issues in the registered social landlord sector* [online] [cit. 2014-01-29] Available on: <http://www.arcom.ac.uk/-docs/proceedings/ar2009-0251-0260_Kempton.pdf>

[5] Modern methods of construction [online] [cit. 2020-01-28] Available on: <http://www.concretecentre.com/technical_information/building_solutions/modern_methods_of_construction.aspx>

[6] Haron N A, Rahman H A, Hanid M 2011 *A literature review of the advantages and barriers to the implementation of industrialised building system (IBS) in construction industry* Editorial Advisory Board 10

[7] Baghchesaraei A, Kaptan M V, Baghchesaraei O R 2015 *Using prefabrication systems in building construction* International Journal of Applied Engineering Research 10(24) 44258-44262

[8] Tam V W, Tam C M, Zeng S X, Ng W C 2007 *Towards adoption of prefabrication in construction* Building and environment 42(10) 3642-3654

Acknowledgments
VEGA project- 1/0336/22 Research on the effects of Lean Production/Lean Construction methods on increasing the efficiency of on-site and off-site construction technologies.

Quantification of acoustic risk from the tram ride in track curves

E Panulinova[1] and S Harabinova[2]

[1,2] Institute of Structural Engineering, Faculty of Civil Engineering, Technical University of Kosice, Vysokoskolska 4, 040 23 Kosice, Slovakia

E-mail: eva.panulinova@tuke.sk; slavka.harabinova@tuke.sk

Abstract. In urban agglomerations, even in our conditions, typical environmental risks are acoustic risks caused by traffic of all kinds. In the case of tram transport, a high-amplitude tonal noise was identified when passing a tram vehicle in a directional curve, which is perceived depending on the emerging frequency as a short-term sound, which emotionally manifests itself as an unpleasant squealing. As a result, the health of the population living in the vicinity of the tram lines is endangered. The described problem forced the authors to look for suitable methods for determining the acoustic risk. The "Risk Management" system was chosen as a suitable tool, the logical step of which is to quantify risk. The results used in the risk assessment process were based on specific measurements at selected locations around the tram line in the city of Kosice. The aim of the risk quantification process is to design the basis for selecting the optimal way to reduce / eliminate acoustic risk to a socially acceptable level.

1. Introduction

Noise is any unwanted sound that produces an unpleasant sensation or harmful effect. Any negative event in the environment acts as an environmental hazard to the surroundings. As a negative activity, environmental noise is classified as an acoustic risk. As a stressor, it alters the natural environment and, if it exceeds limit values, has a detrimental effect on the environment and the human population. The World Health Organization (WHO) has ranked noise as the second most serious cause of harm to public health, the first being air pollution [1].

Residents in urban areas are mainly bothered by noise from car traffic and in some places also from tram traffic [2]. Tramcars are particularly disruptive when riding in a track curve with small radii [3].

The issue of environmental noise is becoming more and more topical and requires targeted measures to be taken, especially in larger urban settlements, to achieve or maintain a suitable living environment for the inhabitants. The described problem forces experts to look for suitable methods to determine the risk of excessive noise also from tram traffic.

2. Risk management

The theory and practice of risk management opens up a new angle of view to achieve the reliability of buildings and the protection of human health. Risk management methods have established procedures that, on the one hand, provide a comprehensive and systematic assessment of what can harm a person or a building and, on the other hand, reveal the most serious deficiencies that pose the greatest risks. The methodology is versatile and applicable in a variety of fields. It consists of steps that help to identify existing but also potential stressors, reveal their possible occurrence, assess and quantify their undesirable consequences, provide the basis for the design of measures oriented to reduce the negative impact of stressors [4].

When any risk is identified, we look for answers to the questions:

"Why and when does a negative phenomenon occur?"

"What can happen as a result?"

The steps of the acoustic risk (AR) management process are shown in Figure 1.

I. Identifying AR:
- Analysing available information on potentially threatened values, stressors or environmental consequences,
- Problem formulation.

II. Analysing AR:
- Identification of hazards to humans and the environment,
- Methodical evaluation of measurements through different criteria and indicators,
- Analysis of the action-effect relationship,
- Establishing the basis for quantifying/characterising the risk.

III. Evaluation of AR:
- Determination of the probability of the occurrence of a negative phenomenon,
- Determining the boundaries - defining the acceptability of the risk,
- Characterisation of the risk - determination of the consequences and their severity.
- Decision making process - acceptance/non-acceptance of risk,

IV. Management/control of AR
- Selection of measures,
- Design of the optimal way to reduce risk,
- Monitoring and control of the implementation of measures.

Figure 1. The acoustic risk management process.

Risk quantification is the risk assessment phase. This phase includes an assessment of the likelihood, consequences and significance of negative effects caused by exposure to the stressor [5].

3. Identifying and analysing the negative phenomenon

The dominant source of noise at speeds around 40 km/h [6], is the wheel-rail contact. When a tram travels through a track curve with a radius of less than 50 m, a high amplitude tonal noise is generated at this point. It is induced by exceeding the voltage by a certain value, which is caused by a sharp jerk and impact release. This phenomenon repeats periodically and leads to oscillation of the whole system [7], [8]. The noise emitted in this way is perceived as a short-term sound, which is felt as an unpleasant squealing, whining sound - the Q-sound. A telling criterion for characterizing Q-sound published in [9] is the difference ΔL between the maximum (L_{AMax}) and the equivalent (L_{Aeq}) sound pressure level A.

IOP Conf. Series: Materials Science and Engineering 1252 (2022) 012023 doi:10.1088/1757-899X/1252/1/012023

To investigate the influence of the directional arc on the noise emission, measurement sites were selected at which the noise emission syndrome was measured in the straight and in the track curves. A total of 282 measurements were recorded. The entire set was subjected to analysis. One of the results of the analysis is the percentage of A sound pressure levels exceeding the limit values [10] at a distance of 2 m from the façade of a residential building. At sites with directional lines in a track curve with r up to 50 m, the presence of Q-sound was detected in 78 % of the total number of measurements, at sites with radius r above 50 m in 17 %, and at straight sections Q-sound occurred only exceptionally (3 %) [11].

4. Quantification of acoustic risk - risk of Q-sound

The quantification of AR is based on the generally accepted definition of risk, where "risk" is characterised by "the combination of the probability of an undesirable Q-sound occurrence and the consequences causing health problems in humans". In order to quantify the risk, it is necessary to estimate and define two supporting parameters, the probability/frequency of the occurrence of the risk event (Q-sound occurrence) and the severity/consequence of the undesirable impact on human health and well-being.

To classify the probability of Q-sound occurrence, a scoring method with 3 probability levels was chosen (Table 1) based on defined intervals based on the numerical expression of the difference $L = L_{AMax} - L_{Aeq}$: Probability of occurrence of the risk event - Q-sound occurrence.

Table 1. Probability of occurrence of a risk event - the occurrence of A-sound.

Degree	Probability	Characteristics	$\Delta L = L_{AMax} - L_{Aeq}$
I. minimal	P1 = 1	there is a Q-sound	ΔL > 8 dB
II. mild	P2 = 2	there is a light Q-sound	ΔL = 5 - 8 dB
III. significant	P3 = 3	no Q-sound	ΔL < 5 dB

A sign that an annoying wailing sound and a form of squealing has been identified in the measurement is the occurrence of a tonal sound that is observable on the spectrogram of the measurements. In the present analysis, the resulting impact of the undesired phenomenon is expressed from the third-octave spectrogram of the individual measurements as an exceedance of L_{Aeq} at a certain frequency by more than 5 dB. The unexpected tonal sound identified in this way is particularly disturbing during sleep and induces hypertension [12].

Table 2 expresses the consequence of the effect of the negative phenomenon as the number of identified distinct peaks on the spectrogram in relation to neighbouring values with a difference of more than 5 dB(A).

It is defined separately for each level of occurrence of the risk event as:

- D1 - insignificant - the number of L_{Aeq} spikes in stage 1 of the event,
- D2 - health-threatening - number of L_{Aeq} surges in the 2nd occurrence level,
- D3 - harmful to health - number of L_{Aeq} spikes in level 3 occurrence.

Table 2. Relative frequency of expression of consequence [%].

Track section	D1	D2	D3
straight	0	0	0
radius up to 70 meters	0	24	36
radius up to 50 meters	0	42	39

The risk that the noise level at a particular frequency will rise above other levels by more than 5 dB and cause an unpleasant disturbance with consequences is shown in Table 3. According to [13], the acoustic risk value is an expression of the functional dependence $R_i = f(P_i \, x \, D_i)$.

Table 3. Acoustic risk quantification.

Route guidance	Degree of occurrence 1	Degree of occurrence 2	Degree of occurrence 3
straight	0	0	0
radius up to 70 meters	0	48	**108**
radius up to 50 meters	0	**84**	**117**

In the next step, the risk is assessed and a risk scale score is determined, which also provides an indication of the urgency to take action to reduce/eliminate the risk. The risk scale and its significance (Table 4) was constructed based on the current state of knowledge in risk science and available information on constructing a risk matrix as suggested by Klaus Winterling [14].

Table 4. Scale and significance of acoustic risk.

Risk level	Risk acceptability	Limits	Significance of acoustic risk
1	acceptable	1 - 20	tonal sounds present in low numbers, slight impact on health, no action, need to monitor the situation
2	undesirable	21 - 50	*light Q-sound present, some measures need to be taken to reduce the undesirable phenomenon, risk of worsening of the condition, need to monitor and control the situation*
3	unacceptable	**more than 50**	***there is a strong Q-sound, widespread, severe impact, necessary measures must be taken to eliminate the risk***

The assessment of the acoustic situation for the different measurement locations showed that the highest resulting risk of Q-sound was recorded for a tram vehicle riding along a route with a radius of up to 50 m. A high degree of risk is also documented in Table 3. for track curve radii up to 70 m, especially when strong Q-sound is present.

5. Conclusion

In conclusion, the measurements demonstrated the presence of a louder and, based on tonal diversity, annoying sub-noise designated as Q-noise and this was classified as an environmental risk. Its assessment is necessary because of the potential health hazard to people moving in the vicinity of the source of the negative phenomenon. The aim of quantifying the risk is to highlight the possible locations of occurrence and the risk of harm to human health. Addressing excessive noise from tram traffic is a solution to a seemingly contradictory problem. On the one hand, the aim is to bring the line and the tram network as close as possible to the area to be served; on the other hand, it is essential to reduce the noise nuisance to the lowest possible level. The information summarised in the study can help planners and managers to manage tram transport in an environmentally friendly way that promotes its sustainability.

6. References

[1] WHO 2018 *Environmental Noise Guidelines for the European Region* [online] [cit. 2021-12-09] Available: http://www.euro.who.int/__data/assets/pdf_file/0008/383921/noise-guidelines-eng.pdf?ua=1

[2] Salomons EM and Janssen SA 2011 Practical ranges of loudness levels of various types of environmental noise, including traffic noise, aircraft noise, and industrial noise *Intentional Journal of Environmental Research and Public Health* **8(6)** 1847-64

[3] Panulinová E, Harabinová S and Argalášová Ľ 2016 Tram squealing noise and its impact on human health *Noise and Health* **18(85)** 329-337

[4] Procházková D 2011 *Risk analysis and management* (Praha: ČVUT)
[5] Zeleňáková M and Zvijáková L 2017 *Using Risk Analysis for Flood Protection Assessment* (Berlin: Springer)
[6] Panulinová E 2007 Curve squealing of tramway wheels *Scientific conference with international learning* (Sofia: VSU' 200 Tom 1)
[7] Thompson DJ, Monk-Steel AD and Jones CJC 2003 *Railway noise Curve squeal, roughness growth, friction and wear Prepared for The railway safety and standards board* (The university of Southampton England)
[8] Heckl MA and Abrahams I. 2000 Curve squeal of train wheels, part 1 Mathematical model for its generation *Journal of Sound and Vibration England 2000* **229(3)** 669-693
[9] Mőhler U et al 1998 Noise emissions from commuter railroads *Journal for Noise Abatement* **6/1998** 209-215
[10] Government Regulation No. 43/2005 Coll., laying down details of strategic noise maps and noise protection action plans
[11] Panulinová E 2010 Analysis of input parameters for the calculation of levels of tramway noise *Selected Scientific Papers Journal of Civil Engineering* **5(3)** 75-82
[12] Sobotová Ľ et al 2010 Community response to environmental noise and the impact on cardiovascular risk score *Science of the Total Environment* **408(6)** 1264-1270
[13] Šimák L 2006 *Risk Management* (Žilina: FŠI ŽU) [online] [cit. 2021-06-06] Available: http://fsi.uniza.sk/kkm/files/publikacie/mn_rizik.pdf
[14] Winterling K 1992 *How crisis management is carried out* (Praha: Babtext) p 38

Acknowledgment
This research was supported by the Scientific Grant Agency of the Ministry of Education of the Slovak Republic under Projects VEGA 1/0374/19.

IOP Conf. Series: Materials Science and Engineering 1252 (2022) 012024 doi:10.1088/1757-899X/1252/1/012024

Applications of a phase-field fracture model to materials with inclusions

R Vodička[1]

[1] Institute of Technology, Economics and Management in Construction, Faculty of Civil Engineering, Technical university of Košice, Vysokoškolská 4, 042 00 Košice, Slovakia

E-mail: roman.vodicka@tuke.sk

Abstract. Fracture in materials with inclusions is considered in the computational model of this contribution which allows prediction of crack propagation. The inclusions cause that cracks may appear inside the materials or along matrix-inclusion interfaces. The presented model can treat them both using two internal variables in a fashion of damage mechanics so that fracture is a consequence of the damage. The first variable is defined at the interface represented by a thin degradable adhesive layer so that an appropriate stress-strain relationship can be obtained as in cohesive zone models. The second variable is defined in the structural domains as a phase-field fracture variable which causes elastic properties degradation in a narrow material strip that forms a smeared crack. Both these damaging schemes are expressed in a unique quasi-static energy evolution process. The numerical solution approach is thus rendered from a variational form with a staggered time stepping procedure related to a separation of deformation variables from the damage ones and using non-linear programming algorithms implemented together within an own MATLAB finite element code. The numerical simulations with the model include simplified structural and material elements.

1. Introduction

Many contemporary structural materials are composite, consisting of fibres or other inclusions more or less randomly distributed in a matrix material. The abundance of loading may cause crucial changes in material structure ensued by a loss of intended functionality of pertinent structural element. Such material deterioration is then macroscopically evident by appearing of cracks in material or between material components, of debonding of structural components etc. One of the mechanical models which describes such issues introduces damage of materials leading subsequently to fracture, see [1], [2]. Successful computational approaches which are capable of described material degradation analysis are of high importance. The present one can naturally capture both in-material and inter-material cracks.

The computational model of material damage based on cumulation of microscopic faults is described by internal variables as introduced in [3]. These variables express the current state of material or of an adhesive along materials interfaces. A limit value of such a variable reflects total degradation of the material or of the joint. This is interpreted as a crack.

Damage and fracture in materials can be formulated using energies, allowing for variational solution techniques. The interface issues can be seen in theory of interface damage or adhesive contact as developed in [4],[5] and followed also by [6] to account also for the cracks in a manner of cohesive zone models (CZM) whose classical interpretations can be found e.g. in [7], [8]. These formulations introduce an internal parameter of damage along the interface, advantageously it then leads to a crack with prescribed path.

The same variational philosophy can be maintained for a material crack, whose path need not be known, as it was studied in [9] trying to extend the classical energy-based Griffith criterion [10]. The relation to damage-based interface models was obtained by introducing another internal variable in fashion of phase-field models (PFM), see e.g., [11], [12], with a non-local character. This model incorporates a predefined characteristic crack width parameter, which may be used to identify crack formation process as presented in [13], [14] and coupled with interface issues as in [15]. Such cracks are also called smeared as the theory relates continuous distribution of the internal variable to presence of a crack in the material. The smeared crack, however, can identify direction in which the crack propagates.

In the following sections the model and some calculations are presented. First in Section 2, the computational model in terms of mechanical energies is described and governing relations for its energy state evolution are formulated. The present model is based on previous works of the research group which were initiated in a brittle interface debonding model of [16], subsequently updated to the models with interface cohesive zones in [17], [18], [19], and newly extended for phase-field modelling as described in [20]. Then in Section 3, the behaviour of the presented model is documented in a couple of numerical tests which contain analysed domain with an inclusion. Though the solved problem involves only an academic simulation, it documents how an inclusion and geometric stress concentrator affect each other and provide various scenarios of crack propagation.

2. Computational model

Let us consider a bounded domain Ω which contains at least one inclusion. For one such inclusion Ω^A and a matrix domain Ω^B, it is shown in Figure 1. The respective boundaries are denoted Γ^A and Γ^B, the common part of the boundaries is denoted Γ_{C}. Each of the boundaries (except Γ_{C}) is disjointly split according to boundary conditions: Γ_{D} denotes the part with prescribed displacements u by the relation $u=g(t)$, the time variable is denoted t, Γ_{N} denotes the part with prescribed tractions p. Here, only displacement control of loading is considered, which means that only free parts of boundary lie in Γ_{N}.

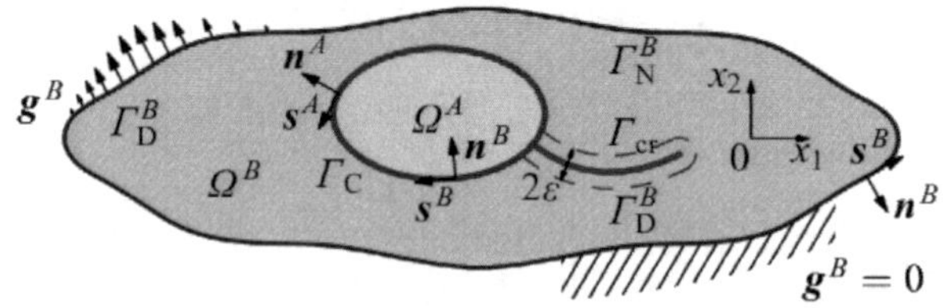

Figure 1. Description of a general domain.

As the boundary conditions change in time, the stress-strain and damage state of the structure evolve. Each state is described by a displacement field u in the interior of the domains (inclusions and matrix), a gap of displacements $[\![u]\!]$ at interfaces, and two internal parameters α and ζ of damage having values between 0 and 1: the former parameter pertains to the damage state derived from the damage PFM in the interior of the domains so that the initial minimal value α=0 means an undamaged state and the maximal value α=1 means total damage and a crack formation. Similarly, the latter damage parameter characterises the damage at the interface in the same manner.

The energy state consists of the free energy represented here by the elastic stored energy E of the domain (in 2D calculations the plane strain conditions are supposed), by the elastic stored energy of the interface Γ_{C}, and energy stored in formed cracks. This free energy may be expressed according to [21] by the following formula:

$$E(t;u,\alpha,\zeta)=\int_{\Omega}\frac{1}{2}\Phi(\alpha)Ce(u){:}e(u)+\frac{3}{8}G_{\mathrm{c}}(e(u))\left(\frac{1}{\varepsilon}\alpha+\varepsilon|\nabla\alpha|^2\right)\mathrm{d}x \qquad (1)$$

$$+\int_{\Gamma_C} \frac{1}{2}\varphi(\zeta)(\kappa[\![u]\!])\cdot[\![u]\!]+G_c^i([\![u]\!])\zeta+\varepsilon_c|\nabla_s\zeta|^2+\frac{1}{2}\kappa_g([\![u]\!]_n^-)^2\mathrm{d}s$$

for an admissible displacement field u, in the sense that it satisfies the displacement boundary conditions on Γ_D, and admissible damage parameters α and ζ, i.e., their values lying in the interval [0,1]. For non-admissible states, the value of E is infinite. The parameters introduced in (1) include initial stiffness matrix C of the materials, initial stiffness κ of the undamaged adhesive layer and the compressive stiffness κ_g for handling possible contact problem by a penalisation energy term, the superscript '-' means that the value $[\![u]\!]_n$ (normal component of the displacement gap) is active only if it is negative. Further, the energy accumulated due to a new crack requires G_c as fracture energy for domain cracks, and G_c^i as fracture energy of the interface cracks. Dependence on the deformation state signalises that their values may be crack mode dependent (different for opening and shearing cracks). Standardly, the term $\int_{\Gamma_{cr}} G_c \mathrm{d}s$ (Γ_{cr} denotes a crack as introduced in Figure 1) is used for accumulating energy of the crack in Griffith-like models. For the domain crack, it is replaced by a regularisation functional (the term containing G_c in equation (1)) controlling the evolution of the damage PFM diffuse crack. Parameter ε is a characteristic length parameter for the PFM (determining the width of the diffuse crack, c.f. Figure 1). Similar regularisation is not needed for the interface crack, nevertheless, to include a non-local character of the interface damage a gradient term with an additional parameter ε_c can be considered (the subscript s in equation (1) stresses that the gradient is in the tangential space of the interface).

The functions $\mathcal{A}(\alpha)$, $\varphi(\zeta)$ are degradation functions of the domain (related to a PFM) and of the interface (related to a CZM), respectively. They determine how the stiffness of the material decreases during the damage process. There are various representations of these functions in literature which are restricted here to the simplest choices: $\mathcal{A}(\alpha)=(\varepsilon_0^2+\alpha^2)$, where the parameter ε_0 adjusts residual material stiffness after total damage to avoid degeneration of the totally damaged material in a numerical solution. The interface degradation function is set to an exponential function $\varphi(\zeta)=\exp(-\gamma^{-1}(\zeta,3))$, with $\gamma(\zeta,3)=\exp(-\zeta)\left(1+\zeta+\frac{1}{2}\zeta^{-2}\right)$ and it modifies the elastic energy of the damageable interface as in the approaches from [17].

Besides, a fraction of energy of the crack formation process is usually dissipated. While in the present model we added this energy to the stored energy, it is still necessary to guarantee a unidirectional character of the damage process at least by the assumption that there is no additional dissipated energy if damage increases. This can be expressed by a trivial dissipation potential $R(\dot{\alpha},\dot{\zeta})=0$, constrained by $\dot{\alpha}\geq 0$, $\dot{\zeta}\geq 0$, otherwise R is infinite to guarantee impossibility of such a state.

The relations which govern the state evolution can be written in a form of nonlinear variational inclusions with initial conditions (corresponding to an undamaged state)

$$\partial_u E(t;u,\alpha,\zeta)\ni 0,\quad \partial_\alpha \mathrm{E}(t;u,\alpha,\zeta)+\partial_{\dot{\alpha}}R(\dot{\alpha},\dot{\zeta})\ni 0,\quad \partial_\zeta E(t;u,\alpha,\zeta)+\partial_{\dot{\zeta}}R(\dot{\alpha},\dot{\zeta})\ni 0,$$
$$u|_{t=0}=u_0,\quad \alpha|_{t=0}=0,\quad \zeta|_{t=0}=0. \tag{2}$$

where ∂ generally denotes a partial subdifferential as the functionals does not have to be smooth, e.g., R jumps from zero to infinity. For smooth functionals, subdifferentials can be replaced by Gateaux differentials and the inclusions by equations.

The numerical solution of equation (2) is approximated by a staggered time stepping procedure. This means that the deformation variables and damage variables are separated and solved one after the other, solution of the evolution is calculated at selected time instants defined by a constant time step. Both the deformation solution and damage solution at each instant are obtained by non-linear programming algorithms in the same way as in [16], implemented together within a MATLAB finite element code.

Numerical simulations with similar models were presented in many works, e.g., [13],[22],[23]. In practical computation there may have some limitations, say, because the crack is modelled as a material with artificially decreased stiffness within the band determined by the length scale parameters ε or ε_c, which subsequently may require very fine meshes to see solution convergence. Nevertheless, such models are usually easy for implementations in modern computational tools.

3. Calculations

Three examples containing at most one inclusion in a square domain, alternatively with grooves, are considered for a numerical solution of equation (2). Their graphical schemes are shown in Figure 2. In fact, they introduce a purely academic structural elements which however may demonstrate various scenarios of crack formation and propagation processes provided by the discussed model.

The parameters of the computational model are the same in all cases: the stiffness matrix C is calculated from the bulk modulus K=52 GPa, the shear modulus μ=29 GPa for the inclusions, and K=3 GPa, μ=1 GPa for the matrix, for the interface $\kappa = \begin{pmatrix} 1 & 0 \\ 0 & 0.5 \end{pmatrix}$ PPam^{-1}, κ_g=100 PPam^{-1}. The minimal element size of the FEM mesh is 0.5 mm. The fracture energy in all domains is G_c=1 kJm^{-2}, and at all interfaces it is G_c^i=1 Jm^{-2}.

The boundary conditions in the solved examples including loading and constraints for all structural elements are also shown in Figure 2.They are loaded by displacement loading so that $g_2(t)=v_0t$, with the velocity v_0=1 mms^{-1}. This load is applied incrementally by the time steps of 20 ms. The constraints include fixed bottom face of the domains.

The behaviour of PFM even in relation to the damage along the interface is documented in a couple of plots. They show either distribution of stress just before a crack appeared, or values of PFM damage parameter. The former, together with (magnified) displacements, reveals actual cracks within the tested structural elements.

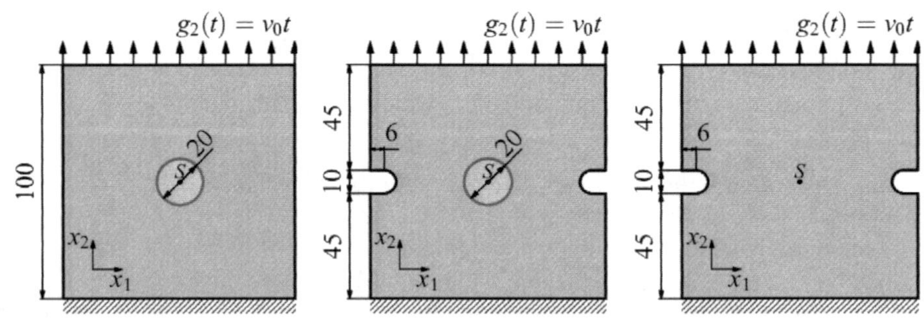

Figure 2. Schemes for calculations, constraints, and loading.

First, the case without a groove is shown. As the crack is more or less in the opening mode, the stress traces σ_{tr} are used for demonstrating the stress state. They are shown in Figure 3. The first instant reflects the situation before an interface crack appeared. The second one documents the stress distribution after the interface crack appeared. The stress concentrations formed near the arisen crack tips. Subsequently, the damage continues in the matrix material which results in a crack. The last plot shows the situation of the separation of top and bottom parts of the corrupted element so that the stresses in the structure vanish.

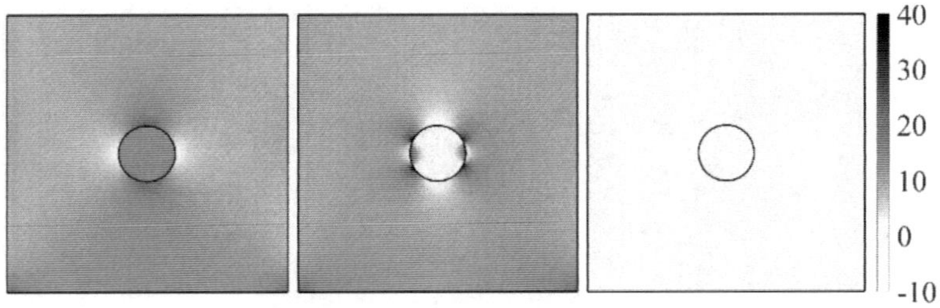

Figure 3. Stress trace σ_{tr} [MPa] distribution for the case without a groove. The time instants used are (left to right): 0.28 s, 0.30 s, and 0.34 s.

These observations can also be seen in the graphs of Figure 4. The PFM damage parameter presents a crack which spreads across the whole element, and it is accompanied by an interface crack, whose position is clear from the magnified displacements.

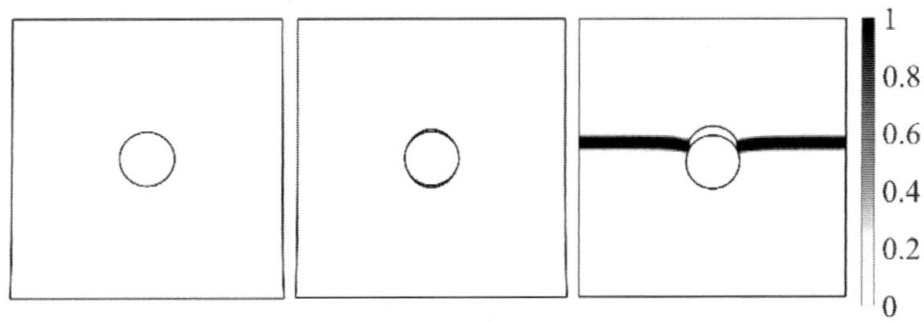

Figure 4. Crack formation process (in terms of PFM parameter α). The same instants as in Figure 3. The deformations are magnified 10 times.

For a better comparison, the instants are selected similarly for the other schemes in Figure 2. The stresses for the second scheme are shown in Figure 5. Comparing them to Figure 3, it leads to conclusion that the interface crack appears first, and slightly different distribution of stress caused by the grooves slightly changes appearing of the domain crack. This crack has two nuclei caused by the grooves. Its final shape is displayed in Figure 6. The cracks which propagate from the inclusion meet those that are generated by the stress concentration at the grooves.

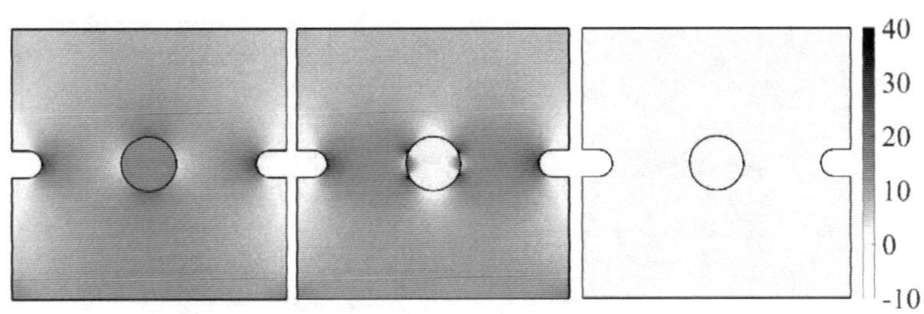

Figure 5. Stress trace σ_{tr} [MPa] distribution for the case with grooves. Time instants used are (left to right): 0.28 s, 0.30 s, and 0.38 s.

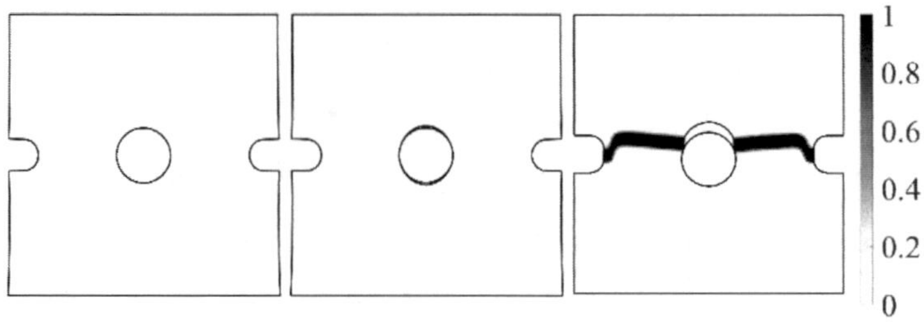

Figure 6. Crack formation process (PFM parameter α) in the grooved domain with an inclusion. The same instants as in Figure 5. The deformations are magnified 10 times.

Nevertheless, adjusting of parameters may lead to a different scenario for the crack propagation. If the parameter G_c^i is changed so that the interface crack formation is postponed, the crack can be purely initiated at the grooves. Such a situation is shown in Figure 7, where the parameter value is changed to 4 Jm^{-2}. No crack is observed at the interface prior to the whole crack formation.

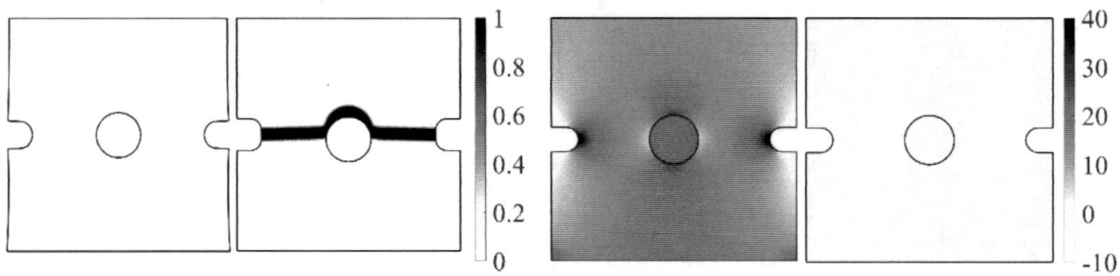

Figure 7. Cracks (PFM parameter α) (left) and stresses σ_{tr} [MPa] (right) for the grooved domain with an inclusion, where G_c^i is modified. The used instants are: 0.4 s, and 0.42 s. The deformations at the plot of FDM damage parameter are magnified 10 times.

Finally, the case without an inclusion is presented. As no interface is here, only two instants were selected: just before the crack formation and right after it. The results are shown in Figure 8. The moment when the crack appears is almost the same as before due to very similar stress distribution near the grooves. Anyhow, parameters of the model can cause substantial changes of the crack formation process.

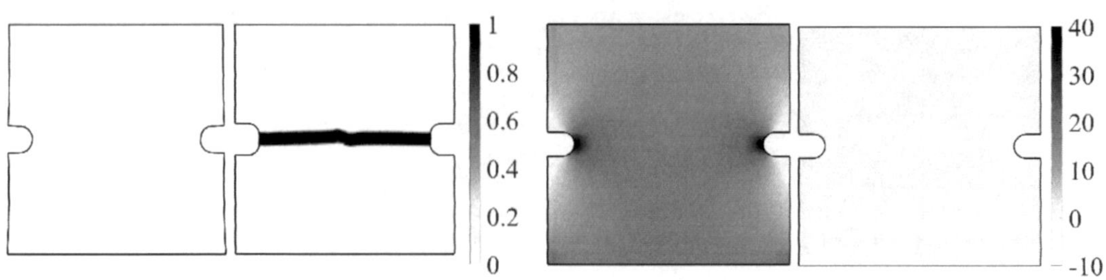

Figure 8. Cracks (PFM parameter α) (left) and stresses σ_{tr} [MPa] (right) for the grooved domain without an inclusion. The used instants are: 0.42 s, and 0.44 s. The deformations at the plot of FDM damage parameters are magnified 10 times.

4. Conclusion

A computational model for studying cracks along interfaces and in bulks has been introduced and examined. The presented academic calculations showed how stress concentrations provoke formation of defects and cracks in materials and along an inclusion interface. They also document how changing of a model parameter may affect what appears first: whether it is an interface crack, or it is a groove-initiated crack. Naturally, the model includes parameters related to a crack formation process which should be appropriately adjusted in order to obtain results comparable to experimental measurements. Such a parametric study has already been planned. Anyhow, the result of the current study can be considered rather promising, and it is supposed that the proposed computational approach may be successfully implemented in more complex engineering calculations, too.

5. References

[1] Besson J et al. 2010 Non-Linear Mechanics of Materials (Dordrecht:Springer).
[2] Lemaitre J and Desmorat, R 2005 Engineering Damage Mechanics (Berlin: Springer).
[3] Frémond M 1985 Dissipation dans l'adherence des solides *Comptes Rendus de l'Académie des Sciences – Series II* **300** 709–714.
[4] Raous M and Cangemi, L and Cocu M 1999 A consistent model coupling adhesion, friction and unilateral contact *Comput. Meth. Appl. Mech. Eng.* **177** 383–399.
[5] Del Piero G 2014 A Variational Approach to Fracture and Other Inelastic Phenomena (Dordrecht: Springer).
[6] Roubíček T and Kružík M and Zeman J 2013 Delamination and adhesive contact models and their mathematical analysis and numerical treatment 2013 .In: Mantič V, ed. *Mathematical methods and models in composites* (London: Imperial College Press) pp 349–400.
[7] Ortiz M and Pandolfi A 1999 Finite-deformation irreversible cohesive elements for three-dimensional crack propagationan alysis *Int. J. Numer. Methods.Eng.* **44** 1267–83.
[8] Park K and Paulino GH 2011 Cohesive zone models: a critical review of traction-separation relationships across fracture surfaces *Appl Mech Rev* **64** 6 061002.
[9] Francfort GA and Marigo JJ 1998 Revisiting brittle fracture as an energy minimization problem *J. Mech. Phys. Solids* **46** 1319–1342.
[10] Griffith AA 1921 The phenomena of rupture and flow in solids *Philos. Trans. R. Soc. Lond. Ser. A* **221** 163–198.
[11] Francfort GA and Bourdin B and Marigo JJ 2008 The variational approach to fracture *J Elasticity* **91** 5–148.
[12] Miehe C and Hofacker M and Welschinger F 2010 A phase field model for rate independent crack propagation: robust algorithmic implementation based on operator splits *Comput. Methods Appl. Mech. Engrg.* **199** 2765–2778.
[13] Tanné E and Li T and Bourdin B and Marigo JJ and Maurini C 2018 Crack nucleation in variational phase-field models of brittle fracture *J Mech Phys Solids* **110** 80–99.
[14] Sargado JM et al.2018 High-accuracy phase-field models for brittle fracture based on a new family of degradation functions J. Mech. Phys. Solids **111** 458–489.
[15] Paggi M and Reinoso J 2017 Revisiting the problem of a crack impinging on an interface: A modeling framework for the interaction between the phase field approach for brittle fracture and the interface cohesive zone model *Comput. Method. Appl. M.* **321** 145–172.
[16] Vodička R and Mantič V and Roubíček T 2014 Energetic versus maximally-dissipative local solutions of a quasistatic rate-independent mixed-mode delamination model *Meccanica* **49** 2933–2963.
[17] Vodička R 2016 A quasi-static interface damage model with cohesive cracks: SQP–SGBEM implementation *Eng Anal Bound Elem* **62** 123–140.
[18] Vodička R and Mantič V 2017 An energy based formulation of a quasi-static interface damage model with a multilinear cohesive law *Discr. Cont. Dyn. Syst. –S* **10** 1539–1561.
[19] Vodička R 2021 A new quasi-static delamination model with rate-dependence in interface damage and its operation under cyclic loading *Int J Solids Struct* **224** 111035.
[20] Vodička R 2019 On coupling of interface and phase-field damage models for quasi-brittle fracture *Acta Mechanica Slovaca* **23** (3) 42–48.
[21] Roubíček T 2019 Models of dynamic damage and phase-field fracture, and their various time discretisations *Topics in Applied Analysis and Optimisation (CIM Series in Math. Sci.)* ed M Hinttermüller and JF Rodrigues (Berlin: Springer) pp 363–396.

[22] Molnár G and Gravouil A 2017 2D and 3D Abaqus implementation of a robust staggered phase-field solution for modeling brittle fracture *Finite Elements in Analysis and Design* **130** 27–38.
[23] Zhou SW and Zhuang XY and Rabczuk T 2019 Phase field modelling of brittle compressive-shear fractures in rock-like materials: a new driving force and a hybrid formulation. *Comput Method Appl Mech Eng* **355** 729–752.

Acknowledgement
The author acknowledges support from The Ministry of Education, Science, Research and Sport of the Slovak Republic by the grants VEGA 1/0374/19 and VEGA 1/0363/21.

IOP Conf. Series: Materials Science and Engineering 1252 (2022) 012025 doi:10.1088/1757-899X/1252/1/012025

The potential of lightweight panel systems in building industry

M Kozlovská[1], D Mokrenko[1], Z Struková[1] and M Špak[1]

[1] Institute of Construction Technology, Economic and Management, Faculty of Civil Engineering, Technical University of Kosice, Vysokoskolska 4, 042 00 Kosice, Slovakia

E-mail: maria.kozlovska@tuke.sk; daria.mokrenko@tuke.sk; zuzana.strukova@tuke.sk; matej.spak@tuke.sk

Abstract. The current construction technologies are often characterized as ineffective, with long construction times, heavy structural weight, increased financial and labour cost, and environmentally unsustainable. Innovative construction methods, including modern lightweight panel systems, are characterized by low weight (this affects not only transport characteristics but also the production, such as easier handling of the panels), faster construction time, the guaranteed quality due to industrial production in a factory, economic efficiency and many other benefits. The presented study examined the potential of efficiency and performance of four variants of lightweight construction panels in the context of the required essential properties of prefabricated construction elements, as well as in comparison with the variant of "heavy" prefabricated concrete panel system.

1. Introduction

Low construction-labour productivity has been a problem in the construction industry for many years [1] what results in the lowest productivity of the sector across all industries. Implementation of new and innovated building structures and technologies, in professional practice referred to as Modern Methods of Construction (MMC), is the way to increase productivity and efficiency of the complex construction process. There is a significant relationship between modern methods of construction and productivity of construction processes. Modern methods of construction are characterized by innovative construction methods, most of which represent off-site construction which moves part of production processes from construction sites to factories (fabrication of more or less completed prefabricated building elements). This relocates some groups of construction works away from the construction site and increases on-site construction-labour productivity [2, 3].

Panel prefabricated structural systems are one from main groups of the modern construction systems, of which lightweight panel systems represent a special group. Based on the several studies [4-6], the benefits of the panel structural systems using can be divided into three main groups:

- Technological:
  - reducing the construction time by moving part of construction works off-site;
  - increasing the quality of constructions and reduction of their defects due to controlled factory environment;
  - less health and safety risks on the construction site as well as in factories;
  - overcoming the skilled labour shortage in the construction sector...
- Economic:
  - increasing the predictability of construction supply;

- reducing the cost of poor quality;
- improving the achievement of the planned profit...

- Environmental:
  - reducing the construction waste in prefab factories as well as directly on;
  - reducing the environmental impact of construction activities due to shorter construction time;
  - reduction of emissions due to need for a lower number of transport cycles...

The panel structural systems are made up of flat prefab panels designed especially for wall constructions. The complexity of a panel structural system is higher if floor and roof panels are also involved in the system. In terms of the complexity of their construction, the panels can be divided into open and closed. In case of open panels, the panel consisting of its basic components (frame, boards, insulation...) is manufactured and transported to construction site. Its surface finishes are made only after its assembly on the construction site, including the assembly of windows or doors. The closed panels are completely ready-made in the factory including finishes, windows, doors, sanitary ware and then are assembled on site, what significantly reduces the on-site construction time [7].

Depending on the level and method of panels completion in factory environment, the figures 1 – 4 present examples of different variants of open or closed panels [8-11]. The Figure 1 reflects a non-insulated open floor panel which is delivered from factory only as a frame with a board on the top and bottom and other finishing works (insulation, installation wiring and distributions, floor finishes, low ceilings) are made on-site. The Figure 2 shows the most common variant of lightweight panel – an insulated open panel without surface treatment. The third example (Figure 3) is an insulated partially closed system with surface treatment on interior or exterior side with filled holes and the Figure 4 demonstrates the variant of a completed insulated closed panel with surface treatment and integrated hole fillings and wiring and distributions installations.

Figure 1. Non-insulated floor panel with board on one side only [9].

Figure 2. Insulated open panel without surface treatment [9].

Figure 3. Insulated closed panel with surface treatment on one side [9].

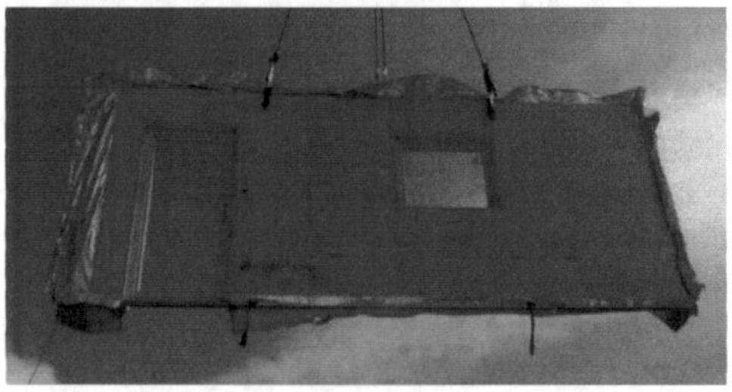

Figure 4. Insulated closed finished panel [9].

The lightweight panel systems are characterized by a sandwich structure consisting of boards or a frame with boards and a panel core which generally has an insulation function. The following products are commonly used as structural boards in lightweight panel systems:

- particle boards such as Oriented Strand Board (OSB), Medium Density Fibreboard (MDF), High-Density Fibreboard (HDF),
- cement-bonded particleboard,
- plasterboards,
- plywood,
- sheets,
- magnesium oxide boards (MGO)

and the insulation layer usually consists of:

- expanded polystyrene (PPU (Polyurethane foam) or PUR (Porous polyurethane)),
- polyurethane foam (PIR (polyisocyanurate)),
- mineral wool.

2. Materials and methods

The current global challenges, focused mainly on environmental sustainability are also challenges for the development of new or innovation of existing panel systems or their components. The innovations aim not only to increase their resilience, durability or efficiency by accumulating their functional properties (thermal insulation, acoustic, fire-fighting), but also by application of renewable materials or secondary raw materials, in order to reduce their carbon footprint.

2.1 Selected technologies of lightweight panel structural systems

Four innovative technologies were selected to analyse the potential of lightweight panel systems.

V(1) - lightweight construction panel HoneyComb ECOCELL (Figure 5) consists of oriented strand boards, using recycled paper as insulation. The insulation core based on the recycled paper has a honeycomb structure, consisting of 90% air. Improved strength parameters, fire resistance and water resistance of the boards are ensured by specially developed impregnating coating based on cement mortar. In the assembly process, the panels can be cut, glued or screwed [12, 13].

Figure 5. HoneyComb panel [12].

V(2) - System Eco-panels (Figure 6) represents innovative panels that consist of oriented strand boards bonded with a special foam core which doesn't contain any formaldehyde, styrene or volatile organic compounds (VOC). In addition to thermal insulation properties, the core withstands temperature around 2.000 °C, what ensures an exceptional fire resistance of the board. The system involves a corner panel, which is applied to insulate the most critical corner points of constructions [14, 15].

Figure 6. Eco-panels [14].

V(3) - Graphite SIP (Structural Insulated Panel) - as an alternative to traditional SIPs with EPS core (Figure 7), where polystyrene is enhanced by graphite powder (GPS – Graphite Polystyrene. Graphite particles more effectively resist thermal radiation, thus improving the insulating capacity of EPS by up to 20%, which allows the application of even smaller panel thicknesses, with the same efficiency. Moreover, the panel has improved strength and acoustic parameters and fire resistance. In case of an open flame, the panels are self-extinguishing due to the presence of flame retardants [16].

Figure 7. Graphite SIP [16].

V(4) - Titan MgO insulated panel – as an alternative to traditional OSBs or plasterboards for construction insulation panels, which applies the unique properties of Magnesium Oxide (MgO). MgO-based boards have improved strength parameters (impact resistance, flexibility), an excellent fire resistance and moisture resistance and in terms of thermal properties, the boards are characterized by a better storage capacity. In addition, the boards are characterized by resistance to mould and pests, which contributes to higher quality of indoor environment of buildings. Thanks to its composition, the Mgo board is among the "green building materials" [17].

Figure 8. Titan MgO panel [17].

The fire test carried out by the producer of the system, where the spread of fire on a magnesium oxide panel system and on a wooden panel system faced with plasterboards was compared. As can be seen in Figure 9, the Titan MgO panels didn't changed significantly, while wooden panels with plasterboards were completely destroyed. Then, to analyse the load bearing capacity of the preserved part of the house, it was loaded with 1.360 kilograms. The building was successful in this load test.

Figure 9. Fire test of Titan MgO panel and wooden panels [17].

In addition to product innovations focused on the material solution of panels, in all of the above mentioned lightweight panel systems, the innovation processes of their assembly on-site are addressed by the producers. The aim is to improve the continuity of the insulation function of panels and minimize time requirements for processes required to assemble panels on site.

2.2 The methodology to select an optimal variant of the structural panel

From the above, the efforts of the construction sector to develop and bring various innovative technologies are clear. However, this brings the need to choose between several technologies with often similar characteristics and parameters. The solutions with the highest efficiency can be most reliably determined with the help of the methods of multi-criteria evaluation of different variants. In a multi-criteria evaluation of variants, it is necessary to [18]:

- create a set of functions – evaluation criteria,
- quantify the values of the individual criteria,
- determine the importance of individual criteria for decision-making problem,
- determine the total performance and efficiency of each variant.

To determine the potential of lightweight panel systems, a technology of "heavy" panel systems was also involved in the multi-criteria evaluation. Specifically, it is about prefabricated reinforced concrete panels of the Benpan system - three-layer insulated ribbed reinforced concrete panel [19], which in terms of strength are undisputed competitors of lightweight prefabricated systems. On the other hand, due to their higher weight, the panels are of higher transport requirements. Thus, the following variants of the panel systems were compared: V(1) HoneyComb ECOCELL, V(2) Eco-panels, V(3) Graphite SIP system Murus, V(4) Titan MgO panels and V(5) panels of the Benpan system.

The following four characteristics were set as evaluation criteria, the values of which were quantified from available technical documentations of the variants [12-20].:

C(1) – density [kg/m^3];

C(2) – compressive strength [N/mm^2];

C(3) – thermal conductivity [W/mK];

C(4) – fire resistance – quantified by scoring, where 1 means better result – the product isn't flammable, and 0.5 means worse result when the product burns in case of open flame.

The unit cost *(C_u)* of the panels in EUR/m^2 is another evaluation criterion needed to determine the efficiency of the variants.

Direct evaluation method was used to determine the importance of the criteria. In the direct evaluation, the evaluator directly determines the order of significance or weight of the criteria. In accordance with the European Commission Regulation nr. 305/2011 and the Directive 89/106/EEC [21], when assessing the parameters of construction products within the ETA (Europe Technical Assessment), lists of essential properties are to be established for individual product groups. Within the product group under code 34 (Building Kits, Units and Prefabricated elements), all the above evaluation criteria correspond to the required assessed properties [22]. Considering the need to meet all the assessed properties at the same, the criteria weight was distributed evenly, thus the weight of 25% was assigned to all of the criteria.

The method of index coefficients, calculated using the values of the base variant, was used to determine the total efficiency of each variant. The index coefficient *(c)* is calculated as the ratio between the input value of the criterion and the value of the given criterion in the base variant. Its value must always be less than or equal to 1 (Equation 1).

$$c_{ij} = \frac{d_{ij}}{b_i}, \qquad \text{or} \qquad c_{ij} = \frac{b_i}{d_{ij}} \tag{1}$$

where c_{ij} – index coefficient of ith criterion of jth variant,
d_{ij} – the value of ith criterion of jth variant,
b_i – the value of ith criterion of the base variant.

The total efficiency of a variant (*E*) is determined as the sum of the products of index coefficients *(c)* and weight (*w)* of the relevant criterion (Equation 2). The best option is the one that has the highest total efficiency value.

$$E_j = \sum(c_{ij} \times w_i) \tag{2}$$

Where E_j – the total efficiency of jth variant,
c_{ij} – index coefficient of ith criterion of jth variant,
w_i – weight of ith criterion.

An extended perception of the optimality of a variant is proven through expression of its efficiency in relation to the cost parameter, which expresses the degree of efficiency of the assessed variants (Equation 3). The variant with the highest degree of efficiency is optimal.

$$DE_j = \frac{E_j}{C_{ui}} \tag{3}$$

where DE_j – the degree of efficiency of jth variant,
E_j – total efficiency of jth variant,
C_{ui} – unit cost of ith variant.

3. Results and discussion

Results of the study are presented in Tables 1 – 3. To determine the index coefficients, it is first necessary to select a base variant based on the individual values of decision criteria (Table 1). The base variant is formed by the best values of criteria. The selection of the best values depends on the type of criterion, indicated by an ascending ↑ arrow (the higher the value the better) or descending ↓ arrow (the lower the value the better). The Table 2 involves the index coefficients of the criteria of individual variants, calculated based on the values of the base variant.

Table 1. Determination of the base variant.

Variant/criterion	C(1) density kg/m^3	C(2) compressive strength N/mm^2	C(3) thermal conductivity W/mK	C(4) fire resistance
V(1) HoneyComb panel	50	1.698	0.22	0.5
V(2) Eco panel	45	0.24	0.21	0.5
V(3) Graphite SIP	20	2.45	0.032	0.5
V(4) Titan MgO panel	25	0.96	0.27	1

V(5) Panel of the Benpan system	240	9.8	0.45	1
Type of the criterion	↓	↑	↓	↑
Base variant	20	9,8	0.032	1

Table 2. Determination of the index coefficients.

Variant/criterion	C(1)	C(2)	C(3)	C(4)
V (1) HoneyComb panel	0.4	0.17	0.14	0.5
V (2) Eco panel	0.44	0.02	0.15	0.5
V (3) Graphite SIP	1	0.25	1	0.5
V (4) Titan MgO panel	0.8	0.09	0.11	1
V(5) Panel of the Benpan system	0.08	1	0.07	1
Base variant	1	1	1	1

The total efficiency of the variants is the sum of the products of index coefficients and weights of individual criteria (Table 3). When the price of the panels is taken into account, a degree of efficiency is obtained based on the overall efficiency. In both cases, the variant with the highest value of the total efficiency is the best one (Table 3).

Table 3. Determination of the total performance and efficiency of the variants.

Variant/weight of the criterion	C(1) 0.25	C(2) 0.25	C(3) 0.25	C(4) 0.25	Total efficiency	C_u EUR/m^2	Degree of efficiency
V(1) HoneyComb panel	0.1	0.0425	0.035	0.125	0.302	45	0.006
V(2) Eco panel	0.11	0.005	0.0375	0.125	0.277	50	0.005
V(3) Graphite SIP	0.25	0.0625	0.25	0.125	0.6875	70	0.009
V(4) Titanium MgO panel	0.2	0.0225	0.0275	0.25	0.5	35	0.014
V(5) Panel of the Benpan system	0.02	0.25	0.017	0.25	0.537	260	0.002

The above results have demonstrated that the graphite SIP is the optimal variant from total efficiency point of view. The result was partly determined by its two best parameters (Table 1) and the same weight of criteria. The Titan MgO panel, from the group of sandwich panels, is a strong competitor of the graphite SIP. In terms of observed parameters, the other two variants of lightweight panels (HoneyComb panel and Eco panel) are of lower efficiency compared to panel of the Benpan system. This was ensured by better properties of the reinforced panel, such as strength, fire resistance and relatively comparable thermal conductivity, achieved by the built-in insulation layer. It follows that even innovations of "heavy" panels, as to their higher insulating function, may be competitive, especially if higher load-bearing capacity is required, for example for multi-storey buildings, where lightweight panels may encounter limitations.

On the other hand, when considering the cost parameter, costly innovations - in this case a panel of the Benpan system in an effort to achieve the performance parameters of lightweight panels - are reflected in higher cost of the solution. Therefore, in terms of the total efficiency, the panel of the Benpan system has reported the worst value compared to all lightweight panels, of which the Titan MgO panel has the highest total efficiency.

Based on the results, the total efficiency of the above variants is in general significantly affected by the cost parameters, what usually affect the result and the customer's opinion when choosing an optimal technology. The presented study demonstrates not only the potential of lightweight panel systems, but also the methodology of their multicriteria evaluation, which can be applied for various number of compared variants, various number of compared criteria and of course various parameter preferences, taking into account the evaluator's view on target optimization requirements.

4. Conclusion

Modern methods of construction, including prefabricated panel construction systems, have an undeniable contribution to increasing the efficiency of the construction industry. Compared to traditional masonry or reinforced concrete monolithic structures, the above methods are clearly characterized by shorter construction times, higher quality, less waste, etc. The construction market brings more and more building technology innovations that bring these benefits [14, 16, 17, 19, 20, 23]. Then, in terms of considering several variants with comparable parameters, it is necessary to decide on the selection of one of them. The paper has presented the study of multicriteria decision-making when choosing an optimal variant of lightweight panel systems, also in comparison with an innovated panel of the Benpan system with a built-in insulation layer.

Using the optimization method based on index coefficients, the total efficiency of different variants was determined, taking into account selected technical parameters of lightweight panel systems. Moreover, the influence of cost parameters was demonstrated. It significantly affects the selection of the optimal variant in terms of its degree of efficiency.

The presented study has demonstrated the potential of lightweight panel systems as a subgroup of modern methods of construction, as well as the constant effort to innovate traditional panel systems based on reinforced concrete. The methodology used is fully applicable to various numbers of compared variants, various number of compared criteria and, of course, various parameter preferences, taking into account the evaluator's view on the target optimization requirements.

5. References

[1] Park H S, Thomas S R and Tucker R L 2005 Benchmarking of construction productivity *Journal of Construction Engineering and Management* vol 131 issue 7
[2] Motwani J, Kumar A and Novakoski M 2015 Measuring construction productivity: A practical approach *Work Study vol 44(8) p 18-20*
[3] Shahzad W, Mbachu J and Domingo N 2015 Marginal productivity gained through prefabrication: Case studies of building projects in Auckland *Buildings* vol 5(1)
[4] Generalova M 2016 Procedia Engineering 153
[5] Gann D and Salter A 2000 Innovation in project-based, service-enhanced firms: the construction of complex products and systems *Research Policy* vol 29 p 955
[6] Davies A 2018 Modern Methods of Construction, A forward-thinking solution to the housing crisis? *Royal Institution of Chartered Surveyors (RICS)*
[7] Ely A 2005 Are Modern Methods of Construction compatible with best practice urban design *Waterfront Conference on MMC*
[8] OECD 2015 The Innovation Imperative: Contributing to Productivity Growth and Well-Being
[9] Construction Industry Development Board (CIDB) 2016 Industrialised building system: the path to enhanced productivity *Heights 3*
[10] Musa M, Mohammad M, Yusof M and Mahbub R 2015 The Way Forward for Industrialised Building System (IBS) in Malaysia *Proceedings of the International Civil and Infrastructure Engineering Conference 2014* vol 14 pp 163
[11] Chai T, Tan C, Chow T, Ling P and Koh H 2018 A Review on Prefab Industrialized Building System Modular Construction in Malaysia: The Perspective of Non-structural Studies
[12] Ecocell official website – Honeycomb 2021 Technical data sheets Available on https://ecocell.ch/en/us-building-system/
[13] Nguyen T and Nguyen H 2021 Lightweight Panel for Building Construction Based on Honeycomb Paper Composite *Nano Hybrids and Composites* vol 32
[14] ECO panels official website 2021 Technical data sheets Available on https://www.eco-panels.com/about/
[15] Mohammadabadi M, Yadama V and Dolan J 2021 Evaluation of Wood Composite Sandwich Panels as a Promising Renewable Building Material *Materials* vol 14
[16] ModernSIP official website – SIP panel 2021 Technical data sheets Available on https://modernsip.hu/sip-panel/
[17] Titan Wall official website –MgO panel 2021 Technical data sheets Available on https://www.titanwall.com/mgo-sips/

[18] Baskova R 2004 *Modelovanie procesov výstavby. Ekonomicko-matematické metódy – časť I. Lineárna optimalizácia a sieťová analýza* Košice SvF TU Košice
[19] Benpan official website - Wall Panel BENPAN Premium 2022 Available on https://benpan.de/
[20] Murus official website – panels 2021 Technical data sheets Available on https://murus.com/product/
[21] EK - Označenie CE stavebných výrobkov. Available on https://www.soi.sk/files/documents/info-verejnost/oznacenie%20ce.pdf
[22] ETA 18/0312 – verzia 01 z 17/09/2018, Technický a skúšobný ústav stavebný Available on https://www.sipeurope.eu/wp-content/uploads/2020/06/ETA-Europske-technicke-posudenie-18_0312-Sipeurope.pdf
[23] Kalús D, Gašparík J, Janík P, Kubica M and Šťastný P 2021 Innovative Building Technology Implemented into Facades with Active Thermal Protection *Sustainability* 13(8)

Acknowledgments
The activity presented in the article is part of VEGA research grant No. 1/0336/22 Research on the effects of Lean Production/Lean Construction methods on increasing the efficiency of on-site and off-site construction technologies.

IOP Conf. Series: Materials Science and Engineering 1252 (2022) 012026 doi:10.1088/1757-899X/1252/1/012026

The effect of manufacturing imperfections in the slip joints

M Mojto[1,2], J Brodnianský[1] and J Brodnianský[1]

[1] Department of Steel and Timber Structures, Faculty of Civil Engineering, Slovak University of Technology in Bratislava, Radlinského 11, 810 05 Bratislava, Slovakia
[2] Corresponding author

E-mail: maros.mojto@stuba.sk

Abstract. The issue of imperfections is mostly related to additional internal forces. However, our understanding of imperfections is associated with the change of structural behaviour. In our case, the most likely indicated effects of imperfections are the reduction of the contact area, and the unreachability of the required overlap length. The behaviour of these slip joints depends on sufficient contact between sliding parts. The contact area is affected by the imperfections of the manufactured steel tubes. This contribution presents the manufacturing tolerances and subsequently, it describes the incorporation of them into the design of the slip joint. Furthermore, the article presents a comparison between the ideal numerical model and model with imperfections involved. Moreover, this numerical model is verified by experimental measurements. The results indicate that imperfections have a significant impact on the slip joints. In our comparison, the effect of imperfections is about 120%.

1. Introduction

The slip joints have many applications in the conical steel columns with circular or polygonal cross-sections. These joints can be determined as the sliding splice of two construction parts together. The lower part supports the upper part. The conical geometry ensures the force transfer from the upper part to the lower. The usage of the slip joints in practise is less frequent than the traditional flange connection despite the indisputable advantages such as saving material, assembling time and additional costs. One of the disadvantages is the lack of a more precise design manual. According to current standards, the minimal overlap length is the only requirement that must be satisfied. This overlap length is related to the diameter and other aspects such as conicity, the thickness of the wall, the magnitude of load, fatigue, or imperfections are not considered.

The issue of slip joint used in construction of masts is presented in these publications [1], [2], [3], [4], [5], [6] and [7]. The similar issue of this contribution is introduced in the thesis [8] in section 3.3 Fabrication tolerances. The used experimental model verified by FEM model was presented in the [9] and [10]. The contribution proceeds on the previous work [11] where the imperfections were introduced.

The overall aim of the contribution is to determine the effect of imperfections. The further aims of the article are the demonstration of the tube manufacturing tolerances and the introduction of the deviation shapes in the slip joints.

In the first part is presented a summary of already reached research and motivation of dealing with imperfection in the slip joint. Chapter 4 introduces the most fluently shape of deviations in slip joints. Chapter 5 includes the presentation and incorporation of manufacturing tolerances into shapes of the deviations. These deviations present maximal allowed imperfections. Based on the comparison between ideal FEM and FEM models with imperfections, Chapter 6 determines the effect of imperfection in the slip joint.

2. A summary of already reached research

This contribution involves the results of the verified FEM model presented in [9] and [10]. The experiments have consisted of two conical polygonal 8-edged steel tubes. Experiment 01 had an overlap length equal to 3.1**d* (diameter of the steel tube in overlap length). After a cut-off of the upper part, experiment 02 has been executed with an overlap length equal to 1.75**d*. The following Table 1 presents the other parameters of specimens and the numerical model.

Table 1. Parameters of numerical and experimental models.

FEM model	experiment	*d* [m]	*h1* [m]	*h2* [m]	*h*	*t* [mm]	*My* [kN*m]	*Material*
01	01	0.115	2.21	2.21	3.1**d*	3	3.0	S275
02	02	0.115	2.21	1.96	1.75**d*	3	2.8	S275

Figure 1 presents the experimental measurement and sensing devices such as strain gauges, displacement transducers and load cells. The results have been collected using hardware Spider8 and evaluated with built-in software Catman. A numerical model based on FEM 3D elements (twenty-node element with possibility of elimination intermediate nodes) has been carried out in the software Dlubal Rfem 5.20. The maximal element size has been equal to 6mm (in the place of overlap length) and 50mm in the positions without the overlap. The contact area between the upper and lower part of the joint is active only in compressive with friction coefficient equal to 0.3. For numerical analysis, the Modified Newton-Raphson method has been used.

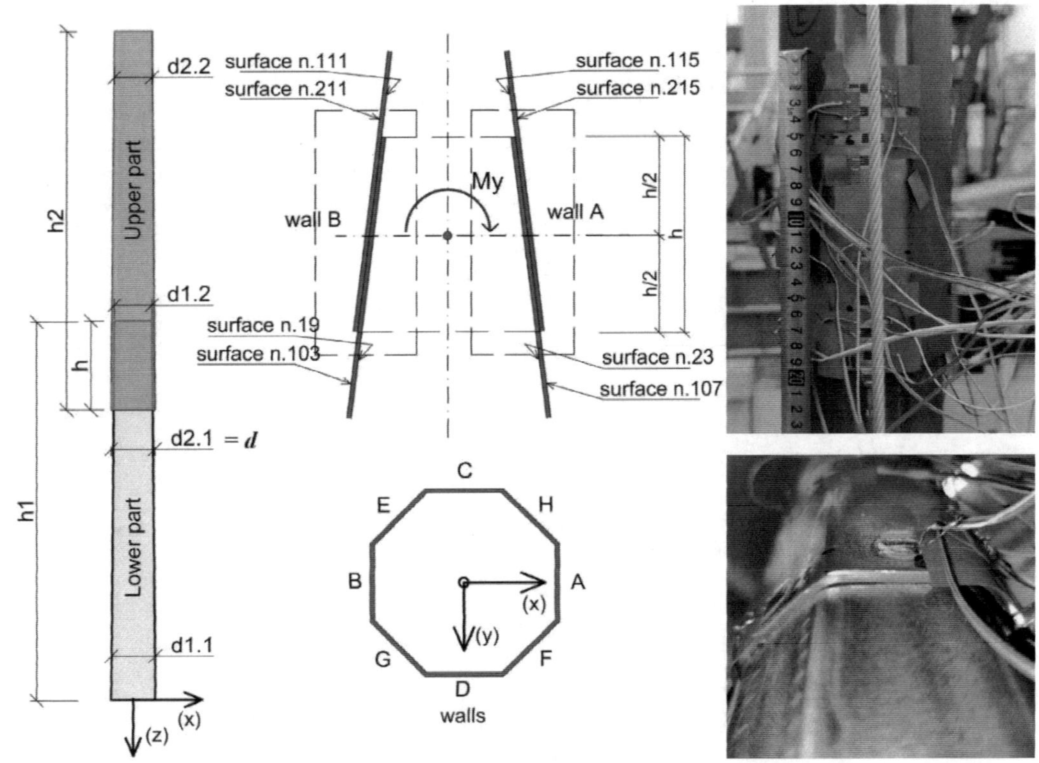

Figure 1. Disposition of FEM model and specimen (left-hand side), experiment (right-hand side).

3. The motivation

The bending moment and vertical force are transferred through the contact area between the upper and lower parts. In Figure 2, the example of the imperfection effect is presented in a slip joint loaded by bending moment. In the middle of the figure is demonstrated the ideal model. On the right-hand side, the model with the imperfection in the shape of the cross-section is presented. While the contact pressure is spread around half of the circumference in the ideal condition, the contact pressure in the imperfect model is concentrated in the small zones on the edges. These geometrical imperfections can reduce the contact area and change behaviour of the slip joint.

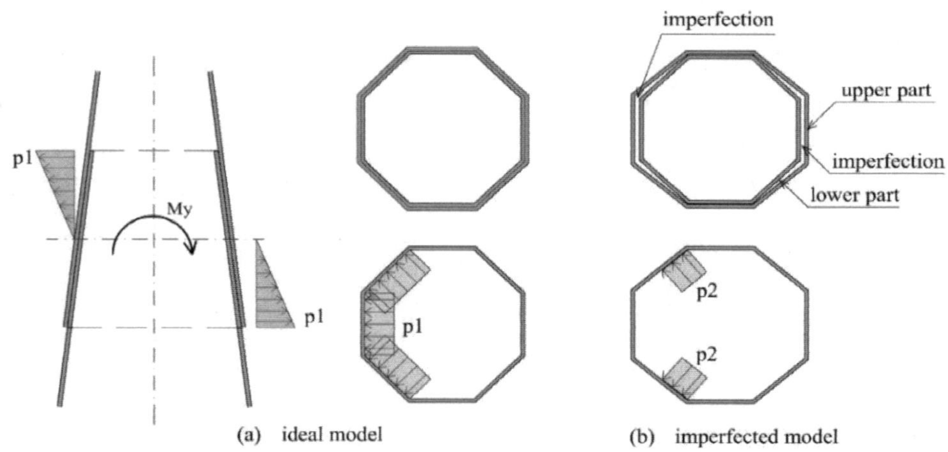

Figure 2. The example of imperfection effect, ideal model (a) and model with imperfection (b).

4. Critical imperfection in slip joint

This paragraph introduces the shape of deviations (imperfections) in the slip joints. The shapes of imperfections have been divided into deviations in the shape of cross-section and deviation in conicity. Note that the studied shapes involve the most probably weighty shapes. Therefore, these shapes of deviations represent one of many other possible deviations.

The deviation in the shape of the cross-sections is presented in Figure 3. These deviations present inaccuracy between the upper and lower parts. These shapes have been carried out by ovalization of the upper and lower part of tubes. Figure 3 also presents the deviation in the conicity where two situations can occur. The first with an upper part with higher conicity to the lower part and the second with lower conicity of the upper part to the lower part.

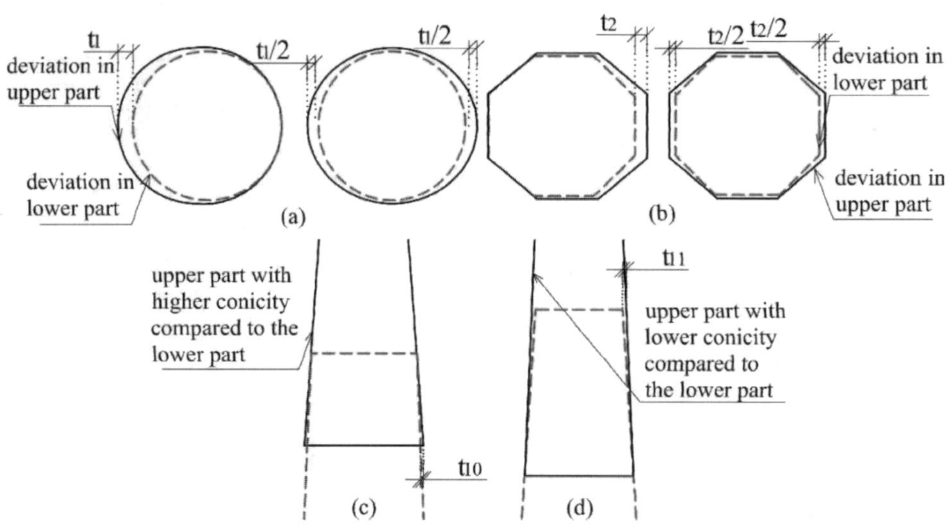

Figure 3. The deviation in cross-section (at the top) and the deviation in the conicity (in the bottom).

5. Definition of the maximal allowed deviations in the slip joint

Table 2 captures the standards which consider the manufacturing tolerances and applicability of the tolerances either in circular or polygonal cross-sections. As an example, in the next table 3 are presented the maximum deviations t_1 and t_2 for tubes with diameters 115 mm and 1250 mm. These maximal tolerances consider the tolerances of both the upper and lower parts of the slip joints.

Table 2. Standards containing the manufacturing tolerances.

		Applicability in cross-section	
Standard	Description	Circular shape	Polygonal shape
STN EN 1090-2 [12] Tab. B.6	Gap Δ between straight edge and surface	No [a]	No [a]
STN EN 1090-2 [12] Tab. B.11	Out-of-roundness (ovality)	Yes	Yes [b]
STN EN 40-2 [13] Section 5.7.1.2	The tolerance for deviation in shape for circular ± 3% related to the diameter	Yes	No
STN EN 40-2 [13] Section 5.7.1.3	The tolerance for deviation in shape for polygonal ± 4% related to the flat of the polygonal	No	Yes
STN EN 40-2 [13] Section 5.7.1.1	The tolerance on the column circumference shall be ± 1%	Yes	Yes
STN EN 10217-7 [14] Tab. B.10	Tolerances on outside diameter and wall thickness	Yes	Yes [b]
STN EN 10217-7 [14] Section 8.8.4.5	Out-of-roundness (ovality)	Yes	Yes [b]

[a] The request is not related to the diameter.
[b] Diameter of described circle has been considered for polygonal shape.

Table 3. Maximal deviations t_1 and t_2 for circular and polygonal cross-section with diameter equal 115 mm and 1250 mm.

	Deviations t_1 in [mm] for circular cross-section		Deviations t_2 in [mm] for polygonal cross-section	
Standard	D = 115 mm	D = 1250 mm	D = 115 mm	D = 1250 mm
STN EN 1090-2 [12] Tab. B.6	2.00/1.00 [a]	2.00/1.00 [a]	2.00/1.00 [a]	2.00/1.00 [a]
STN EN 1090-2 [12] Tab. B.11	A = 3.20 B = 4.60 C = 6.90	A = 17.50 B = 25.00 C = 37.50	A = 3.20 B = 4.60 C = 6.90	A = 17.50 B = 25.00 C = 37.50
STN EN 40-2 [13] Section 5.7.1.2	6.90	75.00	NA	NA
STN EN 40-2 [13] Section 5.7.1.3	NA	NA	8.48	92.39
STN EN 40-2 [13] Section 5.7.1.1	2.30	25.00	2.30	25.00
STN EN 10217-7 [14] Tab. B.10	D3 = 1.73 D4 = 1.15	D2 = 25.00	D3 = 1.73 D4 = 1.15	D2 = 25.00
STN EN 10217-7 [14] Section 8.8.4.5	D3 = 1.73 D4 = 1.15	D2 = 50.00	D3 = 1.73 D4 = 1.15	D2 = 50.00

[a] The request is not related to the diameter.

6. Comparison between experimental measurement and FEM model with imperfections

Figure 4 shows the comparison of the tangential stresses on the walls A and B between experiment 02 and the ideal FEM model 02. The preparation of the experiment and FEM model is visualized in Figure 1. The comparison indicates the discrepancy between curves nevertheless, the trend of curves is similar.

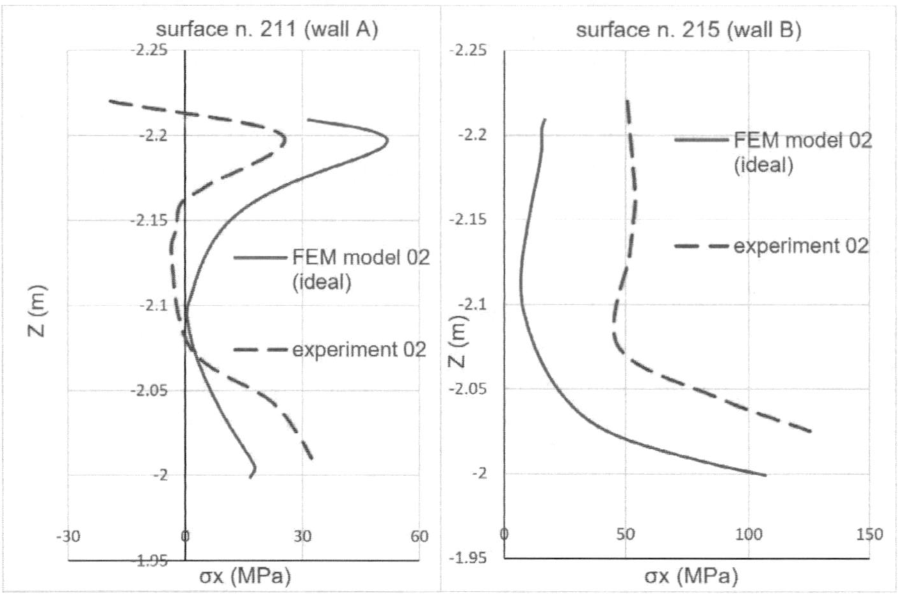

Figure 4. Tangential stresses of experiment 02 and FEM model 02 (ideal).

In the next phase, the manufacturing imperfection of experiment 02 was added to the FEM model. The imperfection between the upper and lower part on walls G and D is visualized in Figure 5. The average distance is about 3 mm (the maximum distance is 5 mm). These imperfections were incorporated by the nonlinear function (see Figure 5 right-hand side). This function ensures that the compressive contact between the upper and lower parts on the walls G and D will be after deformations of 3 mm. Figure 6 shows the comparison between experiment 02 and FEM model 02 with imperfections. The results indicate that the application of imperfections makes the FEM model more accurate, and the FEM model behaves according to the experimental measurements.

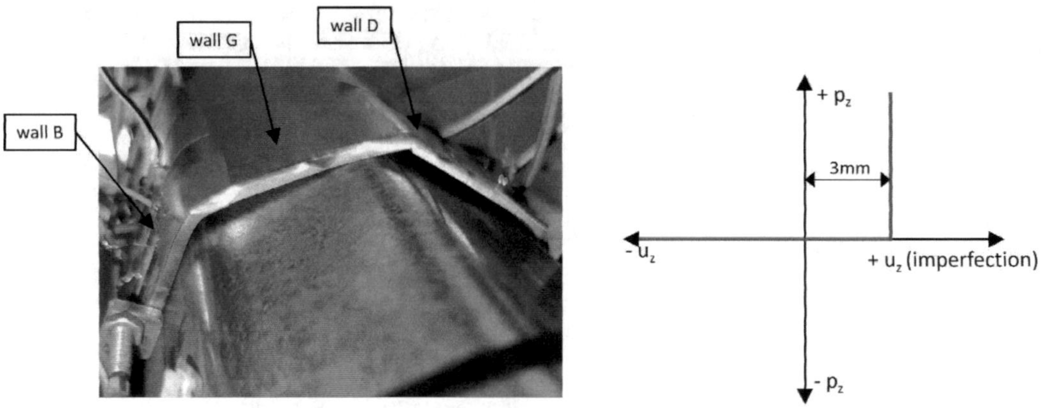

Figure 5. Manufacturing imperfections on experiment 02.

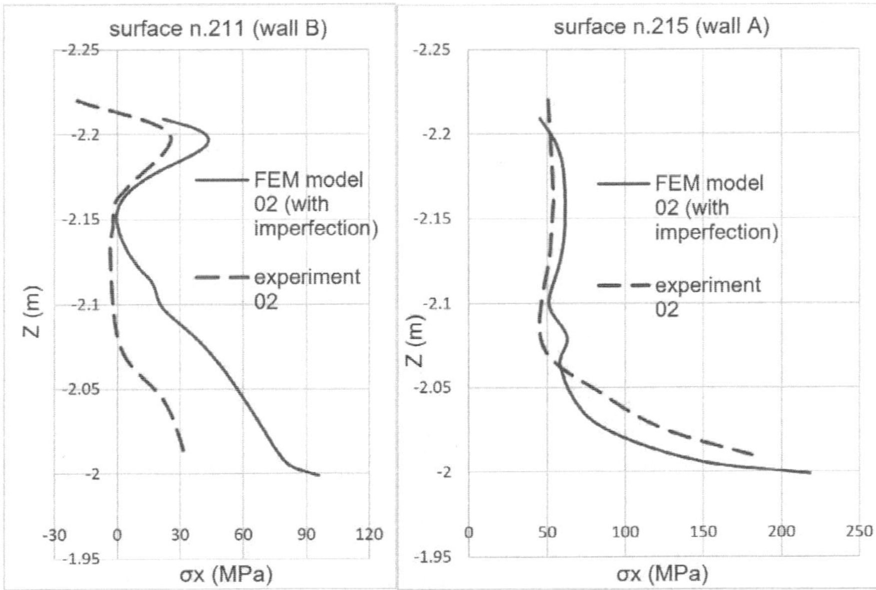

Figure 6. Tangential stresses of experiment 02 and FEM model 02 with imperfections.

6.1. The effect of imperfections in the slip joint

Figure 7 shows the comparison between the ideal FEM model and the FEM model with imperfections. The comparison presents the tangential stresses on walls A and B. The FEM model with imperfections considers the imperfection of 3 mm between the upper and lower part on walls A and B.

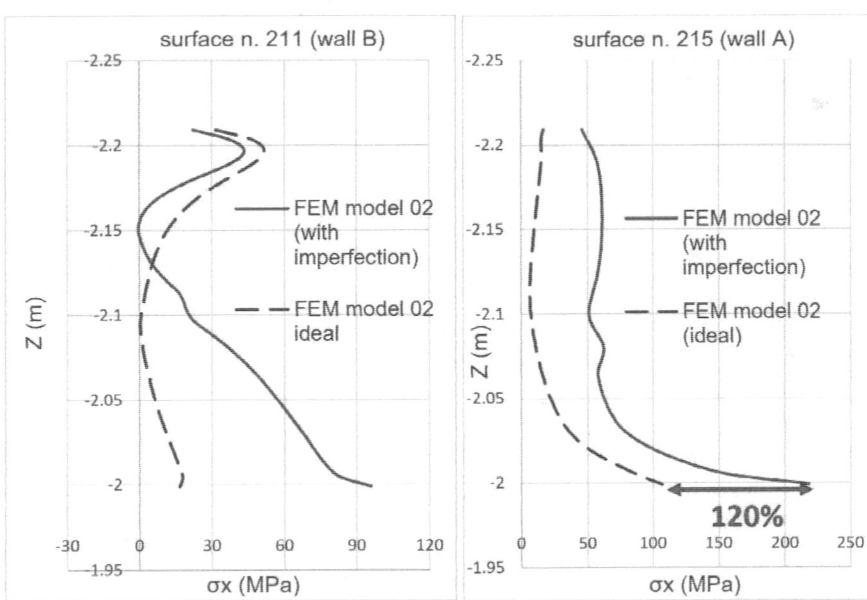

Figure 7. Tangential stresses FEM model 02 (ideal) and FEM model 02 with imperfections.

7. Discussion

The most likely shapes of deviations were presented in the first part of the contribution. Based on the out-of-roundness of the upper part and lower part, the deviations in shapes of cross-sections were carried out. It is important to say that the deviations t_1 and t_2 capture the deviations upper part as well as the lower part of the joint. So, the values presented in Table 3 consider the imperfections of both parts. This section also presented the deviation in the conicity. This deviation has not been examined in the next part of the contribution. The standards do not consider the request for conicity. The conicity deviations t_{10} and t_{11} (see Figure 3) may be calculated by using requirements for cross-section and incorporating them into the deviation in conicity. It seems that the model in Figure 3 (d) has more appropriate behavior than model (c). In a condition where the upper part has higher conicity compared to the lower part, the distance t_{10} between parts is increasing by overlapping.

The standards capturing manufacturing tolerances have been presented in the next part of the contribution. We have considered only with tolerances related to our proposed shape of deviations. The standards contain many other requirements such as wall thickness, the maximum height of the weld seam or the tolerance on exact lengths that are not exactly related to the issue of this article. It is possible to conclude that it is not clear which request is necessary to use in a specific situation. The multiple requirements are related to the similar or the same feature. The tolerances for circular cross-sections are available in a more complex point of view than for polygonal cross-section. However, the requirements for circular cross-sections can be applied also for the polygonal cross-sections in most cases. It can be done by using the diameter of the described circle of polygonal shape as an input to the standards.

The comparison between FEM results was presented in chapter 6. The ideal FEM model 02 presents the geometry of experiment 02 (consisted of tubes with 115 mm diameter) without imperfection. The FEM model 02 with imperfections presents the same geometry but within the imperfections. The results indicate that the effect of imperfection in experiment 02 was about 120%. Note that this effect was reached with imperfection that considers 3 mm distance between upper and lower part on two walls of the slip joint. This value of imperfection lays in the middle of the allowed range of deviations for diameter 115 mm presented in Table 3. It is assumed that the slip-joint consisted of parts that meet stricter requirements mainly according to STN EN 10217-7 [14] will reduce the effect of imperfection more efficiently. It is supposed that individual requirements presented in table 3 could be mutually exclusive. Therefore, it is proposed to verify the possibility of exclusion in the next stage of the research.

8. Conclusion

We conclude that the contribution fulfils all desired aims. The presented contribution provided the shapes and values of maximal limit deviations in slip joints according to the various requirements. In the next approaches, these limit deviations are possible to use in the process of finding the effect of maximal imperfections. One of these approaches was also presented in this contribution where the effect of imperfection was about 120%. It is possible to say that the effect of imperfection should not be neglected.

9. References

[1] Botló M 2017 Slip joint connections of steel conical towers *Disertation Thesis* Faculty of Civil Engineering Slovak University of Technology in Bratislava Svf-13420-50559

[2] Botló M, Brodniansky J and Magura M 2016 Numerical and experimental analysis of structural details of electrical transmission line towers *Proceedings of the IASS Annual Symposium International Association for Shell and Spatial Structures* Tokyo Japan

[3] Brodniansky J, Klas T, Brodniansky J, Magura M, Recký J and Tilinger 2019 Field tests, laboratory evaluation and modelling of slip joints *Proceedings of the IASS Annual Symposium 2019 – Structural Membranes 2019 FORM and FORCE* Barcelona Spain p. 1548–55

[4] Segeren MLA, Lourens EM, Tsouvalas A and Van Der Zee TJJ 2014 Investigation of a slip joint connection between the monopile and the tower of an offshore wind turbine *IET Renew. Power Gener.* **8 4** p. 422–432

[5] Segeren MLA and Hermans KW 2014 Experimental investigation of the dynamic installation of a slip joint connection between the monopile and tower of an offshore wind turbine *J. Phys.: Conf. Ser.* **524**(2014)012080

[6] Cabboi A, Kamphuis T, van Veldhuizen E, Segeren M and Hendrikse H 2021 Vibration-assisted decommissioning of a slip joint: Application to an offshore wind turbine *Marine Structures* **76**(2021)102931
[7] Cabboi A, Segeren M, Hendrikse H and Metrikine A 2020 Vibration-assisted installation and decommissioning of a slip-joint *Engineering Structures* **209**(2020)109949
[8] Kamphuis TPJ 2016 Design, testing and verification of the DOT500 slip-joint support structure *MSc Thesis* Delft University of Technology
[9] Mojto M 2020 Effect of overlap length on slip-joint connection *Advances in Architectural, Civil and Environmental Engineering* Bratislava Spectrum STU p. 477–83
[10] Mojto M 2021 Paramter o overlp length on slip joint connection uniorstav *23rd International Conference of Doctoral Students* Brno ECON publishing, s.r.o p. 255–60
[11] Mojto M 2021 Manufacturing imperfections in slip joints *Advances in Architectural, Civil and Environmental Engineering* Bratislava Spectrum STU p. 463–70
[12] STN EN 1090-2 Execution of steel structures and aluminium structures Part 2: Technical requirements for steel structures
[13] STN EN 40-2 Lighting columns Part 2: General requirements and dimensions
[14] EN-10217-7: European Norm for Welded Steel Tube for Pressure Purposes - Part 7: Stainless steel tube

Acknowledgement
This research is funded by the Scientific Grant Agency of the Ministry of Education, Science, Research and Sports of the Slovak Republic under the program VEGA Nr. 1/0388/19.

Elastic Exploring of Human Lung Parenchyma

B Bošanská[1], M Minárová[2] and K Tvrdá[2]

[1] Lipová 982/7A, 958 01 Partizánske-Veľké Bielice, Slovakia
[2] Slovak University of Technology, Faculty of Civil Engineering, Radlinského 11, 810 05 Bratislava, Slovakia

E-mail: bosanska.bar@gmail.com;maria.minarova@stuba.sk;katarina.tvrda@stuba.sk

Abstract. The paper describes a theoretical study and exploration of authors in the field of biomechanics. The lung parenchyma is focused, and its elastic behaviour under the physiological mechanical load, pressure change during breathing. Anatomical view to the human respiratory system generally is introduced first together with a more detailed description of the alveolus. Then the appropriate theory is elaborated, aimed on the mathematical model establishing, the finite element method is chosen as the resolving tool. The computational implementation is realized in multi physical software based on the finite element method.

1. Introduction

Breathing is one of the essential vital function. The body retrieves the oxygen, which is inevitable for each organ proper functioning. Of course, the carbon dioxide has to be released from the body. The gases pass along the entire air passages, whereas the gas exchange is realized in its smallest parts called alveoli. The alveoli stretch and shrink periodically during breathing, enlarging and decreasing their inner volume. The soft tissue called parenchyma is deformed. As well as in various soft tissue mechanical behaviour, both the geometric and physical nonlinearity is obviously present. This paper focuses on the micrometres level of breathing mechanics, the alveolar level.

2. Human respiratory tract. Anatomy and biophysics

In case of healthy organism, all essential mechanisms operate accurately, aiming to the gaining oxygen and get rid the waste products. Always when the respiratory system is regarded from the medical point of view, it is referred as split to upper and lower part of airways. Upper part includes nose, nasal cavity and nasopharynx, where the inflated air is purified from mechanical dirt, and warmed up to the body temperature. Then the air continues to the lower airways passing through the pharynx, trachea and bronchioles branches down to the lung.

2.1. Lung

Lung is a branched corridor labyrinth with follicles at the ends of each branch. All system is mostly filled with the air. The corridors are symmetrical, full of pores, mutually connected. Approaching to alveoli they turn to alveolar canal. The gas exchange itself is accomplished in the small ending follicles, alveoli through several capillaries – small vessels surrounding each alveolus. The magnitude of alveoli is approximately the same (measured in micrometres) nearby one alveolar canal and moreless even in the entire lung. The tuft of alveoli is called alveolar sac. As the human being grows up, his breathing tract becomes larger and the elasticity of alveoli becoming lower. Also the elasticity of thoracic wall decreases with age, resulting in more effort demand while breathing. The ramification of bronchioles proceeds up to the 23rd level, however, the first 16 levels ensure the air transport uniquely, figure 1. Farther, from the 17th level on we speak about the breathing zone, since this is the level where first alveoli occur, namely the gas exchange takes part there.

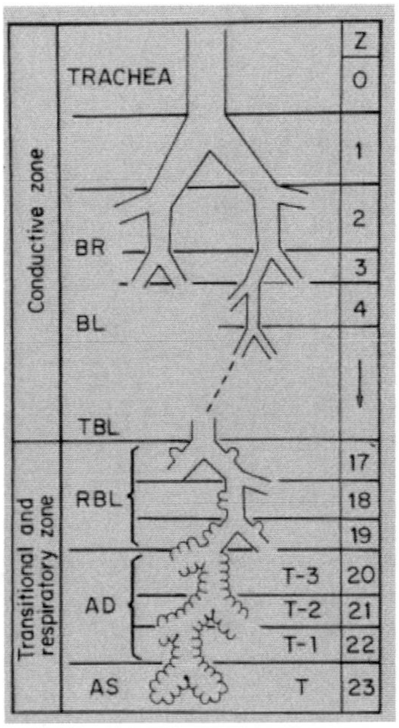

Figure 1. Airway model. BR - bronchus, BL – bronchiole, TBL – terminal bronchiole, RBL – respiratory bronchiole, AD – alveolar duct, AS – alveolar sac [1].

2.2. Alveolus

Alveolus is a terminal part of breathing tract where the gasses get in. Though the exchange between oxygen (O_2) and carbon dioxide (CO_2) are decisive, there are many other gases present, e.g. water vapour, noble gases, etc. There is a certain amount of water in each alveolus spread on its inner surface. Together with inflated air the surface tension arises. Molecules of the liquid interact each with other.

The pressure of the liquid acting inwards is computed by Laplace law quantifying the relation between the surface tension and radius of spherical objects:

$$P = 2\,T\,r^{-1} \qquad (1)$$

where P is pressure in the spherical object, T surface tension and r is radius of the spherical object.

Regarding this law, we can say that with the surface tension increase also the alveolar pressure increases and vice versa, the alveolar pressure decreases with the decrease of the surface tension. For the sake of avoiding the collapse of alveolus – either due to exceeding stretching or exceeding shrinking, the surfactants son the inner surface play an important role here, see figure 2. The surfactant's layer consists of the hydrophobic tails and hydrophilic heads. When alveolus tries to reduce its magnitude too much, surface active matters evolve a pressure against since the hydrophobic tails get to the water film and activate an increase of the magnitude. Vice versa, if the alveolus pumps out too much, the hydrophilic heads find themselves in a too thin water film and indicate the backward shrinking.

In figure 3 the species of human lung is depicted, augmented 240 times. The porosity of tissue is apparent. Particular pores are connected with each other in order to mutual pressure balance regulation. In this manner the tissue is protected against destruction.

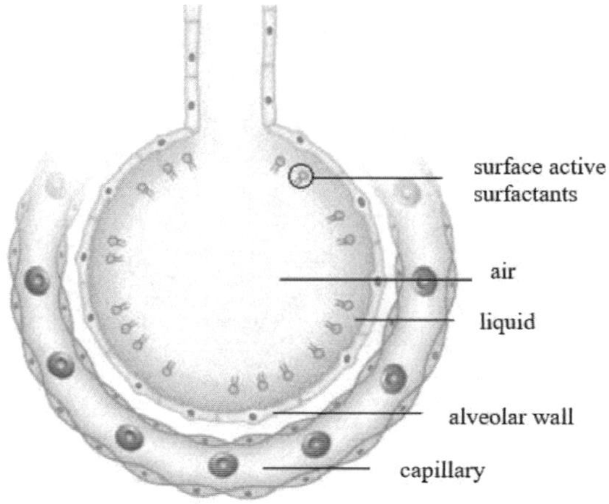

Figure 2. Alveolar anatomy and capillary [2].

The magnitude of alveoli depends on age, weight, height of individual, as well as on the obvious demand of oxygen. Adults' alveolar diameter is about 250 μm, i.e. 0.25 mm. This diameter varies within the same body approximately by 5 – 10 % [3].

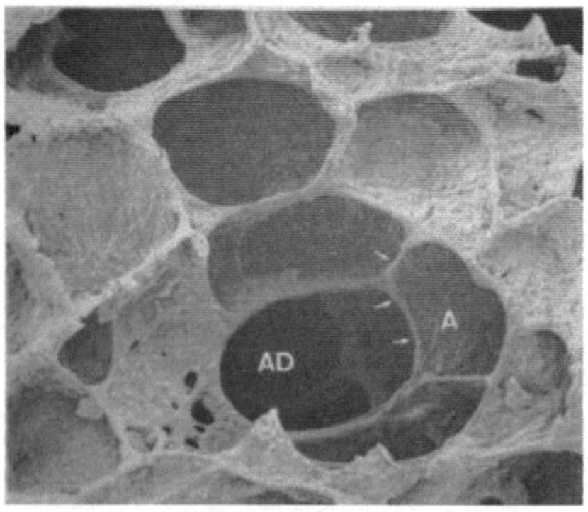

Figure 3. Alveolar tissue, the electron microscope view. Enlarged 240 times.
AD – alveolar duct, A – alveolus [1].

2.3. Biophysics of breathing

During breathing the chest enlarges rhythmically (inflation – inspirium) and relaxes again (expiration – expirium). Inflation is always an active process in which the muscles, mainly diaphragm, are employed. These muscles act against the tissue resistance. Expiration is an inactive process. Due to prevalent elastic property of the tissue, the potential energy saved during inflation is utilized for the reverting process during expiration. During breathing, the pleural pressure, i.e. the pressure surrounding the lung within the pleural space, oscillates, but remains always lower than the atmospheric pressure. It reaches its lowest value in the peak of inspiration, i.e. approximately 1 kPa lower value in comparison to the atmospheric pressure, the maximum differs by –0.3 kPa in comparison to the atmospheric pressure and this happens at the moment of expiration deepness. When the injury of lung results into perforation of pleural space, pneumothorax occurs and the breathing becomes much more difficult unless the original pleural pressure is recovered. In each particular inflation, the mechanical work of breathing has to overcome the elastic and nonelastic tissue resistance, the resistance of chest and the flow resistance of airways. Mathematically this resistance W is the product of pressure and volume change ΔV.

$$W = p\,\Delta V \tag{2}$$

Here p, the pressure, is the sum of all pressures needed to overcoming elastic, nonelastic and flow resistance. In tranquillity the resistance, as well as the same value of mechanical work is decisive for determining the total breathing work, figure 4. In this figure the mechanical breathing work is depicted. In linear case, saved inflation potential energy (represented by the surface of the trapezoid 0ABC on both diagrams) is all utilized during expiration phase. However, the viscoelastic refinement of the model can be anticipated here saying that the inspirium work is indeed a bit greater (by amount corresponding to the surface of canopy ADB) and on the other hand we have to subtract canopy AEB from the energy utilized for the expiration. Finally, sum of both canopies yields the loss of energy – this surface represents the energy that dissipates to heat during one cycle of breathing.

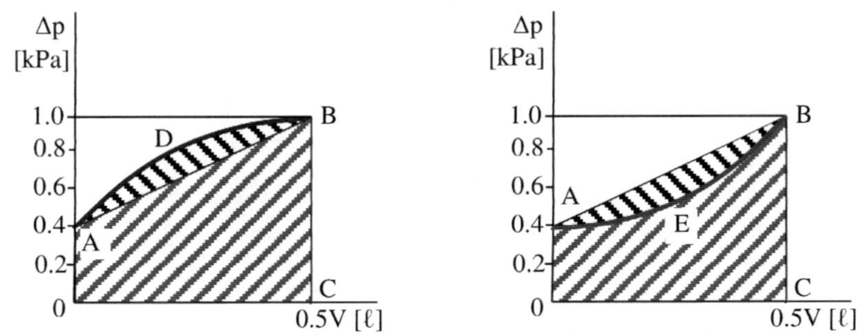

Figure 4. Mechanical work dependence on volume; left – inspirium, right – expiration. Elastic model – with line segment upper edge versus viscoelastic model – curvilinear upper boundary [3].

3. Local deformation of alveoli parenchyma. Mathematical model

From the physiological point of view, it is interesting to explore the local deformation of alveoli. The deformation is affected by the pressure in the surrounding vessels and bronchioles, distribution of the pleural pressure, gravity, diaphragm movement effect and the effect of surface tension. Mathematical model – boundary value problem for this macroscopic task is complex, large deformations and material nonlinearities occur and the accomplishing conditions are very complicated. Accordingly, the finite element method can be employed as a tool herein. Model is of use when the elasticity is characterized as a function of deformation energy and when we can compute the energy needed to deformation of surface.

The first lung deformation analysis utilizing finite element method realized in 1972 pointed out to the importance of gravitation [4, 5]. Another study was carried out in 1978. It was a three-dimensional analysis of finite deformation of lung involving gravitational influences, as well [5, 6]. Also nonlinear elasticity was tackled, however the drawback of the model revealed due to an insufficiency – the surface tension was not involved in the model at all.

3.1. Variation formulation

We proceed in line with thermodynamics which says that each thermodynamical process tends always towards the steady state. Hence we take the stationary value of potential energy considering admissible kinematic deformations. For an elastic system we can express total potential energy Π_E as [7, 8],

$$\Pi_E = \int_{V_0} (W' - \bar{F}_i u_i)\, dV_0 - \int_{S_0} \bar{T}_i u_i \, dS \tag{3}$$

where W' is deformation energy per volume unit, $\bar{F}$ – prescribed volume forces, u_i displacements considered in Cartesian coordinates, $\bar{T}$ prescribed outer loads acting to the particular surfaces, V_0 initial volume, S_0 initial surface, where the load is prescribed.

Problems involving the surface tension is precised by including the additional member representing the energy needed for the new surface creation. If we denote the density of surface energy by Γ then we have the total energy Π quantification in the form

$$\Pi = \Pi_E + \int_{V_0} \Gamma \, dV_0 = \int_{V_0} (W - \bar{F}_i u_i) dV_0 - \int_{S_0} \bar{T}_i u_i \, dS \tag{4}$$

with

$$W = W' + \Gamma \tag{5}$$

where energy of surface tension Γ is a function depending on the surface tension and deformation invariants.

4. Two-dimensional finite element formulation

Let us consider an investigated domain Ω with its boundary S. Let us take the triangle shape finite elements, figure 5, which enable modelling the domains that have various boundaries. We have a finite element Ω_e with nodes in the vertices of triangle i, j, m.

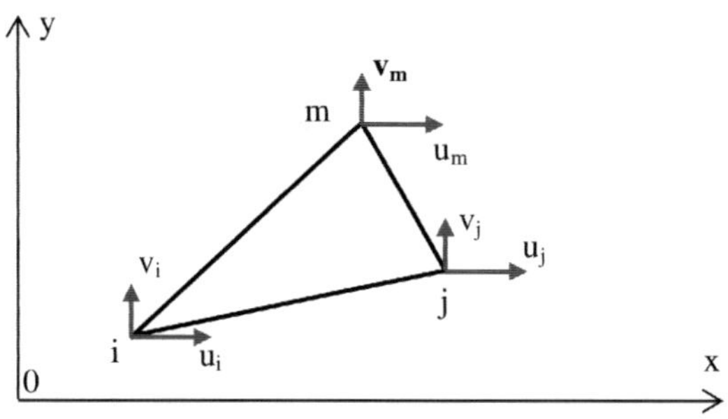

Figure 5. Triangle element.

Real deflection field of the element is approximated by linear polynomials.

$$\begin{aligned} u(x,y) &= \alpha_1 + \alpha_2 x + \alpha_3 y \\ v(x,y) &= \alpha_4 + \alpha_5 x + \alpha_6 y \end{aligned} \tag{6}$$

Six unknown constants α can be expressed by using the primary unknowns, i.e. displacements u, v in vertexes of triangle i, j, m

$$\boldsymbol{\delta}^e = \{\boldsymbol{\delta}_i, \boldsymbol{\delta}_j, \boldsymbol{\delta}_m\}^T = \{u_i, v_i, u_j, v_j, u_m, v_m\}^T \tag{7}$$

with

$$\begin{aligned} u_i &= u(x_i, y_i) = \alpha_1 + \alpha_2 x_i + \alpha_3 y_i \\ u_j &= u(x_j, y_j) = \alpha_1 + \alpha_2 x_j + \alpha_3 y_j \\ u_m &= u(x_m, y_m) = \alpha_1 + \alpha_2 x_m + \alpha_3 y_m \end{aligned} \tag{8}$$

Solving of equation (8) yields

$$\begin{aligned} \alpha_1 &= (a_i u_i + a_j u_j + a_m u_m)(2\Delta)^{-1} \\ \alpha_2 &= (b_i u_i + b_j u_j + b_m u_m)(2\Delta)^{-1} \\ \alpha_3 &= (c_i u_i + c_j u_j + c_m u_m)(2\Delta)^{-1} \end{aligned} \tag{9}$$

with

$$a_i = x_j y_m - x_m y_j, \quad b_i = y_j - y_m, \quad c_i = x_m - x_j. \tag{10}$$

Moreover, the remaining coefficients can be gained by cyclic permutation $i \to j \to m \to i$.

and 2Δ, equation (11), represents twice magnitude area of triangle *ijm*:

$$2\Delta = det\begin{vmatrix} 1 & x_i & y_i \\ 1 & x_j & y_j \\ 1 & x_m & y_m \end{vmatrix} \tag{11}$$

Substituting equation (9) into the first of displacement functions (6) yields

$$u(x,y) = \frac{1}{2\Delta}\left[(a_i + b_i x + c_i y)\, u_i + \left(a_j + b_j x + c_j y\right) u_j + (a_m + b_m x + c_m y)\, u_m\right] \tag{12}$$

Analogously we get

$$v(x,y) = \frac{1}{2\Delta}\left[(a_i + b_i x + c_i y)\, v_i + \left(a_j + b_j x + c_j y\right) v_j + (a_m + b_m x + c_m y)\, v_m\right] \tag{13}$$

Constants $a_i, b_i, c_i, a_j, \ldots c_m$ are known and they depend on vertices coordinates of triangle in sense of equation (10).

In expressing function $u(x,y)$, $v(x,y)$ we have switched from basis 1, *x*, *y* in equation (6) to the basis equation (14).

$$N_k(x,y) = \frac{1}{2\Delta}(a_k + b_k x + c_k y), \quad \text{where } k = i, j, m. \tag{14}$$

The essential property of these basis function is

$$N_k(x_i, y_i) = 1, \quad \text{if } k = i$$
$$N_k(x_i, y_i) = 0, \quad \text{if } k \neq i$$

These basis functions are also called shape functions. Relations (12) and (13) can be rewritten in matrix form.

$$\boldsymbol{u} = \mathbf{N}\, \boldsymbol{\delta}^e \tag{15}$$

We get the components of deformations $\varepsilon_x, \varepsilon_y, \gamma_{xy}$ in a point (x, y) ∈ Ω, if we substitute equation (12) and equation (13) to strain-displacement relations

$$\boldsymbol{\varepsilon} = \left\{\varepsilon_x, \varepsilon_y, \gamma_{xy}\right\}^T = \left\{\frac{\partial u}{\partial x}, \frac{\partial v}{\partial y}, \frac{\partial u}{\partial y} + \frac{\partial v}{\partial x}\right\}^T \tag{16}$$

which yields

$$\begin{Bmatrix} \varepsilon_x \\ \varepsilon_y \\ \gamma_{xy} \end{Bmatrix} = \frac{1}{2\Delta}\begin{bmatrix} b_i & 0 & b_j & 0 & b_m & 0 \\ 0 & c_i & 0 & c_j & 0 & c_m \\ c_i & b_i & c_j & b_j & c_m & b_m \end{bmatrix}\begin{Bmatrix} u_i \\ v_i \\ u_j \\ v_j \\ u_m \\ v_m \end{Bmatrix} \tag{17}$$

We can likewise rewrite equation (16) in the form

$$\boldsymbol{\varepsilon} = \mathbf{B}\, \boldsymbol{\delta}^e \tag{18}$$

Stress values on the element are then

$$\boldsymbol{\sigma} = \mathbf{D}\, \boldsymbol{\varepsilon} = \mathbf{D}\, \mathbf{B}\, \boldsymbol{\delta}^e \tag{19}$$

where **D** stands as the material matrix

$$\mathbf{D} = \frac{E}{1-\nu}\begin{bmatrix} 1 & \nu & 0 \\ \nu & 1 & 0 \\ 0 & 0 & (1-\nu)\, 2^{-1} \end{bmatrix} \tag{20}$$

Moreover, the nodal forces on the element are

$$\boldsymbol{F^e} = \{F_{ix}, F_{iy}, F_{jx}, F_{jy}, F_{mx}, F_{my}\}^T \tag{21}$$

We derive the stiffness matrix of the element from the virtual work of internal and external forces. Let us impose the virtual displacements to the nodes and collect them in one vector:

$$\overline{\boldsymbol{\delta}}^{\boldsymbol{e}} = \{\bar{u}_i, \bar{v}_i, \bar{u}_j, \bar{v}_j, \bar{u}_m, \bar{v}_m\}^T \tag{22}$$

to which virtual functions $\overline{\boldsymbol{u}}$ of displacements in finite element correspond

$$\overline{\boldsymbol{u}}(\boldsymbol{x}, \boldsymbol{y}) = \mathbf{N}(\mathbf{x}, \mathbf{y})\, \overline{\boldsymbol{\delta}}^{\boldsymbol{e}} \tag{23}$$

and virtual deformations

$$\overline{\boldsymbol{\varepsilon}} = \mathbf{B}\, \overline{\boldsymbol{\delta}}^{\boldsymbol{e}}. \tag{24}$$

Virtual work of nodal forces can be evaluated as follows

$$F_{ix}\,\bar{u}_i + F_{iy}\,\bar{v}_i + F_{jx}\,\bar{u}_j + F_{jy}\,\bar{v}_j + F_{mx}\,\bar{u}_m + F_{my}\,\bar{v}_m = \overline{\boldsymbol{\delta}}^{\boldsymbol{e}^{\boldsymbol{T}}} \boldsymbol{F}^{\,e} \tag{25}$$

Virtual work of stress and volume forces is

$$\overline{\boldsymbol{\varepsilon}}^{\boldsymbol{T}}\, \boldsymbol{\sigma} - \overline{\boldsymbol{u}}^{T} . \begin{Bmatrix} \underline{X} \\ \underline{Y} \end{Bmatrix} \tag{26}$$

or by using equations (24), (19) and equation (23) we get

$$\overline{\boldsymbol{\delta}}^{\boldsymbol{e}^{\boldsymbol{T}}} \mathbf{B}^{\boldsymbol{T}} \mathbf{D}\, \mathbf{B}\, \boldsymbol{\delta}^{e} - \overline{\boldsymbol{\delta}}^{\boldsymbol{e}^{\boldsymbol{T}}} \mathbf{N}^{\boldsymbol{T}}\, \underline{\boldsymbol{X}} \tag{27}$$

where $\underline{\boldsymbol{X}} = \{\underline{X}, \underline{Y}\}^T$.

Given boundary load $\underline{p}_x, \underline{p}_y$ per area unit contributes to the virtual work of external forces as follows

$$\overline{\boldsymbol{u}}^{\boldsymbol{T}} \underline{\mathbf{p}} = \overline{\boldsymbol{\delta}}^{\boldsymbol{e}^{\boldsymbol{T}}} \mathbf{N}^{T} \begin{Bmatrix} \underline{p}_x \\ \underline{p}_y \end{Bmatrix} \tag{28}$$

It follows from the condition of zero sum of virtual work of external forces and internal forces in the finite element

$$\overline{\boldsymbol{\delta}}^{\boldsymbol{e}^{\boldsymbol{T}}} \boldsymbol{F}^{e} = \overline{\boldsymbol{\delta}}^{\boldsymbol{e}^{\boldsymbol{T}}} \left(\int_{V_e} \mathbf{B}^{\boldsymbol{T}} \mathbf{D}\, \mathbf{B}\, \boldsymbol{\delta}^{e}\, dV_e - \int_{V_e} \mathbf{N}^{\boldsymbol{T}}\, \underline{\mathbf{X}}\, dV_e - \int_{S_e} \mathbf{N}^{\boldsymbol{T}}\, \underline{\mathbf{p}}\, dS_e \right) \tag{29}$$

From the equation (29) we get the required basic relation of FEM

$$\boldsymbol{F}^{e} = \mathbf{K}^{e}\, \boldsymbol{\delta}^{e} - \int_{V_e} \mathbf{N}^{\boldsymbol{T}}\, \underline{\mathbf{X}}\, dV_e - \int_{S_e} \mathbf{N}^{\boldsymbol{T}}\, \underline{\mathbf{p}}\, dS_e \tag{30}$$

where expression (31) represents the stiffness matrix of finite element Ω_e.

$$\mathbf{K}^{e} = \int_{V_e} \mathbf{B}^{\boldsymbol{T}} \mathbf{D}\, \mathbf{B}\, dV_e \tag{31}$$

Displacements $\boldsymbol{\delta}^{\boldsymbol{e}}$ of nodes for each element are subsets of set of all element's displacements, hence we can rewrite the expression (30) into the form

$$\mathbf{F} = \mathbf{K}\,\boldsymbol{\delta} - \int_{V} \mathbf{N}^{\boldsymbol{T}}\, \underline{\mathbf{X}}\, dV - \int_{S} \mathbf{N}^{\boldsymbol{T}}\, \underline{\mathbf{p}}\, dS \tag{32}$$

with **K** being the global stiffness matrix. For geometric nonlinearity involvement the FEM analysis of lung parenchyma is provided in [8].

5. Linear mechanical response of alveolus subjected to a static load

Computational implementation of the alveolar behaviour under a mechanical load was realized in multiphysical finite element-based software ANSYS.

As far as the material properties concern, the situation is much more complicated. In various publications various properties are provided of various mammals. However, for human lung there it is not easy to find relevant data. Of course, this is reasonable since the material tests on human tissue is subjected to very strict ethical criteria. In our investigation we took the dimensions of alveolus from [9], see figure 6. The geometry was developed directly by using the tools of workbench within the ANSYS.

In linear model we have used Young elastic modulus E = 318080 Pa for bronchiole, E = 101427.8 Pa for alveolus and Poisson number ν = 0.42 in both cases [10, 11]. In figure 7 there is the resulting deformation of alveolus connected to bronchiole under the load corresponding to the inspiration pressure. Because of the difference in elastic modulus, the deformation of alveolar tissue deformation is greater than the bronchiolar one.

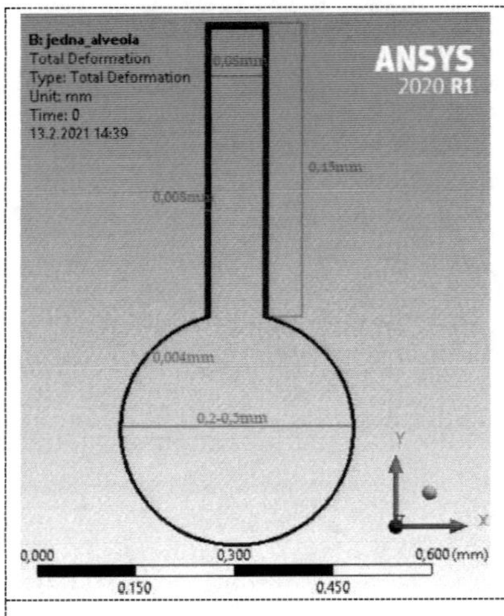

Figure 6. Simple alveolar geometry.

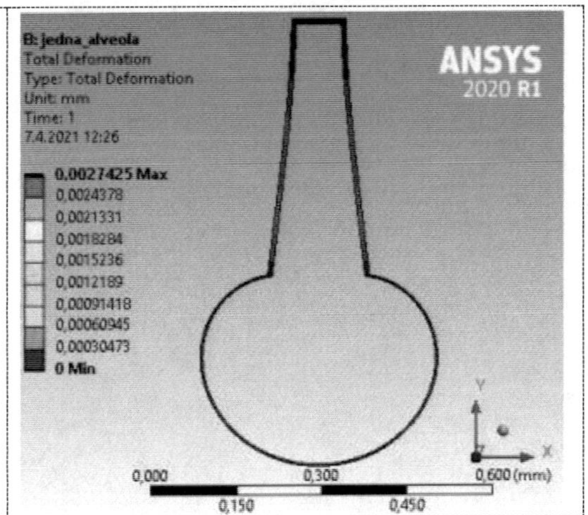

Figure 7. Alveolar deformation.

6. Discussion and conclusion

This paper deals with human respiratory system. The mechanical behaviour of alveoli is of an interest. One single alveolus – bronchiole configuration is studied and its response to the load caused by the pressure change during the inspirium phase is traced. We have dealt with 2D model easy performable graphically. Both geometric and physical linearity were assumed, being simplified in comparison to the reality, whereas indeed, the material properties of bronchioles and alveoli are related to the age and the prevalent physical activity of the person [5]. The mathematical model standing behind this process is arisen, then as the mathematical tool the finite element method is chosen and elaborated. Computational implementation is realized by using the finite element based software ANSYS. The change of the alveolar diameter during inspiration was of our interest among various results yielded from ANSYS. The original alveolar diameter of magnitude 0.36 mm in the state of relaxation increased to the value 0.41 mm evaluated at the peak of inspiration, i.e. 13 %.

Our future research will focus to the physical nonlinearity together with viscoelastic properties involvement. It is worth to creating alveolar sacs, as well and 3D analysis.

7. References

[1] West JB 1977 *Bioengineering Aspects of the Lung*. Lung Biology in Health and Disease. Marcel Dekker Inc, (San Diego)

[2] https://www.aic.cuhk.edu.hk/web8/Surfactant.htm

[3] Hrazdira I and Morstein V 2013 *Medical biophysics and instrumentation*. Textbook, Masaryk University (Brno)

[4] Matthews FL, West JB 1972 Finite Element Displacement Analysis of a Lung. *Journal of Biomechanics*, Vol **5**. (New York)

[5] Vawter DL and Shields WH 1982 Deformation of the Lung: The Role of Interfacial Forces. *Finite Elements in Biomechanics,* John Wiley & Sons (New York)

[6] Liu JT and Lee GC 1978 Static finite deformation analysis of the lung. *Journal of Engineering Mechanics Division*, Vol **104**(1), American Society of Civil Engineers, (Bufallo, New York)

[7] Pian THH and Tong P 1971 Variational formulation of finite displacement analysis. *High Speed Computing of Elastic Structures*, University of Liege, Belgium (Liege)

[8] Karaplan AD, Bienek MP and Skalak R 1982 Finite Element Analysis of Lung Parenchyma. *Finite Elements in Biomechanic,* John Wiley & Sons Interscience publication (Tuscon, Arisona)

[9] Kim J Heise RL, Reynolds AM and Pidaparti RM 2017 Quantification of Age-Related Lung Tissue Mechanics under Mechanical Ventilation. *Medical Sciences* Vol **5**(4)

[10] Polio SR, Kundu AN, Dougan CE, Birch NP, Aurian-Blajeni DE, Schiffman JD, et al. 2018 *Cross-platform mechanical characterization of lung tissue*. PLoS ONE Vol **13**(10)

[11] Weibel ER 1963 Geometry and dimensions of airways of conductive and transitory zones. *Morphometry of the Human Lung*, Springer (Berlin, Heidelberg)

Acknowledgments
This work was supported by grants APVV-17-0066, APVV-18-0052, VEGA 1/0522/20, VEGA 1/0412/18 and VEGA 1/0006/19.

Early-age strains and stresses in bi-axially restrained expansive concrete elements

V V Tur[1] and V H Sannikava[2]

[1] Department of Concrete Technology and Building Materials, Brest State Technical University, Moskovskaya 267, Brest 224017, Belarus
[2] Department of Geotechnics and Transport Communications, Brest State Technical University, Moskovskaya 267, Brest 224017, Belarus

E-mail: profturvic@gmail.com; volhasann@gmail.com

Abstract. Practical application of expansive concretes in reinforced concrete structures needs the rational design model for assessment the early-age stress-strain parameters of self-stressing concrete elements under any restraint conditions during the expansion stage. Various strategies are used to calculate the restrained expansion strains and self-stresses but, in most cases, disregarding the elastic-plastic behaviour of expansive concrete at early ages. It's noticeable that in case of bi-axial restraint conditions both the elastic-plastic parameters of expansive concrete and the influence of restraint in orthogonal directions should be taken into consideration. To achieve a reliable solution a design model for estimation the restrained expansion strains and self-stresses in self-stressed structural elements under bi-axial restraint conditions (2D MSDM) at expansion stage has been developed. The proposed design model is based on the MSDM for the case of uniaxial restraint arrangement. The validation of 2D MSDM was confirmed by the experimental results of expansive concrete elements under 2D confinement carried out by authors as well as obtained by other researchers.

1. Introduction

It is a well-established fact that initial cracking due to shrinkage deformations reduces considerably the serviceability and durability of concrete structures. The most efficient method to prevent this negative phenomenon is prestressing of reinforcement in structure. Conventional methods of prestressing induce the tensile strains in reinforcing elements mainly in longitudinal direction whereas in such structural members as slabs or deep beams the prestressing in two directions meets many technological difficulties.

Development of the self-stressing concrete technology has led to the opportunity not only to eliminate or reduce the drying cracks in concrete structures but also to induce the compressive strains in concrete of a sufficient magnitude for prestressing. The ability to prestress the reinforcing bars in any direction makes the expansive concrete very useful to create bi-axial and tri-axial compressive stresses in concrete.

The most rational field of application of self-stressing concrete involves members with bi-axial or tri-axial confinement [1]. It should be noted that many strategies were developed to assess the strains and stresses in elements where the concrete increase its volume in the presence of bi-axial restraint conditions. According to the basic concept all well-known design approaches for prediction the early stresses and strains in self-stressed members expanded in case of 2D confinements can be divided into the groups: calculation methods founded on the law of conservation of chemical energy [1–5]; empirical calculation methods founded on the exponential function [6, 7]; semi-empirical calculation methods [8] and in paper [9] fuzzy inference system method is presented. Unfortunately, most of the above methods contains the disadvantages that restrict the applicability of these approaches with respect to the wide range of expansive concrete elements. For example, methods [1-8] include coefficients based on the limited quantity of test results therefore erroneous outcomes originate in respect of elements with other boundary conditions. Methods [1–5] founded on the law of conservation of chemical energy consider the design value of concrete self-stress as a basic parameter for calculation whereas a various ratios can occur between the values of free expansion strains and self-stressing degree. Finally, with the exception of the approach in [8], all methods neglect the early-age features of self-stressing concrete behaviour and allow determining the stress-strain parameters towards the end of concrete expansion.

In that way, for the reliable design of bi-axially restrained expansive concrete elements the rational model should consider the elastic-plastic behaviour of material during the expansion period regardless of the geometry of structural element and energy capacity of expansive concrete.

2. Design model used for the calculation of strains and stresses in self-stressed concrete members with bi-axial confinement

2.1. Basic equations applied in case of uniaxial arrangement of restraint

For self-stressed concrete members with uniaxial arrangement of reinforcement the solution was proposed in the researches [10]. The distinctive feature of approach in [10] consists in considering the development of elastic and plastic strains at early ages of expansive concrete as well as acceptance the free expansion strains as a basic parameter. With respect to the above method, the restrained expansion strains are calculated by following an iterative procedure:

$$\varepsilon_{s.x}(t_{i+\frac{1}{2}}) = \sum_{j=1}^{i}\left[(\Delta\sigma_{c.x})_j \cdot J(t_{i+\frac{1}{2}}, t_j)\right] + \varepsilon_{cf}(t_{i+\frac{1}{2}}, t_{\frac{1}{2}}) \tag{1}$$

where $(\Delta\sigma_{c.x})_j$ – is the increment of self-stress in *X*-direction at *j*-th span of time; $\varepsilon_{cf}(t_{i+\frac{1}{2}}, t_{\frac{1}{2}})$ – is the free strain as a result of temperature change and shrinkage or the expansion of concrete in *X*-direction from the period $t_{\frac{1}{2}}$ to the $t_{i+\frac{1}{2}}$; $J(t_{i+\frac{1}{2}}, t_j)$ – function of creep.

It is noticeable that validation of the model [10] was carried out using expansive concrete samples with low energy capacity so a good correlation between the experimental and calculated results occurred. However, experimental investigations presented in the paper [11] have shown that model [10] overestimates the experimental results in case of high energy capacity of expansive concrete especially in respect of elements with high level of restraint. According to the approach [11], this disagreement arises because the additional cumulative force of restraint doesn't taken into account. Such force has the elastic origin and performs the additional restriction to the expansion process in concrete. On that basis, in researches [11] the modified early age strains development model MSDM that is applied in the case of uniaxial arrangement of reinforcement have been developed. In accordance with MSDM the restrained strain $\varepsilon_s(t_{i+\frac{1}{2}})$ at any *i*-th span of time are described by:

$$\varepsilon_s(t_{i+\frac{1}{2}}) = \varepsilon_{cf}(t_{i+\frac{1}{2}}) + \varepsilon_{el}(t_{i+\frac{1}{2}}) + \varepsilon_{pl}(t_{i+\frac{1}{2}}, t_0) + \varepsilon_{cum}(t_{(i-1)+\frac{1}{2}}) \tag{2}$$

In formula (2) $\varepsilon_{cf}(t_{i+\frac{1}{2}})$ is the free expansion strain, $\varepsilon_{el}(t_{i+\frac{1}{2}})$ is the elastic strain at *i*-th span of time, $\varepsilon_{pl}(t_{i+\frac{1}{2}}, t_0)$ is the creep strain at *i*-th span of time in case of constant self-stress applied at t_0

days, $\varepsilon_{cum}(t_{(i-1)+\frac{1}{2}})$ is the additional strain because of the presence of the restrictive force from the restraint at (i-1)-th span of time.

2.2. Design model applied in case of bi-axial arrangement of restraint

Based on the model MSDM [11] for elements with uniaxial restraint arrangement, the new model for calculation of the restrained expansion strains and stresses in self-stressed concrete members with bi-axial arrangement of restraint 2D MSDM has been worked out.

With respect to expansive concrete members where the reinforcing bars are placed in two orthogonal directions, the important remark should be pointed out. When the material is assumed to be elastic, the strains in one direction have influence on strains in another direction. Such interference is considered using the Poison's ratio. Inasmuch as expansive concrete performs the elastic-plastic behaviour and, according to the formula (2), only part of total strain has the elastic origin, so the Poison's ratio should be applied only in respect of the elastic strain. Taking into consideration the results of theoretical and experimental investigations in papers [12–15], in model 2D MSDM the Poison's ratio is accepted as a constant parameter with the value of $\nu = 0.2$.

General equations of proposed design model 2D MSDM were developed on the basis of following assumptions:

a) the presence of restraint in two perpendicular directions is taken into account applying the Poisson's ratio of self-stressing concrete $\nu = 0.2$ in respect of the elastic strains at i-th time interval in X-direction ($\varepsilon_{el.x}(t_{i+\frac{1}{2}})$) and Y-direction ($\varepsilon_{el.y}(t_{i+\frac{1}{2}})$);

b) the restrained expansion strain at i-th span of time in any orthogonal direction is perform the sum (in algebraic form) of free expansion strain ($\varepsilon_{cf}(t_{i+\frac{1}{2}})$), elastic strain at i-th span of time ($\varepsilon_{el.x}(t_{i+\frac{1}{2}})$ in X-direction and $\varepsilon_{el.y}(t_{i+\frac{1}{2}})$ in Y-direction), creep strain at i-th span of time ($\varepsilon_{pl.x}(t_{i+\frac{1}{2}},t_0)$ in X-direction and $\varepsilon_{pl.y}(t_{i+\frac{1}{2}},t_0)$ in Y-direction), additional strain because of the presence the restrictive force from the restraint at (i-1)-th span of time ($\varepsilon_{cum.x}(t_{(i-1)+\frac{1}{2}})$ in X-direction and in Y-direction):

$$\begin{cases} \varepsilon_{s.x}(t_{i+\frac{1}{2}}) = \varepsilon_{cf}(t_{i+\frac{1}{2}}) + \left[\varepsilon_{el.x}(t_{i+\frac{1}{2}}) - \nu\cdot\varepsilon_{el.y}(t_{i+\frac{1}{2}})\right] + \varepsilon_{pl.x}(t_{i+\frac{1}{2}},t_0) + \varepsilon_{cum.x}(t_{(i-1)+\frac{1}{2}}) \\ \varepsilon_{s.y}(t_{i+\frac{1}{2}}) = \varepsilon_{cf}(t_{i+\frac{1}{2}}) + \left[\varepsilon_{el.y}(t_{i+\frac{1}{2}}) - \nu\cdot\varepsilon_{el.x}(t_{i+\frac{1}{2}})\right] + \varepsilon_{pl.y}(t_{i+\frac{1}{2}},t_0) + \varepsilon_{cum.y}(t_{(i-1)+\frac{1}{2}}) \end{cases} \quad (3)$$

c) between the resulting forces in reinforcement and expansive concrete section the equilibrium condition takes place as a result of bond conditions in orthogonal directions at any i-th span of time.

Thus, the restrained expansion strains in orthogonal directions are determined by the formula:

$$\begin{cases} (\Delta\varepsilon_{s.x})_i = (\Delta\varepsilon_{cf})_i - \left[(\Delta\sigma_{c.x})_i - \nu\cdot(\Delta\sigma_{c.y})_i\right]\cdot J(t_{i+\frac{1}{2}},t_i) - \sum_{j=1}^{i-1}\left[(\Delta\sigma_{c.x})_j\cdot\frac{\Delta\varphi(t_i,t_j)}{E_{c.28}}\right] - \frac{\sum_{j=1}^{i-1}(\Delta\sigma_{c.x})_j}{E_c(t_{(i-1)+\frac{1}{2}})} \\ (\Delta\varepsilon_{s.y})_i = (\Delta\varepsilon_{cf})_i - \left[(\Delta\sigma_{c.y})_i - \nu\cdot(\Delta\sigma_{c.x})_i\right]\cdot J(t_{i+\frac{1}{2}},t_i) - \sum_{j=1}^{i-1}\left[(\Delta\sigma_{c.y})_j\cdot\frac{\Delta\varphi(t_i,t_j)}{E_{c.28}}\right] - \frac{\sum_{j=1}^{i-1}(\Delta\sigma_{c.y})_j}{E_c(t_{(i-1)+\frac{1}{2}})} \end{cases} \quad (4)$$

In formula (4) $(\Delta\sigma_{c.x})_i$ and $(\Delta\sigma_{c.y})_i$ describe the increment of self-stress at the *i*-th span of time in *X*-direction and *Y*-direction correspondingly; $(\Delta\sigma_{c.x})_j$ and $(\Delta\sigma_{c.y})_j$ describe the increment of self-stress at the *j*-th span of time in *X*-direction and *Y*-direction correspondingly; $(\Delta\varepsilon_{cf})_i$ represents the increment of the free strain of self-stressing concrete at the *i*-th span of time; $E_{c.28}$ represents the Young's modulus of self-stressing concrete in temperature adjusted concrete age of 28 days; $E_c(t_{(i-1)+\frac{1}{2}})$ is the Young's modulus of expansive concrete at $t_{(i-1)+\frac{1}{2}}$ days in temperature adjusted concrete age; $J(t_{i+\frac{1}{2}},t_i)$ describes the creep function; $\Delta\varphi(t_i,t_j)$ represents the increment of creep coefficient of self-stressing concrete at the *i*-th span of time in case of constant self-stress applied at the *j*-th span of time.

The creep function $J(t_{i+\frac{1}{2}},t_i)$ should be determined by the formula in [10]:

$$J(t_{i+\frac{1}{2}},t_j)=\frac{1}{E_c(t_j)}+\frac{\varphi(t_{i+\frac{1}{2}},t_j)}{E_{c.28}} \tag{5}$$

where $E_c(t_j)$ performs the Young's modulus of self-stressing concrete in temperature adjusted concrete age of t_j days; $\varphi(t_{i+\frac{1}{2}},t_j)$ performs the creep coefficient of self-stressing concrete at the age of $t_{i+\frac{1}{2}}$ days because of self-stress applied at the age of t_j days.

Creep coefficient $\varphi(t,t_0)$ of self-stressing concrete at early ages depends on the notional creep coefficient φ_0 and coefficient $\beta_c(t,t_0)$ and calculated by the formulas in [10]:

$$\varphi(t,t_0)=\varphi_0\cdot\beta_c(t,t_0) \tag{6}$$

$$\varphi_0=5.31\cdot\left(\frac{E_c(t_0)}{E_{c.28}}-1\right)^2+1.11 \tag{7}$$

$$\beta_c(t,t_0)=\left[\frac{t-t_0}{\beta_н+(t-t_0)}\right]^{0.3} \tag{8}$$

In expressions (7) and (8) $\frac{E_c(t_0)}{E_{c.28}}$ is the relative Young's modulus; $\beta_c(t,t_0)$ is the coefficient that performs the creep at the temperature adjusted concrete age t after the temperature adjusted concrete age of load action t_0; $\beta_н$ is the coefficient that describes the effect of loading age on the development of creep:

$$\begin{cases}\beta_н=0.000001,\ \mathit{if}\ 0\le\frac{E_c(t)}{E_{c.28}}<0.346\\ \beta_н=40.5\cdot\left(\frac{E_c(t)}{E_{c.28}}-0.346\right)+0.485,\ \mathit{if}\ 0.346\le\frac{E_c(t)}{E_{c.28}}<1.0\end{cases} \tag{9}$$

The increment of creep coefficient can be calculated as follows:

$$\Delta\varphi(t_i,t_j)=\varphi(t_{i+\frac{1}{2}},t_j)-\varphi(t_{(i-1)+\frac{1}{2}},t_j) \tag{10}$$

The Young's modulus of self-stressing concrete at early age is calculated by the formula in [10]:

$$E_c(t) = E_{c.28} \cdot \exp\left(s\left(1 - \left(\frac{t_{e.28} - a}{t - a} \right)^{0.5} \right) \right) \quad (11)$$

where s and a are the empirical coefficients; t performs the temperature adjusted concrete age:

$$t = \sum_{i=1}^{n} \Delta t_i \cdot \exp\left[13.65 - \frac{4000}{273 + \frac{T(\Delta t_i)}{T_0}} \right] \quad (12)$$

where Δt_i is the number of days with a temperature $T(°C)$; $T(\Delta t_i)$ describes the temperature during the period of time Δt_i (in $°C$); $T_0 = 1°C$.

The increments of self-stress at the i-th span of time in orthogonal directions X and Y are calculated by:

$$\begin{cases} \left(\Delta\sigma_{c.x}\right)_i = \left(\Delta\varepsilon_{s.x}\right)_i \cdot E_{s.x} \cdot \rho_{s.x} \\ \left(\Delta\sigma_{c.y}\right)_i = \left(\Delta\varepsilon_{s.y}\right)_i \cdot E_{s.y} \cdot \rho_{s.y} \end{cases} \quad (13)$$

In expression (13) $E_{s.x}$ and $E_{s.y}$ are the Young's modulus of confinement in X-direction and Y-direction accordingly; $\rho_{s.x}$ and $\rho_{s.y}$ are the reinforcement ratios in X-direction and Y-direction accordingly.

Values of the restrained expansion strains in X-direction and Y-direction at arbitrary span of time are calculated according to the formulas:

$$\begin{cases} \left(\varepsilon_{s.x}\right)_i = \left(\varepsilon_{s.x}\right)_{i-1} + \left(\Delta\varepsilon_{s.x}\right)_i \\ \left(\varepsilon_{s.y}\right)_i = \left(\varepsilon_{s.y}\right)_{i-1} + \left(\Delta\varepsilon_{s.y}\right)_i \end{cases} \quad (14)$$

In turn, the corresponding values of self-stresses in X-direction and Y-direction at arbitrary span of time are defined by:

$$\begin{cases} \left(\sigma_{c.x}\right)_i = \left(\sigma_{c.x}\right)_{i-1} + \left(\Delta\sigma_{c.x}\right)_i \\ \left(\sigma_{c.y}\right)_i = \left(\sigma_{c.y}\right)_{i-1} + \left(\Delta\sigma_{c.y}\right)_i \end{cases} \quad (15)$$

Duration of iterative procedures depends on the law of free expansion strains development.

3. Verification of the model 2D MSDM

For the verification of the proposed model 2D MSDM the three series of the plane self-stressing concrete samples with the orthogonal of reinforcing bars located in the mid-depth of the concrete cross-section were prepared and investigated in laboratory conditions. Each series includes two identical specimens. The reinforcement ratio was accepted as a variable parameter of specimens. Characteristics of reinforcing bars (reinforcement area in X-direction ($A_{s.x}$) and in Y-direction ($A_{s.y}$), reinforcement ratio in X-direction ($\rho_{s.x}$) and in Y-direction ($\rho_{s.y}$)), expansive concrete (free strain at the concrete expansion stabilization (ε_{cf}), Young's modulus at 28 days age ($E_{c.28}$)) and geometry of plane test specimens are shown in the table 1.

The details of reinforcing bars arrangement and measurement device as well as view of prepared concrete sample with fixed gages are presented in the figure 1. In the figure 1b G1 and G2 represent the fixed dial gages, P1-P4 represent the brass pins for mobile dial gages.

Table 1. Characteristics of materials and geometry of plane specimens.

Series	Specimen geometry, mm, bxlxh	Reinforcing bars				Self-stressing concrete	
		X-direction		Y-direction			
		$A_{s.x}$, mm^2	$\rho_{s.x}$, %	$A_{s.y}$, mm^2	$\rho_{s.y}$, %	ε_{cf}, %	$E_{c.28}$, GPa
I	500x500x70	339	0.97	58.9	0.16	0.117	42.66
II		339	0.97	339	0.97	0.437	23.1
III		58.9	0.16	58.9	0.16	0.334	33.3

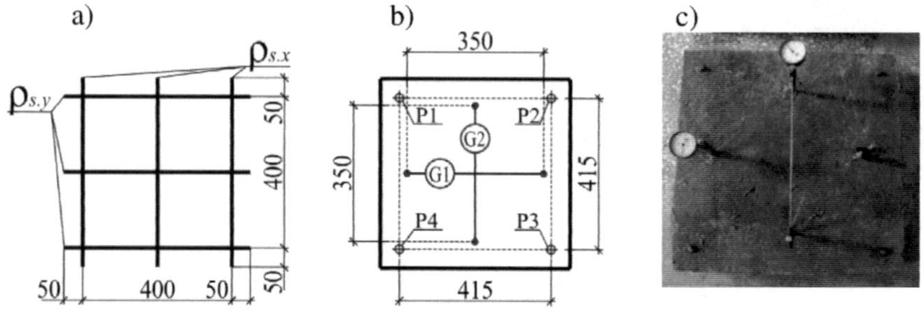

Figure 1. Details of reinforcement (a), gages arrangement (b) and view of prepared sample (c).

Figure 2 demonstrates the development of the experimental and calculated in accordance with the model 2D MSDM values of the restrained expansion strains during the expansion stage. Counting for the Poison's ratio, values of the restrained expansion strains calculated according to the model [10] are also presented in the same figure to show the influence of restrictive force at different boundary conditions.

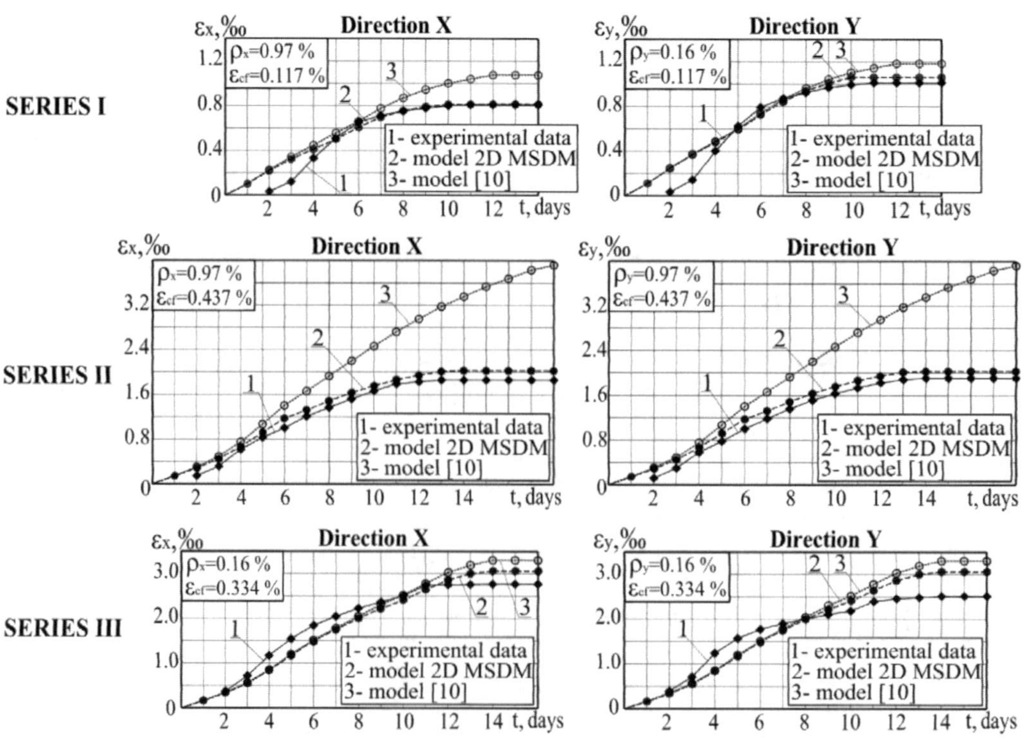

Figure 2. Restrained expansion strains of plane specimens in series I, II, III.

Verification of the model 2D MSDM has also been carried out analysing the results of experimental investigation performed in the papers [4, 5]. Characteristics of materials with respect to the test samples in researches [4, 5] are shown in the table 2. Geometry of test samples in researches [4] is $bxlxh$=100x100x100mm. Geometry of test samples in researches [5] is $bxlxh$=450x450x30mm. In the table 2 concerning the researches [5] test specimens from 1P-1 to 1P-6 perform members with high energy capacity of expansive concrete whereas specimens from 2P-1 to 2P-8 perform members with low energy capacity of expansive material.

Table 2. Geometry and characteristics of materials of specimens in [4] and [5].

Source	Specimen marking	Reinforcing bars				Self-stressing concrete	
		X-direction		Y-direction			
		$\rho_{s.x}$, %	$E_{s.x}$, GPa	$\rho_{s.y}$, %	$E_{s.y}$, GPa	ε_{cf}, %	$E_{c.28}$, GPa
[4]	X8Y8	1.020	2.050	206	206	0.02	26.3
	X12Y8	2.310	2.050	206	206	0.02	26.3
	X8Y12	1.020	4.740	206	206	0.02	26.3
	X12Y12	2.310	4.740	206	206	0.02	26.3
[5]	1P-1	0.105	203	0.526	203	0.75	29
	1P -2	0.263	203	0.526	203	0.75	29
	1P-3	0.373	193.3	0.526	203	0.75	29
	1P-4	0.284	203	0.933	193.3	1.34	29
	1P-5	0.726	189.7	0.933	193.3	1.34	29
	1P-6	1.016	189.7	2.033	189.7	1.57	29
	2P-1	0.560	193.3	0.560	193.3	0.08	29
	2P-2	1.120	193.3	0.560	193.3	0.08	29
	2P-3	1.493	193.3	0.560	193.3	0.08	29
	2P-4	2.240	193.3	0.560	193.3	0.08	29
	2P-5	2.981	197.6	0.560	193.3	0.08	29
	2P-6	0.560	193.3	1.120	193.3	0.20	29
	2P-7	2.240	193.3	1.120	193.3	0.20	29
	2P-8	1.120	193.3	1.493	193.3	0.23	29

For all test specimens from the table 2 the restrained strains and self-stresses were calculated in accordance with the proposed model 2D MSDM from the one day to the stabilization. Theoretical and experimental results of the restrained strains for some specimens are shown in the figures 3 and 4. Like figure 2, figures 3 and 4 also contain the calculated values of the restrained strains based on the formulas in model [10].

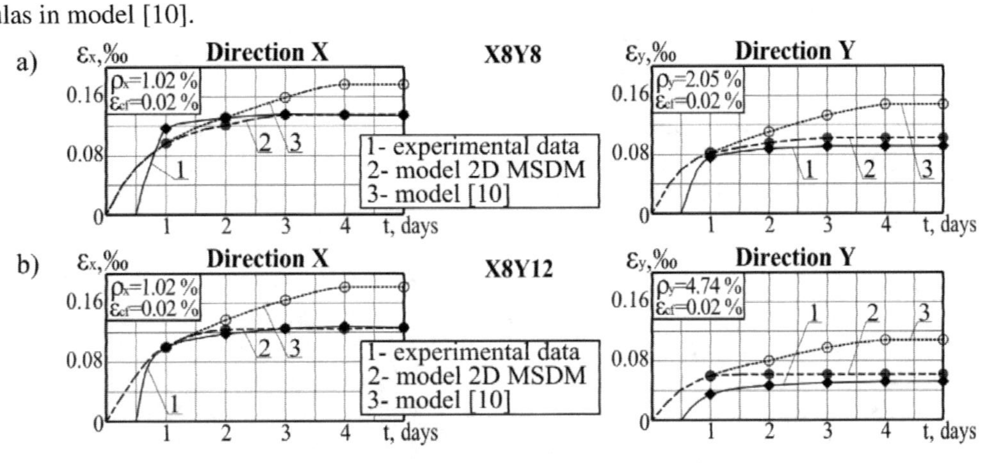

Figure 3. Restrained expansion strains for test specimens X8Y8 (a) and X8Y12 (b) from [4].

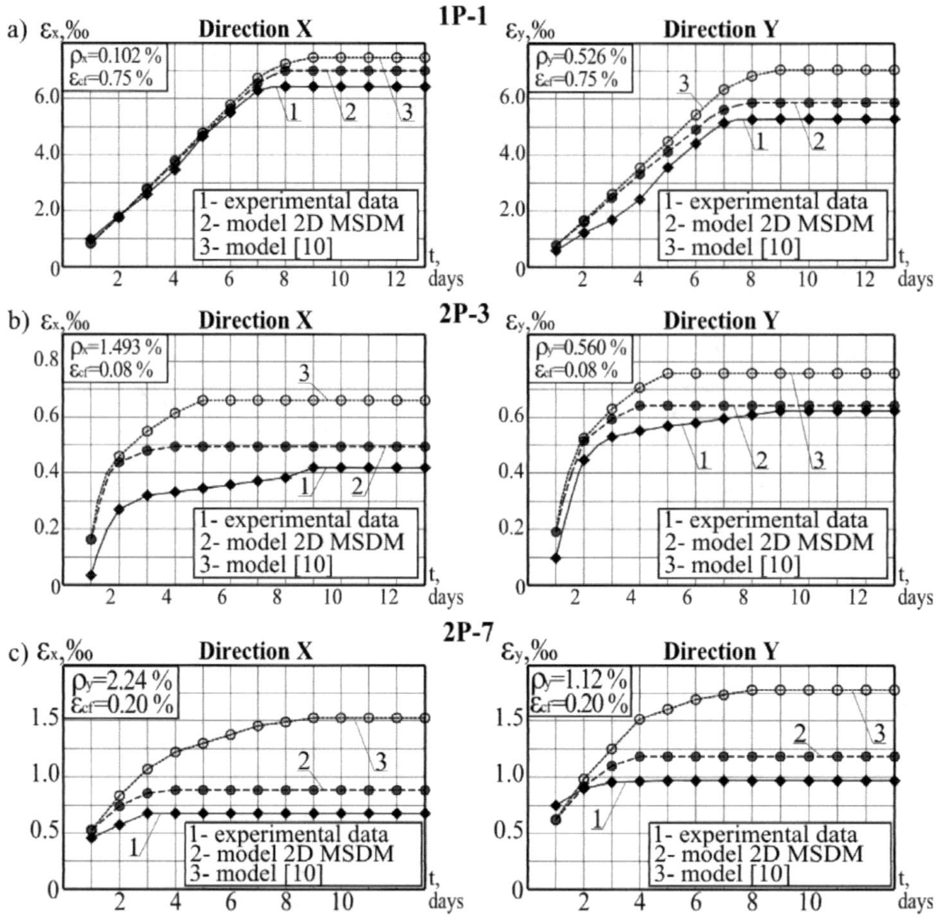

Figure 4. Restrained expansion strains for test specimens 1P-1 (a), 2P-3 (b) and 2P-7 (c) from [5].

It is worth to demonstrate in the figure 5 the comparison between the experimental and calculated in accordance with the model 2D MSDM and model [10] values of the restrained strains by the end of the concrete expansion period for the test specimens described in the table 2 as well as for test specimens from the authors' researches. According to the requirements in [16] the coefficient of variation of the error term was calculated to assess the consistency of model 2D MSDM in general. Such a parameter was also determined in case of the model [10] to compare the divergence between experimental and theoretical results received using two models.

As it could be followed from the figure 5, a good agreement observed with respect to the model 2D MSDM, the coefficient of variation of the error term equals V_δ=14.7 %. In respect of the model [10], that doesn't take into account the additional cumulative force, the coefficient of variation of the error term equals V_δ=24 %.

4. Conclusion

A design model for assessment the restrained strains and stresses in case of bi-axial arrangement of confinement (2D MSDM) is founded on the modified early age strains development model MSDM for the case of uniaxial arrangement of restraint. The model 2D MSDM allows determining the restrained strains and corresponding self-stresses from the one day to the expansive concrete stabilization with taking into account the elastic-plastic development of strains in self-stressing concrete elements and additional restrictive force caused by the confinement.

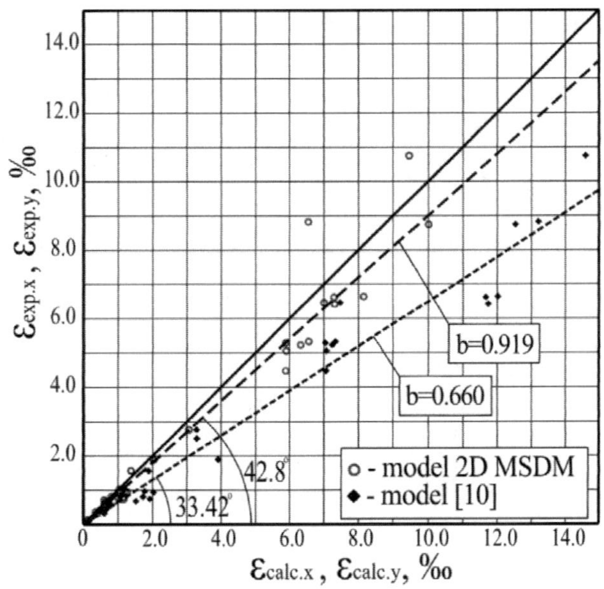

Figure 5. Comparison of experimental and calculated results for all test specimens.

The consistency of developed model 2D MSDM was confirmed based on the results of experimental investigations of bi-axially restrained expansive concrete elements carried out by authors and other researchers. In respect of the model 2D MSDM the coefficient of variation of the error term equals V_δ=14,7 %.

5. References

[1] Mikhailov V and Litver S 1974 *Expansive and self-stressing cements and self-stressed reinforced structures* (Moscow: Stroyizdat) p 312

[2] Mikhailov V and Gershvald V 1984 Plane self-stressed structures *Researches and usage of self-stressing concrete and self-stressed structures* (Moscow: NIIZHB) pp 62–67

[3] TCP 45-5.03-158-2009 2010 *Concrete and reinforced concrete structures from self-stressing concrete: design rules* (Minsk) p 28

[4] Ishikava Y, Shibata K and Tanabe T 2009 *Creep, Shrinkage and Durability Mechanics of Concrete and Concrete Structures* (London: Taylor & Francis Group) pp 437–443

[5] Marchuk V 2002 *Strains and stresses of plane self-stressing concrete elements under nonsymmetrical biaxial restraint at the expansion stage: PhD thesis* (Brest: BSTU) p 176

[6] Litver S and Petukhov A 1984 Stresses under bi-axial restraint conditions *Researches and usage of self-stressing concrete and self-stressed structures* (Moscow: NIIZHB) pp 67–78

[7] Man T, Wang B, Jin H and Zhang X 2016 Expansion behavior of self-stressing concrete confined by glass-fiber composite meshes *Constr. Build. Mater.* **128** pp 38–46

[8] Okada K, Ohta M, Nagafuchi T, Yata A and Tamai S 1983 Characteristics of expansive concrete under bi-axial restraint *J. Soc. Mater. Sci.* **32** pp 182–187

[9] Wang V, Han T and Jin H 2015 Prediction of expansion behaviour of self-stressing concrete by artificial neural networks and fuzzy inference system *Constr. Build. Mater.* **84** pp 184–191

[10] Ito H, Maruyama I, Tanimura M and Sato R 2004 Early age deformation and resultant induced stress in expansive high strength concrete *J. Adv. Concr. Technol.* **2** pp 155–174

[11] Semianiuk V, Tur V, Herrador M F and Paredes G M 2017 Early age strain and self-stresses of expansive concrete members under uniaxial restraint conditions *Constr. Build. Mater.* **131** pp 39–49

[12] Truman K Z, Petrushka D J and Norman C D 1991 Creep, shrinkage and thermal effects on mass concrete structures *J. Eng. Mech.* **117** pp 1274–88
[13] Oluokun F A, Burdette E G and Deatherage J H 1991 Elastic moduli, Poisson's ratio and compressive strength relationships at early ages *ACI Mater. J.* pp 3–10
[14] De Schutter G and Taerwe L 1996 Degree of hydration-based description of mechanical properties of early age concrete *Mater. Struct.* **29** pp 335–344
[15] Byfors J 1980 *Plane concrete at early ages* (Stockholm) p 464
[16] TCP EN 1990-2011 (02250) 2012 *Eurocode. Basis of structural design* (Minsk) p 70

IOP Conf. Series: Materials Science and Engineering 1252 (2022) 012029 doi:10.1088/1757-899X/1252/1/012029

Design and Evaluation of Building Structures Planned for Harsh Conditions

D Katunský[1], E Dolníková[1], M Lopušniak[1], M Vertaľ[1], T Baroš[1], K Hutkai[1], P Jaroš[1], L Kabošová[2], J Katunská[1], V Karľa[2] and M Zozulák[1]

[1] Institute of Architectural Engineering, Faculty of Civil Engineering, Technical University of Košice, Slovakia
[2] Center for Research and Innovation, Faculty of Civil Engineering, Technical University of Košice, Slovakia

E-mail: dusan.katunsky@tuke.sk

Abstract. The paper describes a scientific project that is solved at the authors' workplace. The scope of the project is the analysis and experimental research of fragments and details of building envelope structures designed for demanding conditions. It will be a question of determining the physical properties (thermal, humidity, energy, aerodynamic, light and other) of building structures and the internal environment of buildings, for demanding conditions. In this case, it is the conditions in the design process - the assumption of environmental conditions (climatic conditions of the exterior, conditions of the interior, such as humidity), or the effect of other conditions in the implementation and use of the building. Measurements will be performed in buildings (in situ) and in experimental conditions in the laboratory. After verifying the boundary conditions, the tools leading to virtual reality will be used in the compilation of experimental models with the help of simulation programs for the effects of individual influences. The task of the project is to analyze these influences and solve them in mutual cooperation, i.e. in the interconnection and action of individual influences (heat, light, energy and others).

1. Introduction

Thermal and visual comfort are among the main parameters of the overall comfort of users and the quality of the indoor environment in buildings. The design at an early stage and the selection of the parameters of the building's envelope structures require the necessary consideration of all these aspects of comfort in order to improve the performance, productivity and well-being of the occupants. Several methodologies have been proposed to optimize the thermal and visual state of building users when considering building envelopes, both individually and simultaneously.

The lack of a suitable methodology for the current optimization of thermal and visual properties of the building envelope with the influence of several factors (sun, frost, wind ...) led to a deterioration of the environment inside the green buildings. The aim of this effort is to design a methodology for the current optimization of the thermal and visual behavior of the building envelope in order to minimize the consumption of operational energy and increase the thermal and visual comfort of users.

2. Scientific objectives of the project

The project focuses on meeting three basic goals:
(i) defining the basic characteristics of building structures exposed to harsh environmental conditions
(ii) determining the impact of the basic characteristics of building envelope on their internal environment

(iii) optimization of physical properties of building structures in the interaction of several factors (heat, light, wind)

To achieve the goals, it will be necessary to perform:

1. in situ measurements - measurement of selected physical parameters of building structures and the internal environment of buildings directly in the real environment.
2. experimental measurements - monitoring of selected parameters (physical quantities) of fragments, details of perimeter cladding and the internal environment of buildings in laboratory conditions.
3. dynamic simulations, calculations of the real behavior of buildings and conditions of the internal environment of buildings using simulation tools of building materials, which are exposed to extreme conditions in the building envelope.

The areas of research are different, depending on the professional orientation of the researchers. They can be characterized as follows:

3. Materials and methods

Inputs: Experience gained on the basis of long-term measurements of material parameters and physical characteristics of fragments and construction details in real conditions of exploitation of the construction work - in situ, other findings of the applicant from research in the subject area (projects solved so far).

Outputs: Physical properties of building structures under the interaction of several factors (heat, light, wind) and their impact on the indoor environment of buildings under the influence of demanding conditions.

Methods: Theoretical analysis, synthesis, confrontation, simulations, experiment. In experimental devices, the implementation of measurements on experimental models using monitoring elements in cooperation according to device for monitoring changes in the conditions of harsh outdoor climate: (a / b / c /, d /).

a) mathematical modeling, boundary conditions,
b) models of structural shapes of buildings details and fragments,
c) experimental measurements, data analysis,
d) generalizing conclusions.

4. Researchers, experiences

The research team is composed of experienced experts, associate professors, assistant professors. The team also includes young researchers, doctoral students who, in cooperation with solving project tasks, work on their dissertations. The research team with the designation of the position of each researcher in the team and its specialization is in Table 1. Figure 1 shows the distribution of researchers and staff representation in the project.

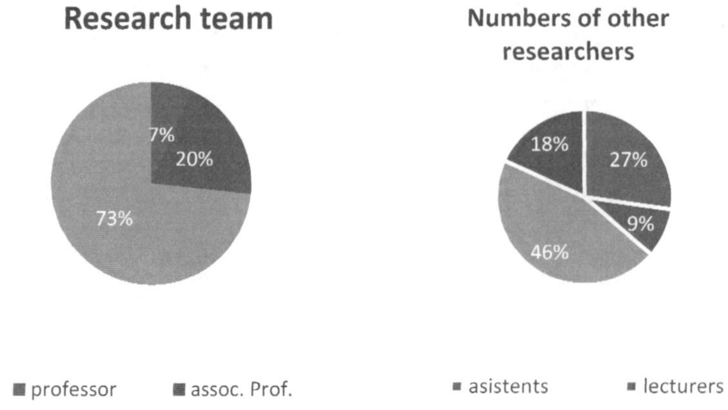

Figure 1. Representation of the number of researchers in the project.

Table 1. Team of researchers.

	position	focus, orientation of research
Katunský D	professor	team leader
Dolníková E	associated professor	deputy leader of the team, daylighting
Lopušniak M	associated professor	energy engineering, fire protection
Vertaľ M	associated professor	water and heat transport material properties
Baroš T	doctoral student	parasitic architecture forms
Blichová Ž	doctoral student	LCA methods
Hutkai K	doctoral student	historical buildings, thermal insulation inside
Jaroš P	doctoral student	historical buildings, sandstone properties
Kabošová L	researcher	wind engineering,
Karľa V	researcher	transparent wood
Katunská J	assistant professor	energy engineering, historical buildings
Merjavá V	teacher	stone family houses, properties of stone masonry
Tóth S	assistant professor	roof constructions
Zozulák M	assistant professor	design detail of building structure, windows
Vargová A	admitted PhD student	green plant roof structures

5. Results

The results in individual areas can be assigned to the researchers and published contributions during the solution period:

5.1. Daylighting

Daylighting is very important in creating an indoor environment and comfort not only in residential buildings, but especially at working places. The authors have published a number of papers on the topic of daylighting in conferences or in magazines. These are contributions concerning, in particular, the lighting of the working environment. Figure 2 shows the results of DF in the room in the conversion of the balcony to the loggia, it is published in the journal Sustainability. One of the most important is an article in the magazine Building and environment [1]. The results of the ratio of Le and Lvz can be seen in Figure 3.

5.2. Energy consumption, building structure details design

Several researchers address the issue of energy efficiency of buildings and the use of non-traditional energy sources. Several publications, textbooks and books on the issue of thermal protection of buildings have been created in this area. Several articles have been published in journals and proceedings contain papers from conferences. These are mainly conferences Thermal protection of buildings. One of the most important publications is an output in the journal Energy and Buildings. [2].

5.3. Thermal and moisture flows

The issue of heat and humidity flows is very closely related to point 2.2, which is monitored mainly in piece building materials (brick, stone, clay blocks). This is the content of some dissertations. Vertaľ, M. and Jaroš, P. are very active in dealing with this issue. Jaroš, P. investigates heat and humidity flows in stone building materials (sandstone) [3].

5.4. Thermal protection inside historical buildings

Thermal protection of historic buildings comes to the fore and is consulted especially with monuments. Labovská, a doctoral student, defended her dissertation on this issue; Katunská, J. as well as Vertaľ, M.

and Jaroš, P. also dealt with historic buildings. At present, PhD student Hutkai, K. solves the problem of installing thermal insulation from the inside of the perimeter wall of old historic buildings. He published some of his first experiences, especially at the Young Scientist conference [4].

5.5. Transparent wood

Collaborator architect Karľa decided to produce transparent wood. After many tests in the laboratories of the Institute of Environmental Engineering in the laboratories of building materials, he succeeded. He made the first small sample of transparent wood and published his experience. He later dealt with theoretical issues, namely the physical and static properties of wood, which demonstrates transparency or translucency [5].

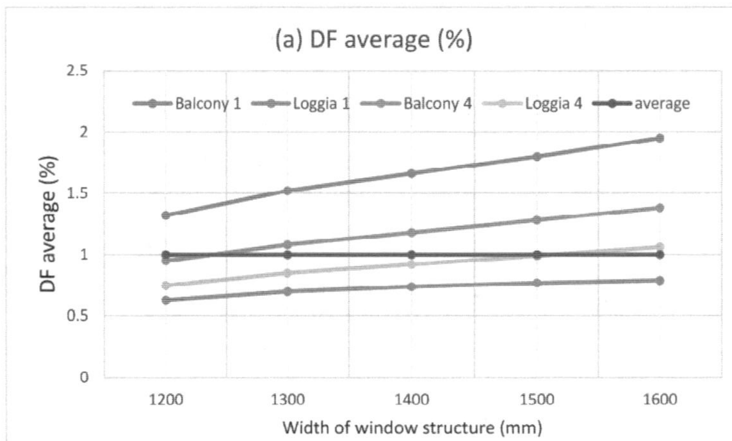

Figure 2. Results of DF average (%) for Rooms before and after reconstruction.

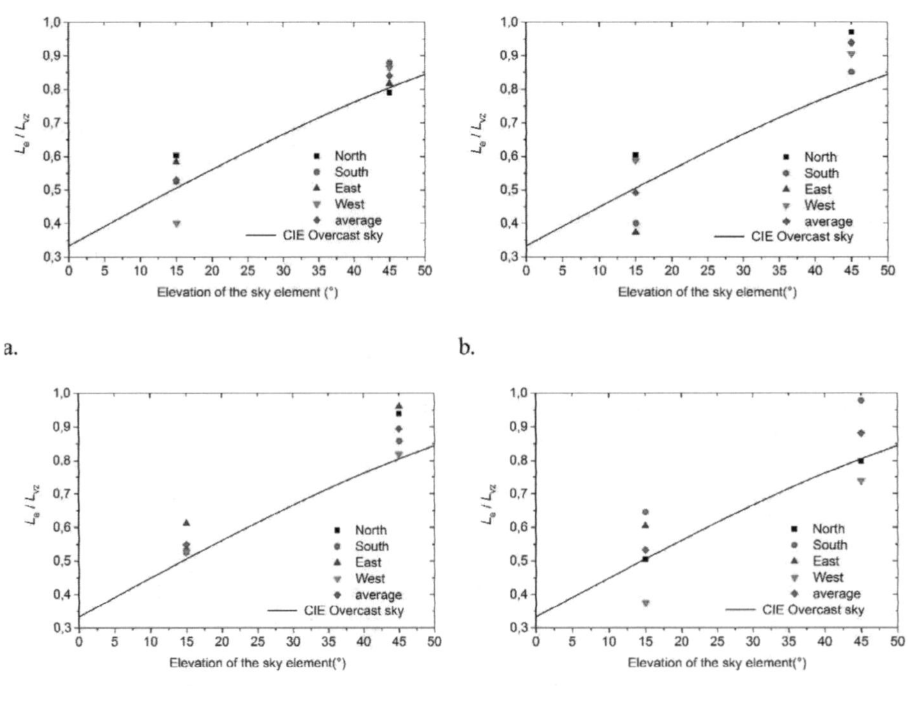

Figure 3. Results of ratio L_e/L_{vz} of sky elements.

Figure 4. Article had a picture on the cover page of Applied Sciences Journal in December issue.

Figure 5a. “Flowbrane” – performance in the westerly winds.

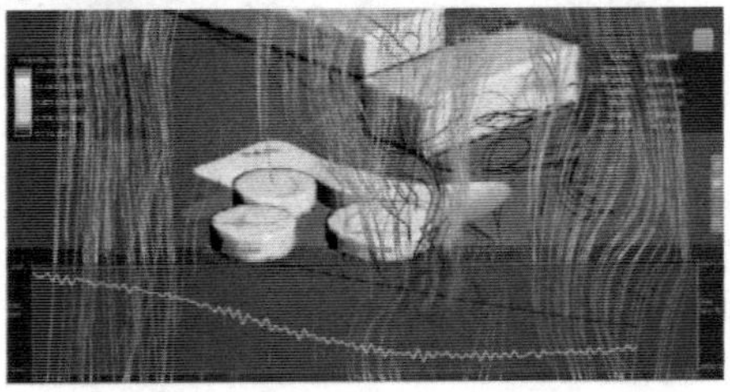

Figure 5b. “Flowbrane” – performance in the southerly winds.

Figure 5. Illustrative pictures for research into the effects of wind on buildings.

Figure 6. Illustration of the essence of Parasitic Architectural Forms by T. Baroš (2020).

5.6. LCA of buildings

Doctoral student Blichová, whose dissertation is devoted to LCA (life cycle assessment), has published many results. One of the most significant results is an article in the CC magazine Building and Environment, international journal of building science and its applications - Amsterdam (NL): Elsevier Vol. 172 (2020), s. 1-13 [print] - ISSN 0360-1323.

5.7. Wind around the buildings

Architect Kabošová handled the topic of her dissertation very well. She has completed many stays abroad at major universities and research parks. She is the author of a number of publications, not only in papers at scientific conferences, but also in recognized journals. An article in Frontiers of Architectural Research [6] is very interesting. Important results are published in Applied Sciences Basel [7]. The editors were so impressed by the editors' results that the schematic results of the research formed the cover of the December issue. See this in Figure 4. Some results from the air flow around the building can be seen in Figure 5.

5.8. Parasitic architecture
Doctoral student Tomáš Baroš chose the issue of Parasitic Architecture. He was to complete an internship abroad, but the pandemic situation of Covid-19 prevented him from doing so. However, he dealt with the topics of research very well. He passed the dissertation exam and established cooperation with experts from the IT environment. His article on intensive authorial cooperation between the issue of building construction, architecture and IT technologies is ready for publication. It is also preparing a case study of the proposed pavilion, as a contribution to a conference in Glasgow DCC'22 (Design Computing and Cognition'22), where the collaboration between architecture and IT is elaborated and visible in more detail. He also published in the journal SSP / JCE and at the Conferences Young Scientist [8]. An illustration of the nature of parasitic architectural forms according to Tomáš Baroš is shown in Figure 6.

5.9. Green plant roofs and walls
Vertaľ, M., Vargová, A. and others deal with the issue of green (vegetation) roofs as well as walls. They created conditions for the application of a case study of a green roof on one of the administrative buildings. They publish the results in magazines. The latest publication from the year 2021 is in the journal Sustainability [9].

6. Conclusion
The project is being worked on on an ongoing basis. The research team also announced the application for a grant for the period from 2022 to 2025. The application is in the process of approval. In the case of a positive statement, the authors will work on ´other cases concerning demanding conditions not only of the design, but also of the implementation or use of the construction work in demanding conditions.

7. References
[1] Dolnikova E, Katunsky D and Darula S 2020 *Assessment of overcast sky daylight conditions: in the premises of engineering operations considering two types of skylights*: *Building an Environment*, 180, 106976.
[2] Vertaľ M, Zozulák M, Vašková A, and Korjenic A 2018 *Hygrothermal initial condition for simulation process of green building construction*: *Energy and Buildings*, *167*, 166-176.
[3] Jaroš P 2019 *Water transport in materials of historical buildings*: *IOP Conference Series: Materials Science and Engineering* (Vol. 566, No. 1, p. 012011). IOP Publishing.
[4] Hutkai K and Katunsky D 2021 *Insulation of Historic Buildings and Case Study Simulation: IOP Conference Series,* IOP Publishing, 2021. p. 012003.
[5] Karľa V 2020 *Methodology of Research on Transparent Wood in Architectural Constructions Selected Scientific Papers - Journal of Civil Engineering*, 15(2), 29-35.
[6] Kormaníková L, Achten H, Kopřiva M and Kmeť S 2018 *Parametric wind design*: *Frontiers of Architectural Research*, 7(3), 383-394.
[7] Kabošová L, Katunský D and Kmeť, S. 2020 *Wind-Based Parametric Design in the Changing Climate. Applied Sciences,* 10(23), 8603.
[8] Baroš T 2020 *Parasitic Architectural Forms (PAF) S01. E01 "Prologue"*: *IOP Publ. Conference Series: Materials Science and Engineering* Vol. 867, No. 1, p. 012001.
[9] Cakyova K, Vranay F, Vertal M and Vranayova Z 2021 *Determination of Dehumidification Capacity of Water Wall with Controlled Water Temperature Experimental Verification under Laboratory Conditions*: *Sustainability*, 13(10), 5684.

Acknowledgments
This contribution was prepared with the financial support of the research project VEGA 1/0626/22 of the Scientific Grant Agency, the Ministry of Education, Science, Research and Sports of the Slovak Republic and the Slovak Academy of Sciences.

On the load-carrying capacity produced by different load models for road bridges

J Gocal[1], J Odrobinak[1] and J Vican[1]

[1] Department of Structures and Bridges, Faculty of Civil Engineering, University of Zilina, Univerzitna 8215/1, 010 26 Zilina, Slovakia

E-mail: jozef.gocal@uniza.sk; jaroslav.odrobinak@uniza.sk; josef.vican@uniza.sk

Abstract. The evaluation of existing bridges is an important process and source of information within the bridge management system. Except for the load-carrying capacity represents the basic evaluation parameter, it moreover expresses a relevant and decisive parameter for the transition of actual road traffic. The paper presents the continuation of an already published analytical study aimed at comparing the effects of traffic load produced by different load models on road bridges. There are presented calculated load-carrying capacities for model cases of bridges with an upper deck with eight alternative spans and six different cross-sectional arrangements. On each of them, multiple load models simulating the effects of traffic were considered, while a certain predefined part of the considered load represented the required weight of the vehicle defining the load-carrying capacity of the respective bridge. The discussion is mainly focused on the pitfalls and possibilities of correctly defining the dependence of the weight of the vehicle representing the load-carrying capacity from the traffic load consideration schemes themselves, which are mostly defined by design codes.

1. Introduction

The load-carrying capacity is a basic evaluation parameter of existing bridges, on the basis of which decisions on further exploitation of the existing bridges are made. Especially the bridges built after the World War II as a result of the reconstruction of war-destroyed transportation infrastructure, reach the limits of their service life and operability at present. Thereby, there is more discussion about the need for their rehabilitation. The reconstruction needs were also supported by the recent serious bridge accidents in Slovakia [1] and abroad [2]. A common feature of these disasters is the insufficient maintenance of transportation infrastructure [3], [4], overestimation of the properties and quality of materials, poor or unmanaged, respectively, construction technologies including imperfect and wrong projects, as well as ever-increasing traffic intensity together with increasing axle loads.

The need for revision of the Technological Standard No. 104 “Load-carrying capacity of road bridges and footbridges” (hereinafter referred to as TS 104) [5] was an impulse for working up an Analysis Task, whose results are summarised in the report [6]. It was worked out by the collective of the Department of Structures and Bridges from the University of Zilina, in cooperation with Slovak Road Administration (SRA), National Motorway Company (NMC) and other co-operators among academicians and designers. The main aim of the Analysis Task was to develop the methodology for determining the load-carrying capacity of road bridges and footbridges. TS 104 is currently available for the calculation of load-carrying capacity of road bridges in Slovakia. However, the TS 104 does not correspond to the current state of European standardization, especially in terms of the reliability level recommended for the assessment of existing bridges. There is also a non-compliance with the definition of loads prescribed for bridges by the standard STN EN 1991-2 [7]. Therefore, the basic goal of this analysis task was to prepare documents to process new TS 104 for determining load-carrying capacity of road bridges and footbridges respecting the actual state-of-the-art and latest trends in the field of existing bridges evaluation. The general concept and basic assumptions considered in the report [6] as well as the basis of load-carrying capacity determination have already been presented in [8]. This paper presents the results of comparative study dealing with calculation of load-carrying capacities for model cases of road bridges using the proposed methodology and their comparing with load-carrying capacities calculated according to the current TS 104 as well as to the origin Slovak standard [9] used at the time of designing most of the older existing bridges.

2. Assumptions of comparative study

As a representative bridge sample, the simply supported girder superstructure with an upper bridge deck was used. In the first stage, the bridge spans were chosen within a range from 2 m to 80 m, with different road widths varying from 5.4 m to 11.5 m corresponding to various numbers of girders at distances from 1.0 m to 9.0 m. The vertical traffic load effects were expressed by the values of the maximum bending moments determined in the span middles and the maximum shear forces above the supports of the simply supported beams. The considered load models according to the current standard [7] and previously valid standard [9] were always located on the bridge in the most unfavourable position in the transverse and longitudinal direction, with respect to the observed internal force. The comparison of the all load model effects are presented in the paper [8].

Subsequently, we selected from the previous very extensive study three representative width arrangements with the smallest and largest number of girders, according to the table 1.

Table 1. Considered bridge cross section arrangements.

road width	b_c	5.0 m		7.5 m		11.5 m	
number of girders	n	2	6	2	8	2	11
girder distance	b	5.0 m	1.0 m	7.5 m	1.0 m	5.0 m	1.1 m

The load-carrying capacities were calculated for eight selected spans: 2 m, 4 m, 8 m, 15 m, 24 m, 35 m, 50 m and 80 m. In total, more than 2,800 normal load-carrying capacities were quantified for the selected model cases. These load models were taken into account:

- Load models for normal load-carrying capacity according to the standard [9]: load configurations ZZ1 (figure 1) and ZZ2 (figure 2), respectively, with a representative vehicle weight of 32 t;
- Load model for exclusive load-carrying capacity according to [9]: four-axle vehicle 4AV (figure 3) with a representative vehicle weight of 80 t;
- Model approximating traffic jam from heavy vehicles in two lanes (or in one lane for narrow bridges) modelled by three-axle vehicles T815 (28 t) (figure 4) at mutual distances of 10 m;
- Load model for normal load-carrying capacity according to the current TS 104: based on the normal load-carrying capacity factor F_z, including effects of load model 1 (LM1) (figure 5) according to [7], multiplied by the weight of the representative vehicle 32 t. This model is denoted as LM1_F (32 t) in the study;
- Modification of the model LM1_F (32 t): the effects of uniformly distributed load (UDL) of value 2.5 kN/m^2 in the notional lanes No. 2 and No. 3 as well as in the remaining area are independent on the load-carrying capacity. This model is denoted as LM1_F_ind_2.5 (32 t) in the study.
- Various load models for normal load-carrying capacity according to the proposed methodology:
  - LM1_ind_2.5 (60 t) — Model based on the LM1, in which the weights of two-axle vehicles (tandem system) TS1, TS2 and TS3, respectively, in the notional lanes No. 1, No. 2 and No. 3, respectively, as well as the UDL in the

1st notional lane directly depend on the load-carrying capacity. However, the UDL in notional lanes No. 2 and No. 3 as well as in the remaining area (2.5 kN/m²) are independent on the load-carrying capacity. The TS1 with the weight of 60 t is considered as the representative vehicle;

- LM1_ind_2.5 (40 t) Model defined as in previous case but with the representative vehicle weight of 40 t corresponding to the weight of TS2;
- LM1_ind_UDL (60 t) Model based on the LM1, in which only the weights of tandem systems TS1, TS2 and TS3 depend on the load-carrying capacity, while all the UDL systems are independent on the load-carrying capacity. The TS1 with the weight of 60 t is considered as the representative vehicle;
- LM1_ind_UDL (40 t) Model defined as in previous case but with the representative vehicle weight of 40 t corresponding to the weight of TS2;

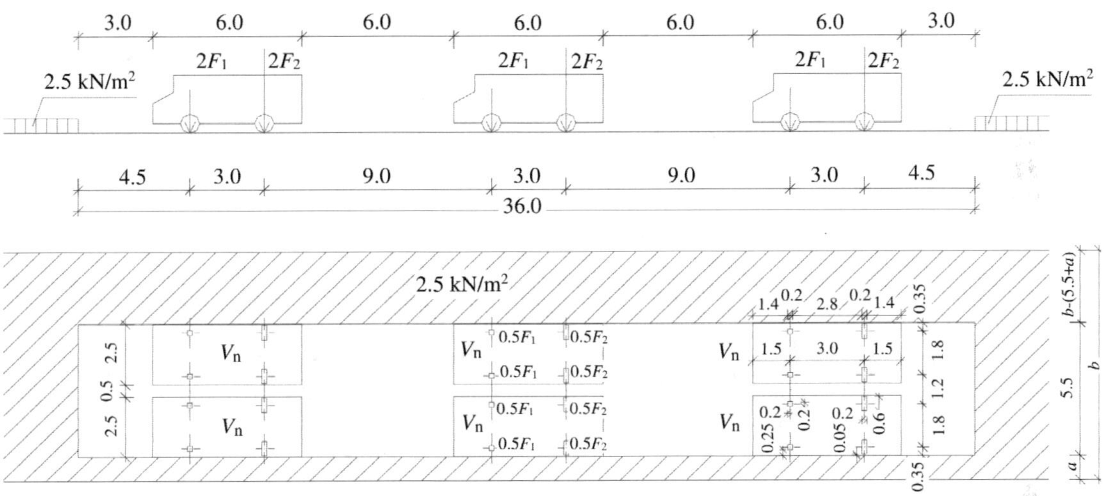

Figure 1. Load configuration ZZ1.

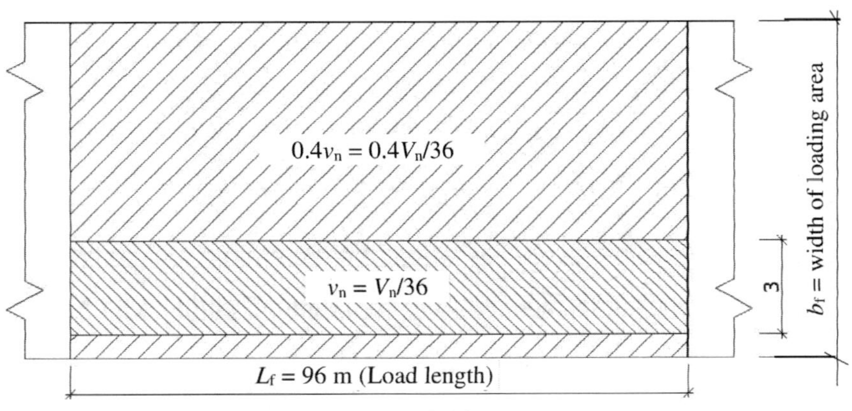

Figure 2. Load configuration ZZ2.

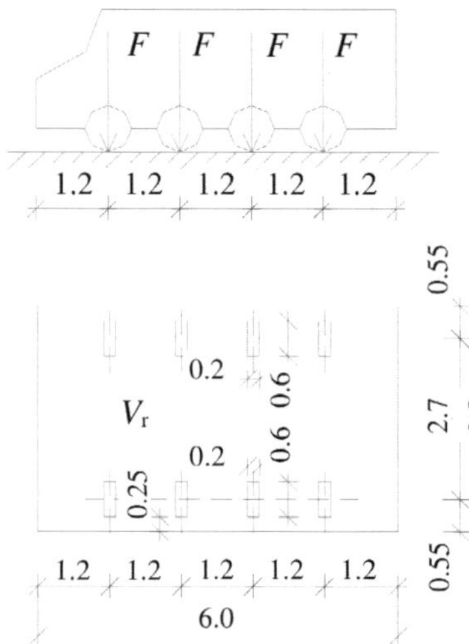

Figure 3. Four axle vehicle.

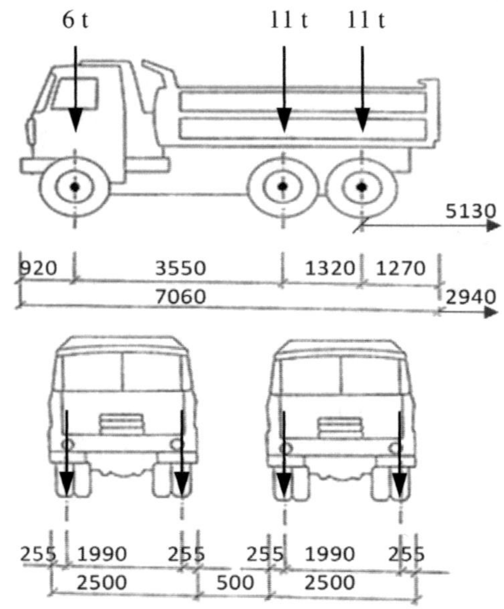

Figure 4. Three-axle vehicle T815.

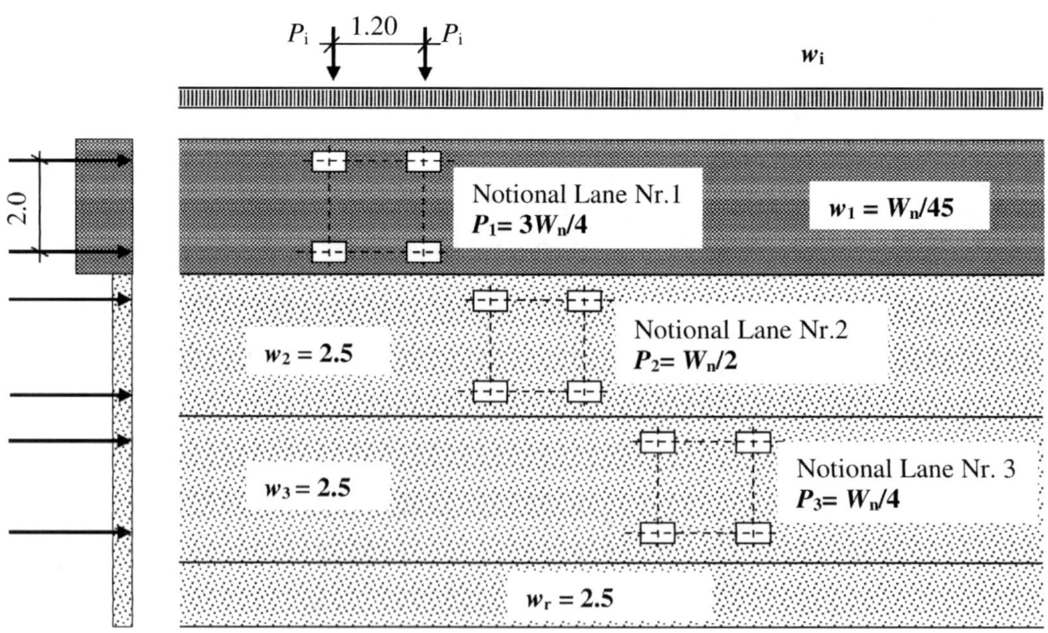

Note: W_n is the vehicle weight in kN, w_i is the value of uniformly distributed load (UDL) in kN/m^2

Figure 5. Load model 1.

The considered load models for normal load-carrying capacity based on the LM1 (figure 5) are summarized in table 2.

Table 2. Considered models for normal load-carrying capacity based on the LM1.

Load model	Loads P_i in (kN) v_i (v_r) in (kNm^{-2})	The whole model	Notional lane 1	Notional lane 2	Notional lane 3	Residual area „r“
LM1_F (32 t)	P_i in TS	V_n	$600F_z$	$400F_z$	$200F_z$	0
	v_i (v_r) in UDL		$9F_z$	$2.5F_z$	$2.5F_z$	$2.5F_z$
LM1_F_ind_2.5 (32 t)	P_i in TS	V_n	$600F_z$	$400F_z$	$200F_z$	0
	v_i (v_r) in UDL		$9F_z$	2.5	2.5	2.5
LM1_ind_2.5 (60 t)	P_i in TS	-	$V_n/2$	$V_n/3$	$V_n/6$	0
	v_i (v_r) in UDL		$3V_n/200$	2.5	2.5	2.5
LM1_ ind_UDL (60 t)	P_i in TS	-	$V_n/2$	$V_n/3$	$V_n/6$	2.5
	v_i (v_r) in UDL		9	2.5	2.5	2.5
LM1_ ind_2.5 (40 t)	P_i in TS	-	$3V_n/4$	$V_n/2$	$1V_n/4$	2.5
	v_i (v_r) in UDL		$9V_n/400$	2.5	2.5	2.5
LM1_ ind_UDL (40 t)	P_i in TS	-	$3V_n/4$	$V_n/2$	$V_n/4$	2.5
	v_i (v_r) in UDL		9	2.5	2.5	2.5

3. Calculation of normal load-carrying capacity

According to the current TS 104, the normal load-carrying capacity factor F_z expresses the bridge ability to transfer load effects caused by the LM1. In general, the relation for the normal load-carrying capacity factor F_z can be expressed as the ratio of the capacity (reserve) of the resistance for the decisive variable load effects to the variable load effect caused by LM1 in the form as follows:

$$F_Z = \frac{R_{d,c}}{E_{LM1,d}} = \frac{R_d - \sum_{i=1}^{n-1} E_{rs,d,i}}{E_{LM1,d}} \tag{1}$$

where F_z is the normal load-carrying capacity factor;
$R_{d,c}$ is the value of the resistance capacity for the decisive variable road traffic load;
R_d is the design value of the resistance of the cross-section or bridge member;
$E_{LM1,d}$ is the value of static quantity due to effects of variable load LM1;
$\sum_{i=1}^{n-1} E_{rs,d,i}$ are the design, combination or group values of the effects of other loads acting simultaneously with the vertical road variable load.

Normal load-carrying capacity is determined as the maximum permissible weight of one vehicle of normal load without limiting the number and position of vehicles passing the bridge. It is determined according to the following formula:

$$W_n = F_Z \cdot W_{n,rep} \tag{2}$$

where W_n is the normal load-carrying capacity in “kN”,
$W_{n,rep}$ is the weight of representative vehicle for normal load-carrying capacity in “kN”.

Since the normal load-carrying capacity is commonly defined as the weight of the vehicle in “t”, its value V_n should be obtained by the formula:

$$V_n = W_{n,rep}/10 \tag{3}$$

According to the report [6], the normal load-carrying capacity V_n in "t" can be calculated directly from the equation

$$V_n = \left[\left(R_d - \sum_{i=1}^{n-1} E_{rs,n,Ed,i}\right) / E_{LM1,Ed}\right] \cdot V_{n,rep} \tag{4}$$

where $V_{n,rep}$ is the weight of representative vehicle for normal load-carrying capacity in "t" and the other quantities are defined as above.

Since we worked with fictitious bridges in the study, it was necessary to estimate their resistance as well as the share of other accompanying loads acting simultaneously with the considered load models in the overall response of the structure to loads. Therefore, we have introduced the assumption that the current technical condition of bridge structures is such that they are able to transfer, after deducting all other loads, a certain share of the effects of the load model ZZ1 according to [9] used in the past. Therefore, for the needs of the study, equation (4) can be modified as follows

$$V_n = \frac{R - (E_{rs} + E_{rs,LM})}{E_{LM}(V_n)} \cdot V_{n,rep} = \frac{(R - E_{rs}) - E_{rs,LM}}{E_{LM}(V_n)} \cdot V_{n,rep} = \frac{(k \cdot E_{ZZ1}) - E_{rs,LM}}{E_{LM}(V_n)} \cdot V_{n,rep} \tag{5}$$

where R is the bending or shear resistance, respectively, of the evaluated cross section;
E_{rs} is the value of effects (bending moment or shear force) of the accompanying non-traffic loads acting together with the considered load model;
$E_{LM}(V_n)$ is the value of load effects (bending moment or shear force) from those parts of the considered load model that are dependent on the normal load-carrying capacity V_n;
$E_{rs,LM}$ is the value of load effects (bending moment or shear force) from the other parts of the considered load model independent on the load-carrying capacity;
$V_{n,rep}$ is the weight of the corresponding representative vehicle;
E_{ZZ1} is the value of effects (bending moment or shear force) of the load model ZZ1;
$R - E_{rs}$ is the resistance reserve (bending or shear) for traffic load effects expressed as a k-fold of the effects of the load model ZZ1, thus $R - E_{rs} = k \cdot E_{ZZ1}$.

The values $R - E_{rs}$ considered in the study are given in table 3, together with the expected load-carrying capacity $V_{n,0}$ of the bridge determined using the load model ZZ1 according to previously valid standard [9], influenced by subtracting the effects of uniform load independent on the load-carrying capacity.

Table 3. Considered resistance reserves $R - E_{rs}$ for determination of load-carrying capacity.

$R - E_{rs}$	$V_{n,0}$ (t)
$0.25 \cdot E_{ZZ1}$	$<$ 8.0
$0.35 \cdot E_{ZZ1}$	$<$ 11.2
$0.50 \cdot E_{ZZ1}$	$\leq$ 16.0
$0.70 \cdot E_{ZZ1}$	$\leq$ 22.4
$1.00 \cdot E_{ZZ1}$	$\approx$ 32.0

For such estimated resistance reserves $R - E_{rs}$ in bending and in shear, respectively, the load-carrying capacities were determined using all considered road traffic load models. From a huge amount of data, only a few outputs are presented. Figure 6 shows the resulting load-carrying capacities calculated from the bending resistance for the case of bridge 5.0 m wide with 2 main girders and for the case of bridge 11.5 m wide with 11 main girders, respectively, when the resistance reserve $R - E_{rs}$ is considered to be 50 % or 100 %, respectively, of the effects of ZZ1 model. Similarly, figure 7 shows the resulting load-carrying capacities determined from the shear resistance for the same model cases.

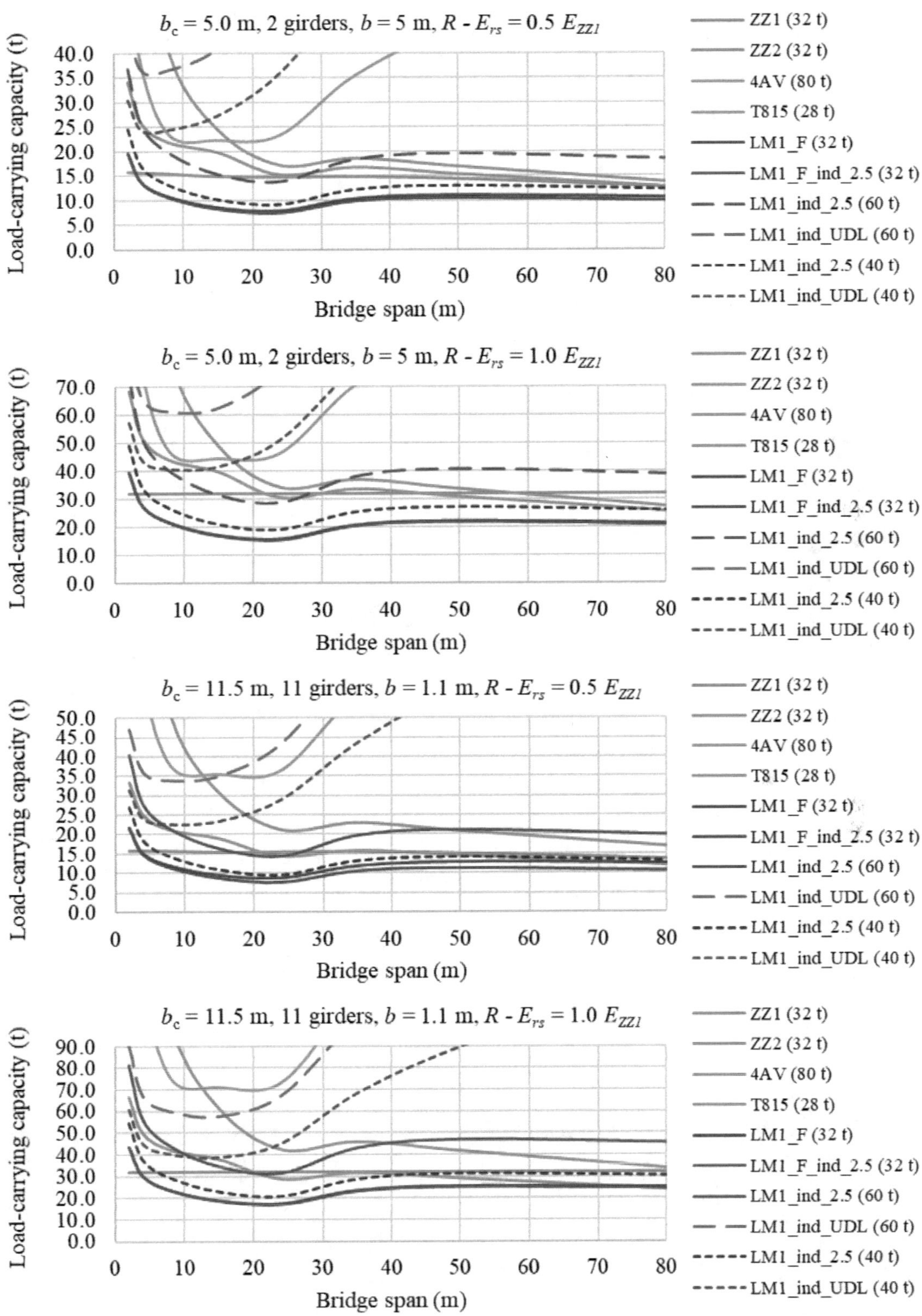

Figure 6. Example of calculated load-carrying capacities from the bending resistance.

IOP Conf. Series: Materials Science and Engineering 1252 (2022) 012030 doi:10.1088/1757-899X/1252/1/012030

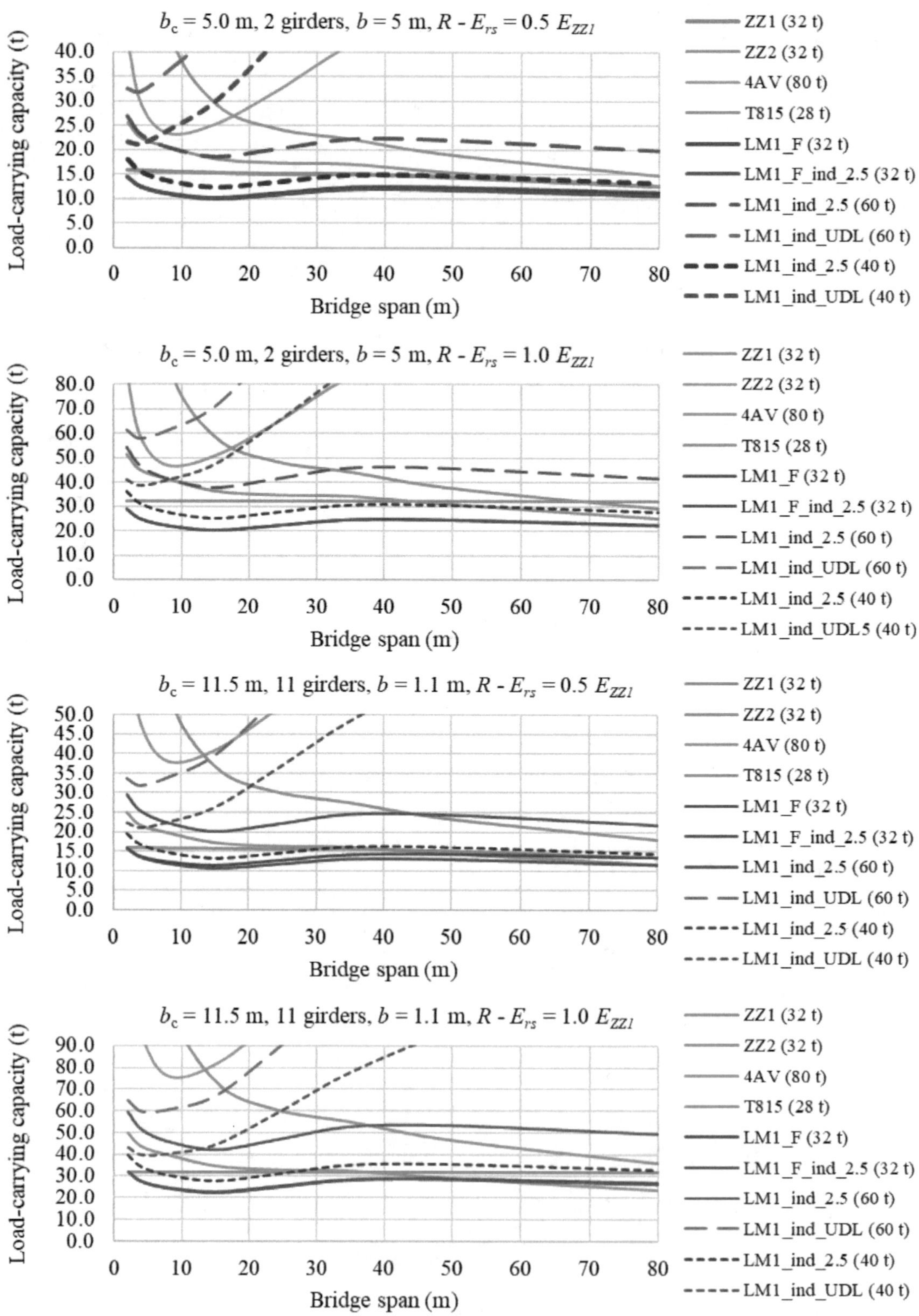

Figure 7. Example of calculated load-carrying capacities from the shear resistance.

4. Discussion of results

From all graphical comparisons, not only the graphs presented here (there are 60 in total), it is clear that the normal load-carrying capacities determined using the load model marked **LM1_ind_2.5 (40 t)** are closest to the load-carrying capacities determined using the ZZ1 load model according to the originally valid standard [9], especially for medium and large spans. It is thus the load model based on the LM1 model according to STN EN 1991-2 [7], in which the tandem system in the notional lane No. 2 with the weight of 40 t is considered to be the representative vehicle for normal load-carrying capacity. As part of the parameterization, other values of the weight of tandem system were also tested, but the value mentioned above corresponded the most to the modelled cases.

All the axle forces of the tandem systems TS1, TS2 and TS3 as well as the uniform load in the notional lane No. 1 depend on the load-carrying capacity. The uniform loads in the other notional lanes No. 2 and No. 3 as well as in the remaining area are not dependent on the load-carrying capacity. They are therefore still considered by the value 2.5 kN/m^2. There is also a practical reason for this fact. Indeed, even if this uniformly distributed load were dependent on the load-carrying capacity V_n, unreasonably low values of the uniform load would be achieved on this loaded area at low load-carrying capacity values V_n. It should be borne in mind that values of uniform load below the value of 2.0 kN/m^2, or even below 1.5 kN/m^2, would not even correspond to the load caused by personal vehicles.

In the case of short spans of 4 m to 20 m, however, the model mentioned above produces obviously the more unfavourable load-carrying capacities compared to the ZZI load model but also, for example, compared to the model case of continuous traffic jam represented by the T815 (28 t) model. This is due to the nature of the LM1 model, in which both discrete (Tandem system) and continuous loads (UDL system) act simultaneously at one point in the loading area (notional lanes No. 1, 2 and 3). Therefore, in the case of short spans, we get to a significantly underestimated load-carrying capacity.

For example, at a span of 10 m, it is unrealistic for more than one truck to stand on a bridge in a single lane. However, within the model LM1 is tandem system TS with a weight of 60 t in the extreme notional lane and, in addition, there is also UDL with an intensity of 9 kN/m^2 in the same notional lane. Together, the load on the 10 m bridge in one 3 m wide lane is up to 87 t. The response of the structure to the load model conceived in this way produces results with significantly lower load-carrying capacity.

Therefore, the report [6] raises, in the case of short bridge spans up to 30 m, the question of a possible modification of the normal load-carrying capacity calculated by the above procedure in order to better capture the nature of the actual traffic even with short bridge spans.

5. Conclusions

This paper presents the results of comparative study, which was worked up by authors as a part of Analysis Task focused on developing the methodology for determining the load-carrying capacity of road bridges and footbridges. The results of a large number of calculations of normal load-carrying capacities of model cases of road bridges using the proposed methodology, which is based on the application of traffic load models according to Eurocodes, are presented. The effects of different traffic load models according to previously valid technical standards, used for design of most of existing road bridges in Slovakia, are compared to the effects of load models based on the load model LM1 according to Eurocode. Based on the results of the study, the parameters of the load model were finally designed for the purpose of determining the normal load-carrying capacity of existing road bridges.

6. References

[1] Moravčík M and Bujňáková P 2018 Collapse analysis and reconstruction process of precast bridge in Slovakia. *10th International conference on Short and medium span bridges*. (Montreal. Canadian Society for Civil Engineering) p 1-6

[2] Vacek V, Kolísko J and Pokorný P 2020 Influence of corrosion of prestressing reinforcement on its properties and possibilities of rehabilitation of such reinforced concrete structures. *iMateriály - portal of the journals Building Materials and Builder for construction professionals*

[3] Kotes P, Brodnan M, Chandoga M, Bacikova T and Kubissa W 2017 The Zlatné bridge - condition survey, temporary remedial works required due to dangerous condition and bridge restoration project. *Roads and Bridges - Drogi i Mosty,* v. 16, n. 3, p 177-90
[4] Kotes P, Brodnan M and Bahleda F 2016 Diagnostics of Corrosion on a Real Bridge Structure, *Advances in Materials Science and Engineering* (2016) 2125604
[5] TS 104: *Load-carrying capacity of road bridges and footbridges*. Ministry of transport, building and regional development of SR. Department of road traffic and roads. 2016
[6] Vican J et al 2020 Analysis Task: Determination of load-carrying capacity of road bridges and footbridges. *Final report.* (University of Žilina)
[7] STN EN 1991-2: 2006 *Eurocode 1. Actions on structures. Part 2: Traffic actions on bridges.* (Slovak Office of Standards, Metrology and Testing)
[8] Vican J, Odrobinak J and Gocal J 2021 Determination of road bridge load-carrying capacity. *Civil and Environmental Engineering No. 17/1* (Zilina, Faculty of Civil Engineering, UNIZA) p 286-297
[9] ČSN 73 6203: 1986 *Action on bridges.* Czechoslovak state standard. (UNM, Prague)

Acknowledgments
This work was supported by the research projects VEGA 1/0630/21 and VEGA 1/0623/21 of the Slovak Grant Agency.

Evaluation of daylight comfort in industrial building

E Dolnikova[1], D Katunsky[1] and M Lopusniak[1]

[1] Institute of Architectural Engineering, Faculty of Civil Engineering, Technical university of Kosice; Vysokoskolska 4, 042 00 Kosice, Slovakia

E-mail: erika.dolnikova@tuke.sk; dusan.katunsky@tuke.sk; martin.lopusniak@tuke.sk

Abstract. Industrial halls have a very high energy consumption due to existing production processes, heating, cooling, lighting. To reduce energy consumption, it is necessary to make a number of technical modifications to the structure itself, or the interior. The industrial sector is one of the largest consumers of energy. One way to save energy is to take advantage of daylight. Natural daylight is an important part of our daily lives at home and at work. Providing natural daylight to the work environment is essential for the personal well-being, work efficiency and safety of the users of this environment. This paper deals with the evaluation of visual comfort in an industrial hall. Visual well-being is assessed on the basis of physical calculations. These were implemented by the RADIANCE simulation program. The lighting calculation was performed for both the cloudy sky and the clear sky. The program also qualified and quantified the brightness distribution. Potential sources of glare have also been identified.

1. Introduction

On average, we spend 90% of our time indoors, mainly at work. Therefore, a healthy workspace is essential for a feeling of well-being. Daytime running lights should be designed to provide adequate levels of lighting in the working plane. Somewhere daylight is the main or only source of light during operation. Changes in the field of view can affect the visual comfort and performance of users [1-3].

Lighting plays a key role in designing not only commercial buildings, shopping malls and retail buildings but also contributes to the visual comfort and quality of the space in these buildings. As artificial lighting can use up to 40% of a building's total energy consumption, there is great potential for energy savings. Skylights can emit up to twice as much light as a normal vertical window. In an industrial building, a large roof area - 15% and 20% - is used to provide daylight. Lighting accounts for up to 40% of energy consumption [4-6]. Transparent fillings in the building affect thermal stability and overheating in summer [7]. This stability in summer is mainly influenced by the thermal and optical properties of the glazing [8]. This problem can be avoided by using a light pipe, that brings light to more distant and inaccessible places [9]. All this of course depends on the settlement of details [10].

This study analyzes the potentials of daylight in a large indoor hall-type building with a capacity of 900 m^2 using an arched skylight located in the northern latitudes and climate of Slovakia. Daylight simulations were performed using Radiance software.

2. Description of the hall and in-situ measurements

The studied industrial building is a single-storey hall 15 m x 60 m x 8.5 m with a total floor area of 900 m^2 in Košice, Slovakia (see Figure 1). In the hall there is a combined lighting, i. top and side lighting. The side lighting consists of windows measuring 5.6 m x 1.8 m and 3.0 m x 1.8 m. The top lighting consists of one arched skylight with dimensions of 2.4 m x 48.0 m with an elevation of 1.0 m (see Figure 2). Neighboring buildings are in the distance, which does not shade the hall. Window systems are made of polycarbonate glass. The optical properties of the interior surfaces of the hall are given in Table 1. The operating hours of the building are from 7:00 to 14:00 from Monday to Friday.

Figure 1. Interior of the hall.

Figure 2. Exterior view on the roof light.

Table 1. Optical properties of the interior surfaces in the test hall.

Surface	Reflectance/Visual transmittance (%)	Coefficient
Wall	reflectance 40	0.40
Floor	reflectance 20	0.20
Ceiling	reflectance 70	0.70
Side windows	visual transmittance 63	0.63
Skylight	visual transmittance 37	0.37

The measurements were realized during two days. The illumination was measured in the interior under the overcast sky. The brightness of the sky was also measured. The following devices were used for the measurements: Ahlborn ALMEMO 2690-10A Datalogger, probe for measuring illuminance Ahlborn FLA623VL, luminance meter LS-110, device for measuring visual transmittance. Interior illuminance was realized at the points in the hall and simultaneously exterior illuminance on the roof in the height 1.5 m. During measurements exterior horizontal illuminance was between 4,500 lx – 9,000 lx. The working plane was 0.85 m above the floor.

The hall is used for medium hard production with work of various kinds and therefore by standard [11] the hall is classified in III-IV lighting classes. With the given combined lighting system in the critical point of functional place on horizontal plane average Daylight Factor = 5-6% [12].

3. Results

Table 2 shows the measured lighting values and the resulting simulations from RADIANCE computer applications. The course of the daylight and the light factor is calculated from the measured values (see Figures 3 and 4). The following figures document the results of the brightness simulation for the indoor environment of the hall. A simulation was performed for the arched skylights. Simulation calculations were performed from two perspectives from the inner hall workplace. Luminance distribution can be seen in Figure 5. The evaluation of visual comfort was performed on the basis of the calculation of the glare index (Guth Visual Comfort Probability), see Figures 6 and 7. Guth's Visual Comfort Probability (Guth VCP) is a metric used to evaluate light scenes. Guth VCP provides an indication of the percentage of people in the area who would consider glare from the lamp acceptable. A minimum VCP of 70% is recommended.

Table 2. Calculated values of Daylight Factor from measured values of illuminance.

	DF_{min} [%]	DF_{max} [%]	$DF_{average}$ [%]
Measured	1.42	9.57	5.40
Calculated	0.22	7.92	4.15

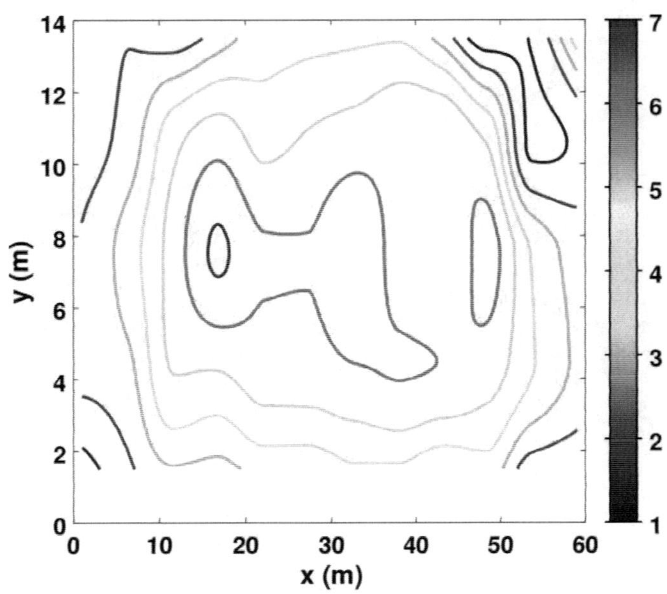

Figure 3. Calculated values of the Daylight Factor (%).

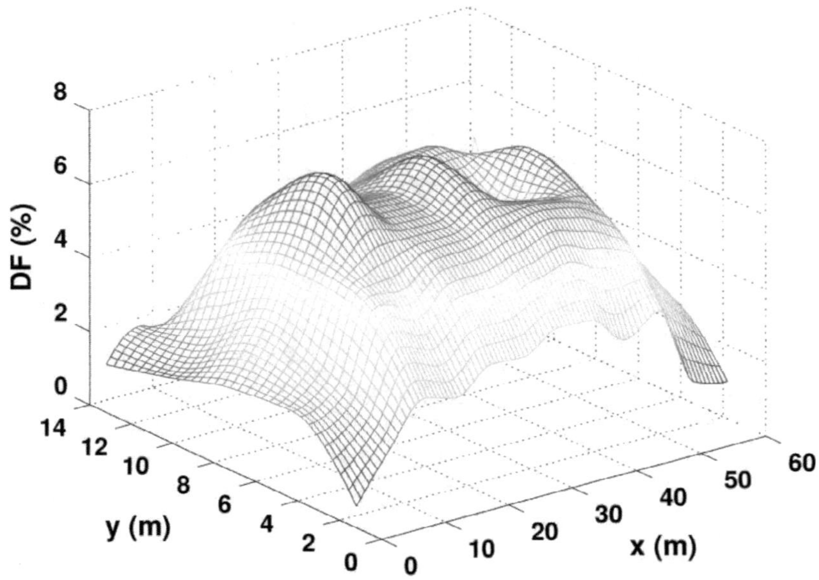

Figure 4. Calculated values of the Daylight Factor – spatial distribution, illumination.

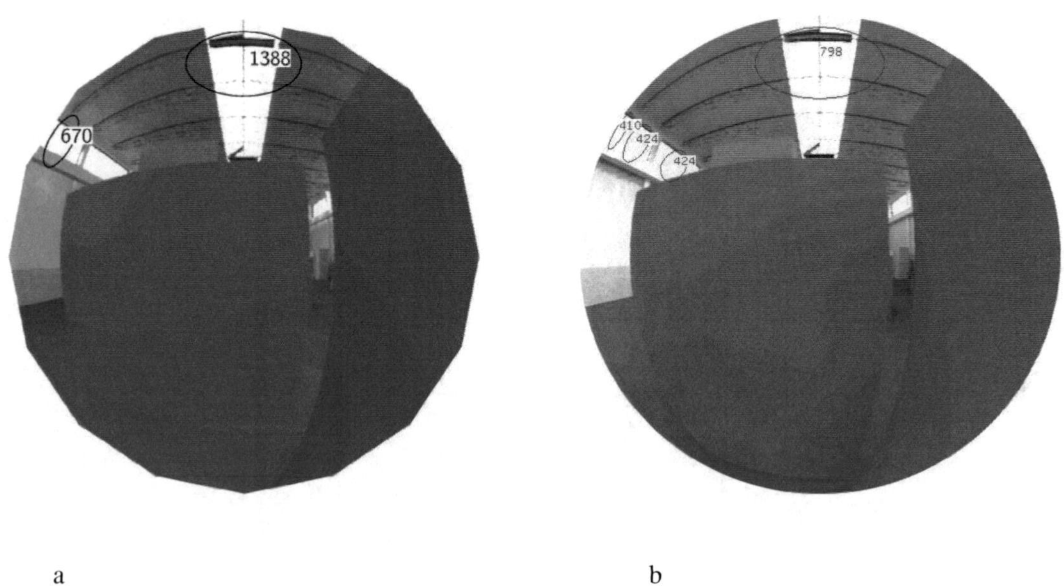

Figure 5. Luminance distribution, a – overcast sky, b – clear sky.

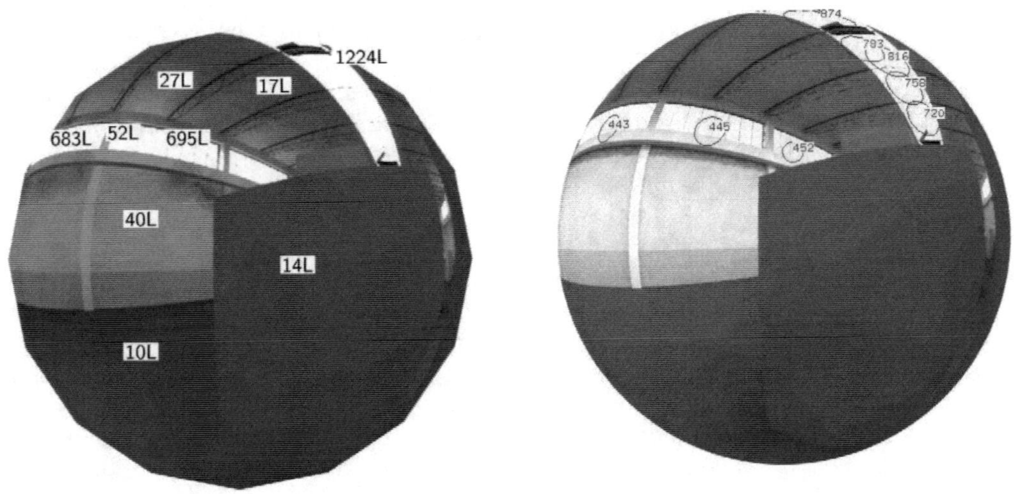

a b

Figure 6. Some luminance values, a – overcast sky, b – clear sky.

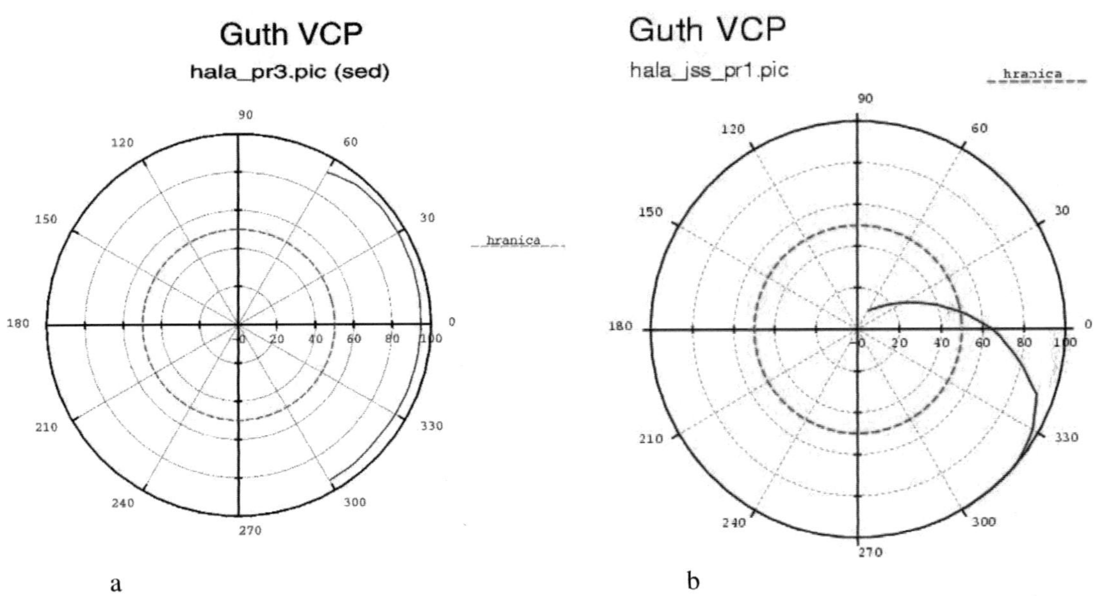

a b

Figure 7. Evaluation of visual comfort - calculated glare index (Guth Visual Comfort Probability). a – overcast sky, b – clear sky.

4. Conclusion

Skylights provide three times more light than the same area of vertical glazing and at the same time provide a much more even distribution of light inside, especially in larger spaces. Of course, using natural light is not without problems. Problems can cause glare, overheating, variability and invade privacy. The most important factor in terms of lighting quality is glare. Glare is a feeling caused by too bright brightness in the field of view. This can result in discomfort, annoyance or reduced productivity. According to the standard STN EN 12464 - 1, users must be protected from glare indoors. Visual comfort is affected for a light ratio of 1:10 and glare for a ratio of 1: 100. To assess visual comfort for a combined daylight and simple wire glass system, it is shown that visual well-being is not affected when the sky is cloudy. In case of clear skies, visual comfort is affected. At the same time, two types of sky are dazzling. The figures (see Figures 5, 6) show the values of the glare sources. Glare contrast occurs for clear skies (see Figures 5b, 6b). The graphs show the percentage of people who would be satisfied (Figure 7a, b). If the cloudy sky would be satisfied with 95% visual comfort (see Figure 7a). In the case of a clear sky, 10-70% of people would be satisfied (see Figure 7b). It can be stated, visual comfort is strongly influenced by the direction of sight.

5. References

[1] Heschong L Wright R L and Okura S 2002 *Daylighting impacts on retail sales performance Journal of the Illuminating Engineering Society* **31**

[2] Reinhart C F and Walkenhorst O 2001 *Validation of dynamic RADIANCE-based daylight simulations for a test office with external blinds. Energy and Buildings.* **33** pp 683-697

[3] Reinhart C F and Mardaljevic J and Rogers Z 2006 *Dynamic daylight performance metrics for sustainable building design. Leukos* **3** pp 7-31

[4] Yu X and Su Y 2015 *Daylight availability assessment and its potential energy saving estimation A literature review. Renewable and Sustainable Energy Reviews* **52**; pp 494-503

[5] Reinhart C F Mardaljevic J and Rogers Z 2006 Dynamic daylight performance metrics for sustainable building design. Leukos. 3 pp 7-31

[6] Knoop M et all 2019 *Methods to describe and measure lighting conditions in experiments on non-image forming aspects Journal of the Illuminating Engineering Society*. 15 (2-3), pp 163-179

[7] Němeček M and Kalousek, M 2015 *Influence of thermal storage mass on summer thermal stability in a passive wooden house in the Czech Republic. Energy and Buildings* **77**, p. 68-75.

[8] Čekon M. Kalousek M Hraška, J and Ingeli R. 2014 *Spectral optical properties and thermodynamic performance of reflective coatings in a mild climate zone. Energy and Buildings* **77**, p. 343-354.

[9] Omishore A Kalousek M and Mohelník P 2019 *Thermal testing of light pipe prototype. Engineering Review* **39**, 3 p. 283-291

[10] Kalousek M and Rybakova A 2014 *Modeling of Details for Single-Layer Wall in a Passive House: The Case of Balcony Slab*. In enviBUILD 2014. Advanced Materials Research (online). 1041. CH: Trans Tech Publications, p. 171-174.

[11] STN 730580; Daylighting in buildings, – 1 Basic requirements, 1986 – 2; Daylighting of residential buildings, 2000, Slovak Republic Office of Standards, Metrology and Testing: Bratislava, Slovakia;

[12] EN 17037; Daylighting in buildings 2018 CEN European Daylight standard, Published: May 2019

Acknowledgments

This work was supported by The Slovak Scientific Grant Agency under the contract No. VEGA 1/0626/22 and No. VEGA 2/0017/20.

Analysis of alternative designs for the reconstruction of an administrative building – case study

A Sedlakova[1]; V Merjava[1]; S Vilcekova[2]; M Gmitter[3] and D Burak[1]

[1] Institute of Architectural Engineering, Faculty of Civil Engineering, Technical University of Kosice, Vysokoskolska 4, 040 23 Kosice, Slovakia
[2] Institute of Environmental Engineering, Faculty of Civil Engineering, Technical University of Kosice, Vysokoskolska 4, 040 23 Kosice, Slovakia
[3] Institute of Technology, Economics and Management in Construction, Faculty of Civil Engineering, Technical University of Kosice, Vysokoskolska 4, 040 23 Kosice, Slovakia

E-mail: anna.sedlakova@tuke.sk; veronika.merjava@tuke.sk; silvia.vilcekova@tuke.sk; marek.gmitter@tuke.sk; dusan.burak@tuke.sk

Abstract. An old office building located on the Garbiarska street, historic part of Kosice city, east of Slovakia was selected because it represents a brownfield and necessarily requires reconstruction. This study is aimed at the analysis of two design alternative solutions to underline the architecture of the reconstruction in terms of sustainability criteria taking into account the significance of the history of the city and its surroundings. Performed SWOT analysis considered strengths and weaknesses of restoration, opportunities and threats, shows that the most suitable is considering a restoration strategy in terms of preserving cultural heritage, at the same time restoring the building in the concept of green architecture. It is more than necessary to prioritize this renovation strategy over an economically viable renovation while maintaining the inappropriate height of the office building. This early analysis confirms the significance of the revitalization of territories in connection to the improvement of quality of life, the enhancement of the historical value of the locality and the creation of new economic opportunities.

1. Introduction

Cultural protection of heritage urban ensembles, the conservation of the environment of individually protected assets, along with centralized and microclimate energy improvements, all happen at the urban scale, that also offers economies of scale and scope [1]. Study [2] presents new methodology for the identification of climate-friendly and cost-effective renovation solutions. This methodology includes the description of uncertain parameters related to the production of the materials and those related to the future building operation, such as climate change, future material replacement and electricity mix, occupancy behaviour and so on. According to study [3] renovation processes mainly of high-rise buildings require a more diverse range of designers, specialists, trades, and workers as well as the increased number of sub-contractors involved in a renovation. For the renovation of buildings, the scheduling method as it is developed and presented in study [4]. This study points out that a scheduling method for office building renovation projects enables the creation of schedule alternatives and suggests optimized schedules in terms of project duration and cost efficiency. Study [5] points out that lots of buildings constructed in the process of urbanization are aged. Countries in the EU have been aware of the problems in terms of the energy consumption and greenhouse gas emissions of these existing buildings. To address these problems, various studies have been conducted to determine an improvement method for the energy performance of aged buildings by adopting the smart city concept. This paper deals with the analysis of a selected administrative building, which was built in the 80s of the last century on Garbiarska Street in Kosice. The building was built to overshadow the surrounding buildings of the historic city centre. The building was purposefully composed on the historical axis of Main Street. The height of this building was given by functionalist realism. At that time, all civic and state's buildings were based on a simple architectural principle, which sought to contrast between the horizontal and vertical appearance of the building. The administrative building on Garbiarska is one of the tallest building in the city and is currently interesting only for its height. The building is currently empty and is being prepared for renovation [6]. The analysis described in this article is focused on the suitability of the design of the reconstruction of the building in terms of cultural heritage in accordance with the green architecture.

2. Materials and methods

Architectural design of the building reconstruction in accordance with the requirements for the design and operation of a low-carbon building and the development of a low-carbon emission built environment was managed as a team project. The architectural design of the reconstruction of an old unused building was presented in the diploma thesis. The building is located on the edge of the Kosice Municipal Monument Reserve and its protection zone. The architectural design of the building and its storey disrupt the important view from the Main Street to the north. This fact led us to the analysis and design of the reconstruction of the building to restore the historic architectural design of the city centre while preserving the heritage.

2.1. History

Garbiarska Street was a typical craft street, after which it was named. The reason for the development of this production activity was the nearby mill drive and the need for quality processed leather and tobacco. A railway line with the dragging of the production block was brought to Strojarenska Street, which later became a tram line. Residential development was gradually formed into a mixture of housing and production, which later dominated - processing of leather, tobacco, later textile and engineering production. Significant development began even after World War II. - apartment buildings, research institute, retirement homes and office buildings. In the 80's, a zoning plan was drawn up, which proposed a complete disruption of the appearance of the street. It also included the administrative building in question, as the germ of the decomposition of the entire street. It was built on the site of three family houses. As already mentioned, the architectural design of the building and its storeys disrupted an important view from the Main Street to the north. There has been an undesirable disruption of the past and present in a very sensitive axis of the city. Figures 1a and 1b depict historical map of the area with the marked place where the investigated building is located. Historic view to the main street of Kosice is shown on figures 2a, 2b, 3a, 3b and 4a.

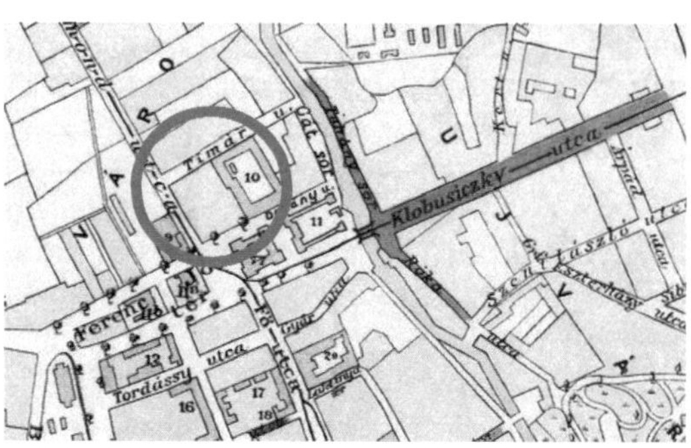

Figure 1a. Historical map of the area from the year 1909 [7].

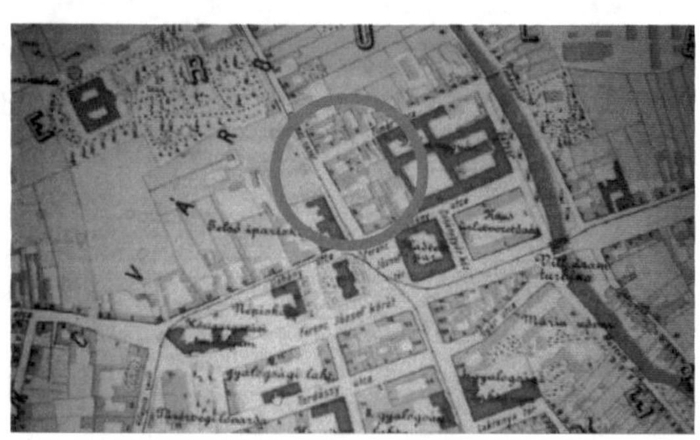

Figure 1b. Historical map of the area from the year 1912 [8].

Figure 2a. Historical view to the Main Street of the centre of the Kosice city from the year 1910 [9].

Figure 2b. Historical view to the northern part of the Main Street of the Kosice city [10].

Figure 3a. Historic view of the Peace Marathon Square before the construction of the administrative building on Garbiarska Street from the year 1964.

Figure 3b. Historic view of the Peace Marathon Square after the construction of the administrative building on Garbiarska Street. Photography from the 80´s.

Figure 4a. Historic view to the northern part of the Main Street of the Kosice city with a view of the historic building of the military garrison headquarters from the 80´s.

Figure 4b. Current view to the northern part of the Main Street of the Kosice city with a view of the Kosice Self-governing Region building. 12/2021.

2.2. Material

The administrative building on Garbiarska Street with thirteen above-ground storeys, the fourteenth retreating service storey and one semibasement storey, is the third tallest building in the city. It is the tallest building in Kosice's Old Town. The state sold the property, and the private owner aimed to convert it into apartments. The construction has been permitted since 2017, but the building is still in disrepair and is in a desolate state. According to the latest plans, the building should be renovated into an office building. Therefore, our study is focused on the architectural design of the reconstruction of a building into the building meeting the current high requirements for environmental, energy, social and economic efficiency. Figures 5a and 5b depicts the view of the building and devastated indoor environment of the building.

Figure 5a. The current view of the building [11].

Figure 5b. The current view to the inside of the building [11].

The plot within the boundary serves as a paved parking space for rent for cars as you can see on the figure 6b.

Figure 6a. Orthophoto map of the analysed area [12].

Figure 6b. The current view of the building and its surrounding [11].

2.3. SWOT analysis

The SWOT analysis as a tool was used in this study to assess the potential for regeneration of the area in terms of its strengths, weaknesses, opportunities, and threats. At the same time, it outlines possible alternatives for the future development of the city, or their solution. A positive scale from 1 to 5 was used to evaluate strengths and opportunities, with 5 indicating the highest satisfaction and 1 the lowest satisfaction. For weaknesses and threats, a negative scale from -1 (lowest dissatisfaction) to -5 (highest dissatisfaction) was used. The weighting criterion was not used due to the asymmetry of the number of individual criteria. The scoring was therefore considered. Based on this SWOT analysis, a suitable alternative to the use of the restoration of the area was considered in accordance with current trends in the design of green buildings in the broader urban - historical context of creating the "color" of the city. Figure 7 shows the parameters considered in the SWOT analysis.

SWOT analysis

	POSITIVE / USEFULL		NEGATIVE	
	STRENGTHS	rating	**WEAKNESSES**	rating
INTERNAL	1 Low-rise urban concept - not disturbing the siruet of the city monument zone	5	1 Failure to preserve the high-rise urban concept of the city during renovation	-5
	2 Green urbanism (interconnection of building and nature)	5	2 Economy prevails over cultural heritage/preservation of historical view	-5
	3 Change of thinking and design in line with the "green" renovation of buildings	5	3 Necessary regular maintenance and care of intensive greenery	-5
	4 Intensive green roof - vegetable growing / beekeeping / biodiversity	5	4 Insufficient experience with design in accordance with historical background of the city	-4
	5 Restoration of greenery with a positive effect on the city's microclimate (heat, noise, dust)	4	5 Increased costs for the irrigation system	-3
	6 Restoration of flora and fauna ecosystems; water retention	4	6 Provision of domestic water for irrigation (not drinking water)	-2
	7 Positive impact of greenery on the physical and mental health of workers	4	7 Seasonality - instability of some greenery; seasonal dependence	-2
	8 Aestheticization of the surroundings / attractiveness of the area	3	8 High trees may reduce the intensity of daylight	-1
	9 Increasing the image/creditworthiness of companies based on the "green building"	2	9 Increased demands for repairs of additional defects of the green roof / facades	-1
	10 Increase in property value	2	10	
	Sum	39	**Sum**	-28
	OPPORTUNITIES	rating	**THREATS**	rating
EXTERNAL	1 Creating space for relaxation zones through a green roof	5	1 Dependence of outdoor greenery on weather	-5
	2 Improving thermal stability in summer through a green roof / facade	4	2 Dependence of outdoor greenery on the quality of the environment	-4
	3 EU support for "green renewal"	3	3 higher incidence of health problems - allergies (pollen, insect bites)	-3
	4 Creating green architecture - new trends/ecological aspect	3	4 Financial complexity of implementation	-3
	5 Sustainability certification of buildings - LEED, BREEAM, WELL, LEVEL(s)	3	5 Roof structure - damaging of waterproofing function of the roof / plaster	-1
	6 Additional protection of waterproofing	2	6	
	7 Attractiveness for new customers - the image of a company based on a "green building"	2	7	
	8 Prioritize the renovation of buildings over the construction of new buildings	2	8	
	9 Legislative support for reconstruction	1	9	
	10 Financial support for reconstruction	1	10	
	Sum	26	**Sum**	-16

Figure 7. SWOT analysis.

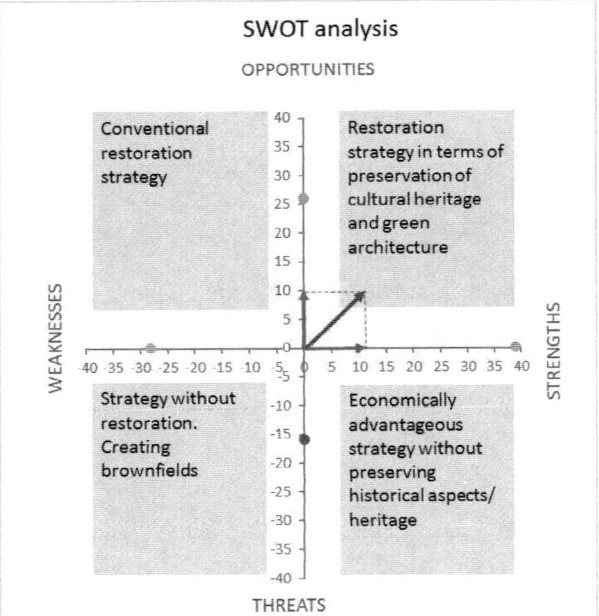

Figure 8. Graphical scheme of SWOT analysis result.

The result of the SWOT analysis evaluation, on the figure 8, is an offensive restoration strategy in terms of preservation of cultural heritage and green architecture, an alternative solution to a variant considering reducing the height of the building. Based on the results, a reference proposal for the reconstruction of the building and an alternative more sustainably advantageous proposal is presented.

3. Design of the building reconstruction

Based on the above-mentioned facts, that the building does not fit into the historic city centre, two different studies of the renovation of this high-rise building were carried out.

3.1. Variant with the preservation of the building height

This proposal considers the restoration of original function of the office building. Therefore, most of the floor space is intended for office space with facilities for employees. The design also uses the basic characteristics of the building, such as its height, through which it provides unmistakable panoramic views of the city of Kosice. It naturally followed the design of the upper floor for the public in the form of a restaurant. This proposal also considers the additional use of a large plot for the extension of a new multifunctional building and the construction of an underground parking house. The newly designed multifunctional building itself would complete the street line and thus consolidate the overall appearance of Garbiarska Street. This building is primarily designed as an office building with facilities for employees and secondarily as a kindergarten. Below the plot, 55% of the area is proposed as an underground car park. The first above-ground floor under the newly designed multifunctional building also serves as a parking area. In the city such as Kosice, the outside air temperature rises from year to year in the summer. One of the reasons is the decreasing number of vegetation and the increasing number of built-up and paved areas. The plot on which the controversial building is situated has an area of 3027 square meters. As this proposal for renovation and extension considers almost 70% of the built-up area of the whole plot, which represents an area of 2104.27 m^2, it was therefore necessary to compensate this percentage by the area of vegetation. We provided compensation in the form of vegetation-extensive roofs and vegetation facade panels with non-invasive vegetation. The total vegetation area in this proposal is 2069.91 m^2, which represents almost 69% of the total plot area. There is no park for the public near the renovated building, so this proposal also considers the use of vegetation areas within the first floor for this purpose.

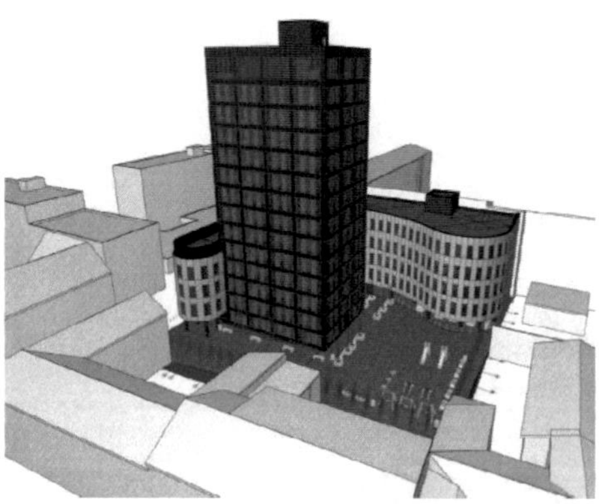

Figure 9a. Model of the building with the preservation of the building height (heigh) and the model of the proposed new building (low).

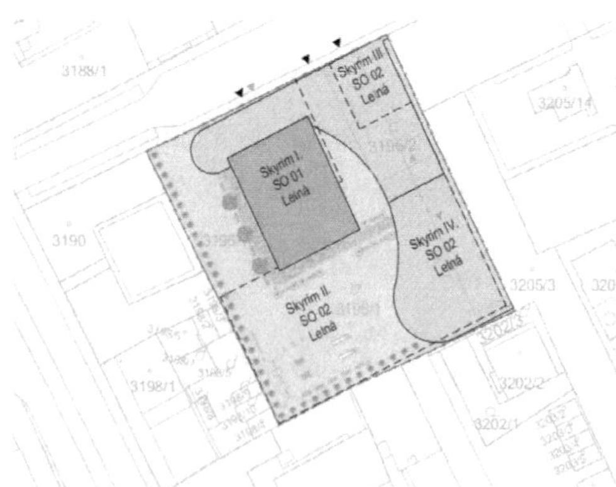

Figure 9b. Situation of the surroundings.

3.2. Variant considering reducing the height of the building

This variant respects the opinion of a group of inhabitants of the Kosice city, who express their negative attitude towards this building. The main reason is the inconsistency of the height of the building with the zoning plan of the city. The high-rise building on Garbiarska Street has 14 above-ground storeys, while the surrounding buildings do not have more than 5 storeys. For this reason, it is necessary to think about the alternative of lowering the building to the height of the surrounding buildings, namely the demolition of 9 floors. Therefore, the design was based on this idea, while preserving the original function of the building. Most of the floor space would be office space with facilities for employees, and the rest would be the technical background of the building. On the ground floor, the possibility of implementing services for the public such as a cafe. As mentioned above, the current problem in big cities is the creation of a heat island. It is therefore necessary to introduce vegetation areas into each new design, thus reducing paved areas. This proposal considers the application of vegetation in the form of a vegetative extensive roof as well as in the form of vegetation facade panels. Irrigation of these areas would be carried out through existing boreholes, which are located on the plot. In case of insufficient water capacity, their number can be increased. The existing asphalt car park will be partially replaced by grass blocks for the implementation of greenery in the largest possible percentage with respect to the total area of the plot. It is proposed to subdivide the lawn area through different height levels of embankments to mimic the natural topography. Planting of trees, shrubs and other flora is being considered. It naturally follows that the parking area would be divided into an area designated for parking and an area intended for the public. The implementation of a public park in a given locality is based on the absence of an area dedicated to this purpose in the area of interest.

Figure 10a. Model of the building with reducing the height of the building.

Figure 10b. Situation of the surroundings.

4. Discussion

The last decades have also been marked by a new view of the process of sustainable development and a new perception of the concept of culture. One of the key, the so-called soft factors of regional development, has become a cultural and social potential of the community. It represents the invaluable capital and unique wealth of the site and its inhabitants. It can be developed and cultivated, but it can also disappear. It is influenced by the business environment and the friendliness of public bodies, but also by the level of use of the uniqueness and uniqueness of the place (genius loci), the potential of its inhabitants and their publicly expressed satisfaction [13].

The attitude of many towards the "heritage of the past" has fundamentally changed. Efforts to protect cultural heritage and preserve it for future generations are often seen as a useless, sometimes even too costly, non-repayable investment [13]. It is therefore necessary to think about what values are important to us and what we want to preserve for future generations when looking at the possible renovation of the office building on Garbiarska Street in Kosice. If we consider that in the current period, when people spend 80-90% of the time in an indoor environment, it is very important to ensure the evaluation of the indoor environmental quality (IEQ), characterized by the effort to create thermal, acoustic and light comfort. At the same time, it is necessary to ensure IEQ with a link to the energy strategy for the renovation of the existing office building and the implementation of green infrastructure in the context of European strategies with the possibility of returning and preserving cultural heritage for future generations in such an intensively perceived part of Kosice.

The current building renovation strategy calls for measures and procedures to support the renovation of the building stock in order to achieve a highly energy efficient and carbon neutral building stock by 2050, and thus a cost-effective transformation of existing buildings into nearly zero energy buildings. A low-carbon study prepared by the World Bank identified the greatest potential for energy savings for Slovakia in the area of building renovation. The major renovation of an existing building must consider the technical, environmental and energy feasibility of high-efficiency alternative systems, as well as healthy climate conditions, the level of fire safety and the risks arising from intensive seismic activity. In the historical parts of the cities, it is also essential and important to evaluate the significant cultural heritage and the construction of green infrastructure, which is pointed out by the proposal of a possible solution of the variant considering the reduction of the building height.

In line with these ideas, it must be remembered that each nation has its own architectural monuments, which it inherited from the past into its heritage. Together with natural conditions and in mutual interaction, they form the characteristic features of individual territorial areas of the country, they become symbols evoking a feeling of homeland and respect for the homeland [14]. Each new generation should be enriched by the history of their surroundings, thus preserving the values of heritage for future generations.

In addition, it should be noted that over the last decade, green infrastructure has become the subject of considerable interest in urban and regional planning. However, the discussion is not entirely new. Significant elements can be traced in the work of Frederick Law Olmsted the Elder in the nineteenth century and to the roots of the professions of urban planning and landscape architecture. As evidence, this article cites three aspects of Olmsted's work in the current theory and practice of green infrastructure: ecosystem services and human well-being; environmental restoration; and comprehensive planning. The article then deals with Olmsted's philosophy regarding the civilizational influence of urbanism and concludes that green infrastructure can be an integral part of the development of the city of the twenty-first century.

Our future research task will be focused on the implementation of this green infrastructure and the evaluation of its importance in terms of reducing climate change [15].

5. Conclusion

The analyzed office building on Garbiarska Street in Kosice is currently in a very poor technical condition and represents brownfield. The insensitive approach to the city's high-rise concept allowed the construction of a 15-storey building in the 1980s that protrudes significantly beyond the surrounding buildings. It was built on the site of three family houses. However, the architectural design of the building and its storeys disrupted an important view from the main street to the north. There was an undesirable comparison of the past and the present in a very sensitive axis of the city. At present, when the restoration of this building due to its devastated condition is a necessity, the "city" is given a new chance to think about the possibilities of its restoration in accordance with the concept of “the height” of the historic city center and its immediate surroundings.

In the decision-making process of a suitable restoration strategy, a SWOT analysis was prepared based on an assessment of strengths and weaknesses of restoration, opportunities and threats, shows that the most suitable for restoring the office building would choose a restoration strategy in terms of preserving cultural heritage. It is necessary to point out the need for reconstruction of the building from the point of view of preserving the character of the historic city center and at the same time implement elements of green and blue infrastructure.

6. References

[1] Villarejo P, Gámez R, Santamaría-L'opez A 2021 Building Renovation Passports in Spain: Integrating exiting instruments for building conservation, renovation and heritage protection. *Energy Policy* **157** 112506

[2] Galimshina A, Moustapha M, Hollberg A, Padey P, Lasvaux S, Sudret B, Habert G 2021 What is the optimal robust environmental and cost-effective solution for building renovation? Not the usual. *Energy & Buildings* **251** 111329

[3] Stopps H, Thorneycroft Ch, Touchie M. F., Zimmermann N, Hamilton I, Kesik T 2021 High-rise residential building makeovers: Improving renovation quality in the United Kingdom and Canada through systemic analysis. *Energy Research & Social Science* **77** 102085

[4] Cho K, Kim T 2021 Optimized scheduling method for office building renovation projects. *Expert Systems with Applications* **168** 114212

[5] Cho K, Yang J, Kim T, Jang W 2021 Influence of building characteristics and renovation techniques on the energy-saving performances of EU smart city projects. *Energy & Buildings* **252** 111477

[6] Košice dnes official website [online] https://kosicednes.sk/zaujimavosti/najvyssie-budovy-v-kosiciach/

[7] Košice online official website [online] https://www.kosiceonline.sk/ako-sa-zrodila-mestska-cast-so-zaujimavym-nazvom-dzungla?fbclid=IwAR1sj8NjFbVmI1VE0LII8fwUnJK9WO5XdOblDkYuB8MvdEYxf-CMicjVHMc

[8] Historické mapy mesta Košice [online] https://www.kosice-estranky.sk/wp-includes/Mapy/HistorickaMapa_Kosice.html

[9] Košice online official website [online] https://www.kosiceonline.sk/historicky-prierez-obchodovania-a-trhovisk-v-kosiciach

[10] Staré Košice [online] https://www.starekosice.sk/hlavna-ulica-kosice.php

[11] Gmiter M 2021 Renovation of an office building. Diploma thesis, TUKE, Faculty of Civil Engineering Košice

[12] Google maps [online] https://www.google.sk/maps/@48.7285679,21.2536007,365m/data=!3m1!1e3?hl=sk

[13] Slušná S, Chomová S 2015 *Súčasné témy v kultúrnej teórii a praxi,* Univerzita Komenského v Bratislave, Available at: https://fphil.uniba.sk/fileadmin/fif/katedry_pracoviska/kkul/SLUSNA_CHOMOVA.pdf

[14] Štibrányi J *Slovensko a architektonické pamiatky,* Available at: https://cdv.uniba.sk/fileadmin/cdv/U3V/studijne-materialy/stiebranyi-slovensko-a--architektonicke-pamiatky.pdf

[15] Eisenman T S 2013 *Frederick Law Olmsted, Green Infrastructure, and the Evolving City,* Volume: 12 Issue: 4, page(s): 287-311. Available at: https://journals.sagepub.com/doi/abs/10.1177/1538513212474227

Acknowledgments
This research was financially supported by the Grant Agency of the Slovak Republic to support projects VEGA 1/0512/20.

IOP Conf. Series: Materials Science and Engineering 1252 (2022) 012033 doi:10.1088/1757-899X/1252/1/012033

Stationary dissipative systems of classical mechanics in the basis of unscientific principles of the theory of creep of reinforced concrete

A Beglov[1], R Sanjarovsky[2] and T Ter-Emmanuilyan[3]

[1] Governor of Saint-Petersburg Administration of St. Petersburg, proezd Smolnyi, d1, lit.1, 91060, Saint-Petersburg, Russia
[2] L N Gumilyov Eurasian National University,11 Kazhymukan str., 010008, Nur-Sultan, Kazakhstan
[3] Russian University of Transport, 9b9 Obrazcova St., 127994, Moscow, Russia

E-mail: gubernator@gov.spb.ru; milasanj@gmail.com; tanya_ter@mail.ru

Abstract. The basis of creation and development of the theory of creep of reinforced concrete is based on unscientific principles taken from systems of classical mechanics that do not correspond to this theory. Linear systems with stationary properties (inertial, dissipative and positional components) are used to describe the properties of aging concrete. The replacement of constant coefficients in classical models is carried out: instead of these constant coefficients ai variable coefficients are inserted, which are functions of time - $a_i(\tau)$, $a_i(\tau)$, i=1, 2, 3, 4. From such the replacement, the integrand function of the theory has lost its physical meaning. Errors in structural calculations are 300-500%.

1. Introduction

The creation and development of the theory of reinforced concrete creep is based on unscientific principles borrowed from systems of classical mechanics that do not correspond to this theory. Linear systems with stationary properties (inertial, dissipative and positional components) are mistakenly used to describe the properties of aging concrete. An enthusiastic and unconscious substitution of constant coefficients in classical models is carried out: instead of these constant coefficients a_i, variable coefficients are inserted that are functions of time $a_i(\tau)$, i = 1, 2, 3, 4. From such a substitution, the integrand function of the theory lost its physical meaning. Errors in structural calculations are 300-500%.

2. Methods

The essence of recording the theory of creep of reinforced concrete

The theory under study is assessed by leading scientists in Europe as, on an international scale ... a new advanced format for calculating creep; it is also emphasized that this theory is generally accepted [1], [2].

We will present the essence of the theory recording and its author's assessments in the original presentation, described by one of its founders. Let us also pay attention to the incorrect assessment of its origins: "... the foundations of this new theory were laid in the theoretical mechanics of continuous media at the beginning of the 20th century by Volterra (1909) under the definition of an elastic aftereffect, however, only at the beginning of the 1940s. this theory has been recognized by design engineers as a suitable mathematical tool for calculating the creep of concrete structures. " The initial provisions of the general theory of reinforced concrete are determined primarily by the creep law, the format of its mathematical recording of short-term and long-term deformations. "This format, first adopted in the preliminary CEB standard under the 1978 Model Code and widely used in the related CEB Manual, is now systematically adopted in the latest guidance documents, recommendations and standards by international standardization organizations [3], [4], [5]. It is based on the introduction of the Volterra integral equation as the basic law for concrete in subsequent equivalent forms, presenting a generalization according to the principle of linear superposition of the concrete reaction to the specific long-term uniaxial superimposed stress or deformation for the time-varying values of these applied actions (we give it in the original notation) [1] ":

$$\varepsilon_\sigma(t)=\sigma(t_0)J(t,t_0)+\int_{t_0}^{t} J(t,t')d\sigma(t') \quad (1)$$

where $\varepsilon_\sigma(t)$ – total deformation due to stress $\sigma(t)$;

$J(t,t')=\dfrac{1}{E_c(t')}+\dfrac{\varphi(t,t')}{E_c(t')}$ – compliance function;

$E_c(t')$ – unsteady modulus of elasticity;

$\varphi(t,t')$ – non-stationary creep characteristic, taking into account aging.

In scientific publications, it is usually integrated into (1) by parts, obtaining

$$\varepsilon_\sigma(t)=\frac{\sigma(t)}{E_c(t)}-\int_{t_0}^{t}\sigma(t')\frac{\partial}{\partial t'}\left[\frac{1}{E_c(t')}+\frac{\varphi(t,t')}{E_c(t')}\right]dt'. \quad (2)$$

Note that the expression $\dfrac{\varphi(t,t')}{E_c(t')}$ is a measure of the concrete creep $C(t, t')$ used in publications in our country, which is preferable to using the creep characteristic when processing experiments.

We emphasize that the aging of concrete is taken into account in $\varphi(t,t')$ and $C(t,t')$ and the modulus of elastic-instantaneous deformation $E_c(t')$ significantly depends on the age of concrete.

"Equations (1), (2)" are substantiated by two fundamental assumptions: the principle of linear relationship between stresses and deformations

$$\varepsilon_\sigma(t,t')=\sigma(t')J(t,t'); \quad (3)$$

the principle of imposition, verbally formulated in various versions of presentation in numerous well-known publications on the theory of concrete creep, reference books.

The unscientific of the considered world theory is estimated below by its following provisions:

a – thinking up the deformation properties of concrete, rejecting the Eurocode;

b – the use of systems of classical mechanics that do not correspond to this theory;

c – substitution of constant coefficients is in the differential equations of mechanics, on the variable coefficients that are functions of time;

d – the presence of mathematical errors in the integrand function of the law (1);

e – gross distortion of experimental data;

f – the application of theory in real design.

3. Results and discussion

a. About linearity of instantaneous deformation and creep deformation

Reputable experimental organizations of the world, as well as Eurocode, especially emphasize the severe nonlinearity of the instantaneous concrete diagram, as well as the highly nonlinearity of creep deformations. In the theory under consideration, this nonlinearity is not even mentioned. The linear theory is presented as a supposedly clear and obvious fact, known to everyone and not requiring explanations [6], [7].

b. The use of linear stationary systems of theoretical mechanics, for aging concrete (non- stationary).

The basic law of motion of a stationary system has the form

$$a_0 \frac{d^2x}{dt^2} + a_1 \frac{dx}{dt} + a_2 x = f(t) \tag{4}$$

where a_0, a_1, a_2 are constants; a_0 is usually taken equal to 1, if it is not equal to 1, all terms of the equation are divided by it.

The kind of the solution essentially depends on the roots of the trinomial $p^2+a_1p+a_2=0$. Below we consider the case given in the textbooks on mathematics (N. Piskunov) for the case $(a_1/2)^2- a_2<0$. The rest of the cases are treated in a similar way.

For the case $x_0=x_0'=0$, the solution has the form

$$x(t) =\int_0^t f(\tau)\psi_1(t-\tau)\, d\tau \tag{5}$$

$$\text{where } \psi_1(t-\tau)=\frac{1}{\sqrt{a_{2}- a_1^2/4}} e^{-\frac{a_1}{2}(t-\tau)} \sin(t-\tau)\sqrt{a_{2}- a_1^2/4} \tag{6}$$

In theoretical mechanics, it is shown that the function ψ_1 is the response of the system "equation (4)" to a unit impulse.

The action $\dot{f}(t)$ in the time interval $(0, t)$ can be represented as a sequential superposition of impulses $\dot{f}(\tau)\, d\tau$; this impulse causes the movement $\dot{f}(\tau)\, \psi_1(t-\tau)\, d\tau$. The quantity $x(t)$ represents the "overlapping motion" produced by the individual pulses. The reaction ψ_1 is obtained after direct integration of the homogeneous equation corresponding to "equation (4)".

Solution "equation (1)" is called in some textbooks the Duhamel integral (French mathematician (1797-1872)).

Integrating "equation (1)" by parts, we find

$$x(t) = f_0\psi_0(t-0) +\int_0^t \dot{f}(\tau)\, \psi_0(t-\tau)\, d\tau. \tag{7}$$

$$\psi_0(t-\tau)=\frac{1}{a_1}(1-e^{-\frac{a_1}{2}(t-\tau)}(\cos\sqrt{a_{2}- a_1^2/4}\,(t-\tau) +\frac{a_1}{2\sqrt{a_{2}- a_1^2/4}}\sin(t-\tau)\sqrt{a_{2}- a_1^2/4})) \tag{8}$$

where $\psi_0(t-\tau)$ is called the response of the system to a single disturbance; there are also other names (for example, creep measure).

"Equation (4)" is also considered in various degenerate forms with other functions ψ_1 and ψ_0 in solutions. For example, neglecting the inertial component (if possible), we have a solution in the form

$$x(t)= \int_0^t f(\tau)\frac{1}{a_1}e^{-(t-\tau)^{a_2}/a_1}\, d\tau \tag{9}$$

or after integration by parts

$$x(t)= f(0)\psi_0(t-0)+\int_0^t \dot{f}(\tau(\frac{1}{a_2}\,(1-e^{-(t-\tau)^{a_2}/a_1}))\, d\tau \tag{10}$$

In some of the simplest cases, the above theory can be used to describe creep deformations. To take into account instantaneous deformations, one must first find them, regardless of the creep deformations. The total deformation is found by summation. If the instantaneous deformations are linear and stationary, then they can be taken into account in "equation (7)" by simple summation with a unit stationary creep deformation:

$$\frac{1}{E_0}+\psi_0(t-\tau)=I(t-\tau) \tag{11}$$

where $I(t-\tau)$ is the coefficient of compliance of total deformations of the system (E_0=const). In the stated theory, it is possible to reveal the presence of generalized deformation (dependence of x on t and τ)

$$x=\dot{f}(\tau)I(t-\tau) \tag{12}$$

conditioned to the properties of linearity and stationarity. With its help, we will find an expression for the speed

$$v=\dot{f}(\tau)I(t-\tau)+f(\tau)\frac{\partial I}{\partial \tau}+ f(\tau)\frac{\partial I}{\partial t} \tag{13}$$

Here, the last two terms are equal in magnitude, and one is negative: this important property will help us in the analysis of errors in the theory of creep of aging (unsteady) concrete.

c. The substitution of constant coefficients a_i = const by variable coefficients $a_i(\tau)$.

In the textbooks of mathematics, it is emphasized that "there are no general methods for finding in the final form the general solution of a linear equation with variable coefficients."

The theory of creep of aging concrete does not notice this, and uses the solution “equations (7), (10), (11)”, by replacing the constants E_0, a_1, a_2 with functions of time.

The theory comes up with data on the existence of generalized deformation “equation (12)” under conditions of nonstationarity (aging ') of concrete properties. However, at the same time, the theory of concrete creep has lost the features of speed “equation (13)”. The existence of a "superposition principle" is invented, under various pretexts, for example, attributing this error to Volterra's theory.

Meanwhile, the condition of the integral “equation (7)” - the "principle of superposition" - is only an expression of the fourth consecutive analytical action in the theory under consideration: first, the law of a linear (stationary dissipative) system “equation (4)” is written; then his solution is found “equation (1)”; third, this solution is integrated by parts “equation (7)”; finally, the features of the velocity structure of this system are revealed, which follow from the properties of its linearity and stationarity: the integral of this velocity structure is given the name of the superposition principle.

d. The availability of a mathematical error in the theory of reinforced concrete creep.

The error arose due to the introduction of an impossible (in the conditions of concrete aging) algebraic linear connection “equation (3)”, by analogy “equation (12)” and also due to imitation in copying the expression of speed, which is present in “equation (7)”. Both the creators and numerous followers, quite famous scientists, have been carried away for eighty years by the verbal formulations of the expression “equation (1)”, called the principle of superposition, carried away by the crafty justifications for it.

Meanwhile, the value of the strain rate corresponding to the assumption “equation (3)” is

$$\mathrm{v}_{\sigma}(t,t')=\dot{\sigma}(t')\cdot J(t,t')+\sigma(t')\frac{\partial J(t,t')}{\partial t}+\sigma(t')\frac{\partial J(t,t')}{\partial t'}$$

From this it can be seen that in the basic law “equation (1)” four terms are lost, caused by the rate of change of the compliance coefficient

$$-\sigma(t')\frac{\dot{E}_c(t')}{E_c^2(t')}+\sigma(t')\frac{1}{E_c(t')}\frac{\partial\varphi(t,t')}{\partial t}+\sigma(t')\frac{1}{E_c(t')}\frac{\partial\varphi(t,t')}{\partial t'}-\sigma(t')\varphi(t,t')\frac{\dot{E}_c(t')}{E_c^2(t')} \tag{14}$$

besides they are comparable in importance with the remaining term. These losses cause significant discrepancies between theory and experiment described in the scientific literature.

Let's take a closer look at this.

The velocity of elastic deformation is

$$\dot{\varepsilon}_{\mathrm{y}}(t')=\dot{\sigma}(t')\frac{1}{E_c(t')}+\sigma(t')\frac{\partial}{\partial t'}\frac{1}{E_c(t')}.$$

Integrating, we have

$$\varepsilon_{\mathrm{y}}(t)-\varepsilon_{\mathrm{y}}(t_0)=\int_{t_0}^{t}\frac{1}{E_c(t')}d\sigma(t')+\int_{t_0}^{t}\sigma(t')\frac{\partial}{\partial t'}\frac{1}{E_c(t')}dt'.$$

Integrating the first term by parts, we find

$$\varepsilon_{\mathrm{y}}(t)-\varepsilon_{\mathrm{y}}(t_0)=\frac{\sigma(t)}{E_c(t)}-\frac{\sigma(t_0)}{E_c(t_0)}-\int_{t_0}^{t}\sigma(t')\frac{\partial}{\partial t'}\frac{1}{E_c(t')}dt'+\int_{t_0}^{t}\sigma(t')\frac{\partial}{\partial t'}\frac{1}{E_c(t')}dt'$$

The short-term deformation is equal to

$$\varepsilon_{\mathrm{y}}(t)=\frac{\sigma(t)}{E_c(t)} \tag{15}$$

It is also seen that the first term under the integral sign (2) is superfluous, and the use in “equations (1), (2)” and of the superposition principle

$$\varepsilon_y(t)=\frac{\sigma(t_0)}{E_c(t_0)}-\int_{t_0}^{t}\frac{1}{E_c(t')}d\sigma(t')=\frac{\sigma(t)}{E_c(t)}-\int_{t_0}^{t}\sigma(t')\frac{\partial}{\partial t'}\frac{1}{E_c(t')}dt', \tag{16}$$

is deeply mistake [8].

The overlay principle mistakenly reconstructs the actual, real elastic linear concrete model with the modulus $E_c(t)$; he attaches to it a non-existent and unrealistic model of a linear viscous fluid with a viscosity coefficient $K_1(t')=\frac{E_c^2(t')}{\dot{E}_c(t')}$, thus forming a Maxwell scheme.

Consider an example, setting in “equations (15), (16)” $\sigma(t)=\sigma_0=const$, we obtain $\varepsilon_y(t)=\frac{\sigma_0}{E_c(t)}$ and $\varepsilon_y(t_0)=\frac{\sigma_0}{E_c(t_0)}=const$. Comparison of unsteady elastic deformations is shown in Figure 1.

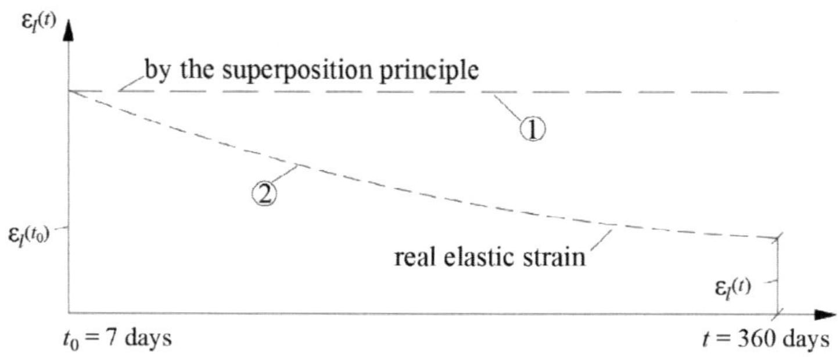

Figure 1. Comparison of $\varepsilon_y(t_0)$ and $\varepsilon_y(t)$.

Curve 2 in Figure 1 corresponds to the VNIIG data on the change in the elastic modulus *Ec (t)* over time. Errors in the value of elastic deformation reach, at $t = 360$ days, ≈ 300%.

Another distortion of the σ-diagram of concrete is caused by the erroneous presence in the creep core of the integral “equation (2)” of instantaneous deformation. This presence introduces a number of well-known scientists into a surprising delusion: "about taking into account the influence of the prehistory of deformation on the modulus of elastically instantaneous deformations."

The basic equation of the theory of creep in these misconceptions takes the form (in the original notation)

$$\varepsilon(t)=\frac{\sigma(t)}{E_c(t,t')}-\int_{t_0}^{t}\sigma(\tau)\frac{\partial}{\partial\tau}\left[\frac{1}{E_c(t,\tau)}+C(t,\tau)\right]d\tau. \tag{17}$$

An "experimentally substantiated" expression appears for the modulus of elastic deformation of concrete $E_{t,\tau}=E_t+a_{n,\tau}\phi_t E_\tau$, where φ_t is the concrete creep characteristic.

Other erroneous forms of the basic law of creep appear.

$$\varepsilon(t)=\frac{\sigma(t)}{E(t)}+\int_{\tau_1}^{t}\sigma(\tau)\frac{\partial}{\partial\tau}\chi(t,\tau)d\tau-\int_{\tau_1}^{t}\sigma(\tau)\frac{\partial}{\partial\tau}C^*(t,\tau)d\tau, \tag{18}$$

where $\frac{\partial}{\partial\tau}C^*(t,\tau)=\frac{\partial}{\partial\tau}\left[\frac{1}{E(\tau)}+C(t,\tau)\right]$;

$\chi(t, \tau)$ - has the name "reducing correction ... to the current specific elastic-instantaneous deformations".

e. The distortions of experimental data.

The creep of concrete, Figure 2, obtained at different loading levels, indicates that "no linear creep exists (see σ=0.1R_{lm})". The error from linear creep application is 200-500%.

Figure 3. shows a diagram of short-term concrete deformations according to the Eurocode, its analytical description [9]. The dummy diagram, from the theory of concrete creep, is shown with a dashed straight line. These data are well known; they are substantiated by numerous experiments. Back in 1931, the «Beton und Eisen» magazine held a discussion on Emperger's suggestion that the theory of elastic concrete was wrong.

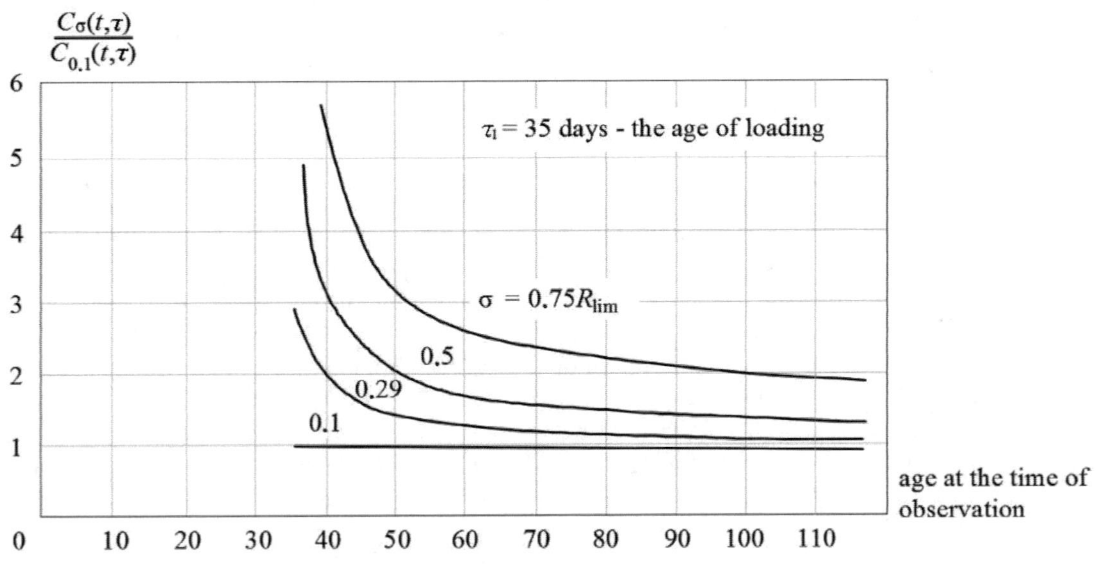

Figure 2. Change in the ratio of specific creep deformations at different initial stress levels $C_{\sigma(t,\tau)}$ to specific creep deformations at initial stress level $C_{0.1(t,\tau)}$.

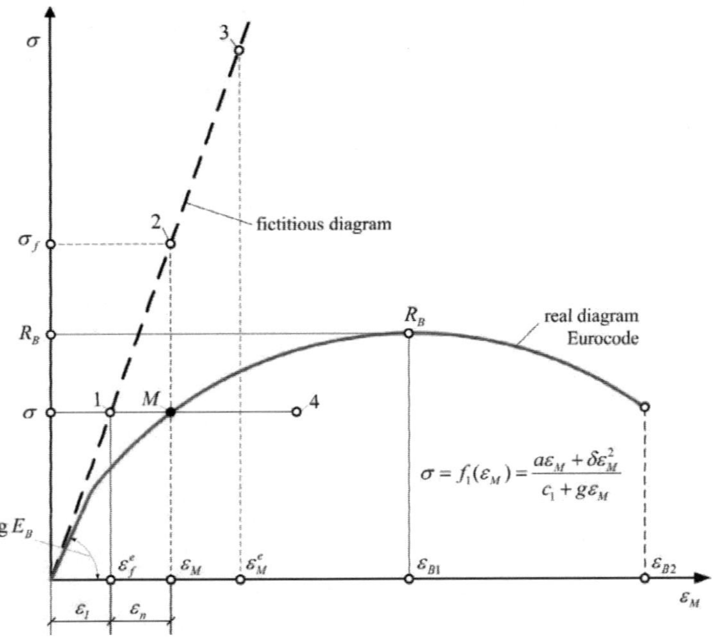

Figure 3. The distortion of the σ-ε diagram of concrete.

4. Conclusion

Application of theory in real design

Gordon Clarke, Director of RAMBOLL, UK, president of fib, points out [2]: "The international experience in accurately predicting the impact of creep ... is highly controversial"; "The above is a warning to designers ... that predicting creep and shrinkage is not an exact science that is still not fully understood to date." It can only be added that in view of the enormous use of reinforced concrete throughout the world, the analyzed unscientific theory of its calculation brings enormous economic damage to world construction. The results of the reinforced concrete calculation theory analysis, as well as the essence of mathematical errors in the theory of concrete creep were reported and discussed at the international symposium in Belgium in 2018 [8], and at the international conference in Moscow in 2014 [10].

5. References

[1] Chiorino M A 2014 Analysis of structural effects of time – dependent behavior of concrete: an internationally harmonized format // Concrete and Reinforced concrete – Glance at Future. III All Russian (International) Conference on Concrete and Reinforced Concrete **7** pp 338-50

[2] Gordon Klark 2014 Challenges for concrete in tall buildings // Concrete and Reinforced concrete – Glance at Future. III All Russian (International) Conference on Concrete and Reinforced Concrete **7** pp 103-12

[3] ACI 209.3R-XX 2011 Analysis of Creep and Shrinkage Effects on Concrete Structures Final Draft Chiorino M A (Chairm. of Edit. Team) ACI Com-mittee 209 pp 228

[4] fib Model Code for Concrete Structures 2010 Ernst & Sohn 2013 402 pp

[5] Yu Q, Bazant Z P and Wendner R 2012 Improved Algorithm for Efficient and realistic Creep Analysis of Large Creep – Sensitive Concrete Structures *ACI Structural Journal* **109** №5 pp 665-75

[6] Sanzharovskij R S, Manchenko M M, Gadzhiev M A, Musabaev T T, Ter-Emmanuil'yan T N and Varenik K A 2019 Sistema nesostoyatel'nosti sovremennoj teorii dlitel'nogo soprotivleniya zhelezobetona i preduprezhdeniya proektirovshchikov // Stroitel'naya mekhanika inzhenernyh konstrukcij i sooruzhenij chapter 15 № 1 pp 3-24

[7] Sanjarovskiy R, Ter-Emmanuilyan, T and Manchenko M 2015 Creep of Concrete and Its Instant Nonlinear Deformation in the Calculation of Structures CONCREEP 10 pp 238-47

[8] Life-Cicle Analysis and Assessment in Civil Engineering Proceedings of the Sixth International Symposium of Life-Cicle Civil Engineering (IALCCE) 2018 Ghent Belgium

[9] EN 1992-2 2004 Eurocode 2: Design of constructions

[10] Concrete and reinforced concrete 2014 look into the future Scientific works of the III All-russian (II International) conference on concrete and reinforced concrete

Quantification of temporal and spatial changes in soil suspension density, specific surface area and parameters of the distribution of the texture of the dispersed soil phase

M Gomboš, D Pavelkova, B Kandra and A Tall

Institute of Hydrology, Slovak Academy of Sciences, Dúbravska cesta 9, 841 04 Bratislava, SR

E-mail: gombos@uh.savba.sk; pavelkova@uh.savba.sk; kandra@uh.savba.sk; tall@uh.savba.sk

Abstract. Number of engineering tasks concerning water management as well as hydrological research require sound knowledge of temporal and spatial changes of soil suspension density, specific surface area and distribution parameters of soil texture in dispersed soil phase. This contribution is based on the hypothesis that there is a dependence between density of soil suspension, specific surface area of soil particles and distribution parameters of soil texture *d(10)*, *d(50)* and *d(90)*. The contribution analyses the results of an experiment investigating the temporal and spatial course of soil suspension density, specific surface area of soil particles and the distribution parameters of soil texture *d(10)*, *d(50)* and *d(90)* in soil suspension. For the research purposes, the methods of laser diffraction, sedimentation analysis and statistical correlation analysis were used. The results of the experiment are displayed graphically. To enhance the visual representation of the course of the investigated parameters, the measured values are approximated by empirical polynomials of various degrees. Close correlations between the investigated parameters are identified.

1. Introduction

The Number of engineering tasks concerning water management as well as hydrological research require sound knowledge of temporal and spatial changes of soil suspension density, specific surface area and distribution parameters of soil texture in dispersed soil phase [1], [2], [3].

Measuring density (bulk density ρ_σ) of soil suspension is standardly used for sedimentation analyses of soil texture. Theoretical basis for sedimentation analyses is Stokes law for spherical particles, valid for laminar flow [4], [5], [6], [7]. Nowadays, there are other state-of-the-art technologies used for measuring particle size distribution. For soil micro-particles, laser diffraction is used with good results. The method is robust and demands less time and effort in comparison to the traditional methods [8], [9], [10], [11], [12]. The analysis enables determination of grain-size distribution in a soil sample. In order to characterize soil texture, basic parameters of distribution function for grain-size distribution *D(10)*, *D(50)* a *D(90)* are determined. The value *D(90)* represents maximum diameter of a soil particle and smaller, which is located with the probability of 90% in the measured sample of soil suspension at certain time point. Analogically, the values *D(10)* and *D(50)* represent the maximum diameter of a soil particle and smaller which is located with the probability of 10% and 50%, respectively, in a sample of soil suspension at certain time point. Determining soil texture is important in hydropedology for analysing hydropedological and hydraulic soil properties. Specific surface area plays an important role in examining the processes which are influenced by surface area, such as adsorption, heat exchange, etc.

This contribution is based on the hypothesis that there is a dependence between density of soil suspension, specific surface area of soil particles and distribution parameters of soil texture *D(10)*, *D(50)* and *D(90)*. The aim is to analyse the results of the experiment investigating the temporal and spatial course of soil suspension density, specific surface area of soil particles and distribution parameters of soil texture in dispersive portion of a soil profile *D(10)*, *D(50)* and *D(90)*. The analysis evaluates the correlations and the possible approximation of the measurements with empirical curves. For the research purposes, the methods of laser diffraction, sedimentation analysis and statistical correlation analysis were used. The results of the experiment are displayed graphically. The correlation between the investigated parameters was assessed by the method of correlation analysis.

Use of sections to divide the text of the paper is optional and left as a decision for the author. Where the author wishes to divide the paper into sections the formatting shown in table 2 should be used.

2. Material and methods

Soil samples were taken in Senné area. This area is characterized by extremely heavy soils where soil particles smaller than 0.002 mm represent more than 77% of soil texture. Content of silt is approx. 18% and sand is 4%, see figure 1.

Clay soil sample was used for creating dispersed phase in water (distilled water) dispersion system. In order to identify the morphological properties of different soil fractions, sedimentation analysis was used. During the process of soil suspension fractionation and subsequent creation of lyofobic soil (composed of clay particles dispersed in water), samples for laser diffraction analysis were taken. The samples were taken at different heights (0.06 m, 0.12 m and 0.18 m) above the bottom of the sedimentation vessel, see figure 2. Grain-size distribution of the soil samples and their specific surface area were measured by MALVERN Instruments device called Mastersizer 2000. The device can work both wet and dry way. The analyses described in this contribution were performed by means of dispersion unit Hydro 2000MU, i.e. wet style. Before each analysis, samples of soil suspension were exposed to ultrasound for 5 minutes in order to enhance dispersion. The measurement range indicated by the manufacturer is between 0.02 and 2000 μm. Soil suspension density was measured by Casagrande densimeter in the course of 6 days, i.e. 8640 minutes, at dispersion system temperature of 20°C. At the end of the measurement, a sedimentation equilibrium began to form in the upper part of the sample (0.18 m above the bottom).

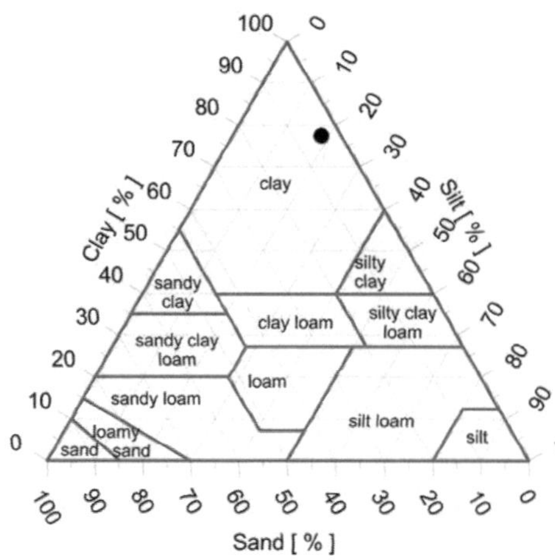

Figure 1. Identification of the soil sample within the USDA triangle.

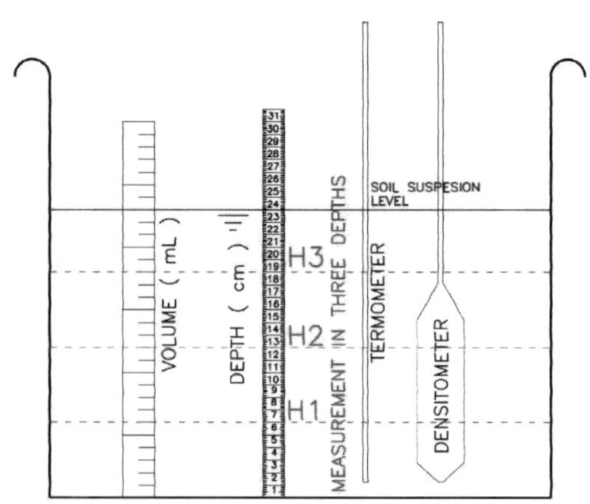

Figure 2. Scheme of the experiment.

The measured density is the value for the whole suspension undergoing sedimentation. Correlation between the investigated parameters was identified. In order to identify the correlations, average values of distribution parameters and specific surface area from all three sampling points (0.06 m, 0.12 m and 0.18 m above the bottom of sedimentation vessel). To enhance the visual representation of the course of the investigated parameters, the measured values are approximated by empirical polynomials of various degrees. Besides, grain-size distribution of dispersed soil phase at all three sampling points were identified at the beginning of the experiment, at the end of density measurement and at the very end after sedimentation equilibrium occurred in the whole soil colloid.

3. Results and discussion

Figure 3 shows grain-size distribution of the dispersed soil in soil suspension at the beginning of the experiment. Soil suspension is homogenous in the whole of the volume. Distribution curves are almost identical at all three sampling heights (see figure 2). This is the starting point of the experiment. Due to subsequent settlement of soil particles, grain-size distribution and the distribution parameters *D(10)*, *D(50)* and *D(90)*, as well as specific surface area and suspension density differ at different sampling height above the settlement vessel.

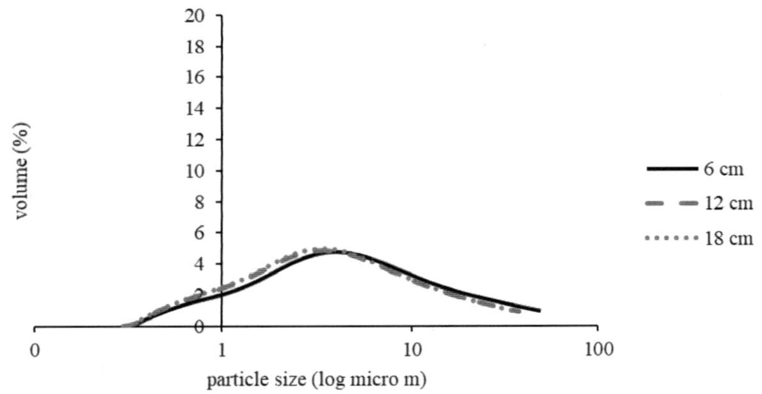

Figure 3. Particle size distribution – 0 minutes after the start of the experiment.

Figure 4, figure 5 and figure 6 show the changes in specific surface area and grain size of soil particles for various measured values of soil suspension density. The figures show that as the suspension density decreases, the size of soil particles in suspension decreases and specific surface area of solid phase increases. Due to the dynamics of settlement, these processes are manifested at the levels higher above the bottom of the settlement vessel.

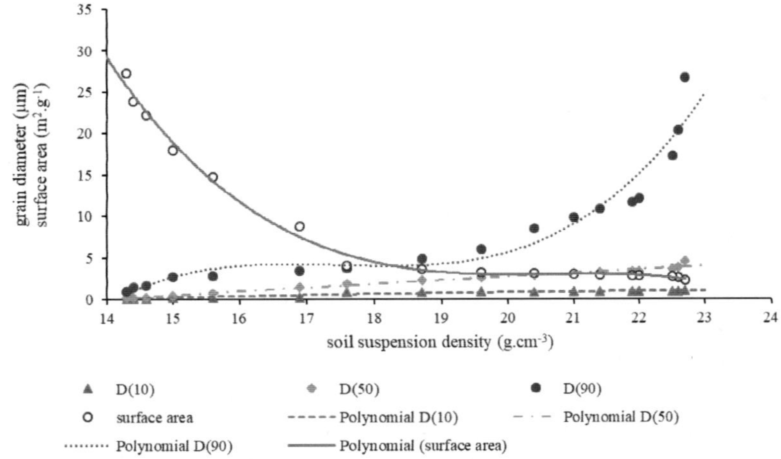

Figure 4. The course of the investigated parameters at 0.06 m above the bottom of the sedimentation vessel.

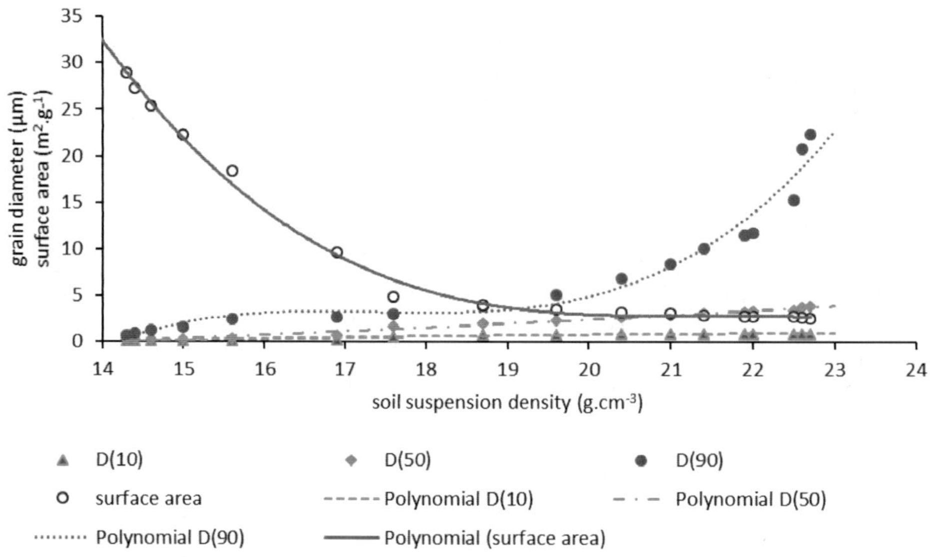

Figure 5. The course of the investigated parameters at 0.12 m above the bottom of the sedimentation vessel.

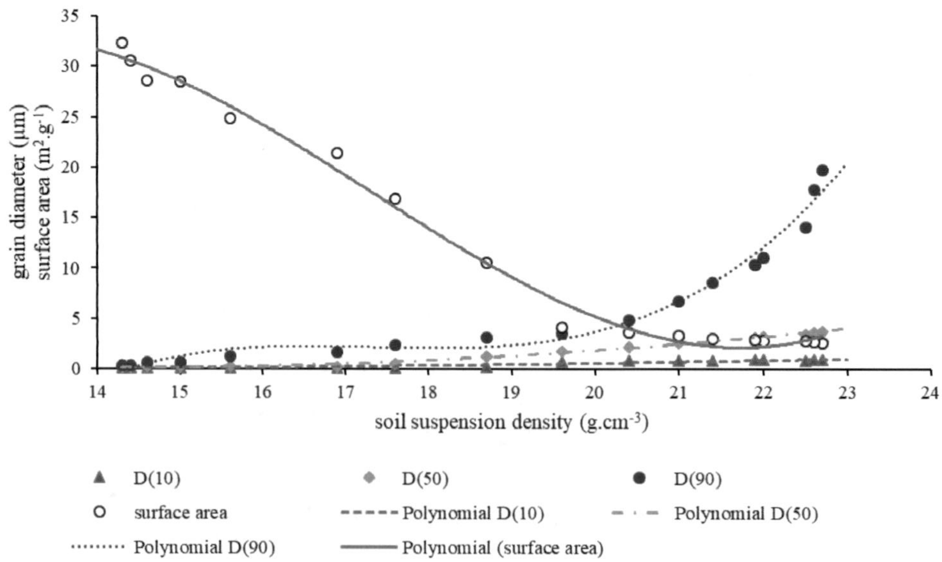

Figure 6. The course of the investigated parameters at 0.18 m above the bottom of the sedimentation vessel.

Analogically, figure 7 shows the courses of average values of the investigated parameters at different heights of the soil suspension. It provides us with the overall picture of the conditions occurring in soil suspension in its whole volume.

Figure 8 describes the course of the investigated parameters in time. It is obvious that as the time increases, both suspension density and particle size for each distribution parameter decreased. On the contrary, specific surface area increased. The identified changes are non-linear.

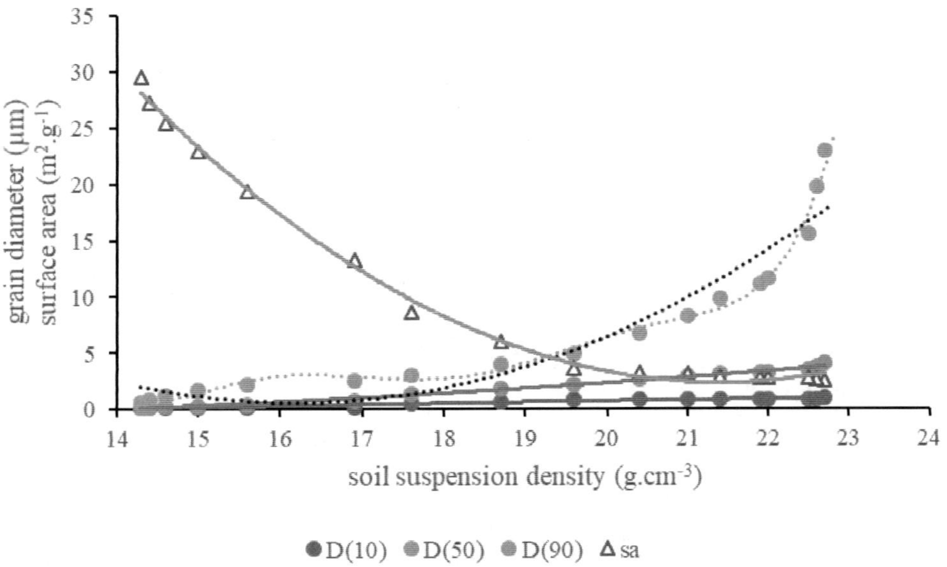

Figure 7. Courses of average values of the investigated parameters in the sedimentation vessel.

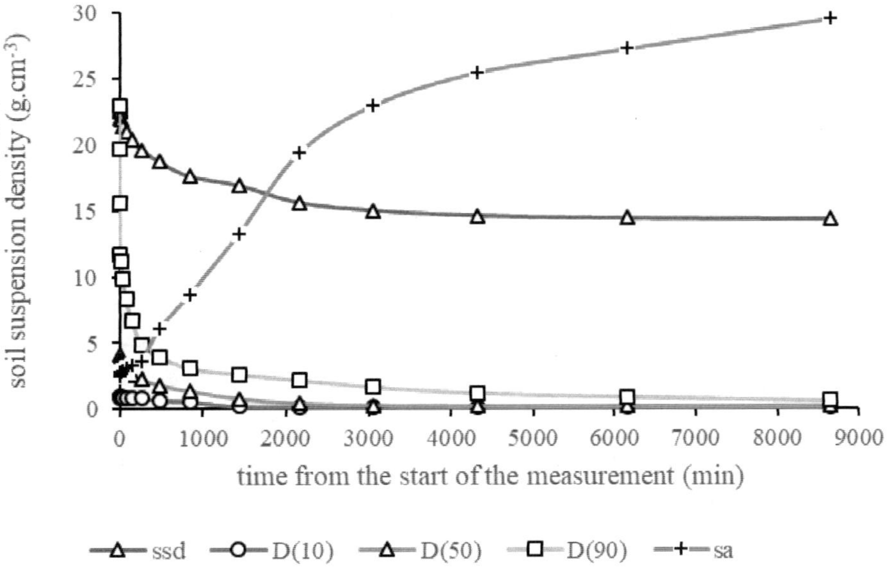

Figure 8. Time course of average values of the investigated parameters in soil suspension.

Figure 9 shows changes in soil particle sizes as a function soil suspension density and grain-size distribution parameters *D(10)*, *D(50)* and *D(90)*. Identified changes are non-linear. The greatest are identified for *D(10)*.

This is visible from the values listed in table 1. Towards the end of the experiment, sedimentation equilibrium starts to be formed at 18 cm above the bottom of the sedimentation vessel.

Towards the end of the experiment, sedimentation equilibrium starts to be formed at 18 cm above the bottom of the sedimentation vessel. Figure 10 shows the formation of sedimentation equilibrium by means of distribution curves, measured 8640 minutes from the start of the experiment.

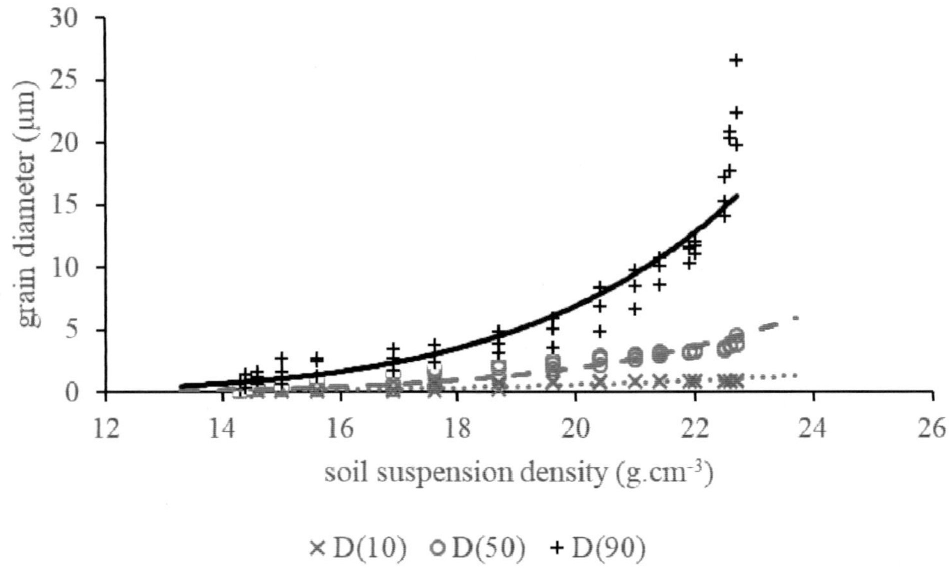

Figure 9. Soil particle sizes as a function of suspension density for grain size distribution parameters *D (10)*, *D (50)* and *D (90)* in the investigated suspension.

Table 1. Time course of the investigated parameters at different heights above the bottom of the sedimentation vessel.

Time from start of measur (min)	Soil suspension density (g cm³)	6 cm Grain diameter *D(10)* (μm)	6 cm *D(50)* (μm)	6 cm *D(90)* (μm)	6 cm Surface area (m^2g^{-1})	12 cm Grain diameter *D(10)* (μm)	12 cm *D(50)* (μm)	12 cm *D(90)* (μm)	12 cm Surface area (m^2g^{-1})	18 cm Grain diameter *D(10)* (μm)	18 cm *D(50)* (μm)	18 cm *D(90)* (μm)	18 cm Surface area (m^2g^{-1})
0	22.7	1.4	4.58	26.61	2.29	0.92	3.87	22.37	2.6	0.91	3.77	19.82	2.65
1	22.6	0.92	3.8	20.37	2.62	0.91	3.76	20.86	2.64	0.9	3.66	17.81	2.7
3	22.5	0.89	3.58	17.25	2.73	0.88	3.48	15.33	2.78	0.86	3.38	14.8	2.85
8	22	0.93	3.31	12.6	2.77	0.92	3.29	11.78	2.78	0.91	3.23	11.11	2.83
18	21.9	0.93	3.27	11.68	2.78	0.92	3.26	11.5	2.81	0.9	3.14	10.32	2.89
40	21.4	0.9	3.2	10.82	2.86	0.89	3.12	10.11	2.91	0.87	2.93	8.62	3.4
80	21	0.89	3.7	9.76	2.94	0.86	2.9	8.46	3.7	0.84	2.61	6.72	3.28
150	20.4	0.87	2.92	8.41	3.5	0.84	2.68	6.88	3.23	0.79	2.23	4.84	3.65
270	19.6	0.88	2.57	5.94	3.24	0.82	2.31	5.1	3.54	0.77	1.82	3.57	4.12
480	18.7	0.81	2.25	4.87	3.6	0.79	1.95	3.85	3.94	0.21	1.21	3.14	10.6
840	17.6	0.78	1.9	3.75	4.2	0.61	1.64	3.3	4.93	0.16	0.49	2.43	16.9
1440	16.9	0.25	1.39	3.43	8.73	0.17	0.69	2.67	9.65	0.15	0.26	1.7	21.4
2160	15.6	0.17	0.74	2.76	14.8	0.16	0.33	2.45	18.4	0.15	0.22	1.3	24.8
3060	15	0.16	0.36	2.7	18	0.15	0.24	1.59	22.3	0.14	0.2	0.65	28.5
4320	14.6	0.15	0.24	1.64	22.2	0.14	0.22	1.26	25.4	0.14	0.2	0.63	28.6
6160	14.4	0.15	0.23	1.43	23.8	0.14	0.21	0.87	27.3	0.14	0.19	0.36	30.5
8640	14.3	0.14	0.21	0.9	27.2	0.14	0.2	0.62	28.9	0.14	0.19	0.28	32.3

The right sides of the distribution curves show that, though minimal, at each sampling height the sedimentation processes still occur. The measurement of density had stopped. The changes in grain-size distribution and specific surface area in soil suspension was further examined until dynamic sedimentation equilibrium occurred. This happened 61860 minutes after the start of the experiment, see table 2. At the state of dynamic sedimentation equilibrium, sedimentation rate of dispersed particles equals diffusion rate in the opposite direction. This is typical especially of colloidal dispersions. Figure 11 shows the distribution curves for all three sampling heights, i.e. 6 cm, 12 cm and 18 cm above the bottom of the sedimentation vessel. All three curves are identical. Soil suspension is homogeneous and reached sedimentation equilibrium.

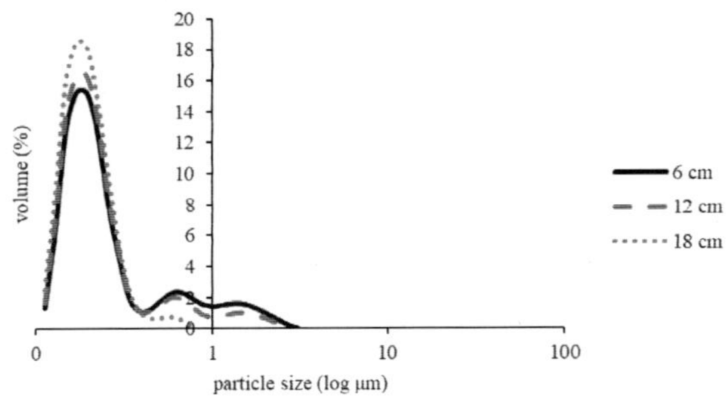

Figure 10. Gain-size distribution – 8640 minutes after the start of the experiment.

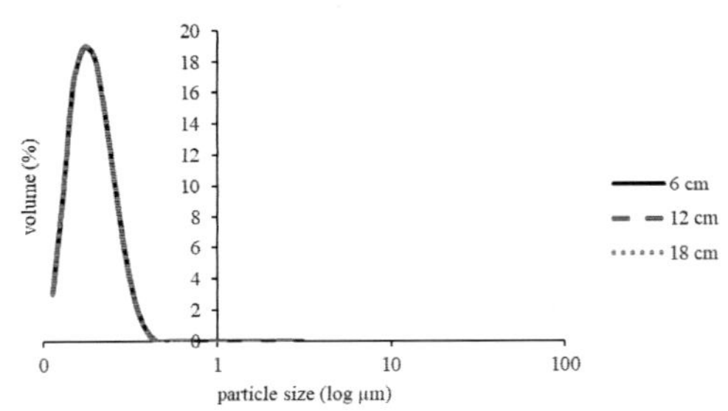

Figure 11. Particle size distribution – 61860 minutes after the start of the experiment.

Table 2. Values of investigated parameters of soil suspension in the state of dynamic sedimentation equilibrium at the end of the experiment.

Height above the bottom	Time from start of measurement	Temperature	Specific surfis area	*D(10)*	*D(50)*	*D(90)*
(cm)	(min)	(°C)	($m^2 g^{-1}$)		(μm)	
6	61860	19.1	33.1	0.14	0.19	0.27
12	61860	19.1	33.2	0.13	0.18	0.27
18	61860	19.1	33.1	0.14	0.18	0.27

Further analysis was performed by statistical correlation analysis. Correlation between soil suspension density, specific surface area and grain-size distribution parameters *D(10)*, *D(50)* and *D(90)* was investigated. The results are described in table 3. The table shows that there is a tight correlation between

Table 3. Table of correlation between the investigated parameters at three heights of the sedimentation tank; *ssd* – soil suspension density, *sa* – specific surface area.

parameters	*ssd*	*D(10)*	*D(50)*	*D(90)*	*sa*
ssd	1				
d(10)	0.94	1			
d(50)	0.97	0.96	1		
d(90)	0.86	0.78	0.90	1	
sa	-0.91	-0.92	-0.90	-0.69	1

the investigated parameters, ranging from $r = 0.69$ to $r = 0.97$. The tightest correlation $r = 0.97$ is between the density of suspension and grain size for the probability distribution parameter *D(50)*. Specific surface area is inversely proportional to grain size for *D(90)* ($r = -0.92$) and suspension density, $r = -0.91$. It is thus effective to express the empirically identified relations by theoretical curves with analytical expression. These are eligible for the assessment of parameters which are difficult to measure (*sa*, *D(10)*, *D(50)*, *D(90))*, on the basis of easily measurable density of soil suspension.

4. Conclusion

This contribution analyses the relations between soil suspension density, specific surface area and grain-size distribution parameters *D(10)*, *D(50)* and *D(90)*. Soil suspension was composed of distilled water and clay sampled in Senné area located on Eastern Slovakia Lowland.

For the research purposes, laser diffraction, sedimentation analysis and statistical correlation analysis were used. The results of the experiment showed important correlation between the investigated parameters. The measurements have shown that it is effective to express the empirically identified relations by theoretical curves with analytical expression. These are eligible for the assessment of parameters which are difficult to measure (*sa*, *D(10)*, *D(50)*, *D(90)*), on the basis of easily measurable density of soil suspension.

5. References

[1] Gomboš M, Tall A and Kandra B 2009 *Creation of Soil Cracks as an Indicator of Soil Drought.* Cereal Research Communications Vol. 37, Supplement 1, pp 367–370

[2] Šoltész A, Zeleňáková M, Čubanová L, Šugareková M and Abd-Elhamid H 2021 *Environmental Impact Assessment and Hydraulic Modelling of Different Flood Protection Measures*. Water; 13(6):786. https://doi.org/10.3390/w13060786

[3] Zhang N, Zhu W, He H and Lv Y 2017 *Experimental study on sedimentation and consolidation of soil particles in dredged slurry.* Journal of Civil Engineering 21: pp 2596–2606. DOI 10.1007/s12205-017-0068-1

[4] Allain C, Cloitre M and Wafra M 1995 *Aggregation and sedimentation in colloidal suspensions*. Physical Review Letters 74:1478-1481. DOI: 10.1103/PhysRevLett.74.1478

[5] Bartovská L and Šišková M 2010 *Fyzikální chemie povrchů a koloidních soustav*. VŠCHT Praha, p. 262 ISBN 978-80-7080-745-3

[6] Miller B A and Schaetzl R J 2012 *Precision of Soil Particle Size Analysis using Laser Diffractometry.* Soil Sci Soc Am J 76:1719-1727. doi:10.2136/sssaj2011.0303

[7] Zaidi A, Tsuji T and Tanaka T 2014 *A new relation of drag force for high Stokes number monodisperse spheres by direct numerical simulation.* Adv Powder Technol 25: pp 1860–71. https://doi.org/10.1016/j.apt.2014.07.019

[8] Arriaga F J, Lowery B and Mays M D 2006 *A fast method for determining soil particle size distribution using a laser instrument.* Soil Sci 171: pp 663–674. doi:10.1097/01.ss.0000228056.92839.88

[9] Everett D H 1988 *Basic Principles of Colloid Science.* London, Royal Society of Chemistry, p 243 ISBN 0–85186–443–0

[10] Moshrefi N 1993 *A new method of sampling soil suspension for particle-size analysis.* Soil Sci 155: pp 245–248 doi:10.1097/00010694-199304000-00002

[11] Storti F and Balsamo F 2010 *Particle size distributions by laser diffraction: sensitivity of granular matter strength to analytical operating procedures.* Solid Earth 1: pp 25–48 https://doi.org/10.5194/se-1-25-2010

[12] Aydin E, Igaz D, Lackóová L and Horák J 2012 *Comparison of methods for the preparation of soil samples and measuring the distribution of grain size fractions of soils in the River Nitra basin* In Acta horticulturae et regiotecturae. ISSN 1335-2563, 2012, vol. 15, extraordinary number, pp 27–30

Acknowledgements

The research was financed by the project VEGA 2/0044/20.

Numerical analysis of shear resistance on composite beams.

M Weissova[1], V Kvocak[1], D Dubecky[1], P Vaňová[1]

[1] Institute of structural engineering, Faculty of Civil Engineering, Technical University of Kosice, Vysokoškolská 4, 040 23 Kosice, Slovakia

E-mail: michala.weissova@tuke.sk, vincent.kvocak@tuke.sk, daniel.dubecky@tuke.sk

Abstract. The part of the preparation of experimental verification of shear connectors is also numerical analysis of the examined problem. The main subject of this research is the FEM analysis of a chosen experiment, the push-out test. For FEM analysis, we have opted for *Abaqus/CAE* software. This piece of work will focus on the modeling of the push-out test via the software stated above.

1. Introduction

This article aims FEM analysis, description of push-out test modeling, and a comparison with an experiment performed in the laboratory. Analysis using the 3D of the finite element method is often a solved issue in software engineering and becomes a common part of scientific experiments in a wide range of expertise. Long-developed software used to model various global problems from design, social or meteorological to epidemiological allows us to look at the solved problem from a different perspective. The correct FEM model allows us to read both the results and the values, which are very difficult to measure in the laboratory. At the level of this work, it is currently necessary to correctly calibrate the individual materials and the sample as a whole to approximate the results from laboratory tests. At the end of this article, I will compare the results from the laboratory with the FEM by analysis.

2. Push-out test

The main goal of the push-out test is to assess the connecting belt's shear resistance. Two concrete blocks having a width of 200 mm, a length of 600 mm, and a height of 600 mm make up the push-out test sample. These two concrete blocks had connection strips with a 400-mm connection length concreted into them. The connection strips are two beams cut from a composite profile with a length of 600 mm that have been concreted into the blocks 100 mm from the block's lower edge and 100 mm above the block's upper edge. A 10 mm thick polystyrene was installed under the beam to assess the actual connection resistance and to prevent the beam from being crushed in the concrete.

The beams were linked to the steel profile to displace the entire central part of the structure with the composite beams. As a result, it was critical to ensure that the joint was rigid enough. This joint is generally done by welding, but it had to be constructed using screws because it was a test of the composite beam connection. As a result, a center section with a total length of 600 mm was built, with steel screws connecting the flanges of two composite beams to the flanges of the steel beam. Because an excessive connection of the screws with the concrete can occur, the mounting foam was used to secure the area surrounding the screws. The connection strip was reinforced with GFRP reinforcement, which made part of the reinforcing bar of the concrete blocks.

The sample had a total of 5 sensors attached to it. During loading, one inductive sensor measured the horizontal displacement of the element's center section. The remaining sensors were installed on the concrete surface in regions where the highest stresses were anticipated. Two tensometers and two inductive sensors were employed. Inductive sensors and tensometers were 150 mm long.
The middle half of the test body was pressed across the distribution surface using an INSTRON which is illustrated in figure1 hydraulic press with a maximum force of 2500 kN. A perforated belt and a ridge belt both were tested as connecting belts.
The sample was loaded in loading cycles, with the sample being loaded to 40% of the anticipated shear resistance and then pulled to 5% of the resistance. Following that, such 25 cycles were carried out with the lowest value of 5% and the greatest value of 40% of the estimated resistance.

Figure 1. Push-out test sample in the laboratory

3. Process of modeling the sample for push-out test

This software can be used for general static as well as dynamic tasks. It solves the whole problem and with the proper modeling, it can find out problems in constructions that are close to reality. This software solves tasks using 3D FEM analysis, which is more accurate than 2D FEM analysis. One of the main inputs which are needed for the right and exact description of the solving problem is the correct and detailed input of material characteristics. [4]
The *Abaqus* software environment works with modules in which the individual inputs for solving the task are gradually defined.

3.1. Modul PART

For the entire geometry of the model, we have used the *AutoCAD* software, in which the entire sample was modeled and then imported into *Abaqus* software in .sat format. Throughout modeling of the whole mass, the individual parts were subtracted from each other so that the volume of masses did not overlap. The reinforcement was modeled as a 2D element – a bar. Its surface and material were assigned in the following module. While modeling the composite part of the sample, it was necessary to define the local coordinate axes due to the anisotropy of this material.

3.2. *Modul PROPERTY*

In this module, the values of material characteristics were entered. They had been tested in detail in the laboratory. Subsequently, a material was assigned to each part of the sample. In the case of reinforcement bars, a cross-sectional area was also assigned. The composite beam material was calibrated according to the results of the tensile test, a yield stress diagram, which was inserted into the *Calibrations* software component. The material characteristics of the middle steel part and the reinforcement were defined by density and elastic properties. [1]

The material of the composite beam is anisotropic in two directions. Therefore, it can be considered as an orthotropic material for which it was necessary to enter material characteristics in all directions. Elasto-plastic analysis was used for the correct definition. To correctly define each material in the 3D FEM analysis, it is necessary to enter the material density. In the case of the composite beam, the density was considered as 1.8 kg/m^3. [3]

Measured values of tensile and shear modulus were used to determine the elastic behavior of the material. The Poisson's constant was not measured. Thus, a value of 0.33 was used. Entering is described in the software manual using the so-called *Engineering Constants*. It includes E12, E13, E23, nu12, nu13, nu23, G12, G13, and G23. [2]

The material characteristics of the concrete were determined in the laboratory by a couple of experiments. The *Concrete Damage Plasticity method* was used to define the characteristics of the concrete. This method consists in defining the material characteristics of concrete using the modules offered by this method, namely *Plasticity*, *Compressive Behaviour,* and *Tensile Behaviour*. [3]

3.3. *Modul ASSEMBLY*

As shown in figure 2, the individual parts of the sample were combined into one unit in this module. In the figure, one part of the concrete is hidden to better represent the geometry.

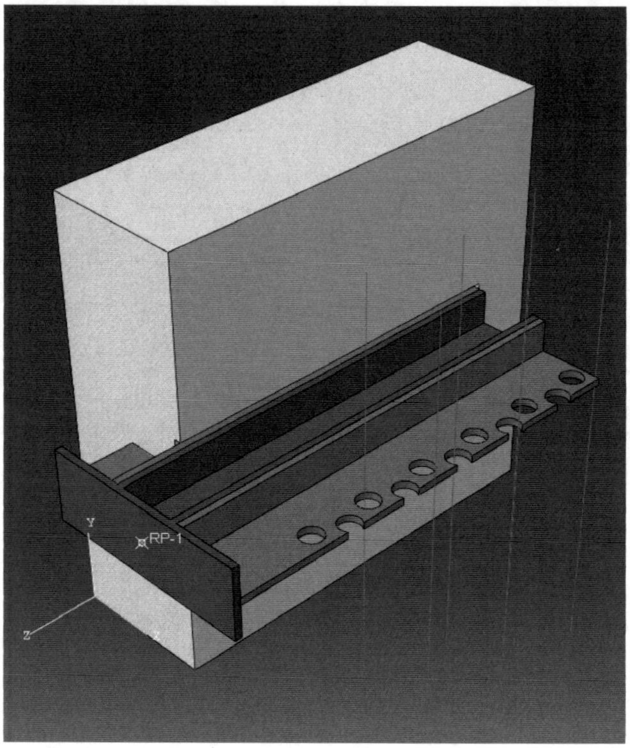

Figure 2. Push-out test sample

3.4. *Modul STEP*

For the calculation, we have used a nonlinear, explicit calculation. The step time was set to 0.1, which reduces the calculation time by ten times, but the density value of the individual elements in the mesh is increased ten times.

Sometimes, very small elements have a slow convergence. Thus, the maximum time to achieve element convergence was set to 1e-7.

This module also determines the number of entries, which was equal to the number of load cycles, as well as the results that were expected, deformation, concrete failure, stresses, and relative deformations. [4]

3.5. *Modul INTERACTION*

The contact between the individual elements had been set so that friction with a coefficient of friction of 0.18 was created between the concrete and the coupling strip. For better convergence of elements and correct input of the contact, we used the *Penalty contact method,* which, in addition to the correct properties of the contact, requires to set the parent and child surface. Since concrete is more represented in the overall model, the parent surfaces were the concrete surfaces and the minor surfaces of the composite.

A perfectly rigid joint was set up between the composite element and the steel, which was secured with screws. It was also necessary to determine the parent and child areas in this case. The steel areas represented the parent areas, and the composite areas were considered as the child areas. [4]

3.6. *Modul LOAD*

The boundary conditions were set so that there was no vertical displacement of the concrete. Since this is storage, this setting is considered before loading the sample.

The load was set in the vertical direction according to the load cycles of the individual coupling belts tested. [4]

3.7. *Modul MESH*
When calculating with FEM, one of the most important things is the correct mesh settings. In the analysis of the push-out test using FEM, there was an effort for a hexagonal 3D mesh, which has one of the basic advantages - there are no faulty elements, and it speeds up the calculation. For a more complicated geometry of the elements of the push-out pattern, it had been necessary to divide the parts into simpler elements (Figure 3) so that it was possible to create a hexagonal 3D mesh of elements. This distribution of matter affects only the elements and not the element as such.
Abaqus also offers the possibility of a mesh in the shape of a quadrilateral pyramid, which is a method enabling the creation of a mesh even on very complicated elements, without the need to create a division. However, the disadvantage is the length of the calculation and the more frequent occurrence of elements that have difficulty converging during the calculation.
For different densities of materials, the density of the elements varied. For concrete, we used the density of elements 15. For the coupling strip, reinforcement, and steel central part, we used the density of elements 8. For individual parts, it was necessary to set the type of elements so that they correctly capture the assumption of the 3D behavior of materials. For reinforcement as a 2D element, it was necessary to set the element type so that they behave like 3D elements and there are no collisions between 3D and 2D analysis. [4]

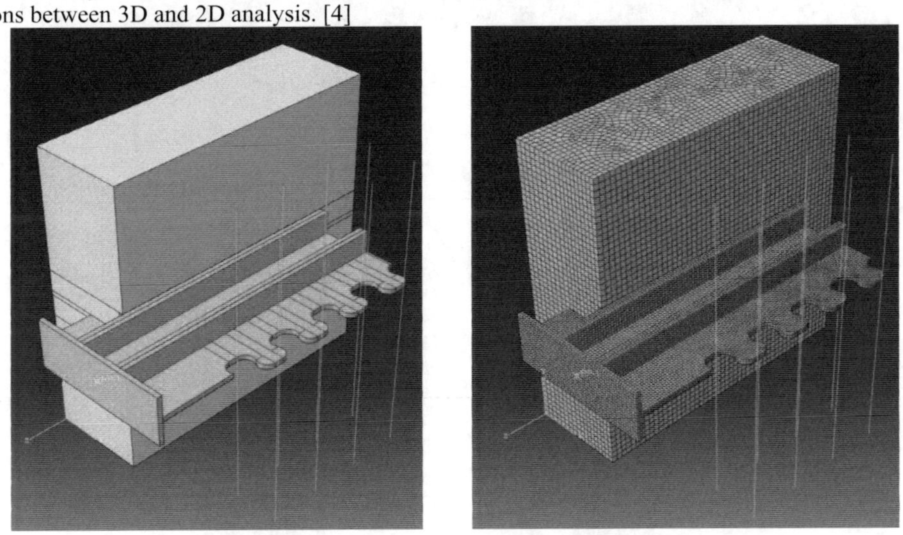

Figure 3. Meshed and not meshed sample

In Figure 3 on the left, the sample is divided into smaller parts. The green and yellow colors indicate the possibility of creating a hexagonal mesh. In the picture on the left is a sample already meshed, where we can see a different size of elements.

4. The FEM model of the sample with ridge strip
For the crodge strip, 130,612 elements, 157,167 nodes were generated, and along with all inputs specified in previous modules, 471,504 variables were generated for analysis. The total calculation time was 1,147, which is approximately 19.1 minutes. The results are illustrated in figure 4, figure 5, figure 6, and figure 7

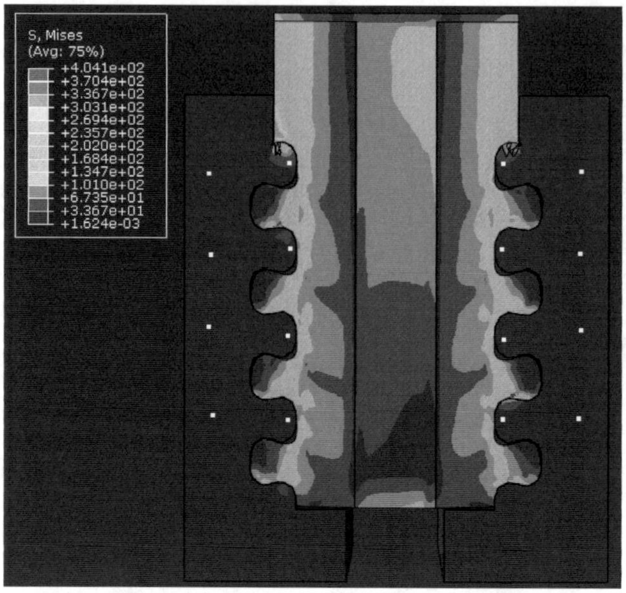

Figure 4. The Von Mises stress in the cross-section of the sample with ridge strip

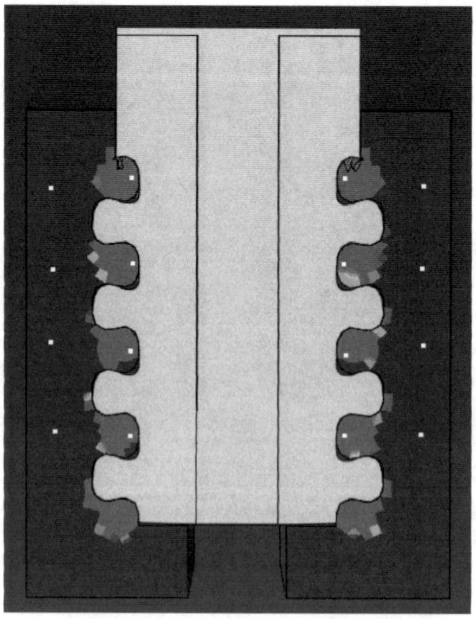

Figure 5. The crush of concrete elements in compression

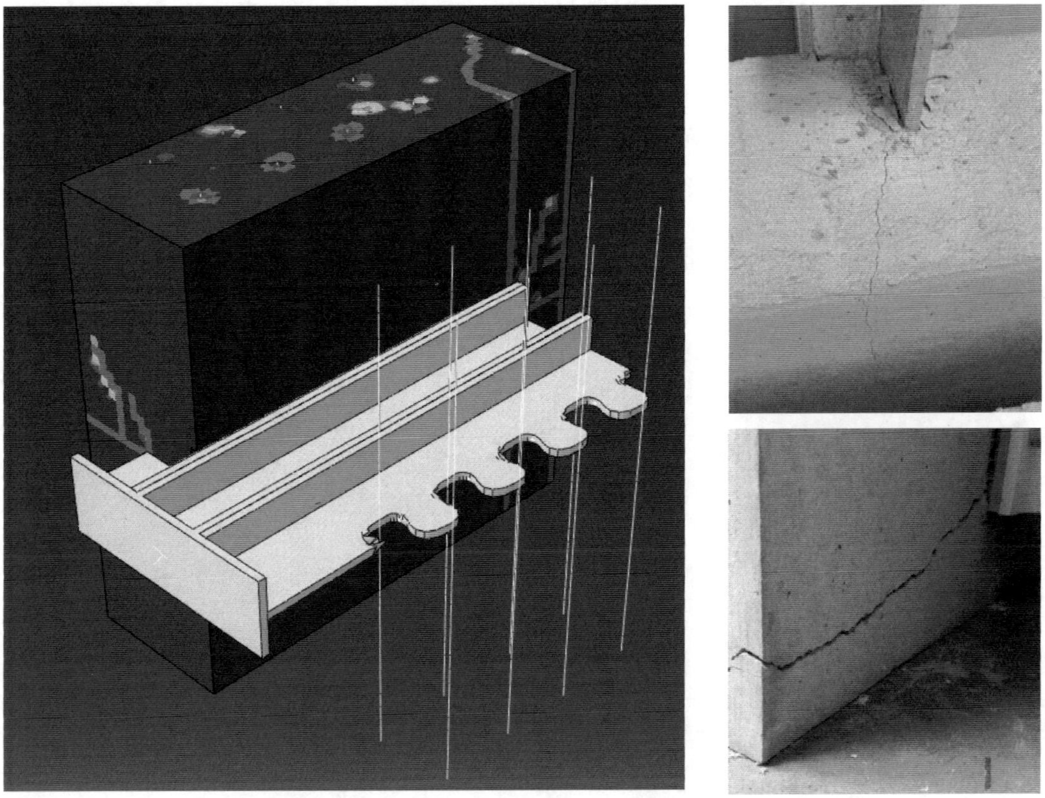

Figure 6. The crush of concrete elements in tension

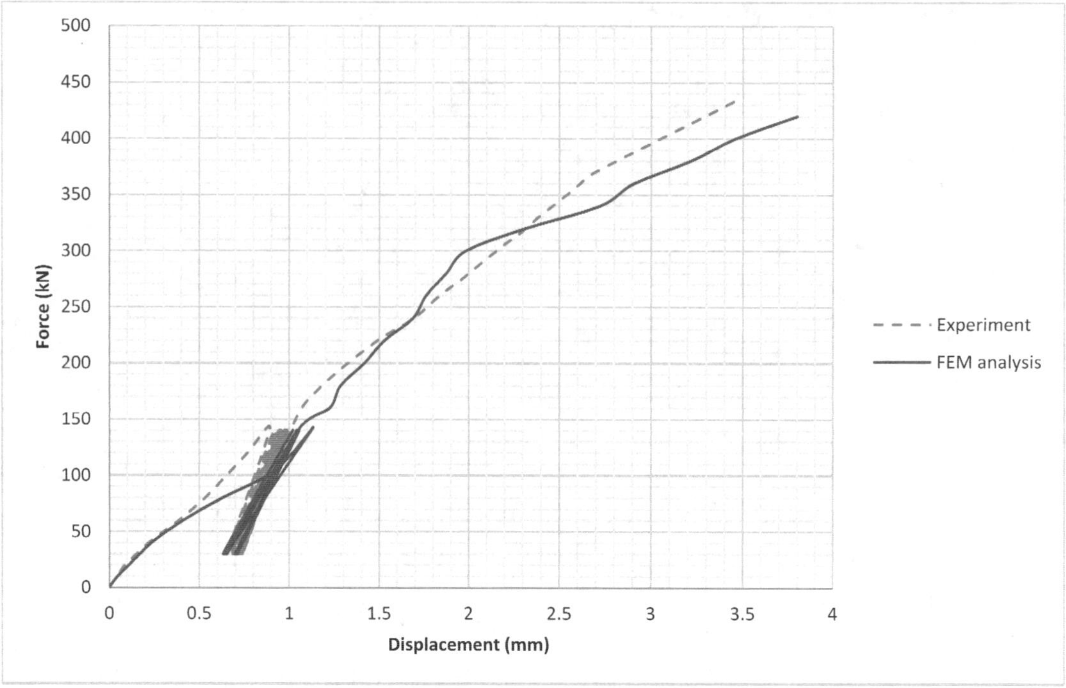

Figure 7. The comparison of laboratory experiment and FEM analysis

According to the results of the FEM analysis compared to the experimental analysis, it is clear that the FEM analysis has similar results to the experimental one, which results is that the use of FEM analysis instead of experimental analysis is possible. In addition to comparing the displacements, it can be concluded that the tension behavior in composite and concrete according to FEM analysis is expected and close to the real behavior of the experiment. In addition to the results of the stress behavior, it is clear that the concrete elements crushed in the analysis are also in places where such a crush was expected even according to a real experiment.

5. Conclusion

The result of the analysis can be evaluated from several points of view. First of all, the shear resistance of the composite material is lower compared to steel. Specifically, according to a previous experimental verification of the coupling shear resistance on steel elements of the same geometry, the resistance of the perforated strip is approximately 725 kN. The resistance of the composite coupling with the comb perforated strip is 569.55 kN.

Based on these results, it can be concluded that the resistance of the perforated strip is reduced by about 20%, which greatly affected the matrix failure during the experiment.

Analysis using FEM creates many open topics. For example, entering material characteristics with a limit value where the sample breaks. Another topic that remains open is the effort to speed up the calculation and its impact on the results. The issue of meshing and creation of elements in FEM analysis is also an important subject, which would also be appropriate to address. It is clear from the results that the creation of the model was correct and can be used to solve similar problems.

6. References

[1] Rewers I 2019 *Numerical Analysis of RC beam with High Strength Steel Reinforcement using CDP model. IOP Conference Series: Materials Science and Engineering* **471** 022025 10.1088/1757-899X/471/2/022025

[2] Demir A 2018 *Effect of viscosity parameter on the numerical simulation of reinforced concrete deep beam behavior*

[3] Hafezolghorani E Milad Hejazi Farzad Vaghei Ramin Jaafar Mohd Karimzade Keyhan 2017 *Simplified Damage Plasticity Model for Concrete. Structural Engineering International* **27** 68-78. 10.2749/101686616X1081

[4] *ABAQUS Analysis User's Manual, Version 6.6 Documentation*

[5] Ivančo V Kubín K Kostolný K 1994 *Metóda konečných prvkov I.* ISBN 80-967131-4-0.

Acknowledgments

The paper presented was supported by the projects: VEGA 1/0172/20 "Stress and deformation analysis of load-bearing components made out of steel, glass and composite materials" of the Scientific Grant Agency of the Ministry of Education, Science, Research and Sport of the Slovak Republic.

Influence of the longitudinal steel ration in RC beams and steel reinforcement ration in RC slabs on the seismic energy dissipation mechanisms for a MR RC frame structure

I Sococol[1*], T C Petrescu[1], P Mihai[1] and D T Babor[1]

[1] Department of Concrete, Materials, Technology and Management, Faculty of Civil Engineering and Building Services, "Gheorghe Asachi" Technical University of Iasi, Romania, 43 "Dimitrie Mangeron" Blvd., 700050, Iasi, Romania

E-mail: ion.sococol@student.tuiasi.ro

Abstract. The MR RC frame structures present a special engineering interest regarding the practical mode of ductile seismic response, through plastic hinges formation in specially designed RC beams end areas and inferior zones of the RC columns from the ground floor. Thus, it was proposed to verify the capacity design concept for this structural system type through seismic response observation of the optimal MR RC frame model. Until this objective is achieved, it was proposed to perform a complex analytical study, which has as finality the current research paper. Thus, in the previous analytical studies the influence of the concrete strength class, the influence of the longitudinal steel reinforcement ratio in RC columns and influence of the RC beams cross sections on the local and global seismic response for a group of RC frame models were analysed. In these conditions, the present paper contains the results obtained in previous research papers and concludes the current analytical study through the observation of the effectiveness of the inferior longitudinal steel reinforcement ratio in the RC beams on the cracking mode, in potentially plastic areas. Finally, it was specified that the K_7 RC frame model represents the optimal structural system for the experimental test.

1. Introduction

The latest earthquake experiences regarding seismic response of the Reinforced Concrete (RC) frame structures - exemplified by 2011 Tohoku earthquake [1], 2010 Chile earthquake [2], 2020 Aegean Sea earthquake [3] etc. - practically demonstrate the inability (incapacity) of this type of structural system to develop the "ductile mechanism" [4-7] present in the current seismic design norms P100-1 [8] and EC 8 [9] (see Figure 1 – Figure 3).

In these conditions, it was proposed to perform an experimental study on the seismic platform in order to analyze the global seismic response of pure Moment Resisting (MR) Reinforced Concrete (RC) frame structures and to observe the local plasticization mechanisms.

Figure 1. Japan 2011 Tohoku earthquake [1].

Figure 2. Chile 2010 earthquake [2].

Figure 3. 2020 Aegean Sea earthquake [3].

In these conditions, it was proposed to perform an experimental study on the seismic platform in order to analyze the global seismic response of pure MR RC frame structures and to observe the local plasticization mechanisms.

This goal cannot be achieved without conducting a preliminary complex analytical study, which should lead to the creation of the optimal experimental model with appropriate theoretical and analytical observations (see Figure 4).

Thus, three analytical studies were performed using the "nonlinear static analysis procedure" [10] with ATENA software [11-13] having this research paper as finality of optimal experimental model establishing process (see Figure 4).

In all analytical studies, as in this research paper, "stress-strain curves for concrete and steel reinforcement with graphical representation of cyclic uniaxial concrete compression/ tension, biaxial compressive failure and tensile failure consideration laws for concrete, bilinear with hardening law for steel reinforcement, cycling steel reinforcement model and steel reinforcement bond model" [12-14], were used, in order to obtain correct analytical results.

Thus, in previous research works, the "optimal concrete strength class" [15], "the optimal longitudinal steel reinforcement ratio in RC columns" [16] and "optimal RC beams cross section" [17] for superior seismic energy dissipation mechanism were established according to the ductile design concept (capacity design concept).

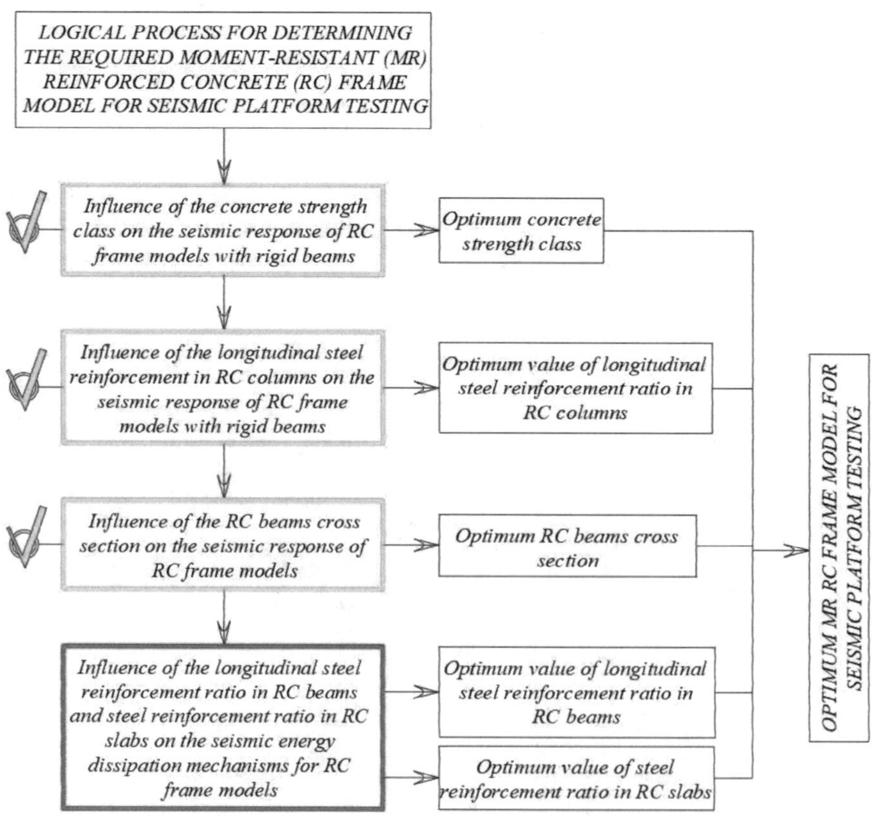

Figure 4. Representation of the research steps regarding the determination process of the optimum MR RC frame model for seismic platform testing. Current research paper represents the study stage in the red square border.

In all research studies, MR RC frame models with the same inter-axis distances and height regime (see Figure 5) were considered, reduced to ½ scale according to ”similarity laws” [18] and pure structural skeleton [19].

Thus, in the present analytical study (ultimate stage of analytical study – see Figure 4), four MR RC frame type models (see Table 1) are considered. Each of these structural RC frame models has a C20/25 concrete strength class (optimal concrete strength class specified in Sococol et al. [15] and the RC columns are reinforced with 4ϕ14 type Bst 500S steel rebars according to the conclusions established in Sococol et al. [16]. Reinforced Concrete beams were developed on 15x20 (cm) cross sections (optimal RC beams cross section specified in research study performed by Sococol et al. [17].

In these conditions, the influence of the longitudinal steel reinforcement ratio in RC beams and steel reinforcement ratio in RC slabs on the global seismic response for MR RC frame models is studied.

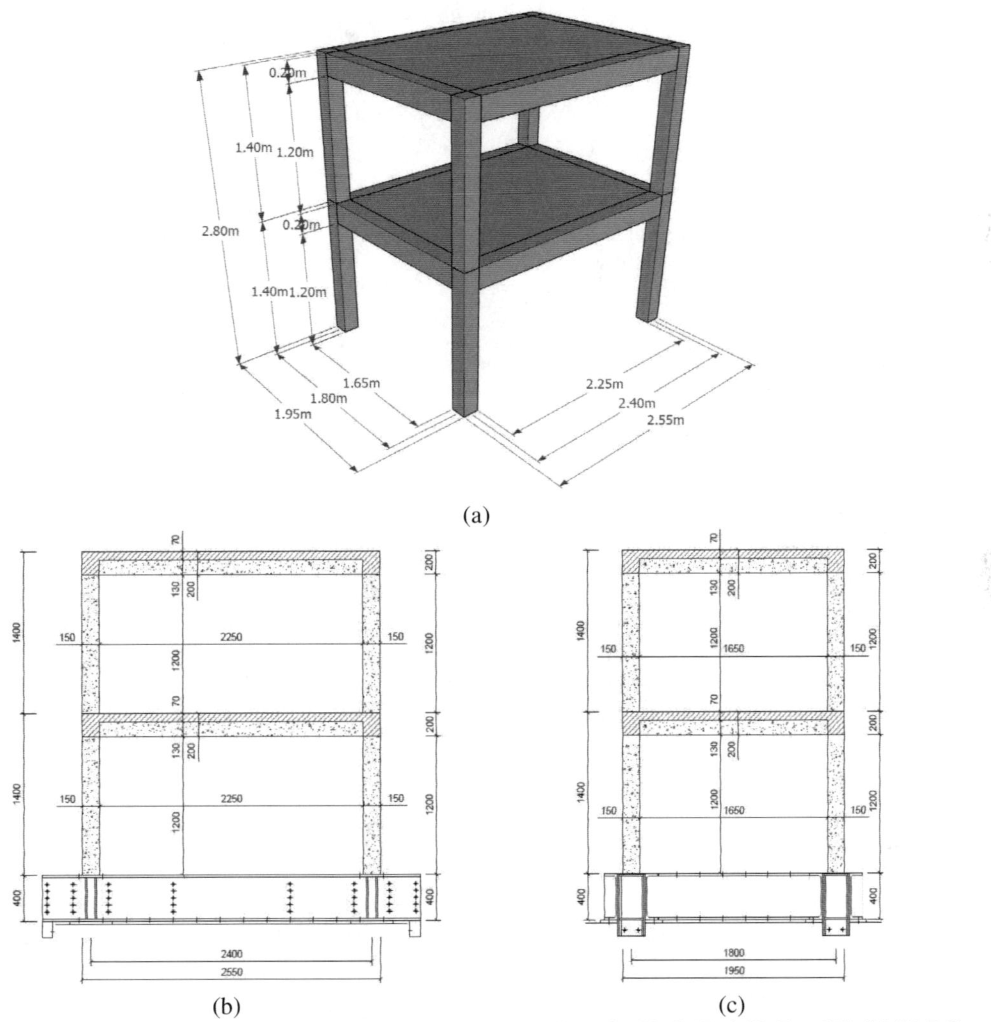

Figure 5. Representation of the: “(a) 3D-structure dimensions for K_5, K_6, K_7 and K_8 MR RC frame model; (b, c) longitudinal and transversal section for K_5, K_6, K_7 and K_8 MR RC frame system” [17].

As output data, the general-structural image of the optimal experimental RC frame model type to be tested on the seismic platform is obtained and specified.

Table 1. Parameters considered in non-linear static analysis for K_5, K_6, K_7 and K_8 MR RC frame models.

CSC	NSC	RC C (15x15 cm)		RC LB (15x20 cm)		RC TB (15x20 cm)		RC S (h_s=7 cm)	
			GR		GR		GR		GR
C20/25	K_5	4ϕ14	Fig. 7 (a)	4ϕ10	Fig. 7 (d)	4ϕ10	Fig. 7 (f)	ϕ6	Fig. 8 (c), (d)
C20/25	K_6	4ϕ14		4ϕ10	Fig. 7 (d)	4ϕ10	Fig. 7 (f)	ϕ5	Fig. 8 (e), (f)
C20/25	K_7	4ϕ14		4ϕ8	Fig. 7 (c)	4ϕ8	Fig. 7 (e)	ϕ6	Fig. 8 (c), (d)
C20/25	K_8	4ϕ14		4ϕ8	Fig. 7 (c)	4ϕ8	Fig. 7 (e)	ϕ5	Fig. 8 (e), (f)

„Note: CSC – **C**oncrete **S**trength **C**lass; NSC – **N**umerical **S**imulation **C**ode; RC – **R**einforced **C**oncrete; C – **C**olumns; GR – **G**raphical **R**epresentation; LB – **L**ongitudinal **B**eams; TB – **T**ransverse **B**eams; S –**S**labs; h_s – RC slabs thickness" [17].

2. Input data required for nonlinear static analysis

General input data required for numerical analysis of the K_5, K_6, K_7 and K_8 MR RC frame models are the same as in research papers of Sococol et al. [13], [15-17] (see Figure 5, Figure 6):

i. the dimensions of the ½ scaled MR RC frame systems are: L=2.4 m, B=1.8 m;
ii. height regime: *GF+1F* (small scale buildings [20];
iii. story height: h_{st}=1.4 m; H_{tot}=2.8 m;
iv. building class importance: III, according to P100-1 norm [8];
v. type of structure: pure MR RC frame (without non-structural components to avoid their negative effects [19];
vi. location: Iasi city;
vii. structural ductility class: *DCH*, according to P100-1 norm [8] and information presented in engineering literature [5-7], [21].

In this research paper, the variables used for MR RC frame models [17] are the RC beams and RC slabs flexural steel rebar ratio. Thus, in nonlinear static analyses, four reinforcement combinations (see Table 1, Table 2, Figure 6 (c), Figure 7, Figure 8) were considered [17]:

i. K_5 MR RC frame model with 4ϕ10 type Bst 500S flexural steel reinforcement for RC beams and standard welded wire type 116GQ283 (6x100/6x100 – welded wire with 6 mm diameter and square mesh) for RC slabs;
ii. K_6 MR RC frame model with 4ϕ10 type Bst 500S flexural steel reinforcement for RC beams and standard welded wire type 111GQ196 (5x100/5x100 – welded wire with 5 mm diameter and square mesh) for RC slabs;
iii. K_7 MR RC frame model with 4ϕ8 type Bst 500S flexural steel reinforcement for RC beams and standard welded wire type 116GQ283 (6x100/6x100 – welded wire with 6 mm diameter and square mesh) for RC slabs;
iv. K_8 MR RC frame model with 4ϕ8 type Bst 500S flexural steel reinforcement for RC beams and standard welded wire type 111GQ196 (5x100/5x100 – welded wire with 5 mm diameter and square mesh) for RC slabs.

Regarding the other cross-sectional characteristics of structural elements, the reinforcement mode and concrete strength class for RC frame models, it can be specified that (see Table 1, Figure 7, Figure 8) [17]:

i. for all RC frame structural models, C20/25 concrete strength class was used [15];
ii. for all RC frame models, 15x15 (cm) cross section dimension of the RC columns was used;
iii. for all RC frame models, 15x20 (cm) (bxh) cross section dimension of the RC beams was used;
iv. for all RC frame models, RC slabs with 7 (cm) thickness was used;
v. for all RC frame structural models, 4ϕ14 Bst 500S flexural (longitudinal) steel reinforcement type in RC columns it was used [16];

vi. cross-sectional steel reinforcement of the RC columns and RC beams it was performed with ϕ4 type Bst 500M stirrups positioned at 5 (cm) in critical zones and 10 (cm) in other areas. Also, it was considered that the RC columns present a critical region on the whole height of the element.

"Lateral loads with implicit horizontal values were considered according to the discrete system represented in Figure 6 (b)" [17].

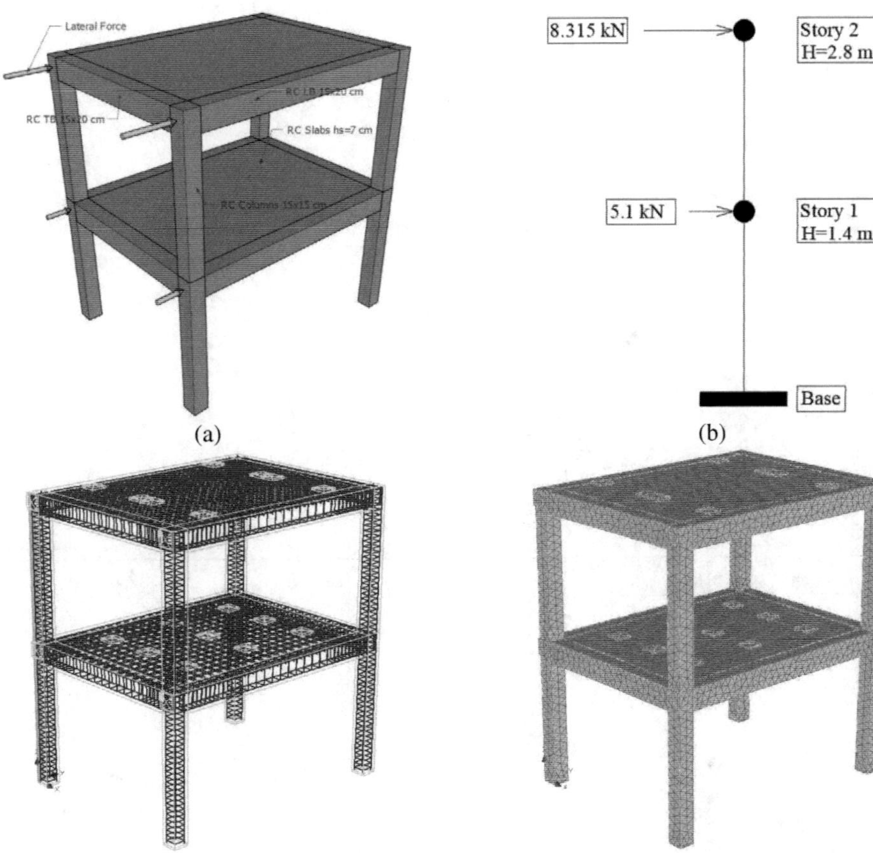

Figure 6. Representation of the: "(a) pushover loading consideration for MR RC frame models; (b) discrete system with lateral forces consideration for MR RC frame models; (c) steel reinforcement framing in MR RC frame models; (d) structural mesh discretization for MR RC frame models" [17], (Note: Sketches in Figure 6 have a general character, being associated with all RC frame models specified in Table 1).

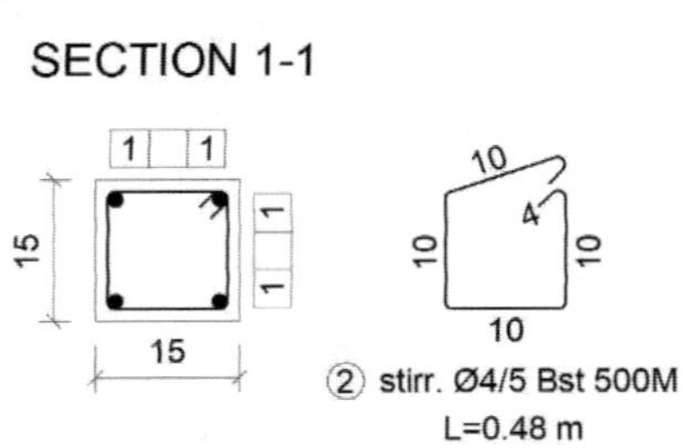

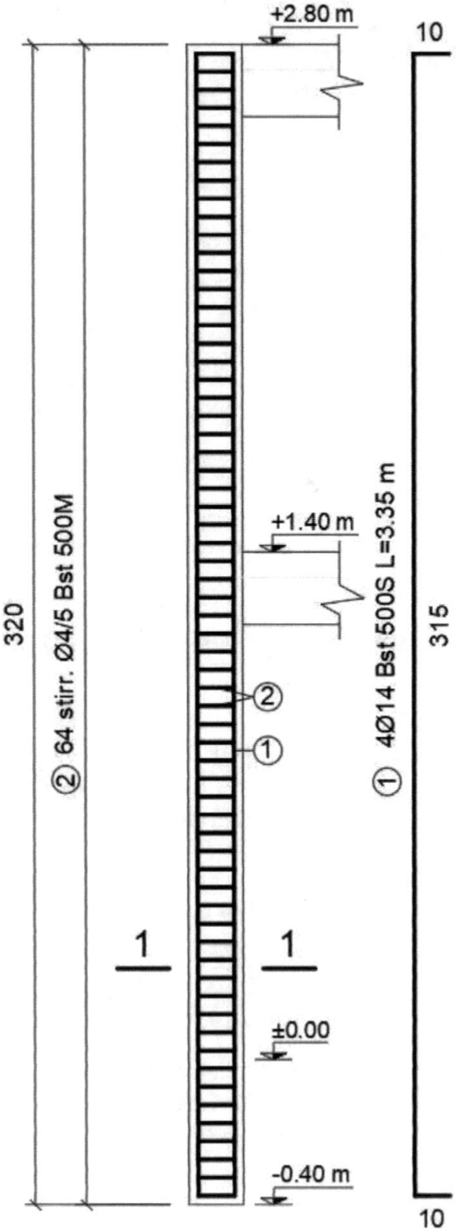
+2.80 m
+1.40 m
±0.00
-0.40 m
320
② 64 stirr. Ø4/5 Bst 500M
① 4Ø14 Bst 500S L=3.35 m
315
10
10
1
1
②
①
(a)

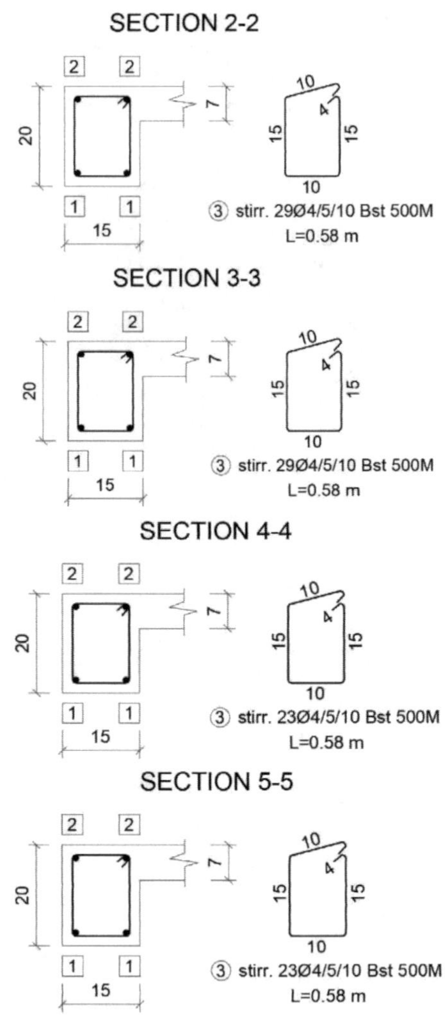
SECTION 2-2
③ stirr. 29Ø4/5/10 Bst 500M
L=0.58 m
SECTION 3-3
③ stirr. 29Ø4/5/10 Bst 500M
L=0.58 m
SECTION 4-4
③ stirr. 23Ø4/5/10 Bst 500M
L=0.58 m
SECTION 5-5
③ stirr. 23Ø4/5/10 Bst 500M
L=0.58 m
(b)

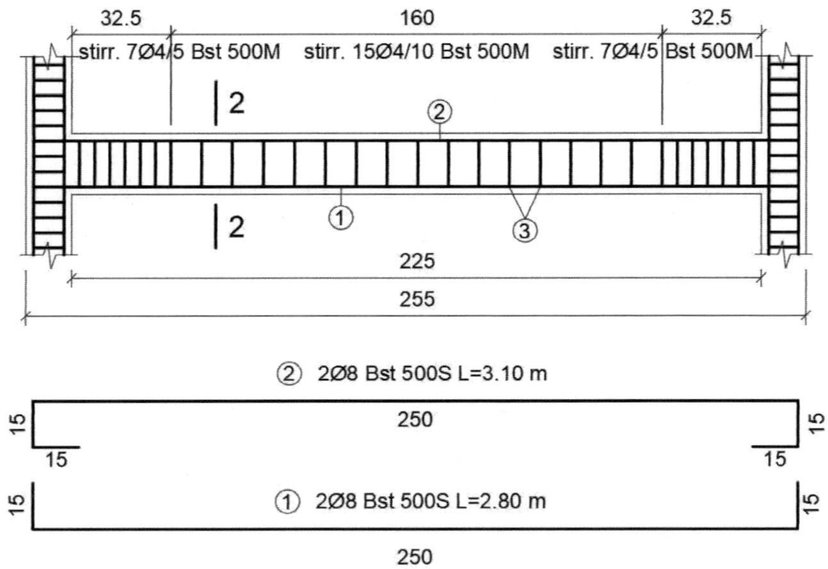
32.5
160
32.5
stirr. 7Ø4/5 Bst 500M
stirr. 15Ø4/10 Bst 500M
stirr. 7Ø4/5 Bst 500M
2
2
②
①
③
225
255
② 2Ø8 Bst 500S L=3.10 m
250
15
15
15
15
① 2Ø8 Bst 500S L=2.80 m
15
15
250

(c)

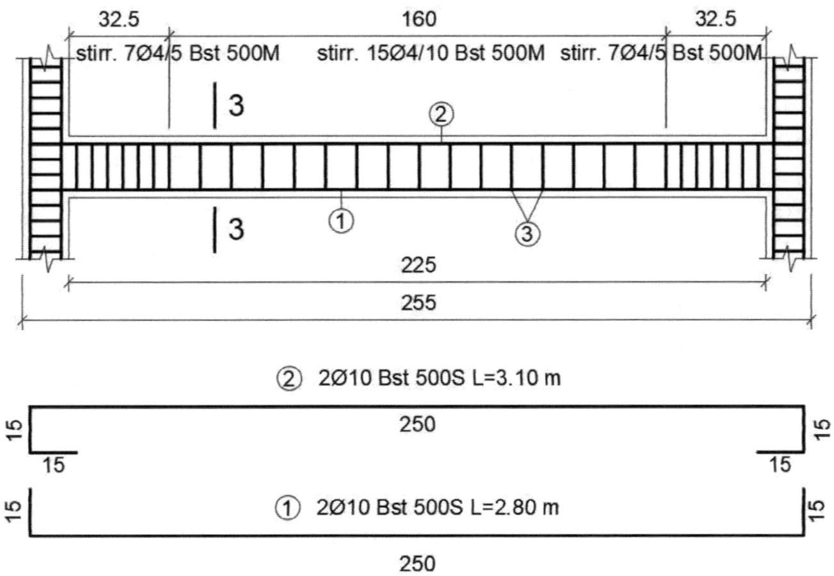
32.5
160
32.5
stirr. 7Ø4/5 Bst 500M
stirr. 15Ø4/10 Bst 500M
stirr. 7Ø4/5 Bst 500M
3
3
②
①
③
225
255
② 2Ø10 Bst 500S L=3.10 m
250
15
15
15
15
① 2Ø10 Bst 500S L=2.80 m
15
15
250

(d)

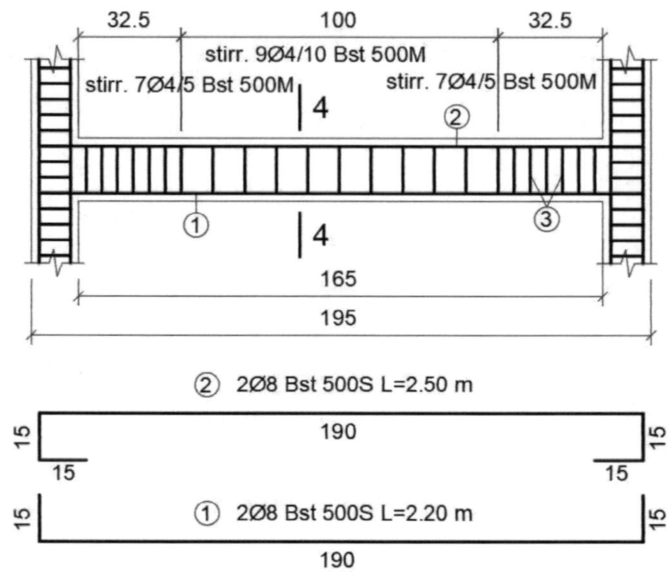

(e)

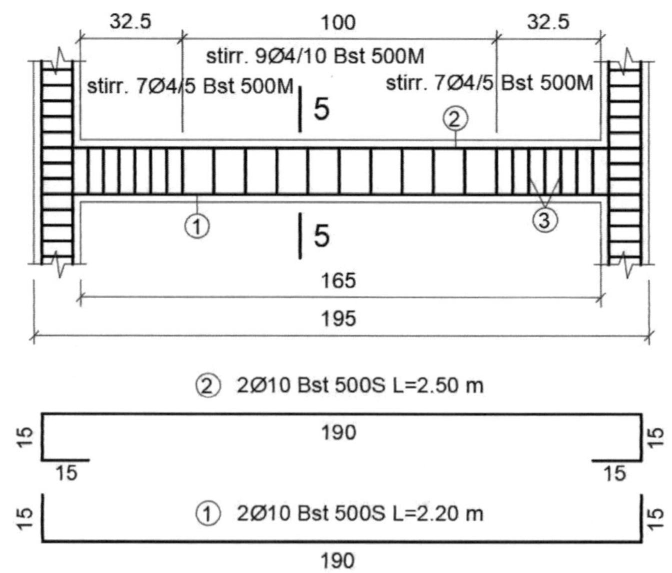

(f)

Figure 7. Representation of the: (a) steel reinforcing method in concrete columns; (b) cross sections for concrete beams and concrete columns; (c) steel reinforcing method in longitudinal concrete beams for K_7 and K_8 MR RC frame models; (d) steel reinforcing method in longitudinal concrete beams for K_5 and K_6 MR RC frame models; (e) steel reinforcing method in transverse concrete beams for K_7 and K_8 MR RC frame models; (f) steel reinforcing method in transverse concrete beams for K_5 and K_6 MR RC frame models.

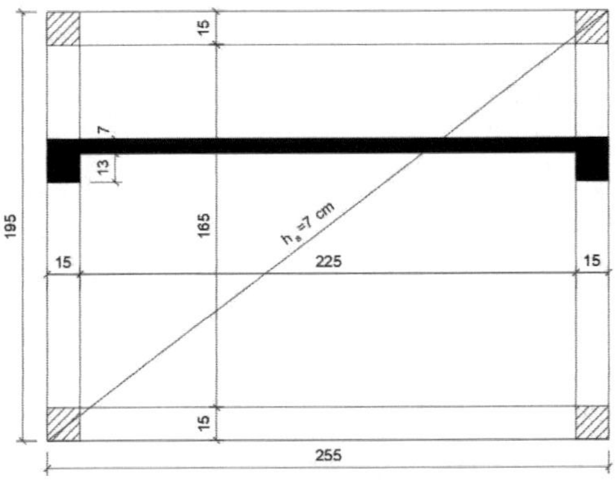
15
7
13
195
165
h_e=7 cm
15
225
15
15
255

(a)

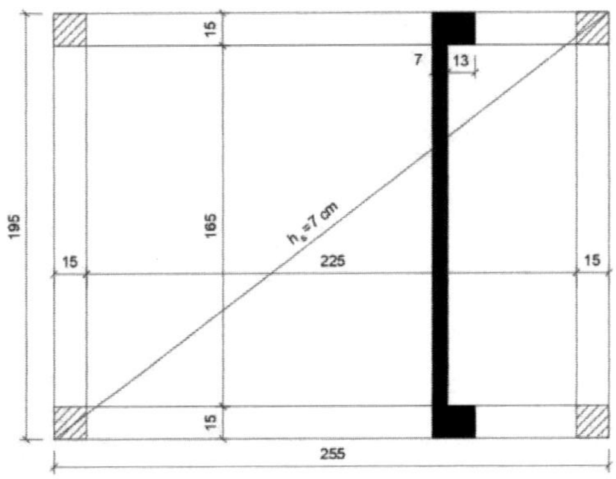
15
7
13
195
165
h_e=7 cm
15
225
15
15
255

(b)

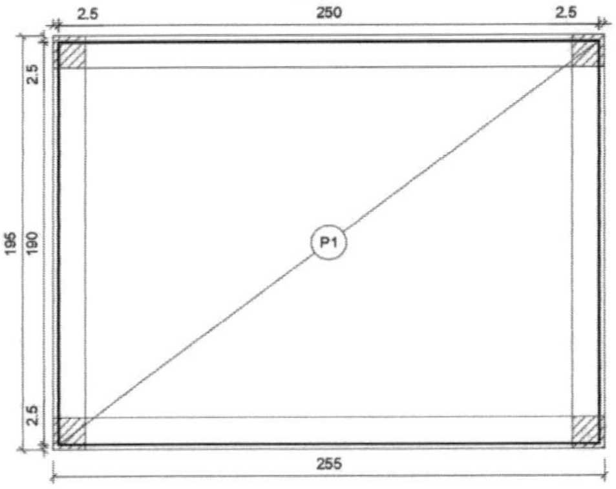
2.5
250
2.5
2.5
195
190
P1
2.5
255

(c)

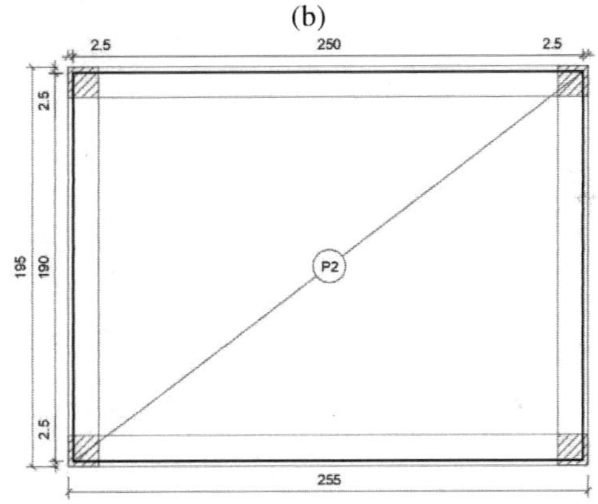
2.5
250
2.5
2.5
195
190
P2
2.5
255

(d)

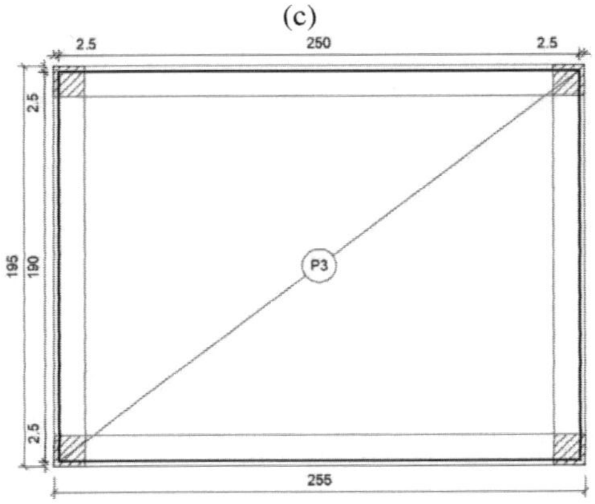
2.5
250
2.5
2.5
195
190
P3
2.5
255

(e)

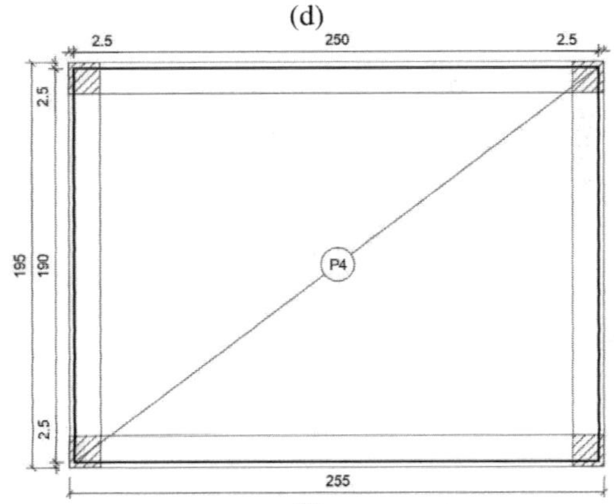
2.5
250
2.5
2.5
195
190
P4
2.5
255

(f)

IOP Conf. Series: Materials Science and Engineering 1252 (2022) 012036 doi:10.1088/1757-899X/1252/1/012036

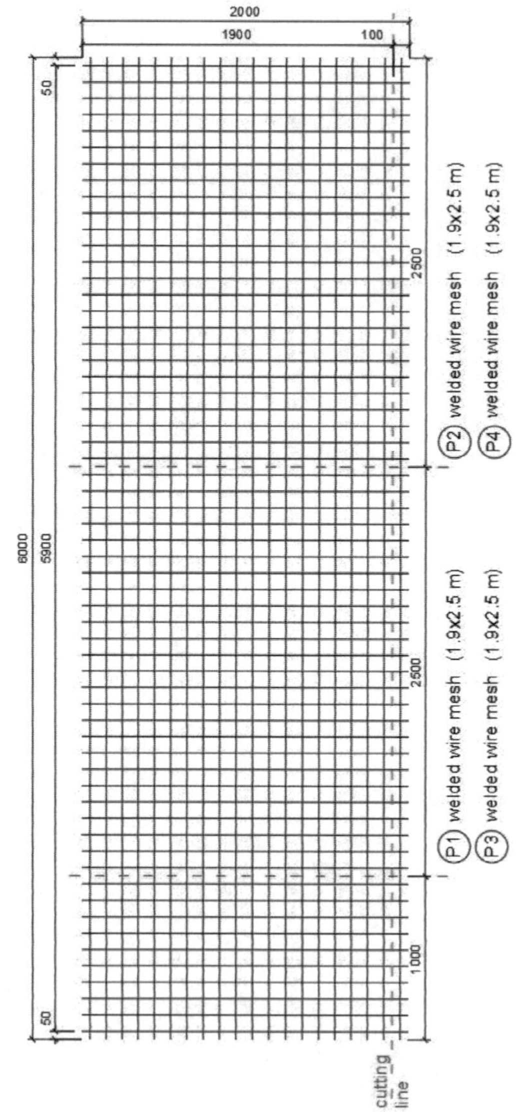

(g)

Figure 8. Representation of the (see Table 2): (a) formwork for concrete beams and concrete slabs in the longitudinal direction of the MR RC frame models; (b) formwork for concrete beams and concrete slabs in the transverse direction of the MR RC frame models; (c) top welded wire mesh reinforcement in concrete slabs for K_5 and K_7 MR RC frame models; (d) bottom welded wire mesh reinforcement in concrete slabs for K_5 and K_7 MR RC frame models; (e) top welded wire mesh reinforcement in concrete slabs for K_6 and K_8 MR RC frame models; (f) bottom welded wire mesh reinforcement in concrete slabs for K_6 and K_8 MR RC frame models; (g) cutting details of welded wire mesh.

Table 2. Details about RC slabs reinforcement with standard welded wire mesh (see Figure 8).

Label	TRS	TSWWM	DRB	SMS (mm)	B (cm)	L (cm)	ST (cm)
P1		116GQ283	ϕ6	100x100			
P2	STNB	116GQ283	ϕ6	100x100	190	250	7
P3		111GQ196	ϕ5	100x100			
P4		111GQ196	ϕ5	100x100			

Note: TRS – **T**ype of **R**einforcing **S**teel; TSWWM – **T**ype of **S**tandard **W**elded **W**ire **M**esh; DRB – **D**iameter of the **R**einforcing **B**ars; SMS – **S**quare **M**esh **S**ize; B – width of the welded wire mesh; L – **L**ength of the welded wire mesh; ST or h_s – **S**labs **T**hickness.

The output parameter data corresponds for all MR RC frame models (see Table 3) in following enumeration form [16]:

i. **U**ltimate **L**ateral **D**isplacements (ULD);
ii. **U**ltimate **L**ateral **F**orces (ULF);
iii. **T**otal **S**trains **E**ps zz (TSE);
iv. **P**rincipal **F**racture **S**trains **M**ax (PFSM);
v. Crack panel for ultimate lateral displacement step.

3. Global analytical results regarding pushover analyses

Following the "non-linear static analyses" [22-26] performed through ATENA software, for the structural MR RC frame models set (see Table 1), the following conclusions were drawn:

The influence of the superior or inferior steel rebars (reinforcement) ratio in RC slabs on the "global dynamic performance of the structures" [27-28], "shear strength" [29-30], and "ductility requirements" [31] through "experimental investigations and analytical calculations" [29] of the "flexural reinforcement type and ratio effect" [32] is insignificant (see Figure 9). However, the rigid diaphragm effect of RC slabs is validated, but in both research situations (with different steel reinforcement ratio), they (RC slabs) present a "non-linear mechanical model" [33].

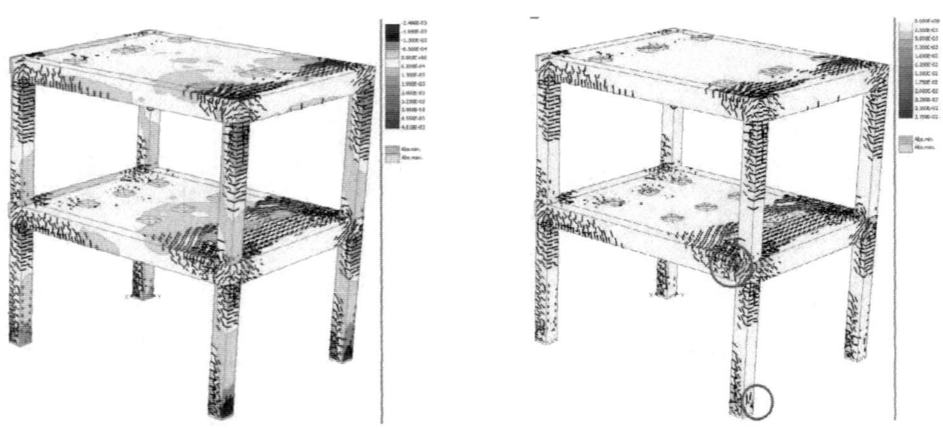

(a) (b)

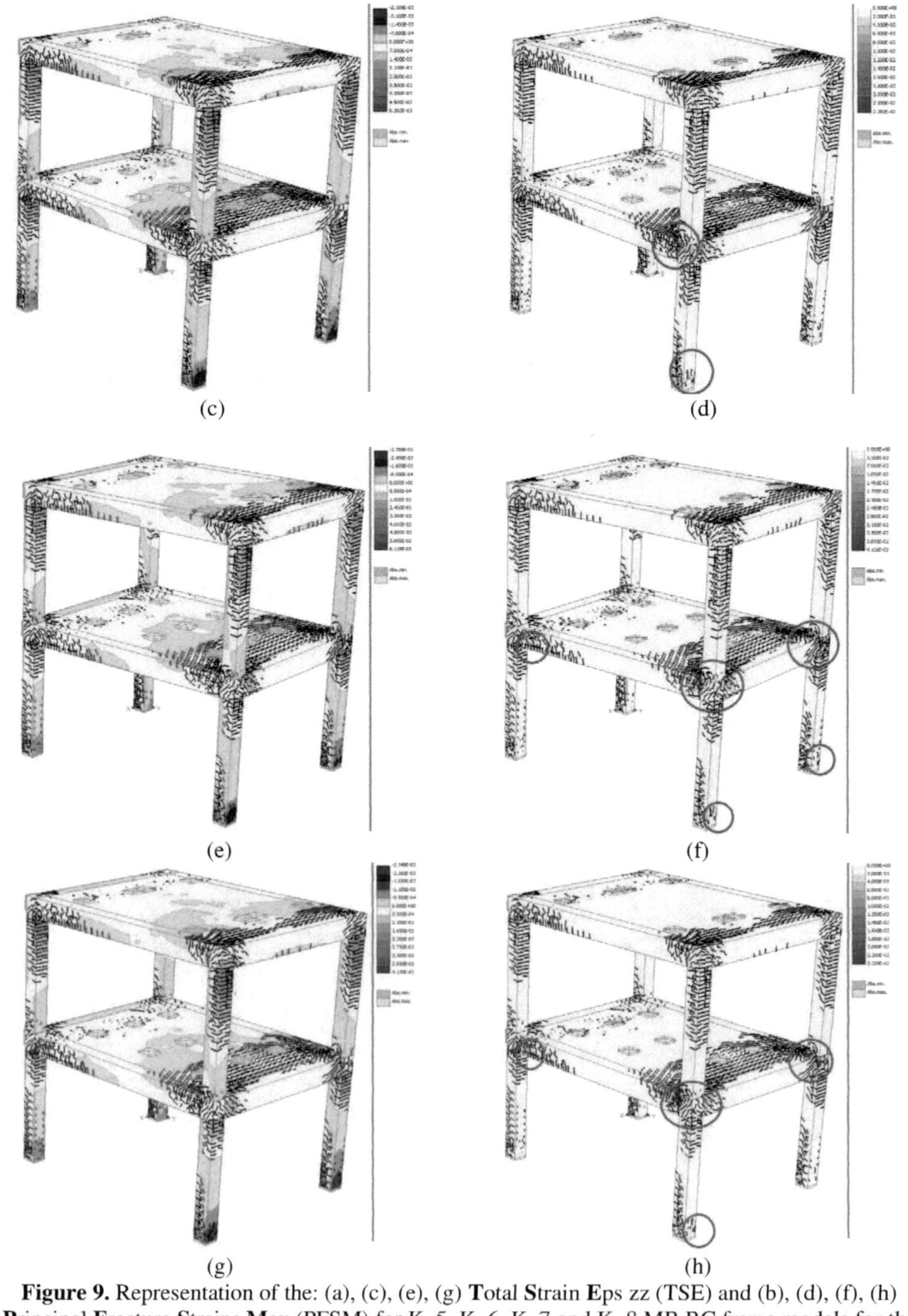

Figure 9. Representation of the: (a), (c), (e), (g) **T**otal **S**train **E**ps zz (TSE) and (b), (d), (f), (h) **P**rincipal **F**racture **S**trains **M**ax (PFSM) for K_5, K_6, K_7 and K_8 MR RC frame models for the ultimate lateral loading stage (see Table 3).

This novel observation ("non-linear mechanical model" [33] raises important issues (problems), regarding the "non-uniform distribution of stresses" [33] in the corner zones of the RC slabs and RC beam-column joint areas. Thus, RC beam-column joints become component parts of the RC beams and RC slab.

Furthermore, an important problem is "crack stabilization" [33] of strongly cracked concrete slabs, so that they (RC slabs) continue to ensure the transfer of inertial masses by means of advantageous mechanisms of the lateral structure.

Another particular effect produced by the RC slab is the development of the "common rigid block (RC beams-slab-node)" [15-17], [34] which can lead to the appearance of the "soft story mechanism" [5], after severe earthquake loading of the structure.

Regarding the influence of the flexural (longitudinal) steel reinforcement (rebar) ratio of RC beams on "structural ductility, shear and flexural performance (behavior), on the moment redistribution and RC beams cracking mode" [35-39], it is possible to observe an increase of the RC beams bending stiffness and decrease of their principal strains for higher ratio (see K_5 and K_6 RC frame models (Figure 9 (a)-(d) and Table 3).

Thus, the K_7 MR RC frame model develops the highest TSE and PFSM (see Table 3) (graphic sketch – see Figure 9 (e)-(f)) in terms of numerical values and highest lateral displacement value for minimum ultimate horizontal force.

Table 3. Principal analytical results for K_5, K_6, K_7 and K_8 MR RC frame models after pushover simulations for the final lateral loading step.

CSC	NSC	ULD (m)	ULF (kN)	TSE (CF)	TSE (TF)	PFSM
C20/25	K_5	0.0276	43.654	0.002486	0.004818	0.0271
C20/25	K_6	0.0288	43.654	0.002589	0.005302	0.0228
C20/25	K_7	0.0329	41.575	0.002789	0.006118	0.0413
C20/25	K_8	0.0290	39.496	0.002349	0.004170	0.0233

„Note: CSC – **C**oncrete **S**trength **C**lass; NSC – **N**umerical **S**imulation **C**ode; ULD – **U**ltimate **L**ateral **D**isplacements; ULF – **U**ltimate **L**ateral **F**orces; TSE – **T**otal **S**train **E**ps zz; CF – **C**ompressive **F**ailure; TF – **T**ensile **F**ailure; PFSM – **P**rincipal **F**racture **S**train **M**ax" [17].

Thus, the concentration mechanism of principal non-linear inelastic strains and concrete/ steel rebars cracking/ yielding process for K_7 RC frame model, develops in a maximum number of areas (see Figure 9 (f)). These plastic areas appear at the end regions of the RC beams and inferior zones of the ground floor RC columns. Thus, the theoretically idealized ductile dissipative mechanism described in "structural engineering literature" [6-9], [21], [40-41] is simulated with many local deficiencies.

These local deficiencies influence the global seismic dissipation mechanism and intensively develop in two categories of structural elements (see Figure 9):

i. RC columns;
ii. RC beam-column joints.

As such, principal plastic strains of the concrete/ steel material in end regions of the RC columns can produce the "global mechanism of progressive collapse" [42-43].

The second problem consists in the non-linear inelastic response of the RC beam-column joint (see Figure 9), with important plastic strains that can lead to the soft story mechanism in current engineering practice. In the research paper of Khan et al. [44],"the need to consider the shear hinge in RC beam-column joints" is confirmed, such that "the model has different stiffness degradation in tension and compression" [14].

However, compressed concrete does not exceed the maximum value of ultimate specific strain limit of compressive concrete ($\varepsilon_{cu,c}$ = 0.0035) [45], as well as the ultimate yielding limit of tensile rebar (reinforcement) (ε_{uk} = 0.075) [45] (see Table 3).

Thus, there is a lateral capacity surplus due to the structural material/ element/ lateral structure over-strength (due to the structural ductility) for the final (ultimate) process of collapse. For this reason, the push-over curve presents the geometric form of Figure 10, stopping at the ultimate lateral force/ displacement values.

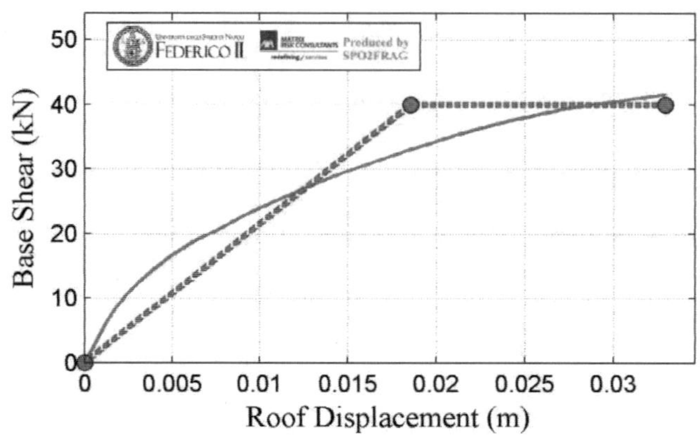

Figure 10. Representation of the **S**tatic **P**ush-**O**ver (SPO) curve (gray line) and bilinearized curve (red line) for K_7 MR RC frame model [46].

4. Conclusions

The static non-linear method proved to be effective in the current analytical study, confirming the conclusions presented in recent specialized publications.

The influence of the flexural steel reinforcement ratio in RC slabs on the seismic response and lateral stiffness response for K_5, K_6, K_7 and K_8 MR RC frame structures is insignificant, producing the inertial masses transfer.

However, non-linear inelastic strains in concrete/ steel material with intense cracking of the RC slabs have been observed. These effects can produce unfavorable mechanisms in RC beams and RC beam-column joints.

The influence of the inferior ratio value for longitudinal (flexural) steel reinforcement in RC beams on global ductility response with principal strains concentration mechanisms in marginal plastic areas (in RC beams), proved to be effective and seismically-favorable.

However, non-linear inelastic strains in RC columns end regions and severe cracking mechanisms in RC beam-column joints were recorded (registered).

Thus, it can be concluded that K_7 Moment Resisting (MR) Reinforced Concrete (RC) frame model represents the optimal seismic-resistant RC frame structure for experimental study on the seismic platform.

5. References

[1] National Centers for Environmental Information (2021) *On This Day: 2011 Tohoku Earthquake and Tsunami* [online] National Centers for Environmental Information (NCEI) Available at: https://www.ncei.noaa.gov/news/day-2011-japan-earthquake-and-tsunami.

[2] Manning K 2012 Quake tech aiming to save lives. *BBC News* [online] 25 Sep.Available at: https://www.bbc.com/news/technology-19589471.

[3] Rahim Z 2020 At least 20 buildings damaged in the Turkish city of Izmir [online] *CNN* Available at: https://edition.cnn.com/europe/live-news/aegean-sea-earthquake-dle-intl/h_e76e639f1d263e9917c8abdc05f699f3

[4] Akiyama H 2002 Collapse modes of structures under strong motions of earthquake *Annals of Geophysics* **45** (**6**) 791-798 doi: 10.4401/ag-3548.

[5] Sococol I, Mihai P and Olteanu-Donţov I 2019 Ductility – Concept for Improving the Seismic Response for Structural Reinforced Concrete Frame Systems, *Bulletin of the Polytechnic*

Institute of Jassy, Construction. Architecture Section **65** (**1**) 17-30.
[6] Budescu M and Ciongradi I 2014 *Inginerie Seismică*, (Iaşi: Politehnium).
[7] Stratan A 2014 *Dinamica structurilor şi inginerie seismică. Note de curs* Timişoara.
[8] P100-1 *"Cod de proiectare seismică", Partea I, Prevederi de proiectare pentru clădiri, Indicativ P100-1, Ministerul Dezvoltării Regionale și Administrației Publice* (2013).
[9] EC 8 *Proiectarea structurilor pentru rezistența la cutremur, Partea 1: Reguli generale, acțiuni seismice și reguli pentru clădiri, Indicativ SR EN 1998-1*, ASRO, București (2004).
[10] Jalilkhani M, Ghasemi S H and Danesh M 2020 A multi-mode adaptive pushover analysis procedure for estimating the seismic demands of RC moment-resisting frames *Engineering Structures* **213** 110528
[11] *ATENA software. http://www.cervenka.cz.*
[12] Cervenka V, Jendele L and Cervenka J 2012 *ATENA Program Documentation. Part 1. Theory* Cervenka Consulting s.r.o. Prague.
[13] Socol I, Mihai P, Toma I O, Olteanu-Donţov I and Venghiac V M 2020 Stress-Strain Relation Laws for Concrete and Steel Reinforcement Used in Non-linear Static Analytical Studies of the Moment Resisting Reinforced Concrete (RC) Frame Models *Bulletin of the Polytechnic Institute of Jassy, Construction. Architecture Section* (In process of publication)
[14] Marzec I, Bobinski J, Tejchman J and Schonnagel J 2021 Finite element analysis on failure of reinforced concrete corner in sewage tank under opening bending moment *Engineering Structures* **228** 111506
[15] Socol I, Mihai P, Toma I O, Venghiac V M and Olteanu-Donţov I 2020 Influence of Concrete Strength Class on the Plastic Hinges Location for a Reinforced Concrete Moment-Resisting Frame Structure with Consideration of the Horizontal Stiffening Effect of the Slab *Bulletin of the Polytechnic Institute of Jassy, Construction. Architecture Section* **66** (**2**) 95-108.
[16] Socol I, Mihai P, Toma I O, Venghiac V M and Olteanu-Donţov I 2020 Static Non-linear analysis of an RC Moment Resisting Frame by Considering Different Values for the Longitudinal Reinforcement Ratio in the Columns *Bulletin of the Polytechnic Institute of Jassy, Construction. Architecture Section* **66** (**3**) 91-106.
[17] Socol I, Mihai P, Toma I O, Olteanu-Donţov I and Venghiac V M 2020 The Influence of the RC Beams Cross Section on the Dissipative Seismic Response of a Moment Resisting RC Frame System *Bulletin of the Polytechnic Institute of Jassy, Construction. Architecture Section* (In process of publication)
[18] El-Attar A G, White R N and Gergely P 1991 Shake Table Test of a 1/6 Scale Two-Story Lightly Reinforced Concrete Building, *Technical Report NCEER-91-0017* Cornell University New York.
[19] Socol I, Mihai P, Pruteanu M and Iftode V 2019 Negative Influences (Effects) of Masonry Infilled RC Frames to Seismic Response of RC Frame Systems and Practical Methods (Solutions) for These Problems (State of the Art) *Bulletin of the Polytechnic Institute of Jassy, Construction. Architecture Section* **65** (**4**) 21-34.
[20] Ishiyama Y 2011 *Introduction to Earthquake Engineering and Seismic Codes in the World* Hokkaido University.
[21] Postelnicu T 2012 *Proiectarea structurilor de beton armat în zone seismice* (București: MarLink,).
[22] Zameeruddin M and Sangle K K 2021 Performance-based Seismic Assessment of Reinforced Concrete Moment Resisting Frame *Journal of King Saud University – Engineering Sciences* **33** (**3**) 153-165
[23] Zameeruddin M and Sangle K K 2021 Damage assessment of reinforced concrete moment resisting frames using performance-based seismic evaluation procedure *Journal of King Saud University – Engineering Sciences* **33** (**4**) 227-239
[24] Praveen O and Gopikrishna K 2020 Assessment of three-dimensional RC moment-resisting frames under repeated earthquakes *Structures* **26** 6-23
[25] Naderpour H and Mirrashid M 2020 A novel definition of damage states for structural elements in framed reinforced concrete buildings *Journal of Building Engineering* **32** 101479

[26] Ismail A 2014 Non linear static analysis of a retrofitted reinforced concrete building *HBRC Journal* **10** (**1**) 100-107
[27] Soltani H, Khaloo A and Sadraie H 2020 Dynamic performance enhancement of RC slabs by steel fibers vs. externally bonded GFRP sheets under impact loading *Engineering Structures* **213** 110539
[28] Sadraie H, Khaloo A and Soltani H 2019 Dynamic performance of concrete slabs reinforced with steel and GFRP bars under impact loading *Engineering Structures* **191** 62-81
[29] Limam S, Nana W S, Bui T T, Limam A and Abouri S 2017 Experimental investigation and analytical calculations on shear strength of full-scale RC slabs with shear reinforcement for nuclear power plants *Nuclear Engineering and Design* **324** 143-157
[30] Schmidt P, Kueres D and Hegger J 2020 Contribution of concrete and shear reinforcement to the punching shear resistance of flat slabs *Engineering Structures* **203** 109872
[31] Santos J and Henriques A A 2021 Span-to-depth ratio limits for RC continuous beams and slabs based on MC2010 and EC2 ductility and deflection requirements *Engineering Structures* **228** 111565
[32] El-Gendy M and El-Salakawy E 2019 Effect of flexural reinforcement type and ratio on the punching behavior of RC slab-column edge connections subjected to reversed-cyclic lateral loads *Engineering Structures* **200** 109703
[33] Gomes J, Carvalho R, Sousa C, Granja J, Faria R, Schlicke D and Azenha M 2020 3D numerical simulation of the cracking behaviour of a RC one-way slab under the combined effect of thermal, shrinkage and external loads *Engineering Structures* **212** 110493.
[34] Sococol I, Mihai P, Iftode V I and Olteanu-Donțov I (2019) Study Regarding the Stiffness Influence of Slab to Beams for a Plan Structural Reinforced Concrete Frame System in Seismic Zones *Computational Civil Engineering Conference (CCE2019)* IOP Conf. Series: Materials Science and Engineering **586** 012013
[35] Li Y and Aoude H 2020 Influence of steel fibers on the static and blast response of beams built with high-strength concrete and high-strength reinforcement *Engineering Structures* **221** 111031
[36] Karimipour A and Edalati M 2020 Shear and flexural performance of low, normal and high-strength concrete beams reinforced with longitudinal SMA, GFRP and steel rebars *Engineering Structures* **221** 111086
[37] Su J, Li Z, Wang J and Dhakal R P 2020 Numerical simulation and damage analysis of RC bridge piers reinforced with varying yield strength steel reinforcement, *Journal Soil Dynamics and Earthquake Engineering* **130** 106007
[38] Yoo D-Y and Moon D-Y 2018 Effect of steel fibers on the flexural behavior of RC beams with very low reinforcement ratios *Construction and Building Materials* **188** 237-254
[39] Diab H M A, Abdelaleem T and Rashwan M M M 2020 Moment redistribution and flexural performance of RC continuous T-beams strengthened with NSM FRP or steel bars *Structures* **28** 1516-1538
[40] Park R and Paulay T 1975 *Reinforced concrete structures* (USA: John Wiley & Sons).
[41] Paulay T and Priestley M J N 1992 *Seismic design of reinforced concrete and masonry buidings* (USA: John Wiley & Sons).
[42] Gowtham S, Prakash M, Parthasarathi N, Satyanarayanan K S and Thamilarasu V 2018 2D-Linear static and non-linear dynamic progressive collapse analysis of reinforced concrete building *Materials Today: Proceedings* **5** (**2**) 8775-8783
[43] Alshaikh I M H, Abu Bakar B H, Alwesabi E A H and Md Akil H 2020 Experimental investigation of the progressive collapse of reinforced concrete structures: An overview, *Structures* **25** 881-900
[44] Khan M S, Basit A and Ahmad N 2021 A simplified model for inelastic seismic analysis of RC frame have shear hinge in beam-column joints *Structures* **29** 771-784
[45] EC 2 *Proiectarea structurilor de beton, Partea 1-1: Reguli generale și reguli pentru clădiri, Indicativ SR EN 1992-1-1*, ASRO, București (2006).
[46] *SPO2FRAG software, http://wpage.unina.it/iuniervo/doc_en/SPO2FRAG.htm.*

Acknowledgments
This research paper is sponsored by "Gheorghe Asachi" Technical University of Iasi with grant number GI/R16_Drd/2021.

Sewage and coronavirus; Maribor case study

M Nekrep Perc[1] and J Kramer Stajnko[2]

[1] Faculty of Civil Engineering, Transportation Engineering and Architecture University of Maribor, Smetanova ulica 17, 2000 Maribor, Slovenia
[2] Faculty of Civil Engineering, Transportation Engineering and Architecture University of Maribor, Smetanova ulica 17, 2000 Maribor, Slovenia

E-mail: matjaz.nekrep@um.si; janja.kramer@um.si

Abstract. Wastewater can be an important indicator of surface water contamination by pathogens, of which SARS-CoV-2 is one, so monitoring of wastewater at treatment plants is important. The virus has been shown to be present in the faeces of infected people, which can be detected much earlier than with medical tests. This will allow prediction of trends in coronavirus outbreaks in advance and may complement current measures to determine prevalence. In Slovenia, pilot monitoring was established in seven WWTPs in October 2020. This paper presents the results and analysis of the Maribor WWTP, where one to three 100 ml samples are collected weekly to detect the presence of SARS-CoV-2. A correlation can be observed between the evolution of the virus in the wastewater and the number of active cases detected by the National Institute of Public Health in the region covered by the WWTP. The presence of the virus in the wastewater is detected about two weeks before the increase in active cases. Some shortcomings and further issues of the wastewater monitoring are analysed.

1. Introduction

Wastewater treatment plants collect wastewater from a large areas, and systematic monitoring can be very useful in detecting various contamination events. Recently, it is particularly important to track the potential presence of pathogens in wastewater, as this can be a useful tool for early warning of pandemics. This information is very important for governments and better preparation of both utilities and public health.

Wastewater research has been used in the past to detect diseases, drug residues, microplastics, hormones, drugs, and pesticides, and to evaluate the success of poliovirus vaccination campaigns. Many chemicals can be found in wastewater that provide more insight into the health and lifestyle of people living in a particular area. Monitoring of wastewater systems has been used in the past, e.g., for early warning of pathogens used as biological weapons [1]. Evidence showed that bioterrorist contamination events result in the pathogen entering wastewater before an epidemic breaks out at the community level. Moreover, the occurrence of enteric viruses, which represent high public health burden, was studied in the effluent of the wastewater treatment plant [2,3].

Recently, sewage research has been gaining importance as a complement to the other investigations conducted by the National Institutes of Public Health for coronavirus surveillance, such as human testing for COVID-19. The main advantage of sewage surveillance over testing people who are already ill is that the virus particles can be detected in the faeces of people who are infected, with or without symptoms. In addition, sewage could act like a COVID-19 early warning system since many people with the virus are asymptomatic and not tested for the disease. If developed further, wastewater research could allow early detection of the resurgence of the virus at an earlier stage, allowing governments and local authorities to develop early action plans to contain the pandemic.

Several studies have been published examining the potential of effluent monitoring to monitor COVID-19 in wastewater. SARS-CoV-2 has been detected in wastewater in several countries (Italy, Netherlands, Spain, France, ...), and good correlations with the number of COVID -19 cases in the population have been found, suggesting that wastewater-based epidemiology could be an efficient adjunct to viral outbreak assessment and containment [4]. However, wastewater-based epidemiology is still not widely accepted as a tool that is adopted by epidemiologists or public health officials as well as governments [5]. In addition, the susceptible-exposed-infectious-recovered (SEIR) model was developed to predict the number of infected individuals. This model provides a framework for quantifying the relationship between the mass rate of virus release in a wastewater catchment and the number of infected individuals present [6]. Wastewater monitoring of SARS-CoV-2 from sewers may be an important adjunct to human clinical testing. However, further research is needed to fill some gaps, e.g., sampling protocols, prediction of case prevalence from virus count by modelling, etc., to determine whether models can be effectively used to back-calculate virion copies to predict case prevalence in specific wastewater catchments/subcatchments [7]. The use of nanofiber filters as a wastewater pretreatment routine and upgrade of existing wastewater assessment and treatment systems to serve as a useful monitoring tool. was proposed in [8], since there is high likelihood that severe acute respiratory syndrome coronavirus 2 may also survive in wastewater.

In Slovenia, pilot monitoring was introduced in October 2020 in seven wastewater treatment plants (Kranj, Ljubljana, Domžale, Šoštanj, Celje, Maribor, Koper). One to three samples are taken every week. The increase of SARS-CoV-2 in wastewater was observed at the beginning of October 2020. The method is efficient to track the trend of the epidemic, the efficiency of the measures and the analysis of the presence of SARS-CoV-2 in facilities such as nursing homes, hospitals, schools and businesses.

In this paper, the results of the pilot monitoring in the Maribor WWTP are presented and the correlation with the number of new cases among people in the region covered by the WWTP is analysed. Some possible uses of the results obtained from monitoring are pointed out and some shortcomings of the method are addressed.

The results are already transmitted to the expert group for the control of the pandemic in Slovenia and are published daily on the COVID-19 tracker.

2. Materials and methods

The Maribor WWTP was built in 2002 and is one of the seven plants where pilot monitoring of wastewater for SARS-CoV-12 was established in October 2020. The sources of wastewater treated in the plant are households, food industry, chemical industry, textile industry, crafts, institutions and hospitals. The share of industry is 20%. The capacity of the sewage treatment plant is 195,000 PE (population equivalent), where 1 PE is the pollution of one citizen per day, the discharge is 109 m^3/s. Currently, the treatment plant covers an area of 258.2 km^2 with 148,000 inhabitants and the following municipalities. (Figure 1):

Table 1. Catchment of the WWTP Maribor [11].

Settlement	Population	Area [km^2]
1. Maribor	113,000	148
2. Miklavž	7,000	12
3. Duplek	7,000	40
4. Hoče-Slivnica	11,000	53.7

Most of the sewer systems are built in combined type.

To detect the presence of pathogens such as SARS-CoV-2 in wastewater, one to three 100 ml samples are collected weekly (24-hour composite sample). The samples are concentrated in the laboratory. Then, the nucleic acid of SARS-CoV-2 (RNA) is extracted from a concentrated sample using a series of chemicals.

IOP Conf. Series: Materials Science and Engineering 1252 (2022) 012037 doi:10.1088/1757-899X/1252/1/012037

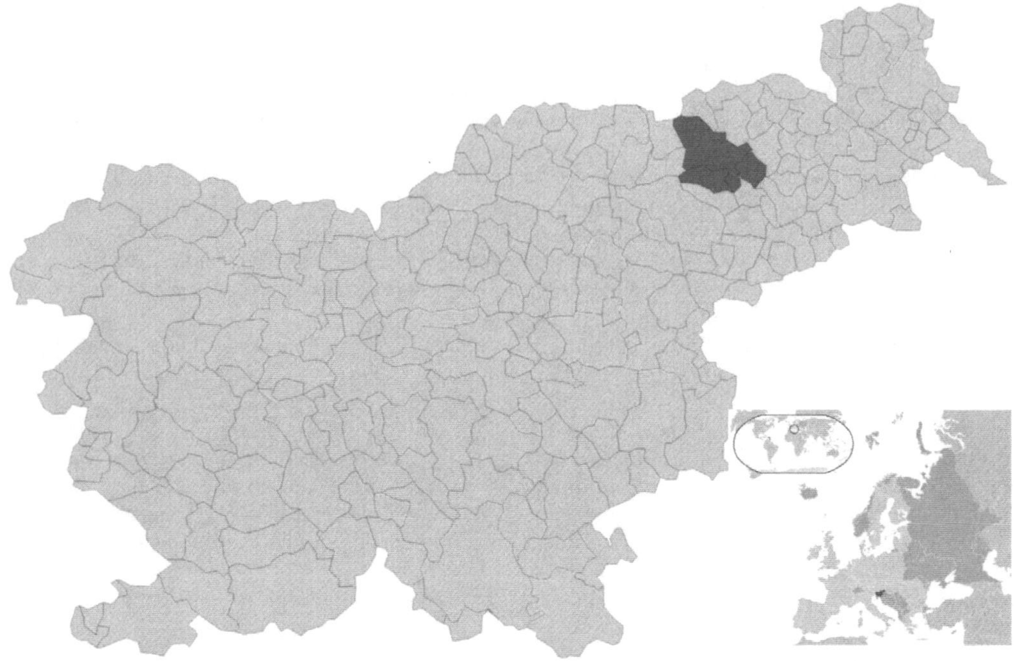

Figure 1. Contribution area of the Wastewater treatment plant Maribor. [9]

3. Results and analysis of current monitoring

Figures 2 and 3 show the results of monitoring a wastewater treatment plant in Maribor from November 2020 to January 2022. The yellow dashed line shows the currently active cases of infected individuals identified by the medical tests and purple dots show the ratio of SARS-CoV-2 gene N2 and PMMoV, respectively.

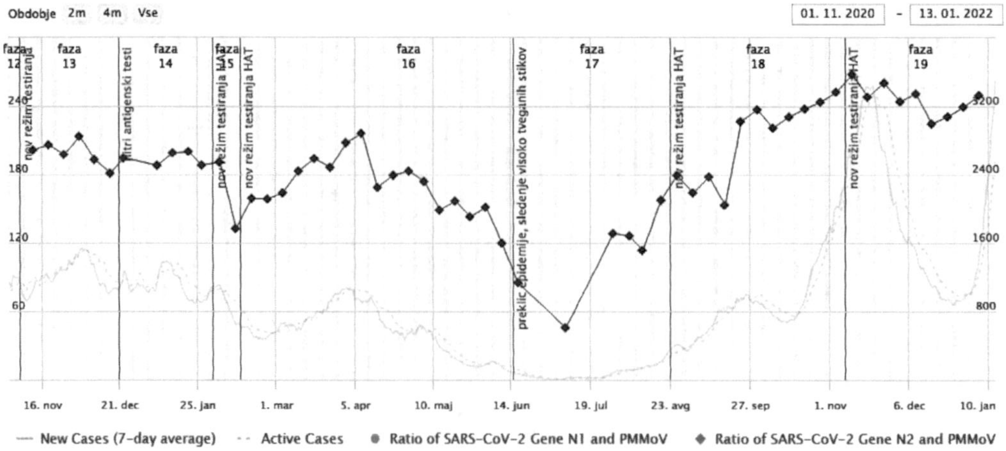

Figure 2. Results of monitoring on a wastewater treatment plant Maribor from November 2020 until January 2022. [10]

A correlation can be observed between the number of active cases and the measured concentration of the virus in the wastewater. Based on the surveillance results, an increase in the pandemic could be predicted, e.g., an increase of virus in the wastewater can be observed in July, two weeks before the increase of actual cases and the opposite trend is observed in early September. In addition, the virus concentration in wastewater increased significantly in the period from late December to January, while the number of active cases began to increase about two weeks later. The presence of the virus in the faeces of infected individuals occurs about two weeks before they become ill and contagious. Therefore, monitoring of sewage monitoring can be an important tool for predicting the outbreak of the next epidemic wave. However, further studies would be needed for a more accurate analysis and prediction of actual trends.

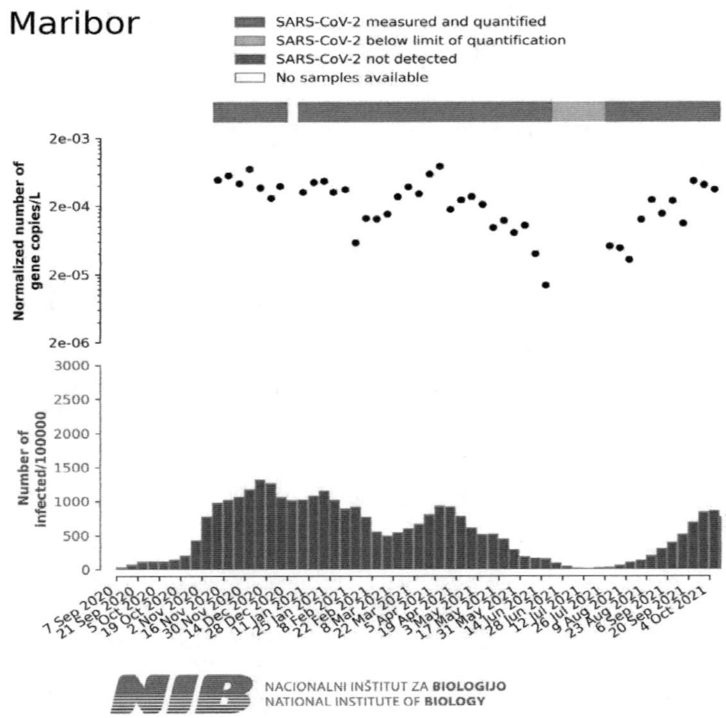

Figure 3. Normalized SARS-CoV-2 RNA concentration in wastewater and active cases for Maribor during September 2020 to October 2021. [11]

Based on the results during the first, second and third waves of the pandemic in 2020 and 2021, it is evident that the wastewater surveillance numbers indicated an advance increase in the virus before it was observed based on indicators such as positive COVID-19 tests and hospitalizations. Therefore, the results of wastewater surveillance could be used as an early warning system for the next wave of the pandemic.

4. Conclusion and recommendations

The results of the monitoring of wastewater in the Maribor wastewater treatment plant, carried out from November 2020, show that this monitoring could be an important early warning tool for the next pandemic wave COVID-19. Due to the good correlation between the number of active cases and the measured concentration of the virus in the wastewater, monitoring of the effluent can be used to detect arrival and decline of the SARS-CoV-2. However, there are still some issues that should be addressed, e.g., the influence of the different types of wastewaters collected in the treatment plant (various industrial and craft enterprises), as well as the impact of heavy rainfall. Indeed, the wastewater system in the municipalities connected to the treatment plant is a mixed system, so rainfall can have a decisive influence on the virus concentrations in the samples and therefore on the reliability of the results. One of the measures to prevent the amount of precipitation from influencing the analytical results could be a special sampling method in which larger samples are taken in proportion to the intensity of precipitation. A representative sample is obtained based on the average volume of wastewater that has flowed through the plant in a day. However, the influence of rainfall is still an open issue that should be studied in more details.

The results of separate wastewater systems are more reliable and accurate, which is an additional advantage over combined wastewater systems. On the other hand, consistent wastewater management is important to prevent a new outbreak of a pandemic.

It remains an open question how safe it is to treat wastewater and human waste contaminated with the SARS-CoV-2 coronavirus. However, if the wastewater is treated consistently, the potential spread of the virus through the wastewater is minimal because the virus has only been detected in the influent, not the effluent, of the facility. Therefore, it is extremely important to prevent wastewater contaminated with the virus from finding a way into municipal water supplies and causing outbreaks of the disease, which poses an additional problem for engineers in the future.

5. References

[1] Sinclair G R, Choi Y C, Riley R M and Gerba P C 2008 Pathogen Surveillance Through Monitoring of Sewer Systems *Adv Appl Microbiol.* **65** 249–269

[2] Prado T, Silva M D, Guilayn C W, Rose L T, Gaspar C A M and Miagostovich P M 2011 Quantification and molecular characterization of enteric viruses detected in effluents from two hospital wastewater treatment plants *Water Research* **45** 1287–97

[3] Steyer A, Gutiérrez-Aguirre I, Rački N, Beigot G S, Brajer H B, Stražar M, Škrjanc I, Poljšak-Prijatelj M, Ravnikar M and Rupnik M 2015 The Detection Rate of Enteric Viruses and Clostridium difficile in a Waste Water Treatment Plant Effluent *Food Environ. Virol.* **7** 164–172

[4] Farkas K, Hillary S L, Malham K S, McDonald E J and Jones L D 2020 Wastewater and public health: the potential of wastewater surveillance for monitoring COVID-19 *Current Opinion in Environmental Science & Health* **17** 14–20

[5] Daughton G C 2020 Wastewater surveillance for population-wide Covid-19: The present and future *Science of The Total Environment* **736** 1–9

[6] McMahan S C et al. 2021 COVID-19 wastewater epidemiology: a model to estimate infected populations *The Lancet Planetary Health* **5** e874–e881

[7] Donner E 2021 Wastewater monitoring for SARS-CoV-2 Coronavirus disease 2019 : COVID-19 WM of SARS-CoV-2 in Australia *Microbiol. Aust.* **42** 18–22

[8] Venugopal A 2020 Novel wastewater surveillance strategy for early detection of coronavirus disease 2019 hotspots *Environmental Science & Health* **17** 8–13

[9] Municipalities in numbers: GOV.SI [online] https://www.gov.si/en/topics/municipalities-in-numbers/

[10] Covid 19 sledilnik official website [online] https://covid-19.sledilnik.org/sl/stats

[11] Milavec M 2021 Merjenje SARS-CoV-2 v odpadnih vodah, Svetovni dan meroslovja – Merjenje za zdravje *Nacionalni inštitut za Biologijo*

IOP Conf. Series: Materials Science and Engineering 1252 (2022) 012038 doi:10.1088/1757-899X/1252/1/012038

Water vapor transmission parameters of the Kežmarok sandstone

P Jaroš[1], M Vertal'[1]

[1] Institute of Architectural Engineering, Faculty of Civil Engineering, Technical University of Kosice, Vysokoskolska 4, 040 23 Kosice, Slovakia

E-mail: pavol.jaros@tuke.sk, marian.vertal@tuke.sk

Abstract. Dynamic complex heat and water transport in building constructions can be calculated using modern simulation software such as Wufi, Delphin, Math, Comsol Multiphysics and other. This software is suitable for the evaluation of the thermal and hygric processes in constructions of historical buildings. Because it includes many important factors such as dynamic boundary conditions, solar radiation, driven rain, ground water, diffusion and others. The heat and water transport parameters of historical materials are needed for simulation of coupled heat and water transport in building materials. The Kežmarok sandstone was chosen for analysis. One of the important material parameters for evaluation of complex heat and water transport in building constructions are water vapor transmission parameters. Vapor transport can be described as water vapor diffusion resistance factor, water vapor diffusion coefficient and diffusion equivalent air layer thickness. The dry cup and wet cup method were used to determine the water vapor transmission parameters of Kežmarok sandstone.

1. Introduction

Water vapor transmission parameters are important parameters, which are needed for simulation of complex heat and water transport in building materials [1], [2], [3]. They describe, how fast water molecules can move through a given material between two spaces with different boundary conditions. Simulation of heat and water transport is suitable for assessment of historical building constructions. In Slovakia, sandstone was the most used building material in past [4], [5]. The water transfer parameters of Slovak sandstones have not yet been presented in any scientific study. Analysis of these parameters and their implementation in the simulation software will allow modelling of the dynamic transfer of water through this material. Subsequently, these simulations can be used in the design of building renovations. There are several scientific studies that describe the use of simulation software to evaluate sandstone structures [6], [7]. In Slovakia, sandstone was used, because strength, appearance and workability. For this paper, Kežmarok sandstone was selected for analysis of water vapor transport. Samples were taken from original abandon quarry in eastern part of Kežmarok city. Coefficients, which describe water vapor transport, were determined using dry cup and wet cup method.

2. Investigated material

In Gothic and Renaissance periods, stone was the main building material. It was used for load-bearing and non-bearing structures such as vertical walls, railings, window frame, cornice and others. Sandstone was very favoured stone. It has a good compressive strength, good workability and good appearance. Kežmarok is one of the oldest town in eastern Slovakia. On the eastern side of the town is an abandon quarry for sandstone mining. It is fine-grained and it has light to gray color. This stone was probably used in the construction of the Basilica of the Holy Cross, the church of the Visitation of the Virgin Mary, for smaller chapels and historic bell towers [8], [9]. Specimens of this stone were taken from an abandoned quarry for analysis of water vapor transmission parameters.

3. Experimental method

The water vapor transmission through a building material can be expressed by the water vapor diffusion coefficient δ and by the water vapor diffusion resistance factor μ. The steady-state dry cup and wet cup methods were used to analyse the water vapor transmission parameters of selected sandstone. It is one of the most performed methods in the field of complex heat and water transport and this method has been used by many researchers [10], [11]. It is described in a European standard [12]. The cup method is based on the transfer of water vapor at steady state between two environments with different boundary conditions. This measurement is based on measuring flow of water vapor passing through the sample either from or into the cup. During the experiment, the weight gain or weight loss of the cup is recorded at time intervals depending on the type of material. Dry or wet cup method can be used. Both are recommended. During experiment of dry cup, specimen is attached to the lid of the cup and the desiccant is stored in the cup. In the case of dry cup, silica gel or C_aCl_2 solution can be used as desiccant, which provide the relative humidity of 0-3%, in wet cup method, the sealed cup can obtain a saturated K_2SO_4 solution, which provides a relative humidity of 97.42%. Also distilled water can be used, which provides a relative humidity of 100%. Principle of cup method is shown in Figure 1.

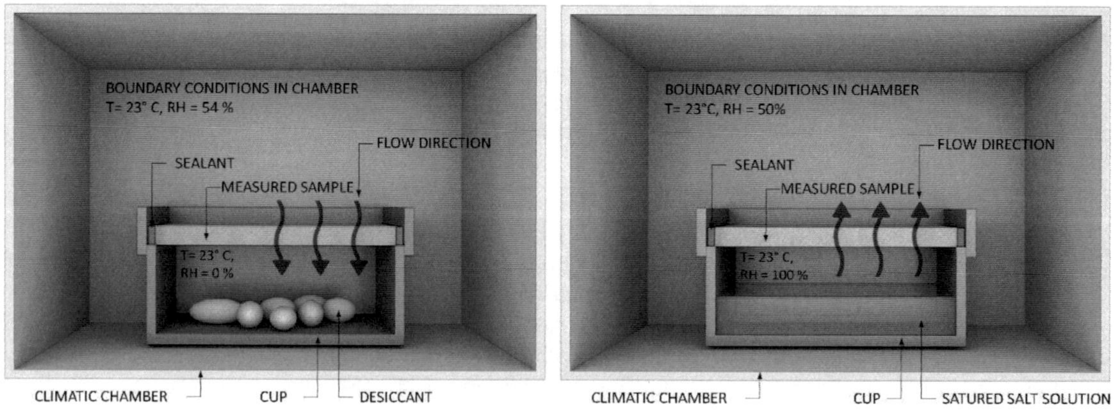

Figure 1. Theoretical principle of the dry (left) and wet cup method (right).

When a steady state occurs in the sample corresponding to the surrounding boundary conditions, the weight increase, decrease of the sample will be constant in a certain time interval. From the average weight gain can be calculated water vapor diffusion coefficient δ according to the equation**Chyba! Nenašiel sa žiaden zdroj odkazov.**:

$$\delta_m = \frac{\Delta m d R T}{S t M \Delta p_p} \tag{1}$$

where Δm is the amount of water vapor, which passes through the specimen (kg), d is the specimen thickness (m), R is the universal gas constant, T is the absolute temperature (°C), S specimen surface (m^2), t the period of the time corresponding to the transport of mass of water (s), p_p is the difference between two different partial pressures above and below the sample (P_a). From the following equation can be determined water vapor diffusion resistance factor**Chyba! Nenašiel sa žiaden zdroj odkazov.**:

$$\mu = \frac{\delta_{air}}{\delta m} \tag{2}$$

where δ_{air} is the coefficient of vapor diffusion in free air (kg/msP$_a$), δ_m is the coefficient of vapor diffusion in measured material.

4. Preparation of the experiment

To ensure only one-dimensional water vapor transport through the material, aluminium cups with high diffusion resistance were made in the shape of a cube with a base of 50x50 mm and a height of 60 mm. For the experiment, samples with block shape were prepared with cross-section of 50 mm and height of 30 mm. Three samples were made to include material variability. Before the dry cup test, specimens were dried in electric oven at 105°C and lateral sides of the specimens were insulated with epoxy resin. The samples were then placed on the cup lid and insulated with self-adhesive tape on the sides of the aluminium cups and samples. They are shown in Figure 2. Silica gel ensured constant relative humidity of 0-3 % in the case of a dry cup, distilled water provided constant relative humidity of 100 % in the case of a wet cup. The air space between desiccant/salt solution and sample was 20 mm. The cups thus prepared were placed in a chamber. Chamber is shown in Figure 3. The relative humidity in the chamber was maintained at 53.49% by salt solution of magnesium nitrate $Mg(NO_3)_2$ at temperature 23°C. Before testing, all samples were stored in desiccator at 23°C and 53.49 % relative humidity to stabilize their weight.

Figure 2. Preparation of the experiment.

In both cases, dry cup, wet cup, three specimens of the same samples were used. They were marked K1, K2, K3. Weight gain and weight loss were recorded by laboratory balance with an accuracy of 0.01 g and resolution 0.001 g. The weighing interval was 48 hours. The average increment per unit time was calculated by linear regression from recorded weights over five weeks.

Figure 3. Chamber during process of measuring.

During the experiment, temperature in chamber was maintained by air conditioning at 23°C. The chamber was airtight, the door was opened only when weighing the cups. The temperature and relative humidity were monitored by sensors.

5. Results

The weight gain in the dry cup method and the weight loss in the wet cup method is shown in Figure 4, the cups were weighed at 48 hours intervals. In the graphs of the dry cup method, it can be observed that the weight increase is not linear in the first period, in this part the material has not yet reached an equilibrium state corresponding to the boundary conditions. After reaching this state, the weight gain is linear for the rest of measurement. A similar situation can be observed in the case of a wet cup method. In the first phase stabilization occurs, then the weight loss is linear. The average increments were calculated from these linear courses.

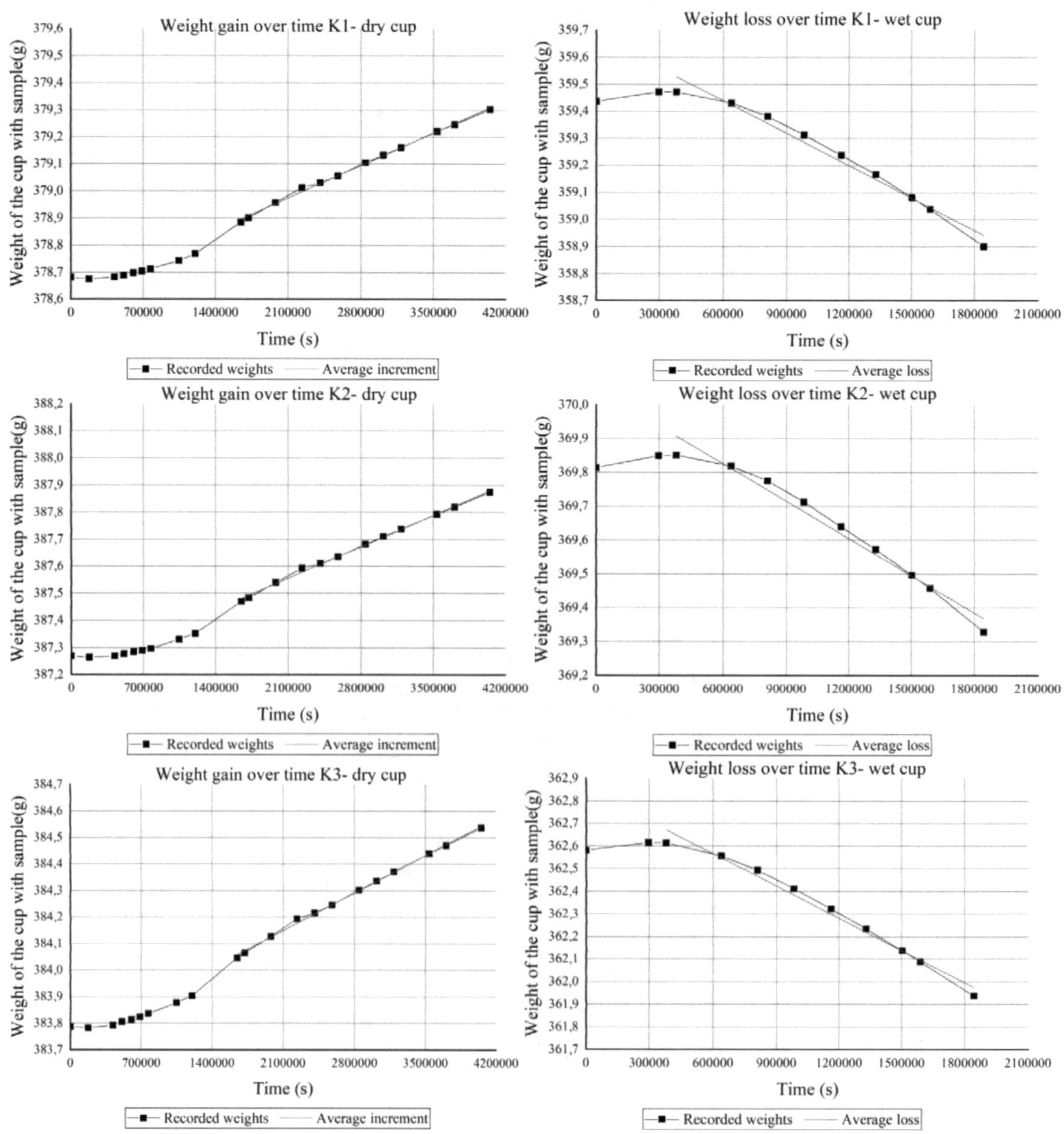

Figure 4. Weight gain in the dry cup method and weight loss in the wet cup method.

The measured water vapor transport parameters are summarized in Table 1 for dry cup and Table 2 for wet cup. The results are not presented as an arithmetic mean because it is not suitable for this type of material. This natural material has an open porosity in the range of 8 to 12%, which causes differences in the resulting parameters. Therefore, it is more appropriate to report results as a range. The water vapor diffusion resistance factor of measured sandstone obtained from the cup method is much higher as compared with the most used building material such as brick, aerated concrete and other. Also, this is not very common with this type of natural material. Studies of the thermal and water transport properties of sandstones from Czech Republic and Germany have shown much lower values of the water vapor transmission properties [1], [6], [13]. However, such high values are in good agreement with the low open porosity of this historical building material. The differences in values between the individual samples are due to the variability of the material. The measured values of water vapor diffusion resistance factor using the dry cup method (3/53.49%) were systematically higher than the values obtained in the case of a wet cup (100/53.49%). It can be attributed to the partial transport of capillary condensed in pore structure of the sandstones [14]. Water vapor permeability parameters of Slovak sandstone are not mentioned in any scientific study. In the national standard, the value of the diffusion resistance factor is 23 for sandstone with bulk densities from 1800 to 2600 kg/m3. The measured values of the investigated sandstone are much higher, therefore it is necessary to analyse these parameters for each sandstone separately.

Table 1. Water vapor transport parameters measured by dry cup method.

Sandstone	δ (kg/ms.Pa)/s	μ (-)
K1	1.52E-12	129.9
K2	1.48E-12	133.7
K3	1.79E-12	110.4

Table 2. Water vapor transport parameters measured by wet cup method.

Sandstone	δ (kg/ms.Pa)/s	μ (-)
K1	3.67E-12	53.9
K2	3.38E-12	58.6
K3	4.37E-12	45.3

6. Conclusion

The coefficients of water vapor transport parameters of the selected sandstone were determined by the cup methods. Determined values of parameters in the range of 45.3 to 53.9 in the dry cup and values from 110.4 to 133.7 in the wet cup are higher in comparison with commonly used materials such as brick or aerated concrete. These values are in a good agreement with open porosity, which is relatively low. These values are not favourable for the moisture regime in building constructions made of this sandstone, because they reduce the process of evaporating water from the porous material. This can cause continual increase of water content, which adversely affects the building constructions. The measured parameters are important for modelling the thermal and hygric behaviour of building constructions, because errors will occur if incorrect values are used. A more detailed explanation of the high diffusion resistance factor of this sandstone can provide mercury intrusion porosimetry, which can be used to obtain a pore size distribution. This part of the research will be realized in another study of thermal and hygric properties of the sandstones. Experimentally determined parameters in comparison with the standard value indicate that it is necessary to analyse the moisture properties of sandstone separately. Obtained parameters of the Kežmarok sandstone can be imported into a simulation software for the analysis of the dynamic transfer of water vapor through a building structure built of sandstone.

7. References

[1] Krus M 1996 *Moisture Transport and Storage Coefficients of Porous Mineral Building Materials: Theoretical Principles and New Test Methods*. Fraunhofer-IRB-Verlag.

[2] Jaroš P and Vertal' M 2020 Coupled Heat and Moisture Transport in Building Material – Water Absorption Coefficient and Capillary Water Content *IOP Conference Series. Materials Science and Engineering* 867 (1): 012012.

[3] Vertal′ M and Ďurica P 2009 Bestimmung von feuchtetechnischen Materialparametern ausgewählter Baumaterialien für die rechnerische Bewertung von Baukonstruktionen unter den Bedingungen in der Slowakei. *Bauphysik 31*(6) 386-392

[4] Čabalová D 2013 *Krása kameňa v živote človeka*

[5] Jaroš P and Vertal' M 2019 Characterization of Sandstones of Eastern Slovakia *Selected Scientific Papers - Journal of Civil Engineering* 14 (2): 19–30

[6] Kočí V Maděra J Fořt J Žumár J Pavlíkov M Pavlík Z and Černý R 2014 Service Life Assessment of Historical Building Envelopes Constructed Using Different Types of Sandstone: A Computational Analysis Based on Experimental Input Data *TheScientificWorldJournal* 2014 (July): 802509

[7] Zhao J and Plagge R 2015 Characterization of Hygrothermal Properties of sandstones—Impact of Anisotropy on Their Thermal and Moisture Behaviors *Energy and Buildings* 107 (November): 479–94

[8] Inžinierskogeologický atlas hornín Slovenska official website [online]: http://apl.geology.sk/igatlasg/

[9] Historic quarries official website [online]: http://hq.chc.sbg.ac.at/

[10] Haba B Boudjemaa A Boudenne A and Benzarti K 2017 Hygric Properties and Thermal Conductivity of a New Insulation Material for Building Based on Date Palm Concrete *Construction and Building Materials* 154 (November): 963–71

[11] Janssen H 2018 A Discussion of 'Characterization of Hygrothermal Properties of Wood-Based Products – Impact of Moisture Content and Temperature *Construction and Building Materials* https://doi.org/10.1016/j.conbuildmat.2018.07.018.

[12] ISO 12572, Hygrothermal Performance of Building Materials and Products—Determination of Water Vapour Transmission Properties, International Standards Organization, Geneva, Switzerland 2001

[13] Pavlík Z Michálek P Pavlíková M Kopecká I Maxová I and Černý R 2008 Water and Salt Transport and Storage Properties of Mšené Sandstone *Construction and Building Materials* 22 (8): 1736–48

[14] Jerman M Keppert M Výborný J and Černý R 2013 Hygric, Thermal and Durability Properties of Autoclaved Aerated Concrete *Construction and Building Materials* 41 (April): 352–59

Acknowledgments

This work was supported by the VEGA: 1/0674/18 Theoretical and experimental analysis of architectural-structural shapes and fragments of building envelope structures designed for harsh conditions.

Mechanical fracture parameters of hemp fibre reinforced cement-based composites with recycled aggregate

H Simonova[1], B Kucharczykova[2], B Poletanovic[3], I Merta[3] and Z Kersner[1]

[1] Institute of Structural Mechanics, Faculty of Civil Engineering, Brno University of Technology, Veveri 331/95, 602 00 Brno, Czech Republic
[2] Institute of Building Testing, Faculty of Civil Engineering, Brno University of Technology, Veveri 331/95, 602 00 Brno, Czech Republic
[3] Institute of Material Technology, Building Physics, and Building Ecology, Faculty of Civil Engineering, TU Wien, Karlsplatz 13/E207-2, 1040 Vienna, Austria

E-mail: kersner.z@vutbr.cz

Abstract. This paper aims to evaluate the effect of replacement of aggregate by recycled one on fracture response of selected composites. These are hemp fibre reinforced cement-based composites. Four different mixtures were prepared: in reference one, the only natural aggregate was used whereas in the other three mixtures, the natural aggregate was replaced by recycled one in the amount of 10, 25 and 50 %, respectively. All mixtures contained hemp fibres with a length of 10 mm in the dosage of 1 % by volume. In order to determine the selected mechanical fracture characteristics of the investigated composites, three-point bending fracture tests were carried out on prismatic specimens with nominal dimensions 40 × 40 × 160 mm^3 provided with an initial notch. During the whole course of the testing, the loading process was governed by a constant displacement increment of 0.02 mm/min. The vertical displacement (midspan deflection) was measured using the inductive sensor mounted on a special measurement frame placed on the specimens. The fracture tests were terminated when the midspan deflection reached 0.5 mm. The mechanical fracture parameters were obtained through a direct evaluation of the fracture test data via the effective crack model and the work-of-fracture method. The paper shows that composites with a higher dosage of recycled aggregate (25 and 50 wt%) achieve similar values of mechanical fracture properties, so this replacement can be recommended.

1. Introduction

The construction industry accounts for a vast quantity of the world's anthropogenic greenhouse gas emissions whereas only concrete is responsible for 8–9 % [1]. Concrete is mostly used man-made material in the world and its consumption is estimated at 4.7 tons per person per year. The main components of the concrete are cement, aggregate and water, whereas the aggregate dominates (around 70 %) with its volume. To foster sustainability in the construction industry, the replacement of the natural aggregate with recycled one should be one of the priorities. Nedeljkovic et al. [2] point out that the most important aspects of replacement of natural aggregate with recycled one are: i) reducing CO_2 emission significantly, ii) reducing the waste landfilling and cutting waste disposal cost and iii) lack of raw materials. However, there are serious drawbacks that prevent the replacement of natural aggregate with a higher quantity of recycled aggregate. Recycled aggregate has a high porosity and water absorption and consequently lower density [3]. Furthermore, recycled aggregate contains hardened cement paste particles that may lead to an alkali-silica reaction [3]. The problem becomes even more pronounced in the case of fine aggregates. After the concrete is crushed, the small particles of hardened old cement paste are mixed with the crushed aggregate and the resulting fine recycled aggregate becomes very heterogenous [2–3].

Since concrete is a brittle material usually it is reinforced with short, randomly dispersed steel or synthetic fibres [4–7]. Recently natural fibres are becoming an attractive replacement for traditional fibres due to their sustainability, worldwide availability, biodegradability, low density and favourable mechanical properties. The objective of this study was to investigate the mechanical fracture parameters of hemp fibre reinforced cement-based mortars with recycled aggregate (up to 50 % of natural aggregate replacement).

2. Experimental part

2.1 Materials, specimens and curing

The mortar matrix used in this study was designed of cement, sand and water with a mass ratio of 1:1:0.4. CEM II/B 32.5R and natural aggregate with a grain size of sand 0–4 mm were used. The recycled concrete aggregate of the company ContraCon (Austria) of type RS III 0/4 U-A according to the Austrian norm [8] was used for the replacement of natural sand.

In total 4 mixtures were produced, i.e., one with natural aggregate (specimens notation MH0: MH0_1, MH0_2,..., MH0_6) and three mixtures where the natural aggregate was substituted with the recycled one in dosages of 10 wt% (MH10), 25 wt% (MH25), and 50 wt% (MH50). For the matrix reinforcement, industrial hemp (*Cannabis sativa L*) fibres were used. The fibres have a diameter of 8–60 μm, a density of 1500 kg/m^3 and, according to the literature [9], the tensile strength of 270–900 N/mm^2. The fibre bundles have been cut to the length of 10 mm and added to the matrix in the volume percentages of 1.0 vol%. Since natural fibres are highly hydrophilic, they are added to the matrix in water-saturated surface dry conditions. The mortar was mixed in a laboratory drum-mixer, the mixture was poured into the moulds 40 × 40 × 160 mm^3, demoulded 24 hours after casting and stored in water for 28 days and then in a closed box with RH > 95 % up to the date of the particular tests were performed.

2.2 Fracture tests

The fracture experiments in the three-point bending configuration were conducted at the age of 35 and 260 days. The above-mentioned prismatic specimens were provided with an initial central edge notch one day before testing. The notch length was approximately one-third of the specimen's height. The span length was set to 120 mm. The stiff multi-purpose mechanical testing machine LabTest 6.250 with a load range of 0–250 kN was used for fracture tests. The loading process was controlled by a constant increment of displacement of 0.02 mm/min in the whole course of the experiment. See the illustration photo in figure 1.

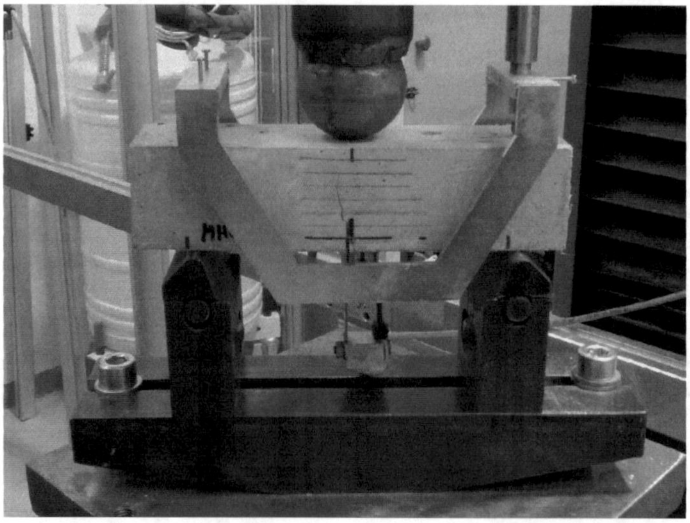

Figure 1. Illustration of the fracture test configuration.

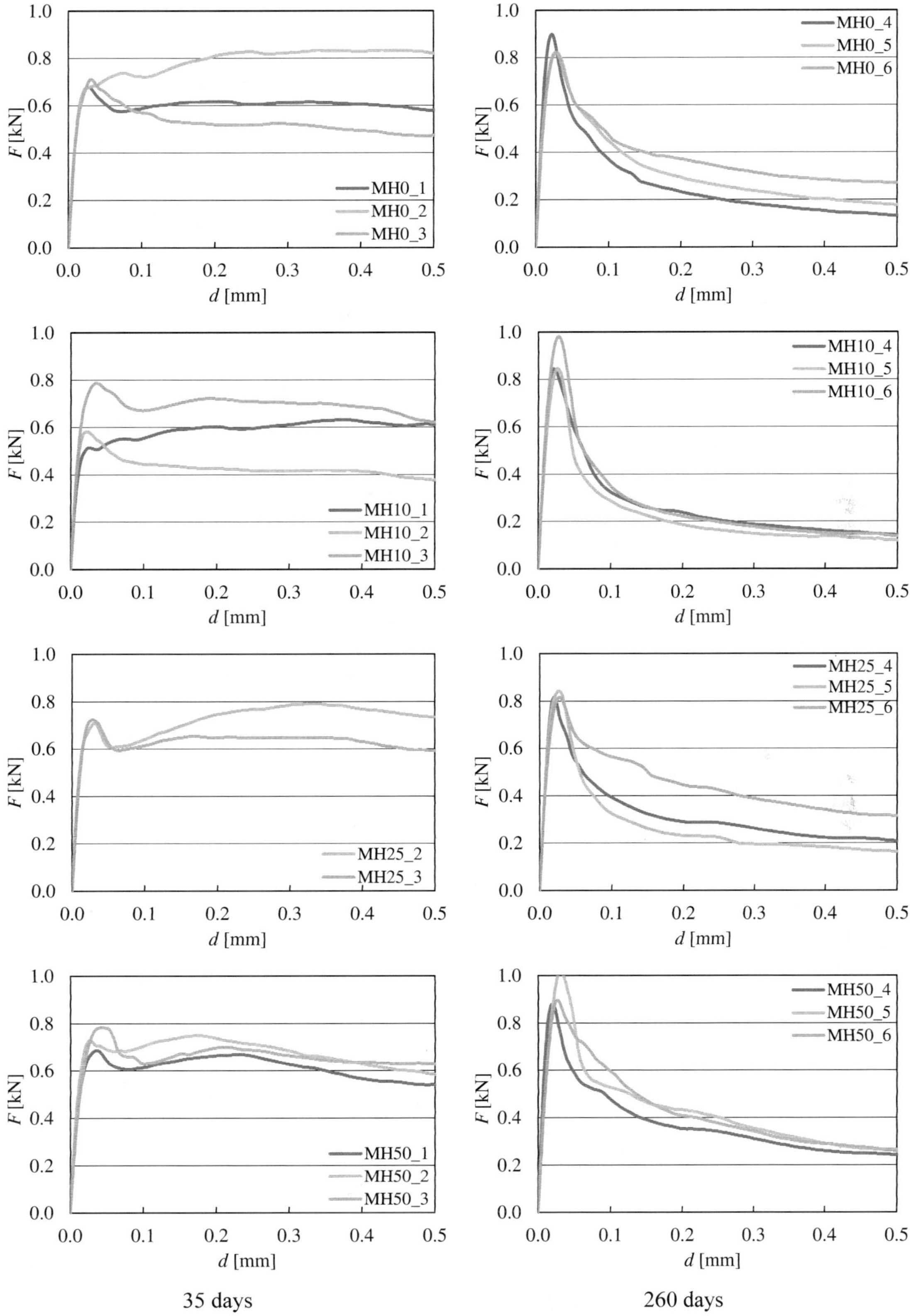

Figure 2. Load vs. displacement diagrams for a particular age of specimens: 35 (left) and 260 days.

The mechanical fracture parameters were determined based on the direct evaluation of the *F–d* diagrams using the chosen fracture model. The modulus of elasticity was assessed from the initial parts of *F–d* diagrams according to Karihaloo [11]. The effective fracture toughness was determined from *F–d* diagrams using the effective crack model introduced by Karihaloo [11]. This approximative model combines the linear elastic fracture mechanics and crack length approaches.

The *F–d* diagrams were further used for the calculation of work of fracture value according to the RILEM method [12]. Then, the specific fracture energy is assessed by dividing the work of fracture by the ligament area (i.e. the projection of the original uncracked area).

Supplementary, a non-destructive test based on the resonance method was employed to monitor the development of the dynamic modulus of elasticity of investigated composites during ageing. For more details about the principle of measurement refer to [13].

Note that the informative compressive strength was determined according to ČSN EN 196-1 [14] on the fragments remaining after the fracture experiments had been conducted.

3. Results

The average value (determined based on 3 independent measurements) and standard deviation represented by error bars are presented in figure 3. The **bulk density** of the specimens does not change significantly, when the natural aggregate is substituted by recycled one (independent of the substitution rate, figure 3), neither at the age of 35 days nor after 260 days.

The **modulus of the elasticity** increases when the specimens tested at the age of 35 days are compared to their counterparts tested at the age of 260 days. The reason could be due to the further cement hydration, which lead to the stronger, but more brittle material (figure 2). The substitution of the natural aggregate with the recycled aggregate does not provide any further increase of the modulus of elasticity when the specimens are tested at the age of 35 days. However, at the age of 260 days, the specimens' modulus of elasticity slightly increases in case of higher aggregate substitution (50 wt%), on the other hand, the variability of values is higher.

Due to the further cement hydration process, the **compressive strength** of the specimens increases with time. The higher substitution of natural aggregate with recycled aggregate leads to higher compressive strength (figure 3), which is in good correlation with other works on recycled aggregate concrete [15]. Only the higher substitution rate (50 wt%) of the aggregate increases the compressive strength of the composite (irrelevant of the age of the specimens).

Due to the significant drop of the post-peak part of the *F–d* diagram (figure 2) the specimens' **fracture energy** with ageing decreases by 50%. However, it can be seen (figure 3) that when the amount of recycled aggregate increases, the drop of the fracture energy becomes lower. In addition to this, it is noticed that the increase in the recycled aggregate results generally in the increase of the specimens' fracture energy, at both ages (35 and 260 days). It is believed that with increased recycled aggregate replacement the heterogeneity and overall porosity of the matrix increases which consequently could prolong/hinder the crack propagation.

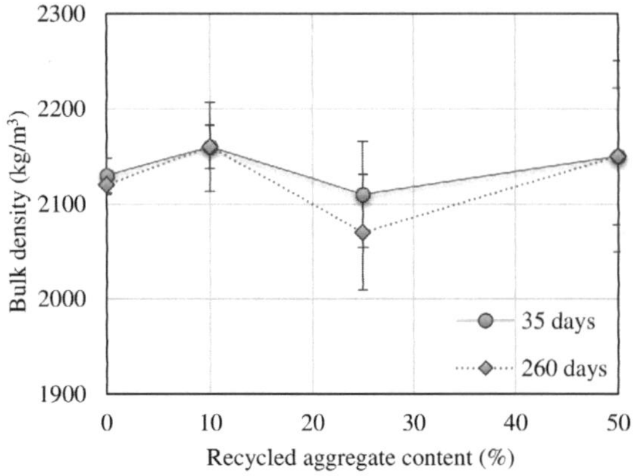

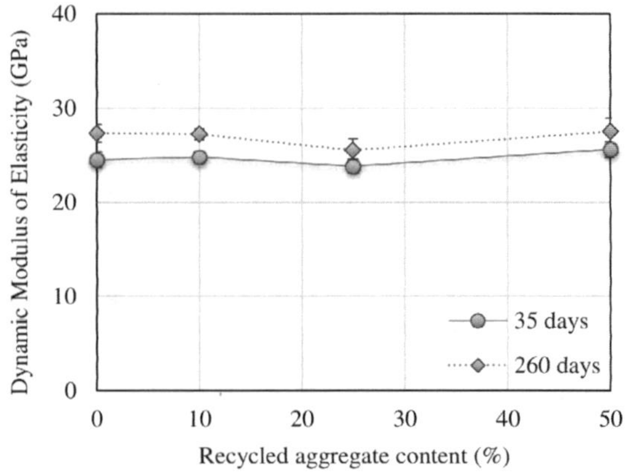

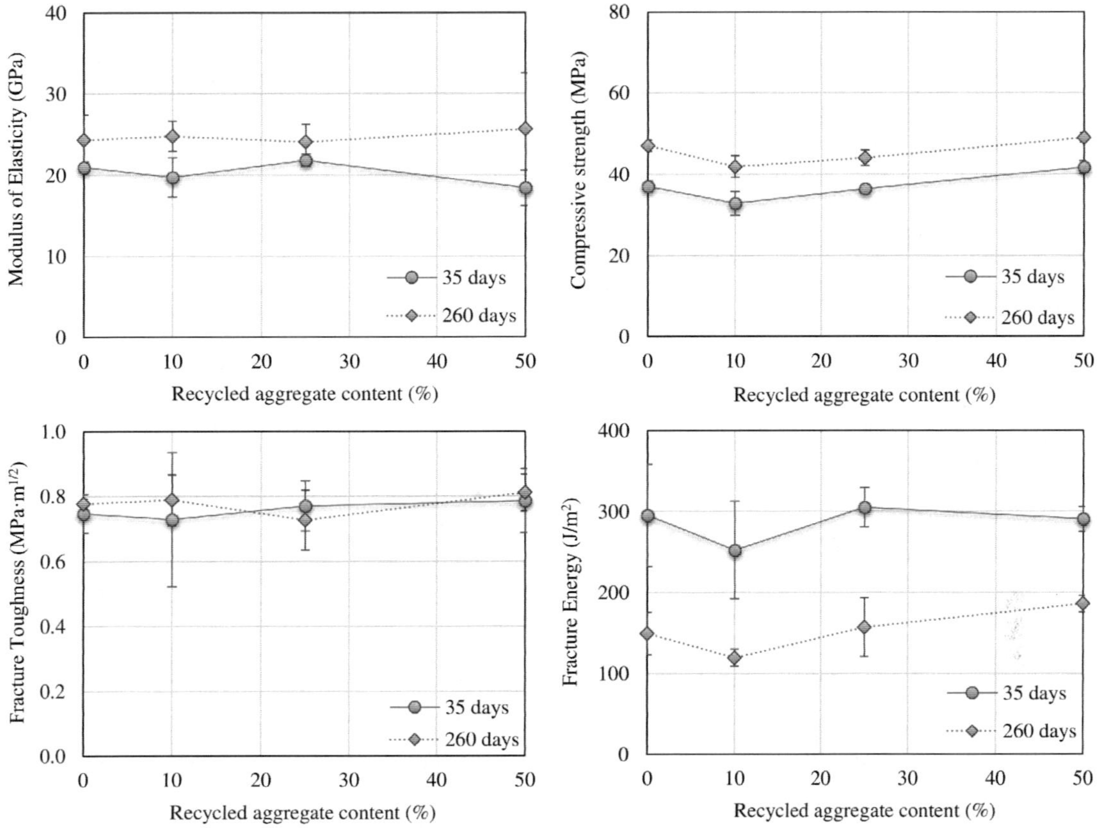

Figure 3. Monitored physical, and mechanical fracture parameters of composites.

4. Conclusions

In the presented work the effect of replacement of natural aggregate with the recycled one (0, 10, 25, and 50 wt%) on the mechanical fracture properties of selected composites. Cement-based composites with dispersed reinforcement (natural fibres of industrial hemp in a dosage of 1 vol%) during the specimen's ageing was studied. Fracture tests of the specimens with a central notch, loaded in a three-point bending configuration, were used for evaluation. It is clear from the results that the values of physical, and mechanical parameters increased with increasing age. Regarding the fracture parameters, the values of effective fracture toughness remained practically constant over time, while the values of fracture energy decreased with age of specimens. It has been shown that composites with a higher dosage of recycled aggregate (25 and 50 wt%) achieve similar values of mechanical fracture properties, so this replacement can be recommended.

5. References

[1] ISO/TC 0712016 *Strategic Business Plan ISO/TC 071* (Tokyo: ISO Technical Committee for concrete, reinforced concrete and prestressed concrete).
[2] Nedeljkovic M, Visser J, Savija B, Valcle S, Schlangen E 2021 *J. Build. Eng.* **38** 102196
[3] Le HB and Bui QB 2020 *Constr. Build. Mater.* **257** 119522
[4] Bentur A and Mindess S 1990 *Fibre Reinforced Cementitious Composites* (New York: E & FN Spon)
[5] Tolêdo Filho RD, Ghavami K, England GL, Scrivener K 2003 *Cem. Concr. Compos.* **25** 185–196
[6] Poletanovic B, Dragas J, Ignjatovic I, Komljenovic M, Merta I 2020 *Constr. Build. Mater.* **259** 119677
[7] Poletanovic B, Janotka I, Janek M, Bacuvcik M, Merta I 2021 *Constr. Build. Mater.* **309** 125072

[8] ÖNORM B 3140 2020 *Recycled aggregates for unbound and hydraulically bound applications as well as for concrete* (Austria)
[9] Yan L, Kasal B, Huang L 2016 *Compos. B. Eng.* **92** 94–132
[10] Frantik P and Masek J 2015 GTDiPS software. http://gtdips.kitnarf.cz/
[11] Karihaloo BL 1995 *Fracture Mechanics and Structural Concrete* (New York: Longman Scientific & Technical)
[12] RILEM TC – 50 FMC 1985 *Mater. Struct.* **18** 287–290
[13] Kocáb D, Žítt P, Daněk P, Čaněk M, Alexa M, Hanuš P 2019 *IOP Conf. Ser.: Mater. Sci. Eng.* **583** 012028
[14] ČSN EN 196-1 2016 *Methods of testing cement – Prat 1: Determination of strength* (Prague: ÚNMZ)
[15] Behera M, Bhattacharyya SK, Minocha AK, Deoliya R, Maiti S 2014 *Constr. Build. Mater.* **68** 501–516

Acknowledgments
The research was carried out within the framework of the project from project No. 8J20AT013 – Integrity and durability aspects of recycled aggregates composites (InDuRAC), with the support of the Czech Science Foundation under project 19-09491S (MUFRAS).

IOP Conf. Series: Materials Science and Engineering 1252 (2022) 012040 doi:10.1088/1757-899X/1252/1/012040

Digital preservation of the Jesuitas Church in Valencia (Spain) using 3D laser scanning

J Moreno-Puchalt[1], A Almerich-Chulia[1] and X Laumain[2]

[1] Universitat Politècnica de València, School of Architecture, Department of Continuum Mechanics and Theory of Structures, Camino de Vera s/n, 46022 Valencia, España
[2] ARAE Patrimonio y Restauración, S.L.P.

E-mail: jemopuc@mes.upv.es; analchu@mes.upv.es; xavier@arae.es

Abstract. The urgent need to improve the quality in the refurbishment of historical buildings has led to the adoption of many innovative technologies. 3D laser scanning is a non-destructive technique used in the study of architectural heritage. It consists of producing millions of accurate 3D points with a very high point density in a short time. To get the complete 3D model, multiple shots must be taken from different directions that provide data from all sides of the building. These scans are integrated into a common reference system so that through a process of aligning the information obtained in all the stations, a complete model is achieved in a single file. This model faithfully reproduces the current volume of the temple, including its deformations and collapses, and provides very precise information from which to make its geometric survey. The refurbishment project of the Jesuitas Church in Valencia (Spain) has been realized thanks to the help of 3D laser scanning carried out three times during the years 2018, 2019 and 2020. The results obtained have not only served to make a very precise graphic survey but have also been applied in the necessary structural calculation.

1. Introduction

Today and increasingly, we have the urgent need to improve the quality in the refurbishment of traditional buildings. The continuous advance in the field of electronics allows us to access a wide range of solutions and tools that were once unimaginable. This reason has led to the adoption of many innovative technologies, specifically 3D laser scanning. The 3D laser scanner is a clear example of innovation and advancement, being a powerful and incomparable measurement tool that arises from the field of topography and it is more frequently present in the building, refurbishment, and architectural heritage sector [1]. Its operation consists of generating a point cloud that reproduces the building under study and its immediate surroundings. To get the complete model of the building, multiple shots must be taken from different directions that provide data from all sides of the building, generating multiple files containing millions of accurate 3D points with a very high point density in a short time. These scans are integrated into a common reference system so that a complete model is achieved in a single file through a process of aligning the information obtained in all the stations. This model faithfully reproduces the current volume of the building, including its deformations, and provides very precise information from which to make its geometric survey.

As a result of the scan, a point cloud is obtained that represents a copy of reality at a 1:1 scale from which the most diverse information about the generated virtual model can be obtained, such as high-resolution orthophotos, measurements in true magnitude, integration of the point cloud in BIM software, virtual tour of the model or even 3D printing [2].

It offers multiple advantages in measurement or data collection. These include precision, quality and speed, in addition to its versatility, simplicity and scope. It is a highly efficient tool as it reduces fieldwork time.

The building under study is the Jesuitas Church located in Valencia (Spain). The church was consecrated on October 12, 1887. It has a Neobyzantine style with some Neoromanesque touches. The church is made up of three naves with seven bays and an altar (figure 1). It also has a two-sided roof cover and a barrel vault with lunettes [3]. Although the state of conservation is relatively good, the building has numerous pathologies that have affected the whole of the Church. To carry out the pertinent refurbishment project, it is necessary to previously have an exhaustive graphic survey. Therefore a 3D laser scan is performed [4].

Picture from the altar

Side nave

Figure 1. State of the church during 3D laser scanning in 2018.

2. Methodology of laser scanning

The scanning process can be divided into three stages: scan plan, data collection and data processing.

2.1. Scan plan

A scan plan is a set of information that describes the scope and approach that will be taken to capture the data at the site. A scan plan should be made after setting the project goals.

Many times a scan plan begins with a detailed analysis of what elements need to be captured. In the case of scanning a building for the first time, you will want to capture the position of all the elements. In the case of continuing with a work started previously, it is possible to set specific work areas on which more information needs to be collected.

Identifying the exact target of the items to be analyzed helps the field team to prioritize their efforts and reduce the time spent capturing unnecessary data. With a clear goal in mind, a document can be created that identifies the optimal location for each scanning station.

In the case of scanning both the interior and exterior of the building, it is important to study the optimal location of each station so that the subsequent joining of the point clouds is carried out correctly.

IOP Conf. Series: Materials Science and Engineering 1252 (2022) 012040 doi:10.1088/1757-899X/1252/1/012040

2.2. Data collection
Scan parameters can be changed from settings at any time during scanning. The main parameters include creating a scanning project, selecting a scan profile indoor and outdoor and setting resolution and quality.

It must be taken into account for the location of the reference points that mathematically three common references are needed between two consecutive scans. However, a higher number of common references per scan will improve the registration results, making it easier and less likely to be errors.

Figure 2. Location of the spheres in the lateral nave.

The scanner used is Faro Laser Scanner Focus 150. Scene, which is the software that Faro technology uses, allows the union of point clouds by points, by planes or by spheres [5]. The spheres used as markers should not be positioned symmetrically with the building. They must form a polygon around the scanner and have varying distances to it. The ideal thing is to place them at different heights, distances and planes as shown in figure 2 and figure 3.

Figure 3. Faro Laser Scanner Focus 150 in the central nave.

IOP Conf. Series: Materials Science and Engineering 1252 (2022) 012040 doi:10.1088/1757-899X/1252/1/012040

The number of stations depends on the size and shape of the building. The scanning time depends on the number of stations, the quality and resolution, the complexity of the scanned building…

2.3. Data processing
The immediate result obtained from the laser scan is a series of point clouds. Each point is represented by coordinates (x,y,z). All point clouds are recorded together in a common coordinate system, resulting in a single point cloud.

Once the complete point cloud is obtained, unnecessary information is eliminated such as people, equipment, surrounding buildings, trees or noise, and then the color is applied (in the case of having scanned in color and not in black and white). Once all the previous steps have been carried out, the project documentation can be prepared by starting the preparation of plans [6].

3. Case study: The Jesuitas Church in Valencia (Spain)
The Jesuitas Church in Valencia requires a comprehensive restoration intervention. This action will range from structural aspects, through decorative aspects, to the adaptation of facilities. Due to this, it is necessary to carry out a graphic survey as exact as possible [7].

A first scan of the church was carried out indoors in 2018. The main objective was to draw the interior of the church in an accurate way. It was also important to check the layout of the vaults. Regarding scan parameters, in our case, we chose indoor to 10 meters, resolution 1/5 and quality 4x which took us an approximate scan time of 8 minutes. Figure 4 shows the indoor stations.

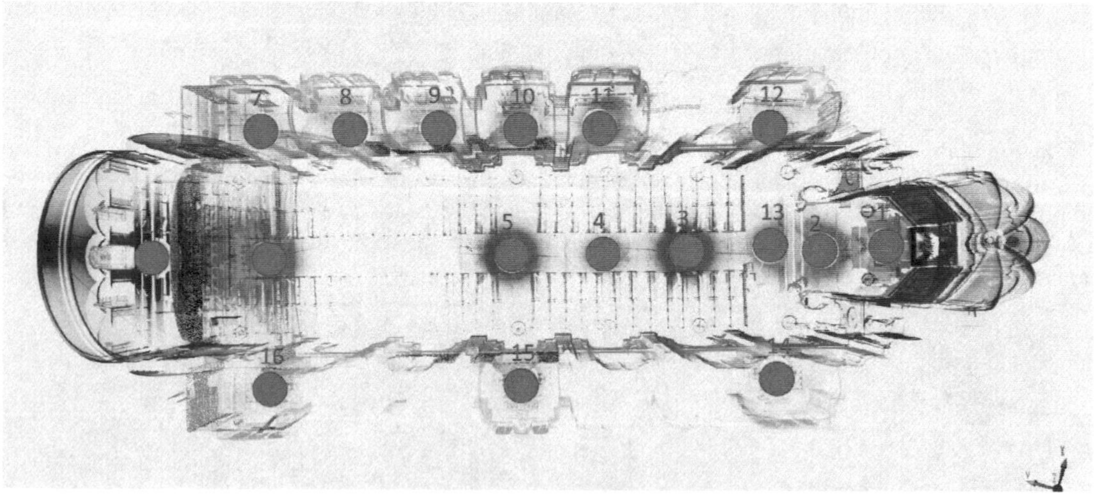

Figure 4. Seventeen positions of the laser scanner.

We decided that the first position of the scanner was at the altar (positions 1 and 2). We continued through the central nave towards the foot of the church (positions 3 to 6), then the right and left sides aisles (positions 7 to 16) and finally went up to the choir (position 17). The total scan time was six hours. This time consisted of installing the scanner, deployment of reference spheres and also changes in positions of the scanner and markers. The positions of the reference points were often different because we had available just five pieces of reference sphere. Due to this, in the union of point clouds, we had to also use planes and not only spheres. In this case, the union defined by planes also provides us with great precision since the marble cladding of the church bases defined perfect planes.

It is important to keep some considerations into account. Very dark surfaces absorb most of the visible spectrum and the reflected signal will be very weak. In contrast, very light surfaces are highly reflective and produce strong reflections. However, if the reflectivity of the object is too high, the laser ray is fully reflected in the direction of the reflection and will either hit another surface or be scattered. This is why shiny surfaces like gold are not easy to register. It can be scanned through the glass, but not with high precision as the glass will refract the ray and give inaccurate data.

We must do the office work once the fieldwork is finished. This work consists of the union of point clouds using Scene by FARO [5]. Each point cloud is joined with the next one by selecting common references that can be spheres, planes or points, as we can observe in figure 5 and figure 6.

Figure 5. Joining stations 5 and 6.

Figure 6. Interconnection between central nave and lateral nave.

This program offers simple measurements, coloring the point clouds, deleting unnecessary points and export the complete point cloud to various formats. Figure 7 shows the finished work.. We use the RCP format to import the point cloud into a CAD program. In this case, the program we work with is Autocad.

This software allows us to make countless sections in the complete point cloud and draw the 2D plans that interest us in generating them in a short time. Façade, floor plans, sections and details can be elaborated simply by making cuts to the complete cloud and drawing on top of them using them as a template (figure 8).

Figure 7. Complete cloud of the interior of the church.

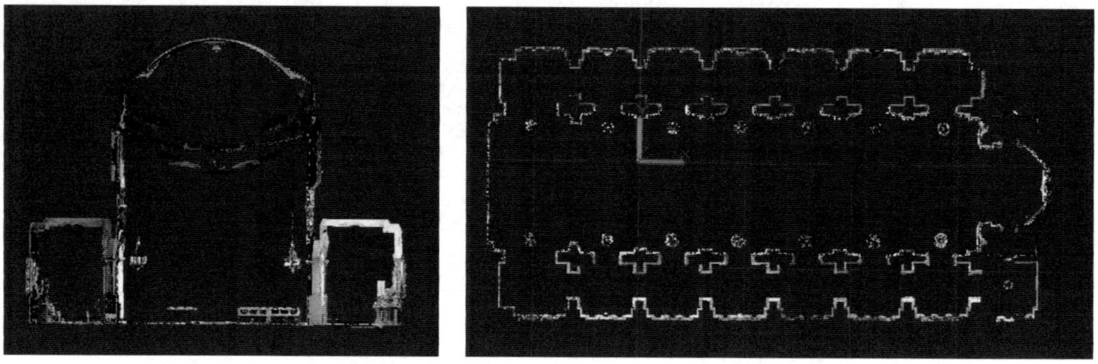

Figure 8. Preparation of sections and floor plan.

4. Conclusion

For the drafting of the project that was presented to obtain the appropriate licenses, this 3D scan of the building has been carried out in order to obtain a real-size and reliable image to refine the intervention in it and adhere to the ideal works according to the requirements.

This paper shows 3D laser technology as a state-of-the-art tool for drawing with great precision plans of traditional buildings. It has been described step by step how to prepare the graphic documentation of Jesuitas Church in Valencia. There is no doubt that the 3D laser scanner is a very effective tool to carry out graphic surveys of the current state of historic buildings.

5. References

[1] Pukanská K 2012 *3D visualisation of cultural heritage : by using laser scanning and digital photogrammetry : monograph* (Vysoká škola báňská - Technická univerzita Ostrava)

[2] Mesároš P and Mandičák T 2017 *Exploitation and Benefits of BIM in Construction Project Management, IOP Conf Ser Mater Sci Eng* **245** (6) 062056

[3] Iglesia Jesuitas Valencia – Proyecto de restauración from https://iglesiajesuitasvalencia.org/

[4] Spišáková M, Mésároš P and Mandičák T 2021 *Construction Waste Audit in the Framework of Sustainable Waste Management in Construction Projects—Case Study, Buildings* **11** (2) 61

[5] FARO Laser Scanner Focus3D 2019

[6] Tkáč M, Mesároš P and Mandičák T 2018 *Terrestrial laser scanning-effective technology for creating building information models, POLLACK Period An Int J Eng Inf Sci* **13** (3) 61–72

[7] Hrozek F, Sobota B and Szabó C 2012 *Digital preservation of historical buildings using virtual reality technologies, Open Comput Sci* **2** (3) 272–82

Acknowledgments

This work would not have been possible without the collaboration and support of the Society of Jesus and those responsible for the Arrupe Center. Thanks to them for allowing us to show the 3D scanning process carried out in the church.

Renewable energy in a dense urban fabric: solar gain estimation in the case of a turn of the century district

V Sugár [1,2] and Zs Fáczányi [1]

[1] Óbuda University, Ybl Miklós Faculty of Architecture and Civil Engineering, 74 Thököly út, 1146 Budapest, Hungary
[2] Eötvös Lóránd Research Network, Centre for Energy Research, 29-33 Konkoly-Thege Miklós út, 1121, Budapest, Hungary

E-mail: sugar.viktoria@uni-obuda.hu; faczanyi.zsuzsanna@uni-obuda.hu

Abstract. Buildings are responsible for significant amount of energy consumption. The new constructions already obliged to use decreased amount of energy, but the older buildings are ineffective compared to them. The new building energetic prescriptions in force since 2021 in Hungary require considerable share of renewable energy produced on-site. In case of a densely built urban fabric, the above renewable ratio can mostly be reached by using solar technologies. The author surveys the solar characteristics of a Budapest downtown area. Considerable ratio of the stock is residential, traditional apartment house type built around the turn of the 19th-20th century. The retrofit solutions and renewable power generation are restricted by the dense fabric and the heritage protection rules as well. In present paper the estimation of solar energy potential of the above characteristic building type is introduced. The European Union conform Hungarian calculations used system is used to define the values. The methodology contains 3D modelling and irradiation software simulation. As result, the connection between the footprint and irradiation data of the characteristic heritage buildings are introduced, as well as the solar power generation potential of the building type in question. Also, the heritage protection compatibility of the solar systems are studied.

1. Introduction

Renewable energy utilization is one of the key elements confronting climate change. Its importance is underlined in several international agreement concerning energy and greenhouse gas (GHG) mitigation aims, including the 2030 Climate and Energy Policy Framework of the European Union [1]. The policy states that at least 40% domestic reduction in GHG emissions to be reached by 2030 compared to the amount of 1990, as well as at least 27% is set for the share of renewable energy. The same mitigation target value was declared in the National Energy and Climate Plan of Hungary [2], with a 21% renewable energy share. In 2020, Hungary's energy mix showed 13,2 % renewable energy share in the gross final energy consumption [2], which is to be increased in the future.

As other aspect, Hungary imports a significant part of its primary energy demand. There is a relatively large difference between energy production and usage (only the 35% of the total energy demand is produced domestically), thus the country depends considerably on the imported energy. 80% of the domestic crude oil demand, 83% of the natural gas demand originates from the former Commonwealth of Independent States, which import is predicted to increase even further. Thus, to reduce the import dependency, decreasing the energy usage and increasing the ratio of domestic produced energy of renewable source is an important aim of the country [2].

The solar sectors' inbuilt electricity generation capacity in 2020 was 1170 MW, which is far the largest value compared to the other renewable sources. The same year the largest renewable source in electricity consumption was the biomass and renewable waste with 2332 GWh compared to the 1335 GWh solar source. This trend is to be inverted until 2030, aiming for the solar source energy consumption to be 6579 GWh compared to the 3328 GWh biomass and waste ratio [2].
The National Energy and Climate Plan of Hungary [2] also highlights that buildings are responsible for around 40% of the primary energy consumption and GHG emission of the country (around 440 PJ of the total 1107,6 PJ), from which the households represent 60%, mainly utilized for heating and domestic hot water generation. The above ratio is also confirmed to be an international trend [3].

Although the new buildings are under strict energetic regulations in the country (as of 1 January 2021, new buildings will only be authorised to usage if the near-zero energy requirements are met), their number is insignificant compared to the older, ineffective buildings. The regulation requires high efficiency and minimum 25% on-site renewable energy generation of the total energy usage [4].

The above requirements are not yet mandatory for the existing building stock, however in light of every stricter rules and tendencies, the nearly-zero energy retrofit methodology is worth investigating. The nearly-zero energy requirements are already in force in case of government supported retrofits for existing buildings.

The on-site renewable energy generation however has its difficulties in the case of the historical city centres, where not only the densely built urban fabric decreases the renewable energy options but the heritage protection principles also hinder the possibilities. Utilizing hydro or wind-power, ground source heat pump or water-based heat pump in the dense urban fabric is difficult or even impossible. The construction activity itself is mostly non-accomplishable, and the required space for the above technologies is scarcely available. The biomass heat production in such an area also should be excluded, not only because of the transportation and storage difficulties, but the amount of dust pollution. The generally applicable solutions seem to be the solar power utilization or air source heat pump. Almost all buildings have suitable roof surfaces to install collectors or photovoltaic panels, the question is its effectiveness and feasibility.

The urban fabric and its compactness in connection with the solar characteristics had been investigated before in multiple researches: Mohajeri et al. [5] surveyed the effect of urban compactness on solar potential, by calculating the annual solar irradiation for various chosen neighbourhood. In the study the authors use geometry indicators, for example site coverage and building density to assess the urban compactness and its effect. Hofierka and Kanuk [6] use 3D model and GIS database to be able to assess the solar potential of the urban areas. They conclude that the assessment of solar potential requires detailed analysis of buildings and surfaces, which underlines the necessity of investigating real, existing neighbourhoods with their irregularities. Chatzipoulka et al. [7] also use the existing stock of London. In their study, also the geometry, the urban layout is in the spotlight.

Alla et al. [8] investigate the local solar energy potential in an Italian case study. They introduce a usable methodology to create GIS maps to identify the areas at best potential. Phap et al. [9] also deal with solar energy potential, their case study is Hanoi city. Their methodology contains remote sensing and a comparism between theoretical and actual data because of the actual photovoltaic system is more affected by environmental condition than previously planned. López and Frontini [10] introduce a system to evaluate the PV panels for heritage buildings based on various criteria like shape, visibility, aesthetics. They underline, that even small interventions can give important results in reduction of energy consumption. They introduce heritage guideline compatible PV technologies. Lingfors et al. [11] introduce a novel approach for visibility in the case of heritage buildings, which is a main question when applying PV panels. The review paper of Cabeza et al. [12] deal with renewable technologies in the case of heritage buildings comparing the acceptable solar panel solutions. Verso et al. [13] use GIS methods to evaluate PV potential in settlements, also comparing PV technologies and their applicability. Wegertseder et al. [14] also deal with solar potential assessment, estimating the energy demand covered by solar energy. Moran and Natarajan [15] examine the potential for roof mounted photovoltaic system at a protected heritage case study area. In this case, the buildings are semi-detached, detached and terrace houses in a conservation area, where also the visibility and the acceptance of the panels arise as a significant question. They argue that the historic buildings already met the new demands and changes like electrification, bathrooms and central heating, which suggests the measures to reduce emissions may become accepted in time. They conclude that the energy usage and the CO2 reduction is significant, especially if the energy use patterns is synchronized. Martinopoulos [16] introduces a life-cycle analysis of solar conversion systems in the case of Net Zero Energy retrofitted buildings. This case study is introducing Greek detached houses with more optimal solar irradiation characteristics than Hungary (and more contemporary houses), stating that the integration of solar systems can provide solutions towards both net zero and nearly zero energy buildings.

The aim of the current study is to be able to estimate the solar irradiation, the amount of the solar energy generation and its ratio compared to the energy utilization of the houses in case of Budapest downtown urban fabric, considering the limitations of the heritage protection.

2. Case Study – Budapest district from the turn of the 19th and 20th century

The 19th century was an outstanding period in the history of Budapest causing fast expansion of the city. The quick urbanisation and the demand for housing gave birth to the traditional apartment house type, called 'bérház', or house for rent in Hungarian.

During this era, the population of Pest had grown rapidly, bringing an unprecedented wave of constructions. The evolution of the aforementioned characteristic apartment building type is closely linked to this population growth, which resulted a dire need of mass construction. As the newly settling population was of less wealthy social state, thus individual constructions were not an option. The few wealthy citizens built larger apartment buildings for rent. The so emerging new buildings were larger, with unique façades, but complying to strict construction regulations, thanks to which the streetscapes remained unified (Figure 1) [17].

This condominium is equipped with differently sized and variously equipped flats. The buildings are mostly built around a courtyard. The street front wing is more decorated, containing larger flats. The courtyard wings on the other hand have simpler flats, often only with a kitchen and a room. The courtyard can be accessed by using the gate on the street front. The flats on the upper stories can be entered from the hanging corridors running parallel the walls. The buildings are mostly built in an unbroken row along the narrow streets of Pest, connecting to each other with firewalls on three sides.

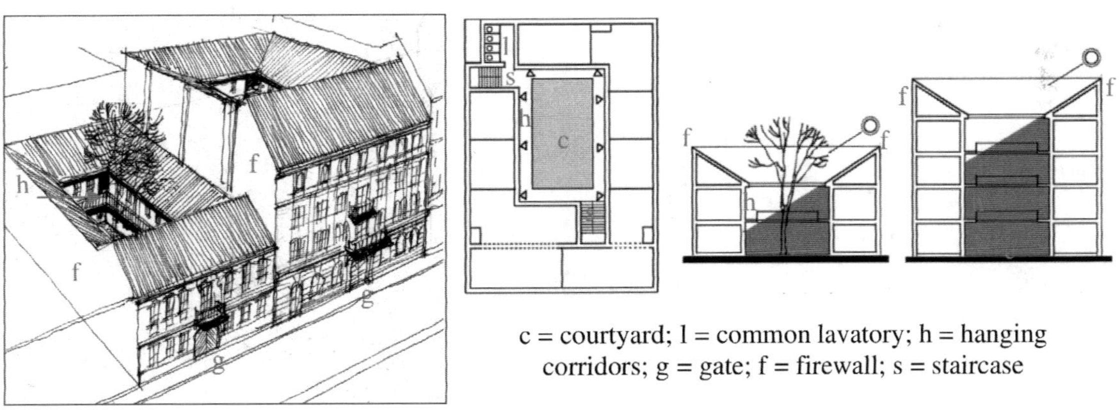

Figure 1. Characteristic traditional apartment houses from the end of the 19th century, built in a row with an enclosed courtyard. Nearing the turn of the 19th and 20th century, they became higher with darker courtyards [18].

The proportions of the buildings (height, ratio of the courtyard), structures and materials were strictly regulated at the time, but due to the diverse architectural forming and ornaments each and every building is unique [19]. The general type of building itself with some differences is widespread in Middle-Europe, in the towns of the former Austro-Hungarian Monarchy. Today, around 10 000 buildings of this type exist in Hungary, with more than 242 000 flats, adding up to a total 14 million m2 footprint. 88% of the buildings is situated in Budapest [20].

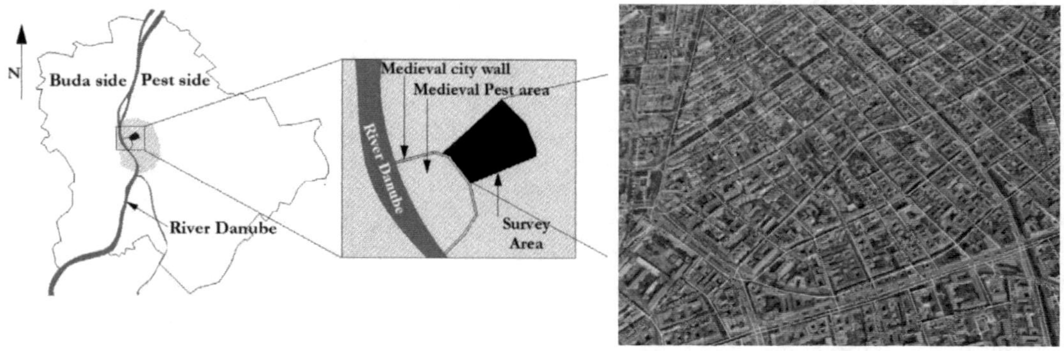

Figure 2. The case study area is situated in central Budapest. The slashed area on the city map represents the area the traditional apartment houses (Authors' own figure).

In the chosen case study area the most characteristic buildings are part of the above mentioned type. The area is situated in Budapest 7th District, bordered by Király Street, Erzsébet Boulevard, Rákóczi Street and Károly Boulevard, named Belső-Erzsébetváros (Inner-Elizabethtown) (Figure 2). Today, 477 plots and 441 buildings are situated on 0,6 km2. They were mostly built around the turn of the 19th and 20th century. Compared to its size, the population is high, its density is almost 25 000 person/km2 [21] making it the most densely populated district of Budapest and Hungary. As explained above, the buildings are built in an unbroken row along the streets, and the footprint layout is most commonly a "frame", where the street front and inner wings enclose a rectangular courtyard (Figure 3) (Frame type (F, 40%).

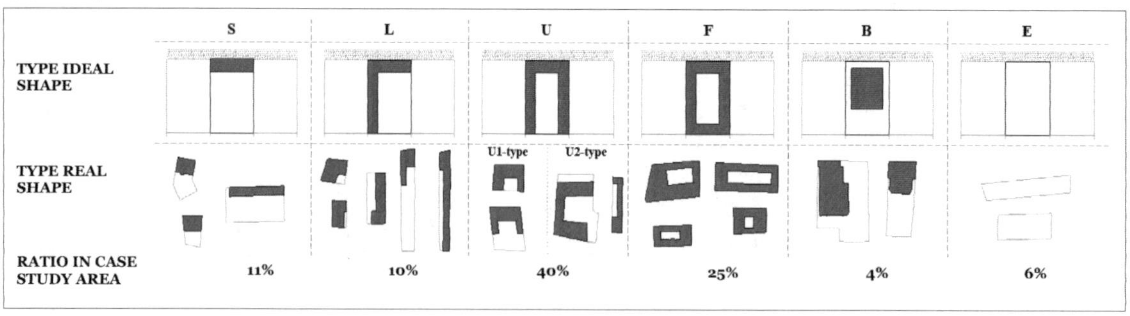

Figure 3. Layout types of traditional apartment houses in the case study area.

Currently, most of the buildings in the case study area are in a poor condition. Their physical characteristics do not comply with today's health, economic or ecological requirements, also, their poor energetic state decreases their value. Apart from some protected ones, most of the traditional stock is not sheltered from demolitions by law. The unprotected buildings are often destroyed to be rebuilt as contemporary apartment houses or modified to the point of losing their original values.

The above facts highlight the importance of a heritage respecting modernization, to be able to protect the values and character of the buildings, but at the same time increasing the living standard of the residents.

It is highly important thus, to offer feasible rehabilitation possibilities to protect this architectural and historical heritage of the Hungarian capital. The author's previous works explain the energetic rehabilitation options [17, 22]. In present paper the focus is on renewable energy utilization.

3. Methodology and calculations

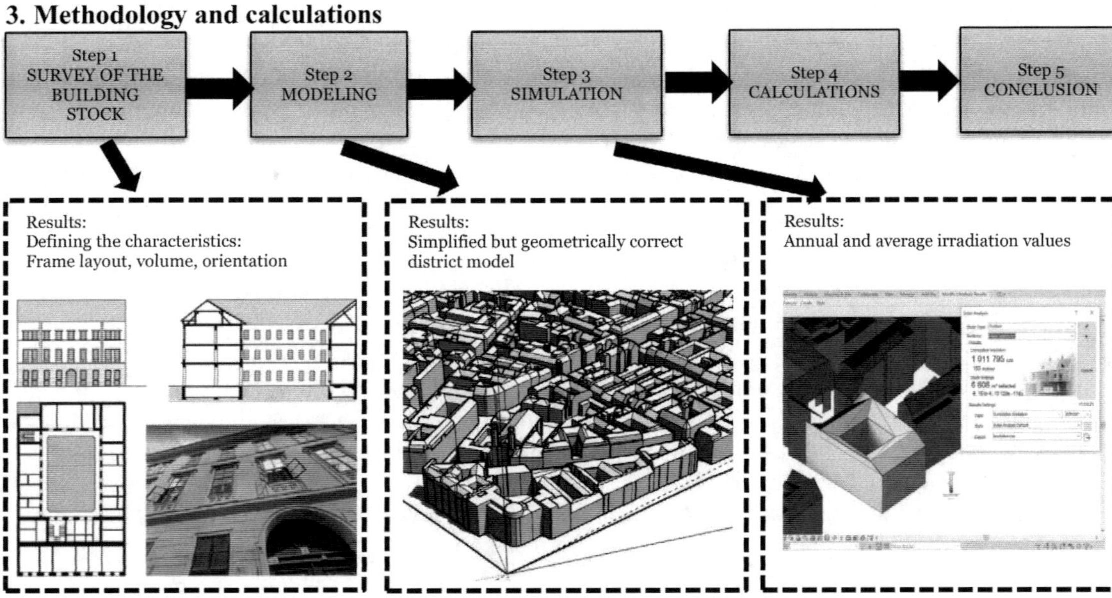

Figure 4. Methodology of the research.

First, the downtown districts of Budapest were surveyed to find the characteristics of the building stock. It was determined that the most common building type is the traditional apartment house, which is a residential, multi-storey building, built around the turn of the 19th and 20th century. Most commonly the 3-4 storeys are covered with pitched roof – saddle roof on the street front, half-saddle roof on the courtyard wings, sloping to the courtyard. The buildings are closely connected together via firewalls on three sides of the plot (Figure 4).

As second step, a characteristic part of the urban fabric (see Section 2) was chosen and modelled in Revit software (see example picture in Figure 4) [23]. The modelling was needed to investigate the solar potential of an existing stock with its irregularities and aims to find connection between geometry and solar potential. The model was based on geometry measurement, base maps acquired from the local government. In the case of unreachable parts, online satellite images were used to complete the existing information. Apart from the 3D model, an extensive database was created containing geometry and other detailed data of each building in question.

Next (Step 3), the irradiation data was acquired via software simulation. As for the solar characteristics of Hungary, it can be stated that the possibilities offered by the solar features are more favourable than other European countries: the total annual number of sunny hours is averagely 1800–2000, the total solar irradiation is 1200-1300 kWh/m2. Due to the geographical situation of Hungary, there are significant differences in winter and summer radiation values, thus the solar thermal energy usage for heating during winter is limited [24].

The software used for simulations is the REVIT 2018 [23], with Insight plugin. The roof areas were simulated separately according to orientation. The software uses the most nearby weather station data after the exact location of the model is added (in this case: 2 km south from the case study area). Weather stations include 'actual year' virtual weather stations and typical year weather stations based on 30-year averages [25].

The simulation results were used as input data for calculations. The average annual irradiation (kWh/m2a), the cumulative annual irradiation (kWh/a), the roof surfaces (m2), their orientation, the footprint of the building (m2) were acquired from the model. The net heated area of the building (m2) was calculated using the geometry input of the database. Two blocks of buildings were used in the detailed simulation.

To determine the electricity generated by the solar panels, the number of the applicable solar panels was calculated first. 1.6 m2 general solar panel size was considered. The roof surfaces of the buildings faced South-East, South, South-West (North facing areas not considered, the East and West facing surfaces being minimal) were used, reduced by 15% to consider the unusable areas around chimneys, drains and edges. Thus, the acquired solar panel number and surface were multiplied by the average annual irradiation values in case of each roof surfaces. The exact model has a great advantage here from the viewpoint of accuracy, as all the overshadowed areas (either by the adjacent building or self-shadowing) are included in the simulation. The efficiency of the solar power system was considered as 15% based on previous studies measuring solar panels [26]. Although the technological advancement is continuous, resulting higher efficiency solar panels, considering the today available, most commonly used polycrystal panels, the value is justified.

To assess the ratio of energy obtained by solar panels compared to the buildings' energy consumption, the Hungarian building energetics calculation system was used [4]. The total primary energy consumption [kWh/m2a] of each building was calculated, which contains the factors of the thermal characteristics of the enveloping structures, the geometry of the building and incorporates the energy consumption of the engineering systems annually in primary energy. In the case of residential buildings, the heating energy and the domestic hot water energy consumption should be summed, as they are the predominant form of energy usage.

Two scenarios of the building energy consumption were investigated. The original state, and a hypothetical upgraded state (See Figure 5 for structural and thermal characteristics of the scenarios).

Enveloping Structure	Original State	Retrofit Scenario
Wall	Brick U = 1.2 W/m^{2}K	*none*
Closing slab under pitched roof	Full timber U = 0.62 W/m^{2}K	Outdoor insulation (e.g. mineral wool) 20 cm U = 0.16 W/m^{2}K
Fenestration	Box-type U = 2.28 W/m^{2}K	Fitting + Low E glass U = 1.45 W/m^{2}K
Floor on cellar	Prussian vault U = 0.56 W/m^{2}K	Underside insulation (e.g. mineral wool) 5 cm + mortar 1.5 cm U = 0.21 W/m^{2}K
Uncovered firewall	Brick U = 1.2 W/m^{2}K	Outdoor insulation (e.g. mineral wool) 20 cm U = 0.16 W/m^{2}K
Heating	individual convector or gas boiler	Heat pump

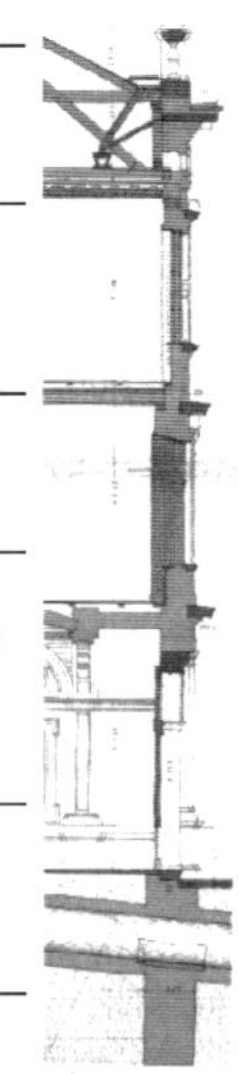

Figure 5. Original and upgraded enveloping structures with their thermal transmittance (U) values.

In case of the upgrade, the following retrofit measures were included: because of the heritage protection guidelines, only the necessary, less visible surfaces were insulated. The decorated façade walls, cellar walls, arcades were left intact. The roof and cellar slabs were insulated as well as the uncovered firewalls(average 20 cm mineral wool insulation on the above surfaces). The windows were upgraded with the full heritage compatible solution (glass low-e glazing). For the engineering system air source heat pump was considered, also produces domestic hot water. The system can be used in summer, cooling the indoor temperature. With this upgrade, the original energy usage of the buildings can be reduced by average 69%. See the author's previous papers for more details on the building stock and its energy characteristics [22].

The simple payback time of the PV system application is calculated by using the household savings on electricity utility costs with solar panels installed compared the construction and application cost of the system, resulting the payback time in years.

The CO_2 mitigation potential (kg/year) is calculated using the data of the Budapest power supplier, stating that the specific CO_2 emission of the supplied electricity of the area, replaced by the renewable option of the calculation is 0.328 t/MWh [27].

4. Results and Discussion

Table 1 shows the average numerical results of the calculations. The table is divided to two main parts. First, the average values per buildings are shown, followed by the average values per net heated m2.

Table 1. The result values of the surveyed area per building (average) and for the case study area's 65 building.

Average data of the surveyed buildings	**Value per buildings**
Average cumulative irradiation on the usable roof surfaces	385.6 MWh/a
Average cumulative energy generated by solar panels on the usable roof surfaces	57.7 MWh/a
Average original energy consumption of the buildings	790 MWh/a
Average retrofitted consumption of the buildings	247 MWh/a
Average ratio of energy consumption covered by the solar panel energy generation in primary energy, original state	8.3 %
Average ratio of energy consumption covered by the solar panel energy generation in primary energy, retrofitted state	25.6 %
Average simple payback time of the solar panel system, original state	13.4 years
Average CO_2 mitigation	18. 9 t/year

Average data of the surveyed buildings	**Value per net heated m^2**
Average cumulative irradiation on the usable roof surfaces	146.8 kWh/m^{2}a
Average cumulative energy generated by solar panels on the usable roof surfaces	22 kWh/m^{2}a
Average original energy consumption of the buildings	267.4 kWh/m^{2}a
Average retrofitted consumption of the buildings	53.2 kWh/m^{2}a
Average CO_2 mitigation	7.2 kg/year

The average cumulative irradiation on the usable roof surfaces is 385.6 MWh yearly on average. This value is decreased when the effectiveness of the solar panel system is also considered, making the generated energy 57.7 MWh yearly per building. If the original energetic state of the building is considered this amount can only cover 8.3% of the building's energy consumption. However, if the retrofitted state is regarded, 25.6% of the energy intake can be covered by the solar panels. Totally 37% of the investigated buildings above renewable ratio reaches the 25% threshold in retrofitted state (meaning that the building could comply to the current national nearly-zero energy prescriptions). Another 32% of the buildings reach at least 20% of the renewable ratio.

Annually, 18.9 t CO2 emission could be avoided if using solar technologies on the surveyed buildings averagely.

IOP Conf. Series: Materials Science and Engineering 1252 (2022) 012041 doi:10.1088/1757-899X/1252/1/012041

The annual solar irradiation on the usable roof surfaces were show a direct connection to the footprint of the buildings (Figure 6 and 7). The result can be explained by the roof geometry following the footprint with saddle and half-saddle forming. The usability of the results lies in the possibility of using Figure 6 diagram for estimation purposes.

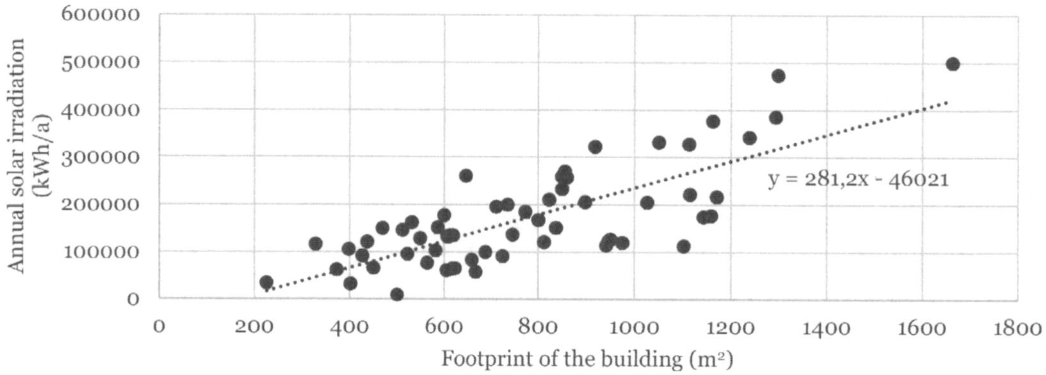

Figure 6. Connection between the footprint of the buildings and the annual solar irradiation on the usable roof surfaces.

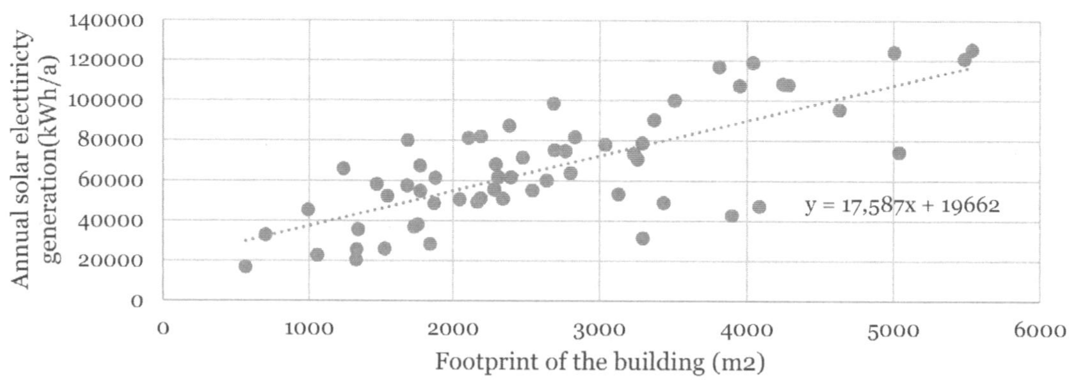

Figure 7. Connection between the footprint of the buildings and the annual energy generation on the usable surfaces.

Concerning the application of the solar panels, the heritage protection guidelines should also be reviewed.

In Hungary, the monument protection system consists of multiple levels: national monument protection, local protection, conservation area protection and monument neighbourhood, their enforcement is, however, mostly weak. It is important to note, that part of the chosen area is classified as an UNESCO Word Heritage Zone, being a Buffer Zone of the Andrássy Boulevard [28]. Thus, visibility of the panels is an important issue.

Figure 8. A wider streetscape with different building heights (left). Example of courtyard wings covered with solar panels (middle). Example of solar tiles [29] (right).

Figure 8 shows a wider streetscape with different building heights (left). The photo indicates that even in case of wider street and different roof heights, the visibility of the applied panels is not an issue. Even in case of stricter regulated areas the courtyard wings can always be utilized for solar pover generation (Figure 8 middle). Also, there is an alternate solution of using solar tiles (Figure 8 right), which blend in with the geometry of the roof.

5. Conclusions

On-site renewable energy generation is an important part of building energy efficiency measures. In Hungary, the national regulations state that the nearly-zero energy buildings require to produce 25% of their energy consumption from renewable sources, on-site.

In the case of dense urban fabric the above prescription can frequently only be satisfied by using solar power. When dealing with dense urban fabric of downtown districts, not only the limitations of space, but the heritage protection guidelines should also be taken into account when planning a retrofit.

The major architectural heritage of Budapest is the area built of the traditional apartment buildings from the 19th and 20th century. They internationally recognised as historical, cultural and architectural heritage. Apart from some protected buildings, most of the traditional stock is endangered by demolitions. The reasoning is often the poor energetic state of the house and the sustainability problems.

In present paper, the Author aimed to determine the solar potential of the characteristic building type of Budapest downtown. Based on the results, it can be concluded, that using solar panels, averagely 57.7 MWh electricity can be generated annually by building. This ratio covers 8.3% of the building's energy consumption. If the hypothetical retrofitted state is regarded, 25.6% of the energy intake can be covered by the solar panels. Totally 37% of the investigated buildings above renewable ratio reaches the 25% threshold in retrofitted state and another 32% of the buildings reach at least 20% of the renewable ratio. The average simple payback time of the solar panel system compared to the original state is 13.4 years. Averagely 18.9 t/year CO2 mitigation could be reached by using panels in each building.

Also, an annual solar irradiation and solar power generation estimation method for the traditional apartment building type is introduced, where the footprint of the building (easily acquirable from open-access satellite images) can be used to approximate the values.

6. References

[1] EUCO 169/14 *European Council 2030 Climate and Energy Policy Framework.* (2014) [online] https://www.consilium.europa.eu/uedocs/cms_data/docs/pressdata/en/ec/145397.pdf Last accessed: 25.01.2021

[2] *Ministry for Innovation and Technology, National Energy and Climate Plan of Hungary* (2020) [online] https://ec.europa.eu/energy/sites/ener/files/documents/hu_final_necp_main_hu.pdf

Last accessed: 20.01.2021
[3] *International Energy Agency, Tracking Buildings. (*2020) [online]
https://www.iea.org/reports/tracking-buildings-2020
Last accessed: 20.01.2021
[4] *Minister without Portfolio Decree 7/2006. (V. 24.) Determining the energetic characteristics of buildings. (2006)* [online]
https://net.jogtar.hu/jogszabaly?docid=a0600007.tnm
Last accessed: 25.01.2021
[5] Mohajeri N Upadhyay G Gudmundsson A Assouline D Kämpf J Scartezzini J L 2016 *Effects of urban compactness on solar energy potential, Renewable,* **93** 469-482
[6] Hofierka J Kanuk J 2009 *Assessment of photovoltaic potential in urban areas using open-source solar radiation tools, Renewable Energy* **34** 2206–2214
[7] Chatzipoulka C Compagnon R Nikolopoulou M 2016 *Urban geometry and solar availability on façades and ground of real urban forms: using London as a case study, Solar Energy* **138** 53–66
[8] Alla S A Bianco Tagliafico V L A Scarpa F 2020 *An innovative approach to local solar energy planning in Riva Trigoso, Italy, Journal of Building Engineering* **27** 100968
[9] Phap V M Huong N T T Hanh P T Duy P V Binh D V 2020 *Assessment of rooftop solar power technical potential in Hanoi city, Vietnam, Journal of Building Engineering* **32** 101528
[10] Lópeza C S P Frontinia F 2014 *Energy efficiency and renewable solar energy integration in heritage historic buildings, Energy Procedia* **48** 1493 – 1502
[11] Lingfors D Johansson T Widén J Broström T 2019 *Target-based visibility assessment on building envelopes: Applications to PV and cultural-heritage values, Energy and Buildings* **204** 109483
[12] Cabeza L F Gracia A Pisello A L 2018 *Integration of renewable technologies in historical and heritage buildings: A review, Energy and Buildings* **177** 96-111
[13] Verso A Martin A Amador J Dominguez J 2015 *GIS-based method to evaluate the photovoltaic potential in the urban environments: The particular case of Miraflores de la Sierra, Solar Energy* **117** 236-245
[14] Wegertseder P Lund P Mikkola J Alvarado R G 2016 *Combining solar resource mapping and energy system integration methods for realistic valuation of urban solar energy potential, Solar Energy* **135** 325-336
[15] Moran F Natarajan S 2015 *PV in historic dwellings: The potential to reduce domestic CO2 emissions, Journal of Building Engineering* **3** 70-78
[16] Martinopoulos G 2018 *Life Cycle Assessment of solar energy conversion systems in energetic retrofitted buildings, Journal of Building Engineering* **20** 256-263
[17] Sugár V 2020 *Rehabilitation of Heritage Buildings: Energetic Retrofit in the Historical District of the Budapest Old Jewish Quarter, PhD Dissertation*
[18] Bitó J 2003 *Lakóházak tervezése (*Budapest: B+V Medical+Technical Lap és Könyvkiadó, Budapest)
[19] Edvi I A 2005 *Budapest műszaki útmutatója 1896* (Budapest: Terc Kiadó)
[20] ÉMI Kft. *Nemzeti Épületenergetikai Stratégia (National Building Energetics Strategy)* (2015)
[21] Hungarian Central Statistical Office *Regional Statistical Yearbook of Hungary (*2015)
[22] Sugár V Talamon A Horkai A Kita M 2020 *Energy saving retrofit in a heritage district: The case of the Budapest, Journal of Building Engineering* **27** 100982
[23] *Autodesk Revit software webpage* [online]
https://www.autodesk.com/products/revit/overview?plc=RVT&term=1-YEAR&support=ADVANCED&quantity=1#
Last accessed: 15.12.2020
[24] *Hungarian Meteorological Agency Database* (2020) [online]
https://www.met.hu/eghajlat/magyarorszag_eghajlata/altalanos_eghajlati_jellemzes/sugarzas/
Last accessed: 15.12.2020
[25] *Autodesk Building Performance Analysis Data* (2020) [online]
http://help.autodesk.com/view/BUILDING_PERFORMANCE_ANALYSIS/ENU/?guid=GUID-DDF449B0-058E-465C-AFF4-7E45917B08EF

Last accessed: 15.12.2020
[26] Török I 2017 *Yields of Solar Energy Systems and Operational Efficiency PhD Dissertation* [online] https://dea.lib.unideb.hu/dea/bitstream/handle/2437/240505/ertekezes_torok_imre.pdf?sequence=2&isAllowed=y
Last accessed: 25.01.2021
[27] E.ON Energiaszolgáltató Kft. *Data of source composition of electricity sold to users website* [online] https://www.eon.hu/content/dam/eon/eon-hungary/documents/lablec/Kotelezo-tajekoztatas/Forr%C3%A1s%C3%B6sszet%C3%A9tel_alapt%C3%A1bla_2018_VP.pdf Last accessed: 25.01.2021
[28] UNESCO World Hetitage website [online] https://whc.unesco.org/en/list/400 Last accessed: 25.01.2021
[29] Solar tile website [online] https://terran-generon.hu/ Last accessed: 25.01.2021

Acknowledgements

This work was supported by the Hungarian Ministry of National Resources and National Resources Support Management in the boundaries of National Talent Program [NTP-NFTÖ-19-B-0207]

Nemzeti Tehetség Program

Emberi Erőforrások Minisztériuma

Corrosion Acceleration Test on Weathering Steel and Paint-coated Steel for Bridge Structural Members

J Liu[1], K Ojima[1], F Jiang[1], M Hirohata[1], T Nakayama[2] and T Sakamoto[3]

[1]Graduate School of Engineering, Osaka University, 2-1 Yamadaoka Suita Osaka 565-0871 Japan
[2]Nisso Industries Co, Ltd, 1-15-2 Nishinakajima Yodogawa Osaka 532-0011 Japan
[3]Railway Technical Research Institute, 2-8-38 Hikari Kokubunji Tokyo 185-8540 Japan

E-mail: j-liu@civil.eng.osaka-u.ac.jp; k-ojima@civil.eng.osaka-u.ac.jp; f-jiang@civil.eng.osaka-u.ac.jp; hirohata@civil.eng.osaka-u.ac.jp; taishi_nakayama@nisso-sangyo.co.jp; sakamoto.tatsuro.51@rtri.or.jp

Abstract. A new accelerated corrosion test authorized as ISO16539 Method B in 2013 has shown a good correlation with corrosion behavior in an actual corrosive atmospheric environment. This study aims to extend the application of this method to steel bridge members. The corrosion acceleration tests by the ISO 16539 Method B to weathering steel without anti-corrosive coating and paint-coated steel were conducted through 6 months of the test periods. A series of fundamental studies were conducted to clarify the characteristics of this test in the durability evaluation of the steel used for bridge structural members by comparing the results of the atmospheric exposure test. The accelerated test of 168 days produced the thickness reduction by 10 to 14 times larger than the exposure test of 365 days to the weathering steel. The blister area of the paint-coating as a measure of the paint degradation by the accelerated test was 11 to 14 times larger than the exposure test of 365 days to paint-coated steel. These results showed the effectiveness of the accelerated test by ISO16539 Method B to evaluate the degradation characteristics of the weathering steel and the paint-coated steel.

1. Introduction

Corrosion is one of the main causes of deterioration in steel structures [1]. Weathering steel and paint-coated steel are used to prevent corrosion in steel bridges. In order to maintain steel structures properly, it is essential to understand the durability of weathering steel and paint-coated steel. Exposure tests are generally conducted as a method for this purpose. Although the exposure tests provide the data on degradation in real environments, it is often difficult to be performed because of the long time required and environmental constraints. As an alternative method, corrosion acceleration tests have been conducted mainly in the field of materials, which enables to verify of the degradation and performance in a short period [2]. Until now, the corrosion acceleration tests have been mainly used as a relative evaluation method for the quality of the new coatings and the new corrosion prevention methods. Although standardized test methods such as Salt Spray Test [3] and Combined Cycle Test [4,5] have been used, it has been pointed out that some test methods may not be able to reproduce the actual phenomena because the corrosion resistance order and corrosion form of materials may differ from the actual environment [6]. Therefore, it is necessary to use appropriate test methods according to the target material and environment. As one of the countermeasures to this problem, ISO16539 Method B has been developed as a new corrosion acceleration test. This test was developed through research on the actual corrosion of surface-treated steel sheets in home appliances and was standardized in 2013 [7]. ISO16539 Method B is a test in which the wetting rate in the process of repeated drying and wetting is made close to the actual environment, and it is said to be able to reproduce the corrosion degradation behavior of surface-treated steel sheets of home appliances. It is expected that this test will be applied to steel structures with weathering steel and paint-coated steel in the future. However, it is unclear whether this test can be applied to the degradation evaluation of the weathering steel and the paint-coated steel due to the lack of basic data. In this study, a series of fundamental investigations on the application of ISO16539 Method B to the weathering steel and the paint-coated steel was conducted based on the comparison with the atmospheric exposure tests.

2. Methods

2.1. Corrosion acceleration test

ISO 16539 Method B [8] was adopted as a corrosion acceleration test. Details of the experimental conditions are shown in figure 1. Artificial seawater with a concentration of 3.5 % was applied to the surface using a spraying device [9], and a salt deposition of 28.0 ± 2.8 g/m^2 was obtained. The repeated drying and wetting process consisted of three hours of dryness (60 °C, 35 %RH) and three hours of wetness (40 °C, 95 %RH), and the transition time from dry to wet and from wet to dry was one hour, which is 8 hours per cycle. This is performed alternately in 8 cycles (3 days) and 11 cycles (4 days).

The artificial seawater spraying apparatus [10] used in this experiment is shown in figure 2, and the environmental test apparatus is shown in figure 3. These samples were washed with water before each corrosion acceleration test to wash off the seawater components adhering to the surface of the steel (rinse treatment), and the surface was dried with a dryer for 60 seconds. Later a certain amount of seawater was evenly sprayed on the surface of the specimens by the artificial seawater spraying apparatus, and then placed in the environmental test apparatus for corrosion acceleration test. To reduce the variation in the corrosion and the paint degradation progresses due to the arrangement of the specimens in the test apparatus, the specimens were rearranged for each spraying process.

2.2. Exposure test

The atmospheric exposure tests were conducted at the Choshi where tests can be conducted under typical Japanese climatic conditions [11] (35°43′ N, 140°45′ E., the distance from the nearest sea side is 4 km, the height is 53.6 m from sea level), Exposure Test Site of the Japan Weathering Test Center. The exposure angle of the specimens was 0 degrees to the horizontal. Figure 4 shows the situation of the atmospheric exposure test.

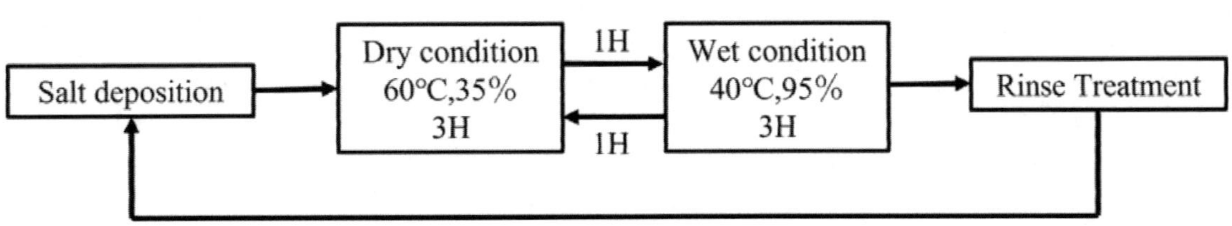

Figure 1. Details of the experimental conditions.

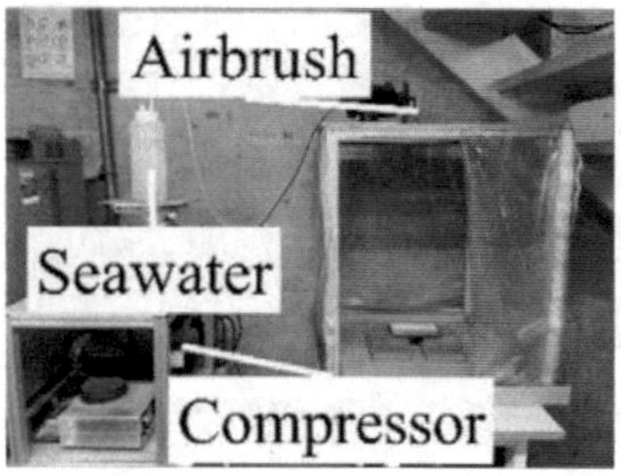

Figure 2. The artificial seawater spraying apparatus.

Figure 3. The environmental test apparatus.

Figure 4. Situation of the atmospheric exposure test.

Table 1. Chemical compositions of weathering steel.

	C	Si	Mn	P	S	Cu	Ni	Cr
SMA400AW (wt%)	0.12	0.20	0.67	0.015	0.004	0.31	0.09	0.49
SMA490AW (wt%)	0.12	0.22	1.14	0.010	0.002	0.31	0.09	0.49

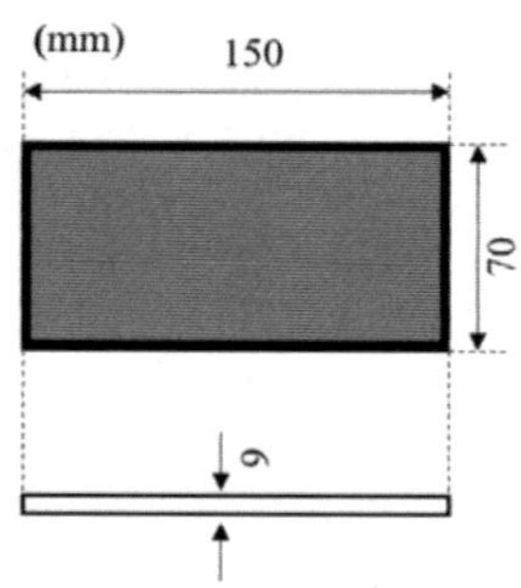

Figure 5. Weathering steel.

2.3. Weathering steel

Weathering steel has been used in the bridges for more than 50 years in Japan. When the structural details are properly designed and the bridge is exposed in a suitable environment, the weathering steel forms a protective oxide coating and should require no coating. In this study, two types of weathering steel (SMA400AW, SMA490AW), with the chemical composition shown in table 1, were cut into coupons with a size of 150 x 70 x 9 mm (see figure 5). The back and sides of the specimens were protected with anti-corrosion tape to evaluate the corrosion condition of one side of the specimen. The number of weathering steels of the same type was three each in the 28, 84, and 168-day corrosion acceleration test and the one-year exposure test.

2.4. Paint-coated steel

In this study, two types of paint-coated steel (SS400) were used as test specimens (hereafter called Paint-coated steel A and Paint-coated steel B). Paint-coated steel A was used for the girder of a roadway bridge in Osaka city, which was built in 1969 and demolished in 2002. Paint-coated steel B was used for the girder of a railway bridge near the seaside in Shimane prefecture, which was built in 1966 and demolished in 2019. As shown in figure 6, Paint-coated steel A and Paint-coated steel B were cut into 150 mm × 70 mm pieces, respectively, and initial defects in the coating were introduced by machining. The number of test specimens was six each, with three specimens used for corrosion acceleration test and three specimens for exposure test. Figure 7 shows the appearance of the coating and initial defects of the specimens. Figure 8 shows the coating thickness of A and B specimens. The back and sides of the specimens were protected with anti-corrosion tape to evaluate only the deterioration caused by the initial defects.

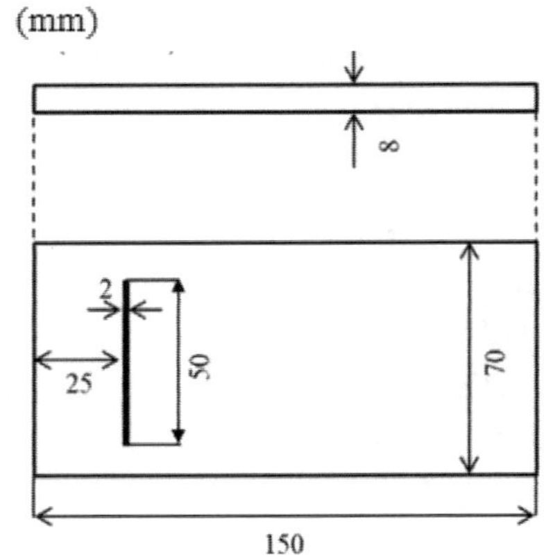

Figure 6. Paint-coated steel.

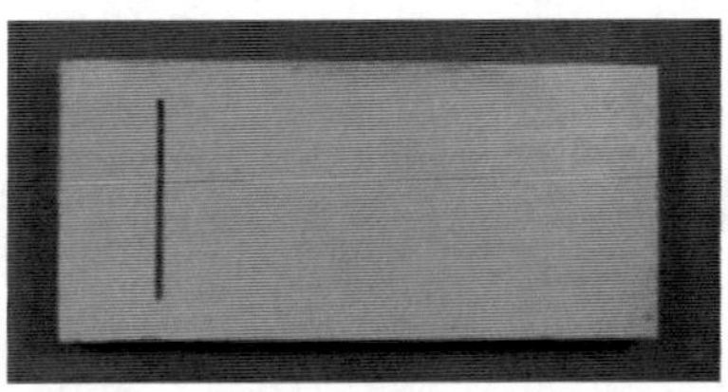

(a) Paint-coated steel A

(b) Paint-coated steel B

Figure 7. Appearance.

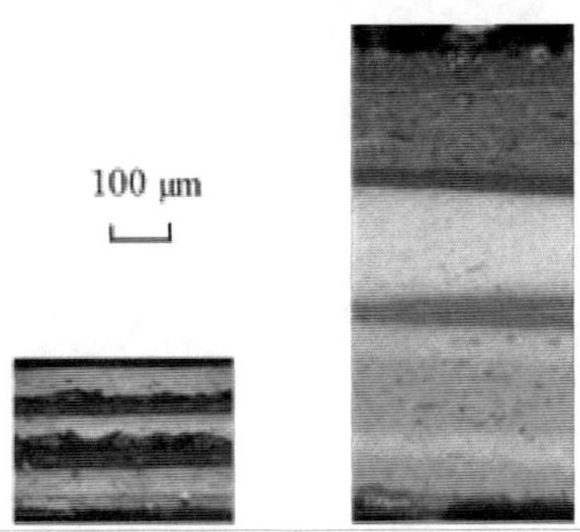

Figure 8. Coating thickness.

3. Results and Discussion

3.1 Weathering steel

Figure 9 shows the appearance of the rust layer formed on the weathering steel as a function of the days of the test. In both cases, rust was observed on the surface of the specimen. It was evident that the surface appearance of the rust layer changed from red-brown to dark-brown and then converted to light-brown with increasing the cycle of test in the corrosion acceleration test, but it appeared yellow-brown in the one-year exposure test. For the corrosion acceleration test, cracks could be observed in the rust formed on the surface area of samples near the anti-corrosion tape at the time of passing 168 days test. Moreover, the rust layer on the surface of the specimens was peeled off as a whole in the 84-day and 168-day corrosion acceleration test, whereas it was not observed in the one-year exposure test.

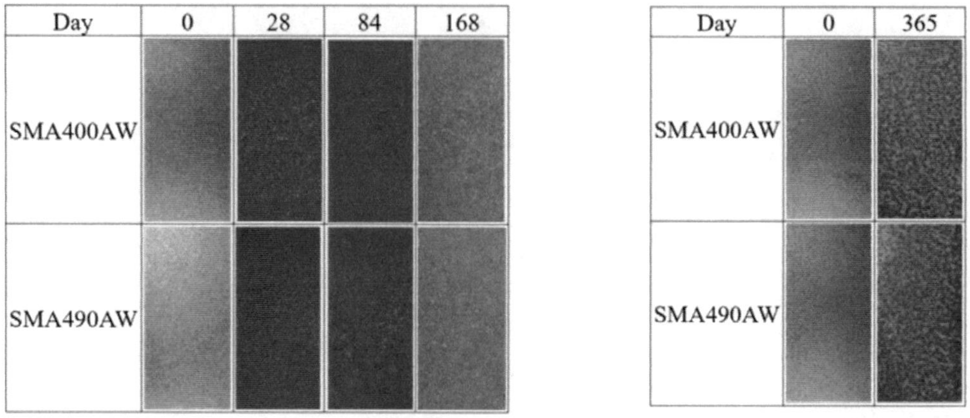

(a) Corrosion acceleration test (b) Exposure test

Figure 9. Appearance of the specimens.

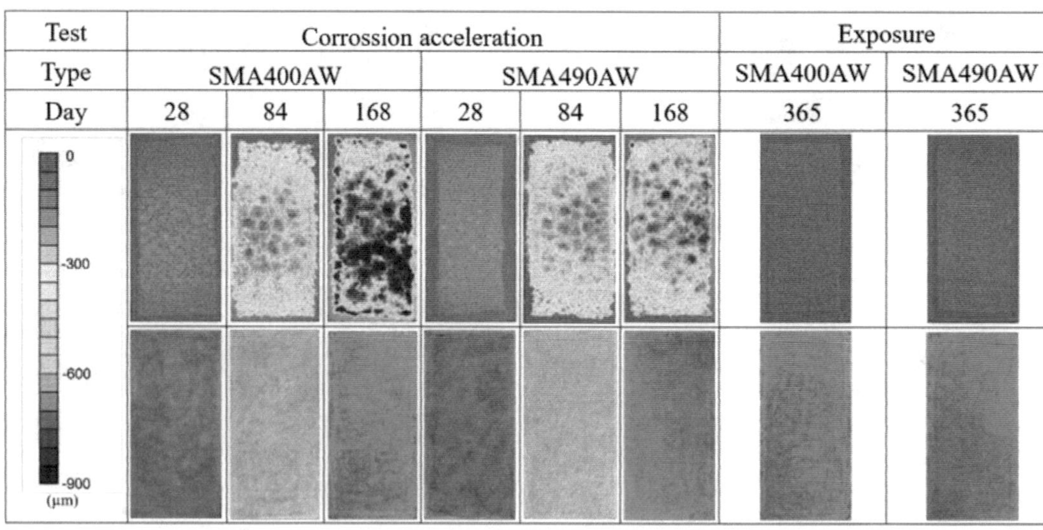

Figure 10. Measurement results of the corrosion distribution on the surface.

To obtain the corrosion condition and corrosion distribution of the steel surface, the surface of specimens after the rust removal process was measured at intervals of 0.3 mm horizontally and vertically using a laser displacement meter. Figure 10 shows the corrosion distribution on the surface of the two types of steel after the 28, 84, and 168-day corrosion acceleration test and one-year exposure test. It could be seen that the corrosion depth of the steel became larger as the cycle of the test increased, and it was much greater than the corrosion depth after one-year exposure test.

In this study, the laser displacement meter was used to calculate the thickness loss of the steel plates. In the measurement by the laser displacement meter, the average of the thickness loss measured in the 50 x 60 mm central area of the three specimens (see figure 11) at the same experimental period, was calculated as the thickness loss. Table 2 shows the average values, standard deviations, and coefficient of variation of the three specimens. It could be observed that for the two types of steel, SMA400AW, and SMA490AW, the thickness losses of the specimens were basically the same in each same experimental period. In other words, there was not much difference in corrosion resistance between SMA400AW and SMA490AW. Figure 12 shows the thickness losses of two types of weathering steel for the corrosion acceleration test. It could be observed that in this test the corrosion rate from 28 days to 84 days, became faster than the results of the previous 28 days, and it slowed down from 84 days to

168 days. The enhanced corrosion resistance of weathering steel in relation to plain carbon steel is due to the formation in low aggressive atmospheres of a compact and well-adhering corrosion product layer known as patina [12]. The study, carried out in Japan on 41 bridges through long-term exposure tests (9 years) between 1981 and 1993 by three organizations: the Public Works Research Institute of the Construction Ministry, the Kozai Club, and the Japan Association of Steel Bridge Construction, confirmed that there was a critical air-borne salinity concentration of around 0.05 mg NaCl/dm^2/day [13]. When the steel is in an environment in which the air-borne salinity concentration exceeds this value, a non-protective rust layer will be formed on the surface of the steel, which will weaken the corrosion resistance of weathering steel. However, it is considered that the protective rust might have formed on the surface of the specimen in the corrosion acceleration test from 84 to 168 days, which might inhibit the corrosion rate of the weathering steel. Although the test conditions were limited, the accelerated test of 168 days produced the thickness reduction by 10 to 14 times larger than the exposure test of 365 days to the weathering steel.

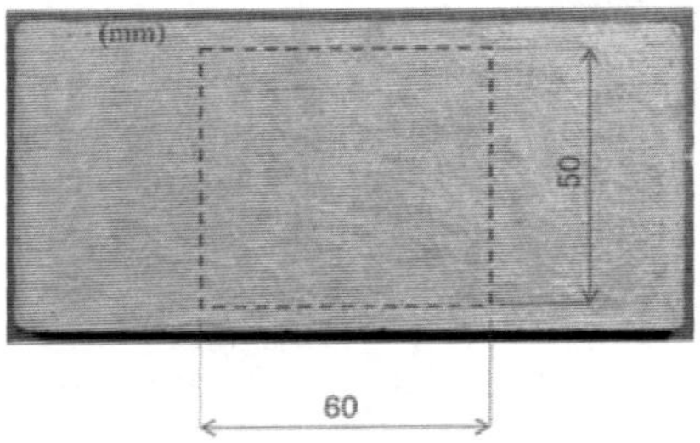

Figure 11. Measurement area of thickness loss by the laser displacement meter.

Table 2. Thickness loss of weathering steel (mm) for each test period.

Type of Steel	SMA400AW				SMA490AW			
Text	Corrosion Acceleration			Exposure	Corrosion Acceleration			Exposure
Days of Text	28	84	168	365	28	84	168	365
Average (M)	0.135	0.591	0.671	0.048	0.152	0.597	0.684	0.069
Standard Deviation (S)	0.014	0.008	0.055	0.006	0.007	0.005	0.043	0.011
Coefficient of Variation (%)	10.22	1.30	8.21	13.13	4.61	0.84	6.29	15.36

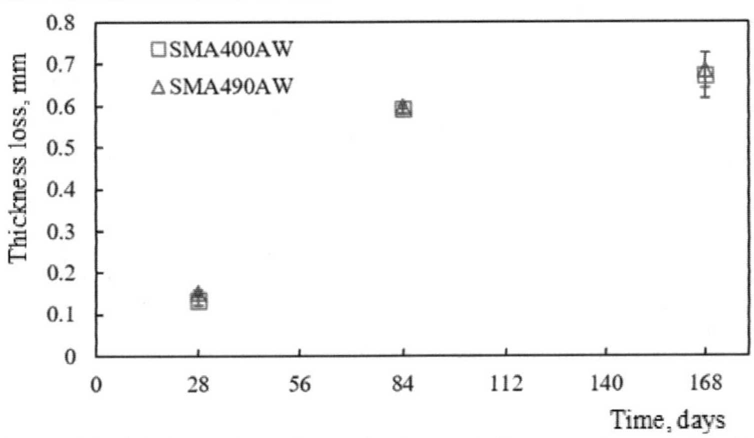

Figure 12. Thickness loss of weathering steel for corrosion acceleration test.

3.2 Paint-coated Steel

When coated steel is exposed to water or oxygen, rust is generated under the coating, and this volume expansion pushes the coating up (figure 13) [14]. The blister area [15] is defined as the area of the coating that is more than 50 μm above the sound part of the coating. Based on this mechanism, the blister area was used as an evaluation index of coating degradation in this study. To calculate the blister area, the surface profile of a 50 mm × 70 mm area centered on the initial defect was measured at intervals of 0.1 mm horizontally and vertically using a laser displacement meter after each cycle of testing.

The photography and the measurements were conducted every 28 days for the corrosion acceleration test and every 91 days for the exposure test. The measurement results of the surface profile near the initial defect area are shown in figure 14, and the transition of the blister area is shown in figure 15. The graphs show the mean values and standard deviations of the three specimens. For the corrosion acceleration test, the blister area of steel A was 298 mm^2, and that of steel B was 431 mm^2 at the time of passing 168 days. For the exposure test, the blister area of steel A was 21.6 mm^2, and that of the steel B was 38.6 mm^2 at the time of passing 184 days. Using the values above, we get the degradation scale factor was about 13.8 times for steel A and about 11.2 times for steel B. ISO16539 Method B shows the characteristics of the actual environment where the speed of corrosion progression of the steel B is faster than that of steel A. This may be a useful method for paint-coated steel, although the results were obtained with a limited number of specimen types, test periods, and test conditions.

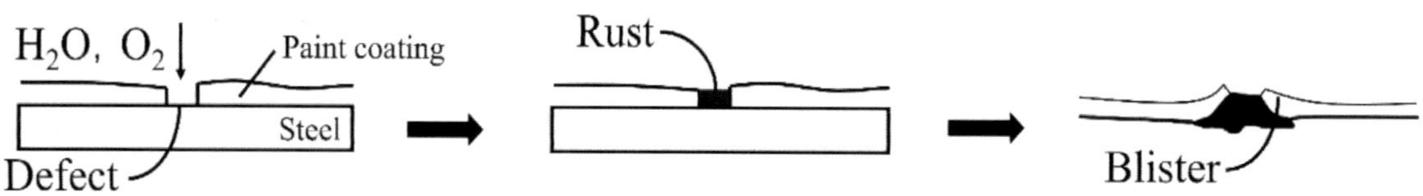

Figure 13. Mechanism of blister.

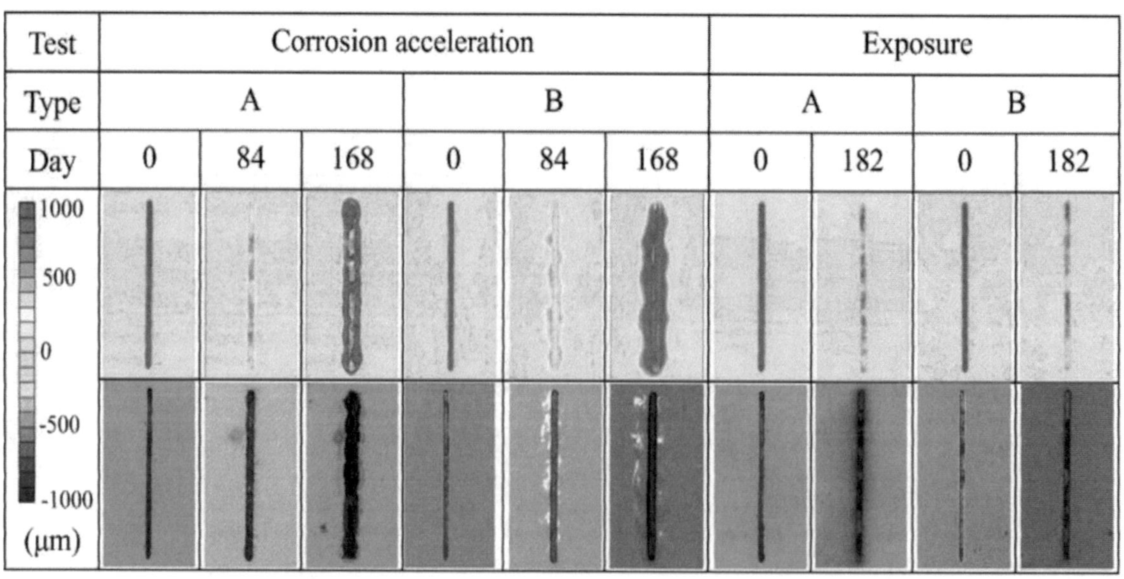

Figure 14. Measurement results of the surface profile.

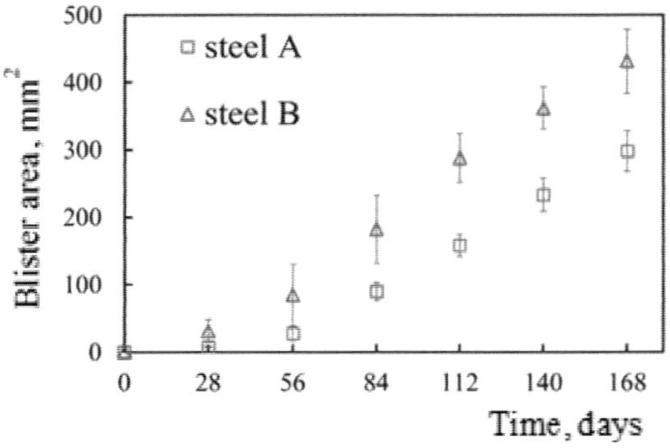

(a) Corrosion acceleration test

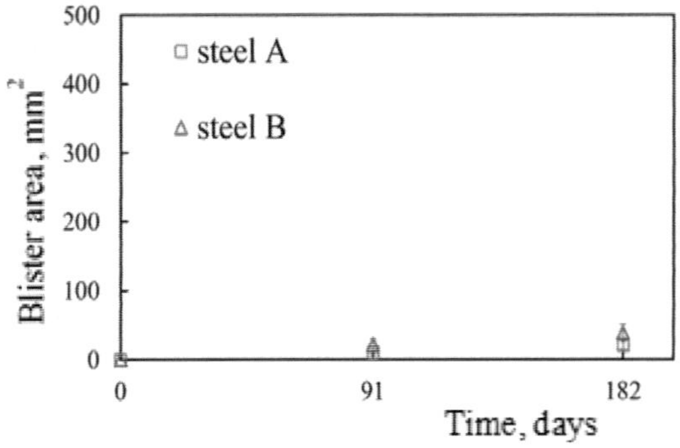

(b) Exposure test

Figure 15. Transition of the blister area.

4. Conclusion

Two types of corrosion tests were conducted to clarify the characteristics of the accelerated corrosion test by ISO16539 Method B in the durability evaluation of steel used for bridge structural members. For the weathering steel, the thickness losses of all samples increased with corrosion acceleration test time. For the paint-coated steel, ISO16539 Method B showed the characteristics of the actual environment where the speed of corrosion progression of steel B was faster than that of steel A. Within the scope of this experiment, both weathering steel and paint-coated steel were found to have an accelerated deterioration effect over 10 times by the acceleration test of 6 months compared to that of the one-year actual environment exposure test. Although continuous data collection and expansion of experimental conditions are necessary, the usefulness of ISO16539 Method B for evaluating durability of the steel members used in the bridge structures was indicated.

5. References

[1] M V Biezma and J R San Cristóbal 2005 *Methodology to study cost of corrosion: Corrosion Engineering, Science and Technology* **40** 344-352
[2] I T Kim and Y Itoh 2007 *Accelerated exposure tests as an evaluation tool for estimating of organic coatings on steel bridges: Corrosion Engineering, Science and Technology* **42** 242-252
[3] ISO9227 2017 *Corrosion tests in artificial atmospheres–Salt spray tests*
[4] ISO14993 2018 *Corrosion of metals and alloys–Accelerated testing involving cyclic exposure to salt mist, dry and wet conditions*
[5] SAE J 2334 2016 *Laboratory cyclic corrosion test*
[6] T Shinohara 2014 *Present State and Visions of Evaluation Methods for Atmospheric Corrosion: Zairyo to Kankyo* **68** 116-120 (in Japanese)
[7] H Kajiyama, S Fujita, K Fuji and M Sakai 2006 *Problem of Conventional Accelerated Corrosion Tests, and Development of New Accelerated Corrosion Test: Zairyo-to-Kankyo* **55** 356-363 (in Japanese)
[8] ISO16539 2013 *Corrosion of metals and alloys-Accelerated cyclic corrosion tests with exposure to synthetic ocean water salt-deposition process-"Dry" and "wet" conditions at constant absolute humidity*
[9] ASTM International West Conshohocken D1141-98 2013 *Standard Practice for the Preparation of Substitute Ocean Water*
[10] K Fujii, K Takahashi and M Sugawara 2015 *Comparison of Salt Deposition Methods in Accelerated Cyclic Corrosion Test ISO 16539: Zairyo-to-Kankyo* **64** 458-461 (in Japanese)
[11] N Enomoto and K Tanaka 2010 *Proposal of test method for prediction of weatherability of construction sealants based on short term out door exposure: Journal of structural and construction engineering* **75** 1199-1207 (in Japanese)

[12] M Morcillo, B Chico, I Díaz, H Cano and D De la Fuente 2013 *Atmospheric corrosion data of weathering steels. A review: Corrosion Science* **77** 6-24
[13] Kozai Club (Ed.) Tokyo 1993 *Technical Report: Guideline for designing and construction of bridges by weathering steel* (in Japanese)
[14] W Funke 1997 *Problems and progress in organic coatings science and technology, Progress in Organic coatings* **31** 5-9
[15] T J Chuang, T Nguyen and S Lee 1999 *Micro-mechanic model for cathodic blister growth in painted steel Journal of Coatings Technology* **71** 75-85

An Investigation on Combined Section by Steels with Different Strengths for Bridge Girder

M Hirohata[1], K Jármai[2]

[1] Department of Civil Engineering, Division of Global Architecture, Graduate School of Engineering, Osaka University, 2-1, Yamada-oka, Suita, 5650871, Japan
[2] Institute of Energy and Chemical Machinery, Faculty of Mechanical Engineering and Informatics, University of Miskolc, Egyetemváros, Miskolc, H-3515, Hungary

E-mail: hirohata@civil.eng.osaka-u.ac.jp; karoly.jarmai@uni-miskolc.hu

Abstract. A basic investigation was carried out in this study to propose a new structure for steel bridge girders by using a combined section composed of steels with different strengths. SM400 as an ordinal strength level steel was selected as the base material for the I-section girder model. It was assumed that the lower flange of the I-section was replaced with an inverted T-shaped member by welding. This member was provided by the different steel with SM400A. The dimension of the inverted T-shaped member was selected to decrease the cross-sectional area to the original section. Furthermore, the moment of inertia and the yielding moment were aimed to increase from those of the original section. With satisfying these conditions, the possibility of cost-saving by using high-strength steel such as SBHS500 was examined. A series of investigations showed the optimum dimensions simultaneously satisfying these mechanical and economic conditions despite being limited.

1. Introduction

Longitudinal girders of steel bridge structures are mainly designed to resist the bending moment. The bending moment applies the tensile or compressive stresses in the flanges. An I-shaped section is one of the typical section types with high bending stiffness. The thickness and the width of the flanges primarily affect the magnitude of the moment of inertia of the section. The load-carrying capacity of the I-section is determined by the yielding of the upper or the lower flanges under the bending loads. Although the lateral-torsional buckling and the shear buckling should be considered in the ultimate state of the whole structure, the initial yielding of the flanges is the origin of the plastic deformation of the section.

Therefore, the concept of the hybrid girder, which has the flanges with high-strength steel and the web with normal steel has been proposed as an economical design [1]. In this concept, high-strength steel is applied to the upper and the lower flanges to save the high-strength steel in the web. Several computing methods have been studied for determining the optimum dimensions of the flanges and the webs of the hybrid girders [2, 3]. Then, the limitation of the difference in the strength of the flanges and the web had been examined from the viewpoint of the premature yielding of the web [4].
Furthermore, the application of the higher strength steels such as HPS 70W in the hybrid girder had been examined with their development [5]. The researches have been continuously performed in the experimental and analytical approaches for the newly developed high-performance steels [6-10]. Recently, the concept of the hybrid girder was applied to not only the normal plate girder but also the corrugated web girder [11].

Although the concept of the hybrid girder has been applied to many cases, it is based on the combination of the flanges by the high strength steel and the web by the normal steel. The authors showed a different concept of the combined I-section of the high-strength steel and the normal steel for improving the mechanical performance of the I-section girder in this study. A series of fundamental investigations were performed on the combined I-section under the bending load. The possibility of improving the bending stiffness and the initial yielding moment was examined. Furthermore, the optimum dimension of the section was searched from the mechanical and the economic viewpoints based on the amount and the price of the materials.

2. Concept of combined I-section

2.1. Assumption of investigation

Figure 1 shows the base section of the investigation in this study. The base section is composed of SM400. The dimensions of the base section satisfy the specifications for highway bridges in Japan [12]. It is assumed that the section is subjected to a uniform bending moment. The influences of a shear loading or the combination of the bending and the shear are not considered. Only the initial yielding of the section under the bending moment is treated as the strength of the I-section. The hardening of the yielded part is not considered for the load-carrying capacity of the I-section. The buckling of the upper flange is restrained by the existence of the floor slab.

2.2. Concept of combined I-section

Figure 2 shows the concept of the combined I-section in this study. The lower flange of the base section is replaced with an inverted T-shaped member of the high-strength steel. The high-strength steel is assumed by SBHS500. Table 1 shows the specified mechanical properties of SM400 by JIS G 3106 [13] and SBHS500 by JIS G 3140 [14].

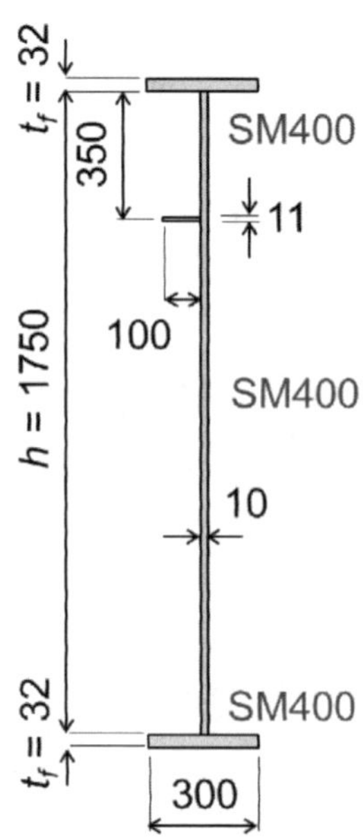

Figure 1. Base I-section.

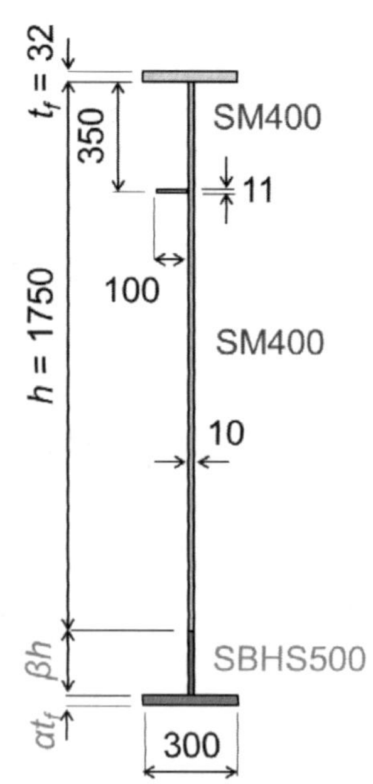

Figure 2. Combined I-section.

Table 1. Specified mechanical properties of SM400 and SBHS500.

	Yield strength (N/mm^2)	Tensile strength (N/mm^2)	Elongation (%)
SM400	Over 235	400 - 510	Over 22
SBHS500	Over 500	570 - 720	Over 20 [a]

[a] A thickness of 35 mm is assumed.

The inverted T-shaped member of the high strength steel is replaced with the lower flange because the additional web height of the inverted T-shaped member increases the moment of inertia of the section. In the case of the base section, the initial yielding occurs at the lower flange. Even if the lower flange of the base section by SM400 is replaced with the steel plate of the same dimension by SBHS500, the moment of inertia is not changed. The concept of the combined section in this study aims to improve the yielding moment and the bending stiffness. If the replaced lower flange of the inverted T-shaped member is the same dimension as the base section, the additional web will increase the cross-sectional area corresponding to the weight. Therefore, the thickness of the lower flange of the inverted T-shaped member is determined so that the cross-sectional area is not increased from that of the base section. The coefficients of α and β are arranged to satisfy the following conditions.

$$A_C/A_B \leq 1 \tag{1}$$

$$I_C/I_B \geq 1 \tag{2}$$

$$M_{YC}/M_{YB} \geq 1 \tag{3}$$

Here, A_C and A_B are the cross-sectional areas of the combined and the base sections. I_C and I_B are the values of the moment of inertia of the combined and the base sections. M_{YC} and M_{YB} are the values of the yielding moment of the combined and the base sections. The values of the yielding moment are calculated by the following equations (4) and (5).

$$M_{YB} = \frac{\sigma_Y}{y_B} I_B \tag{4}$$

$$M_{YC} = \frac{\sigma_{Yi}}{y_C} I_C \tag{5}$$

Here, σ_Y is the yield stress of SM400A applied to the lower flange of the base section. y_B is the distance from the neutral axis to the edge of the lower flange in the base section. σ_{Yi} is the yield stress of the initial yielding part in the combined section. y_C is the distance from the neutral axis to the initial yielding part in the combined section. In the case of the combined section, the position of the initial yielding varies depending on the dimension of the section. When the combined web is relatively high, the initial yielding will occur at the flanges. When the combined web is relatively low, the initial yielding will occur at the boundary between SM400 and SBHS500.

The width of the lower flange and the thickness of the additional web of the inverted T-shaped member may be the parameters influencing the cross-sectional area, the moment of inertia, and the yielding moment of the combined section. However, these dimensions are not considered in this investigation for simplifying the calculation.

The replacement of the original lower flange in the base section with the inverted T-shaped member is assumed to be applied by welding. Although the welding of this replacement may induce deformation and residual stress, their influence on the load-carrying capacity of the combined section is not considered in this investigation.

The first author initially proposed this concept for a repair and reinforcement method on an I-section girder with a corroded lower flange [15]. In the previous proposal, the combination of the steel materials in the section was different from this study. Furthermore, the economic optimization was not performed because of the lack of information related to the material cost.

3. Calculation of sectional properties

The values of the cross-sectional area, the moment of inertia, and the yielding moment of the base and the combined sections were calculated by changing the coefficients of α and β. Figure 3, figure 4, and figure 5 show the improving effects of the cross-sectional area, the moment of inertia, and the yielding moment by fixing α at the specified values and changing β. The cross-sectional area and the moment of inertia of the combined section simply change with the value of β under the fixed α. The tendency of the yielding moment of the combined section is composed of two lines. This is because of the difference in the initial yielding part, as explained. The combinations of α and β satisfying the conditions by the equations (1), (2), and (3) are shown by the striped region in figure 6.

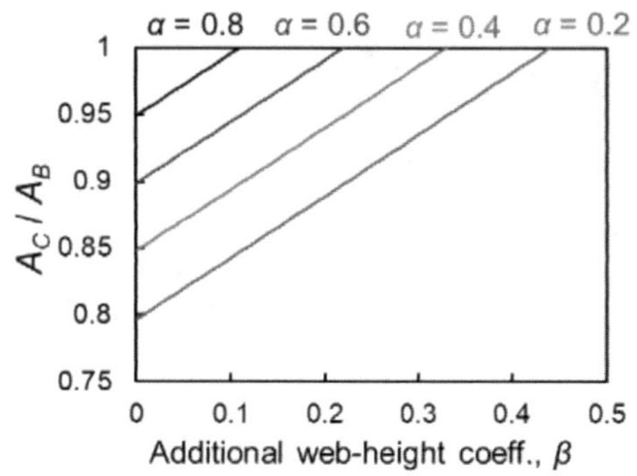

Figure 3. Reduction effect of the cross-sectional area.

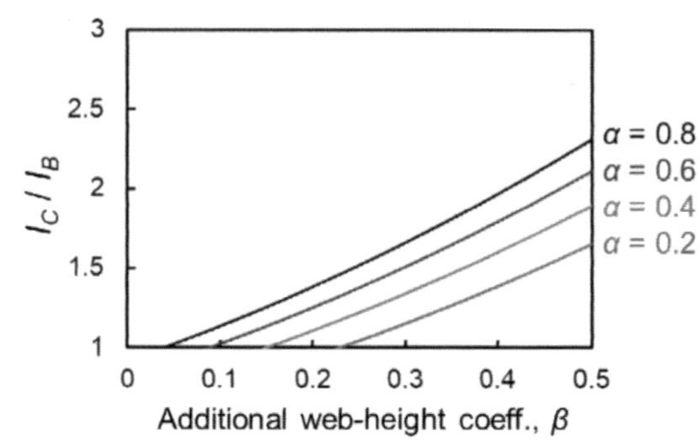

Figure 4. Improving effect of the moment of inertia.

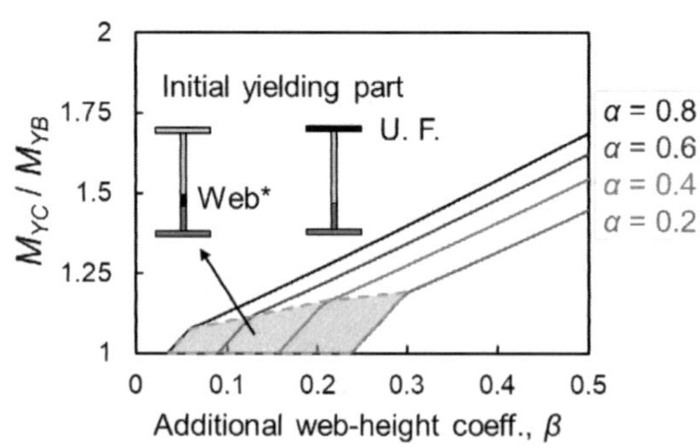

Figure 5. Improving effect of the yielding moment.

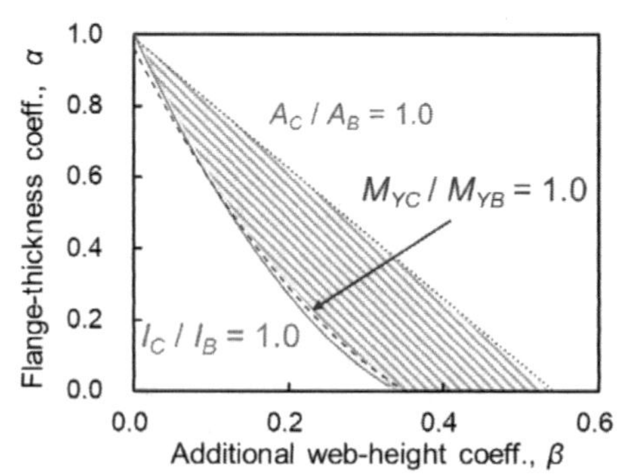

Figure 6. Region of α and β for improving the mechanical performance.

4. Mechanically and economically optimum section

SBHS500 is the high-strength steel for bridge structures developed in Japan [16]. The thermomechanical controlling process ensures a high yield strength over 500 N/mm2, toughness, and weldability. Table 2 shows the material prices of several types of steel in Japan. These prices were estimated by the construction handbook in Japan in August 2021. The actual steel prices vary with the thickness, width, and length. The extra costs for transportation and primer are not considered in this price estimation. Although the application of SBHS in the bridge structures is extending, the price of the material is higher than that of the normal steel such as SM400 or SM490.

The cost coefficient of SM400, γ_{SM400}, is set as 1.00. The cost coefficient of SBHS500, $\gamma_{SBHS500}$, is obtained by dividing the price of SBHS500 by that of SM400. $\gamma_{SBHS500}$ is 1.47. The cost ratio of the base and the combined sections are calculated by the following equation (6).

$$C_C/C_B = \frac{(A_C - A_T)\gamma_{SM400} + A_T\gamma_{SBHS500}}{A_B\gamma_{SM400}} \leq 1 \quad (6)$$

Here, C_C and C_B are the costs of the combined and the base sections. A_T is the cross-sectional area of the inverted T-shaped member of the combined section. The welding costs of assembling the sections are not considered.

Table 2. Material prices of several types of steel in Japan.

	Yield strength [a] (N/mm^2)	Tensile strength (N/mm^2)	Price [a, b] (JPY/t)	Normalized cost by SM400
SM400	Over 235	400 - 510	127,500	1.00
SM490	Over 315	490 – 610	136,000	1.07
SM520	Over 355	520 – 640	144,000	1.13
SM570	Over 450	570 – 720	173,000	1.36
SBHS400	Over 400	490 - 640	152,000	1.19
SBHS500	Over 500	570 - 720	187,000	1.47

[a] A thickness of 35 mm is assumed.
[b] Extra costs for transportation and primer are not considered.

Figure 7 shows the relationship between the cost ratio, C_C/C_B, and the coefficient of the additional web height, β, under the several specified values of α, which is the coefficient of the flange thickness of the inverted T-shaped member. For reducing the cost of the combined section compared to that of the base section, the flange thickness and the additional web height of the inverted T-shaped member should be relatively small. The checked region in figure 8 satisfies the equations (1), (2), (3), and (6). The combinations of α and β in that region may show the possibility of improving the mechanical performance and the cost-saving. It can be said that the mechanically and economically optimum section may be selected from the combinations of α and β in that region.

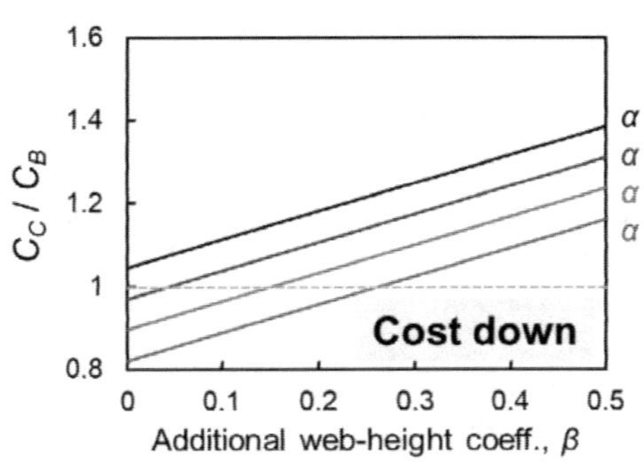

Figure 7. Cost-saving effect.

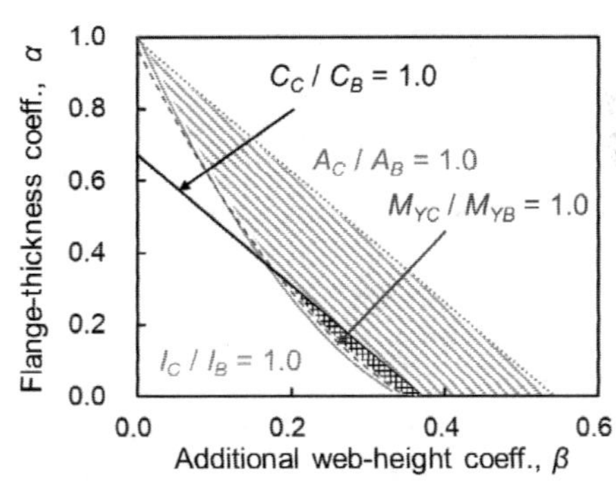

Figure 8. Region of α and β for the mechanical performance improvement and the cost-saving.

5. Conclusions

An approach for searching the mechanically and economically optimum I-section combined by the normal steel and the high strength steel was shown in this study. Even though the basic assumption considering the initial yielding under the bending moment, the possibility for improving the mechanical performance was confirmed by selecting the dimensions of the inverted T-shaped member with the high strength steel welded at the lower part of the I-section. Because the mechanical investigation in this study was only for the yielding moment as the strength of the section, the influence of the welding and the ultimate strength through the elastic-plastic behaviour should be examined in future work. The cost estimation in this study was based on the amount of the material only. The fabrication and the construction cost of welding should also be considered in the economic investigation in the future work in the precise optimization of the combined I-section of the normal steel and the high strength steel in the mechanical and economic viewpoints. More combinations of the different steels should be examined by precise optimization.

6. References

[1] Richard P L 1984 Economical steel plate girder bridges *Engineering Journal* **21** 89-93

[2] Adeli H and Mak K Y 1990 Interactive optimization of plate girder bridges subjected to moving loads *Computer-aided design* **22 (6)** 368-70

[3] Hendawi S and Frangopol D M 1994 Design of composite hybrid plate girder bridges based on reliability and optimization *Structural safety* **15** 149-65

[4] Veljkovic M and Johansson B 2004 Design of hybrid steel girders *Journal of Constructional Steel Research* **60** 535-47

[5] Felkel J P, Rizos D C and Ziehl P H 2007 Structural performance and design evaluation of HPS 70W bridge girders *Journal of Constructional Steel Research* **63** 909-21

[6] Chacón R, Mirambell E and Real E 2010 Hybrid steel plate girders subjected to patch loading, Part 1: Numerical study *Journal of Constructional Steel Research* **66** 695-708

[7] Chacón R, Mirambell E and Real E 2010 Hybrid steel plate girders subjected to patch loading, Part 2: Design proposal *Journal of Constructional Steel Research* **66** 709-15

[8] Shokouhian M and Shi Y 2015 Flexural strength of hybrid steel I-beams based on slenderness *Engineering Structures* **93** 114-28

[9] Wang C S, Duan L, Chen Y F and Wang S C 2016 Flexural behavior and ductility of hybrid high performance steel I-girders *Journal of Constructional Steel Research* **125** 1-14

[10] Kulkarni A S and Gupta L M 2018 Experimental investigation on flexural response of hybrid steel plate girder *KSCE Journal of Civil Engineering* **22 (7)** 2502-18

[11] Shao Y B, Zhang Y M and Hassanein M F 2020 Strength and behaviour of laterally-unrestrained S690 high-strength steel hybrid girders with corrugated webs *Thin-Walled Structures* **150** 106688

[12] Japan Road Association 2017 *Specifications for highway bridges Part II: steel bridges ver 2012* ed JRA (Tokyo: Maruzen) chapter 11 pp 210-241

[13] Japanese Standards Association 2015 *JIS G 3106 Rolled steels for welded structure* ed JSA (Tokyo: JSA)

[14] Japanese Standards Association 2021 *JIS G 3140 Higher yield strength steel plates for bridges* ed JSA (Tokyo: JSA)

[15] Hirohata M, Andoh H and Kim Y C 2008 A proposal and soundness diagnosis of repair/reinforcement by high strength steel for existing steel I-girder *Journal of applied mechanics* **11** 911-8 (in Japanese)

[16] Nishioka K and Ichikawa K 2012 Progress in thermomechanical control of steel plates and their commercialization *Science and Technology of Advanced Materials* **13 (2)** 023001

Acknowledgements

A part of this research was funded by JSPS KAKENHI, grant number 19KK0366. The research was partially supported by the Hungarian National Research, Development and Innovation Office—NKFIH under the project number K 134358.

Influence of Takt Time Planning on Construction Efficiency

M Kozlovská[1], D Klosova[1]

[1] Department of Technology, Economics and Management in Civil Engineering, Faculty of Civil Engineering, Technical university of Kosice, Vysokoskolska 4, Kosice 042 00, Slovakia

E-mail: daria.klosova@tuke.sk; maria.kozlovska@tuke.sk

Abstract. Creating continuous workflow and reducing variability of construction processes is one of the essential principles of Takt Time system. The Takt Time Planning is a powerful lean tool that creates a continuous flow of people, materials, and resources, therefore it is recognized as a technology with great potential for process innovation in construction. A takt is a demand-driven pace or rhythm of work, which ensures that professionals work at the same speed to free up workspace within standard time frames, leading to a more optimized production process. The main purpose of this article is to explore the practical application and influence of Takt Time Planning on the productivity and efficiency of construction projects by providing an overview of selected case studies. Results show that Tact Time helps to establish a continuous flow and, as a result, reduce waste, reduce the duration and number of inventory buffers, and therefore reduce costs of construction projects.

1. Introduction

Construction projects require an end-to-end production system with improved processes on-site that goes beyond outdated methods like the critical path method [1]. The traditional Critical Path Method (CPM) is argued to be disruptive to flow as it only simulates activities and duration. Consequently, a lot of waste is generated during the construction process [1][2]. Moreover, location is valuable resource that must be modelled in all planning and scheduling methods [4]. There are alternative scheduling methods appropriate for construction management such as Takt Time, Flow Line, and Point-to-Point Precedence Relation (PTPPR) [2]. This research mainly focuses on takt time planning as it has received increased attention both in the construction industry and in academia [5].

Takt time originates from the German word "Taktzeit", meaning tempo or rhythm. This is the desired time between unit releases that needs to be synchronized with customer demand [6]. The use of takt means the alignment of the rates of production of various processes [7][8]. Toyota Production System (TPS) is a well-known production method in industrial manufacturing that applies takt time management to set the pace for product flow. TPS plans, schedules and controls its resources and raw materials according to the required Takt [6][9].

The first mention of takt production in construction dates back to the time the Empire State Building was built in the 1930s, when takt time was called a “pacemaker" [10]. A more systematic use of takt time emerged in the early 2010s [5]. Since construction works are classified by tasks and phases, knowing the required tasks and available resources, we can calculate the time it takes for workers to complete a specific task. Here, the takt time is the maximum time allowed to complete the task at any stage of the project [11]. Therefore, the goal of takt time to identify repetitive production processes and balance them to provide an even, stable flow that maintains takt beat across all activities [11][12].

In the manufacturing industry, products move along the production line, while the operators remain in their workplaces, in the construction industry, workers and equipment move around, and the product (building) remains stationary [5]. In this way, it is possible to avoid the situation when the execution of faster tasks is disrupted due to slower tasks, because all tasks included in it must begin and end within takt cycles of equal duration [8][13]. Thus, workers move from one workplace to another at regular intervals which is called takt time [8][14]. If the workers move faster than takt time, the job buffers will grow until they are considered excess inventory and, therefore waste. However, if workers move slower than takt time, activities will take longer than the optimal completion time and delay the successor task, resulting in insufficient productivity that is unable to meet the client's needs [11]. Since location serves as both workstation and product in construction, site management is an important part of construction flow [5][15]. The takt time determines how long each trade reserves individual workplace [6]. This is the best solution when the goal is to get workers to work at consistent productivity [14]. In the manufacturing industry, takt times can be measured in seconds, minutes, hours, days, or weeks. However, in the construction industry takt times can be specified in hours, days, or weeks [3].

Another study describes takt time as the amount of time a product must be produced to match the speed at which the product is required [3]. However, this time is not related to peak performance for the task being performed alone, but rather to time that reflects customer demand [16][17]. That is, the supply rate is equal to the demand rate. However, in terms of construction, on-site production is not largely influenced by changes in market demand. The projects have a contractual term backed by a calendar that specifies the duration of the main processes. Therefore, each process takes a specific takt-time, the calculation of which consists of the relationship between the production time available for the process and the number of repetitive units of the process (eg, floors, departments) [17][18]. The example of tact time schedule is presented in figure 1.

TAKT-TIME 4 ZONES															
DAYS	Mon 1	Tue 2	Wed 3	Thu 4	Fri 5	Mon 8	Tue 9	Wed 10	Thu 11	Fri 12	Mon 15	Tue 16	Wed 17	Thu 18	Fri 19
TASKS															
Structuring Phase															
Vertical Rebar	Z1-S1	Z2-S1	Z3-S1	Z4-S1	Z1-S2	Z2-S2	Z3-S2	Z4-S2	Z1-S3	Z2-S3	Z3-S3	Z4-S3			
Vertical Piping Installation	Z1-S1	Z2-S1	Z3-S1	Z4-S1	Z1-S2	Z2-S2	Z3-S2	Z4-S2	Z1-S3	Z2-S3	Z3-S3	Z4-S3			
Vertical Electrical Installation	Z1-S1	Z2-S1	Z3-S1	Z4-S1	Z1-S2	Z2-S2	Z3-S2	Z4-S2	Z1-S3	Z2-S3	Z3-S3	Z4-S3			
Vertical Formwork	Z1-S1	Z2-S1	Z3-S1	Z4-S1	Z1-S2	Z2-S2	Z3-S2	Z4-S2	Z1-S3	Z2-S3	Z3-S3	Z4-S3			
Vertical Concrete Pouring	Z1-S1	Z2-S1	Z3-S1	Z4-S1	Z1-S2	Z2-S2	Z3-S2	Z4-S2	Z1-S3	Z2-S3	Z3-S3	Z4-S3			
Horizontal Formwork		Z1-S1	Z2-S1	Z3-S1	Z4-S1	Z1-S2	Z2-S2	Z3-S2	Z4-S2	Z1-S3	Z2-S3	Z3-S3	Z4-S3		
Horizontal Rebar		Z1-S1	Z2-S1	Z3-S1	Z4-S1	Z1-S2	Z2-S2	Z3-S2	Z4-S2	Z1-S3	Z2-S3	Z3-S3	Z4-S3		
Horizonal Piping Installation			Z1-S1	Z2-S1	Z3-S1	Z4-S1	Z1-S2	Z2-S2	Z3-S2	Z4-S2	Z1-S3	Z2-S3	Z3-S3	Z4-S3	
Horizontal Electrical Installation			Z1-S1	Z2-S1	Z3-S1	Z4-S1	Z1-S2	Z2-S2	Z3-S2	Z4-S2	Z1-S3	Z2-S3	Z3-S3	Z4-S3	
Horizontal Concrete Puring				Z1-S1	Z2-S1	Z3-S1	Z4-S1	Z1-S2	Z2-S2	Z3-S2	Z4-S2	Z1-S3	Z2-S3	Z3-S3	Z4-S3

Figure 1. Takt-time schedule of structural work (Z1- zone; S1- story 1) [3].

Tact time planning (TTP) can be described as a work structuring technique that can help answer the following questions [19][20][21]:

- What work areas will the project be divided into?
- How will work areas be sequenced?
- How will work be transferred from one working group to another?
- Where do you need decoupling buffers and what are their sizes?
- When will the various stages of work be completed?
- Will successive production units perform work in a continuous flow, or will their work be divided?

Little information has been published about this topic due to the novelty of the application of takt production in construction [22]. However, there are several evidence of implementation of that technology in construction companies. In that case, this study aims to investigate how takt production affects production flow in construction. To achieve the goal of this paper, a qualitative study was conducted based on a literature review and case study analysis.

2. Research methodology

Literary review with detailed analysis of case studies was chosen as the main research method. This method can help direct your research and assess trends within a research topic. In addition, case studies provide a deeper understanding of a specific group or context. Various sources of information have been used to write this article with up-to-date information such as Web of Science, SCOPUS, Google Scholar, and Google Open Search. In the process of selecting literature sources for research, only keyword search and access (open access document) were used as search criteria due to the limited number of articles related to this topic. The following phrases were used as keywords for the search: "Takt time + manufacturing", "Takt time + construction", "Takt time planning", "Construction scheduling methods". Then the content of the articles was assessed by reading the title and analysing the abstract to exclude irrelevance and duplication.

3. Case study analysis

For the purposes of this article, the authors have collected and analysed several case studies on this topic. In should be noted that there are no definitive guidelines for application of takt time planning due to the small number of published articles related to that subject. This statement is supported by numerous publications carried out in this area [8][23].

Lehtovaara et al. conducted a qualitative multiple-case study to validate the hypothesis based on a literature review [5]. Three residential buildings (1,2,3 case study respectively), commercial building (4 case study), hospital / laboratory building (5 case study) and automotive plant (6 case study) were selected as case studies. Data was collected through 27 semi-structured interviews with staff, experts on takt production and other project stakeholders. Data collection also included analysis of projects documentation, site visits, and observation of production meetings where possible. In cases 1, 2, 4 and 5, there was a reduction in unnecessary operations in the production process, but it should be noted that detailed tactical planning required significant efforts and resources. Moreover, in cases with high maturity, such as 3 and 6, the benefits were greater than the costs, in cases with low maturity - 1 and 2, the overall effect was negative and led to resource fluctuations, since excess capacity was not effectively balanced. The same problem was partially found in cases 4 and 5.

In another case study, which analysed five high-rise student dormitories with a kindergarten on the ground floor of each building, the takt time plan worked well. Logistics was highlighted as a key aspect of the takt plan's success, especially given the small amount of on-site storage in the project. Due to the large amount of repetitive work, many crews have shown productivity gains, even though some work packages were cut from five foremen to two before completing the first high rise. Furthermore, there were problems with late completion of work packages due to problems with the delivery of doors from the manufacturer. This problem was solved by installing temporary doors. In addition, foremen received good salaries thanks to the piece pay system. As a result, the use of the tactical plan increased the cost of the project by 18%. It found that it was necessary to rethink the pay system into hourly wages in order to get the most out of the TTP [23].

Yassin et al. [11] applied takt time for construction project consisted with 300 apartments and 170 retail stores with a total area of 3,000 m^2, which was about two years behind schedule due to the large time wasted in terms of excess time buffers. In addition, an unbalanced resource distribution curve was noted. This study focused on the laying of the substructure of pipelines and sewerage systems. Construction tasks in this project were highly repetitive therefore in was considered as a perfect case study for implementation of a takt time. The total duration of the project with the amendments have become 54 days instead of previous 105 days, which means a reduction in the duration of the project by about 50%. The results of the case study show that takt time helps to establish a continuous flow, provide smoother resource distribution and, as a result, reduce waste, shorter duration, and therefore lower costs. In addition, Takt time can help reduce wasteful time and inventory buffers by using flowline visualization techniques that simplify takt timing and provide high level of buffer representation.

Keskiniva et al. [8] has carried out a case study based on a renovation project for a student apartment building in Finland. Keskiniva participated in the planning and design of the takt planning for the case study in collaboration with the main contractor's foremen. The results of this study shoved, that subcontracting model based on progress for small work that require specialized knowledge may be incompatible with tack production because it is difficult for a subcontractor to slow down such work to bring it in line with overall productivity. On the other hand, it can become costly for the main contractor if the subcontractor is required to perform small tasks on a daily basis in order to follow the takt cycle. This finding contradicts the conclusion from [23] about the potential reduction in project costs due to the transition from piece-rate wages to hourly wages. In this case, production was monitored using manual data collection and automated spreadsheets. However, it was not possible to maintain a constant picture of the situation at the worker level. The time schedule must also be in digital format so that the updated data can be processed automatically, that allow comparison of as-planned to as-built data. Therefore, takt planning must also be done digitally, but there is no such software yet [8].

Lehtovaara et al. stated that there has been little effort to comprehensively investigate the impact of tact production on workflow in construction, as tact production research has focused on documenting individual success case studies [5]. Furthermore, control over construction production is often carried out by walks around the construction site and the so-called "vignettes". While walking around the site, the foreman observes the progress of various tasks. The monitoring vignette is created in the form of a table based on the schedule, and, thus, filling it out allows to compare the production plan and implementation. It is very important to make sure that documenting and monitoring the situation in the construction site on a daily basis can be carried out with minimal effort, because it still requires a significant part of the foremen's time [22].

4. Discussion

Takt time planning increases productivity by reducing waste, which can be achieved by optimizing work packages and team sizes according to the desired production rate. This optimization helps workers to produce more in less time, therefore reduces construction costs. On the other hand, implementing a TTP requires extensive plans to be developed in collaboration with subcontractors to make the production process as smooth as possible. All of this can be costly, especially when combined with the large amount of time required to complete plans during the construction phase [23].

Keskiniva et al. offered five clarifications to the existing methods of producing takt:

- Short cycle production requires very detailed task definition to avoid ambiguity.
- Some tasks performed in the takt zones may need to be excluded from takt production.
- Takt buffers should be added according to the sensitivity of the created takt schedule.
- Particular attention should be paid to the coordination takted and non-takted tasks.
- Changing the takt plan during production requires efficient procedures as there is little react time [8].

The TTP works well when the project can be divided into clear zones with repetitive work, this makes it possible to schedule work for one zone and copy the plan to all other zones [5][15], but this is not a requirement for takt time to work [23]. The flow of activities performed in one location can be considered as a primary construction process called the location flow. The workflow is formed by the actions performed by a single trade across different locations called trade flow. The results indicated in [5] show that while tact planning is the best way to increase production flow, it requires major effort in terms of control, planning, and continual improvement. While positively influencing the location flow, the impact on the operations flow or trade flow is contradictory.

Benefits of takt-time planning were collected and summarized below:

- Tact time provides a project with a realistic rhythm and workflow that is tailored to the customer's requirements. Beat denotes activities of the correct size and sequence, and also gives a clear idea of the work ahead [21].
- Increased focus and standardization of the forecasting process. The production process is simplified to standardized and clear work series [21]. The standardization of work in various production industries leads to a reduction in lead times, cycle times and set-up times, as well as to an increase in product quality [24][25].
- Increased Common Understanding. Takt time planning enables the entire production team to develop a common understanding of the overall production strategy. Since the workers, in addition to the daily goal, have an idea of where they will work next time [21][23].

Likewise, it was notice that there is no broader and more comprehensive work on monitoring and control of takt production, as most of the publications related to this topic are conference reports with scattered information. As was mentioned at [22], Alhava et al. [26] have proposed that the collection of the takt production realization data should be carried out by the trade partners' foremen and workers with a mobile application. Data collection is recommended to be carried out by the foremen of the main contractor. In this way, foremen are motivated to visit the construction site frequently and closely follow and documenting progress. It also helps foremen to detect inaccuracies in the takt plan. For example, in the study of Keskiniva et al. [22], the foreman walks the site at the end of each workday and documents the situation in a digital application. The application automatically compares records and schedules and provides analysed data for production management and production development in a user-friendly format. The application contains the steps of work at a high level of detail, in accordance with the time schedule so that it provides an opportunity for a uniform analysis. Nevertheless, this study indicates that it is still important that foremen walk around the site, even if the realization control process is automated. Production monitoring can show where production problems are occurring, but it cannot automatically diagnose problems. The foreman cannot monitor the quality of work without being in the workplace and without talking to those who do the work. However, even with a fully automated system, the presence of a foreman is still required.

5. Conclusion

The purpose of this study is to analyse the practical application of takt time planning in various construction projects and assess the impact of this method on construction productivity. It was found that, tact time is the effective tool for schedule and resource planning in construction that helps to establish continuous workflow. Another benefit for the comfort of workers obtained with the use of tact time is predictable working hours. Nevertheless, successful takt production requires more effort in planning, controlling, continually improving, and managing the supporting conditions. The question also remains which payment system is most appropriate to use with takt time planning.

In addition, it is very important to provide appropriate method for collecting and analysing takt time data, because manual analysis of task collision is time consuming. There is a need of automatic task collision analysis for reduction of problems that are unexpected and require rapid reactions during production. Therefore, site monitoring and control data should be collected and processed automatically using a mobile device.

6. References

[1] Koskela L 1992 Application of the New Production Philosophy to Construction *CIFE Technical Report #72* (Stanford: Stanford University)

[2] Brioso X, Murguia D and Urbina A 2017 Teaching Takt-Time, Flowline, and Point-to-Point Precedence Relations: A Peruvian Case Study *Procedia Engineering* **196** 666 – 73

[3] Brioso X, Murguia D and Urbina A 2017 Comparing three scheduling methods using BIM models in the Last Planner System *Organization, Technology and Management in Construction* **9** 1604–14

[4] Dave B, Hämäläinen J, Kemmer S, Koskela L and Koskenvesa A 2015 Suggestions to Improve Lean
Construction Planning *23rd Annual Conference of the International Group for Lean Construction (Perth)*

[5] Lehtovaara J, Seppänen O, Peltokorpi A, Kujansuu P and Grönvall M 2021 How takt production contributes to construction production flow: a theoretical model *Construction Management and Economics* **39** 1 73–95

[6] Munavalli R J, Vasudeva Rao S, Srinivasan A, Manjunath U and van Merode G G A 2017 Robust Predictive Resource Planning under Demand Uncertainty to Improve Waiting Times in Outpatient Clinics *Journal of Health Management* **19** 4 563–83

[7] Frandson A, Berghede K, Tommelein I D 2013 Takt time planning for construction of exterior cladding *Proceedings of the 21st Annual Conference of the International Group for Lean Construction 31 July–2 August (Fortaleza)*

[8] Keskiniva K, Saari A and Junnonen J-M 2020 Takt Planning in Apartment Building Renovation Projects *Buildings* **10** 226

[9] Eswaramoorthi M, Kathiresan G R, Jayasudhan T J, Prasad P S S and Mohanram P V 2012 Flow index based line balancing: A tool to improve the leanness of assembly line design *International Journal of Production Research* **50** 12 3345–58

[10] Sacks R and Partouche R 2010 Empire state building project: archetype of "mass construction" *Journal of construction engineering and management* **136** 6 702–10

[11] Yassine T, Bacha M B Sa, Fayek F and Hamzeh F 2014 Implementing Takt-Time Planning in Construction to Improve Work Flow *Proceedings IGLC-22 Production Planning and Control (Oslo)* pp 787–798

[12] Dlouhy J et al 2016 Three-level method of takt planning and takt control – a new approach for designing production systems in construction *24th annual conference of the international group for lean construction, 20–22 July (Boston)* pp 13–22

[13] Yassine T, Bacha M B S, Fayek F and Hamzeh F 2014 Implementing takt-time planning in construction to improve work flow *Proceedings of the 22nd Annual Conference of the International Group for Lean Construction 25–27 June (Oslo)*

[14] Haghsheno S, Binninger M, Dlouhy J, Sterlike S 2016 History and theoretical foundations of takt planning and takt control *Proceedings of the 24th Annual Conference of the International Group for Lean Construction 20–22 July (Boston)*

[15] Kenley R and Seppanen O 2010 *Location-based management for construction: planning, scheduling*
and control (New York: Routledge)

[16] Liker J K and Meier D 2005 *Toyota Way: a pratical guide for implementing Toyota's 4Ps* (USA: McGraw Hill)

[17] Fazinga W, Saffaro F, Isatto E, Lantelme E 2019 Implementation of standard work in the construction
industry *Revista Ingeniería de Construcción RIC* **34** 3 288 – 98

[18] Bulhões I R, Picchi F A and Folch A T 2006 Actions to Implement Continuous Flow in the Assembly

of Pre-Fabricated Concrete Structure *Proc. of the 14th Annual Conference of the International Group for Lean Construction* (Santiago: International Group for Lean Construction) pp 407 – 419
[19] Ballard G 1999 Work structuring *Lean Construction Institute, White Paper 4*
[20] Tsao C Y, Tommelein I D, Swanlund E and Howell G A 2000 Case study for work structuring: installation of metal door frames *Proc. 8th Annual Conf. of the Int'l. Group for Lean Constr. (Brighton)*
[21] Frandson A, Berghede K and Tommelein D I 2014 Takt-Time Planning and the Last Planner *Proceedings IGLC-22 Production Planning and Control (Oslo)* pp 571–80
[22] Keskiniva K, Saari A and Junnonen J-M 2021 Takt Production Monitoring and Control in Apartment
Renovation Projects *Buildings* **11** 92
[23] Vatne M E and Drevland F 2016 Practical Benefits of Using Takt Time Planning: A Case Study *Proc. 24th Ann. Conf. of the Int'l. Group for Lean Construction* (Boston) pp 173 – 182
[24] Nallusamy S 2016 Efficiency enhancement in CNC industry using value stream mapping, work standardization and line balancing *International Journal of Performability Engineering* **12** 5 pp 413 – 422
[25] Suganthini Rekha R, Periyasamy P, Nallusamy S 2017 Manufacturing Enhancement through Reduction of Cycle Time using Different Lean Techniques *IOP Conf. Ser.: Mater. Sci. Eng.* **225** 012282
[26] Alhava O, Rinne V, Laine E and Koskela L 2019 Can a Takt Plan Ever Survive Beyond the First Contact with the Trades On-Site? *Proc. of the Annual Conference of the International Group for Lean Construction* 3–5 July *(*Dublin)

Acknowledgments
The activity presented in the article is part of the VEGA research grant No. 1/0336/22 "Research on the effects of Lean Production/Lean Construction methods on increasing the efficiency of on-site and off-site construction technologies" and APVV project No.17-0549 "Research of knowledge and virtual technologies for intelligent designing and realization of building projects with an emphasis on their economic sustainability".

Renovation of dormitory campus using the principles of sustainable development

E Luzay[1], D Shafieva[1], M Zeleňáková[2*] and Š Petkanič[2]

[1]Institute of Economics, Management and Information Systems in Civil Engineering and Real Estate, Moscow State University of Civil Engineering, Yaroslavskoe shosse, 26, Moscow, 129337, Russia

[2]Faculty of Civil Engineering, Technical University of Kosice, Vysokoskolska 4, 042 00 Košice, Slovakia

*E-mail: martina.zelenakova@tuke.sk

Abstract. This study presents the new look on the dormitory's territory, which is located in Jedlikova 5, Kosice, Slovakia. Landscaping design has a significant impact on people, and touches both aesthetic and physical needs, environmental integrity of the area and has an impact on nature habitants. Good landscaping helps to provide a conducive living environment for students by creating needful places for recreation, physical education and studying, as well as producing the natural and healthy ambience to the surrounding by using green roofs, creating a pond for collection of rainwater and planting a lot of greenery. The study framework covers aspects, which are related to the ecological state of the territory, existing environmental problems, natural features and needs of the residents of the dormitory. The aim of the study is to create such a sustainable space, where both students and nature will be in harmony.

1. Introduction

People from the very beginning of their existence have been changing the surrounding space by creating a habitat. After many years, the appearance of their housing has changed fundamentally, but one thing has remained the same – the dwelling should provide a human with a sense of safety and security. In the realities of the modern world, the following criterias have also been added: the house should reflect people's interests and tastes, be visually pleasant, built of high-quality, durable materials, be physically and psychologically comfortable, and should also be located close to transport and social infrastructure facilities. This is very important because the environment can create or destroy interactions among people, can influence people' behavior, activity, mood and effect on their health.

One of the most important stages of a person's life is going to university. According to [1], the number of students who lives in the dormitory is 25-41% of the total number. The dormitory provides students with not only a place to live, but also creates space for socialization, atmosphere of learning, affects their mood, forms habits, creates the basis for further life. It is revealed, that architecture and interior design that meets the student's desires and requirements has a positive impact on the well-being and health. Moreover, with a properly organized space, students improve their study performance [2]. Students are now looking for housing with security, community spaces, and privacy [3]. If universities are unable to provide housing for students, the lack of affordable accomondation may become a significant problem, so many universities are renovating existing dormitories all over Europe now, for example, Comenius University in Bratislava [4], Kaunas University of Technology in Lithuania [5], University of Split in Croatia [6].

The role of landscape design is increasing and expanding, and in this article we pay attention to the improvement of dormitories, because here the student spends most of the day. Therefore, the main task of architectural and planning design and landscaping of the dormitory territory is to create optimal environmental conditions that have a beneficial effect on the aesthetic, moral and mental development of the student. Such a place will encourage them to study and establish contacts with each other, provide them with a place to relax, play sports, and should be as effective and harmless to the environment as possible. The most beneficial way to achieve this goal is to use the green infrastructure: creating green roofs, planting greenery, rainwater harvesting, waste recycling and using energy efficient technologies. With the help of green spaces in the yard, we can not only improve microclimatic conditions, but also create shaded areas for rest and quiet activities. Planting allows you to isolate different territories from each other for their intended purpose and fence off areas from adjacent streets and harmful emissions of vehicles.

The aim of this work is to develop a landscaping project of the TUKE's dormitory campus in Jedlikova 5, Kosice, Slovakia with an increase of green areas that improve aesthetic, quality and microclimate conditions of the object, as well as create favourable conditions for the residents.

To achieve this goal, the following tasks were set:
- create places for entertainment, recreation, sports and studying
- design plant seating plan
- improve lighting (with low energy consumption)
- create convenient pedestrian paths
- "green up" the building with green roofs
- make the field in front of the dorms completely pedestrian

2. Methods

In this research the methods of analysis of the existing environment and landscaping of the territory of the dormitory were applied. The research also provides a comparison of green roofs. Based on the results of the analyses, the most optimal ways to solve the problems were chosen. The research methodology is presented on flowchart (Figure 1).

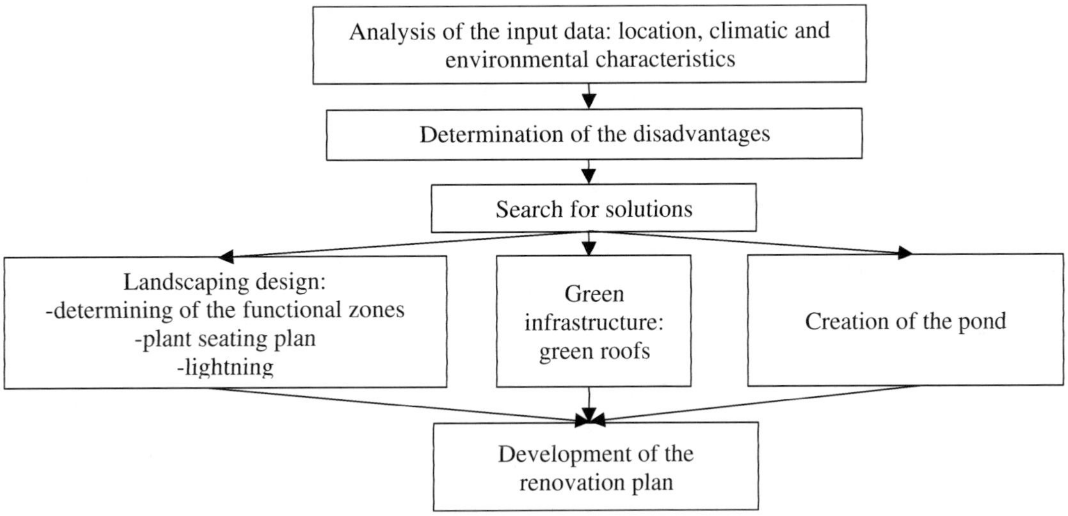

Figure 1. Methodology of research.

2.1 Input data
The subject of our research is located in the Košice - West district, close to the Optima shopping center. It is a 7 - storey dormitory consisting of 4 blocks interconnected by corridors. The dormitory provides accommodation for 1316 people.

Speaking of the environmental conditions of this territory, air quality in Kosice is polluted by nearby industrial complexes (coke, iron, steel, and cement manufacturing). Relatively favourable circumstance is prevailing convection from north directions [7]. This region has a temperate continental climate, which is characterized by warm, humid summers and mild winters. The amount of precipitation in Kosice is significant, with precipitation even in the driest month. There is east wind mostly appeared in Kosice.

2.2 Determination of disadvantages
Before developing new landscaping plan, existing situation was considered and disadvantages were identified. Firstly, the complete absence of green building technologies. In the modern world, this is one of the most developing trends. Many universities have already implemented this technology in the construction of new dormitories. Next, the complete absence of lighting in the territory in front of the building. When trees were cut down, it was very difficult to navigate in complete darkness and not to fall. From a visual and aesthetic point of view, the improvement of the dormitory can't be positively assessed because of absence of vegetation, benches and garbage cans.

2.3 Proposed improvements
After that, we propose the following features how to improve the current situation. For the proper facility arrangement of the main parts of the territory, it was necessary to analyse pedestrian routes, requirements and needs of a modern student. For the comfortable living we decided to create place for playing sports, for entertainment, relaxation, studying. Another important step to replace and update the lighting near the dormitories to increase the safety in areas that people use, create aids in geographic orientation, as people can use well-lit focal points as landmarks to help them find their way.

For the improving of the ecological situation, we updated the dormitory with the green roofs. Green roof has many positive qualities, there are ability to improve the microclimate, helps to filter out dust, smog particles, absorb nitrates, increase rainwater retention by 50–90%, improve noise protection by reducing reflective sound by up to 3 dB, improve sound insulation by up to 8 dB and protect the waterproofing [8-13].

Furthermore, we decided to create a pond in this area because ponds are one of the single best ways of bringing more wildlife into an area. Ponds are a key tool in improving biodiversity because they provide an essential habitat for plants, invertebrates, fish, birds, amphibians and mammals.

3. Results and discussions
After the analysis of the incoming data, identification the disadvantages and ways to further change for the territory to the better, we created the most suitable landscape plan for the territory of our dormitory and its building. We have included data and explanations of our choice about the using of green roofs, the creation of a pond and the installation of a new park light.

3.1 Landscaping design

3.1.1 Determination of the functional zones. Firstly, we have analysed the location (Figure 2), pedestrian routes to the neighbourhoods (to Kaufland, to bus stop "Stodolova', to bus stop near "OPTIMA') and the needs of students and divided territory into 5 semantic parts (Figure 3).

Figure 2. Pedestrian paths.

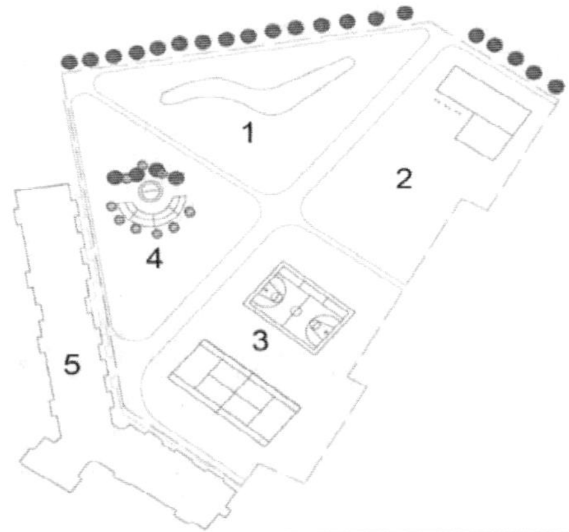

Figure 3. Zones of landscape. Reference.

1. Place to relax, to interact

An open, welcoming gathering space provides a place to make new friends, a place to share experiences, a place to develop a sense of belonging. Main elements: pond, branches, trees, benches

2. Place for studying, co-working and nutrition

Today's students are frequently required to engage in group projects and collaborative work as part of their regular course of study. However, many campuses suffer from a lack of collaborative learning space to support this group work. Spaces like this can fill this need, supporting the academic development of its students and fostering a culture of learning. Moreover, students can grab a coffee and delicious sandwich here, also it is an opportunity to get a job. Main element: cafe

3. Place for playing sports

Like we say in Russia "in a healthy body - healthy mind", so create places where students can go for sports. It is important to take a break from studying and improve yourself. Main element: sportsgrounds

4. Place for entertainment

Modern students are really artistic and imaginative, it is important to create a place where they could arrange concerts, festivals. Main element: stage

5. Dormitory by itself

3.1.2. Plants selection. We have analysed the plants that grow on the territory of the Slovak Republic and decided which trees should be used to improve the park in front of the dormitories. According to the [14], the following tree species (Common beech, European hornbeam, Sessile oak, Sycamore, European ash, Goat willow, European larch, Rowan, Common aspen, Norway spruce, Scots pine) are mainly spread in the Slovak Republic.

There is a grass covering on the territory in front of the dormitory now, but during renovation it will first be completely demolished for laying walking paths, creating cafe and stage, and then planted again, since the grass covering is an ecologically valuable element of the urbanized landscape. The advantage of such a surfacing is the drainage of rainwater directly into the soil. It is also called a plant covering. Grass covering will support ecological diversity in this area.

3.1.3. Lightning. For the lightning we have chosen ES-SYSTEM luminaires (Figure 4) for the territory of the dormitory, because this system is characterized by exceptional durability and resistance to damaging environmental conditions, and with the development of LED technology, they have become more energy-saving, reliable and functional. LED light sources guarantee low operating costs and many years of trouble-free operation. Aside from appropriate lighting settings, the design plays an essential role, blending in quietly with its surroundings during the day and offering a sense of security at night, giving lit areas a distinct character and aesthetic appeal. Each luminaire has a local controller that controls and monitors its operation. The luminous flux of individual luminaires or entire pre-programmed luminaire groups can be controlled via any web browser. The operating parameters of the luminaires, such as their power, power supply voltage, reactive power and the amount of emitted CO_2 can be checked using the interface [15].

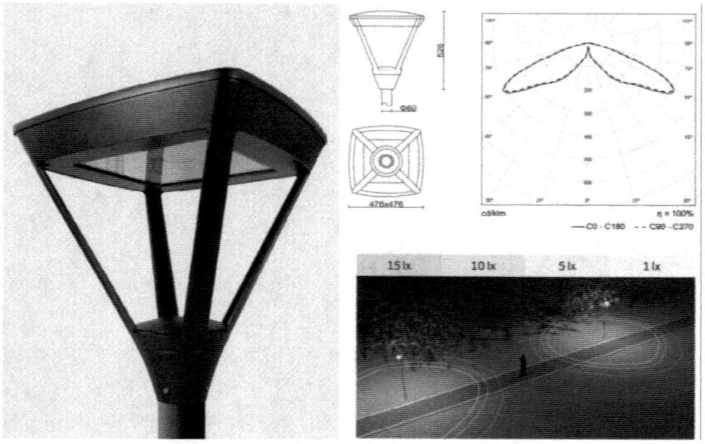

Figure 4. Type of the luminaire.

3.2 Green infrastructure

Next, we have chosen a new greening of the dormitory. There are many types of green roofs, but we settled on an extensive green roof for a number of reasons (Table 1).

Table 1. Comparison of different types of green roofs [16].

	Extensive Green Roofs	Intensive Green Roofs
Maintenance	Minimal maintenance required - inspection 1–2 x / year - supply of water and nutrients mostly by natural processes	Regular maintenance required - garden maintenance such as mowing, fertilizing, watering, weeding etc.
Adapted plant communities	- undemanding, drought-tolerant - self-regenerating	- substrate with higher amount of organic material, with depth > 150 mm
Weight	Little weight and shallow build-up height - mainly mineral substrate with depth up to 120 mm	Weight and build-up height depending on plant selection - substrate with higher amount of organic material, with depth > 150 mm - weight >150 kg/m²
Surface protection	Surface protection with ecological functions	Surface protection with ecological functions

Because irrigation takes place beneath the substrate, the Irrigated Extensive Green Roof (Figure 5) will assist to achieve a harmony between a cost-effective solution and the long-term correct functioning of green roofs in dry areas. As a result, water is available where it is needed most by plants: in the root zone. The system, on the other hand, just provides the quantity of water that the plants require for optimum development. Because of the lesser substrate depth, the green roof build-up is not only lighter, but it also allows for the utilization of different plant communities [17].

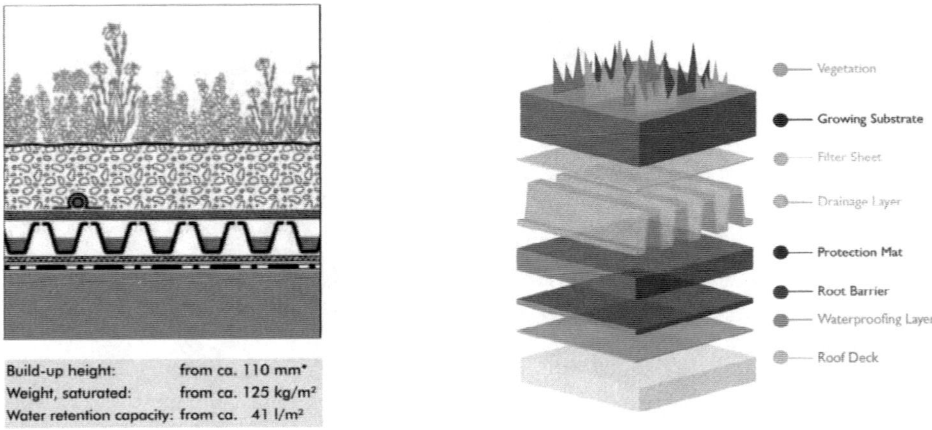

Figure 5. The composition of the green roof.

3.3 Pond

Wet Ponds (Figure 6) are permanent pools of water and provide both quality and quantity control in regards to stormwater. Water quality is controlled when the fast-moving stormwater exits and outfalls and meets with the stationary water of the wet pond. As the water slows down, debris and sediments can settle to the bottom of the pond. The wet pond is designed to slowly release the stormwater over the next 24 - 48 hours to reduce erosion and to delay causing flooding.

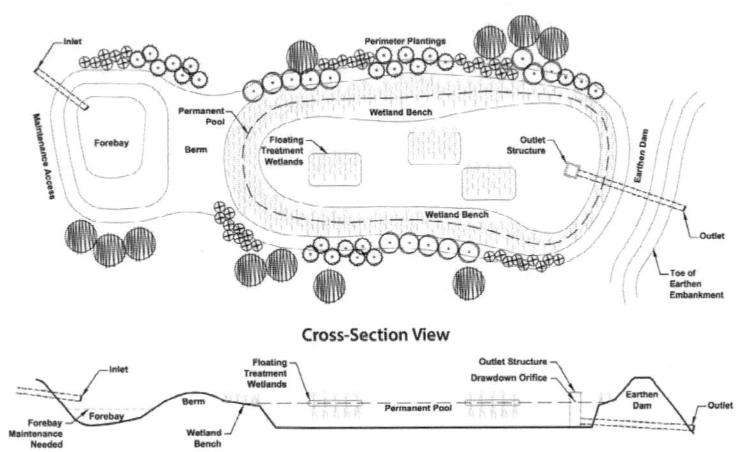

Figure 6. Plan of the pond.

3.4 Renovation plan

As a result of the analysis, we managed to create a new landscape plan for the improvement of the territory in front of the dormitory (Figure 7), as well as to improve the characteristics of the building itself. This image shows the location of new sports fields, a stage, a pond and other entertainment. All these innovations will help to improve unused land, improve the environmental situation around the dormitory and make students feel at home.

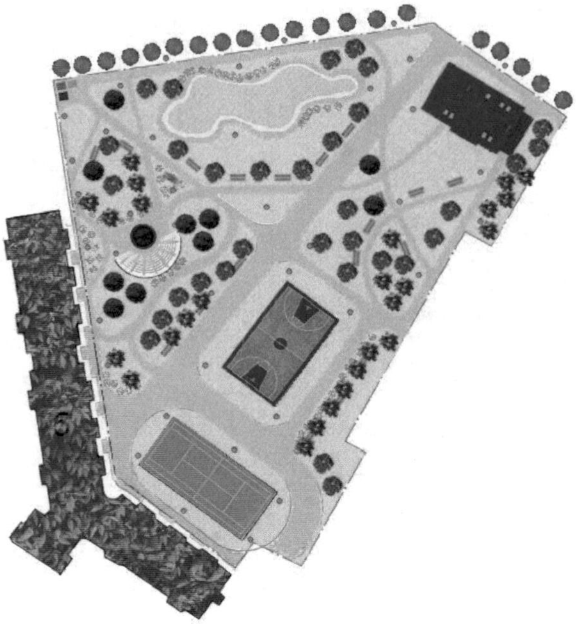

Figure 7. Final result of renovation of the territory.

4. Conclusion

The current territory of the dormitory, although rich in plants, still does not meet the requirements of sustainable development, students living in the absence of space for recreation, sports and study. Our task was to design this place in such a way that answers all the needs of a contemporary young person. The living conditions of a student affects not only their well-being but also their study achievements. Studies, preparation time for lectures and assessments is stressful enough, therefore, the living environment must be pleasant and cosy. Taking into account the needs and requirements of a contemporary student, the area will be divided into several functional zones, such as physical education, relaxation zone and other spaces.

During the project, we placed a basketball and tennis court, established a summer stage for the performances, built a café with the place for studying, dig out a pond for rainwater runoff, placed benches, think over the lightning, plant greenery and placed a green roof on it.

5. References

[1] Eurostudent vi 2016-2018 [online] https://www.eurostudent.eu/download_files/documents/EUROSTUDENT_VI_Synopsis_of_Indicators.pdf

[2] Araujo P and Murray J 2010 *Estimating the effects of dormitory living on student performance* Economics Bulletin AccessEcon vol 30 pages 866-878

[3] Peck S, Callaghan C, Bass B and Kuhn M *Greenbacks from green roofs: forging a new industry in Canada*

[4] Comenius university is planning an extensive dormitory renovation official website [online] https://uniba.sk/en/news-detail/back_to_page/aktuality-1/article/comenius-university-is-planning-an-extensive-dormitory-renovation/

[5] KTU dormitory renovation plan: 5 buildings in 5 years official website [online] https://en.ktu.edu/news/ktu-dormitory-renovation-plan-5-buildings-in-5-years

[6] Reconstruction of Student Dormitory "Bruno Bušić" official website [online] http://www.unist.hr/en/science-and-innovation/projects-office/university-projects/reconstruction-of-student-dormitory-bruno-busic

[7] Air pollution in the slovak republic [online] https://www.shmu.sk/File/oko/rocenky/SHMU_Air_Pollution_in_the_SR_v1.pdf
[8] Bevilacqua P, Mazzeo D and Arcuri N 2018 *Thermal inertia assessment of an experimental extensive green roof in summer conditions* Building and Environment 264-276
[9] Bevilacqua P, Mazzeo D, Bruno R and Arcuri N 2017 *Surface Temperature Analysis of an Extensive Green Roof for the Mitigation of Urban Heat Island in Southern Mediterranean Climate* Energy and Buildings 318-327
[10] Dunichkin I, Naboni E, Korobeinikova A and Poddaeva O 2019 *Issues of wind mode visualization and pedestrian comfort assessment when designing residential buildings on sloping territories in the Arctic zone* SHS Web of Conferences 64
[11] Korol E and Shushunova N 2016 *Benefits of a Modular Green Roof Technology* Procedia Engineering 1820-1826
[12] Korol E, Shushunova N and Rerikh S 2019 *New green roof and green wall systems for implementation in the coverings* E3S Web of Conferences 97
[13] Korol O, Shushunova N, Lopatkin D, Zanin A and Shushunova T 2018 *Application of High-tech Solutions in Ecodevelopment* MATEC Web of Conferences
[14] Pajtík J, Konôpka B and Šeben V 2018 *Mathematical biomass models for young individuals of forest tree species in the region of the western carpathians*
[15] Park lighting [online] https://www.essystem.pl/en/download
[16] Smith J 2000 *Great expectations* Buildings 94 28–32
[17] Nophadrain Extensive Green Roof Systems [online] https://www.nophadrain.com/media/231163/02-design-and-installation-guide-for-extensive-green-roof-systems_en_02-2020_web.pdf

Acknowledgments
This work has been supported by project of the Ministry of Education of the Slovak Republic ID 20210812131460210: Tvorba zelených zón na TUKE a zvyšovanie povedomia o zmene klímy.
This work was supported by the Slovak Research and Development Agency under the Contract no. APVV-20-0281, and a project funded by the Ministry of Education of the Slovak Republic VEGA1/0308/20 "Mitigation of hydrological hazards, floods, and droughts by exploring extreme hydroclimatic phenomena in river basins".

Production of polymer-modified-asphalt mixture as an alternative method to the conventional polymer-modified-bitumen mixtures

R Kovács[1], Ľ Polakovič[2], A Czímerová[2] and J Mandula[1]

[1] Department of Geotechnics and Traffic Engineering, Technical University of Košice Faculty of Civil Engineering
[2] VUIS - CESTY, Ltd. Bratislava

E-mail: robert.kovacs@student.tuke.sk

Abstract. The transfer of new materials and technologies that are necessary for the development of new practices will lead to the development of innovation. These options are considered as (1) innovation, which means a significant improvement in the currently required parameters, mostly in parallel with the price increase, (2) innovation will meet the current parameters, but at a lower product / technology price, (3) innovation will achieve compliance with the applicable parameters, but at a higher price of the product / technology. In most cases, this is the second option, i. e. help the customer to enter the market at a lower price while maintaining the required product quality. This case also concerns the application of mixtures of the polymer-modified asphalt mixture type. The main goal of this work was the production and real laying on the verification section of the polymer-modified asphalt mixture. Based on the results obtained from this work, we can conclude that the values of the observed properties prescribed for a given class of asphalt mixtures in the catalogue sheets of asphalt mixtures were achieved without serious problems. Due to the trouble-free production of such mixtures in the plant, spreading and compaction of mixtures, we can express a clear conclusion that such mixtures are an ideal alternative to conventionally used polymer-modified-bitumen mixtures.

1. Introduction – State of Art

According to EAPA [1], 90 % of the more than 5 million roads in Europe are constructed using asphalt. The representation of asphalt roads is also significantly dominant in Slovakia compared to other types. The production of asphalt mixtures in Slovakia in recent years has ranged from 2 to 2.3 mil. tones per year [2]. The positive aspects of roads with an asphalt cover compared to cement concrete can be found in several factors, such as suitability for low traffic intensities (I. class roads), more advantageous for the influence of the subsoil, application in urban areas, application on bridges, independence of technology from weather, lower noise level of new roads, ease of local and continuous repairs and higher number of producers. However, in addition to the above advantages, pavements with an asphalt cover also have some undesirable properties. Temperature and chemical composition are governing parameters for the mechanical behavior of asphalt binders. During extreme low and high temperatures, asphalt binders can suffer thermal cracking as well as permanent deformation, respectively [3]. The combination of the effects of high temperatures and high-intensity heavy traffic can significantly impair the road's operational capability over several days, and thus road safety. The sensitivity of asphalt mixtures to temperatures (plus and minus - frost cracks) is generally known, and therefore ways to eliminate it are still being sought. Several options are possible, such as: use of crushed coarse aggregate, adjustment of the grain line of aggregate, optimization of the amount of asphalt binder, use of harder asphalts, use of polymer modified asphalts, use of various additives (adhesive, modifying and stabilizing). Polymer-modified-bitumen mixtures derive their technological and conceptual origin from the need for enhancing the performance and durability of asphaltic materials as well as their adhesion to mineral aggregates. The properties of mineral aggregates used for asphalt mixtures are well described in the relevant technical specifications [4]. There are basically two methods of modification of asphalt mixtures. The first one is the modification of the asphalt binder with polymers (PMB) and an alternative way is the modification of the asphalt mixture with polymers (PMAM) [5]. The first technology is generally applied in Slovakia several years. The basic requirements for the used materials and technical conditions for preparation and layering conditions are clearly described in relevant documents [6, 7]. In this case the used PMB is purchased form the company because there is no production of polymer-modified-bitumen. There plants of asphalt mixtures in Slovakia are not equipped with a homogenization station (blender). An alternative way is the direct production of PMAM, which is possible in every production plant of asphalt mixtures in Slovakia. Generally, PMAMs are produced by direct mixing of polymer (usually 3 – 7 % by weight) with mineral aggregates and asphalt [3].

The main goal of this work was to prepare the PMAM directly in the production plant and to compare their properties with an asphalt mixture prepared by a conventionally produced asphalt mixture with PMB. In this work, we have tried to prove that the preparation of a PMAM while meeting all the requirements for asphalt mixtures can be a suitable and economical alternative to the technology of preparation of asphalt mixtures using a modified binder.

2. Theoretical background

Although the technology of production of road modified asphalts was also known in our country, until the beginning of the 1990s there was a problem with the availability of a suitable modifier. Therefore, in the early 1970s, experiments began with modification with polyethylene from domestic sources. These were products delivered in barrels and their great negativity was the inhomogeneity. The first PMB for asphalt mixtures were produced by company JCP Štúrovo (1992) and were supplied under the trade name MOAS in several quality classes. After their successful application on the market, Slovnaft (1993) started the production of polymer-modified asphalts in a modification station in Bratislava. The products were supplied to the market under the trade name Apollobit (after the closure of the modification station, production was moved to Hungary). From the above brief overview, the production of polymer-modified mixtures for road construction is not new in Slovakia. The availability of industrially produced polymer modified asphalts suitable for road construction has dampened interest in PMAM technology.

PMAM have been applied in Slovakia for almost 40 years, so it is good to find a suitable place for this technology to apply today. When choosing a polymer suitable for modifying asphalts, it is important to be aware of the climatic conditions for which it will be used. From the group of plastics, only thermoplastics are suitable for modification [8]. They are more economically advantageous than elastomers, but the resulting products are particularly suitable for improving the resistance to permanent deformations (high value of the KG softening point, insufficient value of the Fraass breaking point). A typical representative of this group is polyethylene (PE). It can be combined with elastomers to achieve efficiency over a wider temperature range. Modifiers from the group of elastomers are more suitable for our climatic conditions. These make it possible to adjust the properties of the asphalt so that they are suitable for use in colder and warmer areas. Typical representatives are SBS and synthetic rubbers, EVA copolymers, synthetic latexes.

In the production of PMAM mixtures, the procedure is in principle the same as in the production of mixtures with PMB. The fundamental difference between conventional asphalt mixtures and PMAM is that in their production additives are always used which are dosed directly into the mixing drum [9]. The additives can be granules or pellets (Figure 1).

Figure 1. Different forms of additives (granules and pellets) to produce PMAM.

The additives used significantly affect the process of dosing the individual components into the mixer. Each of these types requires a different manufacturing process. In general, while granules (pure PE) should be dosed in parallel with or just after asphalt to avoid deterioration on contact with hot aggregate, the pellets are dosed on hot aggregate and only after a few seconds (when to mechanically break it) asphalt is dosed. Caution should be taken in the case of multi-component additives; it must be considered that the mixing of the additive with asphalt must also be extended!

3. Experimental part

To assess the functional properties of asphalt mixtures and to verify the preparation and suitability for use of the polymer-modified asphalt mixture, three model asphalt mixtures were prepared:

1. Asphalt mixture with binder type CA 50/70 (without modification).
2. Asphalt mixture using a commercially available polymer modified binder (PMB 45/80-75).
3. Polymer-modified-asphalt mixture with an additive S CEL-7G®K60 (Improcel K) from CIUR Ltd., Brandýs nad Labem. It is an amide wax-based additive with a polymer content (SBS) of 60 %. We used a 5 % modifier by weight of asphalt to prepare the asphalt mixture. The other components were the same in each mixture due to the comparison.

All mixtures were prepared according to the relevant catalogue sheets and technological standards valid in Slovak republic. The first and the second model mixtures were prepared in the laboratory, and they were used as a blank sample for comparison to the third mixture which was prepared in the asphalt mixture plant and layered to the real construction site. There were no problems with the blank sample mixtures prepared at the laboratory conditions.

Production of asphalt mixture in the asphalt mixing plant:

The asphalt mixture was produced in Pittel + Brauswetter, GmbH, which has a modern production of AMMANN Global 120 asphalt mixture equipment with an hourly capacity of 120 t (Figure 2). The additive was added by hand directly into the mixer in the form of pellets, using cellulose fibers as a carrier.

In the production of PMAM with additive S CEL-7G®K60, the instructions of the additive manufacturer were followed:

- the additive was dosed to hot aggregates,
- mixing of aggregate with additive was extended by 10 s, due to the achieving of perfectly mixed matter
- asphalt was added to the mixture of aggregate and additive,
- mixing of the mixture with the additive and asphalt was extended by 5 s (the time was chosen according to the additive´s manufacturer instructions).
- the mixture was placed in a hot mixture tank (mixture temperature 155 °C).

A total of 96 t of asphalt mixture was produced after visual and satisfactory organoleptic inspection, the mixture was taken to the construction site.

Figure 2. Asphalt mixing plant of the company Pittel + Brausewetter GmbH in Bratislava.

PMAM processing on site:

The mixture was transported to the construction site: "Karloveská street" in Bratislava, which is being implemented as part of the project "Modernization of the Dúbravsko-Karloveská tram radial - Bratislava". The delivery distance was about 18 km. PMAM was laid on a base asphalt layer, 70 mm thick and 3.5 m wide. Spreading with a finisher and roller compaction was without major problems (Figure 3). No special procedures were required for spreading and compaction during the layering. The laying temperatures and mixture compaction conditions were chosen as in case of commonly used PMB´s.

Figure 3. Processing of prepared PMAM on the construction site: Karloveská street, Bratislava.

4. Results and discussion

The main part of this work was the verification of PMAM technology and comparison of the commonly measured parameters of the asphalt mixtures. For the comparison two model mixtures were used:
asphalt mixture which was produced at laboratory condition using basic bitumen 50/70 without additive.
asphalt mixture which was produce at laboratory condition tasing common PMB (PMB 45/80-75). It should be noted that the input materials, as mineral aggregates, and stone flour were the same in all cases The achieved results of individual mixtures are given in Table 1 together with the requirements of KLAZ for asphalt concrete for the loading layer, quality class II with max. grain 22 mm.

The table also shows the results of tests of resistance of asphalt mixtures to permanent deformations together with the KLAZ criterion for the same mixture of quality class I.

Table 1. Achieved results for the various asphalt mixtures.

Particle size distribution				
Sieve size	*Asphalt mixture with AC 50/70*	*PMAM*	*Asphalt mixture with PMB*	*KLAZ [7]*
31.5	100	100	100	100
22.4	100	99	98	90-100
11.2	73	68	67	55-80
2	35	35	35	18-40
0.5	15	15	14	7-25
0.063	8.7	8.2	8.4	3-10
Properties of the mixtures				
Soluble binder content [%]	4.1	4.1	4.2	-
Air void content [%]	5.8	5.7	5.5	4.0-7.0
Voids filled with bitumen (VFB) [%]	62	63	64	
Indirect tensile strength ratio [%]	83	90	89	min. 70
Proportional rut depth in air [%]	3.2	2.4	2.3	max. 3.0[1)]
Wheel tracking slope in air	0.034	0.03	0.029	max. 0.07[1)]

[1)]valid for asphalt concrete for the load-bearing layer of quality class of asphalt mixture I with a maximum grain size of 22 mm.

It is clear evidence from the achieved results that the values of the observed properties prescribed for the given class of asphalt mixtures in KLAZ were achieved for all mixtures. The values obtained from the results of tests of resistance to permanent deformations can also be considered very good for all mixtures. The results of these tests confirmed the benefits of using polymers to produce asphalt mixtures. Compared to the CA 50/70 mixture, an improvement of almost 30 % was achieved. The results obtained with PMAM and with the mixture with PMB are practically the same. The relatively high content of SBS in PMAM also contributed to this to some extent, but in general we can state that the possibility of producing PMAM with properties comparable to mixtures produced with PMB has been practically confirmed.

5. Final evaluation

Based on our knowledge gained in the production and processing of PMAM as well as based on comparison of achieved values of monitored properties of PMAM, mixture with CA and mixture with PMB we can state that PMAM meets all criteria for asphalt mixtures. In addition, it has been shown in the verification section that the production, transport, spreading and compaction of such a mixture is not a problem and does not require special procedures. The only criterion regarding the potential risk of inhomogeneity of the mixture, we strongly recommend the discussion the conditions of the use on the construction sites with the investor for each construction separately.

6. Conclusion

Road pavement start experiencing functional degradation with heavy traffic or freezing of groundwater during the cold season. These undesirably features over the service life of road pavement can improve the detection of the following problems, as fatigue or cracks, potholing stripping and rutting. Modification of the road pavement's structure using additives, as polymers or waxes have the great potential for increasing the service life of the pavement. Moreover, there is a possibility to apply a waste plastic to the asphalt pavement structures. According to the preliminary results reported in this paper an asphalt modifier can be successfully added to the asphalt mixture to improve its properties. From the environmental point of view, in a future the method should be useful for re-use of plastic waste as an additive to the asphalt pavement and thus help to partially preserve the natural fossil resources.

7. References

[1] *Asphalt in figures* 2017, EAPA 2018
[2] Kovaľ P 2019 Activities of the Slovak Association of Asphalt Producers and an overview of the production of asphalt mixtures in Slovakia, *In: Výstavba a rehabilitácia asfaltových vozoviek, 18th Conference of SAAV*, Podbanské
[3] Kakar M, Refaa Z, Worlitschek J, Stamatiou A, Partl M, Bueno M 2019 Effects of aging of asphalt binders modified with microencapsulated phase change material. *Composites Part B: Engineering*, **173**, 107007
[4] KLK 1/2012 *Aggregate Catalogue List*, MDV SR
[5] Porto M, Caputo P, Loise V, Eskandarsefat S, Teltayev B, Oliviero C 2019 Bitumen and bitumen modification: A review on latest advances. *Applied Sciences*, **9**, 742
[6] KLAZ 1/2019 *Asphalt Mixtures Catalogue List*, MDV SR
[7] KLA 1/2019 *Asphalt Catalogue List*, MDV SR
[8] Fakirov S 2017 *Fundamentals of polymer science for engineers*. Wiley-VCH Verlag GmbH&Co. KGaA, ISBN: 9783527802180
[9] Polacco G Filippi S Merusi and Stastna G 2015 A review of the fundamentals of polymer-modified asphalts: Asphalt/polymer interactions and principles of compatibility. *Advances in Colloid and Interface Science*, **224**, 72 – 112

Acknowledgments

This work was supported by the Slovak Research and Development Agency under the contract No. APVV-19-0520 and Grant Agency VEGA under the contract No. 1/0646/19.

Studying the potential microbial risk of construction cementitious materials in the urban environment

M Hološová[1], A Eštoková[1] and A Luptáková[2]

[1]Departement of Material Engineering, Faculty of Civil Engineering, Technical University of Košice, Vysokoškolská 4, 042 00 Košice, Slovakia
[2]Institute of Geotechnics, Slovak Academic of Science, Watsonova 45, Košice, Slovakia

E-mail: miriama.holosova@tuke.sk; adriana.estokova@tuke.sk; luptakal@saske.sk

Abstract. The presented work aimed at analysis of elements in the soil and cementitious materials from the selected locations to study the presence of various microorganisms on the surface of buildings. The methodology of the sampling was focused on the interface of the construction containing cement and the soil around the construction in foundation. The chemical analysis by performing X-ray fluorescence analysis confirmed leaching of the element cations of the clinkers. During the carrying out the investigation of the soil, the results additionally showed exceeded limited values of some of the metals elements in the soil which is considered as an environment burden. Moreover, in this research some of the cultivated microorganisms swabbed from the surfaces of the buildings from 14 selected locations are mentioned. These outputs gradually generate the image of the environmental conditions for cementitious materials of selected location in Košice.

1. Introduction

In several studies, microorganisms in term of both positive and negative influence on cementitious materials are mentioned (*Bacillus sphaericus, B. megaterium, B. cereus, Penicillium chrysogenum, Neurospora crassa, Acidithiobacillus thiooxidans*, etc.) [1], [2], [3], [4], [5]. From the perspective of the durability and life service of materials, it is necessary to inspect the presence of the predominant microorganisms and their impact in investigated area.

In general, the algae influence the concrete containing the Portland cement under the action of microbiologically induced corrosion and its consequent deterioration. The impact of the algae on cementitious materials is caused by several reasons. For example, algae are drying out the concrete due to absorption substance necessary for their life, resulting in the cracks and voids. Another reason is biofilm formation on the surfaces of the materials which can attract more organisms influencing the concrete and can locally generate high concentrations of metabolites chemically aggressive to concrete. [6]. Metabolites as a product of metabolism and have the most significant fungi influence on materials.

During fungi metabolic process, the production of organic acids can lead to the destroying the mechanical integrity of the concrete. On the other hand, the production of the alkaline conditions can protect the concrete [7]. The algae are not the only microorganism which can degrade the cement materials. Fungi's hyphae can absorb nutrients from the environment. The other side of the issue reports the information about so-called bioconcrete using biocement as an environmentally friendly alternative applying mainly *Bacillus* genus or filamentous fungi. The main idea of the healing concrete by microorganisms is to reduce impact of cement production to the environment. The point is in conversion of calcium salts, the CO_2 is consumed in the $CaCO_3$ mineralisation, and these minerals would improve the mechanical properties and durability of the cementitious materials [5].

This study presents an overview of some bacteria, fungi and algae occurring on the surfaces of the constructions. In addition, it analyses the presence of chemical elements in the soil and cementitious materials from the selected locations in order to understand the presence of various microorganisms on the surface of buildings.

2. Methods

2.1. Sampling

For the first step, the various method was chosen to identify the presence of predominant microorganisms on the constructions surfaces in selected location thought the Košice (Figure 1). Three types of the swabbing methods were used – wet (1), moist (2) and dry (3). Wet sampling started with dry swab wiping the surface and subsequently pouring the swab to the sterile water. The moist method was performing by using dry swab immersed to the sterile water and providing the swabbing. Third dry method was performed by sterile scalpel scraping the surface of construction to the lab-tube. Each sample was cultivated on appropriate basal media and provided different results.

The microorganisms were swabbed from the surfaces of the buildings both panel houses coated by plaster and concrete constructions. Fourteen objects were considered under biocorrosion screening study including panel flat house, stairs of a panel flat building, garage, fence, historical brick building, and upper bridge. According to the cultivation methods it was possible to determine fungi, bacteria, and algae. So far, the fungi were identified.

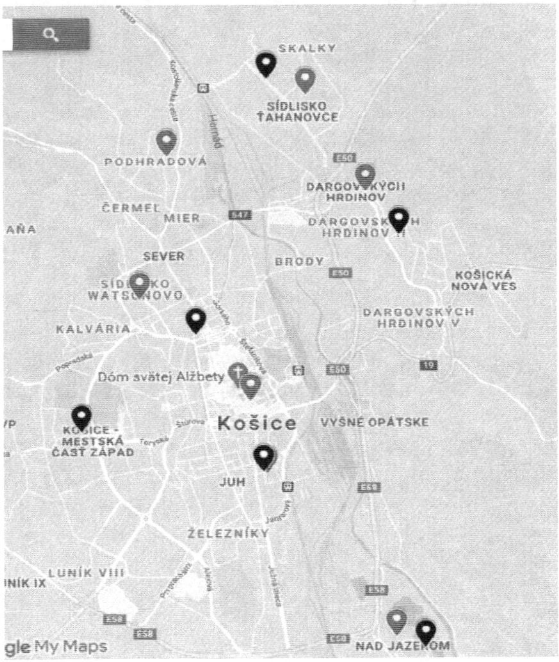

Figure 1. Location of the sampling points of the microorganisms in Košice city map: Adlerova, Maršala Koneva - Dargovských hrdinov; Americká trieda, Berlínska – Ťahaovce; Cesta pod Hradom – Podhradová; Boženy Nmecovej – Sever; Oštepová, Košťova – Juh; Ružínska – Západ; Strojárenská, Zvonárska – Staré mesto; Jenisejská, Amurská – Nad Jazerom.

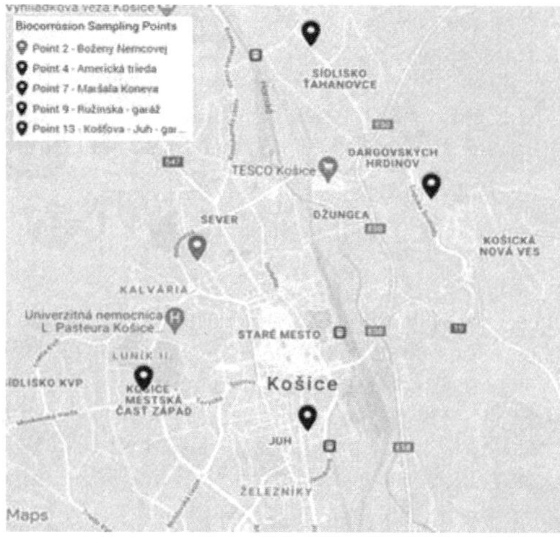

Figure 2. Sampling points of the soil in Košice city map: Maršala Koneva – Dargovských hrdinov; Americká trieda – Ťahaovce; Boženy Nmecovej – Sever; Košťova – Juh; Ružínska – Západ.

It is necessary to know and identify conditions of the environment where the material will be exposed when we want to fight against biocorrosion by using suitable materials. Therefore, the soil material, which was in interaction with the construction studied was investigated as well. The solid materials samples were taken from the interface of the construction and soil and in a proximity of the construction from the depth of 10 cm as well. The soil was taken from locations shown in Figure 2 by hoe toll with fork and spoon to the closable plastic bags and stored for further chemical analysis. Chemical analyses were performed to detect conditions for concrete foundations in the soil.

2.2. Biological analysis

Basal media of Sabouraud Dextrose Agar (M063 – SDA, HiMedia Laboratories, India) for cultivating fungi were prepared by using 16 g of the powder dissolved in 400 ml sterile water in Erlenmeyer flask. The followed process was running according to producent's instructions. Cooled Agar on Petri dishes were inoculated by swabs with sampled microorganisms spread on the surface on agar and letting to flourish.

2.3. Chemical analysis

Prior to the chemical analysis, the soil samples were poured on the glass plates and left to dry on the air. After 5 days, the dried samples were homogenised and crushed in the grinding chamber and sift through the test laboratory sieve with nominal aperture size of 0.063 mm. After sifting the powdered soil samples (with size $\leq$ 0.063 mm) were stored in the closable plastic bags and consequently analysed for the chemical elements by performing the X-ray fluorescence (XRF) analysis. Ametec XRF analyser with silicon detector, working at two currents and voltages (0.24 V and 1 V) was applied. The percentage concentrations of the elements in the samples were detected according to the calibration curve for powder materials.

3. Results

3.1. Fungi identification

According to visual inspection, the predominant fungi occurred on the Petri dishes were observed as illustrated in the Figures 3 - 6. The predominant fungi were identified according to [8-12]. Fungi are divided by division and genus in the Table 1.

IOP Conf. Series: Materials Science and Engineering 1252 (2022) 012047 doi:10.1088/1757-899X/1252/1/012047

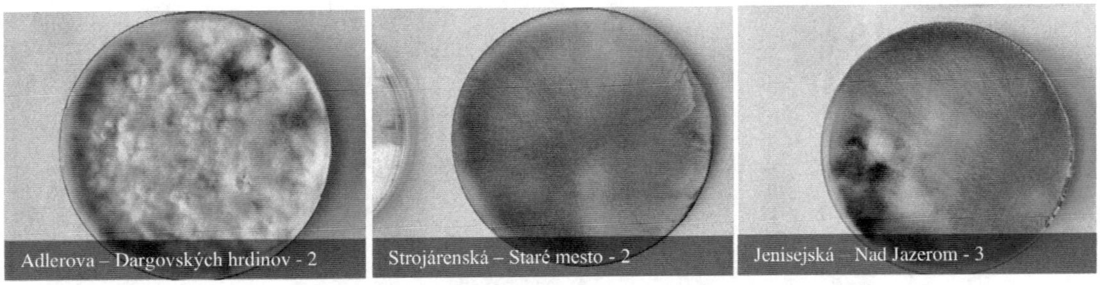

Figure 3. Examples of the present Zygomycota on the surfaces of the selected constructions.

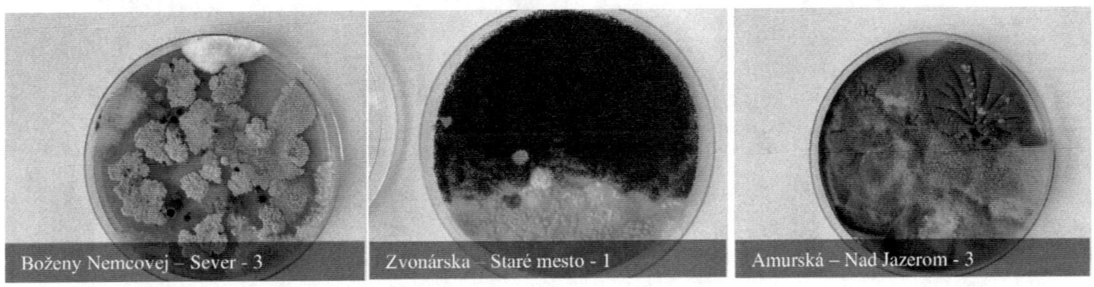

Figure 4. Examples of the present Ascomycetes on the surfaces of the selected constructions.

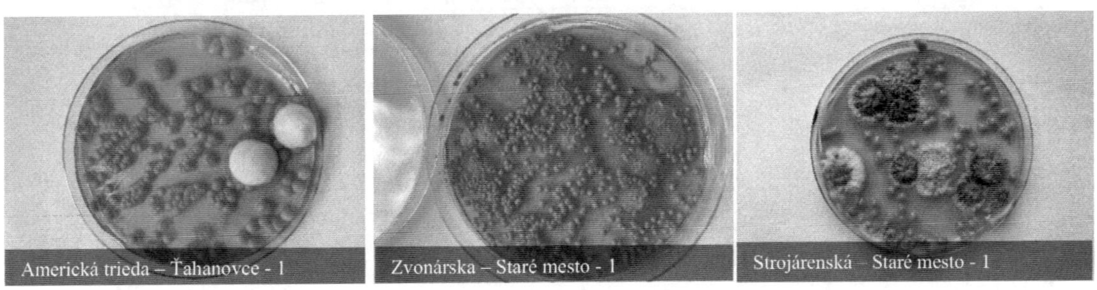

Figure 5. Examples of the present Basidiomycota on the surfaces of the selected constructions.

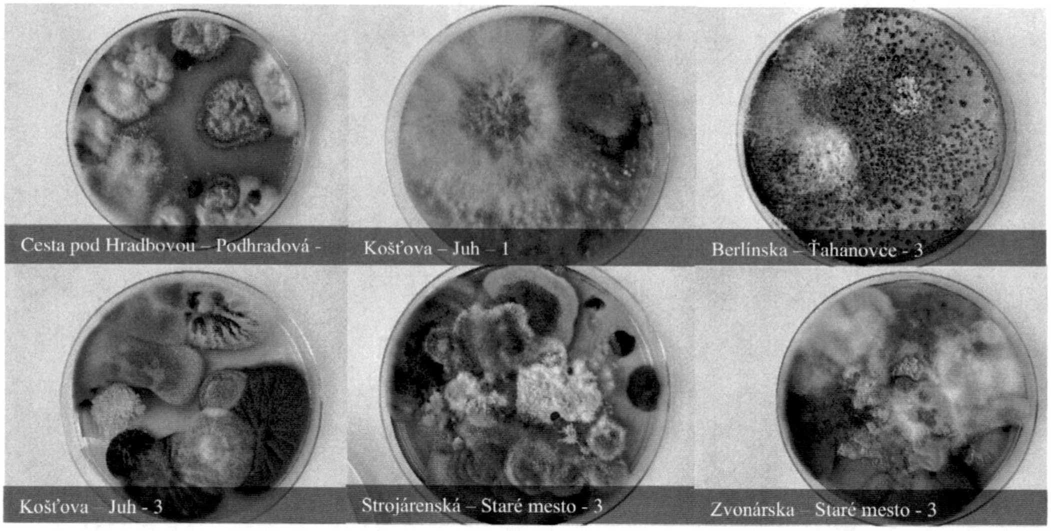

Figure 6. Examples of other present microscopic fungi on the surfaces of the selected constructions.

Table 1. Predominant fungi on surfaces of cementitious materials in selected locations in Košice.

	Division		
	Ascomycetes	**Basidiomycota**	**Zygomycota**
Genus	Acremonium	Rhodotorula	Mucor
	Aspergillus	Pleurotus	Rhizopus
	Botrytis		
	Cladosporium		
	Eutypella		
	Fusarium		
	Paecilomyces		
	Penicillium		
	Sporothrix		
	Trichoderma		

3.2. Chemical analysis of soil

Chemical analysis of the powdered soil sampled from selected location was performed by X-ray fluorescence analysis which results for individual elements are presented in Table 2. Table provides information of some elements concentrations of the sampled soil from the proximity (P) and interface (I) of the construction. The last column shows indication criteria for metal elements [13] which are used as a limited value compared with measured values of XRF analysis displayed in Figure 9.

Table 2. Elemental XRF analysis of soil samples.

	Ťahanovce - Americká trieda		Sever - Boženy Nemcovej		Dargovských hrdinov - Maršala Koneva		Juh - Košťova		Západ - Ružínska		Indication criteria mg/kg of dry matter [13]
	1P	1I	2P	2I	3P	3I	4P	4I	5P	5I	
Z	Concentration [%]										

11	Na	0.535	0.516	0.719	0.722	0.552	0.571	0.524	0.569	0.625	0.564	
12	Mg	0.879	0.841	1.422	1.342	0.904	0.975	5.731	5.310	1.188	1.328	
13	Al	6.311	6.274	5.740	5.253	6.101	6.005	5.094	5.377	4.525	3.862	
14	Si	28.46	28.05	27.01	25.890	21.770	22.450	16.740	18.150	21.390	22.360	
15	P	0.078	0.080	0.127	0.117	0.105	0.099	0.105	0.113	0.143	0.130	
16	S	0.166	0.106	0.081	0.083	0.091	0.093	0.109	0.103	0.244	0.186	*0.1-1%
17	Cl	0.018	0.010	0.020	0.021	0.006	0.010	0.011	0.011	0.018	0.020	
19	K	1.607	1.610	1.490	1.375	1.397	1.358	1.138	1.235	1.115	0.967	
20	Ca	1.382	1.679	3.338	3.766	5.876	6.549	2.978	2.757	4.824	4.943	
22	Ti	0.599	0.608	0.530	0.471	0.428	0.414	0.304	0.312	0.501	0.358	
23	V	0.011	0.009	0.009	0.008	0.010	0.010	0.007	0.008	0.011	0.009	0.034
24	Cr	0.015	0.007	0.008	0.016	0.020	0.040	0.020	0.021	0.035	0.031	0.045
25	Mn	0.109	0.104	0.089	0.084	0.064	0.059	0.147	0.142	0.095	0.091	
26	Fe	3.209	3.104	3.300	3.120	3.035	2.894	3.724	3.699	3.039	2.458	
27	Co	0.002	0.004	0.003	0.003	0.003	0.004	0.002	0.003	0.002	0.002	0.018
28	Ni	0.008	0.008	0.009	0.009	0.009	0.009	0.009	0.009	0.009	0.008	0.018
29	Cu	0.004	0.003	0.005	0.005	0.004	0.004	0.004	0.004	0.005	0.003	0.05
30	Zn	0.019	0.019	0.091	0.078	0.016	0.015	0.054	0.048	0.068	0.053	0.15
33	As	0.002	0.002	0.002	0.001	0.001	0.001	0.001	0.001	0.001	0.001	0.0065
42	Mo	0.000	0.000	0.000	0.000	0.000	0.000	0.000	0.000	0.000	0.000	0.005
48	Cd	0.005	0.005	0.005	0.005	0.005	0.005	0.005	0.005	0.005	0.005	0.001
50	Sn	0.006	0.006	0.007	0.007	0.007	0.007	0.007	0.007	0.007	0.006	0.02
51	Sb	0.010	0.010	0.010	0.010	0.009	0.010	0.010	0.010	0.010	0.008	0.0025
56	Ba	0.086	0.081	0.100	0.090	0.060	0.058	0.056	0.058	0.083	0.056	0.09
80	Hg	0.013	0.006	0.001	0.001	0.001	0.001	0.000	0.000	0.000	0.000	0.00025
82	Pb	0.003	0.005	0.010	0.010	0.004	0.004	0.009	0.009	0.011	0.008	0.025

*Value range according to [14]

Figures 7 and 8 provide information about concentration of the sulphur and phosphorus as a protentional nutrients for microorganisms. The highest concentration of S a P were found in the fifth sampling point. The lowest value for P is first sampling point and for S is the second one. The concentration of sulphur is in a proper range from 0.01 – 1% (Figure 7) [14].

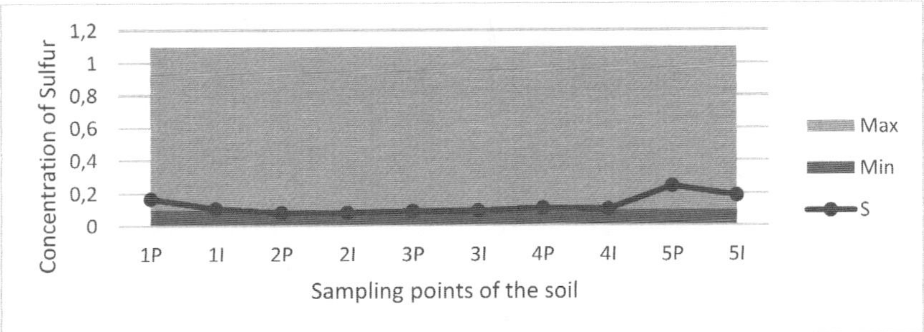

Figure 7. Measured concentration of sulphur in the examined soil in the appropriate range.

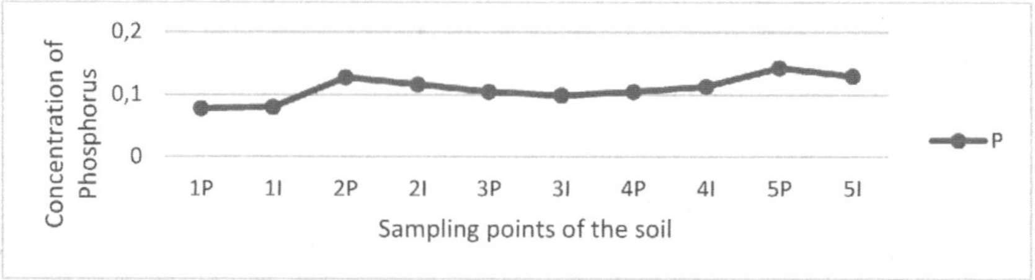

Figure 8. Measured concentration of phosphorus in the examined soil.

In the terms of environment burden, some limited values were exceeded. The guideline [11] provides indication criteria for metal elements in the soil as shown in Table 2. Figure 9 compares criteria with measured concentration of the elements in examined soil. The exceeded values are concerned of barium, cadmium, mercury, and antimony. The Figure 9 represents the limit of the exceedance of mentioned metal elements.

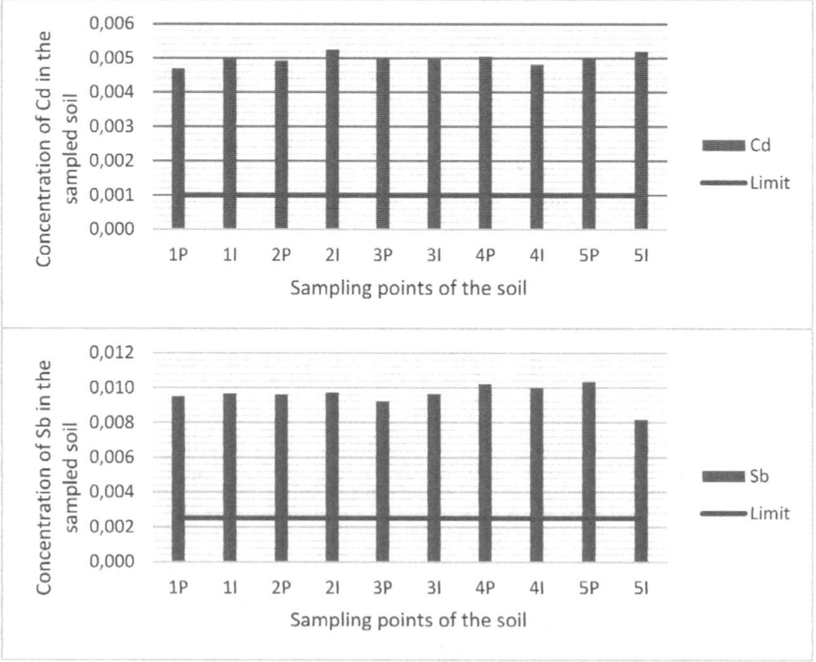

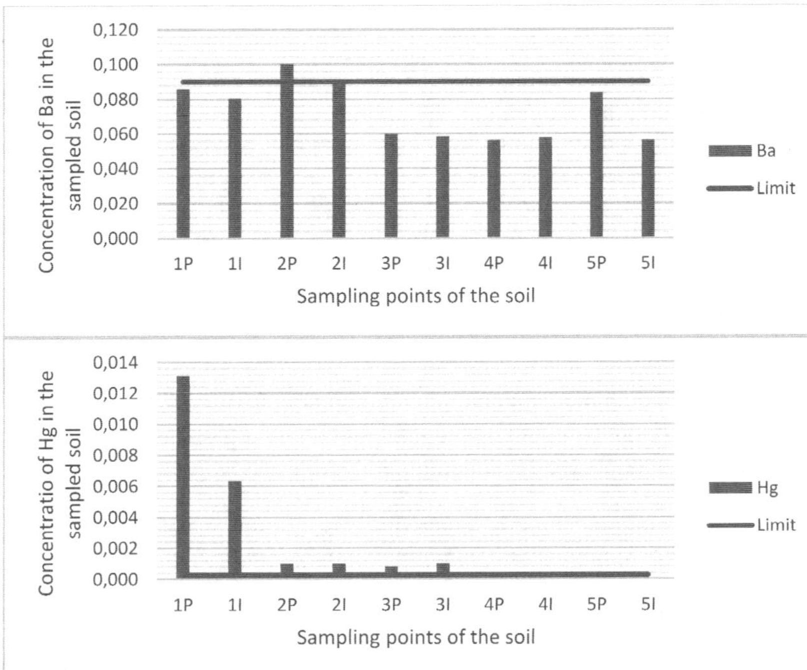

Figure 9. Graphs represent the exceeded limited values of barium, cadmium, mercury, and antimony.

The interesting point of view from the perspective of the leaching the cations of the cement material is shown in the Figure 10 as the results of the XRF analysis. As it can be seen, higher concentrations of calcium, aluminium and silicon were observed in several samples from the interface compared to the samples from the proximity of the constructions. This fact could be caused by an aggressive influence of soil to the cementitious material and subsequent dissolving and leaching of the main cement matrix components, such as portlandite and CSH gel.

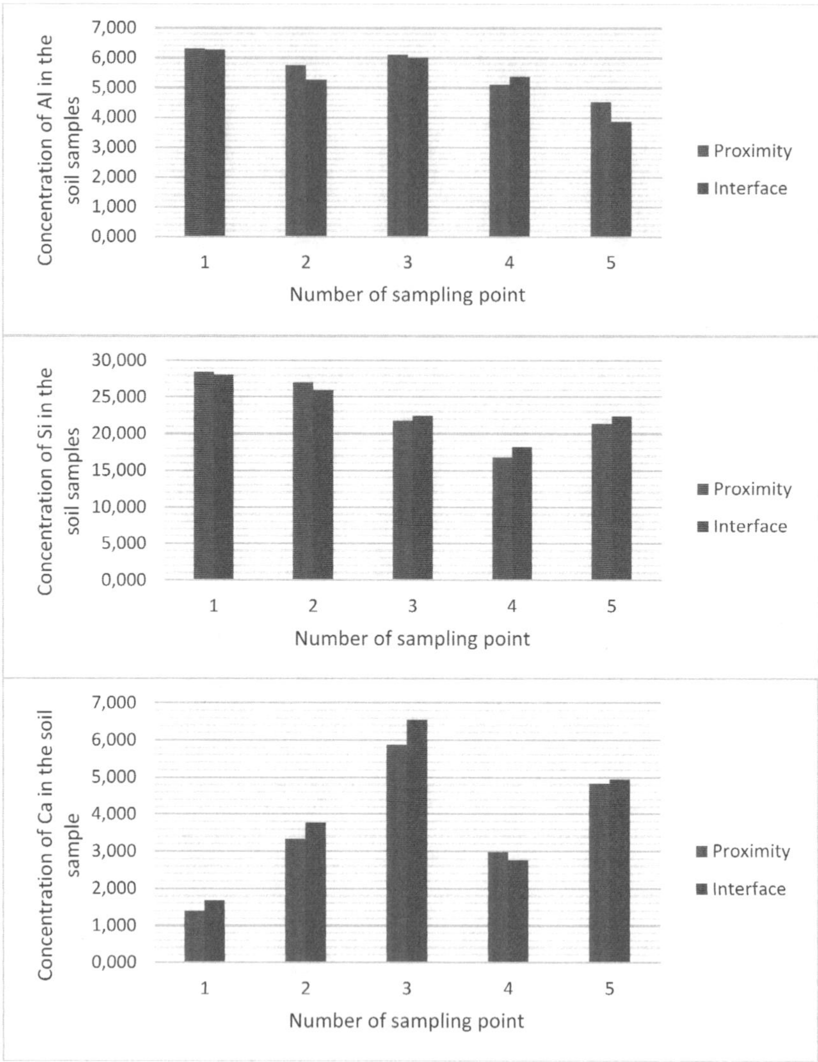

Figure 10. Comparison of the concentrations of selected ions in soil in the interface (I) and in the soil near to the construction (N).

4. Conclusion

Presented work was supposed to contribute for gradually creating an image of the conditions of the construction cementitious materials in Košice in connection to the microbial risk. To these days the present of predominant fungi were identify from all examined microorganisms (bacteria and algae). The pictures of the fungi on the petri dishes provide the visual image of the predominant fungi presented on the surfaces of the selected constructions. Predominant divisions are Ascomycetes. Basidiomycota. Zygomycota.

The other feature for completing the image was examination of soil samples. These samples gave an information of the environmental burden as a secondary effect of the examined parameters. Results of X-ray fluorescence analysis showed exceeded values of elements such as barium, cadmium, mercury, and antimony in comparisons with indication criteria of the soil.

Last investigated feature was comparison of the amount of the cement material cations in the soil on the interface of the construction and in the soils near to the construction. Results pointed to leaching of the calcium cations for 4 in 5 samples, silicon cations in 3/5, and aluminium cations 1/5 samples. Results of the proper concentration of sulphur in soils claimed that the soils environment were not significantly aggressive in terms of sulphate attack.

5. References

[1] Rodin A, Karpushin S and Smirnov V 2020 Cement Composites' Biostability (Trans Tech Publications Ltd) In *Materials Science Forum* Vol **1011** pp 171-178

[2] Stanaszek-Tomal E 2020 Influence of pore structure on humidity parameters of cement-polymer mortars contaminated with filamentous fungi *PloS one* **15**(4) e0231347

[3] Verma R, Kumar M and Saxena S 2013 *Interaction of Fungi with Cement Material And* Properties *There-Of* (Doctoral dissertation)

[4] Irwan J M and Teddy T 2017 An overview of bacterial concrete on concrete durability in aggressive environmen Pertanika J Sci Technol **25** p 259-264

[5] Martuscelli C, Soares C, Camões A and Lima N 2020 Potential of fungi for concrete repair Procedia Manufacturing **46** 180-185

[6] Javaherdashti Reza et al 2009 On the impact of algae on accelerating the biodeterioration/biocorrosion of reinforced concrete: a mechanistic review European Journal of Scientific Research **36** 3 p 394-406

[7] Bertron A 2014 Understanding interactions between cementitious materials and microorganisms: a key to sustainable and safe concrete structures in various contexts *Materials and Structures* **47**(11) p 1787-1806

[8] Hofbauer W K and Gärtner G 2021 Microbial life on Façades (Berlin: Springer-Verlag GmbH)

[9] Læssøe T and Petersen J H 2019 Fungi of Temperate Europe vol **1** (New Jersey: Princeton University Press)

[10] Koul M, Kumar A, Deshidi R et al 2017 Cladosporol A triggers apoptosis sensitivity by ROS-mediated autophagic flux in human breast cancer cells BMC cell biology **18**(1) p 1-15

[11] Nosalj S, Šimonovičová A, Pauditšová E, Hanajík P, Vojtková H and Benková M 2021 Diversity of soil microscopic filamentous fungi in Dystric Cambisol at the Banská Štiavnica – Šobov (Slovakia) locality after application of remediation measures (Biologia) DOI: 10.1007/s11756-021-00774-1

[12] Šimonovičová A, Vojtková H, Nosalj S, Piecková E, Švehláková H, Kraková L, Drahovská H, Stalmachová B, Kučová K and Pangallo D 2021 Aspergillus niger environmental isolates and their specific diversity through metabolite profiling (Front Microbiol 12) 658010; DOI:10.3389/fmicb.2021.658010

[13] Directive of the Ministry of the Environment of the Slovak Republic of 28 January 2015 no 1/2015 - 7 to prepare a risk analysis of the contaminated area

[14] Schnug E 1991 Sulphur nutritional status of European crops and consequences for agriculture Sulphur in agriculture **15** p 7-12

Acknowledgments
This work has been supported by VEGA 2/0142/19.

The influence of inclusive deformation mapping on the kinematic equations of nonlinear elasticity

R Flajs

Faculty of Civil and Geodetic Engineering, University of Ljubljana, Jamova 2 Ljubljana, Slovenia

E-mail: rflajs@fgg.uni-lj.si

Abstract. A model of continuum is often used in statics and dynamics modeling with material and spatial description. In this paper, we describe body deformation using an innovative approach employing embedding. The mapping ϕ maps a domain from the space R^n to the image of space R^m, where the dimensions of the spaces m and n are necessarily different, usually $n < m$. Due to the different dimensions m and n, the determinant of the mapping is not defined, nor is the mapping reversible. This complicates the calculation of the fundamental kinematic quantities, the determinant J of the displacement gradient F, the transposed inverse of F, which occur in the Nanson's equation for deformed surface. In this paper, we derive terms for these quantities. The usage of the Moore-Penrose generalized inverse is essential here. This approach is a generalization of the established approach and thus opens up new possibilities in describing deformation. Displacements often appear as one of the basic unknowns, especially when using the finite element method. In the teaching of nonlinear mechanics, we found that the displacements are not adequately defined. In the paper, we therefore attempt to rectify this shortcoming.

1. Introduction

This paper describes body deformation using insertions. We did not find the mentioned approach in the literature. Mapping $\phi : (\overrightarrow{r_0}, t) \mapsto \vec{r}$ thus maps a domain from the space R^n to an image in space R^m, where the dimensions of spaces m and n are necessarily different, usually $n < m$. At the selected time t and at the selected particle, determined by the material coordinates x_0, y_0 and z_0, the deformation gradient F is presented by a linear mapping from R^n to R^m. Due to the different dimensions m and n, the determinant of the mapping is not defined, nor is the mapping reversible. This complicates the calculation of the basic kinematic quantities such as $J = \det(F)$ and the transposed inverse F^{-T}, which appears in Nanson's equation for change of the deformed surface [1]. In this paper, we derive the terms for these quantities. In our opinion, the displacements [2, p. 96] are not well defined. Otherwise, the authors [2] themselves point out this fact. We therefore attempt to rectify this shortcoming.

At the beginning, we introduce the basic quantities needed to describe body deformations. We then use at a few examples of body deformations to summarize practical conclusions, later going on to generalize them. The Moore-Penrose generalized inverse [3,4,5] proves to be very useful in describing the problem.

1.1. Labels

In general, we will present the deformation of the body by embedding $\phi : R^n \rightarrow R^m$ using a material and spatial description. The particle's material coordinates (from the domain in space R^n) will be equipped with an index of 0, i.e. with (x_0, y_0, z_0), at its spatial coordinates (from the image of the mapping ϕ in space R^m) however, this index will be abandoned and will denote them rather using (x, y, z). Similarly, some other material quantities will be provided with an index of 0. One can express vector $\vec{a}$ with material base vectors

$$\vec{a} = A_x\,\overrightarrow{E_x} + A_y\,\overrightarrow{E_y} + A_z\,\overrightarrow{E_z} = A_x\,\vec{I} + A_y\,\vec{J} + A_z\,\vec{K} \tag{1}$$

or with spatial base vectors

$$\vec{a} = a_x\,\overrightarrow{e_x} + a_y\,\overrightarrow{e_y} + a_z\,\overrightarrow{e_z} = a_x\,\vec{\imath} + a_y\,\vec{\jmath} + a_z\,\vec{k}. \tag{2}$$

Let us denote the scalar product of vectors $\vec{a}$ and $\vec{b}$ with $\vec{a} \cdot \vec{b}$. Linear mappings can be interpreted as elements of the corresponding vector spaces, wherefore they can, like vectors, be written by the sum of the baseline linear mappings of the so-called dyadic or tensor product $\vec{a} \otimes \vec{b}$. By definition the dyadic product $\vec{a} \otimes \vec{b}$ maps vector $\vec{c}$ to vector $(\vec{b} \cdot \vec{c})\,\vec{a}$.

$$\vec{a} \otimes \vec{b} : \vec{c} \mapsto (\vec{b} \cdot \vec{c})\,\vec{a} \tag{3}$$

We will need the transposition

$$(\vec{a} \otimes \vec{b})^T = \vec{b} \otimes \vec{a} \tag{4}$$

and the composition

$$(\vec{a} \otimes \vec{b}) \circ (\vec{c} \otimes \vec{d}) = (\vec{b} \cdot \vec{c})\ \vec{a} \otimes \vec{d} \tag{5}$$

of dyadic products.

2. Embeddings

We will describe the body deformations with embeddings [6], so we will first define some necessary basic terms.

Definition 2.1 Continuous mapping $F : A \subset R^n \to R^m$ is a homeomorphism on $F(A)$ if and only if F is bijective on $F(A)$ and the inverse mapping

$$F^{-1} : F(A) \subset R^m \to R^n \tag{6}$$

is continuous on $F(A)$. The sets A and $F(A)$ are homeomorphic.

Definition 2.2 Differential mapping $\varphi : S \to R^3$ of an abstract surface S in R^3 is immersion if and only if the differential (the tangent map) $\mathrm{d}\varphi_p : T_p \to T_p(R^3)$ is injective. If the space S is metric space with the metric $\mathrm{m}(\cdot,\cdot)$ and the equality $\mathrm{m}(\mathrm{d}\varphi(\vec{v}), \mathrm{d}\varphi(\vec{w})) = \mathrm{m}(\vec{v}, \vec{w})$ is valid for all pairs $\vec{v}, \vec{w}$ in $T_p(S)$ then the mapping φ is called isometrical immersion.

Definition 2.3 Let us S be an abstract surface. Differential mapping $\varphi: S \to R^m$ is embedding if and only if φ is immersion and homeomorphism onto its image.

3. Examples of body deformations

3.1. Wall rotation around vertical axis

Rotation can be described by mapping ϕ

$$\phi : \vec{r_0} = x_0\,\vec{I} + z_0\,\vec{K} \mapsto x\,\vec{\imath} + y\,\vec{\jmath} + z\,\vec{k} = x_0 \cos\alpha\,\vec{\imath} + x_0 \sin\alpha\,\vec{\jmath} + z_0\,\vec{k} = \vec{r}, \alpha = c\,t. \tag{7}$$

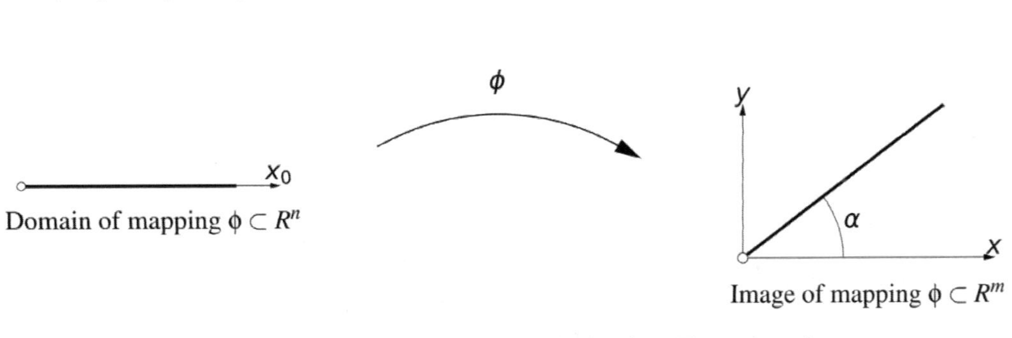

Figure 1. Wall rotation around vertical axis – Floor plan view.

The mapping ϕ, shown in figure 1, is at fixed time t an isometric embedding. The deformation gradient F has the form

$$F = \frac{\partial\phi(\vec{r_0},t)}{\partial\vec{r_0}} = \cos\alpha\ \vec{\imath}\otimes\vec{I} + \sin\alpha\ \vec{\jmath}\otimes\vec{I} + \vec{k}\otimes\vec{K}. \tag{8}$$

One can guess the inverse mapping

$$\phi^{-1}: \vec{r} \mapsto \vec{r_0} = \sqrt{x^2+y^2}\ \vec{I} + z\,\vec{K} \tag{9}$$

and the inverse of deformation gradient

$$\begin{aligned} F^{-1} &= \frac{\partial\phi^{-1}(\vec{r},t)}{\partial\vec{r}} = \frac{x}{x_0}\vec{I}\otimes\vec{\imath} + \frac{y}{y_0}\vec{I}\otimes\vec{\jmath} + \vec{K}\otimes\vec{k} \\ &= \cos\alpha\,\vec{I}\otimes\vec{\imath} + \sin\alpha\,\vec{I}\otimes\vec{\jmath} + \vec{K}\otimes\vec{k}. \end{aligned} \tag{10}$$

3.1.1. Moore Penrose generalized inverse F^+ *of deformation gradient* F. In the chosen coordinate system the deformation gradient F is presented by a rectangular matrix therefore the inverse F^{-1} does not exist so let us try to use a generalized inverse. Under the name generalized inverse, we denote the well-established Moore–Penrose generalized inverse, defined below.

Definition 3.1 If the linear mapping $A \in R^{m\times n}, m \geq n$ has a full rank, the generalized inverse can be defined by equation

$$A^+ = (A^T A)^{-1} A^T \tag{11}$$

otherwise $(m < n)$ by equation

$$A^+ = A^T (A A^T)^{-1}. \tag{12}$$

Definition 3.2 Moore–Penrose inverse A^+ can also be determined by Moore–Penrose conditions:

$$A A^+ A = A,\ \ A^+ A A^+ = A^+,\ (A A^+)^T = A\ A^+,\ (A^+ A)^T = A^+ A. \tag{13}$$

Employing the definition 3.1 one can obtain $F^+ = \cos\alpha\,\vec{I}\otimes\vec{\imath} + \sin\alpha\,\vec{I}\otimes\vec{\jmath} + \vec{K}\otimes\vec{k} = F^{-1}$.
Using the deformation gradient, one can also determine the Green Cauchy tensors C and B.

$$C = F^T F = \vec{I}\otimes\vec{I} + \vec{K}\otimes\vec{K}, \tag{14}$$

$$B = F F^T = \cos^2\alpha\ \vec{\imath}\otimes\vec{\imath} + \sin^2\alpha\ \vec{\jmath}\otimes\vec{\jmath} + \sin\alpha\ \cos\alpha\ (\vec{\imath}\otimes\vec{\jmath} + \vec{\jmath}\otimes\vec{\imath}) + \vec{k}\otimes\vec{k}. \tag{15}$$

Let us determine the polar decompositions $F = R\,U$ and $F = V\,R$ of deformation gradient F. After a brief calculation one can obtain the stretch tensors U and V and the rotation tensor R in the form

$$U = \vec{I}\otimes\vec{I} + \vec{K}\otimes\vec{K} = C, \qquad V = B, \qquad R = F. \tag{16}$$

3.2. Rotating and stretching of the tiny sheet

Deformation can be determined by mapping ϕ

$$\phi\ :\ \vec{r_0} = x_0\,\vec{I} + y_0\,\vec{J}\ \mapsto\ x_0\,\vec{\imath} + y_0\,\vec{\jmath} + c\,t\,x_0^2\,\vec{k} = \vec{r}. \tag{17}$$

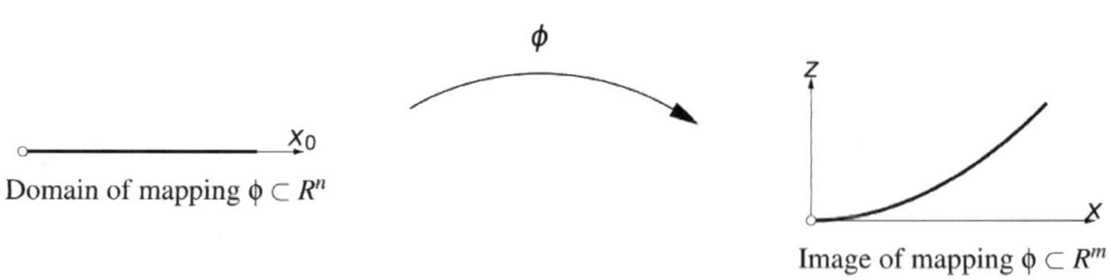

Figure 2. Rotating and stretching of the tiny sheet – The view in the direction $-\vec{\jmath}$.

The mapping ϕ, presented in figure 2, is an embedding at fixed time t. Let us determine the directional derivative

$$\mathrm{D}\phi(\vec{r_0},t)[\vec{u_0}] = \lim_{\Delta\to 0}\frac{\phi(\vec{r_0}+\Delta\vec{u_0})-\phi(\vec{r_0})}{\Delta}, \qquad \vec{u_0} = u_0\,\vec{I} + v_0\,\vec{J}. \tag{18}$$

After a brief calculation one can derive

$$\mathrm{D}\phi(\vec{r_0},t)[\vec{u_0}] = u_0\,\vec{i} + v_0\,\vec{j} + 2\,c\,t\,x_0\,u_0\,\vec{k} = \frac{\partial\phi(\vec{r_0},t)}{\partial\vec{r_0}}:\vec{u_0} = F:\vec{u_0} = F\,\vec{u_0}. \tag{19}$$

The deformation gradient F thus takes the form

$$F = \vec{i}\otimes\vec{I} + \vec{j}\otimes\vec{J} + 2\,c\,t\,x_0\;\vec{k}\otimes\vec{I}. \tag{20}$$

Using the Green Cauchy tensor

$$C = F^T\,F = (1+4\,c^2\,t^2\,x_0^2)\,\vec{I}\otimes\vec{I} + \vec{J}\otimes\vec{J} \tag{21}$$

one can immediately obtain Moore-Penrose generalized inverse

$$F^+ = (F^T\,F)^{-1}\,F^T = C^{-1}\,F^T. \tag{22}$$

4. Basic kinematic quantities

4.1. Deformation gradient

The deformation gradient is an essential mechanical quantity, so let us devote a little attention to it first. It links the differentials of position vectors between spatial and material states.

Theorem 4.1 We have the relations:

$$\mathrm{d}\vec{r} = F\,\mathrm{d}\vec{r_0} \;\Rightarrow\; \mathrm{d}\vec{r_0} = F^+\,\mathrm{d}\vec{r}, \tag{23}$$

$$\mathrm{d}\vec{r_0} = F^{-1}\,\mathrm{d}\vec{r} \;\Rightarrow\; \mathrm{d}\vec{r} = (F^{-1})^+\,\mathrm{d}\vec{r_0}\,. \tag{24}$$

Proof

$$F^+\,\mathrm{d}\vec{r} = F^+\,F\,\mathrm{d}\vec{r_0} = C^{-1}\,F^T\,F\,\mathrm{d}\vec{r_0} = C^{-1}\,C\,\mathrm{d}\vec{r_0} = \mathrm{d}\vec{r_0}, \tag{25}$$

$$F^{-T}\,\mathrm{d}\vec{r_0} = B^{-1}\,\mathrm{d}\vec{r} \Rightarrow B\,F^{-T}\,\mathrm{d}\vec{r_0} = \mathrm{d}\vec{r} \Rightarrow (F^{-1})^+\,\mathrm{d}\vec{r_0} = \mathrm{d}\vec{r}. \tag{26}$$

4.2. Displacements

The displacement vector is usually defined by an equation [2, p. 96]

$$\vec{u} = \vec{r} - \vec{r_0}. \tag{27}$$

However, this definition is misleading, since the vectors $\vec{r}$ and $\vec{r_0}$ are generally located in different spaces. In our cases, the dimensions of these spaces do not match either. The authors [2] point out the fact that the equation (27) is not a tensorial equation.

Definition 4.1 Under the assumption, that the deformation begins at initial time $t = 0$, the displacement vector $\vec{u}$ can be defined by the equation

$$\vec{u} = \phi(\vec{r_0},t) - \phi(\vec{r_0},0) = \vec{r} - \phi(\vec{r_0},0). \tag{28}$$

The difference between vector $\vec{r_0}$ and the image $\phi(\vec{r_0},0)$ of the vector $\vec{r_0}$ at initial time $t = 0$ is clearly seen. The spatial tensor of small strains should be expressed with displacements defined by equation (28).

4.3. The volume change

Let us try to obtain the specific volume change first. The following statement can be helpful in this direction.

Theorem 4.2 The mappings $A^T\,A$ and $A\,A^T$ have the same eigenvalues $\lambda_i^2 > 0$ and the following expression is valid

$$\mathrm{A} = \sum_j \lambda_j \vec{\xi_j} \otimes \vec{\eta_j}, \tag{29}$$

where vectors $\vec{\xi_j}$ and $\vec{\eta_j}$ are the eigenvectors of the mappings $A\,A^T$ and $A^T\,A$, respectively.

Proof

Let us the λ_i^2 and $\vec{\eta_i}$ be the positive eigenvalues and the orthonormal eigenvectors of the mapping $A^T A$, respectively. One can the define the vectors

$$\vec{\xi_j} := \lambda_j^{-1} A \vec{\eta_j}. \tag{30}$$

Let us check first, that the vectors $\vec{\xi_j}$ are orthonormal

$$\vec{\xi_i} \cdot \vec{\xi_j} = \lambda_i^{-1}\lambda_j^{-1}(A \vec{\eta_i}) \cdot (A \vec{\eta_j}) = \vec{\eta_i} \cdot \vec{\eta_j} = \delta_{ij}. \tag{31}$$

Multiplying the relation $A \vec{\eta_j} = \lambda_j \vec{\xi_j}$ with $A A^T$ from the left one can obtain

$$A A^T A \vec{\eta_j} = \lambda_j A A^T \vec{\xi_j} = A \lambda_j^2 \vec{\eta_j} = \lambda_j^3 \vec{\xi_j} \Rightarrow A A^T \vec{\xi_j} = \lambda_j^2 \vec{\xi_j}. \tag{32}$$

Employing the orthogonality of the base vectors $\vec{\xi_j}$ and $\vec{\eta_j}$ and relations $A \vec{\eta_j} = \lambda_j \vec{\xi_j}$ one can derive the desired relation

$$A = \sum_j \lambda_j \vec{\xi_j} \otimes \vec{\eta_j}. \tag{33}$$

4.3.1. Specific volume change. The specific change of the volume can-not be defined by the equation $J = \det(F)$, because the determinant of deformation gradient F is not defined in our case. The specific change of the volume can be defined by equation

$$J := \prod_{\lambda_i > 0} \lambda_i. \tag{34}$$

Using the decomposition $F = \sum_{\lambda_i > 0} \lambda_i \vec{m_i} \otimes \vec{m_i^0}$, where λ_i^2, $\vec{m_i^0}$ and $\vec{m_i}$ are eigenvalues of Cauchy tensors C and B, eigenvectors of tensor C and eigenvectors of tensor B, respectively, one can calculate the volume change

$$\begin{aligned} \mathrm{d}V &= (\mathrm{d}\vec{a}, \mathrm{d}\vec{b}, \mathrm{d}\vec{c}) = (F\, \mathrm{d}\vec{a_0}, F\, \mathrm{d}\vec{b_0}, F\, \mathrm{d}\vec{c_0}) \\ &= \sum_{i,j,k} \lambda_i\lambda_j\lambda_k \, (\vec{m_i}, \vec{m_j}, \vec{m_k}) \, (\mathrm{d}\vec{a_0} \cdot \vec{m_i^0}) \, (\mathrm{d}\vec{b_0} \cdot \vec{m_j^0}) \, (\mathrm{d}\vec{c_0} \cdot \vec{m_k^0}) \\ &= J\, \mathrm{d}V_0. \end{aligned} \tag{35}$$

4.4. The length change

The length change is defined by equation

$$\mathrm{d}s = \sqrt{\mathrm{d}\vec{r} \cdot \mathrm{d}\vec{r}} = |\mathrm{d}\vec{r}| = |U\, \mathrm{d}\vec{r_0}|. \tag{36}$$

4.5. The surface change

With regard to the equality

$$\mathrm{d}V = \mathrm{d}\vec{S} \cdot \mathrm{d}\vec{a} = J\, \mathrm{d}V_0 = J\, \mathrm{d}\vec{S_0} \cdot \mathrm{d}\vec{a_0} = J\, \mathrm{d}\vec{S_0} \cdot F^+ \mathrm{d}\vec{a} = J\, (F^+)^T \mathrm{d}\vec{S_0} \cdot \mathrm{d}\vec{a} \tag{37}$$

Nanson's famous equation $\mathrm{d}\vec{S} = J\, F^{-T} \mathrm{d}\vec{S_0}$ can be replaced by the generalized one

$$\mathrm{d}\vec{S} = J\, (F^+)^T \mathrm{d}\vec{S_0}. \tag{38}$$

4.6. Small strain tensor expressed by displacements

Let us calculate first the directional derivative

$$\begin{aligned} \mathrm{D}F(\phi(\vec{r_0}, t))[\vec{u}] &= \frac{\mathrm{d}}{\mathrm{d}\Delta}\Big|_{\Delta=0} F(\phi(\vec{r_0}, t) + \Delta\, \vec{u}) = \frac{\mathrm{d}}{\mathrm{d}\Delta}\Big|_{\Delta=0} \frac{\partial(\phi(\vec{r_0}, t) + \Delta\, \vec{u})}{\partial \vec{r_0}} \\ &= \frac{\mathrm{d}}{\mathrm{d}\Delta}\Big|_{\Delta=0} \left(\frac{\partial \phi(\vec{r_0}, t)}{\partial \vec{r_0}} + \Delta \frac{\partial \vec{u}}{\partial \vec{r_0}} \right) = \frac{\partial \vec{u}}{\partial \vec{r_0}} = \frac{\partial \vec{u}}{\partial \vec{r}} F = (\nabla \vec{u})\, F. \end{aligned} \tag{39}$$

If the vector $\vec{u}$ in the equation (39) is to represent the displacement vector, then it must be defined according to equation (28), otherwise the quantities in the equation (39) will not be well defined. Expression $\nabla \vec{u}$ denotes spatial displacement gradient. Let us determine the directional derivative of the Green Lagrange tensor of the large strains E [1, p. 120] .

$$\begin{aligned} \mathrm{D}E[\vec{u}] &= \frac{1}{2}(F^T\, \mathrm{D}F[\vec{u}] + \mathrm{D}F^T[\vec{u}]\, F) \\ &= \frac{1}{2}(F^T\, \nabla\vec{u}\, F + F^T\, (\nabla\vec{u})^T\, F) \\ &= \frac{1}{2}F^T(\nabla\vec{u} + (\nabla\vec{u})^T) \; = F^T\, \varepsilon\, F. \end{aligned} \tag{40}$$

From the equation (40) one can extract the spatial tensor of small strains ε in the component form

$$\varepsilon = \frac{1}{2}(\nabla\vec{u} + (\nabla\vec{u})^T) = \frac{1}{2}\sum_{i,j}\left(\frac{\partial u_j}{\partial x_i} + \frac{\partial u_i}{\partial x_j}\right)\vec{\imath}\otimes\vec{\jmath}. \tag{41}$$

4.7. The key roles of Cauchy tensors B and C

These two tensors play an important roles in determining the generalized inverse of the deformation gradient F. According to the definition 3.1 of the generalized Moore-Penrose inverse one can obtain the generalized inverse of deformation gradient by the following definition.

Definition 4.2 If the deformation gradient $F \in R^{m\times n}$, $m \geq n$, has a full rank, the generalized inverse can be defined by equation $F^+ = C^{-1}\, F$, otherwise ($m < n$) we have $F^+ = F^T\, B^{-1}$.

5. Conclusions

In the paper a new way of describing body deformation is presented, based on the concept of inclusive embedded deformation mapping, where the dimensions of domain and codomain are necessarily different. This approach opens up new possibilities in describing body deformations.

As a result of generalizations, the basic kinematic quantities as a determinant of the deformation gradient and the inverse of the deformation gradient do not exist, so some corrections have been made to calculate these essential quantities. The practical examples as well as the theoretical derivations clearly show the importance of the Moore-Penrose generalized inverse in determining these and the other related kinematic quantities.

Since the vector field of displacements as one of the basic unknowns, especially when using the finite element method, was not well defined by the tensorial equation, a proposal for a correct definition of displacements was made, which eliminates this shortcoming.

6. References

[1] Bonet J and Wood R D 2008 *Nonlinear Continuum Mechanics for Finite Element Analysis* (Cambridge: University Press Cambridge)
[2] Marsden J E and Hughes T J R 1994 *Mathematical Fundations of Elasticity* (Dover: Mineola)
[3] Demmel W J 1997 *Applied Linear Algebra* (Philadelphia: Society for Industrial and Applied Mathematics)
[4] Tian Y 2005 The Moore–Penrose inverse for sums of matrices under rank additivity condition *Linear and Multilinear Algebra* **53**:1
[5] Sun L and Zheng B and Bu C and Wei Y 2016 Moore–Penrose inverse of tensors via Einstein product *Linear and Multilinear Algebra* **64**:4
[6] Carmo M P D 1976 *Differential Geometry of Curves and Surfaces* (New Jersey: Prentice-Hall Englewood Cliffs)

Acknowledgement

The work was financially supported by the Slovenian Research Agency (ARRS) under contract P2-0260. The support is gratefully acknowledged.

Effects of Hydro Power Plant Releases on Navigation Conditions in the Lower Lock Approach

P Fošumpaur[1]

[1] Department of Hydraulic Structures, Faculty of Civil Engineering, Czech Technical University in Prague, Thakurova 7, 166 29 Prague 6, Czech Republic

E-mail: fosumpaur@fsv.cvut.cz

Abstract. Waterways are usually designed as multipurpose structures and perform a number of functions in addition to transport. A very common use is the generation of electricity in hydropower plants, water withdrawal for users and flood protection. On canalized rivers, increased attention must be paid to the design of the layout of the individual structures within the barrage. This paper deals with the optimization of barrage layout with respect to the interaction of navigation and hydropower use of the waterway. The navigation conditions below a power plant can be significantly affected by peak operation or by sudden load rejection. These phenomena can generate a negative surge wave with very adverse effects on navigation. In addition, special attention should be paid to the current conditions in the lower lock approach area and the guard wall that separates the lock approach from the weir and the power plant should be optimised. Appropriate design of operational and structural measures can significantly increase the traffic capacity of a waterway. The research was conducted as part of project no. DG18P02OVV004 within a support programme for applied research and experimental development of national and cultural identity, funded by the Ministry of Culture of the Czech Republic.

1. Introduction

Lock approaches (LA) shall be designed to ensure safe navigation conditions for vessels entering the lock and for vessels leaving the lock. Space is also provided in the lock approaches for vessels waiting to pass through the lock. An upper LA is proposed above the weir and a lower LA downstream the weir. The appropriate layout of the lock approaches significantly influences the speed and safety of vessels entering and leaving the lock and the transport capacity. The lock capacity is important for the subsequent cost-benefit assessment of the economic efficiency of the waterway. The traffic capacity is mainly influenced by the opening time of the lock gate, the time required for filling and emptying the lock and the time required for vessels to enter and exit the lock [1]. The lock capacity is a key input in assessing the economic efficiency of new investments in waterways using the cost-benefit method [2]. The effect of the operation of the hydropower plant plays an important role in the assessment of suitable navigation conditions. The operation of the hydropower plant is particularly important for the navigation conditions downstream the weir. Navigable conditions downstream the weir can be significantly affected by peak operation of the hydropower plant or sudden load rejection. These phenomena generally cause negative surge wave which can cause the minimum navigable depth downstrean the lock to be undercut [3, 4]. For this reason, it is advisable to carefully optimise operational measures in advance to eliminate the negative consequences of sudden flow changes in the operation of the hydropower plant. The environmentally safe flow rates in watercourse downstream the weir is also not ensured during a sudden hydro power load rejection [5].

The operation of the hydropower plant may further adversely affect the lower lock approach space through adverse flow conditions. Ships must reduce their speed before entering the LA to allow them to stop safely. The critical area is the entrance to the LA, where the ship must travel at a speed 0.3 to 0.5 $m \cdot s^{-1}$ greater than the water velocity. This requirement arises from the need to ensure that the ship has sufficient response to the rudder to make the ship controllable. It is in this critical area that the ship's trajectory crosses the water current. The water velocity vectors exert a force on the vessel towards the guard wall. The guard wall separates the navigable river from LA. The correct length and type of guard wall significantly influences the navigational conditions in the LA and, in effect, the traffic capacity of the lock. Crosscurrent in the area in front of the guard wall is very unfavourable in terms of navigation conditions and can cause a navigation accident. According to [3], a maximum permissible cross-speed of up to 0.35 $m \cdot s^{-1}$ should be ensured in the area in front of the guard wall. Možiešik et al. [6] dealt with the influence of velocity conditions in the vicinity of the Kolárovo waterworks on the safety of vessels. Currently, methods of automatic shape optimization of hydraulic structures and technological equipment of power plants are beginning to develop intensively [7]. High transverse velocities in the vicinity of the head of the guard wall also often cause scouring of the river bed and the formation of scour holes [8] that threaten the stability of the wall structure. The issue of scouring in the area of hydraulic structures was also studied by Rumann et al. [9].

2. Methods

2.1. Modelling of surge waves caused by sudden flow changes

A standard 1D model based on the approximation of Saint Venant's equations can be used to model sudden flow changes. The 1D model assumes a predominant flow and a surface slope in the direction of the longitudinal axis. Velocity and depth values are approximated for individual cross sections. The relationship between the flow rate Q [$m^3 \cdot s^{-1}$] and the cross-section area A [m^2] is given by the continuity equation (1) and by the momentum equation (2):

$$\frac{\partial A}{\partial t} + \frac{\partial Q}{\partial x} = q \quad (1)$$

$$\frac{\partial Q}{\partial t} + \beta \frac{\partial}{\partial x}\left(\frac{Q^2}{A}\right) + gA\frac{\partial h}{\partial x} = gA(S_0 - S_e - S_f) \quad (2)$$

where q [$m^2 \cdot s^{-1}$] is lateral flow rate (zero in our case), β [-] is the Boussinesq coefficient, h [m] is the water depth, S_0 [-] is the slope of the river bed, S_e [-] is the local loss due to local change in cross-sectional area and S_f [-] is the friction loss.

2.2. Modelling of current conditions in the lower LA

The flow in the lower LA was modelled assuming 3D schematization using Navier-Stokes equations [10]

$$\frac{\partial u_i}{\partial t} + \frac{\partial u_i}{\partial x_j} u_j = -\frac{1}{\rho}\frac{\partial p}{\partial x_i} + \frac{\mu}{\rho}\nabla^2 u_i \quad (3)$$

and the mass conservation (continuity) equation:

$$\nabla \mathbf{u} = 0 \quad (4)$$

where u_i is the velocity component [$m \cdot s^{-1}$],
t … time [s],
ρ … density [$m^3 \cdot s^{-1}$],
p … pressure [Pa],
μ … dynamic viscosity [Pa·s].

To approximate the behavior of a real fluid, the turbulent flow regime must be taken into account. In engineering practice, the k-ε turbulent model is often used, which further introduces turbulent kinetic energy transport and kinetic energy dissipation. The momentum equations go into the form of Reynolds equations:

$$\rho\frac{\partial\overline{\mathbf{u}}}{\partial t}-\nabla.\left[\left(\mu+\rho C_{\mu}\frac{k^{2}}{\varepsilon}\right).\left(\nabla\overline{\mathbf{u}}+(\nabla\overline{\mathbf{u}})^{T}\right)\right]+\rho\,\overline{\mathbf{u}}.\nabla\overline{\mathbf{u}}+\nabla\overline{p}=\mathbf{F}$$
$$\nabla.\overline{\mathbf{u}}=0 \tag{5}$$

The turbulent kinetic energy k is:

$$\rho\frac{\partial k}{\partial t}-\nabla.\left[\left(\mu+\rho\frac{C_{\mu}}{\sigma_{k}}\frac{k^{2}}{\varepsilon}\right)\nabla k\right]+\rho\overline{\mathbf{u}}.\nabla k=\rho C_{\mu}\frac{k^{2}}{\varepsilon}(\nabla\overline{\mathbf{u}})^{2}-\rho\varepsilon \tag{6}$$

The turbulent energy dissipation ε is:

$$\rho\frac{\partial\varepsilon}{\partial t}-\nabla.\left[\left(\mu+\rho\frac{C_{\mu}}{\sigma_{\varepsilon}}\frac{k^{2}}{\varepsilon}\right)\nabla\varepsilon\right]+\rho\overline{\mathbf{u}}.\nabla\varepsilon=\rho C_{\varepsilon 1}C_{\mu}k(\nabla\overline{\mathbf{u}})^{2}-\rho C_{\varepsilon 2}\frac{\varepsilon^{2}}{k} \tag{7}$$

The constants of the turbulent k-ε model were derived by hydraulic research according to Table 1 [10].

Table 1. The k-ε model constants.

C_{μ}	$C_{\varepsilon 1}$	$C_{\varepsilon 2}$	σ_k	σ_{ε}
0.09	0.1256	1.92	1.0	1.3

Similar to the current conditions in the LA region, the current conditions at the inlet of the hydro power plant can be optimized with respect to the operating efficiency of the hydropower plant [11].

3. Study area and material

The České Vrbné barrage is located in the Czech Republic near the town of České Budějovice on the Vltava River. At the beginning of the 20th century, a fixed weir and a hydroelectric canal were built here, which today serves as a slalom canal. In 1968, a new reinforced concrete weir replaced the original weir, creating a stabilisation stage in preparation for a new hydropower station on the left bank. This hydropower plant was built in 1985 and is equipped with two Kaplan turbines with a maximum output of 1.75 MW. The maximum usable flow for the hydropower plant is 38 $m^3\cdot s^{-1}$. The completion of the lock took place between 2009 and 2011, when the weir was modernised and a lock was added as part of the Upper Vltava River navigation. The original hydrostatic reinforced concrete gates were replaced by two steel gates with a width of 22.5 m and a height of 3 m. The situation of the lock is shown in Figure 1. The lock meets the conditions for a Class I waterway with a usable length of 45 m and a width of 6 m. The lock allows ships to overcome a total head of 7.5 m. The maximum navigable flow rate at which navigation can still be carried out has been set at 100 $m^3\cdot s^{-1}$. When this flow is exceeded, navigation is stopped and ships must take shelter in the protective harbour in the upper reservoir (on Figure 1 upstream the weir) or in the harbour at the weir in Hluboká nad Vltavou downstream České Vrbné barrage.

As part of the preparatory work prior to the implementation of the lock, detailed hydraulic research was carried out on mathematical and physical models. This paper deals with the optimization of the interactions between hydropower plant operation and navigation in the lower LA by means of mathematical modelling. The input data were the river morphology from the České Vrbné weir to the lower Hluboká nad Vltavou weir, the geometry of the hydraulic structures, operational data on the hydropower plant and hydrological data.

Figure 1. Layout of the České Vrbné barrage.

4. Results

4.1. Optimization of the weir operation mode during hydropower plant sudden load rejections

The navigation conditions below a power plant can be significantly affected by peak operation or by sudden load rejection. These phenomena can generate a negative surge wave with very adverse effects on navigation. A sudden failure of a hydroelectric power plant will generate a direct negative surge wave, which will cause a rapid drop in the water level downstream the weir and may result in insufficient navigational depth. The minimum navigable depth is given by the sum of the maximum draft of the design vessel and the safety margin between the bottom of the ship and the bed of the waterway. In order to limit the negative effect of a sudden failure of the hydropower plant, an operational measure should be provided, which consists in the prompt replacement of the flow through the hydropower plant by the flow through the weir. Upon identification of the power plant load rejection, the control system of the hydropower plant shall transmit a signal to the weir control system, which shall instruct the automatic opening of the weir gates. The purpose of the mathematical modelling is to determine the necessary speed of response of the control system to a hydropower failure and to verify the effectiveness of this measure in maintaining sufficient depth in the waterway downstream the weir. The following Figure 2 illustrates the implementation of the described operational measure. At time t=0, consider the operation of the hydropower plant with a maximum usable flow rate Q_t=38 m^3·s^{-1}. At time t_1=1 minute, a sudden failure of the hydroelectric power plant occurs, the control system receives information about the failure and issues an instruction to open the weir gates with a delay of Δt_1, then depending on the opening speed of the gates, the flow of 38 m^3·s^{-1} is restored through the weir during the interval Δt_2. The time required to restore the flow rate Δt_2 in this case is equal to 3 minutes. Various control system delay times Δt_1 = 2, 5, 10 minutes and the case where the control system does not react at all (in this case the flow through the weir would be restored only after the level in the upper weir is raised by overflowing the closed weir gate with a considerable delay) were simulated.

Figure 3 shows the water level downstream the weir in the lower LA for different selected control system delays. The simulation is prepared for the assumption of the minimum operating level at the downstream weir Hluboká nad Vltavou. The graph shows a significant influence of the reaction delay Δt_1 on the value of the maximum water level drop. The graph in Figure 4 shows the dependence of the maximum water level drop on the reaction delay time and the hydropower flow. Based on the results of this analysis, the optimum delay time Δt_1 of the control system in which the operation of the hydropower plant is monitored was determined. The value of Δt_1=5 minutes was selected as the optimum because it is fully sufficient to ensure adequate weir gate response to a hydropower plant sudden load rejection with elimination of adverse effects of water level drop in the lower basin.

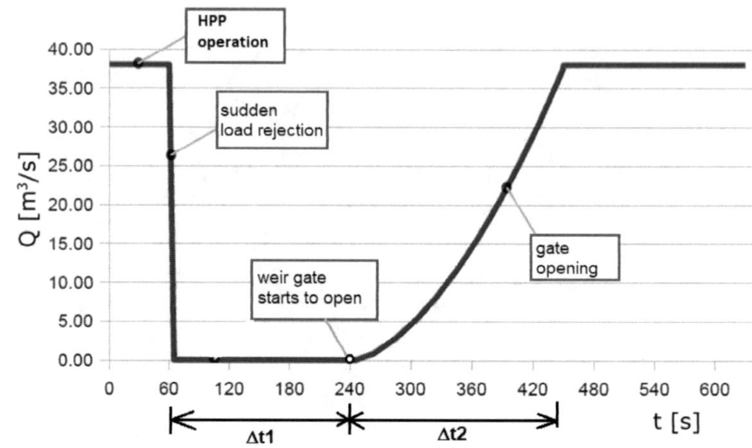

Figure 2. Weir gate opening after hydropower plant (HPP) sudden load rejection.

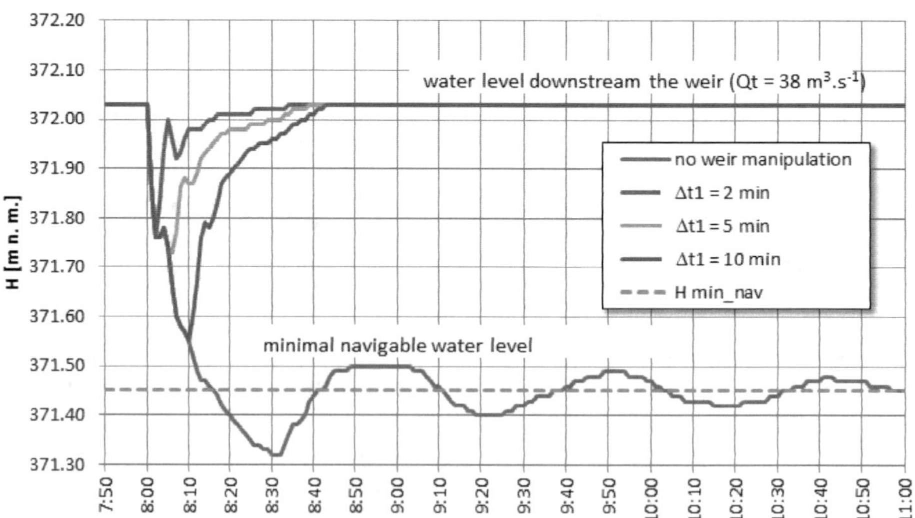

Figure 3. The effect of the reaction delay Δt_1 on the water level drop downstream the weir.

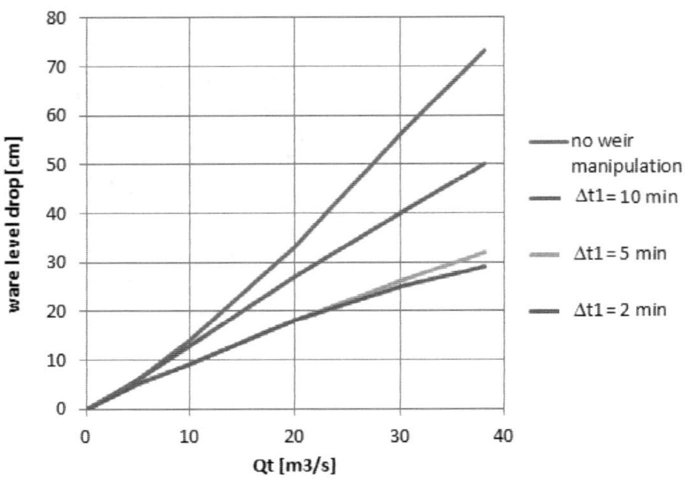

Figure 4. The effect of the reaction delay Δt_1 and power plant flow Q_t on the water level drop downstream the weir.

4.2. Optimization of the guard wall in the lower lock approach

To optimize the nautical conditions, a 3D mathematical hydrodynamic model was used. The model includes the area downstream the barrage including the outflow from both weir gates and the release from the hydroelectric power plant. Figure 5 shows the modelled area.

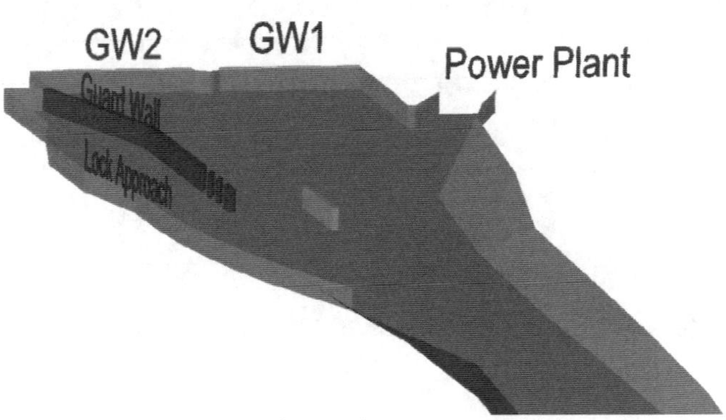

Figure 5. Modelled area downstream the barrage.

In the first phase, the guard wall was optimized with respect to the conveyance of the maximum navigation flow Q_{max}=100 $m^3 \cdot s^{-1}$ through both weir openings and then the conveyance of the maximum available flow through the hydropower plant Q_t=38 $m^3 \cdot s^{-1}$. The results of these simulations are shown in Figure 6. As part of the optimization, a ported guard wall was designed with flow openings in the front. The objective of this design is to minimize the crosscurrent at the entrance area of the lower LA to avoid exceeding the limit transverse velocity of 0.35 $m \cdot s^{-1}$. The intensity of the cross flow depends on the total flow through the weir and the hydropower plant [12]. In order to eliminate the influence of the operation of the hydropower plant on the entrance area of the lower LA, a dividing wall was designed in the next phase of the optimization to divert the release from the hydropower plant. This wall is shown in green in Figure 5 and Figure 7 shows a photograph of the final design of the entrance to the lower LA implemented in 2010. Operational experience has confirmed the conclusions of the model research and demonstrates suitable navigation conditions at all operating stages of the hydropower plant.

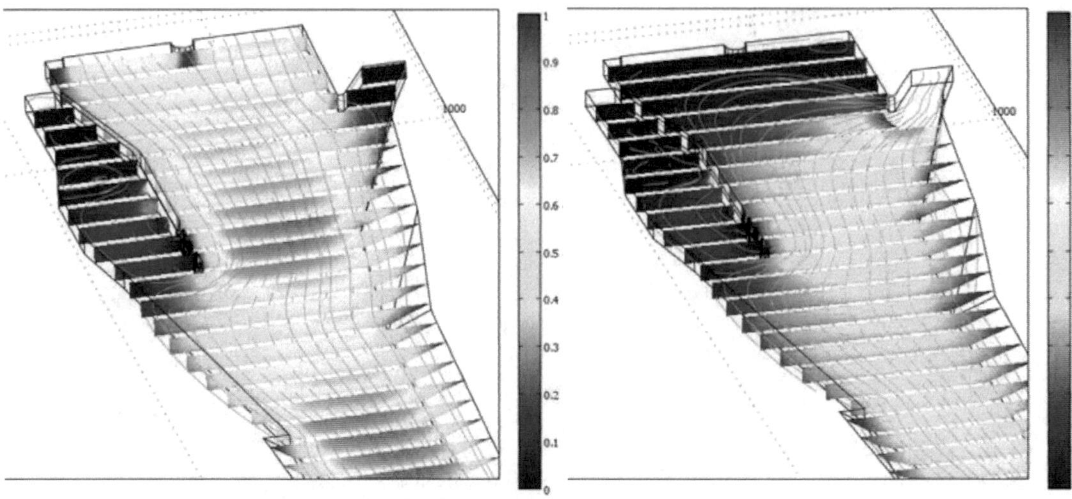

Figure 6. Velocity profile for: (a) maximum navigable flow through weir, (b) maximum power plant capacity.

Figure 7. Final design of lower lock approach with both guard walls.

5. Conclusion

The research focused on the use of mathematical modelling methods to optimise the navigation conditions of the lower lock approach with respect to the operation of the hydropower plant. The research reached the following main conclusions: (1) Operation of a hydropower plant can adversely affect navigation conditions in the waterway by sudden flow changes and the occurrence of adverse flow conditions caused by the discharge from the hydropower plant. (2) The adverse effect of hydropower plant release and the resulting negative surge wave can be eliminated by optimizing the weir control system. The release from the hydropower plant should be compensated by opening the weir gates. This measure has the potential to avoid the risk of undercutting the minimum navigation depth in the downstream area below the weir. (3) Hydraulic modelling methods on a 3D mathematical model can be effectively used to optimise the layout of the lock approaches. The optimization of the flow conditions then allows not only to increase the traffic capacity of the lock, but also to ensure smooth and safe navigation conditions eliminating the risk of traffic accidents. (4) The proposed solution of the ported guard wall application with flow openings below the water surface allows the creation of optimal navigation conditions with minimal impact on the architectural character of the locks and can therefore be used to increase the capacity of historic locks with high cultural value without affecting their external appearance.

6. References

[1] Kooman I C and De Bruijn P A 1975 *Lock capacity and traffic resistance of locks* (Hague: Government Publishing Office)

[2] PIANC 2005 *Economic Aspects of Inland Waterways* (International Navigation Association)

[3] Novak P Moffat A I B and Nalluri C 2007 *Hydraulic structures* (London and New York: Taylor & Francis)

[4] Šulek P and Kinczer T 2019 Expert control system of shipping operation on the Gabcikovo project *Pollack Periodica* **14** 139–50

[5] Balvín P Vizina A Nesládková M Blöcher J Makovcová M Moravec V and Hanel M 2021 Minimum residual flows for catchments in the Czech Republic *Water* **13** 689

[6] Moziesik L Šulek P and Orfánus M 2014 Kolarovo water structure preparation of the research and design proposals *SGEM International Multidisciplinary Scientific GeoConference EXPO Proceedings*

[7] Kantor M Chalupa M Souček J Bílková E and Nowak P 2020 Application of genetic algorithm methods for water turbine blade shape optimization *Manufacturing Technology* **20** 453–8

[8] Hladík M Králík M Ouhel J Sýs V and Zukal M 2021 Effect of energy dissipation on scour hole development downstream of the chute *IOP Conference Series: Earth and Environmental Science* **906**

[9] Rumann J Pavúček M and Dušička P 2021 Model investigations of scouring at the HRIČOV weir using short-range photogrammetry *IOP Conference Series: Materials Science and Engineering* **1203**

[10] Wilcox D C 2010 *Turbulence modeling for CFD* (La Cañada, CA: DCW Industries)

[11] Bytčanková L Rumann J and Dušička P 2021 Analyzing the impact of intake structure on the flow at low pressure Shpp *Pollack Periodica* **16** 114–9

[12] Stockstill R L Park H E Hite J E and Shelton T W 2004 *Design Considerations for Upper Approaches to Navigation Locks* (Vicksburg: US Army Corps of Engineer) pp. 1–46

Acknowledgments

The article was written with support from the project "Documentation and Presentation of Technical Cultural Heritage on the Elbe-Vltava Waterway" No. DG18P02OVV004 solved within the National and Cultural Identity program for 2016 to 2022 (NAKI II) administered by the Ministry of Culture of the Czech Republic.

IOP Conf. Series: Materials Science and Engineering 1252 (2022) 012050 doi:10.1088/1757-899X/1252/1/012050

Analytical Determination of Longitudinal Forces on the Vessels During Filling of the Navigation Lock

T Kašpar[1]

[1] Department of Hydraulic Structures, Faculty of Civil Engineering, Czech Technical University in Prague, Thákurova 7, 166 29 Prague 6, Czech Republic

E-mail: Tomas.Kaspar@cvut.cz

Abstract. The basic parameter for the calculation of the lock filling time is the determination of the forces on the vessels during the filling of the navigation lock. The time required to fill the lock determines the method of filling the lock. The forces on vessels during filling (emptying) directly determine the capacity of the waterway. In the first phase of the lock design, the forces on the vessels during locking are calculated analytically. Most approaches to calculate the forces on vessels are concerned only with finding the maximum value of the forces on vessels. This research deals with the analytical determination of the maximum forces on vessels during the lock filling period. Data from the physical model and the 1D model are used to calculate the maximum forces on the vessels. This simple method can be used to design a new lock. This simple method can be used to design a new lock and to increase the capacity of existing historic locks on the Elbe-Vltava Waterway (Czech Republic) without any structural modifications.

1. Introduction

The determination of the forces on the vessels during the filling and emptying (F/E) of the lock is important for the design of the navigation lock. The F/E time of locks must be reduced to minimize transit time. The excessive flow (inflow or outflow) generates excessive hawser forces, leading to an increased risk of accidents in the lock chambre [1]. Navigational safety is an important criterion in the navigation lock, as well as near the water structures [2] and [3]. The filling (emptying) time depends on the dimension of the lock (length, width, head) and the type of filling (emptying) system. There are a few filling (emptying) systems that are used in inland lock projects: direct system (gate openings, moving gate), lock culverts, wall filling system and floor filling system. In direct filling, there is a concentrated inflow q from one side of the chamber and there are forces due to the flowing water in the longitudinal direction. It is required to design the gate openings to limit the forces on the vessels during F/E process. The longitudinal force acting on the ship during the F/E process is determined by the difference in water levels above the bow and stern of the ship and the flow velocity and friction on the ship (Figure 1). The longitudinal force during filling of the inland lock using the direct filling system can be described by five components (Figure 1): translatory waves F_{transl}, momentum decrease in longitudinal direction of the chamber F_{impulse}, friction F_{frict} and jet action F_{jet} [4] and [5]. The longitudinal force F_{long} is equal to the sum of the individual components:

$$F_{\text{long}} = F_{\text{transl}} + F_{\text{impulse}} + F_{jet} + F_{\text{frict}}. \quad (1)$$

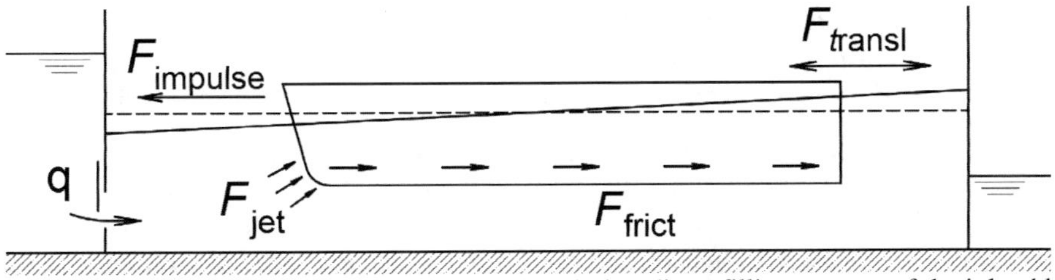

Figure 1. Longitudinal force components occurring during direct filling process of the inland lock.

The translational waves (F_{transl}) caused by the non-constant inflow into the lock chamber are reflected from the ship and the lock gates. This causes oscillation of the water level during the filling process. The filling jet (F_{jet}) and the decrease in momentum ($F_{impulse}$) lower the water level in front of the ship's bow. Translational waves and momentum decrease in the longitudinal direction are the main components of F_{long} [6].

This paper deals with the analytical calculation of the limit of the longitudinal force F_{long} under the dominant action of the force component F_{transl}. Data from previous research on physical and 1D numerical models [7] and [8] were used to obtain the analytical calculation.

2. Experimental data

For analytical determination of the limits of the longitudinal force F_{long}, the investigation of the Děčín lock is used as a basis. On this lock, direct filling of the lock with the Tainter gate is designed. The basic dimensions of the lock are: length of the lock chamber is 200 m, width of the lock chamber is 24 m, maximum head is 5.56 m. This lock was investigated using a physical model at a length scale of 1:20. The research on this physical model is described in detail in [7]. A 1D numerical model of this lock is created and verified with measured data in the physical model. This 1D numerical model is described in [8]. The article [8] also describes a parametric study of the basic design parameters of the lock. Data from the physical model and the parametric study (1D numerical model) are used for this paper to determine the limits of the longitudinal force F_{long}. The course shown in Figure 2 is selected as a characteristic F_{long} course from the researches [7] and [8]. The graph shows the dominant component of the force F_{transl}, which is a damped harmonic motion. The graph also shows the F_{long} course on the verified 1D numerical model.

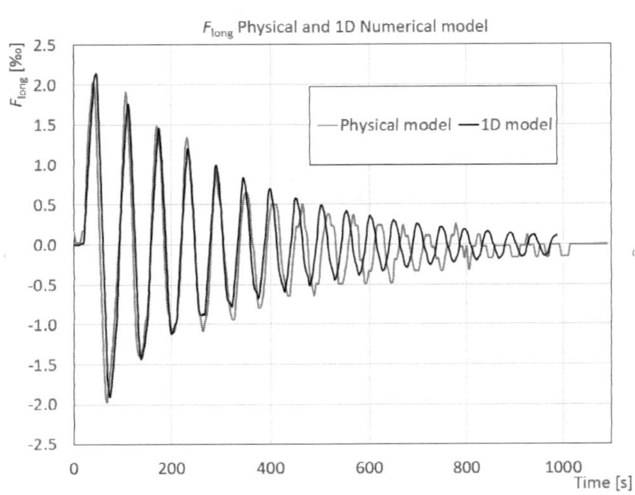

Figure 2. The course of the F_{long} on the physical and 1D numerical model [7], [8].

3. Calculation of the maximum of longitudinal force

The course of the force F_{long} can be determined by physical and numerical models [9], [10] and [11]. The analytical determination of the maximum value (F_{max}) of the F_{long} force can be found in [12], [13] and [14]. Only a few authors have focused on the analytical calculation of the F_{long} course [15]. The advantage of analytical calculations is the simplicity of the calculation. The calculation can be used in the initial design where the lock concept must be determined.

The input for the analytical calculation of the longitudinal force is the calculation of the filling time and the flow into the chamber. In hydraulic terms, this is the discharge of water through the opening from one chamber to another with a variable head. It is the discharge of water through the opening from one chamber to another with a variable head. For the numerical solution of this hydraulic phenomenon, the differential continuity equation for the elementary volume change dV can be used:

$$\mathrm{d}V = \mu \, a_{\mathrm{h}} \sqrt{2\, g\, y}\, \mathrm{d}t, \quad (2)$$

$$\mathrm{d}V = -F\, \mathrm{d}y, \quad (3)$$

where μ expresses the cumulative local loss and the possible friction loss (for culverts) [-], a_h is the area of the filling opening [m^2], F is the surface area in the chamber [m^2], y is the momentary head [m] at time t, t is the time from the start of the filling [s].

For direct filling (or filling with short passes) only the size of the F_{transl} is calculated. The maximum magnitude of the F_{transl} is derived from the literature [12] and depends in the maximum change of inflow into the chamber according to the following:

$$F_{\mathrm{max}} = WS_{max} = \frac{W}{g(b_{\mathrm{k}}h_{\mathrm{k}}-b_{\mathrm{s}}t_{\mathrm{s}})}\left(\frac{\mathrm{d}Q}{\mathrm{d}t}\right)_{\mathrm{max}}, \quad (4)$$

where W is the displacement of the ship [t], S_{max} is the maximum slope of the surface in the chamber [-], b_k is the width of the lock chamber, h_k is the level of water in the lock chamber, b_s is the breadth of the ship and t_s is the draft of the ship.

The maximum value of the change in flow into the chamber is determined by the following equation:

$$\left(\frac{\mathrm{d}Q}{\mathrm{d}t}\right)_{\mathrm{max}} \cong \left(\frac{\Delta Q}{\Delta t}\right)_{\mathrm{max}} = \left(\frac{Q_{\mathrm{i}}-Q_{\mathrm{i-1}}}{\Delta t}\right)_{\mathrm{max}}, \quad (5)$$

The solution of equation (5) is usually discretized into a computational step Δt. Equation (2) is used to calculate Q. Another approach to determine the maximum force F_{max} is given by Peng [14]:

$$F_{\mathrm{max}} = \frac{k_{\mathrm{r}}\, \omega D W \sqrt{2gH}}{t_{\mathrm{v}}(b_{\mathrm{k}}h_{\mathrm{k}}-b_{\mathrm{s}}t_{\mathrm{s}})}, \quad (6)$$

where k_r is the coefficient concerning valve configuration [-], ω is the sectional area of the culvert with valve [m^2], D is the wave force coefficient, H is the maximum lock head [m] and t_v is the opening time of the valve [s].

4. Determination of the limits of the longitudinal force

The F_{transl} can be considered as the dominant component of the F_{long} if the vessel is not positioned too close to the filling opening ($F_{impulse} \doteq 0$) and the energy dissipation under the gate is sufficiently resolved ($F_{jet} \doteq 0$). To determine the F_{long}, the equation:

$$F_{\mathrm{long}} \doteq F_{\mathrm{transl}}. \quad (7)$$

The force component F_{transl} is due to the translational waves that are caused by the non-constant flow rate through the gate openings. Calculation of translational waves (longitudinal slope of the water level) is calculated [11]:

$$F_{\text{transl}} = \frac{h_{\text{bow}} - h_{\text{stern}}}{l_s C_b}, \tag{8}$$

where h_{bow} and h_{stern} are the water levels at the bow and stern, respectively, l_s is the length of the ship and C_b is the block coefficient.

The F_{long} limits can be used from the parametric study [7] the article by [15]. Stockstill [15] states that the F_{long} oscillates with an exponential decrease in amplitude. The limits of the F_{long} in time are according to the equation:

$$F_{\text{lim}} = a\, e^{(b)}, \tag{9}$$

where a and b are exponential coefficients and can be derived from a parametric study [7].

The limits of the F_{lim} can be determined for both the upper limit (positive part of the longitudinal force F_{long}) and the lower limit (negative part of the longitudinal force F_{long}) using the coefficient a. The coefficient a is $\pm F_{\max}$. The coefficient b depends on the calculation time t, the hydrodynamic coefficient C_h and the displacement of the ship W and the width of the chamber b_k. The coefficient b is then given by equation:

$$b = t \cdot -\frac{C_h b_k}{W}. \tag{10}$$

The resulting limits F_{lim} for both the positive and negative parts of the F_{long} can be determined by the resulting equation:

$$F_{\text{lim}} = \pm F_{\max}\, e^{\left(t \cdot -\frac{C_h b_k}{W}\right)}. \tag{11}$$

The hydrodynamic coefficient Ch must be determined experimentally. According to Stockstill [15], the dimensionless hydrodynamic coefficient for locks is around 0.5. Based on a parametric study [7], the hydrodynamic coefficient is determined to be $C_h = 0.53$.

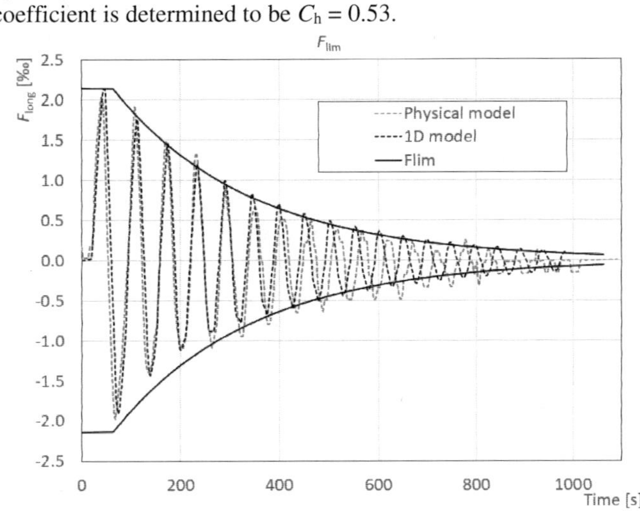

Figure 3. The course of the F_{long} and F_{lim}.

It is clear from the graph (Figure 3) that the F_{lim} force limits must be divided into two intervals. These intervals are given by the length of the wave period T_k in the lock chamber, which is given in equation

(12). The first interval t=0 to t=T_k is $F_{lim} = \pm F_{max}$. The second interval ($t > T_k$) is given by equation (12). The wave period of the course of the longitudinal force T_k can be calculated according to equation [11]:

$$T_k = 2\left(\frac{l_k - l_s}{c_k} + \frac{l_s}{c_s}\right), \tag{12}$$

where c_s is the velocity of the pressure wave in the part of the lock chamber where the ship is; c_k is the velocity of the pressure wave without the ship.

5. Conclusions

In the first phase of lock design, an analytical calculation of the maximum longitudinal filling force is used most often. Research presents the analytical calculation of the limits of the longitudinal force F_{long} during the direct filling of the lock. To perform this calculation, it is necessary to ensure that the dominant force of F_{long} is the F_{transl} component due to translational waves. The force components due to momentum decrease in longitudinal direction of the chamber $F_{impulse}$ jet action F_{jet} must be minimal. This is ensured by sufficient energy dissipation and the position of the ship in the lock. Based on this calculation, it is possible to perform a non-linear opening of the filling gates already in the first lock design. Studies [16] and [17] describe in detail the optimization of nonlinear opening. Proper design will result in faster locking and increase the capacity of the waterway. It is possible to eliminate longitudinal force during direct filling of the lock without structural modifications, making it possible to increase the capacity of historic locks on the Elbe-Vltava Waterway (Czech Republic) without impacting the architectural character of the locks.

6. References

[1] Natale L and Savi F 2000 Minimization of filling and emptying time for navigation locks *Journal of Waterway, Port, Coastal, and Ocean Engineering* **126(6)** 274–80

[2] Pavlíková A and Možiešik Ľ 2016 Optimization of the operation of the Selice water structure on the Váh River with regard to navigational safety *Pollack Periodica* **11(3)** 93–104

[3] Možiešik Ľ, Šulek P and Orfánus M. 2014 Kolárovo water structure – preparation of the research and design proposals *Int. Multidisciplinary Scientific GeoConference Surveying Geology and Mining Ecology Management, SGEM* 447–54

[4] Vrijburcht A 1991 *Forces on Ships in a Navigation Lock Induced by Stratified Flows*, Delft Hydraulics Publication no. 448.

[5] Glerum A and Vrijburcht A 2000 *Design of Locks 1* 459

[6] Nogueira H I S, van der Ven P, O'Mahoney T, De Loor A, van der Hout A and Kortlever W 2018 Effect of density differences on the forces acting on a moored vessel while operating navigation locks *Journal of Hydraulic Engineering* **144(6)**

[7] Kašpar T, Fošumpaur P, Králík M and Zukal M 2021 Parametric study of direct filling system of medium-head navigation locks *Pollack Periodica* **16(2)** 43–9

[8] Kašpar T 2021 *Theory of force action of water flow on vessels using physical and mathematical modelling methods* Ph.D. thesis 121

[9] Fošumpaur P, Králík M and Zukal M 2010 Physical and Numerical Modelling in the Research of Hydraulic Structures *Proceedings of the International Conference on Modelling and Simulation* 109–113

[10] Roux S, Roumieu P, De Mulder T,Vantorre M, De Regge J and Wong J 2010 Determination of hawser forces using numerical and physical models for the third set of Panama locks studies *32nd PIANC Congress, 125th anniversary PIANC* 1–15

[11] Deltares 2016 *Lockfill, user and technical manual* 96

[12] Čábelka J 1976 *Waterways and Navigation* (in Czech) 689

[13] Partenscky H W 1957 *Der Einfluß der Schleusenfüllungen auf den Wasserstand und die Schiffahrt in einer Kanalhaltung*

[14] Peng W 2011 Mooring forces and vessel behaviour in locks - Experience in China *International Workshop, PIANC*

[15] Stockstill R L 2003 Mooring model coefficients for barge tows in a navigation lock *Journal of waterway, port, coastal, and ocean engineering* **129(5)** 233–6

[16] Šulek P, Kinczer T 2018 Expert control system for increase of safety of shipping operation on the Gabcikovo project on the Danube *International Multidisciplinary Scientific GeoConference, Surveying Geology & mining Ecology Management* 297–304

[17] Šulek P, Kinczer T 2019 Expert control system of shipping operation on the Gabcikovo project *Pollack Periodica* **14(1)** 139–50

Acknowledgments

The article was written with support from the project "Documentation and Presentation of Technical Cultural Heritage on the Elbe-Vltava Waterway" No. DG18P02OVV004 solved within the National and Cultural Identity program for 2016 to 2022 (NAKI II) administered by the Ministry of Culture of the Czech Republic.

IOP Conf. Series: Materials Science and Engineering 1252 (2022) 012051 doi:10.1088/1757-899X/1252/1/012051

Mechanical behavior of strengthened reinforced concrete beams an experimental study

G Bou Abdallah[1,2], I Ivanova[1], J Assih[2], C Diagana[2] and D Dontchev[1]

[1] Department of Applied mechanic, University of Chemical Technology and Metallurgy, Kliment Ohridski Boulevard, 8 Sofia, 1756
[2] Civil Engineering Department Sustainable Construction, ITHEMM EA 7548, University of Reims Champagne-Ardenne, Reims 51100

E-mail: i.ivanova@uctm.edu;

Abstract. Sustainability of building structures is of particular importance. In recent decades, the use of composite materials to reinforce reinforced concrete structures has proven to be one of the most promising technologies in the field of structural materials engineering. Composite materials have interesting mechanical properties, light weight and have the ability to make complex shapes. Structural efficiency and the efficiency of strengthening technology is one of the most significant advantages for expanding their industrial application. This experimental study scientifically confirms the effect of strengthening with composite materials on the mechanical properties of reinforced concrete beams. The purpose of this reinforcement technique is to improve the load-bearing capacity of the structure and to increase its rigidity and its service life. The aim is to study the mechanical behavior with different configurations to strengthening the structure. The experimental part includes preparation, fabrication and testing of four reinforced concrete beams with the same geometry (270x30x13cm). One non-reinforced beam, which serves as a reference and 3 strengthened reinforced concrete beams with different types of reinforcement configurations. The results show a significant increase in the ultimate load by more than 203% compared to a non-reinforced beam and a one-third reduction in the sagging of reinforced beams.

1. Introduction

The strengthening technique by external bonding of carbon fiber composite material is used often when a structure performance is reduced. This is necessary when it has internal reinforcement corrosion, freeze-thaw action, or weak original design. In case of accidents such as earthquakes or explosions, in which cracks and damage to the structure occur. When positioning error of the reinforcements, compliance of an existing structure or change of destination of the structure.

The use of the composite materials in the field of civil engineering is remarkable because the RC structure is strengthened or repaired by bonding fiber reinforced polymer (FRP) [1]. The Carbon FRP (CFRP) is used because of their remarkable physical and mechanical characteristics, low density, good resistance to corrosion and fatigue loading. The light weight, the ease of handling and ease of application of composite materials quickly displace the metal plates used in the beginning [2]. The composite material is using also to renovation of old buildings [3].

This strengthening approach increases the structure's the stiffness, the performance and the durability, and also decreases crack propagation. Assih [4] carried out an experimental study on reinforced concrete beams presenting a defect due to an under sizing of the tensioned reinforced concrete steels. The beams were restrengthened by gluing carbon fiber composite plates at the bottom. This study further examined the influence of the main parameters (the thickness and length of the plates, the thickness of the adhesive joint) on the bending behavior of beams. The results showed that a damaged beam can be repaired by bonding carbon fiber composite plates. The bending breaking load of the repaired beam is twice that of an unreinforced beam [4-5].

Another research was carried out to determine the effect of shear reinforcement by carbon composite materials (CRFP) on the ultimate strength and behavior of reinforced concrete beams [6]. The performed 3-point bending test found that the ultimate load of a beam reinforced with composite fabrics at three-quarters of its height is twice that of the control beam. So, this load increases with increasing shear reinforcement area. In addition, when the distance between two frames decreases the effect of reinforcement by composite material fabrics decreases, which indicates the importance of the spacing of the stirrups of the reinforced concrete beam [6]. An increase in the shear contribution of CFRP is noted with increasing stirrup spacing from the beam. The results show that as the stirrup spacing decreases, the angle of the inclined critical crack, which depends on the ratio of the short shear span to the depth, decreases and the number of vertical cracks increases. The number of vertical cracks increases with the increase in reinforced side surfaces. The failure of a beam in short span reinforced sufficiently with a height of three quarters is due to bending [6-8].

One of the problems at the strengthened beams with carbon fiber plate is that the composite plate peeling-off at the ends, discussed in detail by Triantafillou [9]. A solution to prevent the peeling-off effect at the ends has been proposed by Zhou and all [10-11]. They investigate the performance of H-type end anchorage on the flexural strengthening of Fiber Reinforced Polymer (FRP) plated reinforced concrete (RC) beams. The results of the effect of H-type end anchorage on the flexural performance show great improvement as compared to the FRP-plated RC beams without end anchorage [10-11]. Prior to that, in 2011 research was conducted on the strengthening of beams with different numbers of layers of composite laminates bonded to the tensile soffit of the beams [12-13]. The results indicate that the flexural strength of beams was significantly improved as the layers of laminate increased. However, de-bonding of CFRP laminates from concrete surface was observed for the case of multi-layer strengthening of beam [12].

This research studies the mechanical behavior with different configurations to strengthening the reinforced concrete beams. The experimental part includes preparation, fabrication and testing of four reinforced concrete beams with the same geometry (270x30x13cm). One non-reinforced beam, which serves as a reference and 3 strengthened reinforced concrete beams with different types of reinforcement configurations. The objective of this study was to increase the load carrying capacity of the structure. The technique of strengthened concrete structure consists of bonding a composite plate on the beam underside, where the flexion tension is biggest. To prevent peeling-off, composite plate was used fiber fabrics in a U- wraps at both ends of the beam.

2. Experimental programme

2.1. Experimental model

Four rectangular section beams were carried out for this study. The beams have the same dimensions (270x30x13cm) (Figure 1) and the same reinforcement (Φ8 mm for the flexural reinforcement and Φ6 mm for shear reinforcement) and they were tested in 4-point flexion. The flexural reinforcement steel was undersized to investigate the effect of strengthening. The first beam is the reference beam without any strength in composite materials. The other three beams were strengthened with different types and configurations of composite materials. Two beams were reinforced with carbon fiber fabrics in the flexural zone (3 layers of 13 cm width), one with U-wraps fabric with carbon fiber fabrics (2 layers of 15 cm width) at both ends of the beam for shear reinforcement and the other without anchoring. A fourth beam was reinforced with a Carbodur plate (rigid plate) on the lower part of the beam (tensile zone) and with a U-band fabric at (2 layers of 15 cm width) both ends of the beam for shear reinforcement. The configurations of the beams are shown in Figure 3.

The control specimen is labelled as "**B0**," with the letter "**B**" indicating beam and zero **0** indicating no strengthening. The first letter of the strengthened reinforced concrete beam's identification is, as previously said, Beam, while the second letter signifies the type of strengthening (e.g.: **CP** for **C**arbodur

IOP Conf. Series: Materials Science and Engineering 1252 (2022) 012051 doi:10.1088/1757-899X/1252/1/012051

Plate or **FP** for **F**abric **P**late and **UF** for **F**abric in **U**- wraps). The number of layers is then indicated by the numeral (E.g.: 1, 2, 3). Finally, the small letter indicates the type of composite material (e.g., **u** for unidirectional).

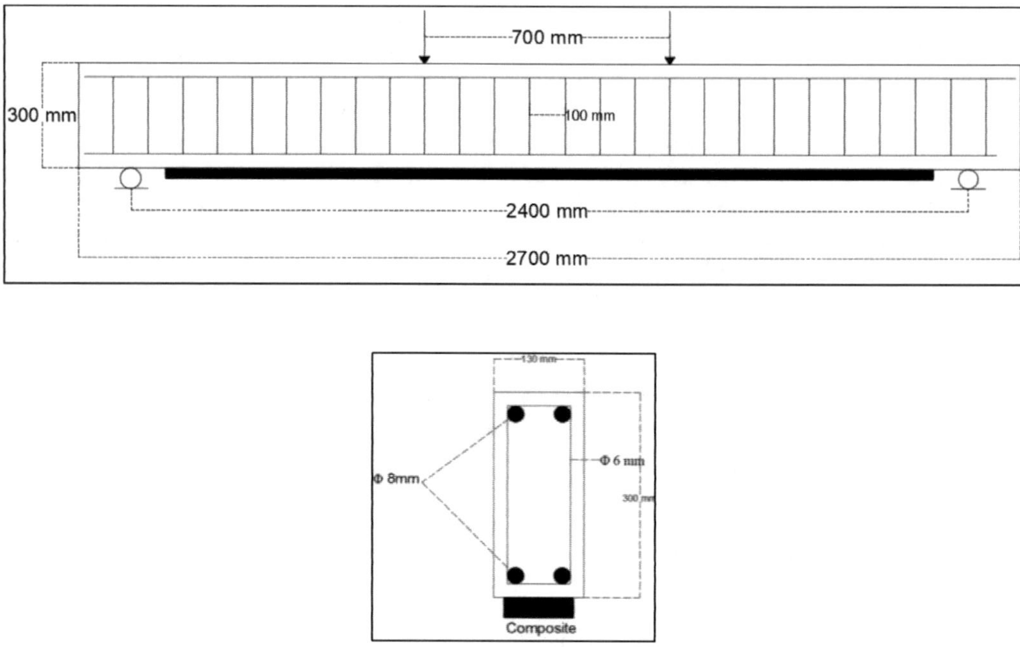

Figure 1. Geometry of reinforced beam.

For investigating the local deformation, the strain electric gauges were used. The resistance of strain electric gauges was 120 Ω and gauge factor of 2.07. They were placed on various position of steel, concrete surfaces and carbon fiber fabrics plate in middle of the beams where the flexural deformations are important (Figure 2).

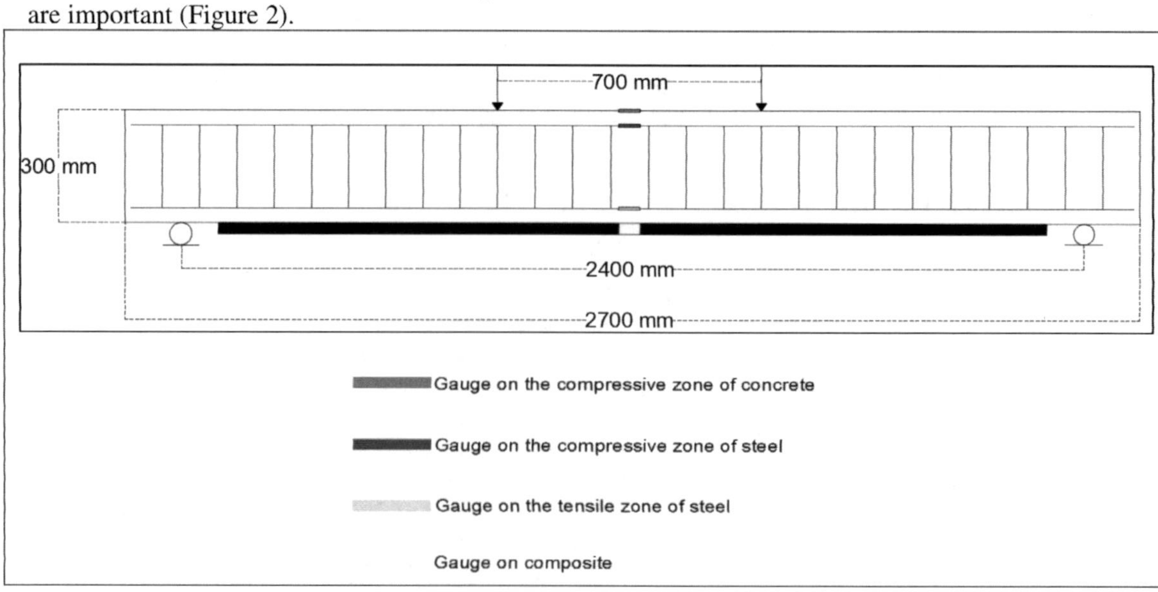

Figure 2. Position of strain gauges on steel bars, on concrete and composite.

Before gluing the composite plate, the beams were prepared and were cleaned by using compressed air just before the bonding procedure to remove any sand or other loose material on the beam surface. After gluing the composite plate, the beams were allowed to dry for 48 hours at room temperature [1].

2.2. Materials

Ordinary Portland cement, gravel, and dry sand are used to make normal strength concrete. Five cylindrical concrete specimens (16x32 cm) and 5 parallelepiped specimens (10x10x40cm) were tested in compression after 28 days using a compression machine Provitec with a maximum load of 3000 KN in order to determine the compressive strength of concrete used in this experiment. The compression loading speed is set to 0.05 MPa/s.

According to the compression test for cylindrical concrete specimens after 28 days, the concrete has a stress of 33±2 MPa and a modulus of elasticity E= 30±2 GPa (Table 1). Steel bars of various diameters are used, including 6, 8 mm. Simple tensile testing is applied to characterize the steel specimens. Table 1 shows the results for stress fu and modulus of elasticity E. In this experiment, unidirectional fibers fabrics and Carbodur plate were used. The properties of carbon composite based on literature [14-15]. The carbon fiber fabric and the Carbodur plate were bonded with a Silkadure-30 adhesive made of two components A and B and an epoxy resin.

Table 1. Properties of Materials [14-15].

Material	Young's Modulus (GPa)	Strength (MPa)	Poisson Ratio
Concrete	30±2	33±2	0.25
Steel bar	200±1	610±10	0,30
Adhesive	11.2	24	
Unidirectional Carbon fiber composite sheet	86±1	1035±63	0,45
Carbodur Plate	165	3100	

3. Results and discussion

3.1. Influence of strengthening on the ultimate load

The experimental results of bending strengthening by adhering directly carbon fiber fabrics or Carbodur plate to the surfaces of reinforced concrete beams are shown in this study. Table 2 show the effect of strengthening on the ultimate load of tested beams in flexural test. $\mathbf{F_u}$ represents the ultimate load for each beam. $\mathbf{F_{uB0}}$ represents the ultimate load of the control beam without strengthening. The results show a significant increase in the ultimate load by more than 203% compared to a non-reinforced beam. For the beam strengthened with carbon fiber fabrics plate in the flexural zone (BFP3u) peeling-off at the ends was observed. The ultimate load for this strengthened beam (BFP3u) was 71 kN. In fact, this value represents 203 % increasing the ultimate load compared with the reference beam B0.

Table 2. Ultimate load of different beams.

	B0	BFP3u	BFP3u+UF2u	BCP3u + UF2u
F_u [kN]	35	71	78	87
F_{ui}/F_{uB0} [%]	1	203	223	249

IOP Conf. Series: Materials Science and Engineering 1252 (2022) 012051 doi:10.1088/1757-899X/1252/1/012051

Design	Designation
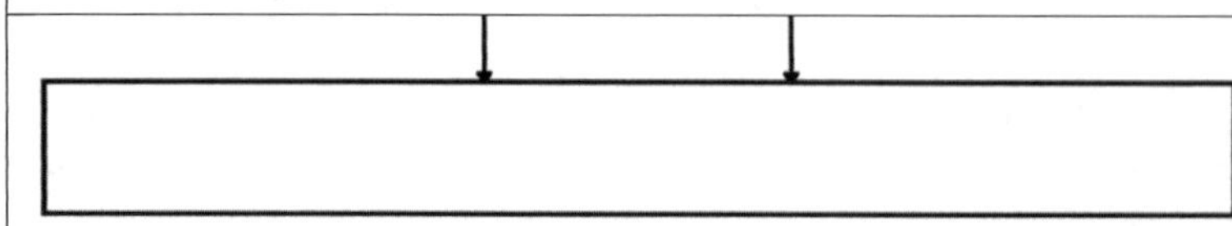	**B0** - Reference beam.
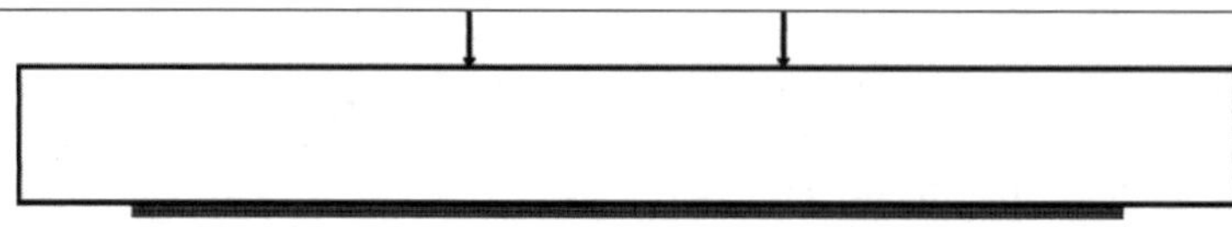	**BFP3u** - Strengthened beam by bonded 3 layers unidirectional carbon fiber Fabrics Plate).
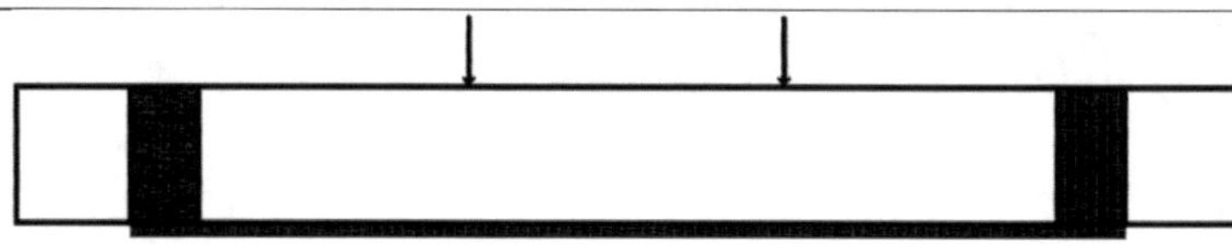	**BFP3u+UF2u** - Strengthened beam by bonded unidirectional carbon fiber fabrics with U2u for shear reinforcement.
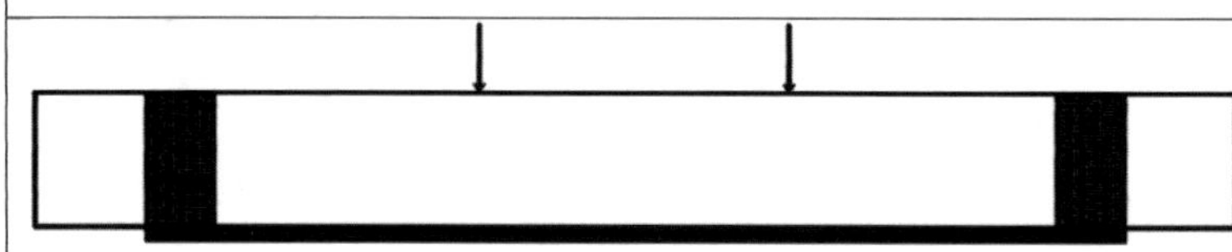	**BCP1u + UF2u** - Strengthened beam by bonded unidirectional Carbodur Plate with shear reinforcement.

Figure 3. Design of Testing Specimen.

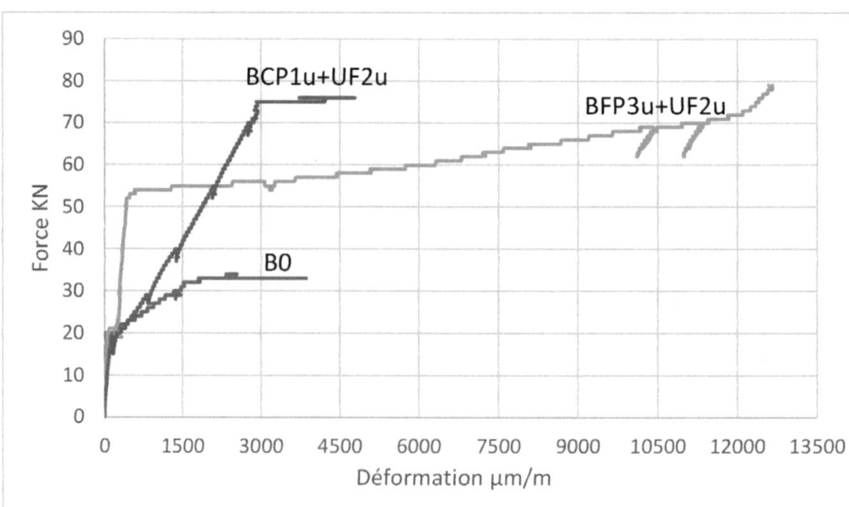

Figure 4. Comparison of Steel Strain of Strengthened beam.

Figure 4 show the result and compares the variation of local deformations on the steel in the same position for the three beams. Up to 20 kN (beginning of flexural cracking) the curves have the same behavior. The horizontal plateau means the beginning of the apparition of the oblique crack due to the shear force. The ultimate load for reference beam (B0) was 35 kN. The ultimate load for the other two reinforced beams (BFP3u+UF2u) and (BCP1u + UF2u) is respectively 78 kN and 87 kN. The local behavior of (BCP1u + UF2u) was explained by the larger rigidity of the Carbodur plate.

3.2. Modes of Failure

The different failure modes and fractures as shown in Figure 5. The results mainly identified three types of failure: flexion and compression failure, fabrics plate peeling-off at the ends and plate delamination and rupture of U-wraps fabrics plates see Figure 5. The results show that the beam reinforced with Carbodur plate is more rigid than the beam reinforced with carbon fiber fabric.

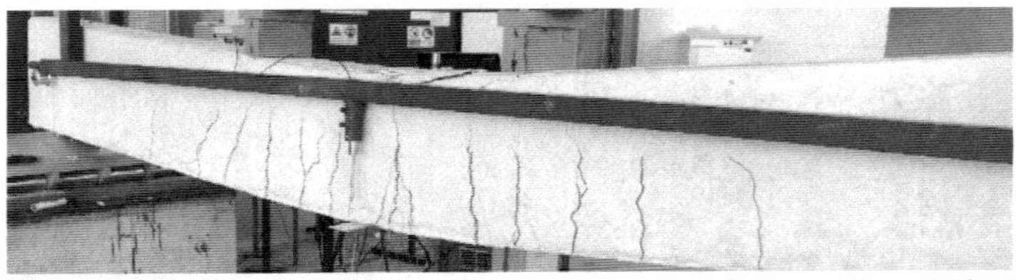

Figure 5a. B0: Flexion failure + compression failure, Fu = 35kN.

Figure 5b. BFP3u: Fabrics plate peeling-off at the ends, Fu = 71kN.

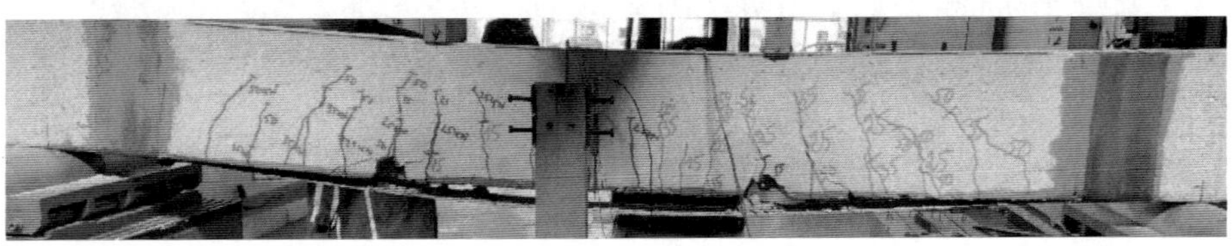

Figure 5c. BFP3u+UF2u: Fabrics plate delamination and rupture of U fabrics plates, Fu = 78kN.

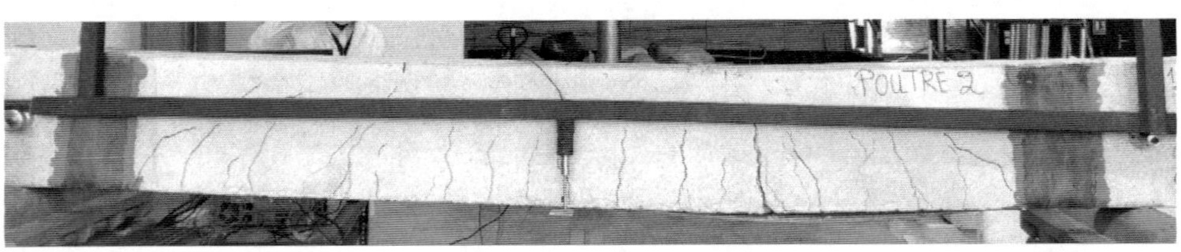

Figure 5d. BCP1u+UF2u: Carbodur plate delamination and rupture of U fabrics plates, Fu = 87kN.

Figure 5. Failure modes of tested specimens.

4. Conclusion

This investigation shows the effect of strengthening with CFRP and the mechanical behavior of beams under flexural test. The paper proves how composite carbon fiber plate can be used to strengthen or repair a reinforced concrete structure. The Bonding carbon fiber fabric and Carbodur plate increase in the ultimate load by more than 203% compared to a non-reinforced beam and a one-third reduction in the sagging of reinforced beams. The use of carbon fiber fabrics in a U-wraps at both ends of the beam substantially influence the performance of strengthened beams. The cracking mechanism was noted and also the different collapses.

5. References

[1] ACI Committee 440 2002 *"Guide for the Design and Construction of Externally Bonded FRP Systems for Strengthening Concrete Structures" American Concrete Institute Michigan USA.*

[2] Raithby K D 1980 *External strengthening of concrete bridges with bonded steel plates TRRL Supp. Report 612, crowthorne Transport & Road Research Laboratory Departement of the Environment.*

[3] Ivanova J, Valeva V, Yanakieva A, Petrova T and Becker W 2016 *Damage of bi-material structures and reinforced composites with different industrial applications Journal of Sustainable Development of Energy Water and Environment Systems* **4**(1) 23-37.

[4] Assih J 1998 *Contribution à l'étude du renforcement de poutres en béton arme par collage de plaques composites en fibres de carbone.* Thèse de doctorat Reims.

[5] Assih T, Li A and Delmas Y 1997 *Flexural strengthening of reinforced concrete beams using externallycarbonfibre composite sheets. EUROMECH 358 Mechanical Behaviour of AdhesiveJoints, Proceeding of the Euromech colloquium 358, edition pluralis* 257–266.

[6] Li A, Diagana C and Delmas Y 2001 *CRFP Contribution to shear capacity of strengthened RC beams. Engineering Structures* **23** 1212–1220.

[7] Ianniruberto U and Imbimbo M 2004 *Role of Fiber Reinforced Plastic Sheet in Shear Response of Reinforces Concrete Beams: Experimental and Analytical Results Journal of composites for construction* **8**(5) 415-424.

[8] Alferjani M B S, Samad A B, Elrawaff B S, Mohamad N B and Ahmad M H B 2014 *Shear Strengthening of Reinforced Concrete Beams Using Carbon Fiber Reinforced Polymer Laminate: A Review American Journal of Civil Engineering* **2**(1) 1-7.

[9] Triantafillou T C and Plevris N 1992 Strengthening of RC beams with epoxy bonded fiber composite materials Materials and Structures **25** 201-211.

[10] Zhou Y, Wang X, Sui L, Xing F, Wu Y and Cheng C 2018 *Flexural performance of FRP-plated RC beams using H-type end anchorage. Composite Structures* **206**, 11-21.

[11] Cheng C, Wang X, Sui L, Xing F, Chen X L and Zhou Y 2019 *Influence of FRP thickness and confining effect on flexural performance of HB-strengthened RC beams Composite Part B* **161** 55-67.

[12] Ahmed E, Sobuz H R and Sutan N M *2011Flexural performance of CFRP strengthened RC beams with different degrees of strengthening schemes International Journal of the Physical Sciences* **6**(9) 2229-2238.

[13] Jayaprakash J, Pournasiri E, Choong K K, Tan C G and De'nan F 2011 *External CFRP repairing of pretested beams reinforced using prestress rebars Journal of Reinforced Plastics and Composites 1–16.*

[14] Ivanova I 2013 *Comportement mécanique de consoles courtes en béton armé renforcées ou réparées par collage de matériaux composites Thèse de doctorat Reims.*

[15] Dontchev D, Ivanova I and Assih J 2018 *Contribution in strengthened reinforced concrete structures by externally bonded carbon fibres fabrics Journal of Chemical Technology and Metallurgy* **53**(4) 765-772.

Acknowledgments

The authors gratefully acknowledge the financial support from, Region Grand Est in France, University of Reims Champagne-Ardenne and University of Chemical Technology and Metallurgy.

IOP Conf. Series: Materials Science and Engineering 1252 (2022) 012052 doi:10.1088/1757-899X/1252/1/012052

Thermodynamic Aspects of Viscoelastic Materials' Linear and Nonlinear Behavior

M Minárová[1] and K Tvrdá[2]

[1] Department of Mathematics and Descriptive Geometry, Faculty of Civil Engineering, Slovak University of Technology Bratislava, Slovakia
[2] Department of Structural Mechanics, Faculty of Civil Engineering, Slovak University of Technology Bratislava, Slovakia

E-mail: maria.minarova@stuba.sk; katarina.tvrda@stuba.sk

Abstract. Materials that exhibit neither purely elastic nor purely viscous behavior, are dealt by viscoelasticity theory. Therein, each material is represented by a viscoelastic model. In this paper a viscoelastic exploration is dealt, focused to the quantification of energy acting during the load – response processes of viscoelastic material. In line with thermodynamic principles, nonnegativity of dissipated energy is supposed within the process, whereby the only reason of energy dissipation is the dashpot, i.e. the viscous member of viscoelastic matter present in the model. The assumption of potential energy storing by and only by the elastic member is taken into account at the same time. Each time we refer a model, the material matching with the model is meant. This work is a continuation, the enhancement of previous research of authors where particular investigations in this field were focused.

1. Introduction

Viscoelasticity is a part of rheology, which studies so called soft materials, i.e. materials that are neither elastic nor viscous. Focused material matched with its attached model is studied together with its response to a mechanical load. Each model of a viscoelastic material consists of several viscous and several elastic elements in various configurations. When loaded, elastic body stores all received energy and disposes with it until the load persist. After ceasing the load, all this energy is used for its total recover. On the other hand, a loaded viscous body dissipates all received energy, nothing remains to recover, [1]. The dissipated energy is released as heat to the ambient. Because of a possible impact of dissipated energy to the farther physical parameters, it is essential to check or control the thermodynamic aspects of the mechanical process, as well. The energy dissipated during mechanical processes – released heat increases the temperature of the ambient, hence the physical parameters of the material operating within this ambient can change. Consequently, if the influence of the temperature gradient to the parameter change is significant, the nonlinear approach has to be taken into account instead of the linear one. Accordingly, the quantification of the dissipated energy is essential within a mechanical behavior exploration of a viscoelastic body, [2]. In [3] the recent exploration is provided, tackling the energy split within the load – response process. The attention is aimed to the models where an elastic member is connected serially to the rest of the model. That enables an instant (partial or total) energy storing. The assumption of "not too big" amount of imposed energy is considered in order to ensure the elastic member to store all this energy initially. As a representative of such models, the Maxwell model (M) was treated and the theoretical material tests were performed under various types of load. The investigation presented in this paper is focused to models that have no freely serially connected elastic element; accordingly they do not store any energy instantly. The simplest representative of such model Kelvin-Voigt model (KV) is dealt herein.

2. Thermodynamics and viscoelasticity

Elementary members of viscoelastic models (H) – Hookean matter and (N) – Newtonian matter, see figure 1, are governed by their particular governing equations

$$\text{(H)}: \sigma(t) = E\varepsilon(t) \tag{1}$$

$$\text{(N)}: \sigma(t) = \eta\dot{\varepsilon}(t) \tag{2}$$

with $\sigma(t)$ and $\varepsilon(t)$ being stress and strain, both functions of time, dot above an item stands as its time derivative; E modulus of elasticity and η coefficient of viscosity. Due to time derivative presence inside each constitutive equation of viscoelastic model, it is apparent that the mechanical response of viscoelastic materials depends not only on the magnitude of the imposed load, but also on its velocity and duration. Having (H) and (N) at hand, we can combine them in parallel or in series, see Kelvin-Voigt model (KV) where one elastic and one viscous element are connected in parallel and Maxwell model (M), where the serial connection of both is used. In this way, by additional members involvement, various configurations of models can be created, that can represent less or more exactly, viscoelastic materials, [4,5, etc.].

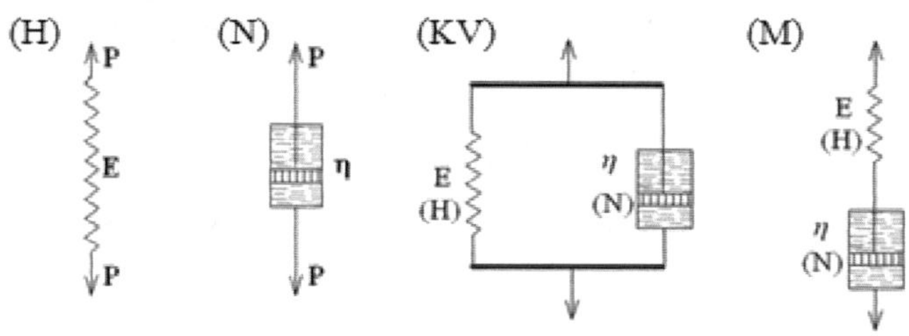

Figure 1. Elementary members: (H) Hookean, (N) Newtonian; simplest configurations: (KV) – parallel connection of (H) and (N), (M) – serial connection of both.

Required thermodynamic consistency of a viscoelastic model states the amount of dissipated energy rate $\dot{U}_d(t)$ during any loading process is nonnegative, [6]:

$$\dot{U}_d(t) = \dot{U}(t) - \dot{U}_p(t) \geq 0 \tag{3}$$

where $U(t)$ is total deformation energy and $U_p(t)$ is potential energy which is stored during the loading maintained, and used further for the total or partial recovery after cease of load. However, since each viscoelastic model consists of elastic and viscous elements and we do not take other energy kinds into account, it is sure that received energy is either stored or dissipated. This energy split still varies in time.

$$U(t) = U_p(t) + U_d(t) \tag{4}$$

Next, we can use the integral expression for energies one by one, following from the relation, [7]:

$$\dot{U}(t) = \sigma(t)\dot{\varepsilon}(t) \tag{5}$$

$$U(t) = U(0) + \int_0^t \dot{U}(z)\,\mathrm{d}z = U(0) + \int_0^t \sigma(z)\dot{\varepsilon}(z)\,\mathrm{d}z \tag{6}$$

Since it is the elastic member, and only itself, which stores the potential energy and only viscous member is responsible for dissipation of energy, the explicit integro differential expressions of potential and dissipated energies follow from the modified, i.e. indexed forms of equation (5): $\dot{U}_\mathrm{p}(t) = \sigma_H(t)\dot{\varepsilon}_\mathrm{H}(t)$ and $\dot{U}_\mathrm{d}(t) = \sigma_N(t)\dot{\varepsilon}_N(t)$:

$$U_p(t) = U_p(0) + \int_0^t E\varepsilon_H(z)\dot{\varepsilon}_H(z)\,\mathrm{d}z = U_p(0) + \int_0^t E^{-1}\sigma_H(z)\dot{\sigma}_H(z)\mathrm{d}z \tag{7}$$

$$U_d(t) = U_d(0) + \int_0^t \eta \dot{\varepsilon}_N{}^2(z)\,\mathrm{d}z = U_d(0) + \int_0^t \eta^{-1} \sigma_N{}^2(z)\mathrm{d}z \quad (8)$$

Remark: It is practical to write energies, equations (7) and (8) in sense of ε or σ. This practical reason is – either the stress is the action and strain the reaction during a material examination, or vice versa. Equation (6) can be expressed in sense of either stress or strain wholly, as well. It can be done so by using so-called time dependent material characteristics – relaxation modulus or compliance modulus, [3].

3. Kelvin-Voigt model, response to a mechanical load; energy quantification

Next and onward, we focus on the Kelvin-Voigt model. We will perform theoretical material tests with various stress functions as the action load used; and strain the reaction. Alongside these tests, we follow up the energies by using forms (6) – (8).

Since the evaluation of energies is the same within all presented examples, we do not see a need to put the exact, often too long and complicated expression of resulting energies. Thus we prefer to provide their graphical performance. For illustration, we put the resulting energies expressions only in Example 2, chapter 3.2, where these expressions are brief and easy to be understood.

The configuration of (KV) – parallel connection, yields the geometric equations

$$\varepsilon(t) = \varepsilon_H(t) = \varepsilon_N(t) \quad (9)$$

$$\sigma(t) = \sigma_H(t) + \sigma_N(t) \quad (10)$$

reflecting the relation among particular and global entities. The incidence with elastic or viscous matter is stated by lower index herein.

From physical equations – equations (1) and (2) and geometric equations – equations (9) and (10), namely by eliminating the indexed variables we derive the global constitutive equation of (KV)

$$\sigma(t) = E\varepsilon(t) + \eta\dot{\varepsilon}(t) \quad (11)$$

which is an ordinary differential equation. Its solution depends on and is tied up with given action.

3.1. Continuous action function

We start the observation in the time instant $t_0 = 0$, and we suppose both action and reaction are zero before. At that time our observation begins, when an immediate increase of load starts, either in a ramped or in a stepped form. We impose the action to a model and study the reaction together with energies functions in this loading time. Hence, with regard to the given action $\sigma(t)$ and given an initial condition $\varepsilon(0)$, we compute the solution to the equation (11) getting resulting reaction function $\varepsilon(t)$. Then we substitute it together with its time derivative, into equations (6) – (8) and evaluate the energies functions. Finally, all results are performed in graphs.

Example 1: Linear action (ramped form): $\sigma(t) = \begin{cases} 0, & t < t_0 \\ at, & t \geq t_0 \end{cases}$

The solution of governing equation (11) with regard to the given action $\sigma(t)$ and with zero initial condition $\varepsilon(0) = 0$, is

$$\varepsilon(t) = \frac{a\eta}{E^2}\left(exp\left(-\frac{E}{\eta}t\right) - 1 + \frac{E}{\eta}t\right) \quad (12)$$

The graphical summarization – action, reaction, as well as and energy split are provided in Figs. 2 and 3. Let us discuss the reaction function and the energy functions split having a look at these pictures: At the beginning of the ramped loading increase, the reaction $\varepsilon(t)$ is delayed, affected by the configuration (KV)=(H)|(N), i.e. elastic and viscous member connected in parallel, which is the phenomenon causing the retardation. Then, after the retardation a linear stabilization appears, figure 2, lower graph.

And what about the energies? Potential energy and total energy perform a delay at the beginning, whereas, after a while they continue upwards linearly; while the dissipated energy rate decreases, resulting in its converging to a constant value finally.

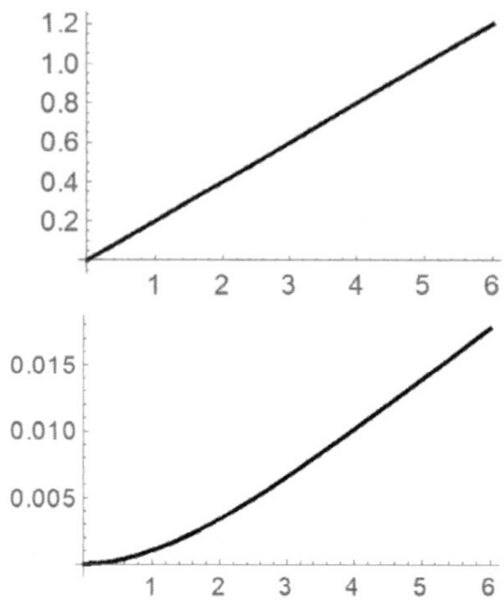

Figure 2. Upper graph: stress load $\sigma(t)$ – linear function with ramped start at $t_0 = 0$, lower graph: reaction function, strain $\varepsilon(t)$.

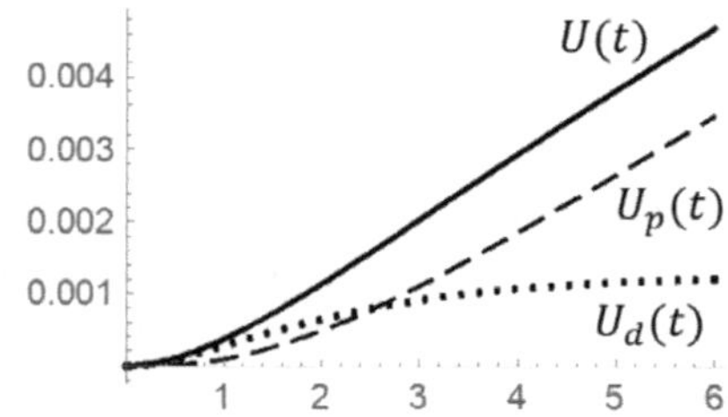

Figure 3. $U(t)$ – Total energy, $U_p(t)$ – potential energy, $U_d(t)$ – dissipated energy.

3.2. Non-continuous action function, discontinuity in $t_0 = 0$

Next, we consider non-continuous action function treated. First, a stepwise constant function jumping from zero to σ^* immediately is taken, then a linear function with an initial jump and exponential action functions are dealt. General considerations about action – reaction dependence is given in [8].

Example 2: Step-wise action function $\sigma(t) = \begin{cases} 0, & t < t_0 \\ \sigma^*, & t \geq t_0 \end{cases}$

The solution of governing equation (11) with regard to the given action $\sigma(t)$ and with zero initial condition $\varepsilon(0) = 0$, is

$$\varepsilon(t) = \sigma^*(\frac{1}{E} exp\left(-\frac{E}{\eta}t\right) - 1) \tag{13}$$

As mentioned before, we provide the energy expressions for the sake of illustration:

$$U(t) = \frac{\sigma^{*2}}{2E}\left[3 - 2exp\left(-\frac{E}{\eta}t\right)\right] \tag{14}$$

$$U_p(t) = \frac{\sigma^{*2}}{2E} exp\left(-\frac{2E}{\eta}t\right)\left[exp\left(\frac{E}{\eta}t\right) - 1\right]^2 \tag{15}$$

$$U_d(t) = \frac{\sigma^{*2}}{2E}\left[2 - exp\left(\frac{2E}{\eta}t\right)\right] \tag{16}$$

Remark: Energies computed by using equations (6) – (8) can be whenever confronted with control sum represented by the equation (4). I.e., the sum of potential and dissipated energy should yield total deformation energy. In this example, the sum of potential energy expressed by the equation (15) and the dissipated energy as expressed in the equation (16) should yield the resulting total energy, performed by the equation (14). And, this is fulfilled. Graphical performance of action, reaction, equation (13), and the graphs of energies functions, equations (14) – (16) can be seen in figures 4 and 5.

In figure 4 it can be seen, how (KV) model copes with a jump. After certain time a constant tendency in responding deformation function is evident. Dissipated energy amount jumps immediately by some value, afterwards continues approaching to a constant value until the load persists. Potential energy first increases slowly, due to dashpot retardation, see the small convex onset of $U_p(t)$. Farther, after filling its capacity it remains at a constant value, see figure 5.

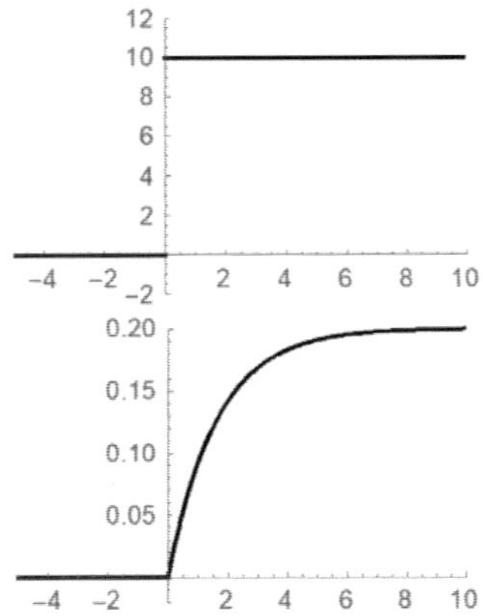

Figure 4. Upper graph: stress action $\sigma(t)$ – step-wisely constant function, initial time $t_0 = 0$, lower graph: reaction, strain function $\varepsilon(t)$.

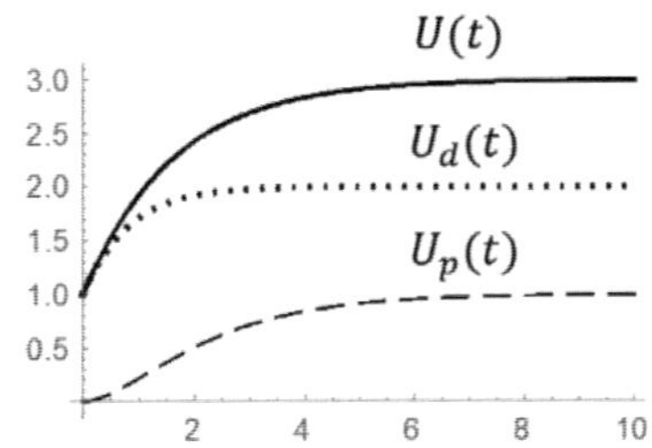

Figure 5. $U(t)$ – total energy, $U_p(t)$ – potential energy, $U_d(t)$ – dissipated energy.

Example 3: Linear action function with nonzero absolute member $\sigma(t) = \begin{cases} 0, & t < t_0 \\ at + b, & b \neq 0, \; t \geq t_0 \end{cases}$ represents the simplest type of load varying with time and involving a jump. The procedure of energy functions computation remains the same. Here, just a coupling of cases from Examples 1 and 2 can be done, hence, the comparison of the results can be even intuitively done.

The solution of governing equation (11) with regard to the given action $\sigma(t)$ and with zero initial condition $\varepsilon(0) = 0$, is

$$\varepsilon(t) = \frac{1}{E^2}\left[bE(1 - exp\left(-\frac{E}{\eta}t\right)) - a\eta(1 - exp\left(-\frac{E}{\eta}t\right)) + aEt\right] \tag{17}$$

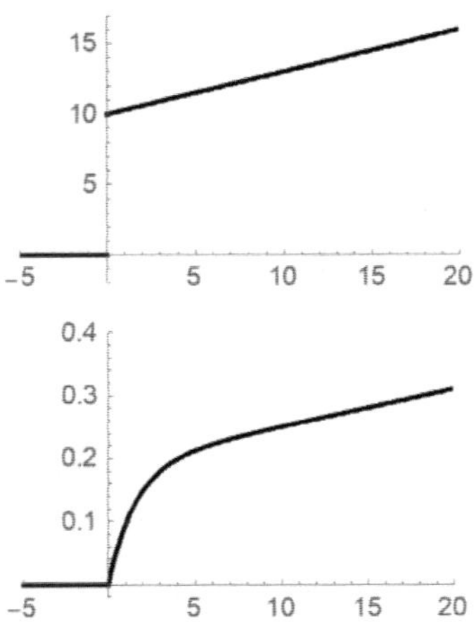

Figure 6. Upper graph: stress action function $\sigma(t)$ – linear function with an immediate change in $t_0 = 0$, lower graph: reaction, strain function $\varepsilon(t)$.

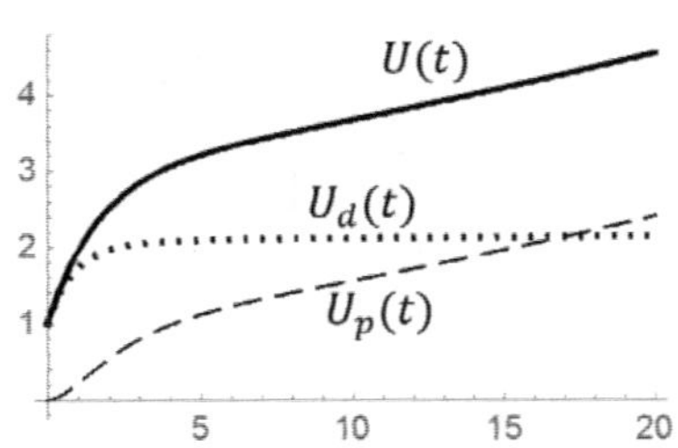

Figure 7. $U(t)$ – total energy, $U_p(t)$ – potential energy, $U_d(t)$ – dissipated energy.

As the energies treated in Example 2 and 3 differ only in a constant, the results are similar. This behavior is totally different in comparison to Maxwell model, [9] and the discussion about the results just couples the consideration in both Example 1 and 2: Logically, the reaction function is retarded, tending to the appropriate linear function, figure 6. An initial jump of dissipated energy function is followed by its continuation towards a linear increasing function, figure 7. The increase of the potential energy is retarded, which is documented by a small convex onset of function $U_p(t)$, likewise in Example 2. Later on, the concave part follows, resulting in a linear shape.

Example 4: Exponential function (involving an initial jump) $\sigma(t) = \begin{cases} 0, & t < t_0 \\ k\, e^{at}, & t \geq t_0 \end{cases}$

The solution of governing equation (10) with regard to the given action $\sigma(t)$ and with zero initial condition $\varepsilon(0) = 0$, is

$$\varepsilon(t) = \frac{k}{E+a\eta} exp\left(-\frac{E}{\eta}t\right)\left[exp\left(at + \frac{E}{\eta}t\right) - 1\right] \tag{18}$$

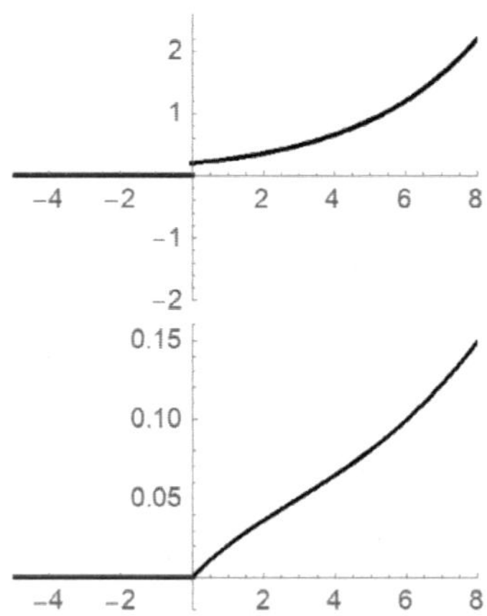

Figure 8. Upper graph: stress action function $\sigma(t)$ – exponential function having an initial jump at $t_0 = 0$, lower graph: reaction, strain function $\varepsilon(t)$.

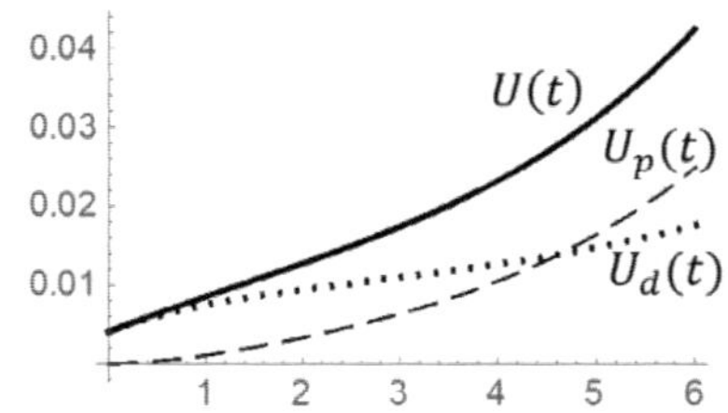

Figure 9. $U(t)$ – total energy, $U_p(t)$ – potential energy, $U_d(t)$ – dissipated energy.

The model again copes with jump while being retarded by the viscous member connected in parallel. That is why we can see a concave part at the beginning of deformation function, in spite of action function being convex all time, figure 8, lower graph. Then, after reaching an inflex point, an overcome of retardation, it continues convexly upwards, even to ∞ theoretically. Dissipated energy, strongly influenced by reaction function $\varepsilon(t)$ starts in a concave, but continues in convex mode, see figure 9.

3.3. Periodic action function

Lastly, we tried to employ a periodic function as an action and study the reaction together with energy split. We chose a nonnegative goniometric function in order to perform the initial retardation.

Example 5: Periodic action $\sigma(t) = \begin{cases} 0, & t < t_0 \\ 1 - \cos(t\,\pi/5), & t \geq t_0 \end{cases}$

The solution of governing equation (10) with regard to the given action $\sigma(t)$ and with zero initial condition $\varepsilon(0) = 0$, is

$$\varepsilon(t) = -25\,E + \frac{\eta^2\pi^2}{E}\left(exp\left(-\frac{E}{\eta}t\right) - 1\right) + 25E\,cos\frac{\pi t}{5} + 5E\eta\pi\,sin\frac{\pi t}{5} \tag{19}$$

As done before, all energies are computed by substituting $\varepsilon(t)$ and its time derivative to the equations (6)-(8), yielding particular energy functions. The graphs are performed in figure 11.

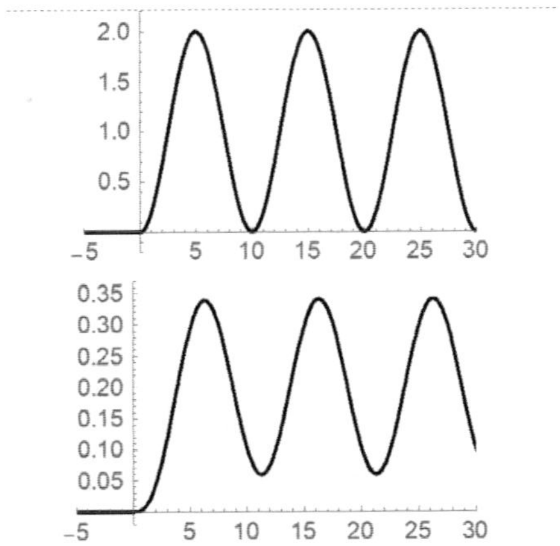

Figure 10. Upper graph: action $\varepsilon(t)$ – periodic function starting from zero value at $t_0 = 0$, lower graph: reaction function $\sigma(t)$.

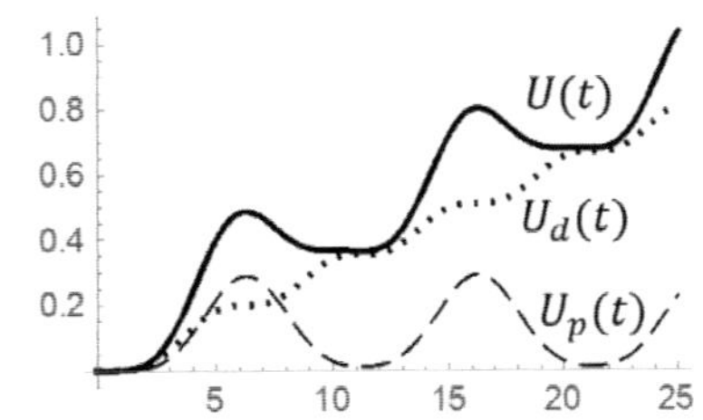

Figure 11. $U(t)$ – total energy, $U_p(t)$ – potential energy, $U_d(t)$ – dissipated energy.

The retardation in reaction $\varepsilon(t)$ is apparent, even the further stabilization in a periodic mode, figure 10, lower graph. As expected, the corresponding potential energy alternates going down almost to zero and then upwards periodically, see figure 11. The dissipation of energy has an increasing tendency, however, a waving appears due to the periodicity of the loading stress function.

4. Conclusion

In this paper an investigation of thermodynamic processes is presented. This exploration is important because of the influence of released heat to farther behavior of the models. According to the type of material and taking the amount of dissipated energy, the linear versus nonlinear modeling is chosen. The amount of total energy delivered to the model is easy to evaluate. Moreover, the amount of potential energy testifies the reversibility of the model (material behind) after load cease.

We are aware that there are a lot of situations within of serviceable life of materials when the load increases in quadratic sense or exponentially, etc. We have executed these theoretical experiments, and a lot of others, and it would be fantastic to compare and validate the results with corresponding lab tests. Nevertheless, in our opinion, prior to this, it is worth having some compact theoretical results at hand.

5. References

[1] Christensen R M 1978 *Theory of Viscoelasticity, An Introduction,* Mir (Moscow)
[2] Sumec J and Malast M 2004 *Influence of dissipative Heating, Proc. New trends in Statics and Dynamics of Buildings* (Bratislava)
[3] Minárová M 2020 *Isothermal viscoelasticity and energy, Proceeding of the 4th International Scientific Conference Structural and Physical Aspects of Construction Engineering (SPACE 2019), MATEC Web Conference,* vol. 310 (Bratislava)
[4] Minárová M 2014 *Mathematical Modelling of Phenomenological Material Properties. Differential Operator Forms of Constitutive Equations, Slovak Journal of Civil Engineering*, vol. 22 (4), (Bratislava)
[5] Sobotka Z 1981 *Rheology of Materials and Structures,* Academia (Prague)
[6] Krejčí P 1996 *Hysteresis, Convexity and Dissipation in Hyperbolic Equations*, Gakuto Intern. Ser. Math. Sci. Appl., vol. 8, Gakkoto-sho (Tokyo)

[7] Tschoegl N W 1989 *The Phenomenological Theory of Linear Viscoelastic Behavior*, Springer-Verlag (New York)
[8] Minárová M 2020 *Duhamel Hereditary Integrals in Uniaxial Viscoelasticity, Proceeding of the International Scientific Conference ESCIM 2020* (Budapest)
[9] Minárová M 2018 *Rheology. Viscoelastic and viscoelastoplastic modelling. Habilitation thesis* (Bratislava)

Acknowledgments

This work was supported by grants APVV-17-0066, APVV-18-0052, VEGA 1/0522/20, VEGA 1/0412/18, VEGA 1/0453/20 and VEGA 1/0006/19.

Use of Building Information Modelling (BIM) in planning and managing the life cycle of buildings

J Smetanková[1], P Mésároš[2], K Krajníková[2], V Biolek [3]

[1] Center of Research and Innovation in Construction, Faculty of Civil Engineering, Technical University of Kosice, Vysokoskolska 4, 040 23 Kosice, Slovakia
[2] Institute of Technology, Economic and Management in Construction, Faculty of Civil Engineering, Technical University of Kosice, Vysokoskolska 4, 040 23 Kosice, Slovakia
[3] Institute of Structural Economics and Management, Faculty of Civil Engineering, Brno University of Technology, Veveří 95, 602 00 Brno, Czech Republic

E-mail: jana.smetankova@tuke.sk, peter.mesaros@tuke.sk, katarina.krajnikova@tuke.sk, biolek.v@fce.vutbr.cz

Abstract. Many studies and research point to a growing trend in industrial digitization. New materials, technologies and advanced automation tools are increasingly being promoted in the construction industry. With the increasing degree of digitization, various ways and methods of planning, management and control across different construction processes are also increasing. BIM is a unified information system that enables the sharing of documentation, supports communication, streamlines individual processes, helps eliminate errors and increases quality. Based on this fact, the paper analyses the current state of use of Building Information Modelling in planning and managing the life cycle of buildings in Slovakia and the Czech Republic. The current state of knowledge and use of BIM, defining the advantages and possible barriers to its use, examining the scope and potential of BIM implementation within the building life cycle and defining tools and applications used in time and financial planning of construction production were analysed. The paper also briefly examines how the implementation of BIM can affect selected key tasks of the preparation and valuation professions.

1. Introduction

Individual aspects of the company have been gradually transformed from manual to digital processes for several years. In the construction sector, the rate of digitization and implementation of innovative tools is increasing. As digitalisation increases, so does the level of productivity, choice, security and the elimination of inefficiencies.

To a greater extent, various elements of robotics are used in individual construction processes, such as drones or elements for improving visualization, such as augmented reality, virtual reality and building information modelling (BIM), which increases efficiency and competitiveness.

The use of BIM tools is constantly increasing. One of the key reasons for adopting BIM is to ensure a balance between project management (quality and function), cost and time. Achieving a balance between these factors is relatively difficult, because of using a number of strategies and solutions, for example innovative technologies, such as building information modelling [1] [2].

Abanda and colleagues described BIM as a global digital technology that will bring a sophisticated structure to the construction industry's processes. STN EN ISO 29481-1: 2016 defines information modelling of buildings as a shared digital representation of the physical and functional characteristics of any constructed object, which is able to provide a reliable basis for decision-making [3] [4].

BIM can be considered as a semantically based and object-oriented approach that allows the management of a comprehensive information system, including 3D visual aids. BIM features allow the evaluation of project activities and the management of all operations within the built environment, while providing a database containing geometric and non-geometric data [5] [6] [7] [8] [9] [10].

The role of BIM technology is to support activities performed across the life cycle of a building by providing information on the geometry of the building and descriptive information about the building and its individual elements. The development of BIM improves construction processes and accelerates the implementation of constructions. Along with the development of technologies such as building information modelling, augmented reality, virtual reality and radio frequency sensors (RFID), new hardware and software tools are also being introduced into processes. These tools have thus increased and accelerated the automation of construction processes [11].

As the implementation of information modelling of buildings increases, so does the number of studies examining the benefits of using BIM across the life cycle of a building. The main advantages of BIM include the ability to collect data in one place, in the BIM model together with the geometric presentation of the building structure, i.e. the 3D model. Data collection across the building life cycle (design phase, use phase) allows to analyse this data, generate new views and simulate various processes and interrelationships by creating new visualization methods using virtual and mixed reality, for example, to improve communication and information at "one place" [12].

The research entitled Dynamics of the level of BIM application in Russia in 2017-2019 examined the extent of BIM utilization in selected construction companies. The rate of active use of BIM in 2017 and 2019 was at the same level - 22%. The slow distribution of technology within Russia has been defined as the main cause of the stagnant implementation of BIM in construction processes. Based on the above facts, there was a decline in interest in the introduction of BIM into individual processes - in 2017 at the level of 14%, in 2019 at the level of 6%.The companies that implemented BIM into their processes defined as the main benefits of use, above all, an increase in competitiveness, quality and efficiency. The main disadvantage of BIM implementation was the low number of qualified staff [13].

Research Assessing BIM Application in the South African Construction Project Management Milieu examined the extent of BIM use in South Africa. The survey was attended mainly by project and construction managers. Approximately 35% of the respondents were familiar with BIM technology and up to 45% of these users actively used it in their projects. 93% of respondents who did not use BIM in their projects assumed that they would implement the technology into their processes in the next 5 years [14].

2. Methodology

2.1. Research aim

The research was focused on the use of Building Information Modelling in the planning and management of the life cycle of buildings with emphasis on selected economic parameters. The aim was to analyze the approach and use of information modelling tools for buildings across the various stages of the construction process. The research analyses general information associated with the use of BIM tools - purpose and rate of use, advantages, disadvantages, and at the same time leads in more detail to the financial and time planning of construction production and the use of this technology in processes.

2.2. Data collection and research sample

Construction companies in Slovakia and the Czech Republic were addressed. The aim of the research was to address the general public - companies which operate in the public or private sector, across the life cycle of construction, companies which involve all participants in construction production -

government, construction and technical supervision, suppliers, designers, planners, construction and technical control. architects, builders and investors. The data were obtained through an online questionnaire. Within Slovakia, about 800 companies were contacted by e-mail and about 700 companies in the Czech Republic. Of the total number of companies and organizations contacted, the return rate of the questionnaire was relatively low, in both countries up to 10%. The reason for the low return was mainly reluctance (high number of parallel online surveys - reduction of personal meetings due to the Covid-19 pandemic), high workload of companies and lower awareness.

2.3. Research step and methodology

In the introduction of the questionnaire, the respondent was acquainted with the purpose of the questionnaire and the approximate time frame for its completion. The questionnaire itself consisted of 33 questions and was divided into 3 sections:

- Section 1 - information about the company / organization - type of organization, sector of operation, focus, size and character of the company,
- Section 2 - information on the use of BIM in the company / organization - use, tools, degree of implementation, advantages and disadvantages of the use and implementation of BIM,
- Section 3 - information on the method of time and financial planning of construction production - used software applications (types, tools), advantages and disadvantages of using BIM in financial and time planning.

3. Results

3.1. Information about the organizations

Information on companies was obtained within this section. The aim of the research was to address the general professional public operating in the private and public construction sector. The research was carried out in the period May - July 2021.

From the point of view of information about the company, the following data were obtained:

- type and size of the organization,
- sector of operation (public private, region),
- focus and character of the company and
- duration of operation on the market.

In terms of the type of organizations, limited liability companies were the most represented in both countries (Slovakia 50%, Czech Republic 42.3%), followed by local governments (Slovakia-20%) and joint-stock companies (Czech Republic -19.2%). A detailed breakdown is shown at Figure 1. Type of company/organization.

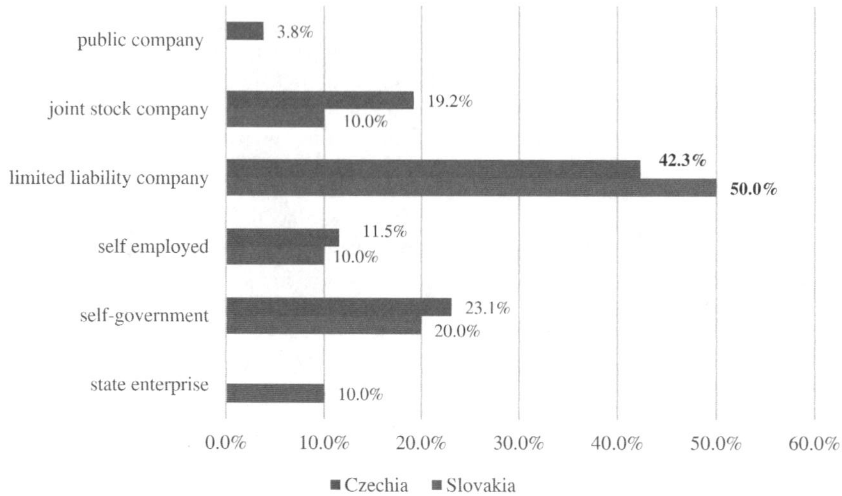

Figure 1. Type of company/organization (data from the period May - July 2021).

The survey was attended mainly by companies operating simultaneously in the public and private sector - Slovakia 65%, Czech Republic 50%, then companies operating mainly in the private sector - Slovakia 25%, Czech Republic 30.8% and the smallest representation were companies operating exclusively in the public sector - Slovakia 10% and the Czech Republic 19.2% of the addressed companies.

From the point of view of operation - geographical division, companies operating in the Prešov (45%), Košice (40%) and Bratislava (35%) regions were mainly represented in Slovakia. In the Czech Republic, companies operating in the South Moravian Region (50%), in the capital Prague (34.6%) and in the Olomouc Region (30.8%) had the largest representation. In terms of focus, or type of construction projects on which the company focuses, in Slovakia were represented mainly companies implementing housing (80%), buildings for administration (65%), other housing (60%) and road, local roads (55%). In the Czech Republic, these were mainly housing construction companies (69.2%), administrative buildings (80.8%), buildings for culture, public entertainment, education and healthcare (76.9%) and buildings for trade and services (61.5%).

From the point of view of the participants in the construction, the companies representing as:

- main construction contractor - Slovakia 50%, a
- designers, architects - Czech Republic 65.4% (see ***Figure 2.*** *Character of the company / organization within the construction activity (type of construction participant)*), operating on the market for more than 30 years (Slovakia 45%) and 21 to 30 years (Czech Republic 46.2%) respectively (see **Table *1.*** Size and length of market presence of the company´s or organization.)

Table 1. Size and length of market presence of the company´s or organization.

Question	Answers	Slovakia	Czechia	Question	Answers	Slovakia	Czechia
The size of the company's organization	micro-company (0 - 9 employees)	10.0%	23.1%	Length of market presence	0 - 10 years	10.0%	15.4%
	small company (10 - 49 employees)	**50.0%**	23.1%		11 - 20 years	25.0%	15.4%
	medium-sized	25.0%	23.1%				

company (50 - 249 employees)			21 - 30 years	20.0%	**46.2%**
large company (250 and more employees)	15.0%	**30.8%**	over 30 years	**45.0%**	23.1%

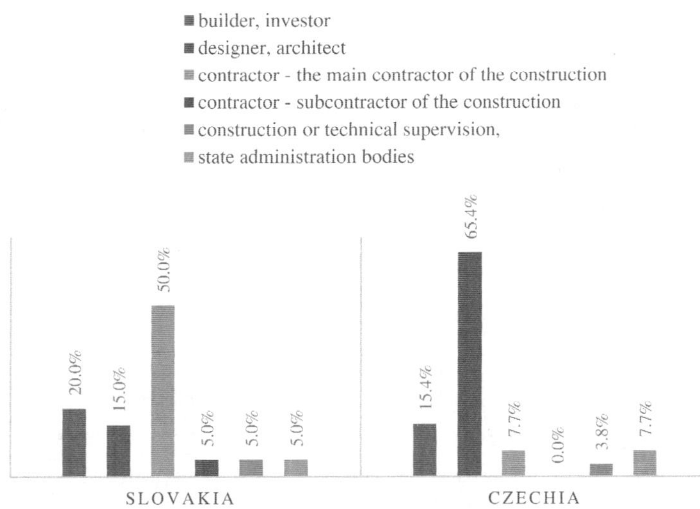

Figure 2. Character of the company / organization within the construction activity (type of construction participant).

3.2. Use of BIM in construction organization

The second part of the research was focused on the degree of knowledge and use of information modelling of buildings within the individual processes of organizations. Respondents answered the questions:

- Do you use BIM in your work?
- Have you heard of this technology?
- What is the purpose for which you most often use BIM applications and tools?
- What is the probability of implementing BIM in your processes?
- What are the reasons not to use BIM in your processes?
- What are the biggest benefits and barriers to implementing BIM in your processes?

The difference in the rate of use of BIM technology within the life cycle processes of buildings differs in Slovakia and the Czech Republic. In the Czech Republic, up to 73.1% of the addressed companies and organizations have already used the tools of BIM technology in their construction projects, while in Slovakia it is only about 35%. Among the respondents who stated that they had no experience with BIM tools, we wondered if they had heard of the technology. In the Czech Republic, all addressed BIM users are familiar with this technology and there is a very high probability that they will implement it in their processes (42.6%). In Slovakia, only 76.9% of the addressed BIM users heard about the mentioned technology, while up to 46.2% of respondents claim that the probability of implementation is low or no probability of implementation (23.1%). More information can be seen at Figure 3. Use and experience with BIM - Slovakia, Czechia.

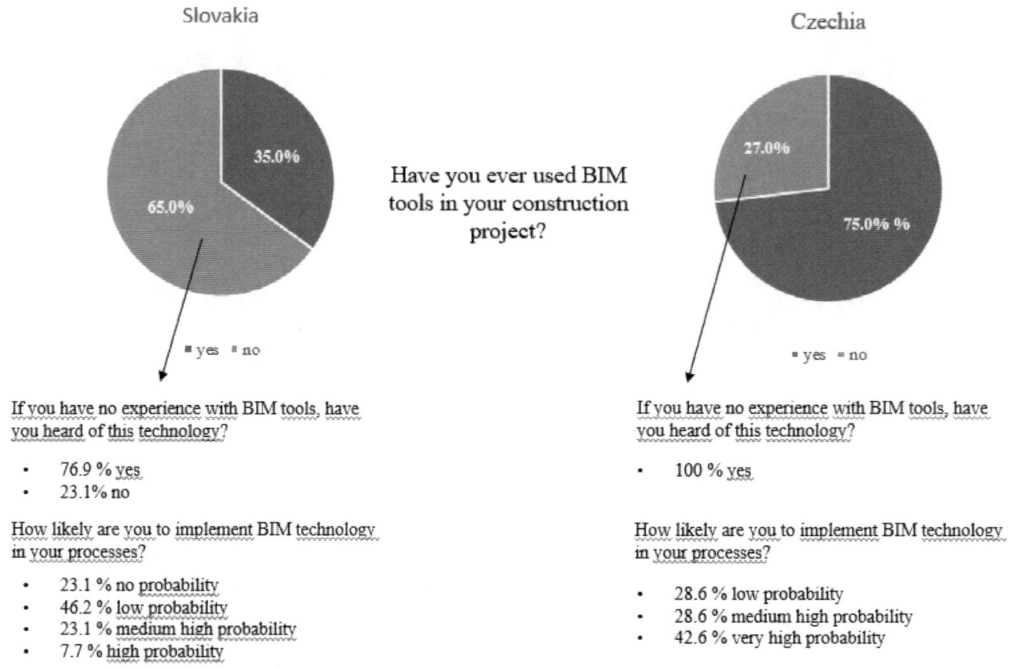

Figure 3. Use and experience with BIM - Slovakia, Czechia.

In the case of companies that stated that they know the mentioned technology and actively use it, we asked what is the current state of use of BIM tools within their implemented projects. In Slovakia, 85.7% of companies also use BIM for current projects and these were mainly projects with a cost of up to EUR 500,000. In the Czech Republic, it was up to 94.7% of companies that currently actively use BIM tools for their projects, especially for projects where the cost was more than EUR 5,000,000 and more.

Slovak and Czech companies most often use BIM technology when designing a design solution and controlling the use of resources. This technology is rarely or not used at all in financial planning, quality control and facility management.

From the point of view of creating a design proposal - a 3D model, the most frequently used software tool in Slovakia is the application Graphisoft ARCHICAD and Autodesk Revit (both approx. 33.3%). In the Czech Republic, it is an Autodesk Revit software tool (approximately 89.9%).

In terms of time and financial planning, the rate of use of BIM tools is low, up to 50% (Slovakia) and 66.7% (Czech Republic) of the addressed companies do not use any BIM tools. In the case of using BIM tools in time and financial planning, the most used tool is the Naviswork Manage application.

The reasons why different organizations do not use BIM technology in their processes are different. The main reasons include in particular:

- no request from the client for its use - 38.9% in Slovakia and 33.3% in Czech Republic,
- satisfaction with existing, already used software tools (not BIM tools) - 11.1% Slovakia, 4.8% Czech Republic,
- too high investment - 16.7% Slovakia, 4.8% Czech Republic,

Reasons were given, such as the considerable time required for the training, the overly complicated use, the lack of information on how to use the BIM tools or the fact that users would like to try the technology first, before purchasing it.

Active users cited the following as the main benefits of using BIM in their processes:

- improving the planning of construction processes,
- reduction of time,
- acceleration of design solutions,
- increase productivity and efficiency,
- identification and elimination of conflicts and collisions of the project,
- improving communication between stakeholders - promoting cooperation and coordination,
- project visualization - operational simulations,
- high degree of adaptation.

Users have defined the following as disadvantages:

- high costs for training, software purchase, BIM tools and learning curve,
- incompatibility with partners - do not use BIM tools,
- companies are not sufficiently familiar with the use of BIM,
- the construction market is not yet ready,
- insufficient demand from clients and other organizations,
- lack of guidelines and standards on how to implement BIM for the construction industry, absence or incomplete national standard for BIM,
- lack of BIM experts,
- social and usual resistance to change.

3.3. Time and financial planning of construction production

Setting goals and creating financial and time plans are among the basic functions of management. These are decision-making processes that are based on a long-term goal, which must accept the development of factors of the internal and external environment on the basis of analysis, anticipation of development and setting of specific goals and procedures for their achievement. Creating a time and financial plan is one of the main aspects that reflect the success of projects.

In this section, the research focused on:

- the most frequently used tools for time and financial planning of construction production,
- the most frequently used functions within BIM applications - time and financial planning,
- advantages of using BIM in time and financial planning,
- the impact of the use of BIM on the work of the preparer and price list of constructions.

The time schedule serves as a guide that informs about the time course of the construction and its overall implementation. The most frequently used tools for time planning of construction production in Slovakia and the Czech Republic include the applications Cenkros / KROS 4 (Slovakia-70%, Czech Republic-11.5%) and MS Project (Slovakia -15%, Czech Republic 53.8%).

Users who use BIM applications for scheduling, such as Nawiswork Manage, most often use the following features:

- creation of time plans,
- detection of 3D and 4D collisions,
- automatic interconnection of graphic and time information,
- rescheduling after change,
- monitoring compliance with deadlines.

Active use of BIM tools in scheduling improves individual activities and processes. Users defined as the main advantages of using BIM applications in time planning, especially improvement of communication between individual participants, improvement of visualization of construction processes, increase of efficiency of construction site planning, improvement of collision detection and identification of problem areas before construction.

Financial planning is an important part of the life cycle. The main task of financial planning is to determine the amount of costs, including profit, as accurately as possible. In Slovakia and the Czech Republic, the most frequently used tool for valuing construction production is the Cenkros / KROS 4 application (Slovakia -90%, Czech Republic -65.4%). In the Czech Republic, the Build Power, Callida and internal budgeting tools of selected organizations - their own tools - are still quite often used.

Users who use BIM applications for financial planning, such as Nawiswork Manage, Synchro Software, most often use the following features:

- determination of costs,
- automated creation of an acreage report,
- use of built-in price databases,
- import of price databases,
- cost management,
- comparison of offers,
- automatic notification of changes.

Users defined as the main advantages of using BIM tools in financial planning, especially increased visualization, support of cooperation between individual participants, increased information quality, better project conceptualization (monitoring costs in individual project phases), more efficient methods and ways of valuing construction production and creating area reports and reducing costs.

The conclusion of the research was focused on the possible impact of BIM implementation on selected professions. We asked the respondents how, according to them, the implementation of BIM will affect the content of the profession of preparer, price list. As many as 50% of respondents in Slovakia and 73.1% of respondents in the Czech Republic stated that the implementation of BIM will significantly change the job description of the profession of preparation and price list of constructions. The rest of the respondents stated that the implementation of BIM will affect the workload of selected professions slightly (Slovakia -40%, Czech Republic 23.1%) or not at all (Slovakia - 10%, Czech Republic 3.8%). According to the respondents, the methods of preparation of tender and contractual documents and the analysis of costs across the life cycle of the project / construction will be most affected. Detailed results are shown in Table 2. Influence of BIM on professions - construction preparer, cost estimator.

Table 2. Influence of BIM on professions - construction preparer, cost estimator.

Question	Answers	Slovakia	Czechia
Influence of BIM on professions - construction preparer, cost estimator	preparation of tender and contractual documents	significant change	significant change
	cost analysis across the life cycle of the project / construction	significant change	significant change
	elaboration of client requirements and feasibility studies,	slight change	significant change
	identification, analysis and development of risk responses	slight change	significant change
	analysis of proposed solutions	slight change	significant change
	invoicing and payment process	slight change	slight change

4. Conclusion

Building Information Modelling (BIM) is an important step towards increasing efficiency, productivity, information and quality within individual processes. The research was carried out in two countries - Slovakia and the Czech Republic. The main goal of the research was to analyse the use of BIM in planning and managing the life cycle of buildings, especially the extent and purpose of using BIM tools, the advantages and disadvantages of using these tools and possible obstacles slowing the implementation of BIM in the processes of valuation and planning of construction production. There was a problem with the reluctance to participate in the online survey in the data acquisition process. In the future, our goal will be to carry out similar surveys in a combined form of data acquisition, ie in the form of personal interviews and an online questionnaire. The main constraint on data collection was the Covid-19 pandemic, many companies were overwhelmed by a number of online surveys and heavily busy, due to various restrictions and problems resulting from the pandemic situation (missing staff, price increases, problems with delivery of materials and services, etc.).

The research was attended mainly by companies operating in the public and private sectors as the main contractors or designers, architects, who mainly carry out the construction of residential buildings, administrative buildings, buildings for culture, public entertainment, education, healthcare and trade and services. A higher rate of use and knowledge of BIM technology was demonstrated in the Czech Republic, while among non-users of BIM in the Czech Republic, there was also a higher degree of probability that they would implement the technology in their processes. BIM is most often used in creating a design proposal and controlling the use of resources. Traditional software applications and tools that are not BIM tools are still the most widely used in time and financial planning processes, these are mainly applications Cenkros/KROS 4, MS Project, Build Power, Callida. The research proved the claim that the implementation of BIM will change the workload of the preparer and the price list of constructions, especially tasks such as the preparation of tender, contract documents and cost analysis across the project life cycle.

5. References

[1] Olawumi T O and Chan D W M 2018 *Beneficial factors of integrating building information modelling (BIM) and sustainability practices in construction projects* Hong Kong International Conference on Engineering and Applied Science (HKICEAS) Higher Education Forum, Hong Kong pp 141-152

[2] Olawumi T O Chan D W M Wong J K W and Chan A P C 2018 *Barriers to the integration of BIM and sustainability practices in construction projects: a Delphi survey of international experts* J. Build Eng. **20** pp 60-71, 10.1016/j.jobe.2018.06.017

[3] International Organization for Standardization (ISO) 2016 *Building Information Models—Information Delivery—Manual—Part 1: Methodology and Format;* STN EN ISO 29481-1:2016 International Organization for Standardization: Geneva, Switzerland

[4] Abanda F H Vidalakis C Oti A H and Tah J H M 2015 *A critical analysis of Building Information Modelling systems used in construction projects* Adv. Eng. Software **90** pp 183-201 10.1016/j.advengsoft.2015.08.009

[5] Marmo R Nicolella M. Polverino F and Tibaut A 2019 A *Methodology for a performance information model to support facility management*. Sustainability 2019 **11** 7007 [Google Scholar] [CrossRef]

[6] Verghote A Al-Haddad S Goodrum P and Van Emelen S 2019 *The effects of information format and spatial cognition on individual wayfinding performance*. Buildings 2019 **9** 29 [Google Scholar] [CrossRef]

[7] Dasovic B Galic M and Klansek U 2019 *Active BIM approach to optimize work facilities and tower crane locations on construction sites with repetitive operations*. Buildings 2019 **9** 21 [Google Scholar] [CrossRef]

[8] Fargnoli M Lleshaj A Lombardi M Sciarretta N and Di Gravio G 2019 *A BIM-based PSS approach for the management of maintenance operations of building equipment*. Buildings 2019 **9** 139

[9] Zhang S Teizer J Lee J and Eastman C M Venugopal M 2013 *Building information modeling (BIM) and safety: Automatic safety checking of construction models and schedules* Autom. Constr. **29** pp 183–195

[10] Issa R Suermann P C and Olbina S 2009 *Use of Building Information Models in Simulations. In Proceedings of the Winter Simulation Conference* Austin TX USA 13–16 December pp 2664–2671

[11] Ratajczak J Riedl M and Matt D T 2019 *BIM-based and AR Application Combined with Location-Based Management System for the Improvement of the Construction Performance* Buildings 2019 **9** 118

[12] Elagiry M Marino V Lasarte N Elguezabal P and Messervey T 2019 B*IM4Ren: Barriers to BIM Implementation in Renovation Processes in the Italian Market* Buildings 2019 **9** 200
[13] Kisel T 2020 *Dynamics of the level of BIM application in Russia in 2017-2019* E3S Web of Conferences 220(2):01025 DOI: 10.1051/e3sconf/202022001025 https://www.researchgate.net/publication/347662637_Dynamics_of_the_level_of_BIM_application_in_Russia_in_2017-2019
[14] Booyens D Beek C 2020 *Assessing BIM Application in the South African Construction Project Management Milieu* In: Journal of Construction 2020 Volume 13 Issue 1 Association of Schools of Construction of Southern Africa https://www.researchgate.net/publication/343231316_Assessing_BIM_Application_in_the_South_African_Construction_Project_Management_Milieu

Acknowledgments

The paper presents partial research results of project "Use of building information model in planning the life cycle of buildings" and of project APVV-17-0549 "Research of knowledge-based and virtual technologies for intelligent designing and realization of building projects with emphasis on economic efficiency and sustainability".

Indoor environmental quality in green certified office buildings

K Harčárová[1] and S Vilčeková[2]

[1] Institute of Architectural Engineering, Faculty of Civil Engineering, Technical University of Kosice, Vysokoskolska 4, 042 00 Kosice, Slovak Republic
[2] Institute of Sustainable and Circular Construction, Faculty of Civil Engineering, Technical University of Kosice, Vysokoskolska 4, 042 00 Kosice, Slovak Republic

E-mail: katarina.harcarova@tuke.sk; silvia.vilcekova@tuke.sk

Abstract. The presented paper deals with the assessment of selected indoor environmental quality (IEQ) factors, specifically the parameters of thermal comfort and indoor air quality (IAQ), in two buildings of a modern office complex. Both buildings are LEED certified. The monitoring of selected IEQ risk factors took place in selected office spaces (cell and open-space offices) representative of all occupied areas under operating conditions. The results of real measurements of IAQ parameters indicate that the monitored office spaces in both buildings did not exceed the required limits set by the LEED certification system. The measured parameters of thermal comfort were within the permissible legislative limits and the results of Predicted Mean Vote (PMV) and Predicted Percentage of Dissatisfied (PPD) indices show that a suitable level of thermal comfort can be expected in the given buildings during their occupancy. Based on the results of real measurements, it can be stated that both office buildings actually meet the LEED criteria for IAQ and guarantee a healthy and comfortable working environment for their occupants.

1. Introduction

In recent decades, the construction of so-called green buildings has become a global trend in the building industry. A building can be considered "green" if its properties are confirmed by certification systems [1]. The aim of green building certification is to achieve sustainable and energy-efficient buildings that are friendly not only to the environment but also to the health of the users themselves [2]. To create a healthy indoor environment, it is necessary to maintain a high quality indoor environment in the building [3]. For this reason, indoor environmental quality (IEQ) control has become an important part of green building rating systems [4]. The factors that define the indoor environmental quality are: thermal comfort, indoor air quality (IAQ), acoustic and visual comfort. One way to control the indoor environmental quality in green building certification systems is to regularly test and evaluate IEQ risk factors [5]. Many green building certification systems set their own proven methods and acceptable limits for assessing indoor environmental quality factors. The LEED and BREEAM certification systems are among the best-known international green building certification systems [6]. In the LEED certification system, IEQ factors are evaluated within the category "Indoor environmental quality" and in the BREEAM system, IEQ is included in the section "Health and well-being".

Up to now, only a few studies have been published on the assessment of IEQ in green certified buildings [3]. Most of them are focused on the subjective evaluation of IEQ in such buildings, while those that would reflect the real state of IEQ in the form of results from real measurements of IEQ risk factors are missing.

The aim of this study is to assess selected IEQ factors in the office spaces of two LEED certified office buildings through real measurements of these factors, in accordance with the requirements of the LEED certification system.

2. Methodology

Monitoring of the selected IEQ factors was carried out in two buildings (A and B) of a modern office complex located in Bratislava (Slovakia). Building A is nine-story and was built in 2006, while building B is eight-story and was built in 2009. Both buildings have received the highest level of LEED certification – LEED Platinum. For the monitoring and evaluation of selected IEQ factors, the methodology prescribed by the LEED certification systems (LEED v4.1 BD+C and LEED v4.1 O+M) was chosen, which was supplemented by the methodology given by the WELL certification system (WELL performance verification guidebook) [7, 8, 9]. Selected IEQ factors were monitored in spaces representative of all occupied areas, during normal hours of use and under typical ventilation conditions. In both buildings, the air was exchanged by natural or mechanical ventilation. The measurements took place in September 2021. In building A, the measurements were carried out in the offices located on the third (A1) and eighth (A2) floor and in building B on the sixth (B1) and eighth (B2) floor. One cell office (*cell*) and one open-space office (*open*) was chosen for IEQ monitoring on each floor. The cell offices were occupied by one person. The number of people in open space offices was variable at the time of the measurement due to the current situation caused by the SARS-CoV-2 virus and the hybrid working regime of many employees working in open space offices.

The thermal-humidity microclimate parameters (temperature, relative humidity and air velocity), as well as carbon dioxide (CO_2) and carbon monoxide (CO) concentrations, were recorded using a TESTO 400 universal instrument with appropriate probes (Testo, Inc.; Germany). Particulate matter concentrations in the 2.5 μm ($PM_{2.5}$) and 10 μm (PM_{10}) fractions were determined using a HANDHELD 3016 IAQ meter (Lighthouse Worldwide Solutions, Inc.; USA), which uses a laser-diode light source and collection optics for particle detection. Total volatile organic compound (TVOC) concentrations were determined with a ppbRAE 3000 UV ionization detector (RAE Systems, Inc.; USA). In the case of TVOC, a two-point calibration, zero and standard reference gas (isobutylene), was performed before the measurement itself. All measured TVOC concentrations were recalculated and are expressed in toluene equivalents (correction factor given by the manufacturer is 0.5). The remaining measuring devices and probes were placed in the breathing zone from 0.9 to 1.8 meters above the floor. The mean radiation temperature was measured at three heights using spherical sensors (d = 0.15 m). Concentrations of all monitored parameters were recorded at 1-minute intervals. The total measurement time in each of the offices was 1 hour (10 min acclimatization + 50 min measurement time), while only the data obtained during the measurement time were taken into the overall evaluation, as required by the WELL methodology.

Thermal comfort was assessed using PMV (Predicted Mean Vote) and PPD (Predicted Percentage of Dissatisfied) indices. PMV and PPD values were calculated using the CBE Thermal Comfort Tool [10]. The CBE Thermal Comfort Tool has been designed in accordance with EN ISO 7730 and EN 16798-1, taking into account the six parameters (metabolic rate, dynamic clothing insulation, air velocity, mean radiant temperature, air temperature and relative humidity) when calculating the PMV indices [11, 12]. From the PMV index, it is possible to determine the predicted percentage of users dissatisfied with a particular thermal environment, the so-called PPD index. At the time of the measurement (autumn), the dynamic clothing insulation was considered to be 0.61 clo (trousers, long-sleeved T-shirt). The metabolic rate value for sitting activities was selected as 1.1 met in the CBE Thermal Comfort Tool (typing).

3. Results and discussion

As part of the IEQ assessment, selected indoor environmental quality factors were monitored, specifically parameters related to thermal comfort and indoor air quality. The obtained results were evaluated in accordance with the requirements of LEED v4.1 certification system.

3.1. Thermal comfort

The results from the monitoring of the thermal-humidity microclimate parameters are summarized in Table 1. These results show that in terms of the measured thermal-humidity microclimate parameters, minimal differences can be observed between selected office spaces. The highest average temperature was recorded in the open-space office located in building A on the third floor. The lowest average temperature was recorded in a cell office in building B on the sixth floor. The same lowest average relative humidity was observed in building B on the eighth floor in a cell and open-space office. The highest average relative humidity was measured in a cell office in building A on the eighth floor. The lowest average air velocity was found in a cell office on the eighth floor of building B and the highest in a cell office on the third floor of building A.

As the LEED v4.1 standard does not define any permissible values for the parameters related to the thermal-humidity microclimate, the obtained results were compared with the requirements for the thermal-humidity microclimate stated in Decree No. 259/2008 Coll. [13]. Requirements of Decree No. 259/2008 Coll. for relative humidity (30–70%) and air velocity (≤ 0.25 m/s) were met in all monitored offices. The resulting values of the operative temperature were within the optimal (23–27 °C) and permissible (20–28 °C) legislative limits relating to the warm period of the year.

Table 1. Results from the monitoring of the thermal-humidity microclimate parameters.

		A1		A2		B1		B2	
		cell	*open*	*cell*	*open*	*cell*	*open*	*cell*	*open*
Temperature [°C]	Average	24.6	25.6	24.0	23.7	23.3	24.0	25.3	24.2
	Min	23.4	25.4	23.7	23.5	23.2	23.4	25	24
	Max	25.2	25.8	24.1	23.9	23.4	24.2	25.7	24.4
	SD [a]	0.66	0.09	0.09	0.09	0.06	0.21	0.19	0.11
Relative humidity [%]	Average	53.65	49.8	57.4	53.2	47.6	48.6	39.3	39.3
	Min	51.2	49.2	56.9	52.6	46.9	47.2	38.4	38.4
	Max	58.2	50.4	58.4	53.7	48.5	51.6	39.6	40.3
	SD [a]	1.84	0.28	0.45	0.28	0.44	1.23	0.35	0.51
Air velocity [m/s]	Average	0.07	0.03	0.03	0.01	0.04	0.02	0.00	0.01
	Min	0.01	0.01	0.01	0.01	0.01	0.01	0	0
	Max	0.17	0.08	0.11	0.08	0.1	0.1	0.01	0.08
	SD [a]	0.04	0.02	0.03	0.01	0.02	0.02	0.00	0.01
Operative temperature [°C]		24.4	24.7	24.1	23.9	23.5	24.0	24.6	24.1

[a] Standard deviation.

The PMV and PPD index values for individual offices calculated using the CBE Thermal Comfort Tool are summarized in Table 2.

Table 2. Calculated values of PMV and PPD indices.

	A1		A2		B1		B2	
	cell	*open*	*cell*	*open*	*cell*	*open*	*cell*	*open*
Mean radiant temperature [°C]	24.5	25.2	24.0	23.6	22.9	23.7	25.1	24.0
PMV	-0.03	0.26	-0.13	-0.27	-0.48	-0.24	0.12	-0.23
PPD [%]	5	6	5	7	10	6	5	6
Category	I	II	I	II	II	II	I	II

Based on ISO 7730 (2005), the indoor environment can be considered comfortable if the value of the PMV index is in the range of -0.5 to +0.5, which represents about 10% dissatisfied [11]. The above results show that a satisfactory level of thermal comfort can be expected in the evaluated office spaces (categories I and II). According to EN 16798-1, the medium level (category II) represents the standard level that users can expect during their stay in the building [12]. Category I includes offices in which a high level of thermal comfort can be expected.

3.2. Indoor air quality

The results of monitoring the IAQ parameters are shown in Table 3. Indoor air quality monitoring was focused on the evaluation of carbon dioxide (CO_2), carbon monoxide (CO), particulate matter in fractions 2.5 and 10 µm ($PM_{2.5}$ and PM_{10}) and total volatile organic compounds (TVOC) concentrations.

The recorded concentrations of carbon monoxide in each of the monitored offices were 0 ppm. The highest average CO_2 concentration of 898 ppm was recorded in a cell office located on the third floor of building A and the lowest 460 ppm in the open-space office on the sixth floor of building B. The low CO_2 concentrations in the offices on the sixth floor of building B were related to the minimum occupancy at the time of the measurement and to the continuous exchange of air through mechanical ventilation. The lowest average concentrations of $PM_{2.5}$ and PM_{10} were observed in the cell office on the sixth floor of building B and the highest in the open-space office on the eighth floor of building A. Regarding TVOC concentrations, the lowest average concentration of 3 µg/m^3 was recorded in building B on the sixth floor and the highest of 312 µg/m^3 in building A on the third floor.

Table 3. Results from monitoring IAQ parameters.

		A1		A2		B1		B2	
		cell	*open*	*cell*	*open*	*cell*	*open*	*cell*	*open*
CO_2 [ppm]	Average	898	738	573	571	490	460	643	520
	Min	821	713	551	553	467	447	568	505
	Max	946	756	608	594	524	469	706	545
	SD [a]	33.96	10.48	15.80	10.46	16.05	5.21	40.05	8.93
$PM_{2.5}$ [µg/m^3]	Average	5.20	7.74	4.07	8.63	0.95	1.07	1.20	1.30
	Min	4.95	7.09	3.80	6.05	0.84	0.90	1.07	1.15
	Max	5.58	8.39	4.69	9.49	1.04	1.20	1.34	1.46
	SD [a]	0.16	0.42	0.23	0.71	0.05	0.07	0.06	0.07
PM_{10} [µg/m^3]	Average	7.94	25.83	5.56	34.77	2.81	3.48	6.86	6.85
	Min	5.87	17.72	4.16	23.98	1.51	1.94	2.93	3.78
	Max	13.52	44.78	9.43	47.87	5.18	5.30	12.41	13.45
	SD [a]	1.66	6.24	1.44	4.28	0.88	1.02	2.69	2.02
TVOC [µg/m^3]	Average	251	312	148	47	22	3	206	205
	Min	202	269	133	0	7	0	186	169
	Max	276	366	161	137	36	21	230	240
	SD [a]	22.68	28.93	5.71	45.13	7.86	5.50	12.22	18.4

[a] Standard deviation.

The LEED v4.1 certification system sets limit values for concentrations of $PM_{2.5}$ ($\leq$ 12 µg/m^3), PM_{10} ($\leq$ 50 µg/m^3) and TVOC ($\leq$ 500 µg/m^3). According to LEED v4.1, CO_2 concentrations should not exceed the recommended value of 1000 ppm. The stated limit values were not exceeded in any of the monitored offices.

4. Conclusion

This study evaluates selected indoor environmental quality factors (thermal comfort and indoor air quality) in two LEED certified office buildings. Both buildings are part of one office complex. The indoor air quality in these buildings was assessed in accordance with the requirements of the LEED certification system. Based on the results of short-term measurements it can be stated that LEED certified office buildings still meet not only the LEED criteria for indoor air quality, but also the limits for thermal comfort set by legislation. Overall it can be stated that both office buildings provide for their users a working environment without increased levels of harmful substances in the indoor air and with a satisfactory level of thermal comfort.

5. References

[1] Mujan I, Anđelković A S, Munćan V, Kljajić M and Ružić, D 2019 *J. Clean. Prod.* **217** 646–57
[2] Wei W, Wargocki P, Zirngibl J, Bendžalová J and Mandin C 2020 *Energy Build.* **209** 109683
[3] Wei W, Ramalho O and Mandin C 2015 *Build Environ.* **92** 10–19
[4] Mattoni B, Guattari C, Evangelisti L, Bisegna F, Gori P and Asdrubali F 2018 *Renew. Sust. Energ. Rev.* **82** 950–960
[5] Lee J Y, Wargocki P, Chan Y H, Chen L and Tham K W 2019 *Indoor Air* **29**(1) 112–129
[6] Altomonte S, Saadouni S and Schiavon S 2016 Occupant satisfaction in LEED and BREEAM-certified office buildings
[7] USGB LEED v4.1 Building Design and Construction Beta Guide official website [online] https://www.usgbc.org/resources/leed-v41-building-design-and-construction-beta-guide-january-2019
[8] USGB LEED v4.1 Interior Design and construction Beta Guide official website [online] https://www.usgbc.org/resources/leed-v41-interior-design-and-construction-rating-system-january2019
[9] IWBI, WELL Performance Verification Guidebook, Applies to WELL v1, WELL v2 pilot and WELL v2 official website [online] https://resources.wellcertified.com/tools/performanceverification-guidebook/
[10] Tartarini F, Schiavon S, Cheung T and Hoyt T 2020 *SoftwareX* **12** 100563
[11] International Organisation for Standardisation 2005 Ergonomics of the thermal environment – analytical determination and interpretation of thermal comfort using calculation of the PMV and PPD indices and local thermal comfort criteria (ISO 7730)
[12] European Standards 2018 Energy performance of buildings - Ventilation for buildings - Part 1: Indoor environmental input parameters for design and assessment of energy performance of buildings addressing indoor air quality, thermal environment, lighting and acoustics - Module M1-6 (EN 16798-1)
[13] Decree of the Ministry of Health, Slovak Republic No 259/2008 Coll on detailed requirements for indoor environment of buildings and minimum requirements for low-standard flats and accommodation facilities

Acknowledgments

The presented contribution was financially supported by the grant agency VEGA with the grant number 1/0512/20 and ITMS 313011T5787.

IOP Conf. Series: Materials Science and Engineering 1252 (2022) 012055 doi:10.1088/1757-899X/1252/1/012055

Moving load effect on a bridge

J Melcer[1] and V Valašková[1]

[1] Department of Structural Mechanics and Applied Mathematics, Faculty of Civil Engineering, University of Žilina, Univerzitná 8215/1, 010 26 Žilina, Slovakia

E-mail: jozef.melcer@uniza.sk; veronika.valaskova@uniza.sk

Abstract. Moving loads represent the dominant component of the load on traffic structures such as bridges and roads. The response of these structures to moving loads can be modeled numerically using a FEM system such as ADYNA or ANSYS. The process requires the creation of a suitable vehicle and bridge computational models. For the purposes of this paper, a spatial computational model of a Tatra T815 heavy truck was created. The modeled bridge is a two span structure with a slab deck. A smooth surface of the bridge is assumed. At the beginning of the solution, the bridge is at rest and the vehicle enters the bridge already vibrated. The dynamic deflections of the bridge and the vehicle over time are monitored. It is possible to display bending moments in the transverse and longitudinal direction and torques at any time. It is also possible to obtain the values of the contact forces between the vehicle wheel and the bridge at any time. The vehicle can move along the bridge at any speed.

1. Introduction

Bridge structures are an important structural element on traffic routes. Moving loads represent their dominant component of the load and excite a dynamic response of the structure. For the correct design of the structure, it is necessary to know the true values of the load and its change over time. The dynamic response can be significantly greater than the static response, so the analysis of the dynamic behavior of the structure should be an integral part of the design and assessment of the bridge. At the time of preparation of project documentation, only numerical analysis of the dynamic response of the bridge is possible. After the bridge is completed, it is possible to perform an in situ dynamic load test. In the case of significant bridge structures, it is also possible to carry out an experiment on a model. The first analysis of the bridge response to the effect of moving loads was performed by R. Willis and G. G. Stokes in 1849 [1, 2]. They solved the movement of mass force along a simply supported beam. Their work prompted a series of works trying to solve the problems of moving loads analytically with varying degrees of applicability. The development of computers stimulated the transition from analytical to numerical methods. The monograph of L. Frýba [3] is known in this direction. Frýba focused mainly on the dynamics of railway bridges [4]. His colleagues at the University of Žilina focused on the dynamics of road bridges and pavements [5-7]. Numerical models can be created in the spirit of classical dynamics or in the spirit of finite methods. Both procedures give good results. However, finite methods provide a wider range of applications. The presented contribution is devoted to the numerical solution of the response of a bridge to the dynamic effects of a moving vehicle in the FEM ADINA program [8].

2. Computational model of a vehicle

The movement of Tatra T815 heavy truck along the bridge is modeled. The numerical model created on the basis of FEM is shown in figure 1. The model is created from 3D solid elements, beam elements and spring elements [7]. Their material properties do not interfere with the calculation and only have a visual function. The model in figure 1 is actually a multi-body model. Equations of motion have the character of ordinary differential equations. The vehicle parameters are listed in table 1.

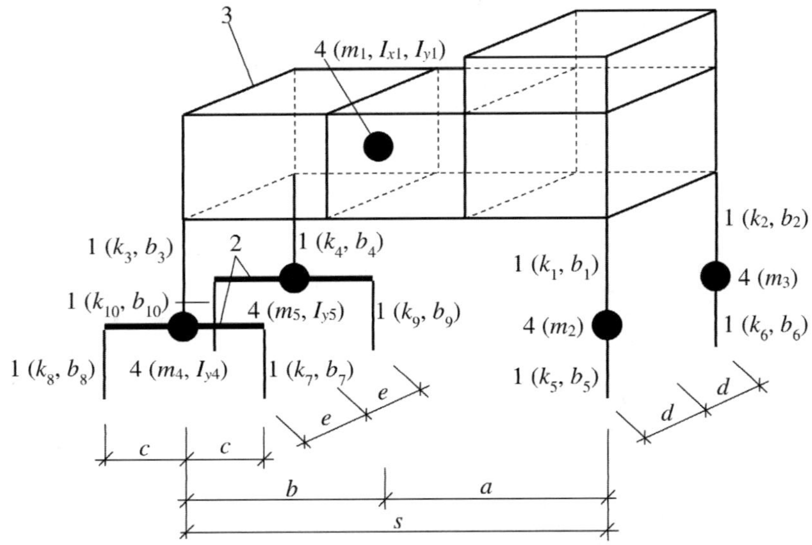

Figure 1. Spatial model of vehicle, 1-linear springs with damping, 2-beam elements, 3-3D solid elements, 4-mass points.

Table 1. Spatial vehicle model characteristics.

Function	Label	Value	Unit
body mass	m_1	22 950	kg
front wheel mass	m_2, m_3	455	kg
rear wheels mass	m_4, m_5	1 070	kg
transversal inertia of the body	I_{y1}	62 298	kgm^2
longitudinal inertia of the body	I_{x1}	22 950	kgm^2
inertia of rear axle	I_{y4}, I_{y5}	466	kgm^2
front suspension	k_1, k_2	143 716.5	N/m
rear suspension	k_3, k_4	761 256	N/m
stiffness of the front tires	k_5, k_6	1 275 300	N/m
stiffness of the rear tires	k_7, k_8, k_9, k_{10}	2 511 360	N/m
front damping	b_1, b_2	9 614	kg/s
rear damping	b_3, b_4	130 098.5	kg/s
front tire damping	b_5, b_6	1 373	kg/s
rear tire damping	b_7, b_8, b_9, b_{10}	2 747	kg/s

It is assumed that the vehicle enters the bridge already vibrated. The initial conditions on the vehicle at the time the vehicle enters the bridge are as follows: r_1 = - 0.021 m; $\dot{r}_1$ = 0 m/s; r_2 = 0.003535 rad; $\dot{r}_2$ = 0 rad/s; r_3 = - 0.00355 m; $\dot{r}_3$ = 0 m/s; r_4 = - 0.00261 m; $\dot{r}_4$ = 0 m/s; r_5 = 0 rad; $\dot{r}_5$ = 0 rad/s. The initial conditions on the vehicle were induced by a force F = 35 000 N acting statically at the center of gravity of the vehicle. The vehicle is assumed to move along the bridge at a constant speed.

3. Computational model of a bridge

The subject of numerical analysis is a highway bridge with two fields. It is a prestressed concrete bridge made of I-73 bridge prefabricated elements, figure 2. The bridge is additionally prestressed, so that due to the moving load it acts as a continuous beam with 2 fields, 2x30 m long.

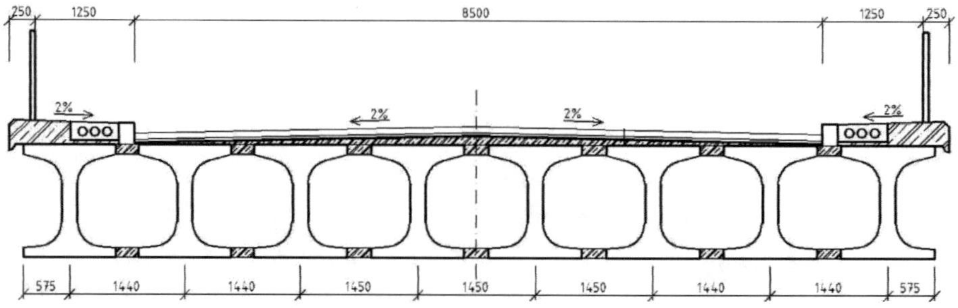

Figure 2. Cross section of the bridge.

The height of I-73 prefabricated beams is 1.4 m. Concrete B 500 (C 35/45), steel 10 425 (B 500B), modulus of elasticity $E = 3.85 \cdot 10^{10}$ N/m^2, Poisson ratio $\nu = 0.15$. The arrangement of the structure in cross section is as follows:

- load-bearing structure, 8 beams I-73, including reinforced concrete between the beams,
- leveling concrete, thickness 2 - 11 cm,
- infiltration coating,
- Balex and Mastix, thickness 1 cm,
- cast asphalt, thickness 3 cm,
- asphalt concrete, thickness 8 cm.

Mass intensity of the bridge μ = 19 680 kg/m. The quadratic moment of the cross section I = 2.409096 m^4.

Fixed bearings are realized as rubber, movable bearings as pot bearings, unidirectionally displaceable. The width between the railings is 11 m, of which the width between the raised curbs is 8.5 m (2 x 3.75 m + 2 x 0.50 m) and the width of the sidewalks is 2 x 1.25 m. The width of the ledge with the railing is 2 x 0.25 m, so the total width of the bridge, including the ledge, is 11.50 m. The angle of intersection with the road under the bridge is 90°. The bridge has a longitudinal slope of 5%.

The substructure consists of 2 outer reinforced concrete gravity supports and 1 intermediate reinforced concrete divided pillar. Everything is based flat on the base concrete. The height of the lower edge of the structure above the ground ranges from 6.85 to 7.80 m.

The bridge is modeled as a slab of constant thickness h according to Kirchhoff's theory from shell elements. Such a model is sufficient to solve the global response of the bridge. The slab parameters are determined so that the model is dynamically equivalent to the original bridge (stiffness and weight). The whole structure is divided into 120·22 = 2 640 finite elements, dimensions 0.5·0.5 m, thickness h = 1.38 m. Cross - section area A = 11·1.38 = 15.18 m^2. Quadratic moment of cross section I = 2.409096 m^4. Mass intensity of the structure μ = 19 680 kg/m. Logarithmic damping decrement ϑ = 0.12. The basic natural frequencies $f_{(1)}$ = 3.75 Hz and $f_{(2)}$ = 5.68 Hz. The bridge is at rest at the beginning of the solution and in a state of static equilibrium. The central flat of the bridge is horizontal and straight. A new bridge with a quality road surface is assumed. Therefore, road unevenness is not taken into account during the calculation. The modeled object consists of three sections, figure 3. The first part is starting. Here, gravity begins to act on the vehicle and contact with the road is established. The second part presents a model of the bridge itself. The third part is used to stop the vehicle.

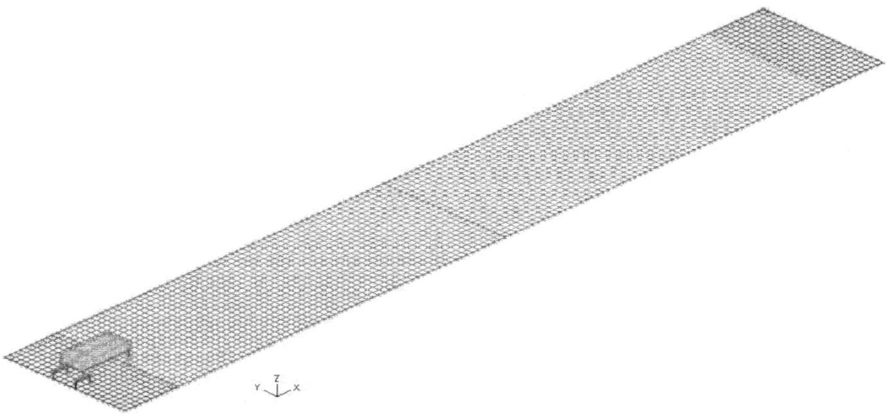

Figure 3. Computational model of the bridge and vehicle.

4. Results of the computation

The results of a numerical simulation when the vehicle moves in the middle of the bridge with the speed $V = 70$ km/h, which is $v = 19.4444$ m/s are presented. Relaxation and making contact of vehicle with the road takes 11 s. The vehicle takes 3.336 s to cross the bridge. The Newmark's method is used for the solution of equations of motion. The output of the solution is done with a time step of 0.001 s. The response of the bridge is monitored in two points, in the middle of the first and the second span. The values of the extreme deflections in the middle of the 1st and 2nd span are given in table 2 and 3.

Table 2. Extreme deflections in the middle of the 1st bridge span.

t [s] 3D	0.96	2.42
w_1 [mm] 3D	-1.1116	0.4736

Table 3. Extreme deflections in the middle of the 2nd bridge span.

t [s] 3D	0.96	2.44
w_2 [mm] 3D	0.4669	-1.0895

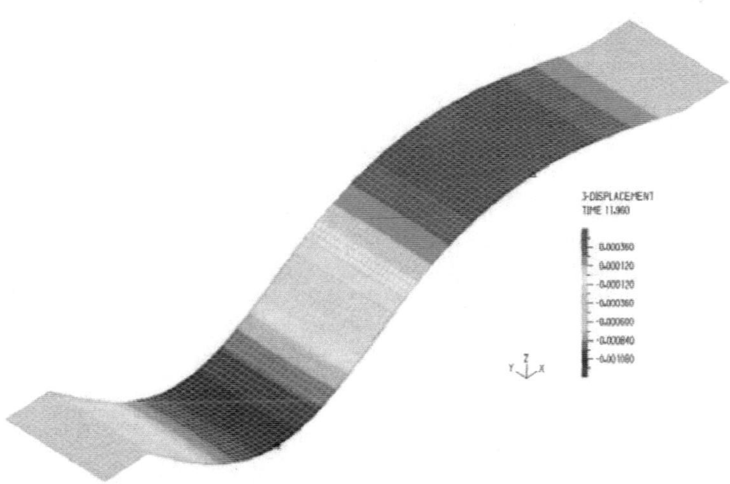

Figure 4. Maximum deflection in the 1st bridge span, $t = 0.96$ s.

The maximal vertical deflections in the middle of the first span when vehicle crossing the bridge occur at time 0.96 s (11.96 s from the beginning of the calculation). They are shown in figure 4.

Another output can be the intensities of bending moments in the longitudinal and transverse directions, figures 5 and 6 and the intensities of torques, figure 7. It is also possible to display the values of contact forces between the vehicle wheel and the bridge.

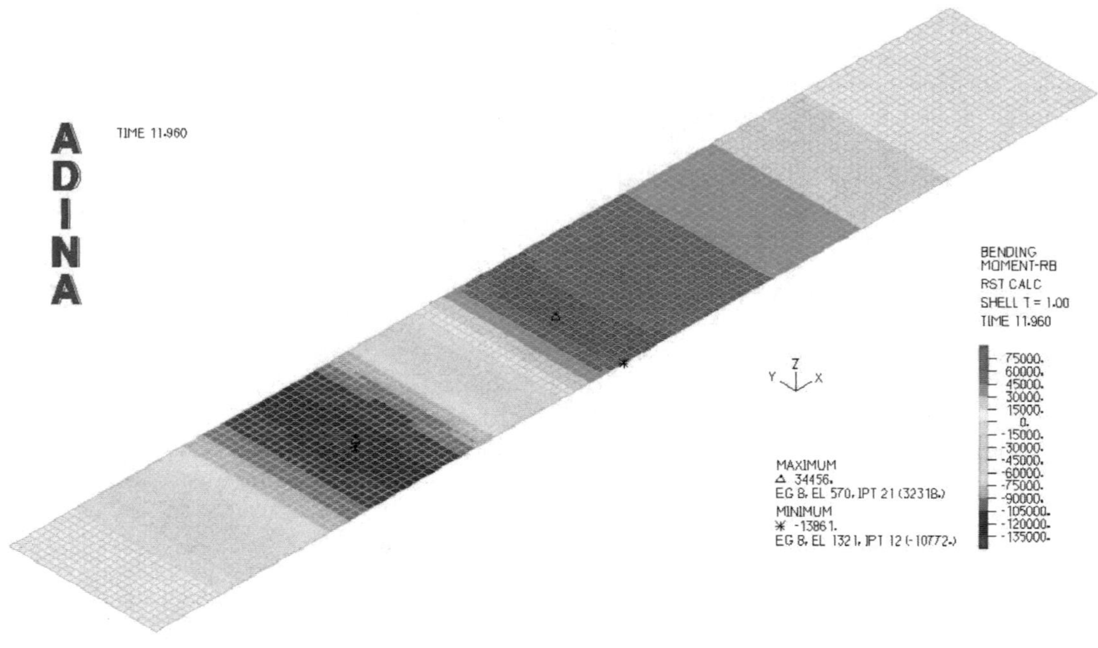

Figure 5. Bending moments in longitudinal directions at time $t = 0.96$ (11.96) s.

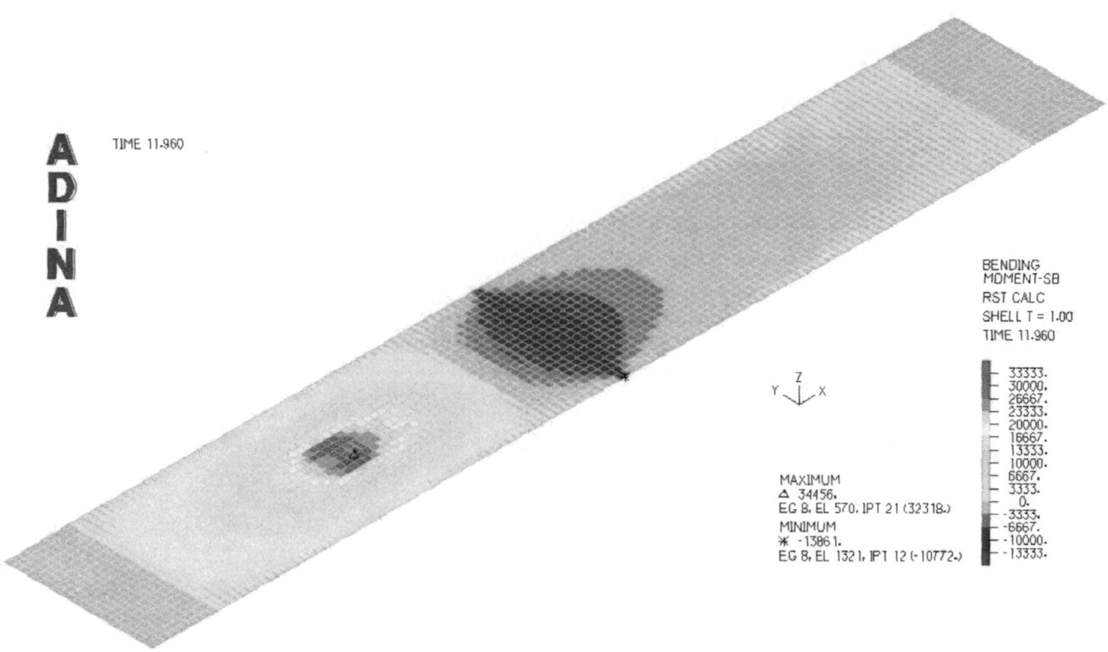

Figure 6. Bending moments in transverse directions at time $t = 0.96$ (11.96) s.

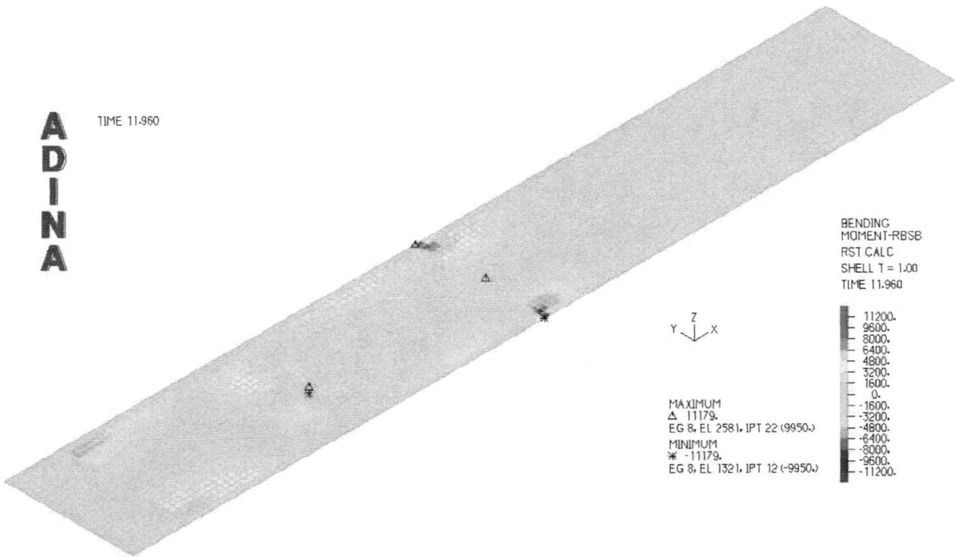

Figure 7. Intensities of torques at time $t = 0.96$ (11.96) s.

5. Conclusion

FEM is an effective tool for the solution of moving loads effect on bridges. It already allows in the process of project preparation to obtain such data that could previously be obtained only from in situ tests. Numerical simulations make it possible to perform interval analysis and monitor the effects of individual parameters. It is possible to observe how the change of one parameter affects the behavior of the whole structure, which is not possible in the in situ test. If the input data are correct, the results correspond to objective reality. The computational model can be created in such a way that it is possible to analyze the behavior of the structure as a whole or to analyze the stress states in individual details.

6. References

[1] Willis R 1849 *Appendix report of the commissioners appointed to inquire into the application of iron to railway structures* (London: Stationary Office)

[2] Stokes G G 1849 Discussion on a differential equation relating to the breaking of railway bridges *Transactions Cambridge Philosophic Society* 8 p 707

[3] Frýba L 1972 *Vibration of solids and structures under moving loads* (Prague: ACADEMIA, Groningen: Noordhoff International Publishing) p 484

[4] Frýba L 1996 *Dynamics of railway bridges* (London: Thomas Telford) p 326

[5] Melcer J 1997 *Dynamické výpočty mostov na pozemných komunikáciách* (Žilina: EDIS, University of Žilina) p 287

[6] Melcer J Lajčáková G Martinická I and Králik J 2016 *Dynamika dopravných stavieb* (Žilina: EDIS, University of Žilina) p 374

[7] Melcer J Lajčáková G and Valašková V 2018 *Moving load effect on concrete pavements* (Kraków: Wydawnictwo Towarzystwa Slowaków w Polsce) p 131

[8] ADINA R & D. 2012 Theory and Modeling Guide. Volume I: ADINA (USA: ADINA R & D, Inc., Watertown, MA)

Acknowledgments

This work was supported by the Grant National Agency VEGA of the Slovak Republic, project No. 1/0006/20.

Natural vibration of a membrane hypar in a steel frame

L Stulerova[1], L Kapolka[1] and S Kmet[1]

[1] Technical university of Kosice, Faculty of Civil Engineering, Vysokoskolska 4, 042 00 Kosice

E-mail: lenka.stulerova@tuke.sk; lukas.kapolka@tuke.sk

Abstract. This paper deals with a subject of a natural vibration. Modal analysis was performed on a simple membrane structure. Examined model is prepared in FEM software Dlubal. The membrane structure has geometry of a hyperbolic paraboloid or so called hypar. This membrane is basically a technical textile with no bending stiffness so therefore to carry different types of external loads it needs to be pre-stressed. This stress is provided by the edge ropes located in the sleeves around the circumference of the membrane. Stated construction is anchored by 2 anchor rods and 2 actuators to the steel frame which is supported with hinged sliding supports. Natural analysis with an add-on module RF DYNAM Pro was performed. The results of the natural frequencies and mode shapes are obtained which represents first phase of a complex dynamic analysis.

1. Introduction

Membrane structures are nowadays very popular between architects, engineers and designers. These thin-walled structures or so called shells are characteristic by their light weight, shape variability and spatial scale.

Kukathasan and Pellegrino focused on the nonlinear vibration of wrinkled membranes. They provided high-fidelity finite element simulation which produces natural mode and frequencies [1]. Different method, particularly h-p version of finite element method, studied Houmat in his paper. He showed results of 10 lowest frequencies of the isosceles triangular membrane, L-shaped membrane [2]. Xian Liu at el published nonlinear vibration analysis of a membrane based on large deflection theory. They compared the transverse displacement of midpoint of the membrane calculate with nonlinear finite element method with assumed mode method and confirmed the accuracy of this proposed method [3].

Natural frequencies of pre-tensioned membrane structure with different boundary geometrical parameters studied Liu et al. For planar membrane, the natural frequencies with circular, L-shape and subdivided arc boundary conditions were calculated [4]. Filipich and Bambill compared again the finite element method with the proposed whole element method. They calculate rectangular and square membranes with multiple intermediate supports and they presented obtained natural frequencies [5]. Borković et al studied first 8 frequencies of singly curved shell and they used isogeometric finite strip method [6].

2. Parameters and properties of the model

As it was mentioned in the abstract program Dlubal RFEM was used for this purpose. This software works on a principle of a finite element method, i.e. FEM. This mathematical approach simply divides every element (beam, cable, surface…) to smaller node elements in which the calculation in running. We obtain more accurate results with finer mesh of the elements.

The researched model consists of 3 important element groups. First group represents membrane surfaces itself. Membrane material is a thin technical textile with a thickness around 1 mm. Membrane is woven in 2 perpendicular directions and according to parameters of this process there are 2 types of resulting membrane: an isotropic or an orthotropic type [7]. The isotropic textile has the same characteristics uniform in all directions and the orthotropic membrane surface differs in 2 perpendicular direction. The directions are called warp (the principal direction) and weft (the secondary direction). In this case the membrane is orthotropic Serge Ferrari Précontraint according to the experimental device which this model represents. It is a polyester textile with PVC coating with thickness of 0.56 mm. As it is already mentioned the parameters vary in directions of the surface, particularly Young's modulus of elasticity in warp X direction and weft Y direction (table 1). Membrane surface is in the shape of hyperbolic parabolic with geometrical parameters listed in the table 1. The surface in numerical model is performed as 2 triangles connected with common edge between lower vertices (figure 1).

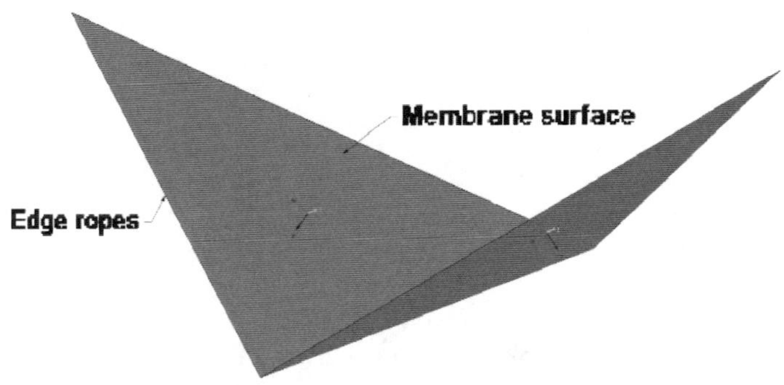

Figure 1. Configuration of the membrane and edge ropes.

Every textile in structural engineering needs to be pre-stressed to carry different external loads, e.g. snow, dynamic wind or impact load of rain drops. The edge ropes are used for this purpose and they represent the second group of elements. The stainless steel ropes of 8 mm diameter are located on each of 4 edges of the hypar. The Young's modulus of elasticity is in table 1. Pre-stress of these cable elements is defined as absolute sag with value of 0.2 m.

The third group of elements is represented by a primary structure. In general, this load-bearing part is able to transfer all of the loads to the foundation. The membrane stabilized with the edge ropes is anchored to a spatial steel frame with axial dimensions of 3.34×3.34×2.5 m (figure 2). The frame is welded of S235 steel with different cross-section:

- 4 edge columns – SHS160×160×8
- intermediate columns – RHS100×80×6, 2×SHS100×80×6
- bottom base – SHS140×140×8
- upper base – SHS140×140×8
- transverse beams of bottom base – SHS80×80×6
- corner strut beams – RHS140×80×6; SHS80×80×6
- bottom tear plate – 6 mm
- upper plate for load piston – 45 mm
- anchorage cross-section (anchor rods/actuator-bottom base) – U135×150×45×50.

In the middle of the both bases and 2 side walls are cross-section doubled and bolted together with 2 rows of bolts. Corner strut beams are welded to 8 edges to all of 3 adjacent sides.

This steel frame is supported to the ground with 4 supports – 4 hinged sliding supports located in every bottom corner of a primary structure.

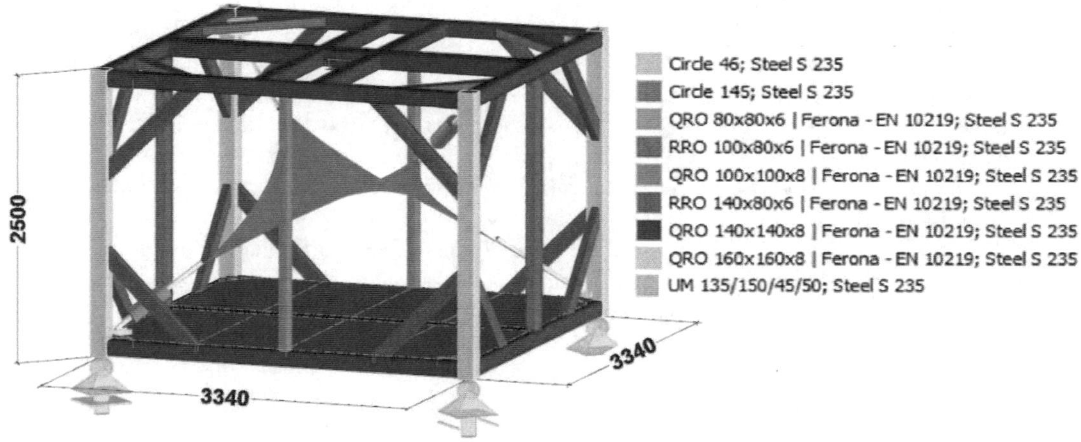

Figure 2. Cross-section of the steel frame and its dimensions.

Membrane surface is anchored to the steel frame through 2 anchor rods and 2 actuators, i.e. hydraulic piston. These elements are not important in this analysis because they were not considered to be used, they will be a subject of a further research. They were modelled just as mass – circular cross-sections with 46 and 145 mm.

Last important characteristics of this model are mesh parameters. Every 1D element was divided into at least 10 parts and 2D elements (membrane surface particularly) were mesh with the size of 50 mm length of each side of triangular finite element. For form-finding it was considered maximum 200 iterations. It was also necessary to change default accuracy and tolerance criteria for convergence: tolerance to 0.5 (in range of 0.01-100) and speed of tolerance to 0.4 (with the same range).

Table 1. Geometrical and material characteristics.

Input parameters	Value
Young's modulus in X direction	1 057 MPa
Young's modulus in Y direction	612 MPa
Young's modulus of a rope	130 GPa
floor plan	2×2 m
height	1 m
thickness of a membrane	0.56 mm
diameter of a rope	8 mm
sag of a rope	0.2 m
pre-stress of a membrane	4 kN/m

3. Analysis of a natural vibration

First step of a modal analysis of a membrane is crucial – it is pretension. As we have already mention it in a previous chapter. Simply said it is a process when we pre-stress surface in its plane and make it stiffer. This specific step defined as 'form-finding' is a routine of equilibrium state between deformation (relocation of a mesh nodes) and stress state (surface pre-stress). There are few different methods for solving this problem: Force Density Method [8], Dynamic Relaxation [9], Finite Element Method – Update Reference Strategy (URS) [10]. The last method is incorporated to the Dlubal RFEM [11] due to its compatibility with finite element method procedures. It is based on the real flat strips that are fabricated and sewed together to one 3-dimensional surface [10]. It is using pre-stress directions and ratios to map the stresses during the procedure of the iteration [10].

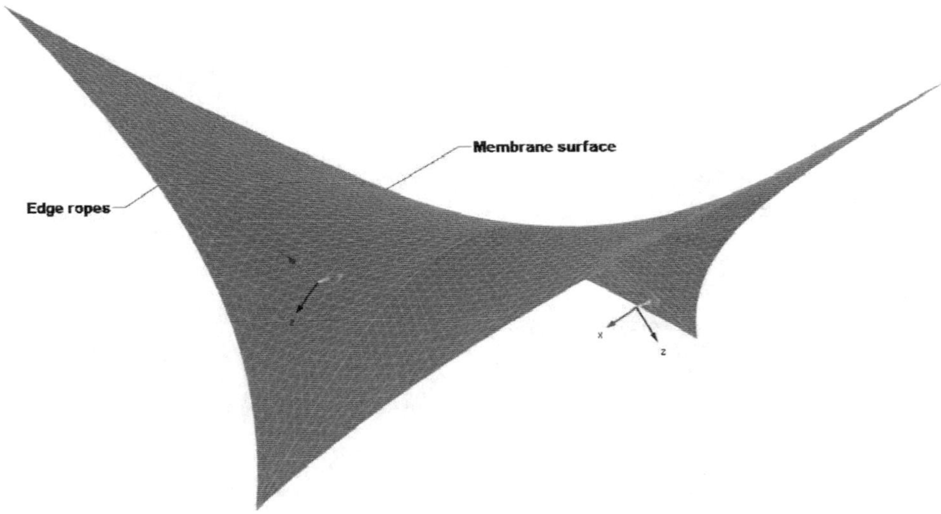

Figure 3. Meshed membrane after the form-finding with LCS.

The initial conditions of this pre-stress are the values in table 1 – sag of a rope and pre-stress of a membrane with values of 0.2 m and 4 kN/m in both of the directions of the orthotropic membrane surface respectively. The self-weight is neglected. With this parameter the calculation URS begins and we received the first shape of a membrane structure (figure 3) which will be statically or dynamically loaded afterwards.

In this case only modal analysis was carried out. 10 modal shapes were considered as sufficient amount of significant eigenvalues. Only 1 mass case was needed – it represents self-weight as permanent load. In table 2 principal adjustment of a natural vibration calculation is listed. Every step was chosen according to the software manual recommendation.

Table 2. Calculation parameters.

Parameters	Value
Acting masses	Self-weight
Mass matrix	Consistent matrix
Method for solving eigenvalue problem	Lanczos
Stiffness modifications	Import axial forces, failing members, surfaces and supports stiffness modifications, extra option and deactivation from: LC Form-finding

4. Results obtained from an analysis

Intended calculation was carried out as geometrically non-linear – large deformation analysis with method for solving system of nonlinear algebraic equations Newton-Rhapson. Monitored quantities are angular and natural frequency, natural period for all of the 10 modes. The results are listed in table 3 below. The angular frequencies are obtaining values from 57.171 to 215.268 rad/s. The natural frequency, as the most important parameter is in range 9.099-34.261 and the natural period raises from 0.11 s for first mode to 0.029 s.

Figure 4 shows first 6 mode shapes of the membrane structure in the steel frame. It is completed with the maximum displacement of its mode shape.

IOP Conf. Series: Materials Science and Engineering 1252 (2022) 012056 doi:10.1088/1757-899X/1252/1/012056

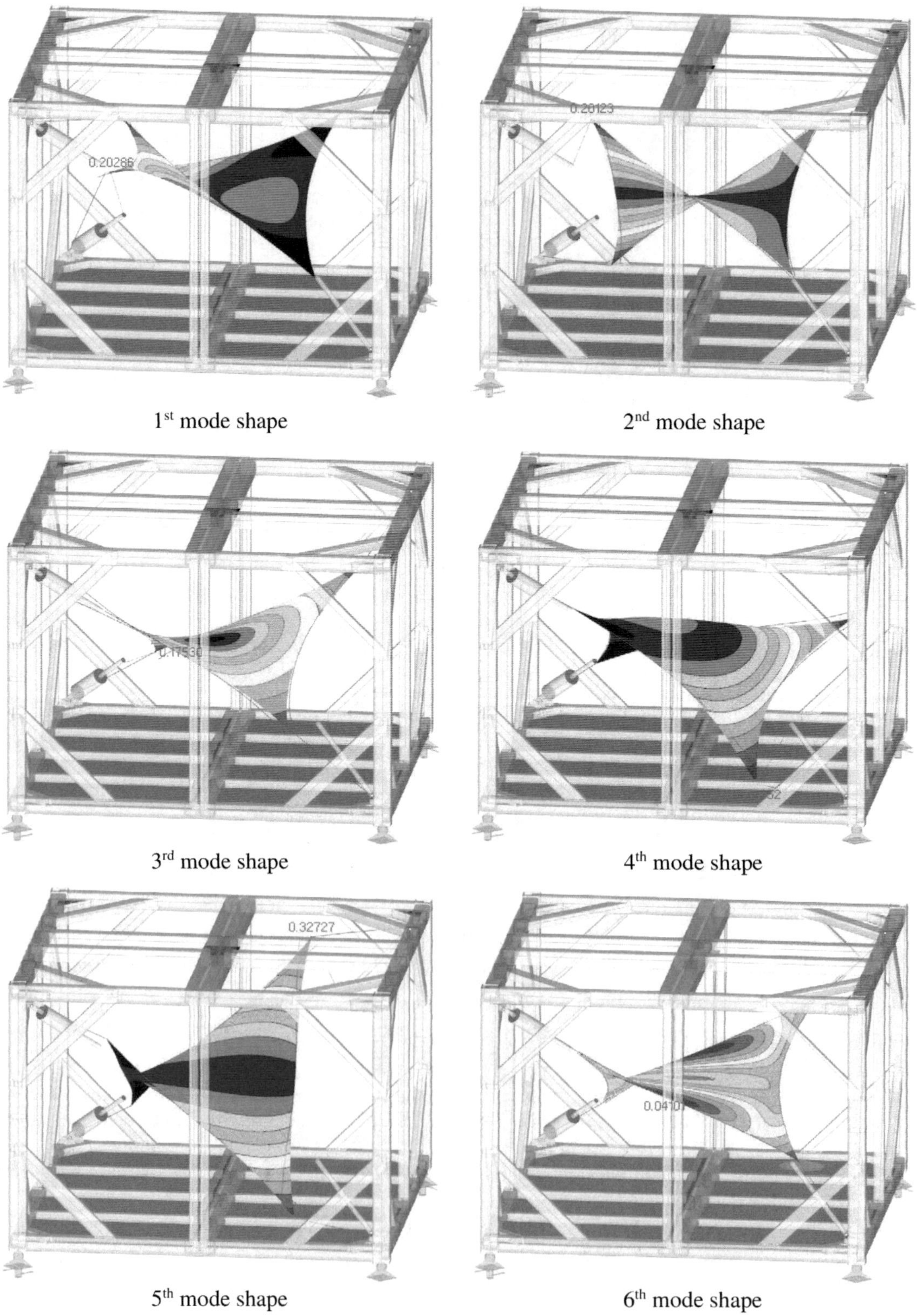

Figure 4. First 6 mode shapes of a membrane structure.

Table 3. Results of the modal analysis.

Mode No.	Angular frequency Ω (rad/s)	Natural frequency f (Hz)	Natural period T (s)
1	57.171	9.099	0.110
2	60.379	9.610	0.104
3	67.270	10.706	0.093
4	92.389	14.704	0.068
5	99.497	15.835	0.063
6	134.851	21.462	0.047
7	140.948	22.433	0.045
8	182.832	29.099	0.034
9	197.459	31.427	0.032
10	215.268	34.261	0.029

5. Conclusion

Main goal of this paper was modal analysis with regard to important parameters of frequencies and periods. Researched model is prepared based on real device and compared to previous basic models [12] it is supplemented with the primary structure – steel frame. This boundary condition influences the stiffness of the model and therefore it has a significant impact on the natural vibration and the results obtain from the analysis.

From the results of the natural vibration analysis it can be stated that with the higher-order mode shape the natural frequencies raises and the natural period decreases. Selected number of 10 modes is not necessary according to the frequencies values: only the mode shapes with the frequencies up to 10 Hz are important for dynamic analysis under the dynamic load – dynamic wind reach these values. It is relevant mostly in the next steps of the research focused on the forced vibration – damped, undamped, harmonic, periodic etc.

6. References

[1] Kukathasan S, Pellegrino S 2003 Nonlinear Vibration of Wrinkled Membranes *44th AIAA/ASME/ASCE/AHS/ASC Structures, Structural Dynamics, and Materials Conference* **1747**

[2] Houmat A 2003 Free vibration analysis of membranes using the h–p version of the finite element method *Journal of Sound and Vibration* **282** 401-10

[3] Xiang L, Guo-ping C, Fu-jun P, Hua Z, Liang-liang L 2018 Nonlinear vibration analysis of a membrane based on large deflection theory *Journal of Vibration and Control* **24** 2418-29

[4] Liu C, Li Y Y, Bao H, Du J L 2015 Natural frequencies of pre-tensioned membrane structure with different boundary geometrical parameters *Journal of Vibration and Shock* **34** 198-202

[5] Filipich C P, Bambill E A 2008 Natural frequencies of rectangular membranes with partial intermediate supports *Journal and Sound and Vibration* **310** 21-37

[6] Borković A, Radenković G, Majstrorović D, Milovanović S, Milašinović D, Cvijić R 2020 Free vibration analysis of singly curved shells using the isogeometric finite strip method Thin-Walled Structures **157**

[7] Seidel M 2007 *Tensile Surface Structures: A Practical Guide to Cable and Membrane Construction* (Berlin: Ernst and Sohn) p 229

[8] Schek H J 1974 The force density method for form finding and computation of general network *Comp. Methods in App. Mech. and Engng.* **2** 115-34

[9] Day A S 1965 An introduction to Dynamic Relaxation *The Engineer* **219** 218-21

[10] Wüchner R, Bletzinger K U 2005 Stress-adapted numerical form finding of pre-stressed surfaces By the updated reference strategy *Int. J. Numer. Meth. Engng.* **64** 143–66

[11] Dlubal 2020 Manual RF-DYNAM Pro
[12] Stulerova L, Kapolka L and Kmeť S 2020 Numerical nonlinear analysis of a membrane structure subjected to dynamic load *IOP Conf. Ser.: Mat. Sci. and Engng.* **867**

Acknowledgments
This work was supported by The Slovak Research and Development Agency under the contract No. APVV-15-0777 and also by The Slovak Scientific Grant Agency under the contract No. VEGA 1/0129/20.

Methods of strengthening the grains of recycled aggregates

A Sicakova[1]

[1] Institute for Sustainable and Circular Construction, Faculty of Civil Engineering, Technical University of Kosice, Vysokoskolska 4, 042 00 Kosice, Slovakia

E-mail: alena.sicakova@tuke.sk

Abstract. Poorer quality indicators than natural aggregates generally characterize recycled aggregates. This is due to the composition of the primary raw material (mostly demolition waste), the method of processing (crushing, sorting) and the typical character of the grains, which, in addition to the original grain of natural aggregate, also contains the original adhered mortar. The resulting grains are heterogeneous, characterized by lower strength, higher porosity, higher absorbency, or the presence of cracks. Improving the quality of these grains is one of the approaches to increase their potential for use in concrete. Several methods of recycled concrete aggregate (RCA) grain strengthening are being studied worldwide, the effect of which on the final quality of recycled aggregates and new concrete is different. The article describes and compares the methods in terms of their impact on selected properties of both. The comparative analysis is based on the collection of results presented in research publications. Based on the evaluation of the results, polymer emulsion treatment can be identified as the most effective way to improve RCA water absorption. Sodium silicate treatment came out the best to improve the compressive strength of concrete mixed with treated RCA.

1. Introduction

The use of recyclates or other alternative aggregates in concrete is a long-term, but still actual topic. On the one hand, there is a critical shortage of natural aggregates for the production of new concrete in some areas, on the other hand, there is a huge amount of demolished material obtained from damaged and old structures, which create serious ecological and environmental problems. The growing production of by-products of industrial activity is leading to ongoing research into their recycling and determination of conditions of use. The limited utilization of by-products may be due to high transportation costs, need for processing, and inadequate data on their geotechnical and environmental properties [1].

The potential for recycling and re-use of Construction and Demolition Waste (C&DW) is considered still underused and varies from one EU country to another. The low quality of the waste is considered the main challenge, which leads to heterogeneity and the possible content of hazardous substances in the waste. This results in uncertainty for potential users of sufficient quality for further use. Lack of confidence reduces the demand for materials recycled from construction and demolition waste [2].

Aggregates from alternative sources usually have worse properties compared to natural aggregates (NA). The size, shape, surface and internal structure of aggregates have a significant effect on the properties of fresh and hardened concrete [3]. In the case of recycled concrete aggregates (RCA), these effects are due to the specific nature of the grains to which a certain part of the original mortar is attached (see figure 1), leading to their porosity, the presence of cracks, the surface roughness and the presence of other substances. RCAs usually have higher absorbency and lower strength than natural aggregates, require more mixing water and cause a faster loss of consistency. This leads to workability problems and overall lower strength and durability of concrete [4, 5, 6].

a) b) c) d)

Figure 1. Comparison of NA and RCA grains. a) NA grain, b) RCA grain: original NA grain having original mortar attached, c) RCA grain: original mortar – conglomerate of cement paste and fine aggregate, d) typical composition of RCA grain with different ITZ [7, 8].

The specific physical-mechanical properties of RCA in comparison with NA can be summarized as follows:

Density: due to the presence of porous mortar on the grain, it is lower by 7-9%, or up to 17%, reaching a value of about 2 350 – 2 900 kg / m^3 [9, 10]

Water absorbency: compared to NA it is higher and is presented with values of 3-12%, compared to 1-5% absorbency of NA, which is 2.5 times more [11, 12]

Surface and shape: The mortar that remains adhered to the RCA grains makes its texture rougher than that of NA. On the other hand, it is stated that the shape index is similar to NA grains [13].

Crush resistance: the value according to Los Angeles abrasion test for RCA can vary in a very wide range of 25-42%, depending on the quantity, quality and phase interface of the original concrete. The resistance of NA (eg basalt) ranges from 13 to 15%. The crush resistance also depends on the grain size; for smaller fractions it is worse [10, 13]. STN EN 12620 [14] requires a value of max. 40% so that RCA can be used for the production of new concrete.

Interphase transition zone: the complex nature of RCA grains also has an impact on the nature of the interphase transition zone (ITZ) in newly produced concrete, which is currently considered a key parameter of its overall quality [15, 16]. It is presented that in ITZ there is a higher porosity, the number of microcracks, as well as the predominance of easily soluble portlandite $Ca\ (OH)_2$, which cause its weakening. The stronger and more durable cement hydration products (mainly CSH) are then at a greater distance from the grain. These facts are due to the higher water/cement ratio around the aggregate grain, as well as the large difference between the modulus of elasticity of the aggregate and the matrix, which leads to higher stresses [17, 18]. Thus, ITZ plays a crucial role in determining the mechanical properties of concrete. Moreover, within the RCA grain, three different types of ITZ can be found unlike the concrete having NA only: between original NA and original mortar, between original NA and new mortar and between original mortar and new mortar, as presented in fig. 1d) [8]. This is why question of ITZ in RCA-based concretes are much more complex.

Therefore, it is possible to record several research activities to increase RCA quality, mainly related to physical-mechanical properties, in order to better meet the requirements for use in concrete. Several methods of grain strengthening are being studied worldwide, the impact of which on the final quality of both RCA and new concrete is different. For the study of the issue there was a need to compare their effectiveness on the basis of the presented results from the research carried out so far. In the article, the methods are listed and briefly described and compared in terms of their effect on improving the water absorption of RCA, as well as on improving the strength properties of concrete. The comparative analysis is based on a set of collected results from available relevant research sources.

2. Methods for strengthening RCA grains

These methods retain adhered mortar on the original aggregate and strengthen it before the aggregate is used for mixing the new concrete. The main goal of quality improvement techniques is to fill the weak areas and develop stronger ITZs in RCA.

2.1. Cement / Pozzolan slurry

Pozzolans are defined as siliceous or alumino-siliceous materials that possess little or no cementitious value but will, in finely divided form and in the presence of moisture, chemically react with calcium hydroxide at ordinary temperatures to form compounds possessing cementitious properties. When pozzolan slurry is applied on the surface of RCA, the products of hydration reaction can modify the surface and strengthen it [19]. The macro-micro pores in attached mortar need to be filled with a fine particle which will not only act as inert filler but also help in the hydration process (secondary hydration). Practically the same principles apply to the composition of pozzolanic slurry for coating RCA grains as to the composition of the binder for concrete - it can range from pure cement slurry [20] to various combinations of cement and supplementary cementitious materials (SCM´s), in such a percentage composition that allows proper hardening and thus obtaining sufficient mechanical performance. These SCM´s include fly ash, silica fume, ground granulated blast furnace slag [21, 22, 23]. However, less common materials are also tested as the effective pozzolanic component of the coating slurry, like Fe_2O_3 [7], nano-silica [24, 25], calcium metasilicate - wollastonite [25], geopolymer paste [20, 26], metakaolin [27] or RCA fines [28].

As for the application of slurry on the RCAs surface, different coating ways were investigated:

- by immersion, spraying or mixing the RCA grains with defined amount of slurry, which means grains are processed before the mixing of new concrete, including defined time and conditions of drying / hardening [7, 20, 29].
- by application of specific mixing approaches, double mixing (DM) or triple mixing (TM), by which the RCA grains are processed during the mixing of new concrete. The principle lies in dividing the mixing process into phases with a precisely defined sequence of adding the individual components of the concrete to the mixer, while in the initial phase the RCA grains are coated with a pozzolanic materials. As part of the sequence of addition of components, there are various coating mechanisms, which can be outlined as follows:
    - initial wetting / soaking of the RCA followed by dosing of dry pozzolans [30]
    - initial mixing of the RCA with dry pozzolans, then addition of a defined amount of water to form a coating paste [31]
    - mixing RCA with the first batch of pozzolans and water (coating with slurry), then optionally the second batch of pozzolans and water (formation of a moist concrete mixture) and finally the last batch of water (ready-mixed concrete) [22]

The authors dealing with these approaches to mixing declare a positive effect on the mechanical properties of concrete, mainly due to the increased quality of ITZ in RCA grain, as well as between the coated RCA and the new cement paste. Authors [32, 30] also state that more accurate TM provides better results than DM. The author attributes the increase in strength to the pozzolanic material used to coat the grains of recycled aggregates, which have the ability to react portlandite (CH) accumulated in the pores and on the surface of the original mortar adhered to RCA grains to form new solid hydration products. These in turn improve the microstructure of the ITZ and strengthen the grain itself, thus contributing to increasing the strength and durability of the new concrete. With further optimization of the mixing process in terms of more detailed water dosing based on RCA absorbency, comes [33, 34]. The procedures contribute to a higher density of the binder in ITZ.

2.2. Polymer emulsions

Polymer emulsions have adhesive properties and can solidify in a short period. If RCA is immersed or sprayed with a polymer emulsion, the polymer molecules can fill the pores of the adhered mortar and seal the surfaces of the RCA. Some may also be water repellent, leading to a reduction in overall RCA water absorption. Materials being tested are Polyvinyl alcohol (PVA) emulsion [35], silan, siloxane [36], mineral oils like paraffin [37], polycarboxilate superplaticizer [5] or acrylic-based waterproofing penetrators [38]. The authors mostly present a positive effect, which is very significant in the case of water absorption of RCA, but the effect on the strength properties of concrete with such a modified RCA is not so great. [39] even concludes, that polymer treatment can improve the quality of RCA and enhance the workability and durability of recycled aggregate concrete as well. However, they cannot enhance but even reduce the compressive strength of the concrete. The reasons might due to the positive polymer groups permeate into the cement paste and make it hydrophobic, which hinders the hydration of the unhydrated cement in the paste. At the same time, the formation of water-repellent film weakens the bonding strength between the aggregate and cement matrix. [16] adds that this method of processing has several disadvantages because it cannot improve the strength of the attached mortar, the process is time consuming, requires specific equipment and economic considerations together with aspects of large scale production also limit its use.

2.3. Sodium silicate solution

The treatment is similar to a polymer emulsion treatment and is also known as a water glass treatment or a pore-blocking surface treatment [40]. Sodium silicate solution might form a continuous layer of sodium silicate on the surfaces of RCA. This can lead to improvement of strength properties, reduction of water absorption and improvement of ITZ. It was found that the concentration and soaking time affected the properties of RCA significantly; with higher concentration and longer soaking time, the compressive strength of the RCA may be reduced [41]. Also, the sodium silicate treatment may introduce alkalis which increase the risk of alkali silica reaction [39, 40]), as well the treatment is not able to improve chloride ions penetration [42]. Therefore, the treatment is not considered to ensure good durability properties of new concrete.

2.4. Carbonatation

When atmospheric CO_2 enters into the concrete through micro-pore and reacts with $Ca(OH)_2$ to form $CaCO_3$, this phenomenon is known as surface carbonation. After carbonation starts within the concrete, the pores start to fill and continue until the blockage of all connected pores [43]. This process can help to get better quality of RCA as attached original mortar is strengthened. For this purpose, improvement technique, where RCA is subject to defined CO_2 conditions (as for concentration and pressure applied) are investigated. Improvements in mechanical properties, density, water absorption, crushing value ITZ in terms of inside the RCA grain, as well as between RCA and new mortar, durability properties like chloride ion permeability or gas permeability are presented [44, 45].

2.5. Calcium carbonate biodeposition

Calcium carbonate biodeposition is based on the ability of bacteria to precipitate calcium carbonate on the surface of the cell wall due to the presence of adequate negative zeta potential. This technology is based on a natural process wherein bacterial cells produce inorganic minerals as part of their basic metabolic activities [46]. This is presented to be beneficial for RCA quality, as leads to better performance by reducing the porosity and moisture absorption inside [47]. The basic advantage of microbial carbonate precipitation is that bacteria are small in size and can enter into the micro pores where cement or fly ash cannot enter. The whole mechanism of deposition depends on the type of bacteria, applied materials and methods, and aggregate porosity [16].

3. Comparison of the methods

3.1. Influence on the quality of RCA grains

The selection of the results given in the relevant references is summarized in table 1, according to the treatment method and the author. For each sample, a value indicating water absorbency of untreated RCA and that of after specific treatment is given, then the percentage change is calculated. The overall improvement in absorbency is determined for each method by the arithmetic mean. A comparison of the effect of the individual methods is shown in the figure 2.

Table 1. Examples of the influence of individual methods of grain consolidation on the absorbency of RCA.

Treatment	Reference	Sample condition	Absorbency [%]
Pozzolanic suspension	[27]	before treatment	7.24
		after treatment (geopolymer with 10% of FA)	4.86
		change [%]	***32.9***
		before treatment	7.24
		after treatment (geopolymer with 10% of metakaolin)	5.47
		change [%]	***24.4***
		before treatment	7.24
		after treatment (geopolymer with 30% of silica fume)	6.59
		change [%]	***9.0***
	[48]	before treatment (20 mm grains)	4.44
		after treatment	3.48
		change [%]	***21.6***
		before treatment (10 mm grains)	5.58
		after treatment	4.48
		change [%]	***19.7***
	[26]	before treatment	8.2
		after treatment (geopolymer with FA)	8.2
		change [%]	***0***
		Overal improvement [%]	**17.9**
Polymer emulsions	[35]	before treatment	6.23
		after treatment – air dried	1.62
		change [%]	***74.0***
		before treatment	6.23
		after treatment –oven dried	2.39
		change [%]	***61.6***
	[38]	before treatment	7.96
		after treatment	2.97
		change [%]	62.7
		Overal improvement [%]	**66.1**
Calcium carbonate biodeposition	[47]	before treatment	5.5
		after treatment	4.7
		change [%]	***14.5***
	[49]	before treatment (20 mm grains)	6.9
		after treatment	6.1
		change [%]	***11.6***

		before treatment (10 mm grains)	9.0
		after treatment	7.9
		change [%]	***12.2***
		Overal improvement [%]	**12.8**
Carbonatation	[50]	before treatment	6.19
		after treatment	4.81
		change [%]	***22.3***
	[51]	before treatment (gravel RCA)	8.6
		after treatment	5.78
		change [%]	***28.7***
		before treatment (limestone RCA)	7.21
		after treatment	6.73
		change [%]	***6.6***
	[44]	before treatment (10 mm grains)	12.25
		after treatment	7.28
		change [%]	***40.6***
		Overal improvement [%]	**24.5**
Sodium silicate solution	[52]	before treatment	5.8
		after treatment (40% concentration, air drying)	3.7
		change [%]	***36.2***
		before treatment	5.8
		after treatment (40% concentration, 60°C drying)	3.8
		change [%]	***34**.5*
		before treatment	5.8
		after treatment (10% concentration, 105°C drying)	5.6
		change [%]	***3.4***
	[27]	before treatment (10 mm grains)	7.24
		after treatment	4.78
		change [%]	***34***
		Overal improvement [%]	**27.0**

3.2. Influence on the quality of new concrete
The selection of the results given in the relevant references is summarized in table 2, according to the treatment method and the author. For each sample, a value indicating strength of concrete having no-treated RCA and that of having RCA treated with a specified method is given. Then the percentage change is calculated. The overall improvement in strength of concrete is determined for each method and each kind of strength by the arithmetic mean. A comparison of the effect of the individual methods is shown in the figure 2. Unlike the table, the figure shows only 28-day strengths.

Table 2. Examples of the influence of individual methods of grain strengthening on the strength properties of recycled aggregate concrete (RAC).

Treatment	Reference	Sample condition	Compressive strength [MPa]		Flexural strength (MPa)	Splitting tensile strength (MPa)	
			7 days	28 days	28 days	7 days	28 days
Pozzolanic suspension	[22]	with no treated RCA	34.5	36.2	4.97		
		with treated RCA (cement)	32.9	43.6	5.16		
		change [%]	***-4.6***	***20.4***	***3.8***		
		with no treated RCA	38.1	43.6	5.69		
		with treated RCA (pozzolanic)	31.5	47.6	7.31		
		change [%]	***-17.3***	***9.2***	***28.5***		
	[7]	with no treated RCA	13.5	25.1		0.8	1.75
		with treated RCA	18.3	32.3		1.4	2.44
		change [%]	***35.6***	***28.7***		***75.0***	***39.4***
	[29]	with no treated RCA	44.0	51.0			
		with treated RCA	48.0	65.0			
		change [%]	***9.1***	***27.5***			
	[20]	with no treated RCA		19.0			
		with treated RCA (cement + fly ash)		24.0			
		change [%]		***26.3***			
		with no treated RCA		19.0			
		with treated RCA (geopolymer)		23.0			
		change [%]		***21.1***			
	[25]	with no treated RCA	37.0	47.0			
		with treated RCA (metasilikate 10%)	47.0	56.0			
		change [%]	***27.0***	***19.1***			
		with no treated RCA	37.0	47.0			
		with treated RCA (nanosilica 0,5%)	37.0	54.0			
		change [%]	***0***	***14.9***			
	Overal improvement [%]		**8.3**	**20.9**	**16.2**	**75.0**	**39.4**
Pozzolanic suspension - mixing methods	[34]	optimised TM - no pozz.	42	52			
		optimised TM with GGBFS	50	62			
		change [%]	***19***	***19.2***			
	[30]	NM (FA)	25.5	35			
		DM (FA)	30	38			

		change [%]	***17.6***	***8.6***			
		NM (FA)	25.5	35			
		TM (FA)	33	46			
		change [%]	***29,4***	***31.4***			
	[32]	NM (SF)	50,29	58.98			
		DM (SF)	50,79	59.26			
		change [%]	***1***	***0.5***			
	[53]	NM (CEM)		24.9			
		TM (CEM)		37			
		change [%]		48.6			
		NM (FA)		34.4			
		TM (FA)		27.7			
		change [%]		***-19.5***			
	Overal improvement [%]		**16,75**	**14,8**			
Polymer emulsions	[35]	with no treated RCA	41.0	49.2		2.6	3.3
		with treated RCA – air dried	39.0	53.0		2.8	3.4
		change [%]	***-4.9***	***7.7***		***6.9***	***3.3***
		with no treated RCA	41.0	49.2		2.62	3.3
		with treated RCA –oven dried	38.0	50.0		2.5	3.1
		change [%]	***-7.3***	***1.6***		***-4.6***	***-6.6***
	[38]	with no treated RCA		33.24			2.91
		with treated RCA		34.03			3.15
		change [%]		***2.4***			***8.2***
		with no treated RCA		34.21			3.09
		with treated RCA		35.05			3.17
		change [%]		***2.5***			***2.6***
	[5]	with no treated RCA	17.5	25.5		1.3	1.9
		with treated RCA	26.1	31.0		1.8	2.5
		change [%]	***49.1***	***21.6***		***38.5***	***31.6***
	Overal improvement [%]		**12.3**	**7.2**		**13.6**	**7.8**
Carbonatation	[50]	with no treated RCA	33.86	38.98			
		with treated RCA	36.63	45.8			
		change [%]	***8.2***	***17.5***			
	[54]	with no treated RCA (from demolition))		46.2	4.08		
		with treated RCA		48.4	4.2		
		change [%]		***4.8***	***2.9***		
		with no treated RCA (from new concrete)		38.2	3.2		

		with treated RCA		43.3	4.1		
		change [%]		***13.4***	***28.1***		
	Overal improvement [%]		**8.2**	**11.9**	**15.5**		
Sodium silicate solution	[55]	with no treated RCA (20 MPa concrete)	8.93	13.14			
		with treated RCA	11.06	16.01			
		change [%]	***23.9***	***21.8***			
		with no treated RCA (25 MPa concrete)	12.01	16.9			
		with treated RCA	17.05	25.02			
		change [%]	***42.0***	***48.0***			
	[56]	with no treated RCA		24.5			2.26
		with treated RCA		26.0			2.4
		change [%]		***6.1***			***6.2***
	[52]	with no treated RCA	29.4	37.9			
		with treated RCA	37.7	48.9			
		change [%]	***28.2***	***29***			
		with no treated RCA	29.4	37.9			
		with treated RCA	33.5	49.5			
		change [%]	***13.9***	***30.6***			
	Overal improvement [%]		**27.0**	**27.1**			**6.2**

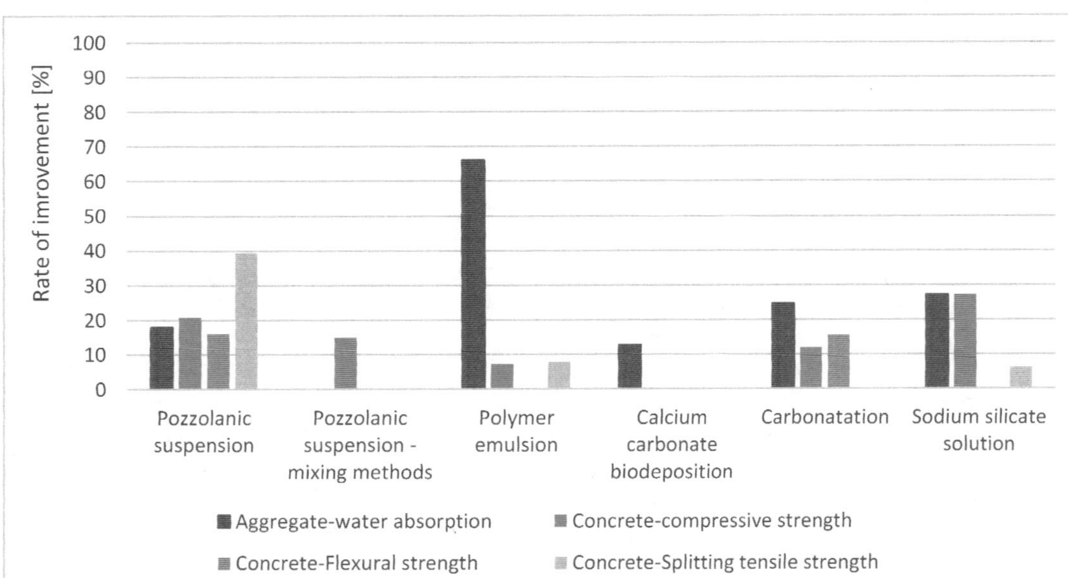

Figure 2. A comparison of the effect of individual methods on water absorbency of RCA and strength characteristics of RAC.

As can be seen in the figure and in the tables, it is difficult to make a complex comparison because the results are not available for some methods or properties (e.g. results for RAC mixed with RCA treated by calcium carbonate biodeposition). However, the effectiveness of methods can be roughly compared to have at least indicative information. Based on the above selection of compared results, the following can be summarized:

- ***polymer emulsion treatment*** can be specified as the most effective method ***to improve RCA water absorption***

The most effective methods for concrete mixed with treated RCA, following can be listed:

- for 28-day ***compressive strength, the sodium silicate treatment*** came out best
- for 28-day ***flexural strength***, the treatments with ***pozzolanic suspension and carbonatation are of comparable effect***
- for 28-day ***splitting tensile strength***, the coating of RCA with a ***pozzolanic suspension*** seems to be of the best solution

4. Conclusion

The article compares the methods for strengthening the grains of recycled concrete aggregate, which are generally presented as beneficial for the functional parameters of both the aggregate itself and the concrete that is produced with the treated aggregate. The methods differ in their principle, strengthening technique, materials used, treatment conditions, etc., so the above comparison of the results should be considered as indicative. It also includes only a limited selection of parameters to be compared (RCA absorbency, RAC strength); for a larger study, it would be appropriate to supplement the data with other, mainly durability, as well as economic aspects. However, it provides information on differences in the effectiveness of individual methods to help develop a basic research in this area, or on parameters for which there is not much information and there is a certain gap in research.

5. References

[1] Sorvari J 2008 Developing environmental legislation to promote recycling of industrial by-products – An endless story? *Waste Management* vol 28 pp 489-501

[2] European Commission 2016: EU Construction & Demolition Waste Management Protocol. Brusel. https://ec.europa.eu/growth/content/eu-construction-and-demolition-waste-protocol-0_en

[3] Jamkar S S, Rao C B K 2004 Index of Aggregate Particle Shape and Texture of coarse aggregate as a parameter for concrete mix proportioning Cem and Concr Res pp 2021-2027

[4] Eguchi K et al 2007 Application of recycled coarse aggregate by mixture to concrete construction Constr and Build Mater Vol 21 pp 1542-1551

[5] Ryou J S, Lee Y S 2014 Characterization of recycled coarse aggregate (RCA) via a surface coating method Int J of Concr Struct and Mater vol. 8 (2) pp 165-172

[6] Tabsh S W, Abdelfatah A 2009 Influence of recycled concrete aggregates on strength properties of concrete Constr Build Mater vol 23 pp 1163-1167

[7] Choi H et al 2016 Evaluation on the Mechanical Performance of Low-Quality Recycled Aggregate Through Interface Enhancement Between Cement Matrix and Coarse Aggregate by Surface Modification Technology Int J of Concr Struct and Mater pp 2234-1315

[8] Zhao Y, Zeng, W, Zhang H 2017 Properties of recycled aggregate concrete with different water control methods. Constr Build Mater vol 152 pp 539–546

[9] Limbachiya M C, Leelawat T, and Dhir R K 2000 Use of recycled concrete aggregate in high-strength concrete Mater and Struct vol 33 pp 574–580

[10] Sagoe-Crentsil K K, Brown T, & Taylor A H 2001 Performance of concrete made with commercially produced coarse recycled concrete aggregate. Cem and Concr Res vol 31 pp 701–712

[11] Shayan A, Xu A 2003 Performance and properties of structural concrete made with recycled concrete aggregate Materials Journal Vol 100 (5) pp 371–380

[12] Katz A 2003 Properties of concrete made with recycled aggregate from partially hydrated old concrete. Cem and Concr Res Vol 33 (5) pp 703–711

[13] Descarrega A 2011 Quality improvement of the recycled aggregates through surface treatment Universitat Politecnica de Catalunya Thesis, Barcelona
[14] STN EN 12620 (2008) Aggregates for concrete Slovak Office of Standards, Metrology and Testing Bratislava Slovakia
[15] Al-Bayati H K A et al 2016 Effect of Different Treatment Methods on the Interfacial Transition Zone Microstructure to Coarse Recycled Concrete Aggregate. TAC 2016: Efficient Transportation - Managing the Demand Conference and Exhibition of the Transportation Association of Canada
[16] Mistri A et al 2020 A review on different treatment methods for enhancing the properties of recycled aggregates for sustainable construction materials Constr Build Mater 233 (117894)
[17] Jawahar J G, Sashidhar C, Reddy I R and Peter J 2013 A Micro and macro level properties of fly ash blended self-compacting concrete Materials & Design 46 pp 696-705
[18] Akçaoğlu T, Tokyay M and Çelik T 2005 Assessing the ITZ microcracking via scanning electron microscope and its effect on the failure behavior of concrete. Cem and Concr Res 35 (2) pp 358-363
[19] Shaban W M et al 2019 Quality improvement techniques for recycled concrete aggregate: A review J of adv concr technol vol 17 pp 151-167
[20] Gupta A, Ghosh S and Mandal S 2012 Coated recycled aggregate concrete exposed to elevated temperature Global Journal of research in Engineering, Civil and Structural Engineering Vol 12 (3) pp 26-32
[21] Kou S C and Poon C S 2012 Enhancing the durability properties of concrete prepared with coarse recycled aggregate Constr Build Mater 35 pp 69–76
[22] Li J, Xiao H and Zhou Y 2009 Influence of coating recycled aggregate surface with pozzolanic powder on properties of recycled aggregate concrete, Constr Build Mater 23 pp 1287–1291
[23] Li W, Long C, Tam V W Y, Poon C and Duan W H 2017 Effects of nano-particles on failure process and microstructural properties of recycled aggregate concrete, Constr Build Mater 142 pp 42–50
[24] Zhang H, Zhao Y, Meng T and Shah SP 2015 The modification effects of a nano-silica slurry on microstructure, strength, and strain development of recycled aggregate concrete applied in an enlarged structural test Constr Build Mater 95 pp 721–735
[25] Ismail S and Ramli M 2014 Influence of surface-treated coarse recycled concrete aggregate on compressive strength of concrete Int. J of Civ Constr and Archit Eng Vol 8 No 8
[26] Sicakova A et al 2016 Changes in technological parameters of selected waste aggregates after the surface treatments for use in concrete IAHS World Congress on Housing pp 1-10
[27] Bui N K, Satomi T and Takahashi H 2018 Enhancement of recycled aggregate concrete properties by a new treatment method Int. J. of GEOMATE Vol 14 Issue 41 pp 68-76
[28] Urban K and Sicakova A 2017 The influence of kind of coating additive on the compressive strength of RCA-based concrete prepared by triple-mixing method IOP Conference Series Earth and Environmental Science: AEE 2017 Bristol: IOP Publishing pp 1-5
[29] Zhao Z. et al 2013 Evaluation of pre-coated recycled aggregate for concrete and mortar. Constr Build Mater 43 pp 191-196
[30] Kong D. et al 2010 Effect and mechanism of surface-coating pozzalanics materials around aggregate on properties and ITZ microstructure of recycled aggregate concrete Constr Build Mater 24 (5) pp 701–708
[31] Lee S H et al 2014 Influence of aggregate coated with modified sulfur on the properties of cement concrete Materials Vol 7 (6) pp 4739–4754
[32] Tam V W Y and Tam C M 2008 Diversifying two-stage mixing approach (TSMA) for recycled aggregate concrete: TSMAs and TSMAsc Constr Build Mater 22 pp 2068–2077
[33] Sicakova A and Urban K 2018 The Influence of Discharge Time, Kind of Additive, and Kind of Aggregate on the Properties of Three-Stage Mixed Concrete Sustainability Vol 10 (11) pp 3862-3862
[34] Zhang W et al 2019 Effect of the optimized triple mixing method on the ITZ microstructure and performance of recycled aggregate concrete. Constr Build Mater 203 pp 601–607

[35] Kou S C and Poon C S 2010 Properties of concrete prepared with PVA-impregnated recycled concrete aggregates Cem Concr Comp Vol 32 Issue 8 pp 649-654
[36] Spaeth V and Tegguer A D 2014 Improvement of recycled concrete aggregate properties by polymer treatments Int J Sustain Built Environ 2 pp 143–152
[37] Tsujino M et al 2007 Application of conventionally recycled coarse aggregate to concrete structure by surface modification treatment J Adv Concr Technol 5 pp 13–25
[38] Atmajayanti A T et al 2018 The effect of recycled coarse aggregate (RCA) with surface treatment on concrete mechanical properties MATEC Web of Conferences 195 (01017)
[39] Shi C 2016 Performance enhancement of recycled concrete aggregate - A review J of Cleaner production 112 pp466-472
[40] Pan X et al 2017 A review on concrete surface treatment Part I: types and mechanisms Constr Build Mater 132 pp 578–590
[41] Chen JJ, Thomas JJ and Jennings HM 2006 Decalcification shrinkage of cement paste Cem Concr Res Vol 36 (5) pp 801-809
[42] Ibrahim M et al 1999 Use of surface treatment materials to improve concrete durability J Mater Civ Eng 11 pp 36–40
[43] Borges PH et al 2010 Carbonation of CH and C-S-H in composite cement pastes containing high amounts of BFS Cem Concr Res 40 pp 284–292
[44] Kou S C, Zhan B J and Poon C S 2014 Use of a CO2 curing step to improve the properties of concrete prepared with recycled aggregates Cem Concr Comp 45 pp 22-28
[45] Xuan D, Zhan B and Poon C S 2017 Durability of recycled aggregate prepared with carbonated recycled concrete aggregate Cem Concr Comp 84 pp 214-221
[46] Dhami N K, Reddy M S and Mukherjee A 2013 Biomineralization of calcium carbonates and their engineered applications: a review Front Microbiol
[47] Grabiec A M et al 2012 Modification of recycled concrete aggregate by calcium carbonate biodeposition Constr Build Mater Vol 34 pp 145-150
[48] Ismail S and Ramli M 2013 Effect surface treatment of recycled concrete aggregate on prioperties of fresh and hardened concrete IEEE Business Engineering and Industrial Colloquium (BEIAC) pp 651-656
[49] Wu Ch R et al 2018 Improving the properties of recycled concrete aggregate with bio-deposition approach. Cem Concr Comp 94 pp 248-254
[50] Li L et al 2017 Effect of carbonated recycled coarse aggregate on the dynamic compressive behavior of recycled aggregate concrete Constr Build Mater Vol 151 pp 52-62
[51] Zhang J. et al 2015 Influence of carbonated recycled concrete aggregate on properties of cement mortar Constr Build Mater Vol 98 pp 1-7
[52] Yang L X et al 2016 Effects of Recycled Coarse Aggregate Reinforcing Treated by Water-Glass on the Performance of Recycled Concrete Advances in Engineering Research, Proceedings of the 3rd International Conference on Material Engineering and Application (ICMEA) Vol 103 pp 181-186
[53] Urban K and Sicakova A 2018 The Effect of Mixing Technique and Prolonged Mixing Time on Strength Characteristics of Concrete Environment, Green Technology and Engineering International Conference: proceedings MDPI Basel Vol 2 pp 1-4
[54] Xuan D, Zhan B and Poon C S 2016 Assessment of mechanical properties of concrete incorporating carbonated recycled concrete aggregates Cem Concr Comp Vol 65 pp 67-74
[55] Misra A K, Kalra M and Bansal S 2017 Influence of polymer treatment on strength and water absorption capacity of recycled aggregate concrete Int J of Sust Build Techn and Urban Dev Vol 8 Issue 2 pp 81-91
[56] Al-Taie M W and Al-Shafi'I N T 2020 Mechanical Properties of Normal Concrete Using Treated Recycled Concrete Aggregate Int J of Innov in Eng Res and Technol Vol 7 (07) pp 94-101

Acknowledgments
This research has been carried out within the project of Slovak Scientific Grant Agency VEGA (Grant No. 1/0336/22) "Research on the effects of Lean Production / Lean Construction for improving the efficiency of on-site and off-site construction technologies" and NFP project No. 313010T578.

IOP Conf. Series: Materials Science and Engineering 1252 (2022) 012058 doi:10.1088/1757-899X/1252/1/012058

VR-based Safety Training Research in Construction

V Chellappa[1], P Mésároš[2], M Špak[3], M Spišáková[4] and P Kaleja[5]

[1] Department of Design, Indian Institute of Technology Guwahati, Guwahati-781039, India
[1,2,3,4,5] Faculty of Civil Engineering, Technical University of Košice, Košice -042 00, Slovakia

E-mail: vigneshkumarchellappa@gmail.com

Abstract. Safety training in the construction industry widely adopted virtual reality (VR) technologies to train and educate employees and workers to enhance the safety performance of projects. In this study, a review of VR-based safety training research in construction was carried out to understand the body of literature. Using keywords such as "construction industry" OR "safety training" OR "safety learning," etc., a total of 52 journal articles published between 2000 and 2021 were collected from the Scopus database after advanced filtering. A two-stage analysis of VR-based safety training methods and application areas was carried out were followed by a review. The results indicate that the immersive safety training method was widely used followed by desktop-based, game-based, and BIM-based training methods. VR-based safety training methods are largely applied to hazard recognition (HR), safe operations (SO), and safety awareness (SA). The study findings could assist both academicians and construction practitioners in choosing the best optimal VR-based safety training methods to improve training performance.

1. Introduction

Globally, the construction sector offers more job opportunities for many individuals [1]. Also, the construction sector corresponds to 13 percent of the gross domestic product (GDP) worldwide [2]. Despite its benefits, the construction industry contributes to poor safety performance. According to global statistics, accident rates in construction were higher than in other sectors [1]. Furthermore, the harmful nature of accidents in construction results in unintended consequences such as loss of work and project delays [3]. In addition, fatalities cause serious financial damage to individuals and society [4].

Previous studies indicated that most construction accidents were caused by a lack of proactive control measures such as identifying hazards, awareness of safety, workforce training, and so on [5]. To avoid accidents at construction workplaces, safety training has been recognized as an effective strategy to improve individuals' safety awareness and risk identification skills [6]. On the other side, many researchers (e.g., [7,8]) indicated that traditional safety methods are not effective in transferring safety knowledge during the training process. Also, due to the limitations of workplace visualizations, it is challenging to visualize the risks involved in the construction work through traditional training methods [9].

With the evolution of innovative technologies, virtual reality (VR) has been recognized in construction as a promising tool for safety training programs due to its effective nature in enhancing such programs [10]. Hence, VR has been hailed as the appropriate solution to the aforementioned shortcomings. VR is a computer-assisted method of providing visualization from immersive and interactive views [11]. The benefits of VR technology include the ability to visualize complex construction environments and the use of a gamification strategy to improve users' involvement and enhance knowledge-based training [12]. Hence, many studies established VR technology as an effective safety training method compared to traditional training methods [12, 13-15].

Despite the growth of VR technology in construction for safety training, no review of studies on the development and execution of VR-based safety training methods in construction has been conducted till now up to the authors' knowledge. According to Chellappa and Srivastava [16], conducting a periodic review of existing studies in a particular domain can help to understand the state of research. Therefore, in this study, a review of the literature was conducted on VR-based safety training methods in construction using a Scopus database. A total number of 52 journal documents were retrieved from 2000 to 2021. The following are some of how this study adds to the knowledge: (1) to visualize the distribution of the paper by publication years; (2) to show the distribution of articles by journal sources; (3) to expose methods used for VR-based safety training in construction.

2. Research method

A literature search was conducted on the Scopus database to retrieve relevant documents for the review. Due to the broader coverage of studies, Scopus was chosen in this study [16]. Also, previous literature review studies [17,18] in the construction management domain used the Scopus database to collect documents.

The keywords used to search and collect suitable articles were:

TITLE-ABS-KEY ("construction industry" OR "occupational health and safety" OR "safety management' OR "safety training" OR "safety learning" OR "safety education" OR "virtual reality" OR "game technology" OR "health and safety training." AND "virtual environment").

The research methodology framework is shown in Figure 1. It can be seen in figure 1 that 323 documents were yielded initially from the database. The identified documents were limited through advanced search rules i.e., 'from 2000 and before 2022" for the timeframe, "English" was the language, source type was "Journal", and "Article" as the document type. Due to reliable sources of knowledge, only journal articles were selected [17]. The articles that do not target VR for safety training but still used the term in the abstract were omitted. Finally, after careful refinement, 52 journal documents were chosen for this study. The MS Excel sheet was used to classify and store the retrieved documents based on the publication year, journal name where it got published, the title of the document, and the author keywords used in the documents. In the next section, the study results are presented based on the number of documents published per year, publications per journal, and the existing studies carried out on VR for safety training.

3. Results and discussion

3.1. Overview of selected documents

The documents published per year in the selected period are shown in figure 2. From figure 2, no significant growth in studies focusing on VR-based safety training was observed during the period 2000 to 2019. It can be seen the highest number of publications was documented in the year 2021 (16 publications). It is anticipated that a greater number of documents could be published in upcoming years in this research domain due to its positive impact on safety performance, although there is a slight deviation in past years.

The number of documents published per journal source is listed in table 1. Table 1 shows the list of journals with at least 2 articles published within the scope of this study. It can be seen that the most influential journal is Automation in Construction (8 articles), followed by the Journal of Information Technology in Construction (5 articles), Advanced Engineering Informatics (4 articles), and Engineering, Construction and Architectural Management (3 articles). Overall, the listed journals in Table 1 published 54% of the documents that were retrieved for this study.

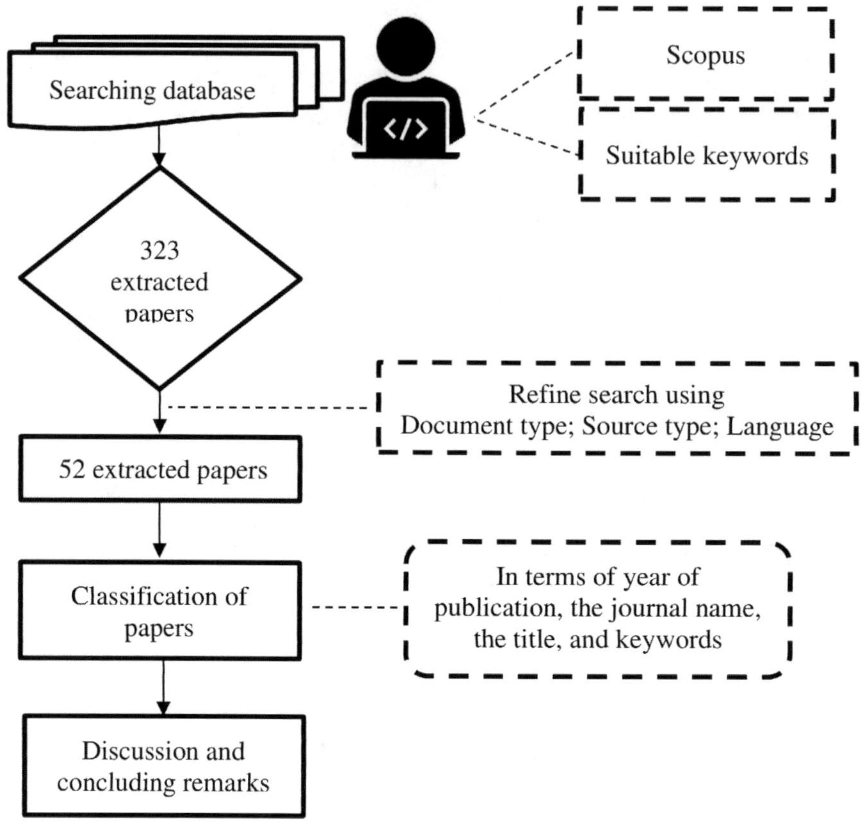

Figure 1. Research methodology framework.

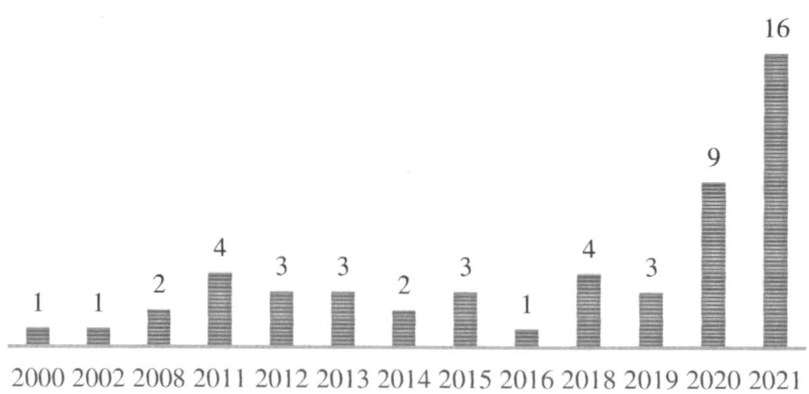

Figure 2. Number of documents published each year.

3.2. Overview of the identified studies

VR-based safety training studies in construction were categorized based on four different methods such as BIM-based training, desktop-based training, game-based training, and immersive training. Figure 3 shows the distribution of the retrieved articles characterized by VR methods. It was observed from figure 3 that immersive training (48%) followed by desktop-based training (34%) were most commonly used for safety training in construction. In terms of growth, the immersive training method gained much attention among the construction safety management research community in the recent decade with 25 publications (48%) from 2011 to 2021, although there is a little growth in the desktop-based training method. Also, BIM-based and game-based safety training methods emerged in the year 2011.

Table 1. Documents published per journal source.

S.No	Source	No of publications
1	Automation in Construction	8
2	Journal of Information Technology in construction	5
3	Advanced Engineering Informatics	4
4	Engineering, Construction and Architectural Management	3
5	Construction Innovation	2
6	Safety Science	2
7	Journal of Computing in Civil Engineering	2
8	Sustainability	2

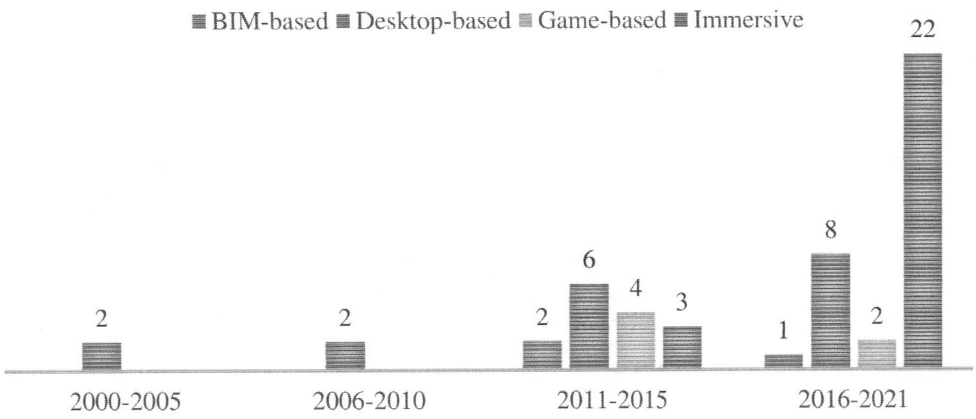

Figure 3. Distribution of documents characterized by publication year and VR methods.

The retrieved documents distribution characterized by research areas is shown in figure 4. Since 2000, the application of VR focused mainly on training and educating about safe operations (SO) (37%) of construction activities. Since 2011, the research focus was also placed on hazard recognition (HR) training (34%) and safety awareness (SA) (21%) in the construction workplace. It can also be observed in figure 4 that a few studies (8%) were focused on combining different areas such as HR and SA [19-22]. The VR methods and their application in different fields are discussed in the following sections.

3.2.1. Immersive training

Among 52 retrieved documents, 25 documents (48%) were focused on safety training through immersive VR. According to Wang et al. [23], immersive VR depends on the adoption of a head-mounted display (HMD for providing an immersive world to the users by withdrawing them from the physical environment. Various sensors and other equipment such as sensor gloves, controllers, and treadmills are also added to the immersive VR setup to provide real-time feedback [24]. Figure 5 shows the example applications related to immersive VR for safety training.

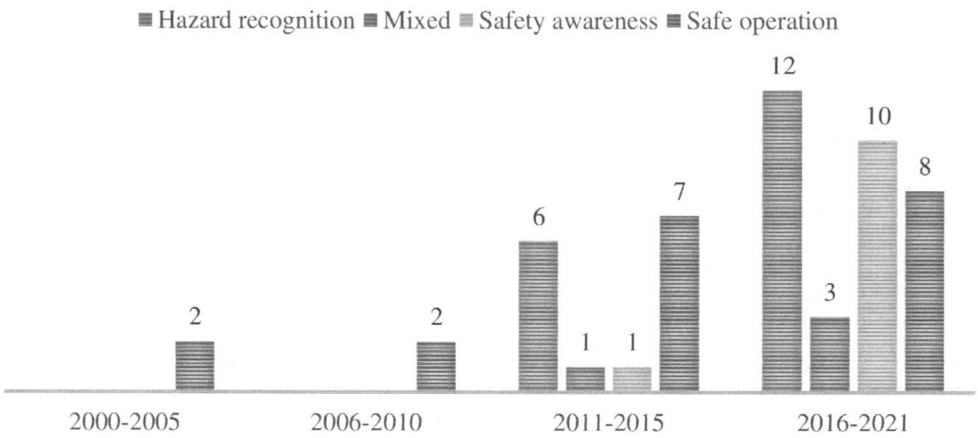

Figure 4. Distribution of documents characterized by research areas for VR safety training.

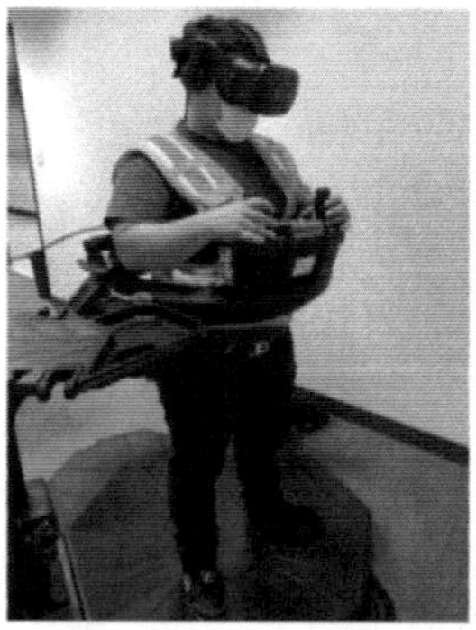

Figure 5. Immersive virtual training [25].

Figure 6 shows the VR safety training methods applied to different research areas. It can be seen in figure 6 that 25 (48%) out of 52 retrieved documents used the immersive VR method for safety training. Among 25 documents, most of the documents (i.e., 10) focused on training for HR. For instance, Eiris et al. [26] conducted an immersive VR safety training with construction management students to recognize fall hazards in the electrical trade of construction projects. The findings of this study indicated that the proposed VR method could reduce the time spent on HR compared to traditional methods. Similarly, several studies (e.g., [27]) adopted the immersive VR method for HR at construction workplaces. The immersive VR method has also been widely used to train and educate about SO (8 documents), followed by SA (5 documents). In addition, some studies adopted the immersive VR method to cover different areas. For example, Nykänen et al. [21] developed a novel safety training method for HR and to create SA among construction workers [22].

3.2.2. Desktop-based training

Desktop-based VR training has been a widely used VR method for safety training in construction since 2000 (see figure 3). Without the use of any tracking equipment, the desktop visualizes the virtual environment (VE) on the computer screen [23]. Mouse and keyboards were used to perform the tasks. This method is considered to be cheap for VR safety training compared to other VR methods. An example application related to desktop-based VR for safety training is shown in figure 7.

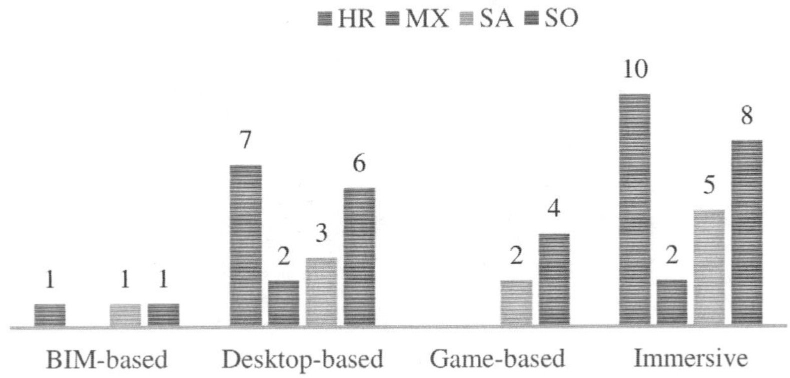

Figure 6. VR methods applications in different research areas.

From figure 6, it can be observed that the desktop-based method has been adopted for HR (7 documents), to perform SO (6 documents), followed by creating SA (3 documents). In terms of HR, Pham et al. [29] developed a prototype to improve the learning skills of educators. The desktop-based method has also been used by some studies to cover different areas. For example, Pedro et al. [30], developed a system to visualize the construction environment for HR and to create SA among workers.

Figure 7. Desktop-based VR safety training [28].

3.2.3. Game-based training

Game-based technology is a computer-based setting that contains multi-user operating technologies, networks, interactive, and so on to improve user interactions [23] and focuses more on interactions of game objects. According to Wang et al. [23], game-based VR can minimize the complexity of complex objects like cranes and excavators in construction. Figure 8 shows the example applications related to game-based VR for safety training.

From figure 3, it was observed that 6 documents (12%) focused on game-based VR. In which, a game-based VR method has been adopted to perform SO and create SA, with 4 and 2 documents, respectively (see figure 6). In terms of SO, Li et al. [31] developed a multi-user safety training system for dismantling tower cranes at construction sites. Similarly, Dickinson et al. [32] adopted safety games with students to enhance their knowledge of trench safety.

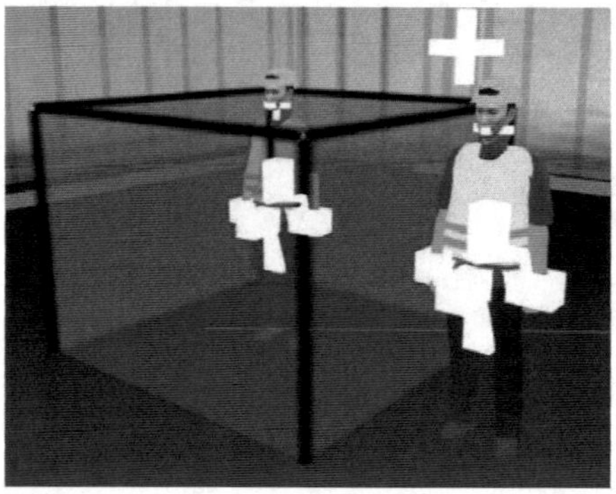

Figure 8. Game-based VR for safety training [31].

3.2.4. BIM-based training
Among other VR methods, BIM-based training use was limited to safety training (3 documents). BIM is concerned with the development and usage of three-dimensional objects [23]. Unlike other VR categories, BIM-based VR relies on the model to replicate construction processes and operations, stressing the data binding and linkages behind it [23]. The example application related to BIM-based VR for safety training is shown in figure 9.

Figure 9. BIM-based VR for safety training [33].

In research areas such as HR, SO, and SA, the BIM-based method has been used. Park et al. [12] developed a BIM-based system prototype for HR and validated it with construction practitioners. Likewise, game-based virtual training was conducted to perform SO [33] among construction workers and create SA [34] among construction university students.

4. Conclusion
This study conducted a literature review on VR-based safety training research in construction to understand the VR-based methods used and their application research areas. The articles for the literature review were retrieved from the Scopus database from the year 2000 to 2021. The findings indicated that although the studies were scattered from 2000 to 2019, the scope of this research got much attention among researchers in 2020 and 2021 with the highest number of publications. It is expected that a greater number of studies could be carried out in upcoming years focusing on VR-based safety training research in construction. The most influential journals under the scope of the study were Automation in Construction, followed by the Journal of Information Technology in construction and Advanced Engineering Informatics.

In terms of methods, the immersive safety training method has been widely used particularly after 2010. The desktop-based training method was found to be a stable method for safety training since 2000 and it is considered to be cheaper. Other methods such as game-based and BIM-based training were used less for safety training. This is because the software for creating VE is expensive. Concerning research areas, most of the VR-based methods were used for HR followed by performing SO and SA. Some of the studies were also found to cover multiple areas. The findings of this study are expected to assist construction practitioners in choosing the best optimal VR-based safety training method to improve training performance.

This study comes up with some limitations. First, this study retrieved articles only from the Scopus database. Next, only journal articles were selected from the Scopus database that is published in English by omitting all other sources such as conference papers, book chapters, and so on. In the future, other databases and other sources of publication could be used to have a broader view of this study's scope.

5. References

[1] Sousa V, Almeida N M and Dias L A 2014 Risk-based management of occupational safety and health in the construction industry – Part 1: Background knowledge *Saf. Sci.* **66** 75–86

[2] Barbosa A, Gambatese J, Das A and Pestana A C 2014 Mapped workflow for safety and reliability assessments of use and re-use of formwork *Proc. Constr. Res. Cong.* (Atlanta) (Georgia/ASCE) p1821–1830

[3] Sunindijo R Y and Zou P X W 2013 Conceptualizing Safety Management in Construction Projects *J. Constr. Eng. Manag.* **139** 1144–1153

[4] Yuan J, Yi W, Miao M and Zhang L 2018 Evaluating the Impacts of Health, Social Network and Capital on Craft Efficiency and Productivity: A Case Study of Construction Workers in China *Int. J. Environ. Res.* **15** 345

[5] Park C S and Kim H J A 2013 framework for construction safety management and visualization system *Autom. Constr.* **33** 95–103

[6] Sawacha E, Naoum S and Fong D 1999 Factors affecting safety performance on construction sites *Int. J. Proj. Manag.* **17** 309–315

[7] Demirkesen S and Arditi D 2015 Construction safety personnel's perceptions of safety training practices *Int. J. Proj. Manag.* **33** 1160–1169

[8] Bhandari S and Hallowell M 2017 Emotional engagement in safety training: impact of naturalistic injury simulations on the emotional state of construction workers *J. Constr. Eng. Manag.* **143** 1–10

[9] Ge L and Kuester F 2015 Integrative simulation environment for conceptual structural analysis *J. Comput. Civ. Eng.* **29** 1–10

[10] Akinlolu M, Theo C, Edwards D J and Simpeh F 2020 A bibliometric review of the status and emerging research trends in construction safety management technologies, *Int. J. Constr. Manag.* https://doi.org/10.1080/15623599.2020.1819584

[11] Li X, Yi W, Chi L, Wang X and Chan A P C 2018 A critical review of virtual and augmented reality (VR/AR) applications in construction safety *Autom. Constr.* **86** 150–162

[12] Zhang M, Shu L, Luo X, Yuan M and Zheng X 2022 Virtual reality technology in construction safety training: Extended technology acceptance model *Autom. Constr.* **135** 1-13

[13] Wang P, Wu P, Wang J, Chi H and Wang X 2018 A critical review of the use of virtual reality in construction engineering education and training *Int. J. Environ. Res. Public Health* **15** 1–18

[14] Goulding J, Nadim W, Petridis P and Alshawi M 2012 Construction industry offsite production: a virtual reality interactive training environment prototype *Adv. Eng. Inform.* **26** 103–116

[15] Goedert J, Cho Y, Subramaniam M, Guo H and Xiao L 2011 A framework for virtual interactive construction education (VICE) *Autom Constr.* **20** 76–87

[16] Chellappa V and Srivastava V 2022 Science mapping approach based review of thermal comfort studies in Indian residential buildings *J. Eng. Des. Technol.* ahead-of-print

[17] Vigneshkumar C and Salve U R 2020 A scientometric analysis and review of fall from height research in construction *Constr. Econ. Build.* **20** 17-35

[18] Chellappa V, Srivastava V and Salve U R 2021 A systematic review of construction workers' health and safety research in India *J. Eng. Des. Technol.* **19** 1488-1504

[19] Pedro A, Pham H C, Kim J U and Park C 2020 Development and evaluation of context-based assessment system for visualization-enhanced construction safety education *Int. J. Occup. Saf. Ergon.* **26** 811-823
[20] Nykänen M, Puro V, Tiikkaja M, Kannisto H, Lantto E, Simpura F, Uusitalo J, Lukander K, Räsänen T, Heikkilä T and Teperi A M 2020 Implementing and evaluating novel safety training methods for construction sector workers: Results of a randomized controlled trial *J Safety Res.* **75** 205–221
[21] Nykänen M, Puro V, Tiikkaja M, Kannisto H, Lantto E, Simpura F, Uusitalo J, Lukander K, Räsänen T and Teperi A M 2020 Evaluation of the efficacy of a virtual reality-based safety training and human factors training method: study protocol for a randomised-controlled trial *Inj. Prev.* **26** 360–369
[22] Le Q T, Pedro A and Park C S 2014 A Social Virtual Reality Based Construction Safety Education System for Experiential Learning *J. Intell. Robot Syst.* **79** 487–506
[23] Wang P, Wu P, Wang J, Chi H L and Wang X 2018 A Critical Review of the Use of Virtual Reality in Construction Engineering Education and Training *Int. J. Environ. Res.* **15** 1204
[24] Kaufmann H, Schmalstieg D and Wagner M 2000 Construct3D: A virtual reality application for mathematics and geometry education *Educ. Inf. Technol.* **5** 263–276
[25] Adami P, Rodrigues P B, Woods P J, Becerik-Gerber B, Soibelman L, Copur-Gencturk Y and Lucas G 2021 Effectiveness of VR-based training on improving construction workers' knowledge, skills, and safety behavior in robotic teleoperation *Adv. Eng. Inform.* **50** 101431
[26] Eiris R, Jain A, Gheisari M and Wehle A 2020 Safety immersive storytelling using narrated 360-degree panoramas: A fall hazard training within the electrical trade context. *Saf. Sci.* **127** 104703
[27] Afzal M and Shafiq M T 2021 Evaluating 4D-BIM and VR for Effective Safety Communication and Training: A Case Study of Multilingual Construction Job-Site Crew. *Buildings* **11** 319
[28] Zhao D and Lucas J 2014 Virtual reality simulation for construction safety promotion. *Int. J. Inj. Contr. Saf. Promot.* **22** 57–67
[29] Pham H C, Dao N N, Cho S, Nguyen P T and Pham-Hang A T 2019 Construction Hazard Investigation Leveraging Object Anatomization on an Augmented Photoreality Platform *Appl. Sci.* **9** 4477
[30] Pedro A, Pham H C, Kim J U and Park C 2020 Development and evaluation of context-based assessment system for visualization-enhanced construction safety education *Int. J. Occup. Saf. Ergon.* **26** 811–823
[31] Li H, Chan G and Skitmore M 2012 Multiuser Virtual Safety Training System for Tower Crane Dismantlement *J. Comput. Civ. Eng.* **26** 638–647
[32] Dickinson J K, Woodard P, Canas R, Ahamed S and Lockston D 2011 Game-based trench safety education: development and lessons learned *ITcon* **16** 119-134
[33] Ahn S, Kim T, Park Y J and Kim J M 2020 Improving Effectiveness of Safety Training at Construction Worksite Using 3D BIM Simulation *Adv. Civ. Eng.* 1–12
[34] Ku K and Mahabaleshwarkar P S 2011 Building interactive modeling for construction education in virtual worlds *ITcon* **16** 189-208

Acknowledgments
This research is supported by the National Scholarship Programme of the Slovak Republic. This research was also supported by Erasmus+ 2019-1-SK01-KA203-060778: "Construction Safety with Education and Training using Immersive Reality" and KEGA 009TUKE-4/2022: "An interactive tool for designing a safe construction site in an immersive virtual reality".

The effect of physicochemical properties of bottom sediments on nitrogen and phosphorus sorption

N Junakova[1], J Junak[1] and M Balintova[1]

[1] Institute of Environmental Engineering, Faculty of Civil Engineering, Technical University of Košice, Vysokoškolska 4, Košice 042 00, Slovakia

E-mail: natalia.junakova@tuke.sk; jozef.junak@tuke.sk; magdalena.balintova@tuke.sk

Abstract. River and reservoir bottom sediments are characterized by their physical and chemical properties, which affect the quality of water and deposited sediments in reservoirs. The grain composition of sediments is their fundamental property that determines the detachment, transport and sedimentation of sediments. In addition to understanding the dynamics of transport and distribution of particles in the watershed, the granularity is closely related to the mineral and chemical composition of the sediment and plays an important role in terms of its ability to bind contaminants. Other important property of sediment includes the specific surface area, which determines the nature of sediment surface reactions and influences ecosystem level, biological processes.

This study investigates sediment properties, including physical properties such as grain-size distribution, specific surface area, organic matter and chemical composition, and their reflections on nitrogen and phosphorus adsorption by the sediments deposited in the small water reservoir. The results clearly indicate the dependence of the content of nutrients (nitrogen and phosphorus) in reservoir sediments on their grain size and specific surface area.

1. Introduction

Sediments are an integral, natural and dynamic part of every waterbody. They represent eroded particles derived from rocks or biological materials from surrounding area that have been transported by the liquid phase. It is estimated that the average annual amount of eroded particles transported to the oceans is in the range of 24 to 30 milliard tons [1]. Some of the transported sediments are stored in water reservoirs [2, 3], which reduces the amount of sediment flowing into the oceans by 1.4 milliard tons yearly [4]. It is generally stated that the sedimentation rate in reservoirs is 0.3-2% per year from the original reservoir storage volume, giving most reservoirs a lifetime of approximately 100 years [5]. In the Slovak Republic, the rate of reservoirs' silting with sediments is at the level of 0.8% per year, which in 2050 assumes a reduction of their storage volume due to sedimentation by almost half.

The permanent process of transport and sedimentation of particles brought by inflows and runoff from the surroundings also affects the change of qualitative characteristics of watercourses and reservoirs. Erosion-washed soil particles bind substances contained in the soil (e.g. heavy metals, specific organic substances, etc.) and biogenic elements (especially nitrogen and phosphorus), making them secondary pollutants of waterbodies [6, 7]. The accumulation of pollutants in sediments poses a potential risk of their remobilization into the aquatic environment.

Sediments deposited at the bottom of reservoirs represent the largest sink of phosphorus in most reservoirs. Therefore one of the most studied mobilization processes at the sediment-water interface is the study of phosphorus sorption by bottom sediments [8, 9]. Sorption interactions between sediment particles and water control the amount of phosphorus present in the water. The most important factors influencing phosphorus mobilization from / to sediment in most types of lakes and reservoirs include the physicochemical properties of the sediment [10], oxidation-reduction conditions, temperature and pH. It is the grain composition of sediments and the specific surface area that belong to the basic characteristics of sediments, which determine the nature of sediment surface reactions and are used in the sorption of nutrients. More studies (e.g. [11]) have shown that sediments with a predominance of fine-grained clay minerals with Fe and Al hydr (-oxides) on their surface, have a greater specific surface area for phosphorus sorption. Similarly, the presence of total nitrogen in sediments is affected by sediment granularity and specific surface area [12]. Some authors [13] also have reported a positive effect of organic matter present in sediment on phosphorus uptake in sediment. On the contrary, this fact was refuted by Galy et al. [14] who found that this dependence does not apply universally.

The paper focuses on the study of the influence of selected physicochemical properties of bottom sediments on nitrogen (N) and phosphorus (P) sorption. The aim is to analyze the properties of sediments and evaluate the relationship between particle size, specific surface area, organic matter content, pH and the presence of selected chemicals.

2. Material and methods

2.1. Material

To study relationship between sediment physicochemical properties and their impact on nitrogen and phosphorus sorption, samples were taken from the Hervartov small water reservoir located in the north-east of Slovakia in the Topľa river basin. The reservoir is situated in an agricultural watershed with no presence of significant point source pollution. Composite sediment samples were collected from nine different places of the reservoir (near the dam, samples S1-S4; and along the reservoir S5-S8; sample S9 in inflow to the reservoir). To determine the physical-chemical and chemical parameters, sediment samples were analyzed as described below.

2.2. Sediment analysis

Particle size distribution (PSD) of sediment samples was performed using Mastersizer 2000 (Malvern Instruments, UK) with a Hydro 2000S wet dispersion unit. Measurement parameters were: pump speed - 2,800 rpm; ultrasonic - turned on for 5 min after sample addition.

Specific surface area (SSA) was estimated by low temperature gas adsorption using nitrogen gas (Quantachrome NOVA 1000e, USA) and subsequent application of Brunauer – Emmett – Teller isotherm.

The organic matter (*OM*) in sediments was determined by rough estimate based on the measuring the weight loss on ignition. First, two grams of dry sediment were transferred into a pre-weighed crucible and sample weight (*A*) was determined using an analytical balance. Then the samples were burned at 550° C for 2 hours in a laboratory muffle furnace (Thermolyne 1400) to remove organic matter. After cooling in a desiccator, samples were reweighed (*B*) and organic matter percentage was determined using equation:

$$OM\ (\%) = \frac{100-(A-B)}{A} * 100 \tag{1}$$

Sediment pH was measured in a solution of 0.01 M $CaCl_2$ in a 1:5 sediment-to-solution ratio at 25 °C according to the corresponding ISO method (ISO 10390). After shaking the mixture for 1 hour, sediment pH was determined using FiveGo™ electrode (Mettler Toledo, Switzerland).

Total nitrogen content was determined using elemental analysis (LECO CHN628, USA) and total phosphorus using inductively coupled plasma-atomic emission spectrometry technique (Agilent 5100 ICP-OES, USA).

The chemical composition of the sediments (oxide analysis) was determined by X-ray fluorescence method using SPECTRO iQ II (Ametek, Germany) with SDD silicon drift detector with resolution of 145 eV at 10,000 pulses.

3. Results and discussion

Representation of individual grain fractions in bottom sediment samples (clay: <0.004 mm, silt: 0.004-0.063 mm, sand: 0.063-2.0 mm, gravel: > 2.0 mm), analyzed by laser diffractometry, is given in table 1.

Table 1. Particle size distribution.

Sample	Clay (%)	Silt (%)	Sand (%)	Gravel (%)
S1	13.87	84.53	1.60	-
S2	15.37	83.61	1.02	-
S3	11.70	87.52	0.78	-
S4	10.99	88.70	0.31	-
S5	7.88	89.86	2.26	-
S6	8.29	90.84	0.87	-
S7	9.64	90.19	0.17	-
S8	8.60	86.89	4.51	-
S9	1.62	60.26	30.33	7.79

The sediment specific surface area (table 2), as a dominant factor controlling its surface reactions, is around 8 m^2 / g in most fine-grained sediments. The fine-grained sediments' specific surface area is in this case almost 5 times larger than the coarse-grained sediments' specific surface area (sample S9), which represents a value of 1.7 m^2 / g. This result confirms the indirect relationship between the specific surface area and the particle size. According to literature [15], the clay particles contribute the most to the specific surface of inorganic sediment.

The pH values of the sediments, measured in the leachate, range from 7.05 to 7.21 (table 2).
The amount of organic matter in sediments (table 2), determined by weight loss on ignition, ranges from 4% to 12.5%. The largest amount of organic matter is contained in sample S3, the smallest content of organic matter is recorded in the coarse-grained sediment (sample S9).

Table 2. Selected physico-chemical characteristics of sediments.

Sample	SSA ($m^2.g^{-1}$)	pH (-)	OM (%)
S1	8.851	7.11	6.86
S2	8.899	7.21	8.23
S3	8.831	7.05	12.46
S4	8.763	7.21	11.60
S5	7.792	7.21	7.27
S6	8.649	7.17	7.40
S7	8.664	7.08	5.29
S8	8.731	7.08	5.13
S9	1.741	7.07	4.01

The contents of the main chemical elements in the sediments together with the total contents of nitrogen and phosphorus are given in table 3.

The results of chemical analyzes (table 3) show that compounds of silicon, aluminum, and iron, predominate in sediment samples. The concentrations of total nitrogen and phosphorus in individual samples of bottom sediments are different. The reason may be the uneven deposition of sediments in the reservoir and thus the uneven granularity of taken sediment samples, which is related to the ability to bind contaminants. Nutrient concentrations are relatively high in sediment samples taken near the dam at the outflow from the reservoir (S1, S2 and S6). As the sampling distance from the dam increases, the proportion of particles smaller than 0.063 mm decreases and at the same time the concentrations of total nitrogen and phosphorus in the bottom sediment samples decrease. A similar trend is recorded for other monitored elements Si, Al, Fe, Mg, K, except calcium.

Table 3. Content of main elements in sediments.

Sample	Si (%)	Al (%)	Fe (%)	Mg (%)	K (%)	Ca (%)	P_{total} (%)	N_{total} (%)
S1	32.05	9.58	3.6	1.45	2.45	0.79	0.112	0.26
S2	28.91	8.06	3.78	1.29	2.23	2.84	0.113	0.24
S3	29.38	6.57	2.62	0.99	1.83	0.87	0.066	0.23
S4	26.63	5.82	2.33	0.97	1.62	3.21	0.066	0.22
S5	26.01	6.05	2.13	1.06	1.52	1.05	0.067	0.22
S6	32.29	8.3	3.66	1.21	2.49	1.21	0.112	0.24
S7	28.13	6.93	3.1	1.09	1.89	0.87	0.061	0.18
S8	26.87	6.67	2.89	1.11	1.68	0.59	0.049	0.17
S9	23.21	2.85	1.06	0.64	0.56	0.41	0.034	0.07

In order to study the linear dependences between the analyzed physicochemical parameters of bottom sediments, an intercorrelation matrix was designed (table 4). The dependence between particle size, specific surface area, organic matter content, pH and the presence of chemicals was evaluated. The evaluation was performed by Pearson correlation analysis, at the significance level $p \leq 0.05$. Correlation dependence was interpreted using the empirical classification according to Evans [16]. Significant correlations are marked in bolt in this table. The results of the correlation analysis indicate a very strong relationship between the specific surface area and the grain size of the sediments ($r \geq |0.817|$, $p \leq 0.05$). This confirms that the specific surface area is inversely proportional to the particle size [15]. The results indicate that the grain size and the associated specific surface area are also responsible for the sorption of nutrients.

Table 4. Pearson correlation coefficients (r) for the study of dependences of individual physico-chemical parameters of sediments.

	Sand	Silt	Clay	OM	SSA	N_{total}	P_{total}	Al	Fe	pH
Sand	1	-0.964	-0.788	-0.524	**-0.985**	**-0.886**	-0.560	**-0.772**	**-0.746**	-0.367
Silt		1	0.597	0.442	**0.936**	**0.790**	0.422	**0.666**	**0.639**	0.343
Clay			1	0.530	**0.817**	**0.845**	**0.705**	**0.800**	**0.782**	0.305
OM				1	0.491	**0.593**	0.232	0.181	0.132	0.272
SSA					1	**0.868**	0.557	**0.804**	**0.790**	0.293
N_{total}						1	0.807	**0.859**	**0.768**	0.472
P_{total}							1	**0.865**	**0.843**	0.478
Al								1	**0.950**	0.219
Fe									1	0.222
pH										1

Sediments with a predominance of fine-grained silt-clay particles show a larger specific surface area for phosphorus and nitrogen sorption, similar to those reported by [11, 12]. Several studies [11, 13, 17] also present that sediment particle size, iron and aluminum content are the main factors influencing sorption. Even in the case of the studied sediments, the content of aluminum and iron, probably bound on the surface of the finest clay particles in the form of hydroxides/oxides [13], contributed to significantly higher phosphorus concentrations in the fine-grained fraction of sediments. Similarly, the presence of total nitrogen in sediments depends on their grain size and larger specific surface area [12], indicating high concentrations of observed elements in sediments deposited near the dam area. The results showed that the total nitrogen and phosphorus concentrations in the sediment samples tended to decrease with increasing distance of the sediment collection site from the dam (figure 1), while the proportion of fine-grained particles smaller than 0.063 mm also decreased.

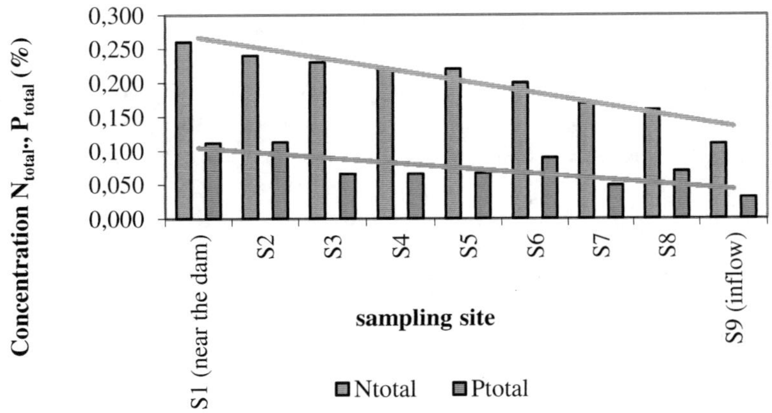

Figure 1. Influence of sediment sampling distance from dam and granularity on nutrient concentrations in sediments.

Total nitrogen concentrations correlated with the presence of organic matter in the sediment [18], what may indicate that most of the nitrogen content in the sediment comes from sediment-bound organic matter [19]. However, the presence of organic matter does not increase phosphorus retention in the studied sediments, as reported by [18], probably because phosphorus comes from inorganic sources from agricultural land in the form of fertilizers [19]. The relationship between nutrient content and pH was observed as moderate.

A moderate positive correlation was also found between the specific surface area of the sediments and the bound organic matter, similar to that reported by [12]. Fine-grained sediments tend to accumulate more organic matter [19], with an increasing proportion of larger particles, the organic matter content in the sediments decreases. However, a strong relationship between the content of organic matter and clay particles was not found [20], the correlation was at a medium level ($r = 0.530$, $p \leq 0.05$). The reason may also be a less accurate method of determining the organic matter content of sediments based on the measurement of weight loss on ignition.

4. Conclusion

The study of reservoir sediments from small water reservoir located in agricultural watershed indicated, that the content of nutrients (nitrogen and phosphorus) in reservoir sediments depends on their grain size and specific surface area. The closer sampling point at the dam, the higher proportion of fine-grained sediments (particles smaller than 0.063 mm) and the higher concentration of nutrients. The phosphorus sorption by reservoir sediments is affected not only by the granularity and specific surface area but also by the chemical composition of the sediments (especially the contents of aluminum and iron, which are probably bound on the surface of the finest clay particles in the form of hydroxides/oxides). The results also showed that the presence of total nitrogen in studied sediments depends on the presence of organic matter in the sediment, which may indicate that most of the nitrogen content in the sediment comes from the organic matter bound to the sediment. The presence of organic matter does not increase the retention of phosphorus in monitored sediments, probably due to the fact that phosphorus comes from inorganic sources from the agricultural area in the form of fertilizers.

5. References

[1] Sedimentation Committee 2009 *Sedimentation and Sustainable Use of Reservoirs and River Systems* Draft ICOLD Bulletin (France: ICOLD).

[2] Wisser D, Frolking S, Hagen S, Bierkens MFP 2013 Beyond peak reservoir storage? A global estimate of declining water storage capacity in large reservoirs *Water Resour Res* **49** 5732–5739

[3] Junakova N, Balintova M 2012 Predicting of soil loss in the Tisovec catchment, Slovakia *Chem Eng Trans* **28** 265-270

[4] Wohl E 2011 *A World of Rivers: Environmental Change on Ten of the World's Greatest Rivers* (USA, Chicago: University of Chicago Press)

[5] Lempérière F 2006 The Role of dams in the XXI century: Achieving a sustainable development target *The International Journal on Hydropower and Dams* **13**(3) 99 – 108

[6] Junakova N, Balintova M 2012 Predicting nutrient loads in chosen catchment *Chem Eng Trans* **26** 591-596

[7] Junakova N, Balintova M 2011 Model for predicting the nutrient content in reservoir bottom sediments *11th International Multidisciplinary Scientific Geoconference and EXPO - Modern Management of Mine Producing, Geology and Environmental Protection, SGEM 2011* **3** 475-482

[8] Lee GF, Jones RA 1992 Water Quality Aspects of Dredging and Dredged Sediment Disposal *Handbook of Dredging Engineering* (USA, Michigan: McGraw-Hill) p 9-23 to 9-59.

[9] Junakova N, Balintova M, Smolakova M 2017 Influence of granularity of sediment from a water reservoir on phosphorus sorption processes *Environmental processes* **4**(1) S239–S249

[10] Singovszka E, Junakova N, Balinova M 2016 The effect of sediment grain size on heavy metal content in different depth in water reservoir Ruzin, Slovakia *Solid State Phenomena* **244** 240-245

[11] Rawlins BG, Turner G, Mounteney I, Wildman G 2010 Estimating specific surface area of fine stream bed sediments from geochemistry *Appl Geoch* **25** 1291–1300

[12] Bergamaschi BA, Tsamakis E, Keil RG, Eglinton T, Montluçon DB, Hedges JI 1997 The effect of grain size and surface area on organic matter, lignin and carbohydrate concentrations and molecular compositions in Peru Margin sediments *Geochim Cosmochim Acta* **61** 1247–1260

[13] Wang Y, Shen Z, Niu J, Liu R 2009 Adsorption of phosphorus on sediments from the Three-Gorges Reservoir (China) and the relation with sediment composition *J Hazard Mater* **162** 92-98

[14] Galy V, France-Lanord C, Lartiges B 2008 Loading and fate of particulate organic carbon from the Himalaya to the Ganga–Brahmaputra delta *Geochim Cosmochim Acta* **72** 1767–1787

[15] Percival JB, Lindsay P 1996 Measurement of physical properties of sediments *Manual of Physico-Chemical Analysis of Aquatic Sediments* eds. A Mudroch, JM Azcue, P Mudroch (USA, Boca Raton: CRC Press) p 7-38

[16] Evans JD 1996 *Straightforward Statistics for the Behavioral Sciences* (USA, California: Brooks/Cole Publishing)

[17] Jalali M, Peikam EN 2013 Phosphorus sorption–desorption behaviour of river bed sediments in the Abshineh river, Hamedan, Iran, related to their composition *Environ Monit Assess* **185** 537–552

[18] Shen Z, Niu J, Wang X, Wang H, Zhao X 2013 *Distribution and Transformation of Nutrients in Large-scale Lakes and Reservoirs: The Three Gorges Reservoir* (Germany, Berlin: Springer-Verlag)

[19] Chandrakiran KS 2013 Assessment of Physico-Chemical Characteristics of Sediments of a Lower Himalayan Lake, Mansar, India *International Research Journal of Environment Sciences* **2**(9) 16-22

[20] Burone L, Muniz P, Pires-Vanin A M, Rodrigues M 2003 Spatial distribution of organic matter in the surface sediments of Ubatuba Bay (Southeastern - Brazil) *An Acad Bras Cienc* **75**(1) 77-90.

Acknowledgments
This research has been supported by the Slovak Grant Agency for Science (Grant No. 1/0419/19) and by the Slovak Research and Development under the contract No. APVV-20-0140.

IOP Conf. Series: Materials Science and Engineering 1252 (2022) 012060 doi:10.1088/1757-899X/1252/1/012060

Renewal of the roof structure of the Winter Stadium in Prešov

J Vojtuš[1]

[1] Institute of Structural Engineering, Faculty of Civil Engineering, Technical university of Kosice, Vysokoskolska 4, 042 00 Kosice, Slovakia

E-mail: jaroslav.vojtus@tuke.sk

Abstract. Winter stadium in Prešov is situated in an exposed area near the city center, in Slovakia but also in the context of world architecture, it is today an exceptional building with a rope roof structure. The author of the construction is Ing. arch. Čejka, structural engineer doc. Ing. Hajdu, CSc. At the time of design and implementation, it was a very progressive design of the supporting structure, which met the parameters of a world format. The project was modified in 1984 by the completion of outbuildings on the west side of the stadium, mainly due to the need to expand the number of players' locker rooms according to the current needs of athletes. The ice rink was renovated more than 10 years ago, including mantinels and safety glazing.

1. Introduction

Ropes, wires and chains are load-bearing elements, characterized in that they can resist predominantly only tensile forces, while they resist forces acting transversely or in compression only very slightly. By idealizing the real elements, the concept of a perfectly flexible fiber is obtained. It is assumed that the perfectly flexible fiber resists only tension, is intangible and that the transverse dimensions of the fiber are negligible compared to the length of the fiber. No cross-section of a perfectly flexible fiber can produce a bending moment *Mx*, a torque *Tx*, a transverse force *Vx* or a negative normal force *Nx* [1]:

$$M_x = 0 \text{ kNm},\ T_x = 0 \text{ kNm},\ V_x = 0 \text{ kN},\ \text{N} > 0 \text{ kN} \tag{1}$$

The resultant of all forces acting on one side of the cross-section passes through the center of gravity of the fiber and has the direction of normal to the cross-section. Every fiber problem is geometrically nonlinear and the principle of superposition does not apply. If, in addition to the forces at the end points, a system of solitary forces acts on the fiber, then the fiber is tensioned in all parts and has the shape of a polygon [1-4].

The aim of the proposed reconstruction of the stadium is primarily a comprehensive replacement of the roof structure due to its emergency condition and creating conditions for comprehensive provision of hockey teams, training, rehabilitation, administrative - operational and representative activities of the hockey club, consisting of several senior, junior and student teams. possible to use the hall for about 5000 visitors. Last but not least, one of the goals of the reconstruction is to improve the thermal parameters of the building, including the roof cladding.

2. Description of the stadium construction

The structural load-bearing system of the Prešov Winter Stadium is a monolithic reinforced concrete, part is made of plain concrete, part of the point of view is a prefabricated construction site reinforced concrete. The winter stadium is located at an altitude of 242,000 m above sea level (Figure 1). The foundations are monolithic concrete footings and foundation strips made of plain concrete for additional vertical structures. Perimeter and internal columns, together with horizontal and inclined beams, form frames for a stepped ceiling and ceilings mounted on horizontal beams. The stairs are monolithic reinforced concrete, later an elevator with a brick shaft was built. The reinforcement of the building is supported by monolithic reinforcing walls inserted into the supporting frames at the level of the 1st floor. The building is one expansion unit.

a b

Figure 1. External view on the Winter Stadium, a – west view, b – east view.

a b

Figure 2. External view, a – view of the perimeter beam, b – detail of the hole for the rope.

The supporting structure of the roof consists of a prestressed rope network with a grid of 1.0 x 1.0 m in the shape of a hyperbolic paraboloid with overall dimensions in the longitudinal direction of 92.00 m and in the transverse direction of 77.88 m. The supporting ropes are in the transverse direction in the number of 77 pieces and stabilizing, tensioning ropes in the longitudinal direction in the number of 74 pieces. The ropes are made of new unprecedented ropes. The ropes are anchored in a circumferential monolithic reinforced concrete beam measuring 3.5 x 1.3 m mounted axially above the circumferential columns. Anchoring is realized by means of steel pipes and washers (Figure 2).

In the longitudinal axis of the hall there are pylons with an upper sloping edge terminated by a reinforced concrete drainage gutter of rainwater from the stadium roof to the shafts at ground level for water / snow capture and drainage to the subsoil / sewer. The footings of the pylons are monolithic stepped with a toothed load joint. There is a hole 4.30 m wide and 1.50 m high to 3.40 m high in the wall of the pylons to ensure the movement of pedestrian and car traffic. The total longitudinal axis of the hall with pylons and rainwater shafts is 135.00 m.

IOP Conf. Series: Materials Science and Engineering 1252 (2022) 012060 doi:10.1088/1757-899X/1252/1/012060

3. Roof structure, existing condition

During the existence of the building and its roofing, uneven deformations of the rope subsystem were observed, which cause geometric imperfections of local and global character (Figure 3). The causes of this condition were caused by both the unstabilized ropes used and the implementation technology, local loads on the rope system, as evidenced by the increase in the thickness of the concrete layer compared to the project, rheological influences, etc. (source: Static-expert opinion, TUKE 2015).

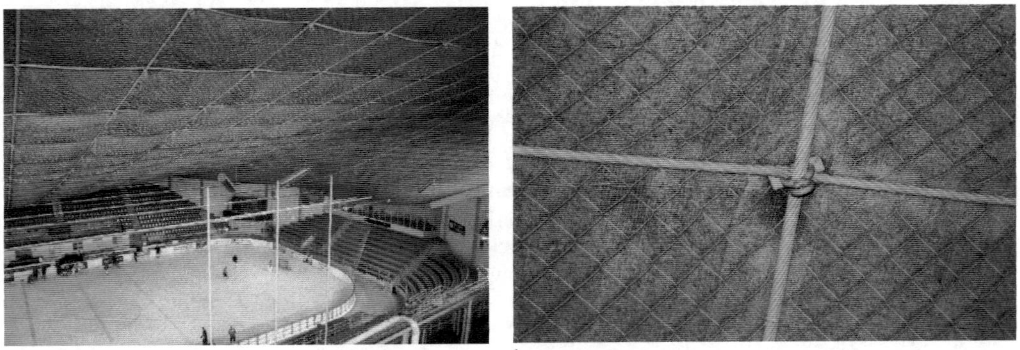

a b

Figure 3. Internal view of the roof cladding, a – view of the spatial shape , b – detail of the shape.

Figure 4. External view of the roof covering.

The composition of the roof cladding determined on the basis of an in situ probe (see Figure 4):

- mesh hr. 5.0 mm on the rope net;
- bag;
- glass wool th. 10 mm;
- lightweight concrete th. 60 mm;
- boards heraklith th. 25 mm;
- asphalt cardboard th. 2 mm;
- plain concrete th. 80 mm;
- sheet metal folded roofing th. 1 mm.

The composition differs from the composition stated in the original project documentation. The inspection reveals corrosive effects at the joints of the sheet metal strips, as well as the degradation of other layers. Due to the condensation of water vapor on the lower surface of the roof cladding, water drips onto the ice surface.

4. Roof structure, solution design

The subject of the reconstruction of the stadium roof is the dismantling of the existing roof cladding, including the supporting rope structure. The Roof Reconstruction Project was developed in 2017 Technical University in Košice, where the author of the design is prof. Ing. S Kmeť and prof. Ing. M Tomko. The content is the replacement of supporting ropes and the implementation of its own roof cladding. The new supporting ropes will be 24 mm in diameter and 12 mm in diameter. Ropes of the so-called open type with a modulus of elasticity of 165,000 MPa are used. The supporting ropes will be terminated with a potted end, which will be passed through the existing ring with rectification using steel sheets. Load cells will be mounted in selected supporting ropes. The prestressing ropes will be terminated on both sides by a prestressing piece, while on one side the rope will have a load cell. Anchoring of these ropes in the wreath will be in two ways, by means of threaded rods and by means of anchored sheets into the reinforced concrete wreath. The joints of the individual ropes will be solved by means of sheets connected by screws. Each joint has four holes in the sheet metal, into which the roof panels will be fastened with four screws and washers. Roof panels must expand separately due to significant deflections of the ropes, which are in different curvatures, so it is necessary to have larger holes in the sheet. After the realization of the rope net, it is necessary to prestress the net by means of prestressing ropes so that there is a force of 30 kN in each prestressing rope with a diameter of 12 mm (Figure 5).

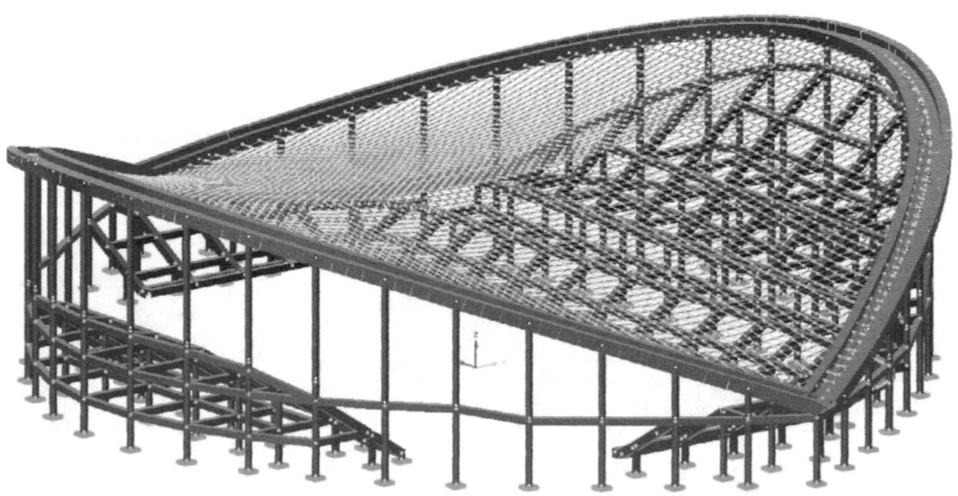

Figure 5. Model of the rope construction of the roof of the Winter Stadium in Prešov.

The roof cladding solution consists of a single-skin roof construction:

- realization of the supporting layer of the roof sheathing on the system of supporting ropes with thermal insulation in the form of prefabricated cassettes;
- realization of waterproofing layer.

The roof cladding will consist of completed panels made of trapezoidal steel sheets with thermal insulation filling of mineral wool with thickness 200 mm, dimensions approx. 1000/1000 mm. The panel will be reinforced with a steel "C" profile glued to the edge of the panel in the direction of the supporting ropes. The joint of the panels will be flexibly filled and covered with a self-adhesive asphalt strip. The waterproofing layer is formed by a TPO foil. The edge of the roof will be solved with XPS insulation on a reinforced concrete perimeter frame with a flap made of OSB3 board, including cladding (see Figure 6). The roof area will be complemented by elements of the lightning protection system. The supporting system of the roof structure, i. rope net, with sufficient reserve to transfer the additional load from the newly designed composition of the roof sheathing.

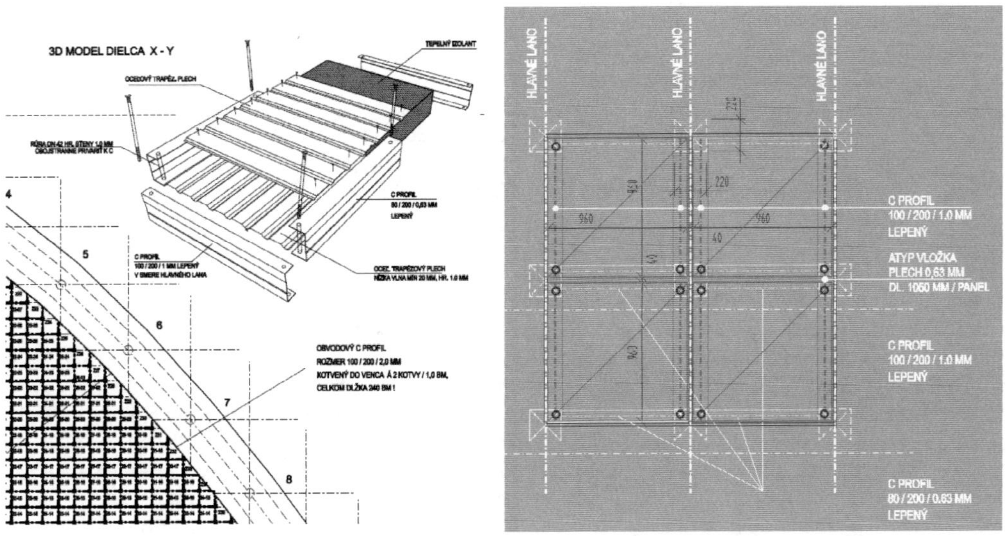

Figure 6. Design of the solution of thermal insulation cassettes of the roof cladding of the Winter Stadium in Prešov.

Boundary conditions [5, 6]:

- calculated outdoor air temperature in winter according to the map of temperature areas depending on the geographical location and altitude for the locality of the town Prešov = 250.00 m above sea level is θ_e = -13 ° C;
- calculated relative humidity of the outside air for the calculated outdoor air temperature in winter φ_e = 84%;
- calculated indoor air temperature for hockey hall θ_i = +12° C /1 meter above the ice surface/;
- aspect θ_i = +20 ° C;
- relative humidity of indoor air φ_i = 50%;
- dew point temperature $\theta_{si,80}$ = +12.62 ° C, safety margin $\Delta\ \theta_{si}$ = 0.5 / 1.0 K for α_i> / < 8 W/m^2. K.

The design of the roof structure affects the phenomenon of water vapor condensation in the roof structure, where in building structures that separate the external environment from the interior, water vapor condenses inside the structure in colder periods of the year, respectively. on the contrary, for this type of construction. According to STN 73 0540-2 [5, 6], roof structures should be designed so that water vapor does not condense in them. If water vapor condensation occurs in the roof, it can be considered a suitable structure if the following condition is met for single skin roofs, that the condensed amount of water vapor in the flat roof structure is less than $Mc < 0,1$ kg/m^{2}a.

Composition of the project proposal:

- waterproofing layer, TPO membrane thickness 1.5 mm, mechanically anchored to the panel;
- separating layer, geotextile Filtek 300 g/m^2;
- completed panel in composition:
  - top cover plate made of trapezoidal plastic coated sheet thickness 0.63 mm with a wave height of 5 mm;
  - thermal insulator made of non-combustible material, e.g. made of mineral fiber boards thickness 200 mm;
  - lower support plate made of trapezoidal plastic coated sheet thickness 1.0 mm with a wave height of 20 mm.
- supporting structure, rope system.

Thermal engineering calculation:
200 mm of insulator with mineral wool $\lambda = 0.041\text{-}0.038$ W/m.K, U = 0.20 W/m^{2}K, R = 4.89 m^2.K/W
Amount of condensed water vapor: 0.7287 kg/m^{2}a
Amount of evaporated water vapor: 0.8319 kg/m^{2}a
The construction has a favorable annual balance of condensed and evaporated water vapour [5].

The proposed composition solves the construction of the roof cladding using a set of prefabricated cassettes - panels in a grid of ropes, i. axis 1.0 x 1.0 m. These panels will be anchored in the corners to a plate measuring 215 x 215 mm, which is part of the so-called rope couplings. Said panels are formed by cassettes measuring 960 x 960 mm, and are made of two shaped sheets - trapezoids, a lower support thickness of 1.0 mm with a wave height of 20 mm and an upper cover thickness of 0.63 mm with a wave height of 5.0 mm, between which will be thermal insulation. It is also possible to use trapezoids with a different height, but the assumption is that the panels can carry at least 300 kg when placed at 4 points on their edge. On the sides parallel to the main ropes, the panel will be reinforced with a thin-walled C profile 1.0 mm thick, which will be glued to the thermal insulation with a vertical wall and attached to the sheets by shooting 100 mm. In the opposite direction, in the direction of the tensioning ropes, the panel will also be reinforced and closed with a sheet metal C profile 0.63 mm thick. This C profile will also be anchored to the sheets by shooting at 100 mm. The panels in contact with the reinforced concrete rim of the stadium will be atypical, and will be anchored to a C profile 2 mm thick by 1 m of glued anchors with a diameter of 12.0 mm.

The joints between the individual cassettes - panels will be glued on both sides with elastic strips. On the underside, a sheet metal washer measuring 500 x 500 x 0.63 mm with a cross groove with the possibility of eliminating movements is placed at the point of intersection of the panel joints. Belts with a vapor barrier function are connected to this pad, e.g. made of asphalt self-adhesive tape of minimum width 250 mm. On the upper side, the joints of the panels are glued also by means of an asphalt self-adhesive tape with the possibility of expansion at the point of contact. A flexible sheet metal omega profile will be inserted into the actual contact of the individual panels, into which a thermal insulation strip made of mineral wool will be inserted.

The roof surface will be provided with waterproofing from a single-layer membrane of the TPO type, resp. made of PVC-P foil, thick 1.5 mm. The waterproofing will be mechanically anchored to the sheets of the supporting panels of the roof sheathing. The end of the waterproofing at the edge of the roof will be complemented by system elements of plastic sheeting. The edge of the roof - the perimeter reinforcement wreath must be insulated, connect the thermal insulation of the roof with the thermal insulation of the perimeter cladding. The coating must be secured against winds by mechanical anchoring and airtight at the edges. In connection with this, plumbing structures will be implemented at the site of rain gutters.

5. Roof construction, implementation

After the demolition of the roof structure, roof sheathing and rope system, there was a multi-criteria reassessment of the overall solution by the builder and the construction company. Due to these facts, the solution of the roof cladding has also changed, especially the solution of the supporting structure of the roof cladding (Figure 7).

Figure 7. Implementation of a new rope network for the roof of the Winter Stadium in Prešov.

The new design solves the construction of the roof cladding by means of a set of boards again in a grid of ropes, i. axially approx. 1.0 x 1.0 m. These boards will be anchored on the sides to the strips in the shape of the letter omega, which will connect to the rope connectors. Trapezoidal sheets became the supporting element of the roof cladding. The composition of the thermal insulation and waterproofing layer was preserved. The mentioned solution is to ensure the elimination of movements from the deflections of the roof cable system (Figure 8).

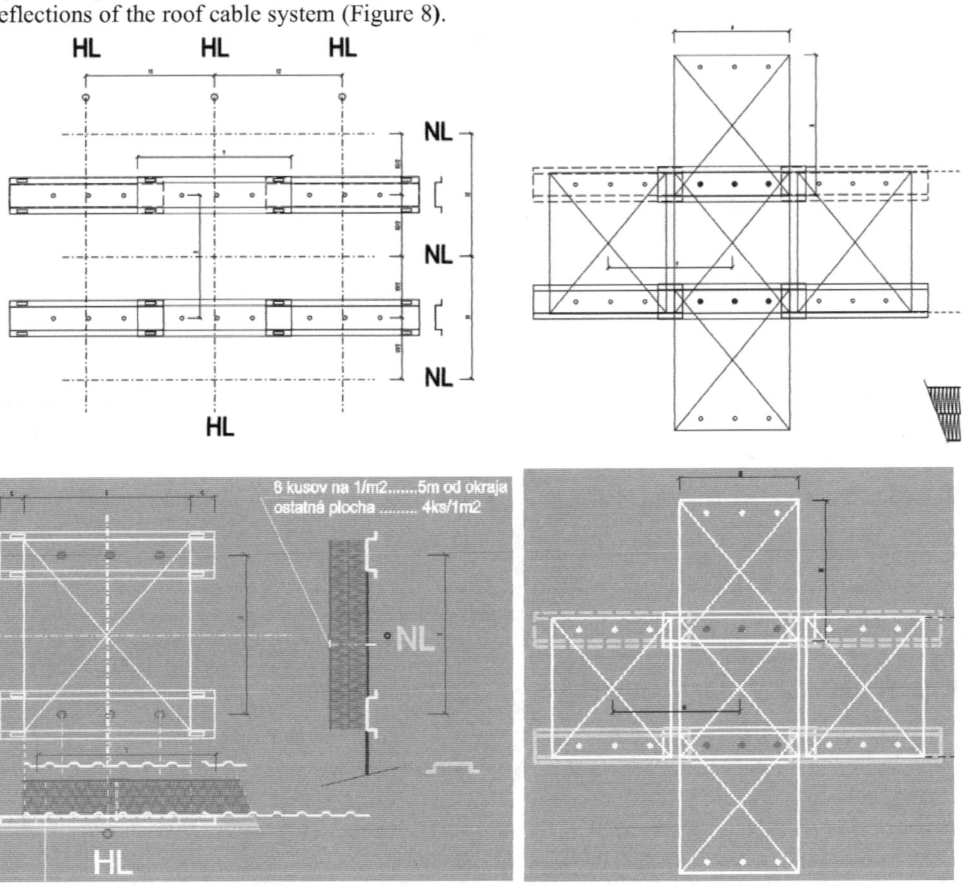

Figure 8. The concept of the solution of the supporting layer of the roof cladding using omega profiles and supporting plates.

Proposed composition of the realized roof cladding [7, 8]:

- waterproofing layer, TPO foil thickness 1.5 mm, mechanically anchored, white color, total area 4,215 m^2, weight 7,713.45 kg,
- thermal insulation layer, non-flammable, 2x mineral wool boards, lambda 0.040 W/m.K, order no. weight 165-200 kg/m^2, mechanically anchored, weight 139,095.00 kg,
- vapor barrier, asphalt self-adhesive tape, hr. 1.5 mm, weight 6,955.00 kg,
- supporting layer of the roof cladding, trapezoidal sheet metal, thickness 1.0 mm, min. 0.75 mm,
- supporting layer of the roof cladding, "omega" profile / "C" molding,
- supporting structure of the roof, rope net.

The last change of the project was the replacement of "omega" profiles with "C" rails (Figure 9,10).

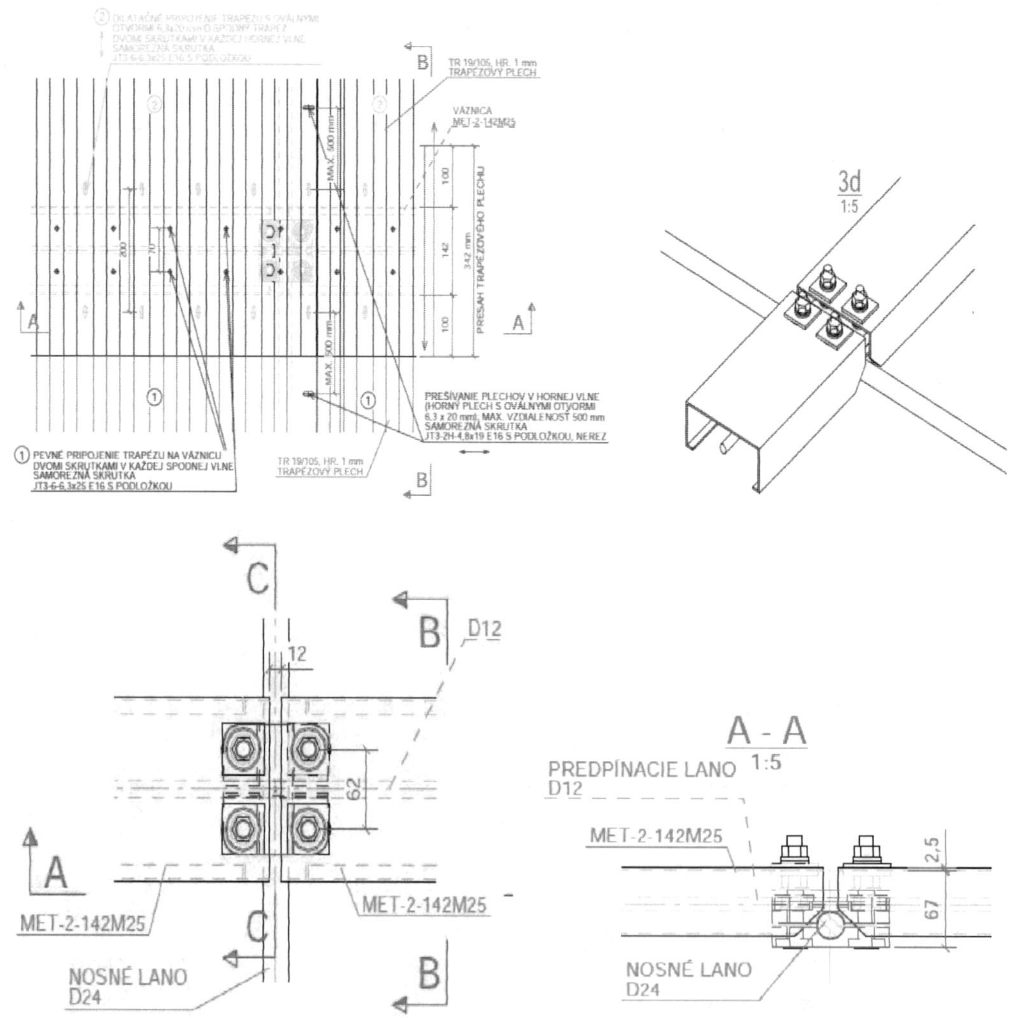

Figure 9. Details from the solution of the supporting structure of the roof cladding.

Figure 10. Interior view of the load-bearing structure of the roof of the Winter Stadium in Prešov.

6. Appendices

The design and implementation of the new roof of the Winter Stadium in Prešov is designed to preserve the uniqueness of the rope structure, ensure further functionality of the building, extend the life of the roof structure, eliminate the faults of the roof, increase the market value of the building, etc (Figure 11). All participants in the implementation are pleased that we managed to preserve a rare architectural-engineering work and reconstruct at a high technical level such a unique structure, with a hyperbolic-parabolic prestressed rope network, which currently ranks 4th in the world [2, 4].

Figure 11. Views on the implementation and completion of the roof cladding of the roof of the Winter Stadium in Prešov.

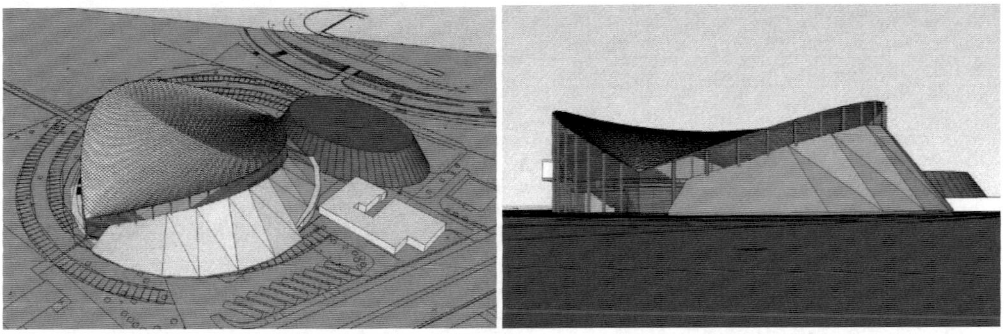

Figure 12. General view of the Winter Stadium in Prešov after the completion of all stages of construction.

7. References

[1] Kmeť S and Platko P and Mojdis M 2009 Navrhovanie a analýza tensegrity sústav, *The International Journal of TRANSPORT & LOGISTICS*, ISSN 1451-107X

[2] Chard C and Mallen C 2013 Renewable Energy Initiatives at Canadian Sport Stadiums: A Content Analysis of Web-Site Communications. *Sustainability*, *5*, 5119-5134.

[3] Ding M. Luo B. Ding S. Shen Y. Huang, L. 2021 Experimental Investigation and Numerical Simulation of a Levy Hinged-Beam Cable Dome. *Buildings*, *11*, 110.

[4] Viskovic, A. 2018 Hemp Cables, a Sustainable Alternative to Harmonic Steel for Cable Nets. *Resources*, *7*, 70.

[5] STN 73 0540 – 1-4/Z1:2016 a Z2:2019 Tepelná ochrana budov. Tepelnotechnické vlastnosti stavebných konštrukcií a budov

[6] STN EN ISO 6946 Stavebné konštrukcie. Tepelný odpor a súčiniteľ prechod tepla. Výpočtová metóda

[7] STN 73 1901, 73 1901/01 Navrhovanie striech. Základné ustanovenia

[8] STN 73 3610, 73 3610/Z1 Klampiarske práce stavebné

Acknowledgments

This work was supported by The Slovak Scientific Grant Agency under the contract No.VEGA 1/0512/20.

IOP Conf. Series: Materials Science and Engineering 1252 (2022) 012061 doi:10.1088/1757-899X/1252/1/012061

Indoor air quality assessment in the apartment – case study

P Kapalo[1], M Kováč[1], R Nagy[1] and Z Baka[1]

[1]Institute of Architectural Engineering, Faculty of Civil Engineering, Technical university of Kosice, Vysokoskolska 4, 042 00 Kosice, Slovakia

E-mail: peter.kapalo@tuke.sk, martin.kovac@tuke.sk, richard.nagy@tuke.sk, zdeno.baka@gmail.com

Abstract. The application of demand-controlled ventilation is one of the ways, how to reduce energy consumption in residential buildings. The main producer of pollutants in these types of buildings is human being and its activities. The most common measurable air parameters that can be characterized the level of air degradation in rooms are the concentration of carbon dioxide (CO2), air temperature and relative humidity. The aim of our research was to monitor and evaluate these air parameters in real conditions of apartment usage. As part of the research, the experimental measurements were carried out in a selected apartment, where the concentration of carbon dioxide CO2, air temperature and relative humidity were monitored in individual rooms. During evaluating the monitored air parameters, their maximum and minimum values were evaluated. The achieved research results can be applied in computer simulation the operation of a ventilation system in the apartment in order to increase the energy efficiency of the HVAC system while maintaining the required indoor air quality that is necessary for healthy living.

1. Introduction

Indoor air quality has a significant impact on human comfort, health and occupant performance. Several studies [1-4] have examined indoor air quality in buildings. In most cases, the effects of energy measures on indoor air quality in buildings were investigated. The study [5] shows that efforts should be made to improve ventilation when modernizing buildings in order to avoid adverse effects on IAQ. The scientific team of Kim et al [6] studied the air quality in the bedroom in relation to the quality of sleep. Among other things, the results showed that people were exposed to various environmental problems during sleep, such as too low or too high air temperatures or relative humidity and high CO_2 concentrations. One of the ways to increase the energy efficiency of buildings while ensuring quality indoor air is to use simulation software that simulates the processes taking place in the apartment. Study [7] XV developed and applied a system for estimating the effects of energy efficient household interventions on energy consumption and health. In this study, the monitoring of selected air parameters is documented in order to expand the database of experimental measurements of dwellings, which can be used in the creation of simulation programs. From the measured air parameters, the volume flow of ventilation air for sedentary activity, sleep and cooking is roughly calculated.

2. Materials and methods

In order to determine the air quality in a standard apartment located in a panel house in the housing estate, the measurement of normally measurable air parameters was performed during normal operation of the apartment. From the courses of the measured air parameters, it is then possible to determine the required air volume flow required to ensure a healthy indoor environment in the apartment.

2.1. Characteristics of the measured room

The experimental measurement was realized in a selected three-room apartment (Figure 1) with a floor area of about 86m2, which is located in an apartment building in the housing estate. In order to examine the quality of the air in the selected apartment during the whole week, physical quantities such as air temperature, relative humidity and CO_2 concentration were measured.

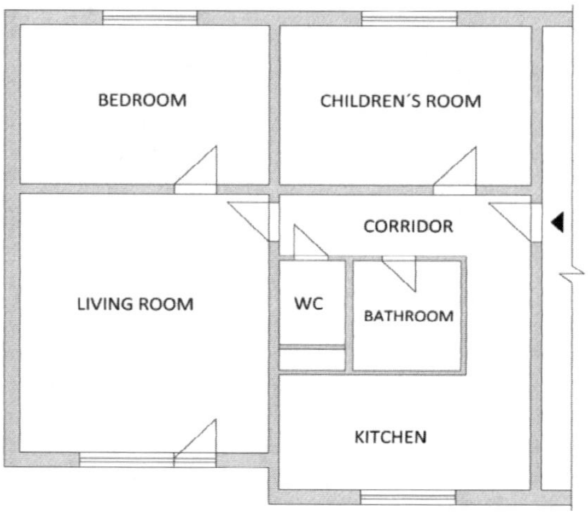

Figure 1. Floor plan of the apartment.

The measurements were performed in the following areas of the three-room apartment: living room (25.18m^2), children's room (15.44m^2), kitchen (11.16m^2) and bathroom (4.61m^2). Indoor air parameters were recorded 24 hours a day, throughout the week in January. During the measurements, standard family activities took place in the apartment.

2.2. Measuring devices

A CO_2 sensor (type C-AQ-0001R) and a thermo-hygrometer S-3541 were used to monitor the indoor air parameters. The measuring principle of the CO_2 sensor C-AQ-0001R is based on variations of infrared radiation with CO_2 concentration. The CO_2 concentration measurement range is from 0 to 5,000ppm and the measurement accuracy is ± 75ppm. The principle of measuring CO_2 concentration is based on the dependence of the attenuation of infrared radiation on the concentration of CO_2 (Non Dispersive Infrared Radiation absorption). The concentration of CO_2 indicates the air quality in the room and this module can also be used to control the ventilation unit providing room ventilation. The S-3541 thermometer and hygrometer is designed to record air temperature and relative humidity. The temperature measurement range is from -30°C to +70°C and the accuracy is ± 0.4°C. The relative humidity measurement range is from 5% to 95% and the accuracy is ± 2.5% relative humidity at 25°C. The recorded values are stored in a non-volatile electronic memory. During the measurements, the thermo-hygrometer was connected and used simultaneously with the CO_2 sensor. Dataloger software developed by Comet System was used to download data to a computer.

3. Results and discussion

In order to determine of airflow volume needed to eliminate the influence of pollutants arising from the breathing of household persons, experimental measurements were performed, during which the following were measured: room air temperature, relative humidity and CO_2 concentration in the room. Subsequently, the mass flow of carbon dioxide and then the airflow volume required for room ventilation were calculated using the gas mixing equation.

3.1. Measurement of air parameters

The experimental measurement was realized in the winter season in the living room, children's room, kitchen and bathroom. The CO_2 concentration in the exterior was in the range of approximately 414 - 436ppm. The recorded air parameters of the individual rooms are documented in Figures 2 to 5.

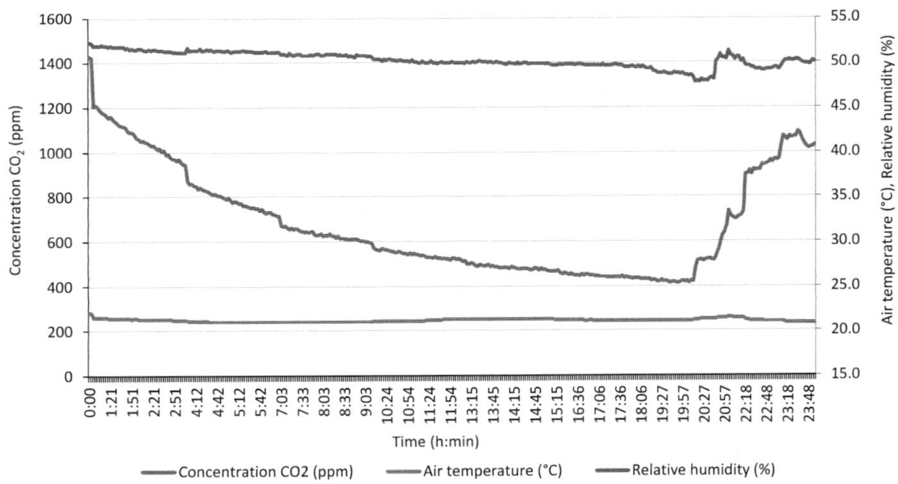

Figure 2. The measured parameters of indoor air quality in the living room.

From Figure 2 we can observe the presence of people in the room according to the course of CO_2 concentration. On this monitored day (Friday), a decrease in CO_2 concentration can be observed from midnight, which indicates the absence of people after midnight until 20:12 (h:min), when the CO_2 concentration began to rise again. During the observed hour, a minimum CO_2 concentration of 414ppm and a maximum of 1428ppm were recorded in the room. Therefore, it can be stated that from the point of view of the CO_2 concentration, the room was ventilated within 20 hours to the level of the background concentration of the outside air. The increase in CO_2 concentration due to the presence of people in the room was 245%. The relative humidity in the room was in the range of 47.8–52.3% RH– the percentage difference was 8.6%. The air temperature in the room was in the range of 21.8–22.0°C - the percentage difference was 5.45%.

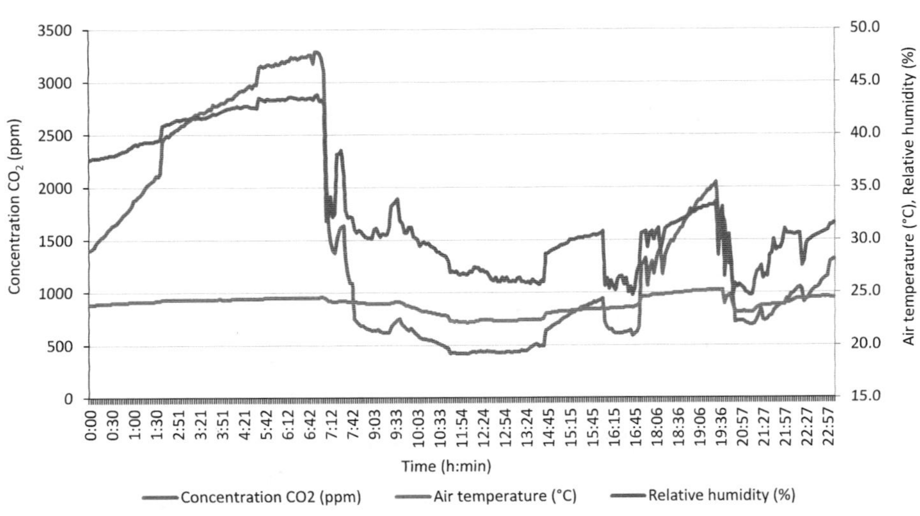

Figure 3. The measured parameters of indoor air quality in the children room.

Similarly, from Figure 3 we can observe an increase in the concentration of CO_2 in the children's room from midnight to morning 7:00, from 13:48 to 16:00 and from 16:48 to 19:30 (h:min). A sharp drop in CO_2 concentration indicates ventilation of the room through a window. During the observed day, a minimum CO_2 concentration of 418ppm and a maximum CO_2 concentration of 3287ppm were recorded in the room. The increase in CO_2 concentration due to the presence of people in the room was 686%. The relative humidity in the room was in the range of 24.7 - 43.8% RH - the percentage difference was 43.8%. The air temperature in the room was in the range of 22.1-25.2°C - the percentage difference was 12.30%.

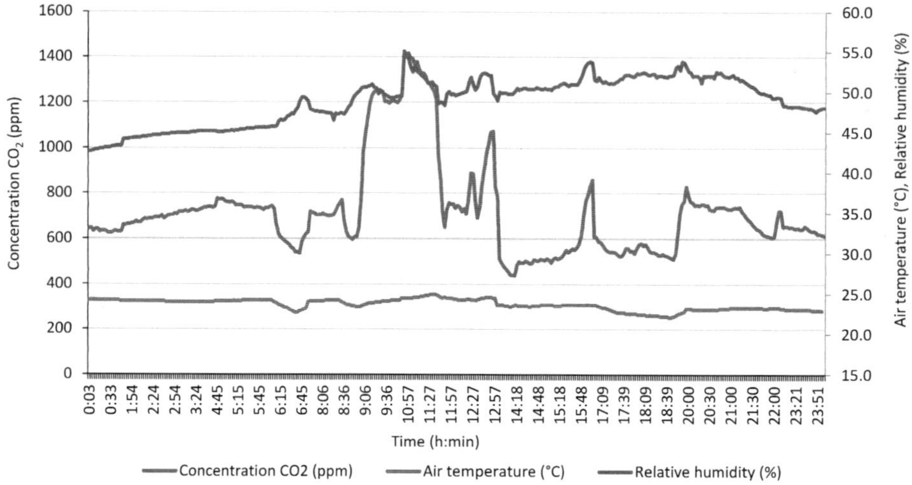

Figure 4. The measured parameters of indoor air quality in the kitchen.

A minimum CO_2 concentration of 436ppm and a maximum of 1425ppm were recorded in the kitchen. The increase in CO_2 concentration due to the presence of people and cooking was 227%. The relative humidity in the room was in the range of 42.7-55.1% RH - the percentage difference was 22.50%. The air temperature in the kitchen was in the range of 22.1-25.0°C - the percentage difference was 11.60%.

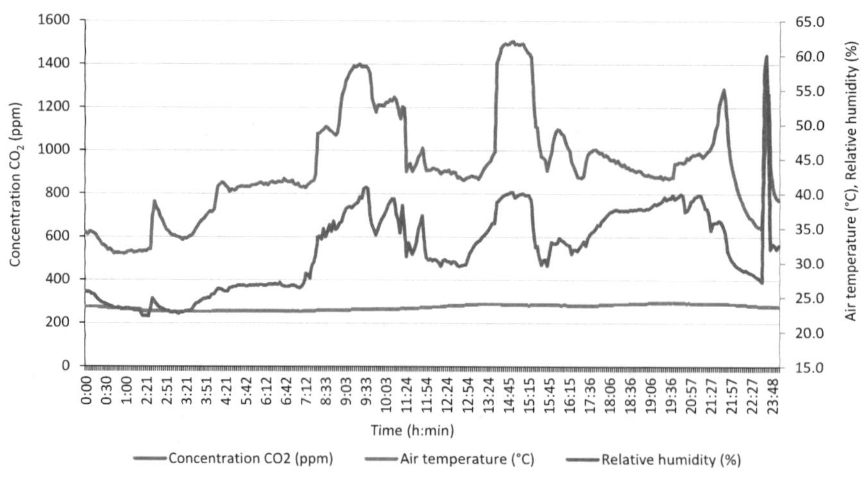

Figure 5. The measured parameters of indoor air quality in the bathroom.

A minimum CO_2 concentration of 522ppm and a maximum of 1506ppm were recorded in the bathroom inside the apartment, which cannot be ventilated directly in a natural way through a window but only by means of shaft ventilation and mechanically by means of a fan. The increase in CO_2 concentration due to the presence of persons was 189%. The relative humidity in the room was in the range of 22.2-60.0% RH - the percentage difference was 63.0%. The air temperature in the bathroom was in the range of 22.9-24.3°C - the percentage difference was 5.76%.

The comparison of individual courses of CO_2 concentration in rooms is documented in the graph in Figure 6.

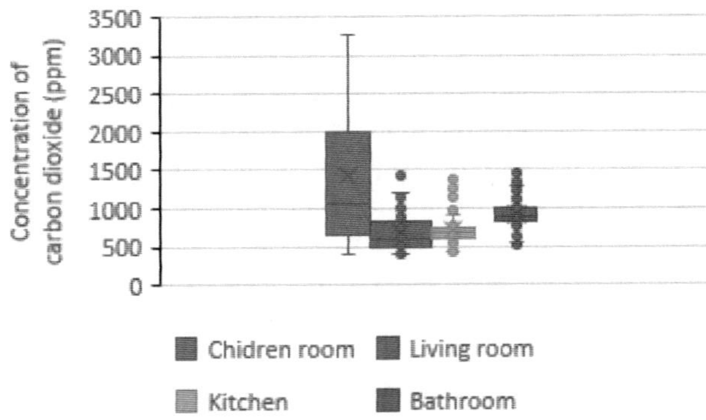

Figure 6. The comparison of CO_2 concentration courses.

From the comparison of the CO_2 concentration in the monitored apartment, it can be stated that the highest concentration was recorded in the children's room, which was inhabited for the longest time. The lowest concentration was recorded in the living room and kitchen. There was a higher concentration in the bathroom, because it is the smallest room.

3.2. Calculation of the air volume flow rate

From the measured course of the CO_2 concentration, it is possible to calculate the airflow volume required to ensure the required indoor air quality. For the purposes of this study, airflow volume (m3/h) and air exchange rate (1/h) were calculated for sedentary activity, sleep and cooking. The calculation of the required ventilation airflow volume was performed according to the published method [8-11]. The values of the results are documented in Table 1.

Table 1. Airflow volume and air ventilation rate.

	Sedentary activity	Sleeping	Cooking
Airflow volume (m^3/h)	27.43	24.31	30.77
Ventilation rate (1/h)	0.78	0.70	1.02

The results shown in Table 1 cannot be generalized. They are valid only for the apartment in which the experimental measurement took place and for the specified measurement days.

4. Conclusion

The aim of this paper was to determine the course of CO_2 concentration, air temperature and relative humidity during normal operation in the apartment by experimental measurements. From the results of experimental measurements, it was found that the highest concentration of CO_2 was achieved in the children's room and the lowest in the living room. It was similar with the air temperature. The highest relative humidity values were achieved in the bathroom for a short time. The highest long-term measured values of relative humidity were achieved in the kitchen and living room, which results from the nature of their operation. The highest value of air volume flow is required in the kitchen and the smallest in the bedroom. The achieved research results can be applied in simulating the operation of a ventilation system in the apartment in order to increase the energy efficiency of the HVAC system and while maintaining the required quality of indoor air needed for healthy living.

5. References

[1] Du L, Prasauskas T, Leivo V, Turunen M, Pekkonen M, Kiviste M andAaltonen A et al 2015 Assessment of indoor environmental quality in existing multi-family buildings in North-East Europe *Environ. Int.***79**pp 74-84

[2] Langer S, Beko G, Bloom E, WidhedenWA and Ekberg L 2015 Indoor air quality in passive and conventional new houses in Sweden *Build. Environ.***93** 92-100

[3] DerbezM, BrunoB, ValerieC, MurielleL, CecileP, JacquesR andSeverineK 2014 Indoor air quality and comfort in seven newly built energy-efficient houses in France*Build. Environ.*72173-87

[4] Kaunelienė V, Prasauskas T, Krugly E, Stasiulaitienė I,Čiužas D, Šeduikytė L andMartuzevičius D 2016 Indoor air quality in low energy residential buildings in Lithuania*Building.Env.*10863-72

[5] Földváry V, Bekö G, Langer S, Arrhenius K and Petráš D 2017 Effect of energy renovation on indoor air quality in multifamily residential buildings in Slovakia *Environ. Int.***122** 363-72

[6] Kim M, Chun C and Han J 2010 A study on bedroom environment and sleep quality in Korea. Indoor and Built Environment. **19**(1) 123-8

[7] Underhill LJ, MilandoChW, Levy JI, Dols WS, Lee SK and Fabian MP 2020 Simulation of indoor and outdoor air quality and health impacts following installation of energy-efficient retrofits in a multifamily housing unit *Environ Int.* 170106507

[8] Kapalo P andSiroczki P 2014Calculating the intensity of ventilation in classrooms on the basis of measured concentrations of carbon dioxide in Slovakia - case study *Int J Vent.***13**(3) 247–58

[9] Persily AK 1997 Evaluating building IAQ and ventilation with indoor carbon dioxide. ASHRAE Transactions **103** Part 2 (Boston, United States)

[10] Persily A 2006 What we think we know about ventilation *Int J Vent.* **5**(3) 275–290

[11] Kapalo P, Domnita F, Bacotiu C and Podolak M 2018 The influence of occupants' body mass on carbon dioxide mass flow rate inside a university classroom – case study *Int J Environ Health Res*28(4)432–447

Acknowledgments

This work was supported the Slovak cultural and educational grant agency under the contract No. KEGA 047TUKE-4/2020.

Overview of meteorological and hydrological conditions in the Hornád river basin

P Nagy[1], M Garaj[1] and M Hlinková[1]

[1] Institute of Environmental Engineering, Faculty of Civil Engineering, Technical University of Kosice, Vysokoškolská 4, 040 23 Košice, Slovakia

E-mail: patrik.nagy@tuke.sk, miroslav.garaj@tuke.sk, maria.hlinkova@tuke.sk

Abstract. The assessment of deficit and surplus volumes is important for assessing the effects of climate change on water management. Climate change needs to be taken into account in water management. At a time of climate change, there is a frequent occurrence of such periods, but also torrential rains, it is necessary to assess the quantity and quality of surface waters. Building structures must be designed in such a way that, during the period of torrential rains, they retain water from the surface runoff and prevent any damage that could arise from the original; sufficient water and dry season must be ensured during the period. Water shortages can also be caused by increased water consumption. The assessment of the volume of water in the Hornád river basin has shown that it is possible to mitigate climate change.

1. Introduction

Climate change is gaining more and more attention around the world than potentially one of the biggest challenges facing modern society. Although much of this attention is focused on how to reduce greenhouse gas emissions into the atmosphere (climate change mitigation), perhaps the most important and uncertain question for civil engineers is how to prepare for a future world with changing climatic and hydrological conditions, and to adapt to climate change. Most scientists believe that many climate-related changes will occur regardless of the level of success in regulating greenhouse gas emissions that are already in the atmosphere and the long-term effects of these gases in the climate system [1,2,3,4,5].

Changes in precipitation and water levels are another consequence of global climate change, which will occur over a longer period of time than the average life of most of the infrastructure components built today. More soil moisture and hydrostatic pressure accumulated in structures such as retaining walls and piles can cause a reassessment of the types of materials used in construction and in dimensions such as e.g. base plate thickness. Consolidation of saturated soils should also be considered in the context of paved areas. Higher groundwater levels could affect the design of column foundations for bridges and other deep foundation-dependent structures [6,7,8].

Increasing the level of precipitation will have a significant effect on drainage outflows. The proposed water flow currently expected for infiltration and drainage systems would probably need to be changed, leading to the introduction of higher capacity systems. Increased precipitation could increase the outflow rates through culverts and the design of the inlets are influenced by the velocity of the water flow entering the outflow [9,10,11].

Floods due to extreme events such as heavier and more frequent storms can affect the way overflow systems are designed and the cross sections under the bridges are designed [12,13,14,15]. The ways in which bridge foundations are protected from increased flows have been investigated by Meyer and Weigel [1]. The opposite of climate change is the opposite of hydrological drought. Currently, the most frequent method of identifying the occurrence of drought is the index method. In this study, deficit volumes and surplus volumes from daily average surface runoff from the studied area are evaluated.

2. Methods and study area

2.1. Study area

The studied area is the Hornád basin, which is located in eastern Slovakia. The Hornád basin is formed by the rivers Hornád, Hnilec, Torysa and Olšava. The evaluated station in the Hornád basin was the Ždaňa station, which is the last station in the Hornád basin in the Slovak Republic (figure 1). The catchment area is 4249 km^2. The average daily flow at the Ždana station is 29.3 m^3 / s and the maximum daily average was 772.3 m^3 / s and the minimum daily flow was 3.94 m^3 / s for the observation period 1961-2010. Data provided by the Slovak Hydrometeorological Institute.

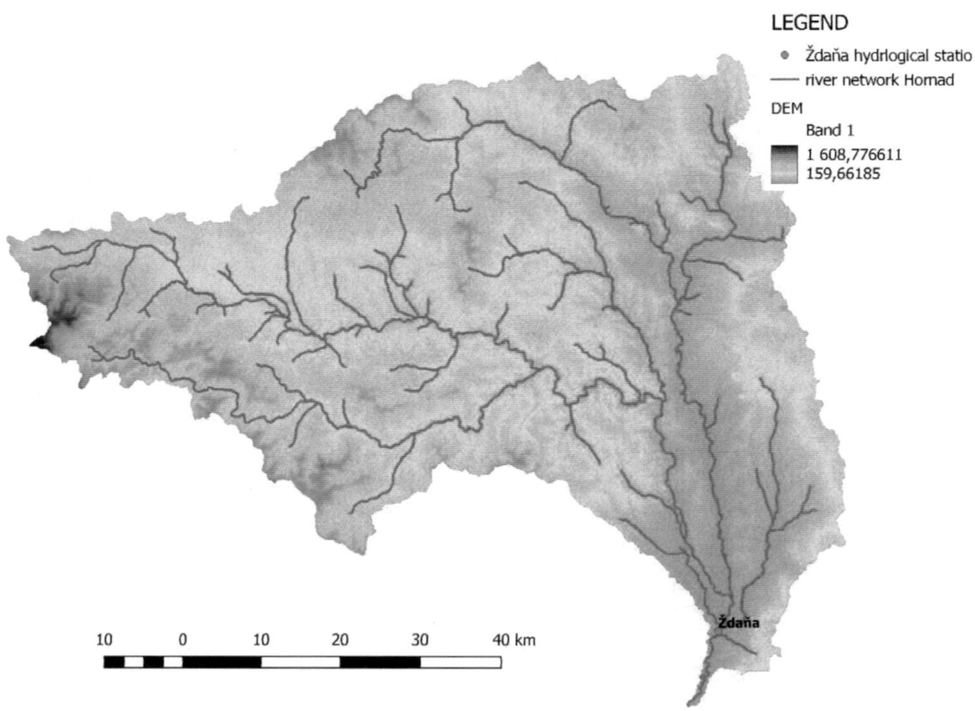

Figure 1. Basin Hornád and hydrologic station Ždaňa.

2.2. Methods

Daily average catchment water outflows R [mm] were used to determine deficit volumes and excess volumes, which were calculated from daily average discharges Q [m^3 / s].

$$R = \frac{Q[m^3/s]\times t[s]}{S[m^2]} = \ [m] = [mm] \tag{1}$$

Deficit volumes in the river basin were determined by the 20% quantile and surplus volumes were determined by the 80% quantile. Quantili, deficit and surplus volumes were processed in the programming and statistical language R. Quantili were calculated by the method of Hanzel, which is popular among hydrologists. Subsequently, the maximum deficit and surplus volumes in a given year for the whole observation period were selected [16].

IOP Conf. Series: Materials Science and Engineering	1252 (2022) 012062	doi:10.1088/1757-899X/1252/1/012062

3. Results

In figure 2 we see the course of the surface outflow quantili Q20 and Q80 in the station Ždaňa for the observation period 1961-2011.

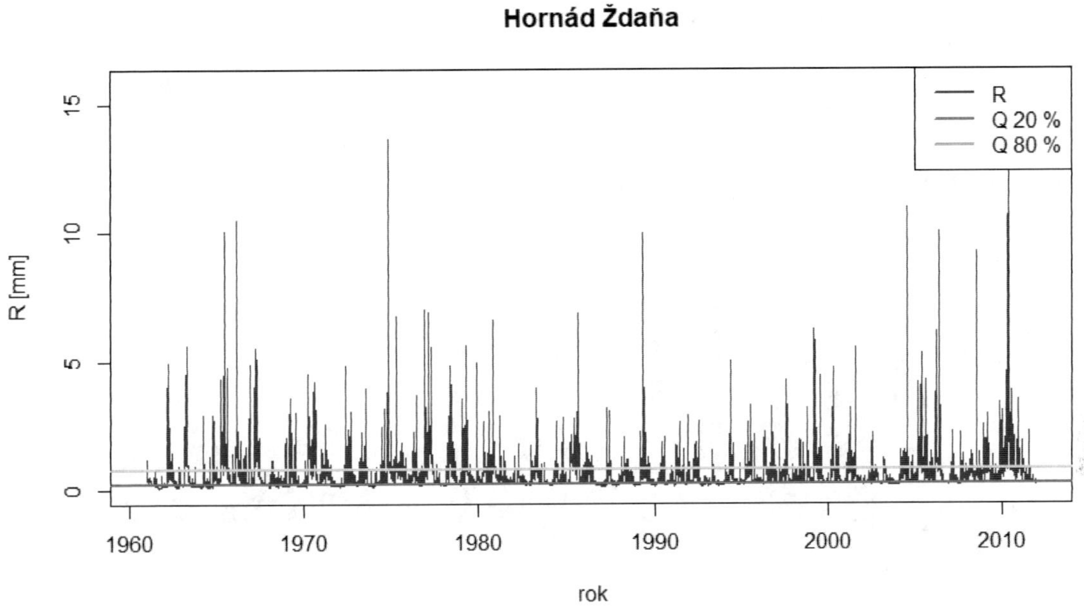

Figure 2. Basin Hornád surface runnof and quantili 20% a 80%.

Deficit volumes at the Ždaňa station were also significant at the beginning of the observation period at the beginning of the 1960s and then at the turn of the 1980s and 1990s (figure 3.).

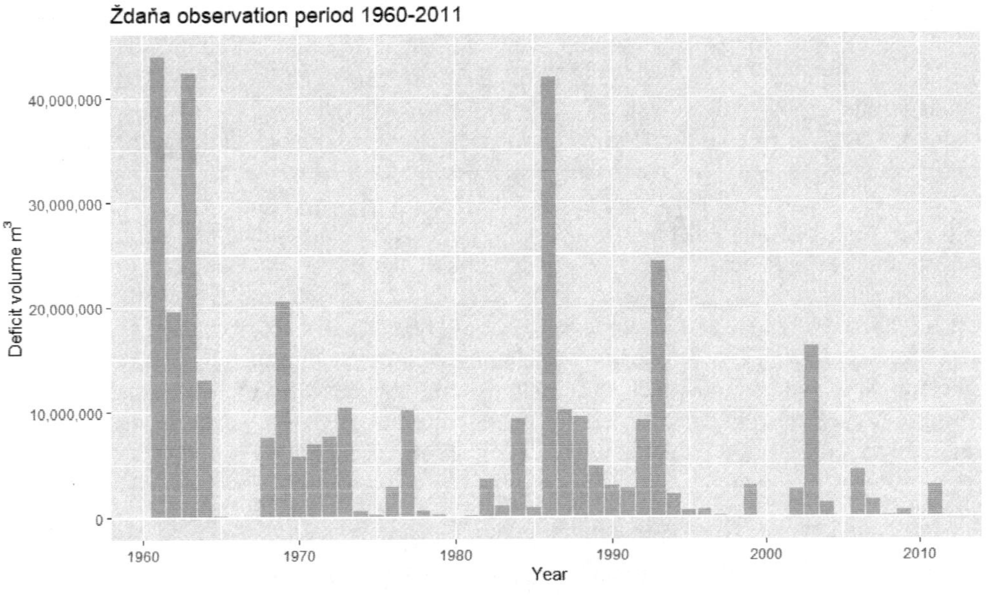

Figure 3. Deficit volume.

Excess volumes assessed at the Ždana station can be seen in figure 4. During the entire observation period, they were recorded, they were more pronounced in 2010 when there were also major floods. In the second half of the 1960s and 1970s, significant surplus volumes were also observed.

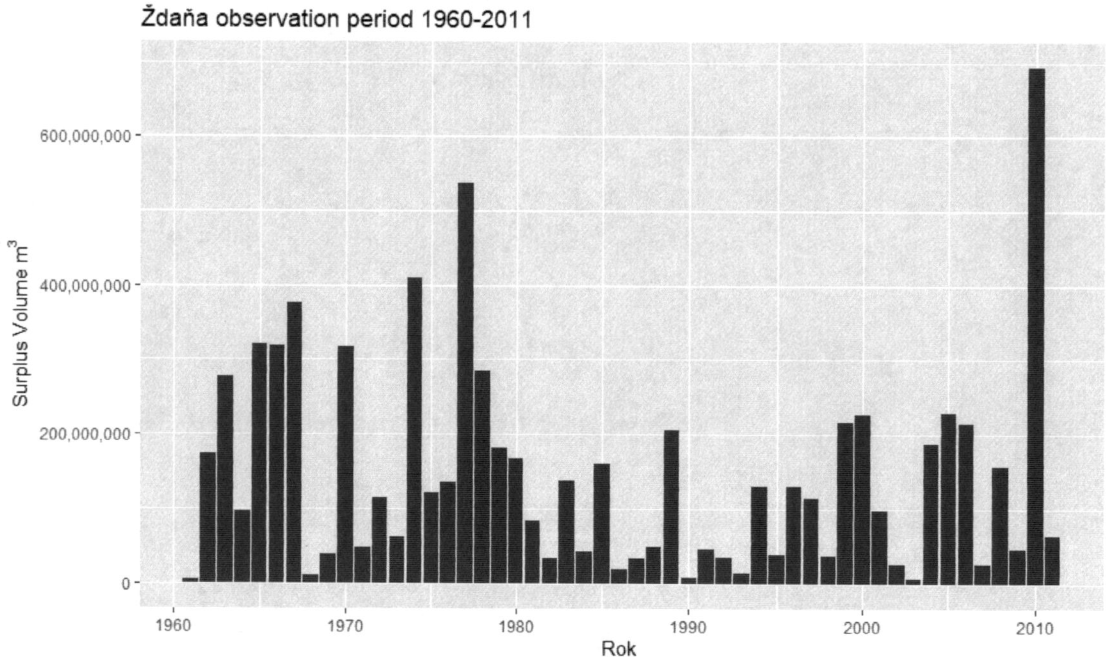

Figure 4. Surplus volume.

From the assesment of volumes in the Hornád river basin, it can be seen that the dry and wet periods alternate. There are periods where there is a significant deficit of water volumes in the river basin, there are periods where there is an excess volume of water in the river basin, which just flows away.

4. Conclusion

Despite the high number of studies dealing with climate change and drought, drought remains a phenomenon that requires further research. To date, many methods have been developed to assess climate change and drought, depending on the type of drought studied. This study was devoted to deficient and surplus volumes in the Hornád basin. It is clear from the results that the study of deficit and surplus volumes of water in the basin is necessary for better water management and water management in the country. Especially during the dry season, it is necessary to overestimate the water in the basin, but here the water must be retained in wet periods by suitable buildings in the basin. In this study, a large enough area was assessed so that specific measures could be identified. For a better assessment of the area, it is necessary to divide the river basin into smaller sub-basins and then it would be possible to determine where and what measures would be needed to mitigate climate change in terms of water management. Water scarcity can be caused by climate change such as increased temperature, lack of precipitation or increased water consumption industry, agriculture but also increased drinking water consumption. Reviewing and assessing water volumes in river basins leads to better water management in the country and reducing the effects of climate change.

5. **Reference**

[1] Meyer M D, Weigel B 2011 *Climate change and transportation engineering: Preparing for a sustainable future. Journal of Transportation Engineering* **137.6** pp 393-403

[2] Zelenakova M et al 2014 *Trends in minimal stream flows at eastern Slovakia. 9th International Conference Environmental Engineering (9th ICEE)* - Selected Papers

[3] Zelenakova, M et al 2016 *Flood occurrence evaluation based on the heights of water level in the rivers. In: 3rd International Conference Water Resources and Wetlands* pp 153-157

[4] Nagy P et al 2020 *Identification of dry and wet years in eastern Slovakia using indices. In: Advances in Environmental Engineering (AEE 2019)* **444** 012041

[5] Kubiak-Wojcicka K et al 2021 *Identification of Extreme Weather Events Using Meteorological and Hydrological Indicators in the Laborec River Catchment, Slovakia. Water* **10** 1413

[6] Mishra A K. Singh V P 2010 *A review of drought concepts. Journal of hydrology* **391.1-2** pp 202-216

[7] Tomlain J 1991 *Charakteristika suchých a vlhkých oblastí Slovenska* (Bratislava, SHMÚ) *Zborník prác, zv*, **33** pp 173-183

[8] Iturbide M et al 2020 *An update of IPCC climate reference regions for subcontinental analysis of climate model data: definition and aggregated datasets. Earth System Science Data* **12.4** pp 2959-2970

[9] Mckee T B et al 1993 *The relationship of drought frequency and duration to time scales. Proceedings of the 8th Conference on Applied Climatology*. 1993. p. 179-183.

[10] Mishra A K, Singh V P 2010 *A review of drought concepts. Journal of hydrology* **391.1-2** pp 202-216

[11] Skalský R Nováková M Mišková M 2012 *Analýza sucha v krajine ako príklad využitia simulačných modelov v geografii. Geografický časopis* **64** pp 55-69

[12] Zelenakova, M Preliminary flood risk assessment in the Hornad watershed. In: *5th International Conference on River Basin Management*, 2009, pp.15-24

[13] Zelenakova, M et al Flood occurrence evaluation based on the heights of water level in the rivers. In: *3rd International Conference Water Resources and Wetlands*, 2016, pp 153-157Nagy, P et al Identification of dry and wet years in eastern Slovakia using indices. In: *Advances in Environmental Engineering* (AEE 2019), 2020, 444: 012041

[14] Kubiak-Wojcicka, K et al Identification of Extreme Weather Events Using Meteorological and Hydrological Indicators in the Laborec River Catchment, Slovakia. *Water*, 2021, **10,** 1413

[15] Hyndman R J, Fan Y 1996 *Sample quantiles in statistical packages. The American Statistician* **50 (4)** pp 361-365

Acknowledgments
This work was supported by the Slovak Research and Development Agency under the Contract no. APVV-20-0281, and a project funded by the Ministry of Education of the Slovak Republic VEGA1/0308/20 “Mitigation of hydrological hazards, floods, and droughts by exploring extreme hydroclimatic phenomena in river basins”.
This work has been supported by the Slovak Research and Development Agency by supporting the projects SK-PT-18-0008, SK-SRB-21-0052 and SK-CN-21-0043.

Nonvisual aspects of daylight in the built environment

Jozef Hraška[1]

[1] Department of Building Construction, Faculty of Civil Engineering, Slovak University of Technology, Radlinského 11, 810 05 Bratislava, Slovakia

E-mail: jozef.hraska@stuba.sk

Abstract. This paper investigates melanopic equivalent daylight vertical and horizontal illuminances within an indoor space as a metric for the nonvisual effects of its lighting. The article is based on the findings of photobiological research and the recommendations of several institutions for the threshold values of melanopic illuminance at the eyes of users of indoor spaces, which is sufficient for the proper functioning of the circadian system. This study uses a computer workplace as an environment to investigate experimentally the applicability of melanopic daylight metrics for the practical evaluation of the circadian potential of the indoor daylight environment. The article points out a number of problems associated with the practical evaluation of vertical melanopic daylight illuminance in a specific space. The paper also outlines practical guidelines for designing facades and color treatment of building surfaces in terms of their circadian potential.

1. Introduction

In 2002, light sensitive cells (ipRGC - Intrinsically Photosensitive Retinal Ganglion) were discovered in the human eye [1], which mediate the direct connection between the eyes and the suprachiasmatic nucleus (SCN), which is part of the hypothalamus that holds our "body clock" also called "circadian clock". This discovery provided a solid basis for research into the nonvisual effects of light. Subsequent extensive research has provided insights into the spectral composition and intensity of light needed to send a signal to the SCN. The "body clock" drives of many hormonal levels (e.g. serotonin, melatonin, cortisol), daily wake-sleep cycles, body temperature, cardiovascular functions, alertness, memory, cognitive performance, blood glucose levels, cell division, etc. [2–4]. Light sensitive ipRGC cells express a unique photopigment melanopsin. Melanopsin is particularly sensitive to the absorption of short-wavelength (mainly blue) visible light, senses light at relatively high intensities, and playes a dominant nonvisual role in the setting of circaan rhythms. However, cones and rods, the image-forming cells, also participate in the circadian photoentrainment because all send information to the SCN [5–7]. The mechanism is complex and is the subject of intensive medical research. Extensive basic and applied research on nonvisual aspects of light on people has led to the gradual formulation of new requirements for building lighting [8–10]. It seems that these requirements for the level and spectral composition of light can be significantly higher compared to the vision requirements given in current standards, especially for artificial lighting. Satisfying these new needs is sought through innovations in artificial lighting; from an environmental and humane point of view, increased use of daylight in buildings is very desirable. Daylight is considered to be "proper light at the right time". In all indoor spaces, in which people stay for a long time during the days, there should be enough daylight not only in terms of its quantity but also in terms of spectral composition. There are currently a number of trends opposed to the need to ensure sufficient daylight in buildings. For examples overcrowded urbanization, increasing of depth of rooms, application of large offices, application of glazing with reduced light transmittance or changing their color, changed way of life of people in technically advanced countries where the traditional way of using buildings is changing during the day and night as well as other impacts. However, the excessive economisation of the creation of the buillt environment should not be at the expense of the availability of natural light in buildings, as natural light has numerous and not negligible positive effects on human health. The nonvisual effects of light on people open up the old question of how much daylight should be in a building so that it does not disturb the oscillation of circadian rhythms. There are many factors determining entrainment of the circadian cycle – light intensity, spectrum and timing, duration, pattern of light received at the eye (directionality), individual history of light exposure and the age of the interior user play a big role. These factors are widely investigated, most often in laboratory conditions. In this paper, we will experimentally determine the levels of photopic and melanopic illumination in a computer workplace with a real light environment consisting of daylight and artificial light and light from computer monitors. We will focus on daylight conditions in cloudy and clear skies and we will compare the measured values of photopic and melanopic illuminances in workplaces with the current recommended criteria and metrics.

2. Methods and requirements of "circadian lighting" evaluation

The term "circadian lighting" is usually used in cases where the lighting is designed to stabilize human day/night rhythms. Several metrics have been proposed to evaluate such lighting systems [11]. The following methods are most commonly used in scientific studies and practice: (I) the Equivalent Melanopic Lux (EML) [9], (II) the Circadian Light (CL_A) and Circadian Stimulus (CS) models [12], and (III) the Melanopic Equivalent Daylight Illuminance (m-EDI) [13]. The EML method is defined with reference to the equi-energy illuminance, the CL_A method is normalized to 2856 blackbody radiation, and m-EDI is based on light source D65. The standard of the Commission Internationale de l'Eclairage (CIE) [13, 14] accepted the upgraded proposal of Lucas et al. [9] to combine the sensitivity of the currently known five ocular photoreceptors to daylight, which is quantified by the α-opic Equivalent Daylight Illuminance (α-EDI) model. The functioning of signals originating from ocular photoreceptors is interconnected and is currently not sufficiently investigated. By [15] is melanopic illuminance sufficient to define nonvisual responses in humans. For the above reasons, this article evaluates the non-visual aspects of daylight using the m-EDI method with peak sensitivity at 490 nm. The International WELL Building Institute [16] uses both EML and m-EDI to evaluate circadian lighting design because they are fundamentally similar and simple. The ratio between m-EDI and EML is

$$\text{m-EDI} = 0.9058\ \text{EML} \quad (1)$$

According [13] the Melanopic Equivalent Daylight Illuminance for D65 standard daylight is represented as

$$E_{V,m}^{D65} = \frac{E_m}{K_{m,V}^{D65}} \quad \text{(lx)} \quad (2)$$

Where E_m is the melanopic irradiance in watts per square meter and melanopic efficacy of luminous $K_{m,V}^{D65}$ is defined as

$$K_{m,V}^{D65} = \frac{\Phi_m^{D65}}{\Phi_V^{D65}} \quad \text{(W / lm)} \quad (3)$$

Where Φ_m is the melanopic radiant flux and Φ_V is the luminous flux.

The state of research into the nonvisual effects of light on humans does not yet make it possible to establish clearly quantified criteria for assessing circadian lighting of indoor spaces in buildings. However, intensive research to date in this area has led several authors and several institutions to propose recommendations for the level, length and timing of eye exposure for indoor users [16]. Most of these recommendations are based on Circadian Action Factor [17], Circadian Stimulus [18] and Equivalent Melanopic Illuminance. Relationships are given between these metrics, e.g. according to [18] a photopic illumination of 500 lx will provide CS = 0.3 in the case of most current light sources. According to [16]

CS = 0.3 is equivalent to m-EDI = 217 lx or EML = 240 lx. It is a vertical illuminance at a height of 1.2 m (usually) or 1.4 m (according to [16]) above the floor, which represents the height of the eyes of a seated person. The circadian illuminance values recommended in [16] for all spaces except dwelling units are given in Table 1. In option 2 in Table 1 projects will receive one points if they meet at least one of the following requirements on each floor:

- 70% of all workstations are within 7.5 m of transparent envelope glazing or atria. Visible light transmittance of transparent glazing is greated than 40 %.
- Window area is not less than 10% of the regularly occupied floor area and visible light transmittance of transparent glazing is greater than 40 %.

One point is also obtained if $sDA_{300,50\%}$ (spatial Daylight Autonomy of the area that has more than 300 lux illuminance for at least 50% of the annual time of use) is achieved for more than 55% of regularly occupied floor area. If a project demonstrates through computer simulations $sDA_{300,50\%}$ for more than 75% of regularly occupied floor area, it will get 2 points (according Feature L05 in [16]). The light levels in Table 1 should be achieved at least between the hours of 9 a.m. and 1 p.m.

Although not all of the above requirements do not guarantee sufficient daylight in the indoor environment, they provide the designer with at least a basic orientation when designing daylighting. The stated requirements of the WELL Building Standards in countries of Central Europe apply to all workplaces in buildings and the requirements for visible light transmittance through glazing are usually higher, e.g. in Slovakia it is 60%. Sufficient daylight in the interior of buildings is the best way to achieve circadian benefits. It seems that the WELL recommendations could support daylighting more strongly.

Table 1. WELL Building Standard (Version 2) recommendations for illuminance levels for circadian lighting for all spaces except dwelling units [16].

Option 1 Electric light only		**Option 2** Electric light and daylight
At least 150 EML which is 136 m–EDI (1 point)	OR	At least 120 EML (109 m–EDI) with electric light and at least 2 points in Feature L05: Enhanced Daylight Access (1 point)
At least 240 EML which is 218 m–EDI (3 points)	OR	At least 180 EML (163 m–EDI) with electric light and at least 2 points in Feature L05: Enhanced Daylight Access (3 points)

3. Melanopic and photopic daylight illuminance of the virtual reality training laboratory

In this study, the light environment of the virtual reality training laboratory at the Faculty of Civil Engineering of Slovak University of Technology in Bratislava is experimentally analyzed, Figure 1 – 2. The classroom is located on the seventh floor. The surrounding buildings obstructed the windows of the laboratory minimally. The floor plan dimensions of the room are 5.9 × 5.9 m and its clear height is 2.67 m. The entire width of the external wall of the room is occupied by windows 1.7 m high with a windowsill 0.9 m high. Clear triple glazing in windows has a visible light transmittance of 70%.

The aim of the experimental measurements is to determine the photopic and melanopic illuminance at individual workplaces in the conditions of cloudy and clear skies. It is obvious that the intensity of light and its spectral composition is constantly changing under daylight conditions. Due to head movements, manipulation of blinds, changing brightness and spectrum of monitors, etc. the light exposure of the users of the space also changes. We are primarily interested in the ratio of horizontal and vertical photopic and melanopic illuminances in the workplaces with a focus on vertical melanopic illuminances. The light environment in the laboratory consists from combination of daylight spectra, artificial illuminants spectra, spectra from computer screens and spectra of reflecting surfaces in space. Artificial lighting was not in operation during the measurements published in this article. The new MSC 15 light spectrum meter from Gigahertz-Optik GmbH (Figure 7) was used for measurement of the photopic and melanopic illuminance and other spectral characteristics of light. MSC15 covers a spectral range between 360 nm and 830 nm with a spectral bandwidth of 10 nm and offering a wide measurement range for illuminance and color between 1 lx and 350,000 lx. The device can be operated manually or measured data can also be stored in a computer running Windows as GOS extension source files, which are exported in ASCII and ExcelTM formats.

Figure 1. View of the room from the windows with the designation of computer workplaces (numers 1, 2, 3).

Figure 2. View of part of the room and part of the windows.

Illuminances were recorded in September 28, 2021 between 9:00 a.m. and 11:30 a.m. and in October 6, 2021 between 9:00 a.m. and 10:34 a.m. (overcast sky) and also were recorded in October 1, 2021 between 10:00 a.m. and 11:40 a.m (clear sky). Typical spectrum of exterior daylight in cloudy sky conditions in September 28, 2021 is in Figure 3. The spectrum of daylight in cloudy sky conditions after passing through glazing and pale gray Venetian blinds is in Figure 4. Typical spectrum of exterior daylight in clear sky conditions in October 1, 2021 is in Figure 5. The spectrum of light from clear sky after the light has passed through the glazing and blinds in the horizontal position is in Figure 6. The measured room is oriented to the northeast and no direct sunlight entered it during the measurements. The spectrum of light reflected from the wall with burgundy paint in conditions of daylight under cloudy sky is shown in Figure. 8. From the measured data documented in Figure 3 to 6 and 8, it is apparent that the photopic and melanopic illuminance ratios vary from case to case. The light reflected from the burgundy coating significantly reduces its melanopic content.

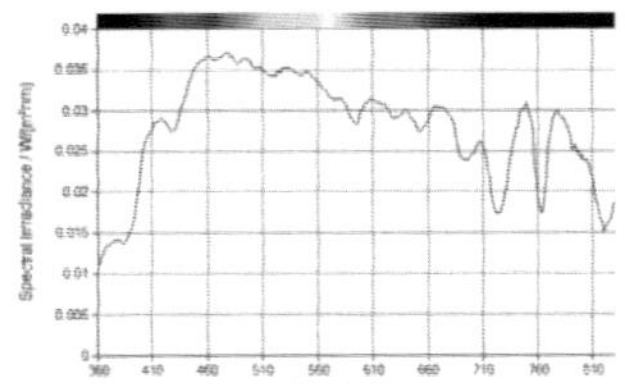

Figure 3. Typical spectrum of exterior daylight in cloudy sky conditions during measurements.

Ratio $E_{v,ph} / E_{v,mel} = 1.035$

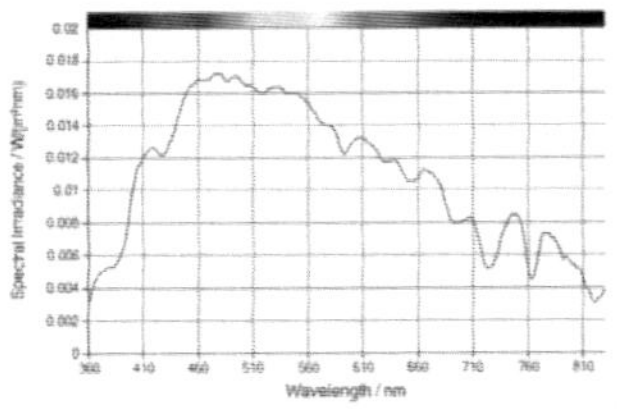

Figure 4. The spectrum of daylight in cloudy sky conditions after passing through glazing and blinds.

Ratio $E_{v,ph} / E_{v,mel} = 1.005$

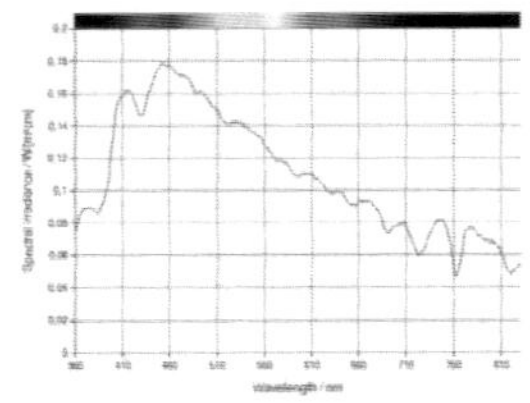

Figure 5. Typical spectrum of exterior daylight in clear sky conditions during measurements.

Ratio $E_{v,ph} / E_{v,mel} = 0.896$

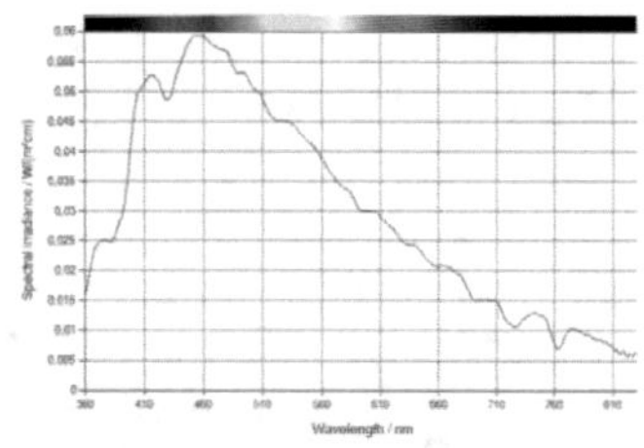

Figure 6. The spectrum of daylight in clear sky conditions after passing through glazing and blinds.

Ratio $E_{v,ph} / E_{v,mel} = 0.820$

Figure 7. The MSC 15 light spectrum meter. The device measures photopic, scotopic and melanopic illuminance, color properties, etc.

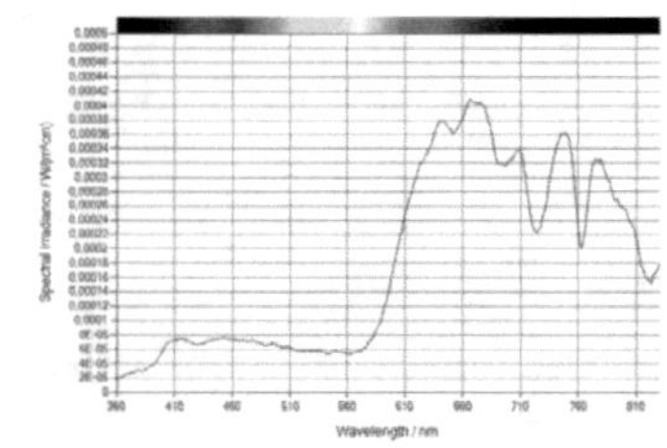

Figure 8. The spectrum of light reflected from the wall with burgundy paint in conditions of daylight under cloudy sky.

Ratio $E_{v,ph} / E_{v,mel} = 1.762$

Due to the limitation of the extent of this article, we deal with photopic and melanopic illuminances only in the workplace marked with the number 3, which is located furthest from the windows.The place where the horizontal and vertical photopic and melanopic illuminances were measured under clear sky conditions is shown in Figure 1. Named arrows in the Figure 1 indicate the directions of measurement. During the measurements under conditions of clear sky, the photopic illuminance on the vertical windows fluctuated in range from 7 964 lx to 9 366 lx. The melanopic exterior illuminance on the windows ranged from 8 512 lx to 10 450 lx. The vertical illuminances of the window under a clear sky was created largely by reflected direct sunlight, which fell on the facades, roofs and terrain below the level of the windows of the room on the seventh floor.

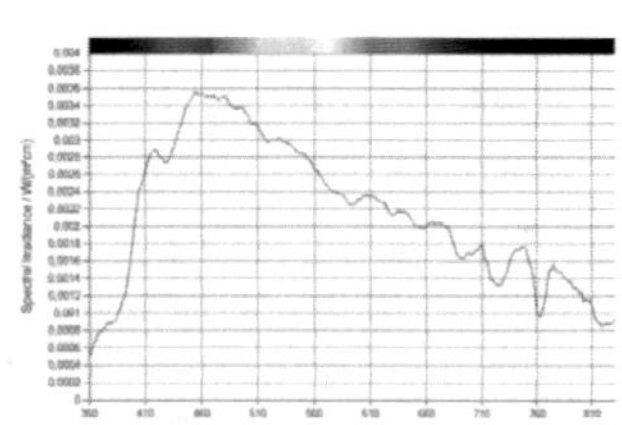

Figure 9. The spectrum of horizontal daylight illuminance on the worktable (indicated in Figure 1 by the letter H); clear sky conditions; pulled out Venetian blinds.

$E_{v,ph} = 196.0$ lx $E_{v,mel} = 214.2$ lx

Ratio $E_{v,ph} / E_{v,mel} = 0.915$

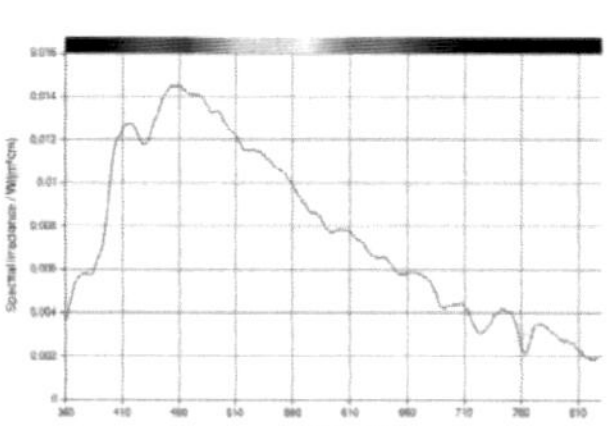

Figure 10. The spectrum of vertical daylight illuminance on the worktable towards the windows (direction V1); clear sky conditions; pulled out Venetian blinds.

$E_{v,ph} = 717.6$ lx $E_{v,mel} = 851.6$ lx

Ratio $E_{v,ph} / E_{v,mel} = 0.843$

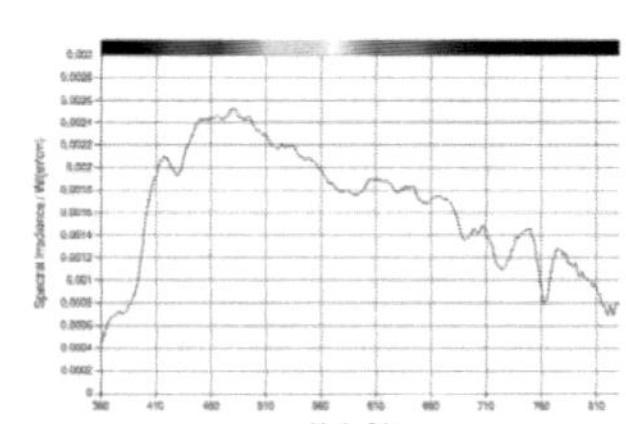

Figure 11. The spectrum of vertical daylight illuminance on the worktable towards the windows (direction V2); clear sky conditions; pulled out Venetian blinds.

$E_{v,ph} = 146.7$ lx $E_{v,mel} = 153.3$ lx

Ratio $E_{v,ph} / E_{v,mel} = 0.957$

Measurements were performed when the Venetian blinds were pulled out (Figure 9 – 11) and when the pale gray blinds were in a horizontal position (Figure 12 – 14). In the case of blinds in the horizontal position, lower levels of illuminance were achieved (compare Figures 9 to 11 with Figures 12 to 14). A slightly larger decrease was observed in the case of melanopic illuminances. This is probably because the light directed by the blinds fell on the burgundy walls and ceiling in a larger amount compared to the situation without blinds. The blinds also partially change the spectral composition of daylight entering the room. Although the effect of burgundy-painted walls in a room intensely illuminated by daylight did not manifest itself significantly on the values of melanopic illuminance, changes depending on the measurement directions are noticeable.

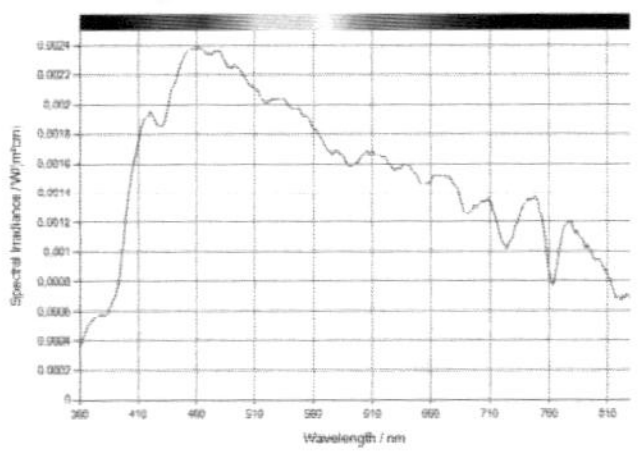

Figure 12. The spectrum of horizontal daylight illuminance on the worktable (indicated in Figure 1 by the letter H); clear sky conditions; pale gray Venetian blinds in a horizontal position.

$E_{v,ph}$ = 135.3 lx $E_{v,mel}$ = 144.1 lx

Ratio $E_{v,ph} / E_{v,mel}$ = 0.940

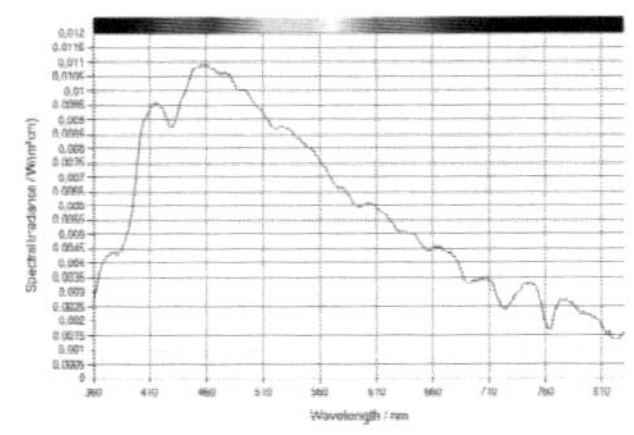

Figure 13. The spectrum of vertical daylight illuminance on the worktable towards the windows (V1); clear sky conditions; pale gray Venetian blinds in a horizontal position.

$E_{v,ph}$ = 548.3 lx $E_{v,mel}$ = 642.4 lx

Ratio $E_{v,ph} / E_{v,mel}$ = 0.854

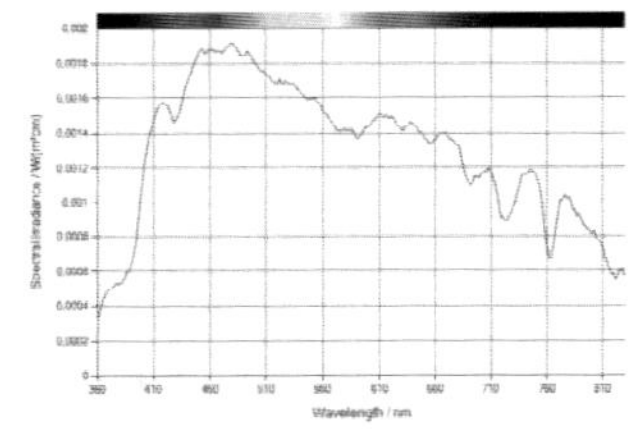

Figure 14. The spectrum of vertical daylight illuminance on the worktable (direction marked in Figure 1 by the letter V2); clear sky conditions; blinds in a horizontal position.

$E_{v,ph}$ = 113.9 lx $E_{v,mel}$ = 117.2 lx

Ratio $E_{v,ph} / E_{v,mel}$ = 0.972

At least the influence of the blinds in the horizontal position on the change of melanopic lighting was manifested in the vertical plane facing the windows. The relatively greater loss of melanopic light was measured in the vertical plane facing the sidewall of the room, although much of the burgundy-painted wall was covered with a white projection screen.

If we compare the measured vertical values of melanopic illuminance at a height of 0.85 m above the floor at the workplace no. 3 (Figures 10 and 13) and the data in Table 1 (option 2), they far exceed the levels specified in the WELL rating system in clear sky conditions (in the direction of the view to the windows). Melanopic illuminance in clear sky conditions meets the m-EDI > 109 lx criterion given in Table 1 also in the case of the direction of the view from workplace no. 3 to the side wall (direction V2, Figures 11 and 14).

Horizontal melanopic illuminance at the reference plane in the room axis at a height of 0.85 m at floor level and vertical melanopic illuminance at 1.2 m and 1.4 m above the floor towards the window under a cloudy sky are documented in Figure 15. At a distance of approximately 3.8 m from the window under cloudy sky conditions, at the time of measurement, the vertical melanopic illuminance values at a height of 1.2 m and 1.4 m, respectively, will fall below the level required by the WELL system. At the time of measurement (October 6, 2021 between 9:00 a.m. and 10:34 a.m.), the vertical melanopic illumination on the outside of the windows ranged from 1,217 lx to 1,663 lx. This means that for some time the vertical melanopic illuminance even in the depth of the room exceeded the WELL requirements (i.e. m-EDI > 109 lx) and for some time lower values were recorded. Even during cloudy days, the level of daylight fluctuates to a large extent and it is difficult to guarantee that between 9 a.m. and 1 p.m. during whole year, the value of vertical melanopic illuminance did not fall below the criteria required by the WELL system, although it is sufficient to meet them only on a part of the regularly occupied floor area. It seems that the criteria given in Table 1 are not best suited for assessing melanopic illuminance in daylight conditions only.

4. Conclusion

Natural daylight is the best light source that provides visual comfort in the indoor environment of buildings, circadian stimulation and has the potential for significant energy savings in their operation. The methodology and initial findings from a case study of photopic and melanopic illuminance of a side-lit room under clear and cloudy skies have been described in this paper. Architectural factors of the evaluated room are standard, with the exception of a part of the walls painted with dark border paint. Very good daylighting of the room significantly reduces the adverse effect of burgundy color on the

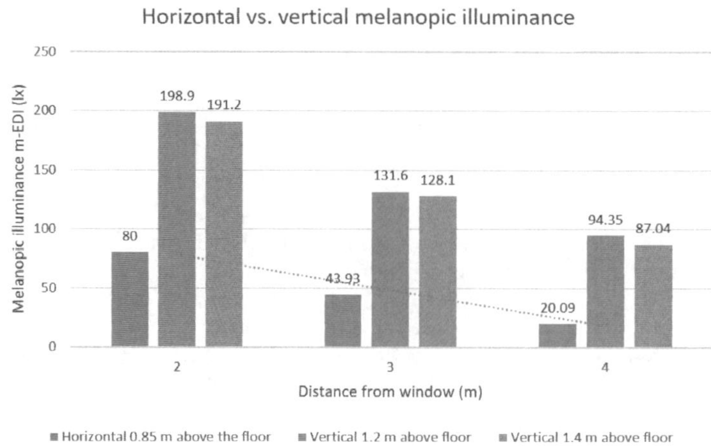

Figure 15. Horizontal melanopic illuminance at the reference plane in the room axis at a height of 0.85 m at floor level and vertical melanopic illuminance at 1.2 m and 1.4 m above the floor towards the window under a cloudy sky.

vertical melanopic illuminance in the gaze direction towards the windows. On the contrary, in the workplace near the burgundy wall and in the direction of view towards this wall, a ratio of photopic and melanopic illuminance of up to 1.7 is achieved. The ratios between photopic and melanopic illuminance change not only depending on the direction of view, but also the state of the sky. Melanopically more efficient daylighting is in clear sky conditions compared to cloudy skies.

There are currently several methods for evaluating the indoor light environment in terms of its circadian quality, but they are still in preliminary stage. In practice, the WELL Building Standard evaluation system is usually used, which is mostly focused on the evaluation of circadian properties of artificial lighting. The requirements and criteria of the WELL rating system do not take sufficient account of the dynamics and specifics of daylight. In daylighting, it is not possible to ensure that during the year between 9:00 a.m. and 1:00 p.m. the vertical melanopic illumination has not fallen below the criteria values of the WELL system, although practically the whole year, in a well-designed daylighting, the melanopic illumination will be several times higher than its required values.

Future research on circadian-efficient indoor spaces of buildings should in addition to architectural and interior aspects also takes into account urbanism, local climatic conditions and the spectral characteristics of glazing, artificial light sources including light from screens and all surfaces involved in the interaction of light in the built environment.

5. References

[1] Berson D M, Dunn F A, Takao M, 2002 Phototransduction By Retinal Ganglion Cells That Set The Circadian Clock, Science 295, pp. 1070-1073

[2] Lockley S W 2009 Circadian Rhythms: Influence of Light in humans. Oxford, UK: Academic Press, pp. 971–988

[3] Spitschan M, Stefani O, Blattner P, Gronfier C, Lockley S W, Lucas R J 2019 How to Report Light Exposure in Human Chronobiology and Sleep Research Experiments. Clock&Sleep, 1, 280–289

[4] Gooley J J, Rajaratnem S M W, Brainard G C, Kronauer R E, Czeisler Ch A, Lockley S W 2010 Spectral responses of the human circadian system depend on the irradiance and duration of exposure to light. Science Translational Medicine, 2(31):31ra33

[5] Dacey D M, Liao H W, Peterson B B, Robinson F R et al. 2005 Melanopsin-expressing ganglion cells in primate retina signal colour and irradiance and project to the LGN. Nature, 433(7027), 749

[6] Altimus C M, Güler A D, Alam N M, Arman A C et al. 2010 Rod photoreceptors drive circadian photoentrainment across a wide range of light intensities. Nature neuroscience, 13(9), 1107
[7] Spitschan M 2019 Melanopsin contributions to non-visual and visual function. Current Opinion in Behavioral Sciences, 30:67-72
[8] Aries M, Aarts M, van Hoof J. 2015 Daylight and health: a review of the evidence and consequences for the built environment. Lighting Research and Technology; 47: 6–27
[9] Lucas R J, Peirson S N, Berson D M, et al. 2014 Measuring and using light in the melanopsin age. Trends in Neurosciences, 37: 1–9
[10] Neberich M, Opferkuch F 2021 Standardizing Melanopic Effects of Ocular Light for Ecological Lighting Design of Nonresidential Buildings—An Overview of Current Legislation and Accompanying Scientific Studies. Sustainability, 13, 5131
[11] Lowry G D 2018 A comparison of metrics proposed for circadian lighting and the criterion adopted in the WELL Building Standard. In CIBSE Technical Symposium, Stretching the Envelope. London South Bank University, 12-13 April, pp. 1–10
[12] Rea M S, Figueiro M G, Bierman A, Bullough J D 2010 Circadian light. Journal of circadian rhythms, 8(1), 2
[13] CIE: S 026/E: 2018 System for Metrology of Optical Radiation for ipRGC-Influenced Responses to Light. CIE, Vienna.
[14] CIE 2020 CIE Toolbox and User Guide to use with CIE S 026:2018. Retrieved September 24, 2021, from https://mailchi.mp/cie/vvg94rcpua-3230141
[15] Brown T M 2020 Melanopic illuminance defines the magnitude of human circadian light responses under a wide range of conditions. Journal of Pineal Research, 69: e12655
[16] International WELL Building Institute 2021 Circadian Lighting Design. Retrieved September 21, 2021, from https://v2.wellcertified.com/v/en/light/feature/3
[17] Gall D, Bieske K 2004 Definition and measurement of circadian radiometric quantities. Light and health - non-visual effects: proceedings of the CIE symposium, 30 September - 2 October 2004, Vienna, Austria, Commission internationale de l'éclairage, 2004, 129–132
[18] Underwriters Laboratory (UL) 2019 Design guideline for promoting circadian entrainment with light for day-active people. UL Design Guideline 24480, Edition 1.

Acknowledgments
The Ministry of Education, Science, Research and Sport of the Slovak Republic under VEGA contract 1/0042/21 and the Slovak Research and Development Agency under contract APVV-18-0174 have supported this article.

Gradual transformation of an administrative building into a green structure

M Vertal[1], Z Vranayova[2], K Cakyova[3], A Vargova[1] and F Vranay[2]

[1] Department of Architectural Engineering, Faculty of Civil Engineering, Technical University of Kosice, Vysokoskolska 4, 04001 Kosice, Slovakia
[2] Department of Building Facilities, Faculty of Civil Engineering, Technical University of Kosice, Vysokoskolska 4, 04001 Kosice, Slovakia
[3] Centre of Research and Innovation in Construction, Faculty of Civil Engineering, Technical University of Kosice, Vysokoskolska 4, 04001 Kosice, Slovakia

E-mail: zuzana.vranayova@tuke.sk

Abstract. The paper describes our vision of a green conversion of selected administrative building in Košice, which is supported by the research project APVV-18-0360 ACHIEve Active Hybrid Infrastructure for a sponge city. The greenIZOLA vision represents a set of complex measures, the result of which is a logically, economically, and environmentally justified design of a green building. Within separate stages, the individual parts of the complex and the building are gradually transformed into structures with a potential for retaining rainwater. At the same time, the proposed architectural structures introduce a potential for the creation of experimental structures designed for multidisciplinary research of synergistic links between the building, the environment and vegetation units. The transformation process uses a multidisciplinary approach (BIM, data analysis, life cycle analysis, water retention modelling in the environment, analysis of vegetation units, energy and multiphysical simulations). By applying these approaches, it is possible to monitor the proposed structures, analyse their effects, predict their behaviour, and then optimise the system.

1. Introduction

Cities will increasingly feel the consequences of climate changes. Different nature based solutions can not only increase climate resilient in cities as well as promote a more sustainable and climate friendly development for their inhabitants. Green roofs and rainwater harvesting systems are constructive measures that are becoming increasingly important in our urban environment. Massive buildings radically change the character of urban space. In this dynamic, a pleasing aspect is the growing interest in the transformation of parts of buildings with an implemented vegetation element. Most often, these are vegetation flat roofs, which create ideal conditions for the creation of a good-quality architectural space for people to stay in. In terms of the ability to accumulate water, the roof represents the area with the highest amount of incidental atmospheric precipitation. Knowledge of the behaviour of a vegetated roof, interaction with the building, users and the environment is a key aspect of its proper design.

In the paper we present a vision of a possible gradual green conversion of the premises of the construction company IZOLA Košice, s.r.o. in cooperation with the Technical University in Košice and is covered by a research project called APVV-18-0360 ACHIEve Active hybrid infrastructure for a sponge town. The vision of greenIZOLA represents a set of complex measures, the result of which is a logically, economically and environmentally justified proposal for the green conversion of the premises.

The project has separate stages, in which the individual parts of the complex and the building are gradually transformed into structures with the potential for retaining rainwater. At the same time, the proposed architectural structures create the potential for the creation of experimental structures designed for multidisciplinary research of synergistic links between the building, the environment and vegetation units. The cooperation of the university with the practice integrates multidisciplinary approaches into the process of transformation of the selected area of interest (modeling of water retention in the environment, analysis of vegetation units, energy and multiphysical simulations, data analysis, life cycle analysis + LCC, BIM, etc.). By applying these approaches, it is possible to monitor the proposed structures, analyze their manifestations, predict their behavior and then optimize these structures. Currently, the first stage of the greenIZOLA project is completed, focusing on the measurement and analysis of the thermal-humidity behavior of vegetation roofs with not only biodiversity potential [1].

The collective from the Faculty of Civil Engineering, Technical University of Kosice organizes the project. The co-authors are from other departments of TUKE as well as from other universities in the region (Botanical Garden of University of Pavol Jozef Safarik, Kosice). The reason for the cooperation is the interdisciplinarity of this research project.

The main goal is to create a smart resilience water and energy system in accordance with the concept of sustainability and durability. In connection with the analysis of green walls or roofs and their influence on the microclimate in the external environment of buildings, new water capturing elements will be simulated and measured as parts of the projects. The novelty is in the synergy between architecture, building services and technologies, thermal technology, water industry through the use of numerical methods, the search for interrelationships, the improvement of existing methods in the combination of aspects to simulate complex phenomena and their intelligent management.

Our task is to draw up recommendations on how to build "green buildings" progressively supplied with water, connecting infrastructures - from existing administrative buildings. We will examine how these measures will affect the quality of life of their users and how they can help to change society's mind towards water-saving culture.

The analysis of the co-called sponge city creation by green infrastructures has not been dealt with so far in Slovakia, and therefore our projects bring a new approach to designing or redesigning buildings in terms of water and energy management. Part of the multivalued hybrid infrastructure will be verified by various research methods (in-situ experiment, questionnaires, environmental and economic assessment). Experiments include real buildings with and without active elements of hybrid infrastructure.

Within the project "ACHIEve" a pilot phase was finished, which consisted of construction of experimental vegetation roof terraces as active elements of functioning sponge infrastructures.

2. Material and methods

Testing of building structures with a vegetation layer is the research ambition of many university workplaces and development centres in the world [1], [2], [3], [4], [5]. The type of vegetation suitable for the given conditions, the influence of the vegetation roof on the building, street or city with regard to the retention of water in the environment, overheating of the buildings or elimination of the so-called heat islands. Using in-situ measurements on buildings, laboratory experiments on scale models and simulations, researchers try to describe the behaviour of building structures with a vegetation layer, optimize its design for conditions of use and predict the behaviour of such structures under different compositions and different boundary conditions. The ACHIEve research project is focused on a long-term monitoring of vegetation roofs in the climatic conditions of the city of Kosice. Test segments with different compositions were implemented on the roof terraces of the administrative building. The installation of the measuring infrastructure in the individual layers of the roof and the monitoring of the relevant climatic parameters make it possible to monitor the thermal-humidity response of the test segments in interaction with the environment. The setup of the experiment assumes the possibility of comparing the measured data from vegetation segments with a referential conventional roof segment with gravel backfill [6], [7], [8], [9].

As part of the project an area of 3926 m^2 in the south industrial zone of the city of Košice, Nad jazerom district, was analyzed. There is an administrative building with an area of 618 m^2 with parking spaces for employees and customers, shelters for a temporary warehouse for temporary storage of building materials and a fire tank. The area of interest is drained by means of an oiled sewer, which passes through an oil separator and then opens into a public sewer. The surface area with vegetation cover makes up approximately 14% of the area of interest.

Green conversion of the IZOLA Košice, s.r.o. counts on the maximum retention of rainwater in the defined area of interest. Priority is given to the direct accumulation of rainwater by building structures with a vegetation layer through the implementation of vegetation roofs and shelters. Excess water from vegetation segments and water from non-absorbent surfaces will be properly treated and subsequently accumulated using infiltration blocks designed in the area of interest.

The administrative building is a four-storey building built as a two-way prefabricated skeleton with reinforced concrete prefabricated ceiling panels and infill masonry. The south terrace (see Figure 1), which is accessible from the last above-ground floor, was chosen to carry out the test vegetation roof experiment. There is an administrative space under the roof terrace.

Figure 1. Experimental building of the company IZOLA, s.r.o. in Kosice [10].

The roof structure has been modified in a several ways [11-13]. The originally flat single-skin roof with the reverse order was transformed into a roof with the classic order of layers, by supplying the main waterproofing layer based on PVC foil (see Figure 2). The original waterproofing layer of the roof subsequently took over the function of a vapour barrier. In accordance with the current requirements for thermal protection of buildings, the original thermal insulation layer made of XPS was extended by a thermal insulation board based on PIR thickness of 60 mm. The heat transfer coefficient of the treated roof is $U = 0.15\ W/(m^2\ K)$.

For the purposes of in-situ experiments, the test roof was divided into three parts/test segments, which are identical to the level of the roof waterproofing (PVC foil) (Figure 2).
Above the level of the waterproofing layer, the experimental roof is divided into:
• extensive vegetation roof with a roof substrate height of 120 mm (Test segment I)
• referential part with gravel backfill (Test segment II)
• extensive vegetation roof with a roof substrate height of 240 mm (Test segment III)

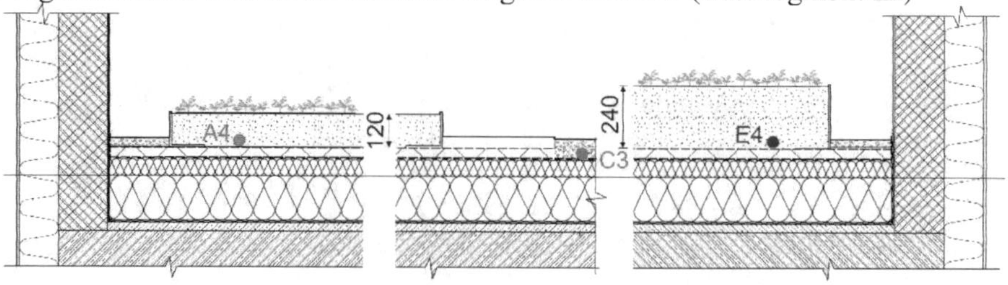

Figure 2. Scheme of three test segments of the experimental roof [13].

The choice of the thickness of vegetation substrates takes into account the requirement to grow higher plants with biodiversity potential (vegetation of herbs and grasses rich in diverse species). In addition to the typical extensive vegetation planted immediately after the implementation of the test roof segments, there is room for experimentation with various types of greenery suitable from the point of view of the city climate. A system solution from BAUDER with a Pflanzerde LBB-E roof substrate (retention potential = 45 l/m^2, at a thickness of 120 mm;) and a BAUDER RE 40 retention element (constant retention potential 13.5 l/m^2) was used to implement the vegetation roof [11], [12], [13].

In the greenIZOLA project, each roof and roof terrace will be transformed into a vegetation system with the potential to retain rainwater. In the individual phases of the project, as first the roof terraces are gradually transformed into experimental segments, which are focused on in situ measurements of parameters of variant solutions of vegetation roofs. In total, the greenIZOLA project provides the potential for the creation of 17 independent experimental in situ roofing sites (see Figure 3).

The application of vegetation to the facades of the building will change the visual of the area of interest, reduce the amount of CO2 emissions and reduce the temperatures of external surfaces through the processes of evapotranspiration (the effect of the so-called heat islands). From an experimental point of view, the vegetation facade will create a space for testing the impact of building structures with variant vegetation layers on the building and its internal environment.

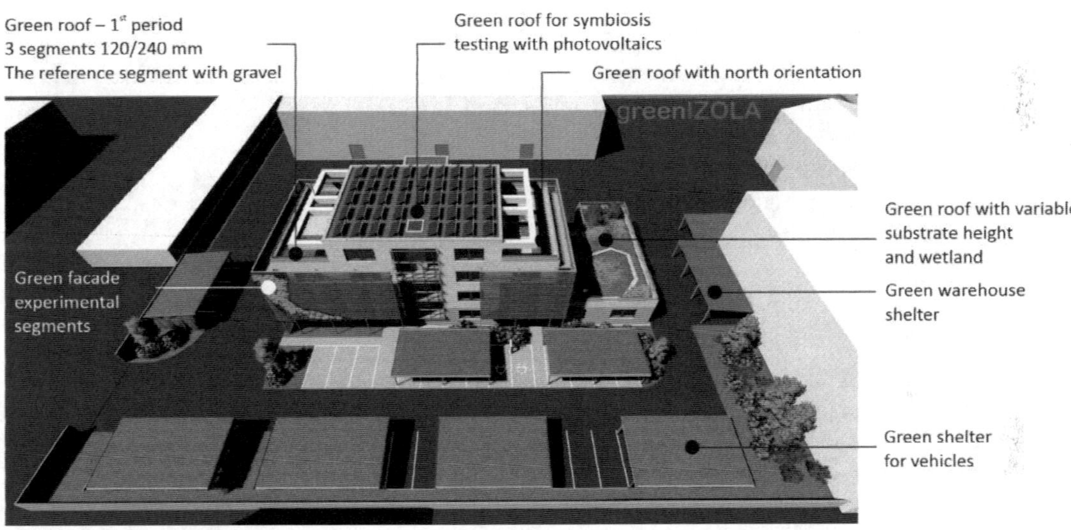

Figure 3. GreenIZOLA vision, SvF research center, TUKE of the administrative building and the area of interest and a possible visual after the implementation of all phases of the greenIZOLA project.

At the beginning of the first phase of the greenIZOLA project, partial research goals were set, which reflect current issues of practice by applying in situ research procedures. In the practice of applicable research of building structures with a vegetation layer, it appears as an optimal concept for considering and creating the restoration of sustainable buildings. In the first phase, the effectiveness of roofs with a vegetation layer is monitored by application of standard compositions [12-16] and is focused on the following sub-objectives:

- ability to accumulate precipitation water and its redistribution in the substrate,
- the effect of rainwater on the temperature in the roof cladding,
- analysis of the effects of shortwave and longwave radiation,
- the influence of the snow layer on the course of temperatures in the roof cladding,
- overheating of roofs in long-term dry weather,
- a set of measured input and validation measurements for simulation analyzes.

IOP Conf. Series: Materials Science and Engineering　1252 (2022) 012064　doi:10.1088/1757-899X/1252/1/012064

2.1. Preparation of an experimental roof

In the pre-implementation phase, project documentation of the roof modification was prepared, which included the project of fitting and tracing the measuring infrastructure, diagnostics of the roof cladding, preparation of perforations, grooves of brackets and stands. The roof was then covered with a new waterproofing layer (PVC foil).

The transformation of the roof terrace into a test vegetation roof was carried out within three working days.

Figure 4. Progress in the green roof with not only biodiversity potential.

2.2. Measurement equipment

Monitoring the behavior of test vegetation roof segments is realized through long-term in-situ measurements.

The measurements are performed in two circuits according to the diagram in Figure 8. The first circuit measures and records relevant climatic parameters (temperature and relative humidity, atmospheric pressure, wind speed and direction, solar radiation intensity, total and intensity of atmospheric precipitation) in this locality of the city of Košice.

The second circuit measures and records selected quantities in selected layers of test roof segments using built-in sensors.

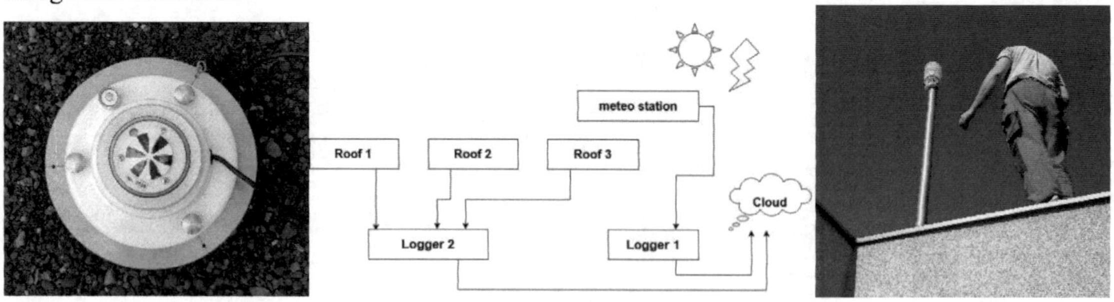

Figure 5. Diagram of measuring infrastructure (middle), pyranometer (left) and weather station (right) for recording climatic parameters in the test site.

2.3. Experimental vegetation roof and its monitoring

The terraces of the south façade, realized in the spring of 2020, unite a fully functional and usable space of a vegetation roof and a fully-fledged experimental segment of research (Figure 6).

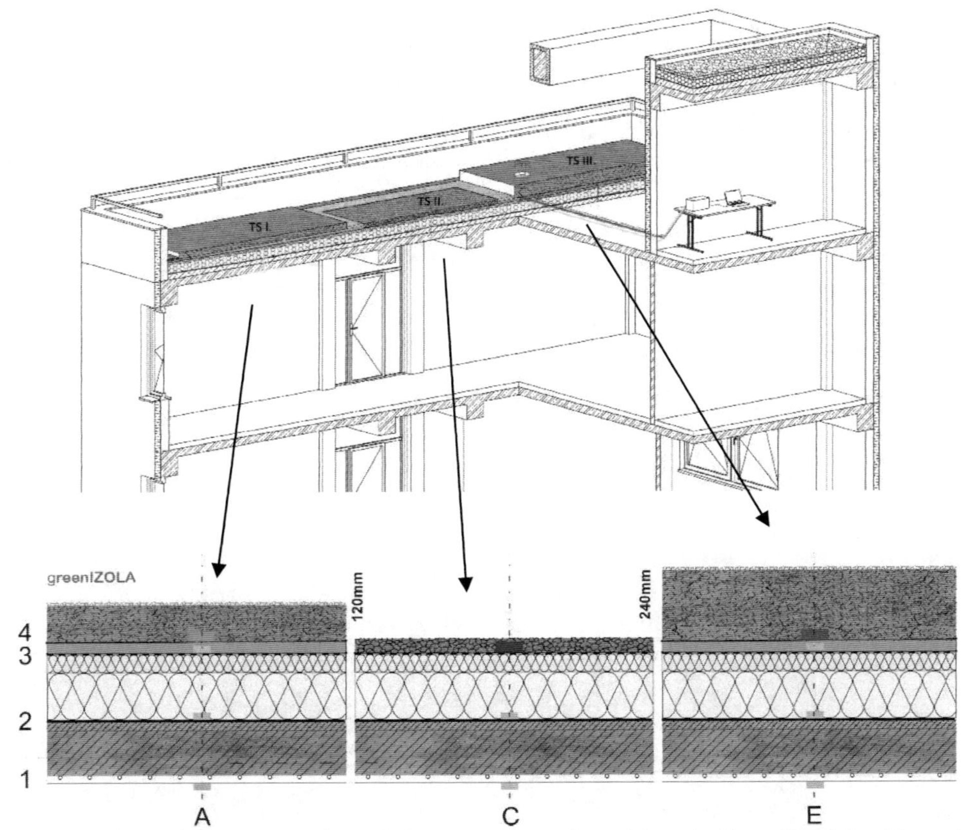

Figure 6. Schematic representation of the experimental roof facade GreenIZOLA - Phase 1 and location of sensors in test segments of roofs with a substrate height of 120 and 240 mm and a reference segment of a gravel roof covering.

By monitoring the temperatures in different layers of the roof with vegetation layer and comparing them with the data measured in a conventional roof (gravel layer), it is possible to analyze the ability of the structure, protect the building from cooling in winter and overheating during summer. Figure 6 also shows the location of selected measuring sensors in the test segments of the experimental roofs.

3. Results and discussion

The measured data from both measuring circuits are stored in a cloud storage are accessible in real time for further analysis. Clear control of measurements and visual display of measured quantities is mediated by an on-line interface, which shows the visualization of measurements. For the processing of measured data, tools for quick evaluation as well as possibilities for more demanding analyses are available. An example of the analysis of the measured data is shown in Figure 7.

The upper part of Figure 7 shows the waveforms of the measured temperatures under the substrate of vegetation segments with a substrate thickness of 120, 240 mm and a reference segment with a gravel layer.

From the curves the influence of the presence of the roof substrate on the course of temperatures in the test segments of the roof structure is clearly visible. The influence of the vegetation substrate and its thickness on the temperature maximum and its time shift during five days can be seen as well. The climatic parameters are shown in the lower part of Figure 7. The subtraction of temperatures under the substrates of thickness 120 and 240 mm with respect to the referential (gravel) roof is in the middle part of Figure 7.

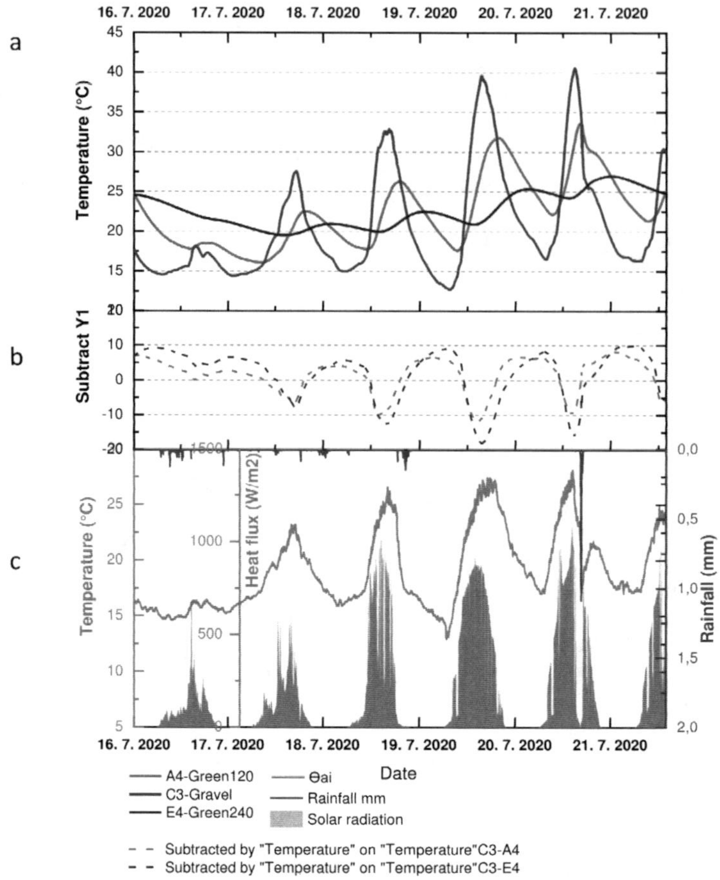

Figure 7. An example of the analysis of the measured data.

a - the course of measured temperatures under the substrate layer thickness 120 mm (green curve), gravel (purple curve) and substrate thickness 240 mm (blue curve),
b - mutual subtraction of measured temperatures (gravel / substrate 120 mm green intermittent and gravel / substrate 240 mm blue intermittent curve),
c - boundary conditions, climatic parameters during measurement.

The measured temperature profiles in the analyzed roof coverings show four characteristic time periods in the period 2020/21 (Figure 8).

The summer week after intense rain and during a long warm period without precipitation is documented in parts of graphs a) and b). The cold period without snowfall and the period immediately after snowfall can be monitored on the course of measured temperatures c) and d).

From the presented data it is possible to observe a significant influence of the vegetation layer on the temperature profiles in the roof cladding. The presence of a layer of substrate with vegetation protects the building from winter cold and summer heat. This "protective effect" is largely influenced by the concentration of water in the roof cladding. Rainwater in the retention layers of the roof increases its thermal storage capacity, which means that the roof is better able to alleviate the manifestations of summer overheating. The process of evapotranspiration in the vegetation layers reduces the temperature of the upper layers of the roof, which significantly affects the overall energy balance of the roof.

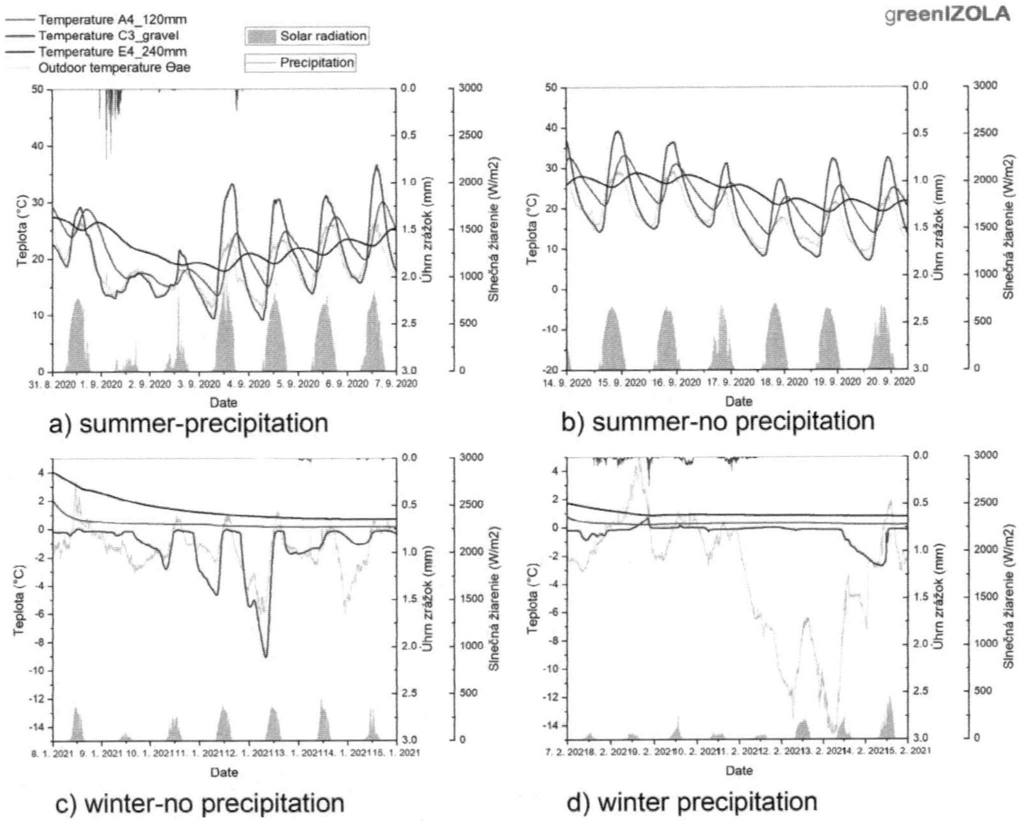

Figure 8. Examples of measured waveforms in the test segments of the roof cladding under the substrate layers 120 and 240 mm thick and under the gravel layer of the reference roof segment.

The increase of direct infiltration of rainwater will be ensured by the use of building structures with a vegetation layer. Transformations of building structures are proposed after a thorough analysis of technical limits, economic parameters and environmental impacts of the project. The greenIZOLA project envisages the construction of 1025 m^2 of vegetation roofs or shelters within the area of interest. According to the vision in Figure 9, the application of green solutions will increase the share of vegetation areas in the area of interest from the original **14% to 41%.**

a b

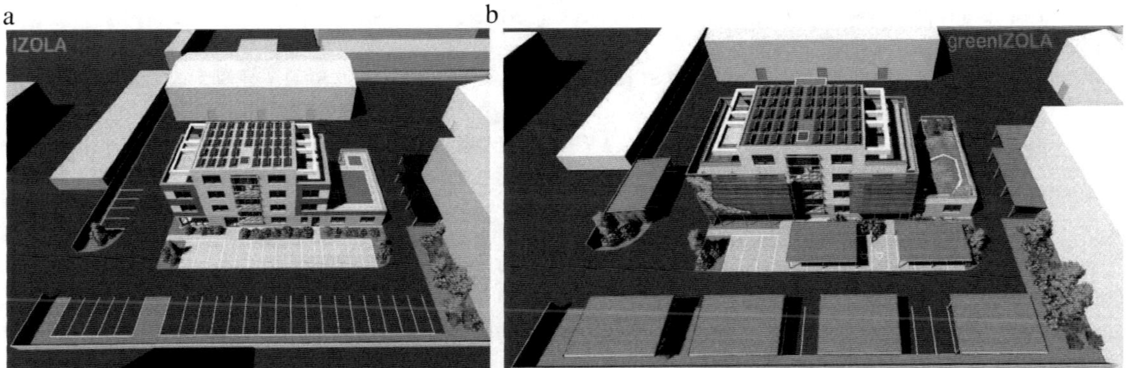

Figure 9. Our area of interest before a) and after b) the implementation of the greenIZOLA green conversion.

The paved areas in the area of interest formed by the asphalt road cover are in a technically satisfactory condition and their replacement does not represent a more fundamental economic or environmental benefit. Rainwater from asphalt roads and parking areas will be collected in shafts around the building, cleaned of oil substances and accumulated directly in the area of interest using infiltration blocks according to the scheme in Figure 10.

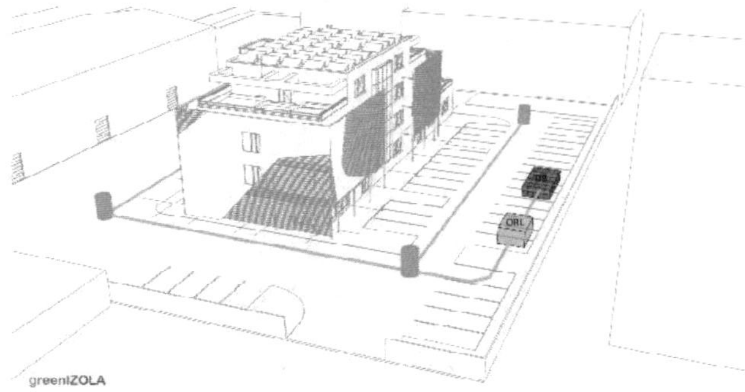

Figure 10. Drainage of paved areas of interest [10].

The retention potential of 39 areas intended for parking motor vehicles is ensured by the realization of vegetation shelters (Fig. 11). The proposed shelters protect vehicles from bad weather, but above all prevent summer overheating of vehicle interiors. The areas of vegetation shelters in car parks and emergency material storage reduce the non-absorbent asphalt area from the original 3308 m^2 to 2765 m^2 and turn it into a functional water retention element.

Figure 11. Visual representation of the proposed vegetation shelters [10].

The added value of vegetation shelters is the protection of motor vehicles against increasingly frequent weather extremes, such as hail. This damaged the company's fleet and employees' motor vehicles from the summer of 2019.

After the implementation of all individual phases, the greenIZOLA project will bring:

• increasing the share of vegetation areas in the area from the original 14 to 41%,
• increase of the amount of infiltrated rainwater in the area of interest by 94%,
• creation of 440 m^2 of new vegetation roofs or terraces,
• reduction of non-absorbent areas by using vegetation shelters with an area of 590 m^2,
• reducing the energy intensity of the building,
• increasing the quality of the indoor microclimate inside the building,
• increasing the attractiveness and aesthetic level of the building,
• creation of potential for the creation of 17 experimental roof vegetation segments and facade elements,
• providing shelter for students and young researchers.

BIM tools were used in the proposals for the renewal of the area of interest, including the administrative building, which also became the general platform for follow-up processes. From visualizations through multidisciplinary analyzes to interactive experimental mapping of the installed measurement infrastructure, the information is accumulated in a complex BIM model (see Figure 12).

Figure 12. BIM representation of greenIZOLA [10].

4. Conclusions

Information from in-situ measurements is an irreplaceable factor for the preparation and debugging of simulation models. Correct modeling of water accumulation and transport in the substrate appears to be key to the accuracy of the simulation calculation of a building structure with a vegetation layer [12], 13]. Simulation models make it possible to effectively investigate, design and optimize solutions for structures with a vegetation layer and analyse their impact on humans, buildings and the environment.

Our research of vegetation roofs is focused on testing vegetation roofs with not only biodiversity potential. The in-situ experiment analyses the temperature behaviour of experimental vegetation roofs with a roof substrate height of 120 and 240 mm and a referential roof with a conventional gravel finish. By long-term monitoring of temperatures in individual layers of the vegetation roof, it is possible to monitor its thermal-humidity response and analyse the effects of various factors.

In-situ experiment is defined as a multidisciplinary project, with emphasis on the technical aspects of the tested structures such as water retention, protection of buildings from overheating, thermal protection in the winter or the so-called effect of heat islands. In addition to the mentioned technical attributes, the project provides space for the implementation of biological, mathematical, sociological or economic analyses, the results of which will be presented in further studies.

Cooperation of the Faculty of Civil Engineering, TUKE with the company IZOLA Košice, s.r.o. resulted in the implementation of an experimental workplace called greenIZOLA, where the latest knowledge in the field of design of building structures with a vegetation layer is gradually applied in practice. This approach is extended by the experimental level, on which each designed and implemented building structure becomes an experimental system designed to verify its effectiveness and operation in real climatic conditions.

5. References

[1] Liu K K Y and Baskaran B A 2004 Green Roof Infrastructure - Technology Demonstration, *Monitoring and Market Expansion Project* (National Research Council of Canada)

[2] Zirkelbach D et al. 2017 A hygrothermal green roof model to simulate moisture and energy performance of building components. *Energy and Buildings* 145 pp 79-91

[3] Global Goal for Water: Synthesis of key findings and recommendations from UN-Water. Executive summary A Post 2015. [online] [cit. 2014-10-15] Available: http://www.unwater.org/fileadmin/user_upload/unwater_new/docs/Topics/UNWater_technica advicepost_2015_global_goal_ES_final_highres.pdf

[4] Tudiwer D et al. 2019 Illustration of the heat-insulating effect of a facade greening system in a simulation. *Bauphysik* 41(3) p 155

[5] Hart S 2011 EcoArchitecture the work of Ken Yeang Wiley. Italy ISBN 978 0 470 72140 7

[6] Poorova Z and Vranayova Z 2019 *Green Roofs and Water Retention in Košice, Slovakia* (Cham Springer Nature 2020) 133 p ISBN 978-3-030-24038-7

[7] Sapkota M, Arora M, Malano, H et al. 2015 An Overview of Hybrid Water Supply Systems in the Context of Urban Water Management: Challenges and Opportunities. *Water* vol. 7 pp 153- 174

[8] Yeang K and Spector A 2011. *Green design. From theory to practice* (Black dog publishing London) ISBN 978 1 907317 125

[9] Vranayova Z. et al., 2019. APVV-18-0360 Active hybrid infrastructure for a sponge city (In Slovak). Project application and reports

[10] Vargova A 2021 Refurbishment of Administrative building, Diploma thesis (TUKE Košice) 65 p. (In Slovak)

[11] Vranayova Z. et al., 2015. Stormwater Management in Compliance with Sustainable Design of Buildings. *Storm Water Management : Examples from Czech Republic, Slovakia and Poland.* (Cham Springer International Publishing) p 113-135 ISBN 978-3-319-25833-1

[12] Vertal M et al. 2020 Experimental vegetation roof with biodiversity potential *Roofs, facades, insulation* Vol 27 No. 11-12 Ostrava 2020 (In Slovak)

[13] Vertal M et al. 2018 Hygrothermal initial condition for simulation process of green building construction. *Energy and Buildings* 167 pp 166-176

[14] Macková D, Peráčková J 2021 Ensuring the Required Potable Water Temperature in Water Pipeline Inside Buildings. *Periodica Polytechnica Mechanical Engineering*, 65(4), pp. 353, https://doi.org/10.3311/PPme.18205

[15] Stanowska P, Dziopak J, Słyś D, Starzec M 2021 An innovative rainwater system as an effective alternative for cubature retention facilities. *Studia Geotechnica et Mechanica*, 43(s1) 532-547. https://doi.org/10.2478/sgem-2021-0037

[16] Zeleňáková M 2021 Urban Rainwater and Flood Management. *Water 13*, 974. https://doi.org/10.3390/w13070974

Acknowledgements
The project uses measuring components from AHLBORN, GmbH installed by AREKO, s.r.o.
The article was created with the support of the project APVV-18-0360 Active hybrid infrastructure for a sponge city.

IOP Conf. Series: Materials Science and Engineering 1252 (2022) 012065 doi:10.1088/1757-899X/1252/1/012065

The possibilities of using digital twin technology in the life cycle of buildings

Z Struková[1], M Kozlovská[1], A Tažiková[1] and J Čabala[1]

[1] Institute of Construction Technology, Economics and Management, Faculty of Civil Engineering, Technical University of Košice, Vysokoškolská 4, 042 00 Košice

E-mail: zuzana.strukova@tuke.sk; maria.kozlovska@tuke.sk; alena.tazikova@tuke.sk; jozef.cabala@tuke.sk

Abstract. The article deals with Digital twin (DT) technology, as one of the latest technologies, which, unlike construction, is already widely used in other industries. The theoretical knowledge basis of this technology in construction industry lies in understanding the characteristics of the digital twin and in using knowledge about it from other areas of application. Based on acquired theoretical knowledge and an example from construction practice, the aim of the article is to demonstrate the development process of a digital twin and to indicate the benefits of using Asset Information Model (AIM) as an essential source of information for development of DT during the phases of the life cycle of buildings. The case study analyses the process of development of a digital twin of the heating system in family house, from the initial design of a heating project, through use of 3D scanning to create a virtual model and the integration of various tools of Internet of Things technology to the creation of its virtual model. Reflexion on the benefits of AIM in various phases of the life cycle of buildings, induced from the study of theoretical knowledge and based on the case study, is an essential part of the article.

1. Introduction

The construction industry is one of the sectors that significantly affects the country's economic growth. However, at the same time, the construction has long struggled to achieve quality and higher productivity and efficiency [1]. With the consistent introduction of digital technologies, these problems have evidently begun to subside. Nevertheless, the speed of adoption of information and communication technologies (ICT) is an important aspect. While industries such as banking, manufacturing and services have long enjoyed the benefits of digitalisation, the construction industry is only very slowly adopting work with digital technologies. The implementation and use of ICT is considered to be the beginning of the fourth industrial revolution.

The increase in large volumes of data and the associated transfer of information were the impetus for the introduction of Industry 4.0, as this revolution is called. Not only in construction, it is actually a matter of connecting technologies that can connect the physical and digital worlds. Building Information Modelling (BIM), 3D scanning, augmented reality, drones, Electronic Data Exchange (EDI), Internet of Things (IoT), Cloud Computing etc. are among favourite ICT with the largest potential to improve efficiency and productivity of architectural, engineering and construction sector. The interaction of digital technologies with smart technologies seems to be eminent. One such technology, which is also an essential element of Industry 4.0, is the Digital Twin (DT) technology. This simulation-based planning and optimisation concept will potentially transform the construction industry just like other industries [2]. The investment in DT has the potential to increase productivity through predictive analytics and minimize the numerous challenges facing the construction industry [3]. However, the development of DT is gradually improving in other industries like the manufacturing and automotive industries, its development within the construction industry is still at its infancy. Even though, some applications of DT have been made in the different phases of the lifecycle of a construction project. The application of DT in the various lifecycle phases tackled a specific problem [4]. Among the applications of a digital twin, Khajavi et al. focused in their research on lowering maintenance cost, increasing tenants' comfort, and lowering the overall management and operational cost of a building. The aim of the article is to demonstrate the development process of a digital twin and to indicate the benefits of using a digital twin during the phases of the life cycle of buildings.

2. The Digital Twin technology – state-of-the art

Some people mistakenly believe that the Digital Twin (DT) is a replica/imitation of a real model. However, a distinction needs to be made between the digital model, simulation and digital twin. The concept of a digital twin is not a static model, but rather a system that is connected between a physical and a digital system. This connection can be achieved using techniques such as machine learning, analytics, streaming data from sensors, etc. So, it could be claimed that the DT is essentially a combination of these techniques [5], [6]. The most widely used and recognized definition of a digital twin is from 2012 by Glaessgen and Stargel: "A digital twin is an integrated multi-physical, multilevel, probabilistic simulation of a complex product and uses the best available physical models, sensor updates, etc. to reflect the life of matching twins." [7]. In essence, the digital twin can be defined as a constantly evolving model that is a virtual replica of the actual and historical behaviour of a physical object or process and ultimately helps companies achieve better success, whether the overall business or individual products, services, processes, etc. [8]. Michael Grieves and John Wickers have described the concept of the digital twin as a whole, which consists of three main parts: i) physical products in real space, ii) virtual products in virtual space, and iii) connected data that brings physical and virtual products together [9]. Setting up the right processes and methodology is very important, because the more information a DT contains, the higher its value. The essence of the DT should be the ability to answer "what if" questions. This can be achieved by simulation, which is therefore considered a key component of digital twin technology [8]. The continuous exchange of data between the virtual and physical worlds can only be ensured with the help of network devices and technologies such as IoT, Big Data, etc. which are extremely helpful in improving the quality of the DT. Thanks to them, the DT receives all available data from its physical twin so that it can analyse and re-evaluate the errors. Based on this, it can anticipate possible unpleasant scenarios, propose solutions for system maintenance and, last but not least, optimize changes. It then sends all this information back to the physical twin, which should respond to these suggestions. The physical twin then re-sends the already updated data to his digital twin. This cycle with regular data exchange takes place throughout the life cycle of the entity [10]. A digital twin can be created at any time during the life of the product. However, it reaches the best value when it is created as soon as possible, preferably right at the beginning, most ideally after the idea was born [11]. In order for a DT to be functional, it is necessary to perform the following six steps in its design and creation:

- Creating a virtual image of a physical product
- Data processing to facilitate the final design decision
- Simulation of product response in a virtual environment
- Instructing a physical product to perform the recommended functioning/behaviour
- Creating a two-way and safe connection between the physical and the virtual product in real time
- Collection of all types of product data from various sources

2.1. The Digital Twin in the construction industry

The market is quickly changing and there is a growing demand for ecological and energy efficient buildings. It is expected that the use of DT in will be equally useful for clients, suppliers, investors, etc. Therefore, the interest of building managers, building owners, and building operators should increase.

The DT mirrors the digital data of an entity. In case of a building, it provides information on each change that occurred, when it occurred and who reported the change, the problems that occurred, how, when and by whom they were addressed, all throughout the construction [12].

The basis of DT technology is the combination of key elements that lead to the digital transformation of buildings [13]:

- Big Data - used for a thorough analysis of the generated economic or operational information that buildings provide us. With this analysis we can achieve a better understanding or obtain the formula of how users behave or how they use the building.
- Internet of Things (IoT) - mobile devices, drones, sensors are important devices with which we achieve the interaction of the real and digital world.
- Artificial intelligence - allows us to achieve building optimization through human knowledge, the right analytical tools and machine learning strategies.
- Cloud based access - provides all parties involved with remote access to the application, which makes work more efficient, speeds up the processing of requests, and eliminates problems.

Despite the initial cost of developing a digital twin, it brings a number of benefits. Data from one digital twin can be used to create new applications without greater costs. For example, by sensing the indoor environment and monitoring whether a room is in use or not, the digital twin allows the building user to turn the air conditioning on or off, thus contributing to energy reduction.

The most significant benefits of DT technology in construction include mainly reducing construction and operating costs, increasing productivity and collaboration, improving occupational safety, and optimizing performance and revenue sustainability [11]. We can store and update information such as project documentation, time plans, material specifications, etc. that the building acquires during its life cycle. If necessary, we can use them in the implementation of other projects and thus prevent possible risks. By transmitting construction data in real time, we can warn workers in advance about areas that are dangerous and the associated potential threats and risks. In case of an emergency, they will be provided with relevant health and safety instructions. Thanks to the monitoring and analysis of the collection of operational and working data in real time we obtain the necessary information on the benefits of the construction and how it is used. From this we can determine where the highest risk occurs during maintenance and also how the construction or part of it will be affected if any of the data changes.

The most common applications of DT technology in construction are presented in Table 1 [7].

Table 1. Applications of DT technology in construction industry.

Scope	Demonstration
Lifecycle management of construction	Transparency and cooperation between project participants (owner, architect, engineer, contractor) in individual phases of the project
Big data	Big data allows the analysis of the necessary data among the amount of information collected during the entire life cycle of the project. On the platform, we can simulate, optimize, verify information, which simplifies the way we work and do processes more effective
Real-time monitoring	Using algorithms, it allows to monitor cracks and material displacement and thus prevent a possible accident. It also monitors the behaviour of construction workers and can send warning signals if they find themselves in a danger zone.
Performance decision/prediction	Knowledge from the analysis of stored data allows to make better decisions and based on constantly up-to-date data we can also predict performance

Efficiency	It is important to deliver project on time. When designing, it is possible to achieve it by taking into account various architectural and conceptual aspects, by making the necessary changes before implementation. During implementation, it monitors the work to see if it is made to specifications and timeline and take corrective action based on the digital model as needed. Efficiency is also achieved in predicting resource requirements in monitoring material movement and equipment utilization. During the operation of the building, it is possible to anticipate possible errors and thus schedule maintenance.
Improved cyber-physical systems	The systems enable two-way flow of information and work closely between the physical product and the virtual one.

2.2. The Building Information Model vs. Digital Twin

The Building Information Modeling (BIM) and the DT may seem to be the same. However, the difference between the technologies lies mainly in the following aspects [12]:

- BIM is designed primarily for the design of buildings/constructions
- BIM is not designed to achieve real-time feedback
- BIM is more focused on buildings/constructions than on the human factor
- DT provides information on subsystems such as air conditioning, heating or lighting and how this may affect the consumer/customer
- DT evolves continuously with each phase in the life cycle and thus brings higher value
- It is possible that DT will replace BIM in the design phase in the future

The BIM with well-managed data is ideal for creating a digital building/structure that then achieves dynamic performance through simulation. In case that any changes occur during design, implementation, operation, reconstruction or demolition, DT will immediately realize, react and adjust the differences. This accurate information will help making better decisions and efficiency. DT can analyse changes in energy supply and draw attention to the need for repairs or modernization. If we compare the BIM and the DT of a building/structure according to who uses them, what they are aimed at and what technology they support, we can conclude that BIM is mostly used in the design, to improve communication between participants, cost control, monitoring the time schedule, etc. Unlike BIM, the DT focuses on optimization through "what if" simulation, resource efficiency, anticipating maintenance time and mainly focuses on the people who will use the building/construction and their comfort. The main difference, however, is that the digital twin works with the data it obtains in real life time [14].

3. The creation of the digital twin of the heating system in the family house – case study

In order to take benefits of the digital twin technology or what this technology can provide, it is necessary to create such a digital twin, whether it is a DT of a whole building/construction or of its structure. A great advantage is the beginning of its creation immediately after the agreement with the investor, already during the processing of the initial documentation of a construction project. The study analyses the creation of the digital twin of the heating system in the family house in Slovakia. The digital twin was created by the company Klima-Teplo designing, s.r.o. which deals with the design of affordable and environmentally friendly solutions for heating systems, air handling, air conditioning, sanitary installations and general technical equipment of various types of buildings.

3.1. Basic information on the design of the heating system

The design of the heating and air conditioning system was prepared on the basis of a mutual agreement of the initial requirements of the investor and the recommendations of the company that made the design. The design of the heating system consists in the calculation of heat losses and balances, in the design of the heat source and the selection of a suitable heating system. The design of air conditioning system includes data such as air volume flow, inlet air temperature in winter and outlet air temperature in summer. The heating system designed for the building consists of the floor heating system PYD ALU® FLOOR and a ground-to-water heat pump from Alpha Innotec. The PYD ALU® FLOOR heating system is specific; it is an aluminium sheet, the shape of which is in the form of small pyramids. The combination of aluminium, which ensures excellent thermal conductivity and the shape of the pyramids, guarantees fast heat transfer.

3.2. The digital model of the heating system

The design of the heating system was preceded by the calculation of heat losses and heat load. These calculations were performed in the design company on the basis of manual entry of all necessary data, i.e. data from what material are the structures, dimensions of the building and individual rooms, which were obtained from the documents from the investor of the building. Then, the creation of a digital model of heating system by Revit® software was conducted (Figure 1). It consisted in modeling the individual elements of the PYD ALU® FLOOR system, including manifolds and a heat pump, which complete the overall system. Since this technology is new and does not yet have a library of 3D elements created by the manufacturer, it was necessary to create them first. In this phase, the 3D model of heating system made in this way is usually compared with projects of other professions to detect possible collisions. All this documentation is the basis for the implementation project, i.e. to create a physical work from which using other technologies, we get a virtual heating model.

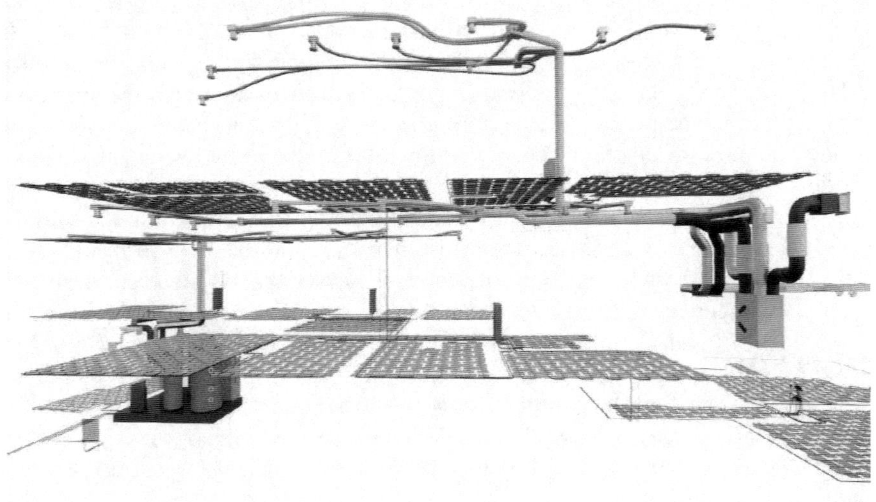

Figure 1. The visualization of 3D model of the heating system. [source: Klima-Teplo designing s.r.o.]

3.3. Construction of the physical model of the heating system at site

During the implementation of heating system at the site, it is necessary to constantly check the work performed, as even with any precise planning, problems can occur anywhere and at any time. In case of larger construction projects, it is possible to use technologies such as virtual reality and augmented reality to get a more realistic view of how it is going, whether everything that has been built is in the right place and whether there are any serious changes. Any change in the construction of the heating system during implementation, whether it is a change in the material used or a change in dimensions or a change in any component documentation used, must be recorded and all these changes must then be updated to the original digital model. After the implementation of the entire heating project, we get a physical model of heating, which forms one part of the digital twin. The subsequent capture of the actual state of the implemented heating project, together with the incorporation of all possible changes and modifications, we get a true virtual model that will represent the actual physical construction/structure. This represents the second part of the digital twin.

3.4. The virtual model of the heating system

To record the exact view of the building, 3D scanning is used for data collection, which consists in creating a point cloud. The family house including the in-built heating system is scanned by the technology of terrestrial laser scanning. Before scanning, it is necessary to decide on a reasonable number of positions, i.e. scanning positions so that individual scans capture the entire building. Unlike the scanning positions around the building with wider mutual distances due to fewer obstacles, i.e. one position can capture a larger scanning area, the scanning positions in the main part of the building are closer, because it is necessary to capture and scan all possible details of the object. This is achieved only by successive moving the scanner from one position to another until the desired result is achieved. After the final scanning of the whole family house, the resulting connection of point clouds is cleaned. All unnecessary data such as cars, people, various materials, etc. are cut out of the point cloud. Then, the cleared point cloud is optimized by adding colors to various elements. This color resolution is especially important in the technical support of buildings. By such a cleaned and optimized cloud points the virtual model is developed. The virtual model is expected to represent the real physical model of the building including the heating system in the virtual environment.

3.5. Steps to link the physical and the virtual model

The essence of a digital twin, or without which a digital twin would not be a digital twin, is to link a virtual and a physical model by receiving various data from the physical model and then real-time sending that virtual model data. After constructing the physical model of the building and after developing the virtual model that reflects the reality of the heating system, it is possible to proceed to the implementation of a measurement and control system using various sensors and IoT technologies. The sensors that are implemented in the case of the heating system include temperature sensors, humidity, pressure and motion detection sensors. Temperature sensors are used to sense the temperature in individual rooms or outdoor temperature, a pressure sensor to measure the tightness of joints, or to identify pressure drops or fluctuations, a humidity sensor similar to a temperature sensor and a motion detection sensor, to determine whether someone is in rooms or not. The information that will be continuously obtained from these sensors will be converted into data and then sent to the virtual model via a cloud system. It is therefore necessary to implement so many smart meters that the result we require or want to achieve results in an optimal smart solution of data collection. The virtual model is expected to analyze the data collected through machine learning technology and to evaluate the acquired knowledge or situations, how the physical product reacts to various changes (influence of ambient air, temperature fluctuations, ...) and if it is possible to influence the improvement of operation, it will propose the optimal solution to achieve this and forward this control optimization proposal to the physical product.

4. The role and benefits of Asset Information Models in the building lifecycle

Based on the previous case study, the digital twin, or especially Asset Information Model, as the source of information for DT, of the heating system can be expected to capture all important operational data and provide an idea or option that will then help decide whether to improve further designs, whether necessary adjustments or future developments. However, it is important to ensure that any changes to data or information that occur during any phase of the building life cycle are documented on an ongoing basis and information about them is entered into the virtual construction model.

Asset Information Model (AIM) supports the creation of a DT. Digital asset information models such as BIM models have been broadly embraced within the design and construction phases. Regardless of this, the adoption of BIM within the maintenance and operation phase has been rather limited. In the PAS 1192-3:2013 standard by British Standard Institution, an Asset Information Model is defined as data and information that relates to asset to a level required to support an organization's asset management system. The scope of a DT is much broader than the scope of an AIM. Moreover, the Project Information Model (PIM) is in the PAS 1192-2:2013 described as the information model developed during the design and construction phase of a project. The PIM is developed progressively, first as a design intent model then a virtual construction model. Once the construction is complete, the PIM is developed into an AIM to be used during the operational phase. In fact, an AIM provides all the data and information that relates to or are required for the operation of an asset. It can provide graphical and non-graphical data and information, documents and metadata, such as information describing the original design intent, 3D models, operational performance information, condition information, etc.

4.1. Usability of the Project Information Model of heating system in the planning phase
The potential use of a Project Information Model (PIM) of heating system, made for one building, in the planning phase of a heating project for other buildings lies in the simulation and consistent design of an optimal solution. It all depends on the input data that is required and important when designing new projects. The required inputs may vary in various ways, which may occur by changing a single element or by combining multiple inputs. Input data that may change in the planning phase are summarized in Table 2.

Table 2. The variable input data of a PIM.

Input data	Explanation
External climatic conditions	External climatic conditions greatly influence the design of the heating system. The change in this temperature data will also change the calculation of heat losses and thus the need for power input for heating. By the simulation, it can be assessed whether the originally designed elements in the PIM meet the requirements of the heating system design or not. If the simulation shows that the design is not satisfactory, the PIM should propose an optimal solution that still needs to be changed in order to achieve the correct heating design for the construction in the given location.
Building orientation and layout solution	Changing the orientation of the building to the world sides and changing the layout of the building can significantly affect the design of the heating system. The layout solution is part of the design of the building and not the design of the heating system, but how the building is designed actually affects the design of the heating system itself. Similarly, when the house has a layout solution identical to the virtual model, but the layout of the building is oriented to another side of the world, the new data is inserted into a virtual model of the PIM, a simulation is performed and a solution is proposed.
Material and structures used and their dimensions	Heat losses due to heat transfer are of course also influenced by the material and construction used. The use of a before developed PIM consists in entering changes in the data of the material used as well as changes in wall and insulation thickness. Of course, it is also necessary to take into account the share of glass/windows in the new building. Only a small change in one of these parameters will affect the final design of a new heating system. As a result, not only the data on the material used or its thickness will change, but also information such as technical sheets or the service life of individual structures will be added for the needs of the updated virtual model.

Sizes of rooms	In any project, it is enough to change only the construction height of the floor and we can say that it is a completely different project. Although the clear height does not directly affect the underfloor heating, it affects the temperature losses through ventilation, which in turn affects the total designed heat input and thus the design of a pump. By simulation in a virtual model of a PIM based on a change (clear height of the floor or the area of rooms) we get optimal design needed for such a new construction.
Type and dimensions of elements	The DT dealt with in Chapter 3 can also be used in the case of another identical family house with the same layout, located in the same location, with the same orientation to the world sides, but the difference may lie in the size of the land, related to the design of another type of pump. In this case, it is necessary to use a virtual model of the heating system to simulate whether a change to a new pump is sufficient to achieve all the necessary values. If not, we will achieve the necessary design by simulation, either in the form of the design of other dimensions of elements, or in the need for several quantities of individual elements.

4.2. Usability of the Project Information Model of heating system in the construction phase

The Project Information Model can be used in the construction phase in:

- Construction time scheduling
- Structure constructing into building
- Ensuring safety and health of workers

If the construction schedule is linked to the construction model, it provides a time-space construction model, providing information not only about the time but also about the space where the individual activities on the construction site are performed. During construction, the PIM can collect data on the progress of construction through various camera systems. It does not only reveal the delay, but it also uses various algorithms to design the required number of workers so that the deadline can be met. The whole calculation will be performed with respect to the input data that will be part of the plan, with respect to the maximum number of workers so as not to interfere with each other in the various parallel activities. The PIM can also evaluate the increase in staff financially, as the costs incurred should be part of it. The digital twin can monitor the real conditions on the construction site using humidity sensors. It analyzes and evaluates these data and, if they are in accordance with the limit data from technological regulations, it can propose a solution, for example in the form of postponing the implementation date to another date. Or, based on the analysis, it finds out that the surrounding moisture may have affected the material in some way, it draws attention to this fact and thus prevents the devaluation of the structure. Based on monitoring of workers as well as machines and equipment on the construction site using various proximity sensors and various wireless accessories such as bracelets, the digital twin can check where the workers are and whether there are unexpected situations. Based on the real data, the digital twin may evaluate such a situation and automatically send a signal of a danger.

4.3. Usability of Asset Information Model of heating system in the maintenance and operation phase

The main use of digital twin technology is in the operation of buildings and their maintenance. For example, a digital twin of heating system is expected to turn the temperature when somebody entered the room even if the owner has set a lower room temperature as he assumed that no one will be there at a given time. The digital twin also analyzes and records this situation and, in the event of such a situation recurring, it automatically adjusts the temperature or sends a message to the building owner. At the same time, the digital twin also senses the outdoor temperature, it analyzes the measured data, evaluates and proposes optimal solutions. The digital twin technology greatly influences the regulation of temperatures in the building. If necessary, the digital twin will also design a control for the heat pump.

The reliable Asset Information Model includes information on the service life of individual elements, as well as information on the method of maintenance and possible replacement of elements. It continuously monitors the time and collects data on how the individual built-in elements are used, for example whether they are not overloaded. If it discovers any anomalies, it proposes a recommendation that the element needs to be replaced due to its premature wear. At the same time, it also draws attention to the dates of regular revision tests and professional inspections of technical equipment. Of course, all these changes need to be recorded and stored in the digital twin so that we have up-to-date information.

4.4. Usability of the digital twin of heating system in the renovation/demolition phase

In this phase, especially information on the service life of individual elements of the heating system is useful. Before the expiration of the warranty period, the digital twin notifies and informs the owner and operator about the current functionality of the elements. If this functionality is insufficient, it is necessary to count on repair or replacement of any of the components. At the same time, it will also propose the method of disposal of these elements in accordance with the applicable regulations. If the digital twin is used for renovation, it is necessary to record all changes, whether dispositional or material. However, if all heating elements are degraded at the same time, the digital twin itself will be destroyed.

5. Conclusion

The rapid adaptation of digital technologies relative to Industry 4.0 benefits the construction industry and their profit is becoming more pronounced and beneficial for all construction participants and for building users. It is important to clarify and explain the technology of the DT, as some of the professional public confuses this term with a virtual representation of a product or service. It is basically a digital representation of an entity, but what the digital twin technology offers is its benefit by combining several different technologies that are commonly used in practice. Thus, the essence and the main difference is in the connection of physical and virtual models through such tools that would allow monitoring the functionality and properties of the physical model and then be able to send this information to the virtual model. For such a connection, the digital twin uses IoT technology, which can monitor and transmit data in real time. Before we focus on creating a digital twin of a building, it is necessary to create digital twins of individual structures. Already these smaller DTs can evaluate the situation and recommend the proposal of new optimal solutions. These solutions or designs are then analysed in conjunction with other digital twins and eventually create one larger digital twin building.

The case study of the DT of heating system has demonstrated the process of the DT creation as we do not know what DT contains, we will not be able to use its benefits in practice. The main benefit of the DT lies in the use of collected real-time information and in the proposal of recommendations for more efficient design, construction and use of similar buildings. The role and benefits of Asset Information Models in the building lifecycle have been discussed in the closing part of the paper. The AIM handles as a single source of validated and approved information that relates to a built asset and is used during the operational phase of a building. It is a term that may relate to a single asset, a system of combined assets, or an organisation's entire portfolio of assets.

The development of systems and components is moving forward, and therefore what we consider to be an advantageous way of heating today may not be effective in a few years. It is all about changing climate conditions and also with regard to renewables. One can only assume that DT technology could help in the design of other buildings, especially by reducing the workload by simply simulating various designs, or by ensuring the protection of safety and health of workers during work on the construction site or achieving an efficient supply of building materials.

6. References

[1] Greco M, Grimaldi M, Locatelli and Serafini M 2021 *How does open innovation enhance productivity? An exploration in the construction ecosystem, Technological Forecasting and Social Change* **168** 120740

[2] Tao F, Zhang H, Liu A and Nee A.Y.C. 2018 *Digital Twin in Industry: State-of-the-Art, IEEE Transactions on Industrial Informatics* **15**(4) 2405-2415

[3] Lee J, Lapira E and Kao Hung-An 2013 *Recent advances and trends in predictive manufacturing systems in big data environment, Manufacturing letters* **1**(1) 38-41

[4] Opoku D J, Perera S, Osei-Kyei R and Rashidi M 2021 *Digital twin application in the construction industry: A literature review, Journal of Building Engineering* **40** 102726
[5] ARUP *Digital twin: towards a meaningful framework*. [Online]
[6] Johnson Controls *Applying Digital Twins to the Built Environment* [Online]
[7] Kan C and Anumba C J 2019 *Digital Twins as the Next Phase of Cyber-Physical Systems in Construction, Computing in Civil Engineering 2019: Data, Sensing, and Analytics*, Atlanta, GA
[8] Parrott A and Warshaw L 2017 *Industry 4.0 and the digital twin,* Deloitte Development LLC, Deloitte University Press
[9] Pradeep N, Venkatesan and N Vivek D *Is Your Organization Ready to Embrace a Digital Twin* [Online]
[10] Barricelli B R, Casiraghi E and Fogli D 2019 *A Survey on Digital Twin: Definitions, Characteristics, Applications, and Design Implications*, IEEE Access 7 167653-167671
[11] Evans S, Savian C, Burns A and Cooper Ch 2019 *Digital Twins for the built environment*, The Institution of Engineering and Technology [Online]
[12] Mateev M 2020 *INDUSTRY 4.0 AND THE DIGITAL TWIN FOR BUILDING INDUSTRY, Industry 4.0* **5** 29-32
[13] Siemens Switzerland Ltd 2018. *Digital twin – Driving business value throughout the building life* cycle [Online]
[14] Khajavi S H, Mothlagh N H, Jaribion A, Werner L and Holmstrom J 2019 Digital Twin: Vision, Benefits, Boundaries, and Creation for Buildings, IEEE Access 7 147406-147419

Acknowledgments
The paper presents a partial research results of project VEGA 1/0336/22 “Research on the effects of Lean Production/Lean Construction methods on increasing the efficiency of on-site and off-site construction technologies”. The article presented a partial research result of projects: KEGA No. 038TUKE-4/2020 “Supporting the platform Construction 4.0 through adoption of digital technologies in the learning process”.

Construction segments applying modern methods of construction in Slovakia

M Spišáková[1], J Galla[1] and T Mandičák[1]

[1] Institute of Technology, Economics and Management in Construction, Faculty of Civil Engineering, Technical University of Košice, Vysokoškolská 4, 042 00 Košice, Slovakia

E-mail: marcela.spisakova@tuke.sk; jozef.galla@student.tuke.sk; tomas.mandicak@tuke.sk

Abstract. Construction industry has a huge innovation potential focused on the development of new building elements or process procedures. Progressive construction processes are often referred to as modern and cover technologies and systems of various material bases. The drivers and barriers associated with modern methods of construction are generally perceived as a perspective of ensuring sustainable construction, which also reflects their potential application in various construction segments. The aim of the paper is to map companies in the Slovak construction market, which focus their activities on modern methods of construction and then identify the material basis of modern methods of construction. The paper is also focused on determining the application measure of individual types of modern methods of construction in specific construction segments. The research was carried out through own survey and identification of construction companies that declare their activities in the field of production of structural elements, construction of buildings within modern methods of construction.

1. Introduction

The construction industry presents one of the most significant industries and provides the infrastructures for economic growth [1] and economic pillar in many countries. An innovation in architecture engineering and construction (AEC) industry have devoted the introduction of new products, processes or services. AEC suffers from labor shortages, low productivity and increased demands for sustainability [2]. These barriers to construction can be minimized by applying modern approaches to production of new products, processes or services. Modern methods of construction (MMC) can be defined as a method in which less or greater extent building blocks (walls, floor slabs, beams, columns and staircases) are mass-produced in the factory under strict supervision with high quality [3]. Other authors [4] argued that the MMC is a continuous system utilization craftsman at each construction activity that uses a manufacturing production to minimize wastage of resources and increased value for end users. The research focused on the perception of the modern methods of construction using in Slovakia has been pointed out the benefits and barriers of MMC in Slovak construction industry and markets [5]. Research has shown (see table 1) that modern approaches can eliminate the disadvantages of construction. On the other side, there is still a distrust in the use of MMC by users as well as contractors. MMC can also be sorted from the material point of view. A range of materials is used for MMC, the most common being wood, steel and concrete [6]. The aspect of construction material was identified as the most important for investors [7].

IOP Conf. Series: Materials Science and Engineering 1252 (2022) 012066 doi:10.1088/1757-899X/1252/1/012066

The MMC usability is possible in various construction segments. Sorting of buildings is possible from several points of view, e.g. size, function, construction, style, design etc. The International Building Code (IBC) and Uniform Building Code (UBC) categorize buildings based on occupancy and use [8]. Criteria from IBC and UBC are logical since they control the design and construction of structures. Ten types of buildings are identified according to the IBC and UBC categorization of buildings: assembly buildings, business buildings, educational buildings, factory and industrial buildings, high hazard buildings, institutional buildings, mercantile buildings, residential buildings, storage buildings, utility and miscellaneous buildings. Slovak legislation classifies buildings through the Building Act [9] as follows: land buildings (residential buildings and non-residential buildings) and engineering buildings. Currently used classification created on the basis of the European standard is Classification of buildings which is used by Slovak statistical office. Like Building act, Classification of buildings divides buildings into buildings and engineering buildings.

Table 1. Benefits and barriers perception of MMC using in Slovakia.

	Rate	Attributes
	1	Reduction on site works
	2	Minimal wastage
	3	Less site materials
	4	Cleaner environment
	5	Safer construction sites
	6	Faster project completion
Benefit	7	Controlled quality
	1	Higher cost
	2	Problem with joints
	3	Non acceptance by client
	4	Problems in design alternation during or after construction
	5	Limited suppliers of prefabricated elements
	6	Preference of traditional methods of construction
Barriers	7	Lack of expertise in implementation and installation

Modern methods of construction are most often used for the design, implementation and occupancy of land buildings [5]. The classification of land buildings was determined for the purpose of a systematic approach to the research solution. Various segments of land buildings were defined in terms of use (see Table 2).

Table 2. Classification of land buildings.

residential buildings	non-residential buildings	
	service buildings	industrial buildings
family houses	housing facilities for short-term stays (hotels, etc.)	industrial buildings and warehouses
flat houses	buildings for administration, (banks, post offices, etc.)	agricultural buildings
other buildings for housing	buildings for education system (schools, etc.)	warehouses, logistics buildings
	buildings for culture and public entertainment (museums, libraries, etc.)	other non-residential buildings
	buildings for shops and services (shopping centres, restaurants, etc.)	
	hospitals, health and social facilities	
	transport and telecommunication buildings (stations, garages, etc.)	
	covered building for sport	

2. Material and methodology of research

The aim of the paper is to map companies in the Slovak construction market, which focus their activities on modern methods of construction and then identify the material basis of modern methods of construction. The paper is also focused on determining the application measure of individual types of modern methods of construction in specific construction segments.

2.1. Research material

Research examines methods modern of construction that are available on the Slovak construction market. Four basic MMC construction systems were analyzed: blockwork system, frame and beam system, panel system, volumetric system (figure 1). Three basic groups of MMC were identified, divided according to the material base - wood, steel and concrete composite.

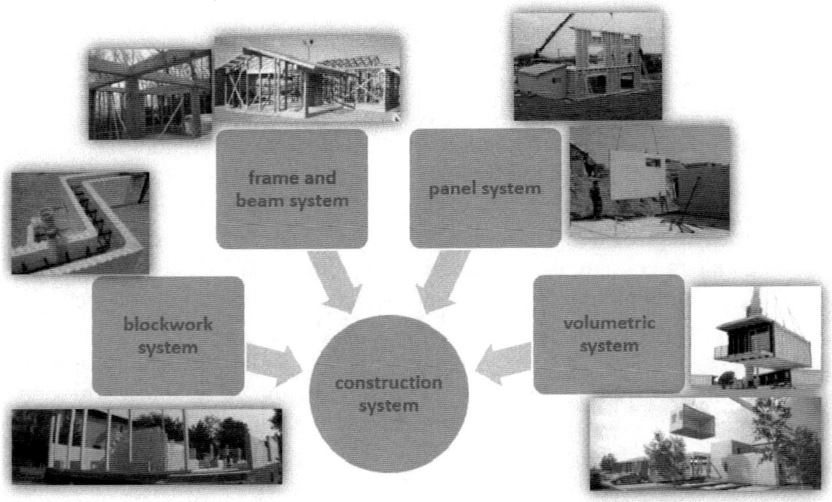

Figure 1. Construction systems of modern methods of construction.

The modern methods of construction are used mainly for the implementation of land buildings. Therefore, the research was carried out only for this construction segment.

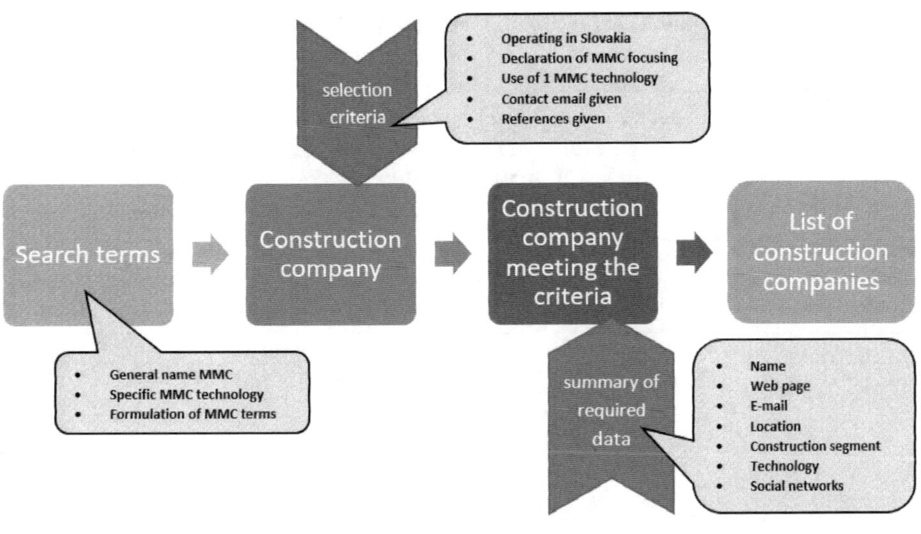

Figure 2. Flow chart of the process of identification of construction companies.

2.2. Research methodology

The mapping of construction companies devoting with modern methods of construction in Slovakia was carried out through the desktop study of companies on the Slovak construction market. The desktop study was conducted through a survey of available information that construction companies presented on the internet. Procedural steps of the process of identification of construction companies are in a flow chart (see figure 2). The result of this research part is list of construction companies devote with modern methods of construction in Slovakia, then the research was extended by the material basis of MMC technologies and types of constructions.

Before the survey of construction companies, the following hypotheses were expressed:
H1: *More than 50% of construction companies focus on modern methods of construction wood-based.*
H2: More than 80% of construction companies focus on modern methods of construction of one material base.

3. Results and discussion

The result of the survey was the identification of 174 companies on Slovak construction market that focus their activities on modern methods of construction - their design, implementation, import or assembly. The companies were summarized on the basis of selection criteria (see figure 2). Subsequently research on these companies determined the material basis and the target segment of buildings.
We can say that the operation of construction companies in Slovakia is not evenly distributed. Most construction companies focused on MMC (86 out of 174) are located in western Slovakia, 58 companies are in central Slovakia and 46 companies operate in eastern Slovakia.

The operation of 174 construction companies in Slovakia reflects the wide representation of the industry of modern construction methods, regardless of the material basis of technologies or systems. The resulting list of 174 construction companies can be divided into two groups according to the material focus of the companies as follows. The first group are construction companies, which have their focus and offer of technologies or construction systems, only on one material basis, e.g. they focus only on buildings made of steel-based technologies. The second group are construction companies that have their focus and offer technologies or construction systems on several material bases, e.g. they focus on buildings made of wood-based technologies but also steel-based or concrete composite-based. Up to 60% of all construction companies dealing with modern methods of construction focus on MMC wood based (110 companies). In addition, up to 102 of them implement only this material base. The next group are MMC steel-based which are focused on 58 construction companies. The 49 companies focus exclusively on the steel base MMC. The material base of the concrete composite has the smallest share, which is the focus of 16 companies (see Table 3.). The research confirmed hypothesis H1: "*More than 50% of construction companies focus on modern methods of construction wood-based.*". Wood is a renewable building material available in Slovakia. Its use takes into account the requirements of sustainability in construction. The hypothesis H2 "*More than 80% of construction companies focus on modern methods of construction of one material base.*" was also confirmed. Changing the material base of a construction company's profile is a demanding process. The individual technologies differ significantly from each other. Focusing the company on only one material base is easier and more efficient for construction companies.

Table 3. Material basis of technologies or systems.

construction company				construction company focusing	
type of material base (MB)	number	share of MB in 100%		1 type of MB number/share (%)	1 type of MB number/share (%)
wood based	110	60%	→	102 / 93%	8 / 7%
steel based	58	31%	→	49 / 84%	9 / 16%
concrete composite based	16	9%	→	13 / 81%	3 / 19%

The construction segments were determined according to the classification of land buildings in Table 2.

Modern methods of construction, regardless of the material base, have the largest representation in the residential segment of buildings and specifically in family houses. According to the survey systems (see Table 4.), 84% (146 out of 174) construction companies focused its activities on the family houses sector. The second building type is the non-residential segment, where construction companies have their focus mainly on service buildings - 106 construction companies. 63 companies focus on other non-residential buildings (mainly buildings like garages, gazebos).

Table 4. Number of construction companies in terms on building segments and material base.

type of material base (MB)	construction segment					
	residential buildings			non-residential buildings		
	family houses	flat houses	other buildings for housing	service buildings	industrial buildings	other buildings
regardless of MB	146	53	28	106	47	63
wood based	109	29	15	50	10	43
steel based	32	21	13	51	39	19
concrete composite based	15	8	4	13	2	5

In generally, residential buildings are built by modern methods of construction based on wood. Implementation of non-residential buildings prefers MMC based on steel or concrete composite. It should be noted that some construction companies do not have a precisely defined target segment of their activities. On the other hand, many companies focus their activities on the design, implementation, import or assembly of wooden family houses. As many as 109 companies out of 174 focused on the construction of wooden family houses. Based on the research, we can say that construction companies devoting on MMC in Slovakia focus their construction activities mainly on the material base of wood in the segment of residential buildings - family houses.

4. Conclusion

The development of new building materials and composites, as well as the basic requirements for buildings, increasingly taking into account their energy and environmental aspects, also determine the development of new construction methods and systems. These methods must take account in addition to aesthetic, architecturally diverse but also socio-economic requirements and customer limitations. The application of new construction methods is in line with the sustainable trend in construction together with increasing productivity and quality of construction. The presented survey confirmed the acceptance of MMC on the Slovak construction market - 174 construction companies dealing with MMC were identified. However, it should be noted that modern methods of construction are applied and adopted mainly in the residential buildings segment, less in the non-residential buildings segment. From the material point of view, construction companies design, implement, import or assemble mainly wood-based MMC. Wood is a renewable building material available in Slovakia. Its use takes into account the requirements of sustainability in construction. 110 construction companies in Slovakia focus on wood-based MMC. This represents 60% of all companies dealing with modern technologies. At the same time, 84% (146 out of 174) construction companies focused its activities on the family houses sector.

5. References

[1] Maki T and Kerosuo H 2015 *Site managers'daily work and the uses of building information modelling in construction site management*, *Construction Management and Economics*, **33** 163-175

[2] Noktehdan M, Shahbazpor M and Wilkinson S 2015 *Driving innovative thinking in the New Zealand construction industry, Buildings*, **5** 297-309
[3] Trikha J 1999 *Industrilised building system Proseeding of world engineering congress* (Kuala Lumpur)
[4] Chen Y et al. 2009 *Sustainable performance criteria for construction method selection in concrete buildings, Automation in Construction,* **19** 235-244
[5] Kozlovská M, Spišáková M and Mačková D 2017 *Modern methods of construction – towards adoption in Slovakia* (Brusel: EuroScientia vzw) p 219
[6] *Modern methods of house building*, 2003, Postnote, United Kingdom
[7] Kozlovská M 2014 *Comparison of low-energy houses selected parameters made by traditional and modern methods of construction, Advance Materials Research*, **1041** 51-54
[8] International building code 2018 pp 757
[9] Act On Land-use Planning and Building Order (Building Act) 50/1976 Coll.

Acknowledgments
This research was supported by the Slovak Scientific Grant Agency (VEGA) under Grant number 1/0336/22: "Research on the effects of Lean Production/Lean Construction methods on increasing the efficiency of on-site and off-site construction technologies" and KEGA 009TUKE-4/2022: "An interactive tool for designing a safe construction site in an immersive virtual reality".

Analysis of stress concentration at the edge of hole in plates with different widths by using FEM

B Safaei[1,2], Z Pezeshki[1], K Kotrasova[3] and E Kormanikova[3]
[1] Department of Mechanical Engineering, Eastern Mediterranean University, Famagusta, North Cyprus via Mersin 10, Turkey
[2] Department of Mechanical Engineering Science, University of Johannesburg, Gauteng 2006, South Africa
[3] Institute of Structural Engineering, Faculty of Civil Engineering, Technical University of Kosice, Vysokoskolska 4, 040 01 Kosice, Slovakia

E-mail: babak.safaei@emu.edu.tr

Abstract. The aim of this work was to investigate the concentration of stress due to circular cut out in sheets under tensile loads. Geometrical changes and loading create stress concentration. Under this condition, stress has increased in the regions of geometrical changes and if its value exceeds yield stress, the material fails in these regions. In this work, the distribution of stress in sheets with circular cut out was numerically investigated to study stress concentration at the edge of hole and the obtained results were compared with analytical solutions. In addition, relationships among stress concentration factors and relative dimensions of hole were determined using the equations acquired from elasticity theory. It is worth mentioning that the obtained results proved the validity of numerical analysis of all parameters by ANSYS.

1. Introduction

Stress concentration is defined as the distribution of stress for all categories of axial, bending, and shear stress due to geometrical irregularities and material structural components such as fillets, holes, grooves, keyways, spline, and scrapes [1]. Stress concentration due to flaws or impurities of brittle material cannot bring about dire consequence of fraction. Contrary fatigue and impact loading are able to leave the irreparable influence on structures and spark off dire damage [2]. Investigators have studied stress distribution in components through experimental methods. Recently, researchers have shown increased interest in studying stress concentration by extracting finite element simulation. For instance, Pisarenko et al. [3] applied governing experimental method to analyse the stress distribution of beams with different dimensions of segment perforation.

Furthermore, Ghuku and Saha [4] studied on deflection and stress distribution behaviour from the generated hole due to the drill along the body of the beam by considering the experimental analysis under various loading situations. Duan at al. [5] investigated the stress behaviour of perforation in glass fiber-reinforced polypropylene structures using experimental and theoretical simulation methods. Zhao et al. [6] added to this body of research by presenting investigation, which proved the effect of various normalized time and passion's ratio in stress behaviour. Moreover, simulated the failure factor of various circular tunnel radii of cylindrical board by governing LS DYNA and modified split Hopkinson pressure bar. Zamanian et al. [7] also investigated the stress concentrations of circular and square perforations through fiber laminate by deriving FEM method utilizing Abaqus software. Prakash and Hithendra [8] conducted a novel study on the influence of increasing the radius of perforation on stress intensity factor of plate due to the pin inside it according to the FEM simulation by governing ANSYS. ACMC et al.

[9] used FEM and governing experimental method to simulate stress distributions of different widths and diameter holes on plate and the effect of the elastic moduli of laminates. Zappino et al. [10] applied localized 3D printed carbon fibre reinforced to simulate the stress and strain behaviour of plates with hole. Interestingly, based on the obtained results, the 3D model experienced noticeable decrease in comparison with other material. A pioneering work on this topic was performed by Dong et al. [11] who investigated the influence of crack growth around the edge of perforation and plate on fatigue based on stress intensity factor. Cabrera-Gonzalez et al. [12] simulated originated perforation from riveted join failure behaviour in composite plates, and the stress concentration around the hole of the riveted joint which is caused by orthotropic behaviour of join and nonlinear connection of rivet, in order to forecast the amount of failure. It is worth mentioning that a pioneering work on this topic was conducted by Ding et al. [13-15] who applied linear elastic FEM method to demonstrate the influences of notch depth, width, and thickness of ground blade on stress distribution.

Finite element method is an approximation method with relatively high precision for the investigation of the value and distribution of stress. In this method, the assumed geometry is meshed. Smaller meshes are used in the regions of geometrical changes in order to obtain more precise stress distribution. The types of elements and meshing have significant effect on the precision of the obtained results. Therefore, when the pattern of distribution is not known, it is necessary to conduct a comprehensive study on effective parameters to make sure the precision of the results. Furthermore, investigating the theories showed increased stress up to 3 times in the boundary of circular cut out region.

2. Formulation and theory

Stress concentration at hole edge in a plate is illustrated in figure 1.

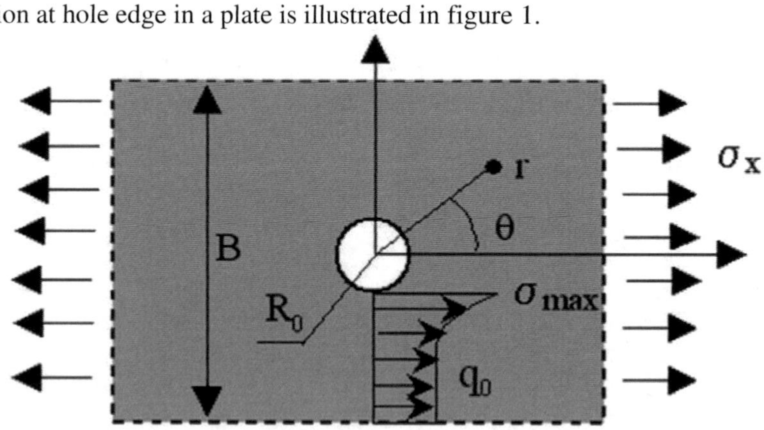

Figure 1. Stress concentration at hole edge in a plate.

Kirsch's solution is linear elastic solution for stress around a hole is for the case of uniaxial tension in an infinite plate, which consists of three prominent analytical solutions as follow [3, 4]:

$$\sigma_r = \frac{q_0}{2}(1 - \frac{R_0^2}{r^2}) - \frac{q_0}{2}\cos 2\theta (1 - \frac{R_0^2}{r^2})(1 - 3\frac{R_0^2}{r^2}) \tag{1}$$

$$\sigma_\theta = \frac{q_0}{2}(1 + \frac{R_0^2}{r^2}) - \frac{q_0}{2}\cos 2\theta (1 - \frac{R_0^2}{r^2})(1 + 3\frac{R_0^4}{r^4}) \tag{2}$$

$$\tau_{\gamma\theta} = \tau_{\theta\gamma} = -\frac{q_0}{2}(1 - \frac{R_0^2}{r^2})\sin 2\theta (1 - \frac{R_0^2}{r^2})(1 + 3\frac{R_0^2}{r^2}) \tag{3}$$

where σ_r, σ_θ, and $\tau_{r\theta}$ are radial, hoop and shear stress, respectively. Furthermore, q_0, θ, R_0, and r are remote stress, remote loading direction, radius of hole, and radial coordinate, respectively.

Using these functions, distribution of σ_θ would be along vertical axis, where $\theta = 90°$:

$$\sigma_\theta(\theta = 90°) = q_0(1 + \frac{1}{2}\frac{R_0^2}{r^2} + \frac{3}{2}\frac{R_0^4}{r^4})$$

$$\sigma_\theta(\theta = 90°, r = R_0) = 3q_0 \tag{4}$$

Furthermore, concentration factor is the ratio of maximum stress at a hole, fillet, or notch (but not a crack) to remote stress. Our case of a hole in an infinite plate, is defined as:

$$K_t = \frac{\sigma_{max}}{q_0} \tag{5}$$

which was considered to be 3 in our case.

3. Finite element model

Plane182 was chosen for element modelling. Selection of Shell 181 was also possible. The formula of this element was chosen based on planar stress. Thickness was assumed to be 1mm. In a wide range of material resistance problems, the results are independent of material properties. Material properties are equivalent Young modulus of 200 GPa and Poisson ratio of 0.3.

The geometrical shape was a sheet with thickness of 1 mm and dimensions of 1200×100 mm. Circular cut-outs with radiuses of 10, 20, 30 and 40 mm were considered. Figure 2 illustrates the geometrical shape of the plate, which the regions around the cut out are separately sectioned to obtain more precise meshing.

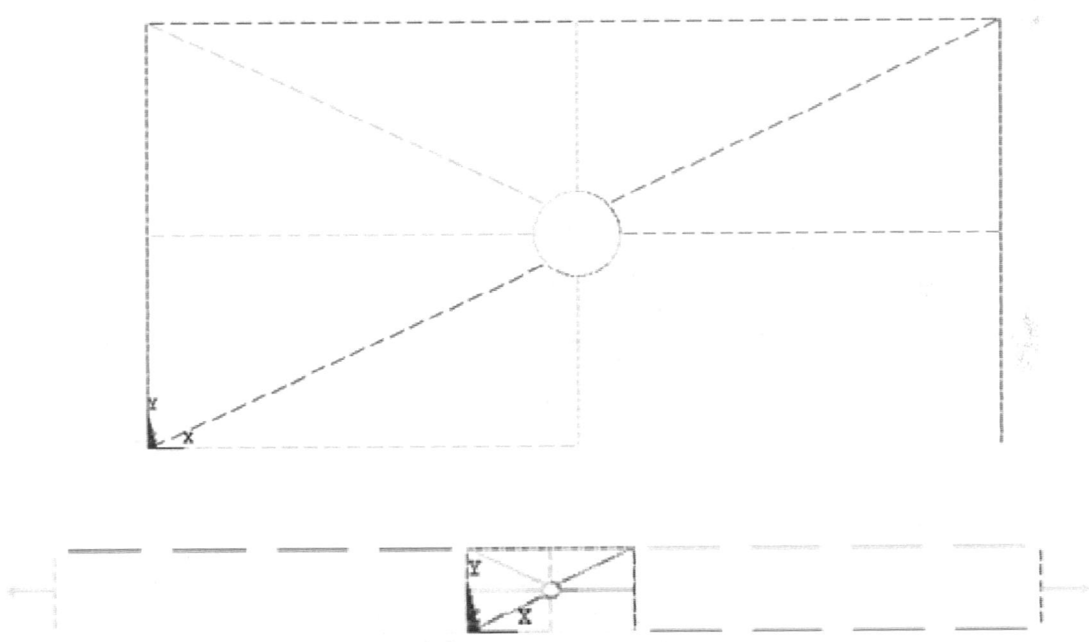

Figure 2. Geometrical shape of the plate.

In the next step, the drawn geometry based on the considered element was meshed. Due to the concentration of the stress near the cut off, more elements and finer meshes were used in these regions. In other regions with less stress concentration, coarser meshes were used. Figure 3 shows meshing patterns in different geometries.

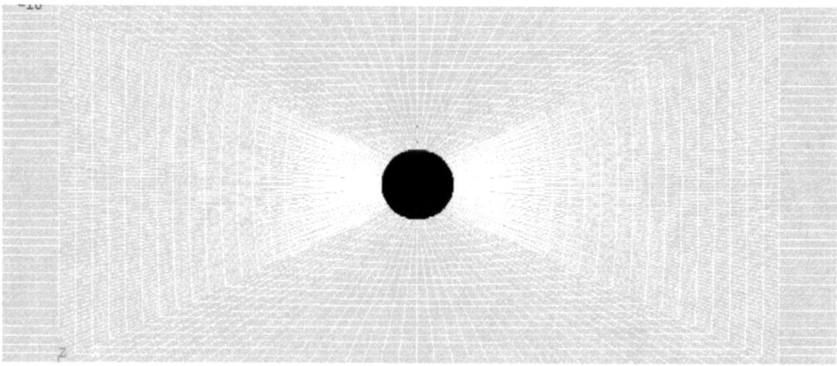

(a)

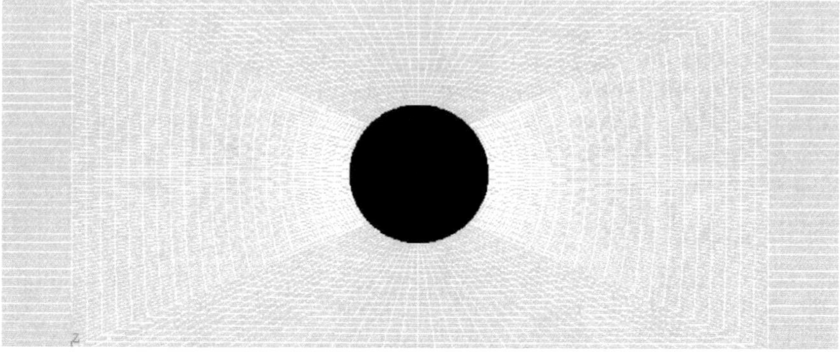

(b)

(c)

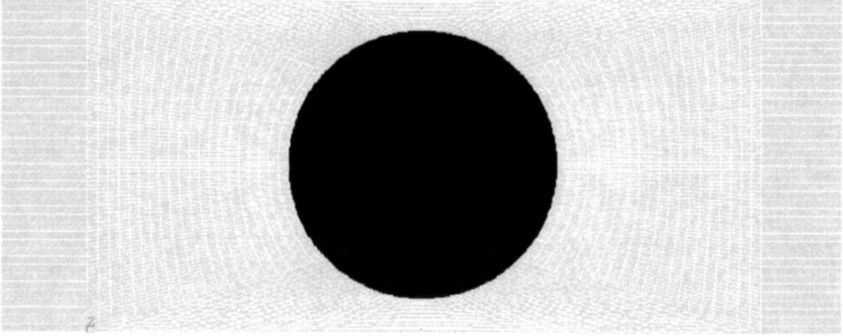

(d)

Figure 3. Meshing patterns in different geometries (a) $R = 10$ mm, (b) $R = 20$ mm, (c) $R = 30$ mm, (d) $R = 40$ mm.

As shown in figure 4, loading was performed by applying 10 MPa stress on the initial and final levels of the sheet (here, on the lines).

Figure 4. Loading diagram.

4. Results and analysis

After applying above items, finite element model was solved. Then, using the menu of data analysis, investigation of stress distribution was performed. Stress distributions and stress distributions diagram for sheets with $R = 10$, 20, 30 and 40 mm cut-outs are shown in figures 5-12, respectively.

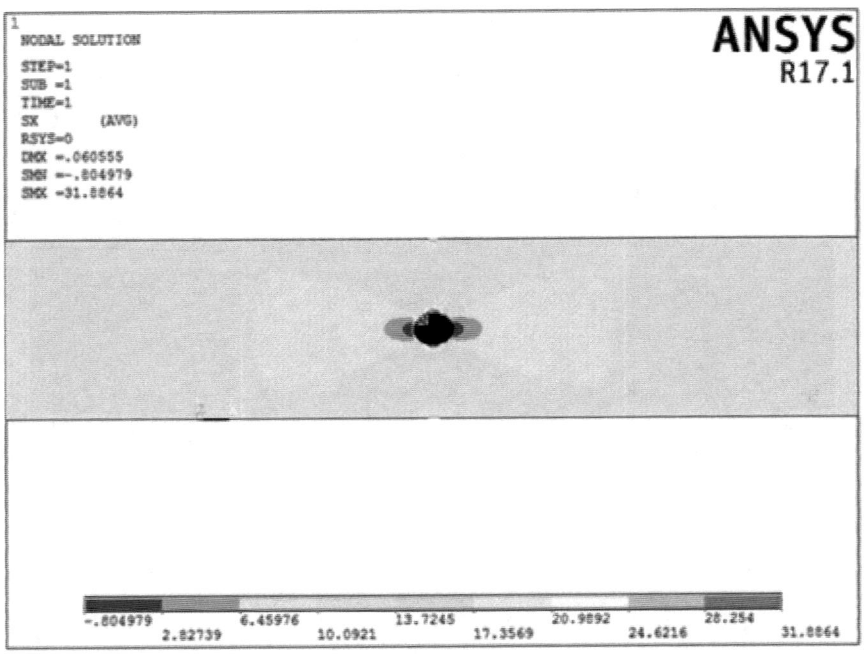

Figure 5. Stress distributions for sheets with $R = 10$ mm.

Under these conditions, stress distribution diagram on vertical line below the cut out is showed in figure 6.

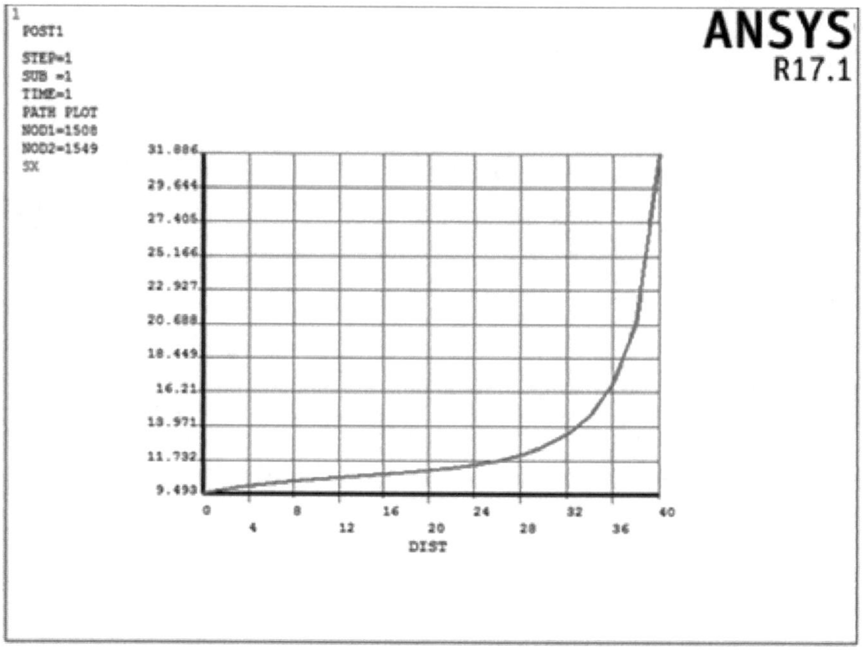

Figure 6. Stress distribution diagram with $R = 10$ mm.

As can be seen in Figure 6, stress in x direction ranged from 31.886 MPa at the edge of the cut out to 9.493 MPa at the end of the line. The obtained maximum value was 3 times the nominal stress of the sheet being 10 MPa. This distribution had a good agreement with stress distribution obtained from analytical solution.

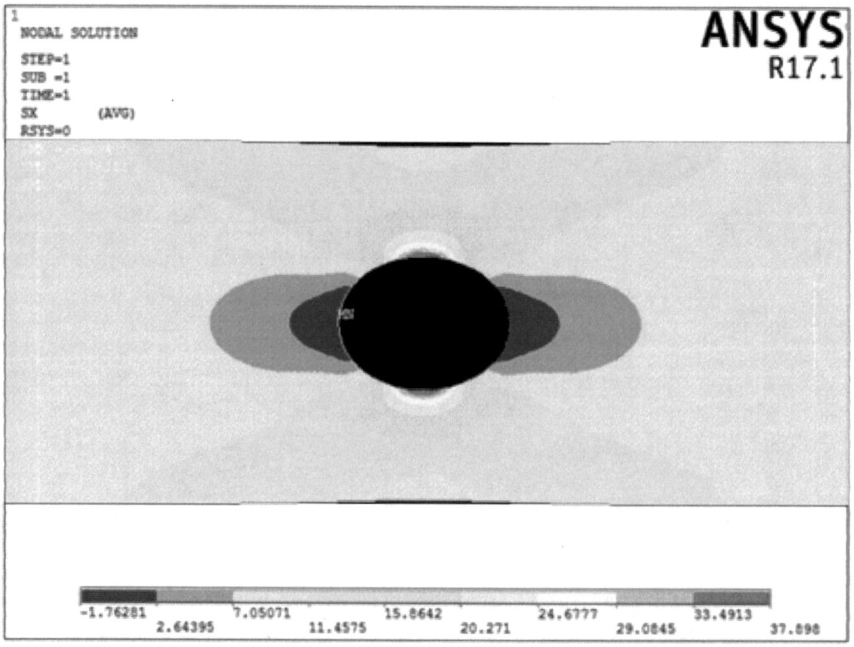

Figure 7. Stress distributions for sheets with $R = 20$ mm.

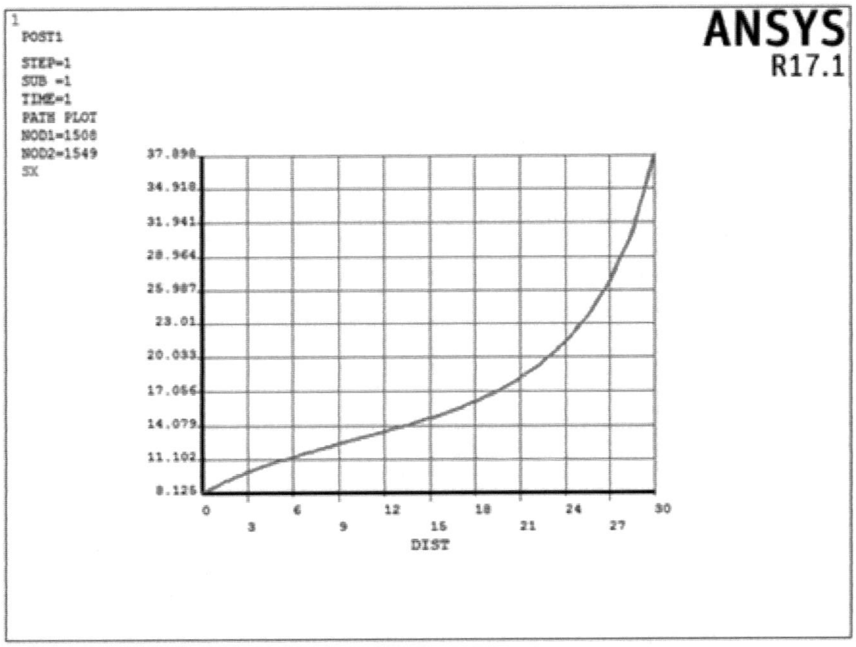

Figure 8. Stress distribution diagram with $R = 20$ mm.

It is noticeable form figure 8 that although the amount of stress around the perforation with 20 mm radius experienced significant increase from 8.125 MPa at the end of the line to 37.898 MPa the at the edge of the cut out, it is slightly bigger than that in hole with 10 mm width.

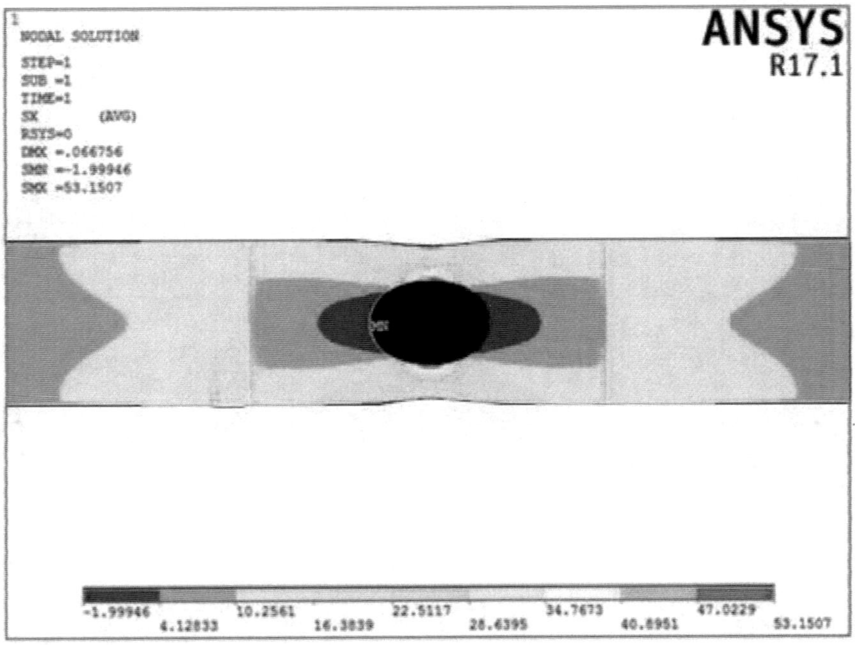

Figure 9. Stress distributions for sheets with $R = 30$ mm.

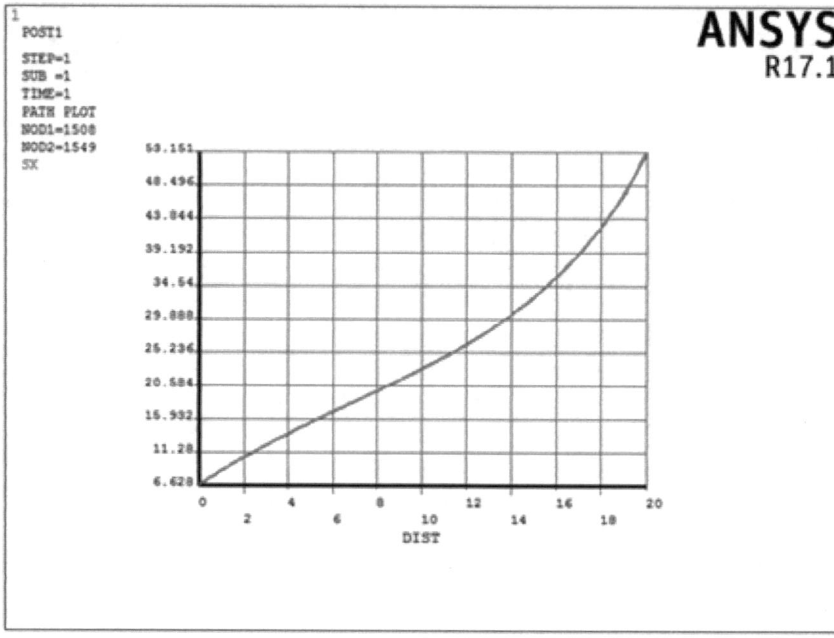

Figure 10. Stress distribution diagram with $R = 30$ mm.

It is apparent from Figure 10 that, stress distribution around 30 mm radius perforation of the plate ranged from 53.151 MPa at the edge of the cut out to 6.628 MPa at the end of the line. Comparing the data obtained from Figure 10, it is clear, the stress distribution of the plate, which is contained the hole with 30 mm radius approximately consists of 1.4 times the stress of the plate with 20 mm radius.

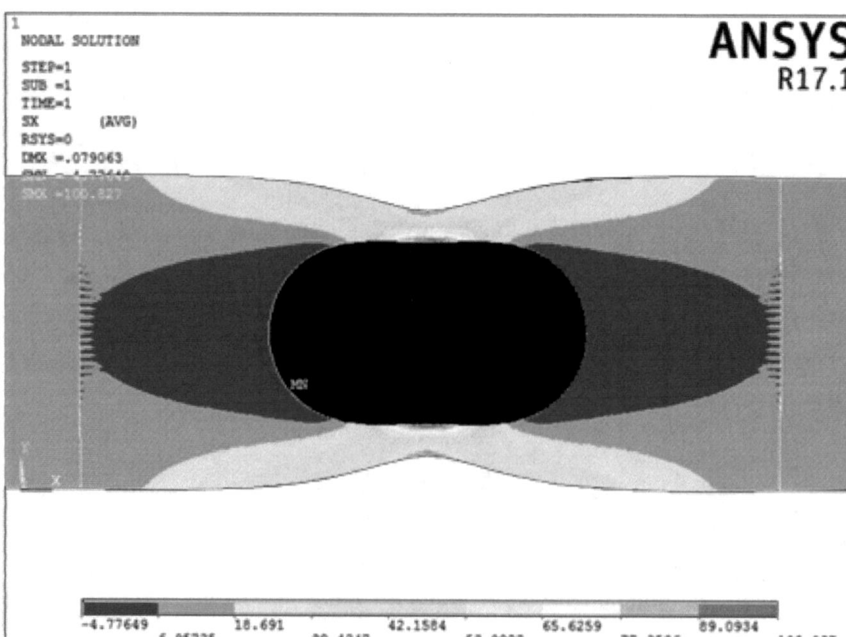

Figure 11. Stress distributions for sheets with $R = 40$ mm.

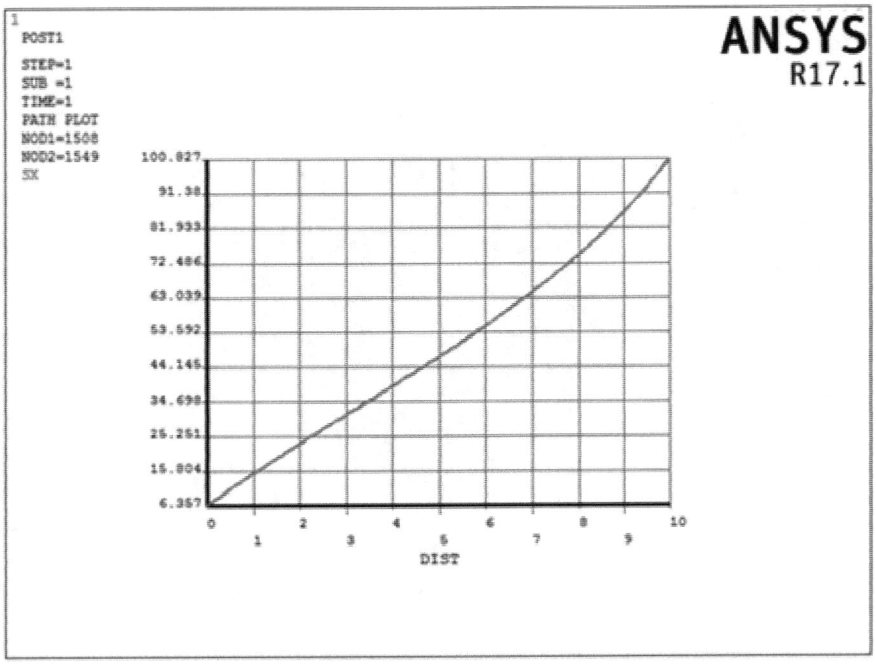

Figure 12. Stress distribution diagram with $R = 40$ mm.

As can be seen in figure 12, stress in x direction ranged from 100.827 MPa at the edge of the cut out to 6.357 MPa at the end of the line. Figure 12 shows that the stress distribution of the plate, which is contained the hole with 40 mm radius approximately consists of doubled amount of this plate with 30 mm radius near the edge (100.827 MPa and 53.151 MPa). In other hand, almost it has the same amount of stress with the 40 mm and 30 mm perforation (6.357 MPa and 6.628 MPa).

Table 1 documents stress distribution range in the narrowest parts of 100 mm wide plates with circle cut-outs with radius R = 10, 20, 30 and 40 mm.

Table 1. Stress distribution interval for variable radius of cut-outs R.

R [mm]	Stress distribution range [MPa]
10	9.493 – 31.886
20	8.125 – 37.898
30	6.628 – 53.151
40	6.357 – 100.827

Furthermore, least figures 13 and 14 illustrate stress distribution in sheets with R = 10 mm cut-outs and 440 mm width.

IOP Conf. Series: Materials Science and Engineering 1252 (2022) 012067 doi:10.1088/1757-899X/1252/1/012067

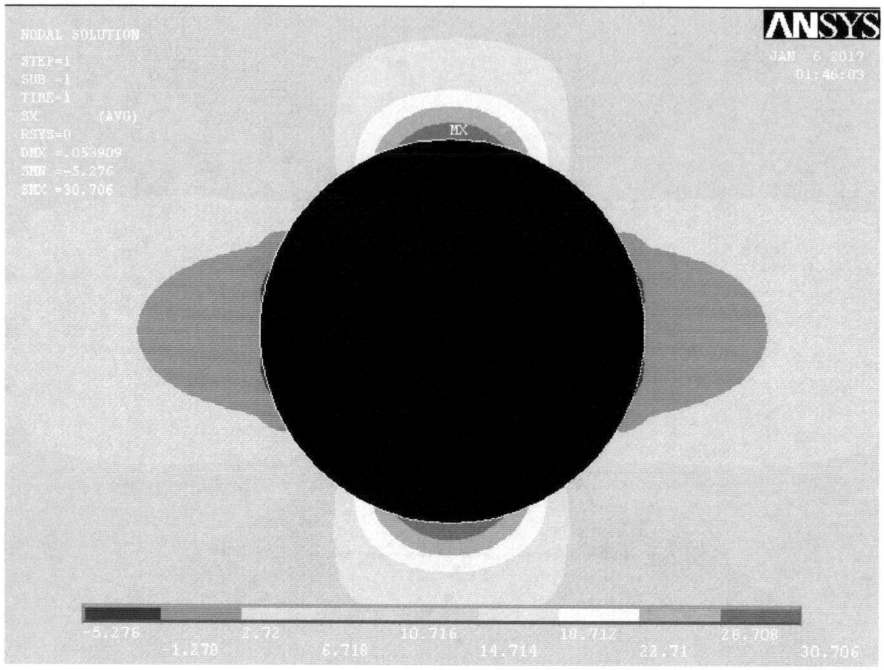

Figure 13. Stress distribution with $R = 10$ mm, $W = 440$ mm.

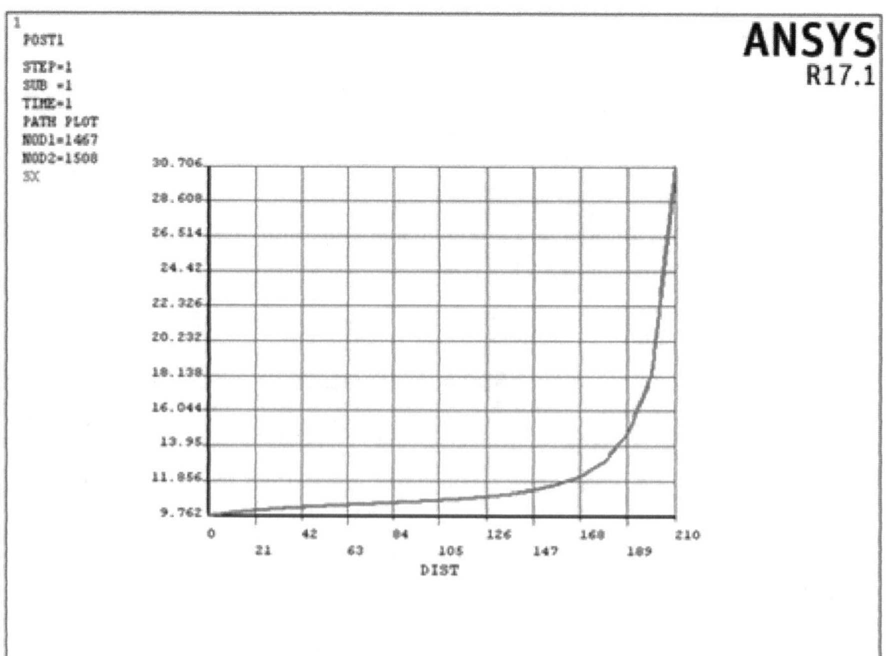

Figure 14. Stress distribution diagram with $R = 10$ mm, $W = 440$ mm.

As can be seen, stress behaviour of perforation along the plate witnessed a smooth increased from the 9.762 MPa to 30.706 MPa. It was clear that, comparing the graph with figure 6, increasing the width of the plate from 100 mm to 440 mm has not influence on stress distribution for the sheets with endless width theory. To be precise, near the hole of plate, falling from 31.886 MPa to 30.706 MPa.

5. Conclusion

Creation of cut-outs in sheets for doors and other structural elements is inevitable. By increasing the radius of cut out compared to its width, the maximum stress increased (table 1). What is more, the analytical equations were obtained for the sheets with infinite width. Stress distribution obtained using the finite element method had good agreement with those obtained from analytical solution methods. Consequently, the study of theories showed that stress at the boundary of the circular shear zone was increased up to 3 times.

6. Reference

[1] Santos A 2013 Determination of stress concentration factors on flat plates of structural steel *: Conf. Ser.: J. Phys.* **466** 12035
[2] Hearn EJ 1997 Theories of elastic failure *Mechanic Material* 1 (Elsevier) pp 401-429
[3] Pisarenko GS, Stetsyuk NA, Khyuvenen VA and Chekulaeva S V 1986 Use of an experimental-numerical method to study the stress concentration in a beam with a hole during bending. *Stren. Mater.* **18(10)** 1346-1350
[4] Ghuku S and Saha KN 2016 An Experimental Study on Stress Concentration Around a Hole Under Combined Bending and Stretching Stress Field *Procedia. Technol.* **23** 20-27.
[5] Duan S, Zhang Z, Wei K, Wang F and Han X 2020 Theoretical study and physical tests of circular hole-edge stress concentration in long glass fiber reinforced polypropylene composite *Compos. Struct.* **236** 111884
[6] Zhao R, Tao M, Zhao H, Cao W, Li X and Wang S 2020 Dynamics fracture characteristics of cylindrically-bored granodiorite rocks under different hole size and initial stress state *Theor. Appl. Fract. Mech.* **109** 102702
[7] Zamanian H, Marzban B, Bagheri P and Gudarzi M 2013 SE Journal Science and Engineering ORICPublications On Stress Concentration Factor for Randomly Oriented Discontinuous Fiber Laminas with Circular/Square Hole *J. Sci. Eng.* **3(1)** 7-18
[8] Prakash RV and Hithendra K 2020 Combined influence of pin-hole interference and additional hole on the stress intensity factor for a pin loaded finite plate *Proc. Struct. Integ.* **28** 1125-1133
[9] ACMC B, SRL T, RS F, SHS N and EMF A 2020 Analytical, experimental and finite element analysis of the width/diameter hole ratio effect in vinylester/carbon hybrid twill weave composites *Compos. Part. C. O. Acc.* **2** 100033
[10] Zappino E, Filippi M, Pagani A, Petiti M and Carrera E 2020 Experimental and numerical analysis of 3D printed open-hole plates reinforced with carbon fibers *Compos. Part. C. O. Acc.* **2** 100007
[11] Dong Y, He X, Sun X and Li Y 2020 Fatigue crack growth of a corner crack at the edge of a hole *Int. J. Fat.* **139** 105699
[12] Cabrera-González JA, Vargas-Silva G and Barroso A 2021 Riveted joints in composites a practical tool to estimate stresses around the rivet hole *Compos. Struct.* **263** 113735.
[13] Ding M, Zhang Y, Lu H and Sun Y 2021 Numerical investigation on stress concentration of surface notch on blades *Eng. Fail. Anal.* **122** 105241
[14] Murakami Y 2019 Stress concentration *Met. Fat.* 13-27
[15] Liu K and Wang Z 2019 Stress distribution around waist hole on finite plate based on mapping function *AIP. Adv.* **9(7)** 075120

Acknowledgments

Financial support from the Scientific Grant Agency of the Ministry of Education of Slovak Republic and the Slovak Academy of Sciences the project VEGA 1/0374/19 is gratefully acknowledged.

Modelling of Stormwater runoff and Wastewater Flow in the Sewer Network

R Wittmanová[1], I Makro[1], J Hrudka[1], M Šutúš[1], Š Stanko[1]

[1] Department of Sanitary and Environmental Engineering, Faculty of Civil Engineering, Slovak University of Technology in Bratislava, Radlinského 11, 810 05 Bratislava, Slovakia

E-mail:reka.wittmanova@stuba.sk; ivana.marko@stuba.sk, jaroslav.hrudka@stuba.sk, marek.sutus@stuba.sk, stefan.stanko@stuba.sk

Abstract. The purpose of mathematical modelling of sewer networks is mainly to assess the hydraulic capacity and monitor its behaviour under different conditions to predict the future state. Sewerage network models are also part of the design process. Their advantage is the possibility of simulating the future state of the network and the precipitation and runoff process in the context of climate change. With the help of simulations, it is possible to anticipate future conditions in urban catchments and thus effectively design new networks. The aim of this paper is to summarize mathematical simulation models that are used to model sewer networks.

1. Introduction

The purpose of mathematical modelling of sewer networks is mainly to assess the hydraulic capacity and monitor its behaviour under different conditions to predict the future state. Sewerage network models are also part of the design process and their advantage is the possibility of simulating the future state of the network and the precipitation and runoff process in the context of climate change. Future conditions can be anticipated in the urban catchment and thus effectively design new networks [1].

Simulation models are a simplified but detailed description of the precipitation-runoff process using mathematical equations that describe the relationships between the inputs and outputs of the model. Mathematical models combine all sub-elements of sewer networks into an integrated system that allows monitoring these elements individually or the entire network as a single system. In addition to hydraulic assessments, some simulation models can model the transport of pollutants that enter the receiving water during overflow or discharges from accumulation tanks during rainfall events.

As many authors have suggested, simulation models can be characterized as a set of equations and algorithms that describe a real system and mimic its behaviour [2]. Urban drainage systems models must be built to show the various hydrological parameters that can subsequently be transformed into hydraulic parameters (e.g. depth, pressure, flow) [3].

Currently, there are many simulation models that scientists have tried to gradually categorize according to different assumptions. Simulation models are most often divided into stochastic and deterministic models. Stochastic models are characterized by random parameters and indicate uncertainty in the simulation. However, in the hydrology of urbanized areas, deterministic models are used primarily, which are based on mathematical relationships between the physical quantities of the precipitation-runoff process [4]. The deterministic model is characterized by the fact that a certain combination of input parameters always provides the same output. In these models, random events are excluded [1; 5].

The aim of this paper is to summarize knowledge about mathematical modelling of flows in the sewer network and the runoff transformation, the basic division of simulation models, and to characterize individual modelling software.

2. Complex Model of a Sewer Network

The urban drainage system is recorded in the models using state and output equations. The continuity equation is used as the equation of state, and the input data represent the rain data. The shapes of the output equations are conditioned by the description of the precipitation-runoff process [6]. Modelling the precipitation-runoff process consists of several phases that are directly connected. Most models include all relevant precipitation and runoff processes. In general, the models consist of two basic components, namely: a hydrological model (surface runoff model) and a hydraulic model (rainwater inflow and flow transformation model).

The complex sewer network model combines a precipitation runoff model, including all physical processes that contribute to the formation of surface runoff and efficient rain, and a hydraulic model that considers the transformation of rainwater into the network and the gradual transformation of rainwater due to unsteady flow in the network. The software for creating precipitation-runoff models is based on the idea of dividing the model into sub-models. These sub-models are based on the time sequence of surface runoff formation. The hydrological model also includes four modules that describe surface runoff, namely: rainfall event, surface runoff, surface runoff concentration. The result of this model is a surface runoff hydrograph, which serves as an entrance to another sub - model. The hydraulic sub - model already deals more comprehensively with the hydraulic flow conditions in the sewers and the transformation of the rain inflow into the sewer network [3]. The quality of the simulation model is greatly affected by the accuracy of the boundary conditions, as their values are constant throughout the simulation and may not correspond exactly to reality. Boundary conditions include, for example, elevation and topographic data, urban catchment topology (runoff coefficient), drainage characteristics (roughness, average), and design rainfall [6].

3. Runoff Transformation in the Sewer Network

From a hydraulic point of view, we can define the flow of urban wastewater as an unsteady flow. Its characteristics change in time and space. The wastewater flow changes daily depending on the current water consumption and the rainwater inflow into the network. Another important prerequisite is the flow of wastewater with a free surface. However, this assumption may not appear to be correct in torrential rains, as overloading results in capacitive flows and pressure flow may occur. With unsteady flow in partially filled sewer pipes, the relationship between water depth and actual flow is more complex than steady flow. The shape of the stormwater discharge wave in the sewer also changes. The wave flattens and, thus, the maximum value of the peak flow decreases. Without considering the transformation of the rain wave, it is not possible to accurately predict the relationship between flow and time. The flow of stormwater in sewers is simulated using empirical hydrological equations, Saint-Venant (S-V) equations, or their approximations according to kinematic and diffusion waves, or by simple hydraulic methods [7]. The solution of Saint-Venant equations is relatively complicated and time consuming, so their simplified forms are often used, which are obtained by approximation by diffusion wave, kinetic wave, or dynamic wave [1; 7]. The most used solution method is the finite difference method (FDM), which is based on the conservative form of determining equations in the network of current field nodes and involves the division of length and time into small separate steps [8].

4. Simulation Models

The stochastic and deterministic models mentioned above are divided into conceptual and empirical models. The difference between these models is whether they are based on physical laws or empirical relationships. According to the spatial variability of the parameters, lumped and distributed models are distinguished. Lumped models do not consider the spatial variability of the input parameter, while distributed models consider the simulation of hydrological processes in special stages of the precipitation and runoff process [5]. Elliott and Trowsdale distinguish up to six different simulation time intervals in their study, namely: lumped average annual steps, lumped daily time steps (or by rain events), distributed daily time steps, lumped hour or minute steps, distributed hour steps, and distributed minute steps (or second) steps. In general, time-resolution simulations divide models into short time-step simulations, long-term series simulations, and long-term continuous simulations [9]. Stochastic and empirical models belong to the group of focused models (models with concentrated parameters) and are based on the measured parameters of observed events [5]. An example of an empirical model is the rational method. These methods were also used in a study by Trudeau and Richardson (2016) to assess the impact of urbanization on the hydrology of the Great Lakes-St. Lawrence, Canada [10].

The deterministic models used to model the rain flow in urban drainage system sewers are classified as hydrological and hydraulic methods. Hydrological models solve only the continuity equation, while hydraulic models solve the continuity and momentum equations as a connected system [11].

Butler and Davies describe sewer network simulations that combine simulations of one-dimensional flow in a pipeline with one-dimensional surface runoff flow and flow from a spillway to an urban catchment. The flood flow created is then solved using the S-V equations. They also define a 1D-2D model that combines one-dimensional flow in pipes with two-dimensional surface flow. Both models share unknown variables, namely flow rate and water level (depth). In a one-dimensional model, velocity is a one-component parameter, and in a two-dimensional model, it is a two-component parameter. The aim of these models is to predict more accurately the development of floodplains and the depth of the water during floods. They are the most accurate presentation of flooding in urban catchments [1].

Depending on the precipitation runoff module, the models of urban drainage systems are further divided into semi-distributed and fully-distributed. Semi-distributed models are based on predefined sub-catchments, which are limited to analyses of surfaces around the point of discharge. This point represents the node of the assessed sewerage system. All sub-catchments have assigned morphological and hydrological properties, from which the runoff hydrograph for a given area are derived. Fully distributed models are determined by discretization of a two-dimensional surface network, and the collision itself is applied to each two-dimensional element. The resulting surface runoff is then simulated by a two-dimensional surface flow. These models are used in 1D-2D simulations. The study by Pina et al. (2016) provides a complete comparison of these models with the design of new concepts for building models and connecting modules semi-distributed and fully-distributed models [12].

Table 1. Comparison of hydrological models [12] [13] [14] [15] [16] *.

Model	Land use	Purpose	Elements	Time scale	Water quality	Availability
HSPF	U, R	P, D	SR, P, OCH	1/C	no	free
HEC-1 (HEC-HSM)	R	P, O	SR	1/C	no	free
HEC-5Q	R	P, O	R, NF	C	no	free
WinTR-20	R	P,O	SR, RE, OCH	1	no	free
WinTR-55	U, R	P, O	SR, RE, OCH	1	no	free
HydroCAD	U	P, O, D	SR, P, RE	1	yes	commercial
SUSTAIN	U	P, O, D	SR, P, OCH	1/C	yes	free
EPA NSWC	U	P, O	SR	1/C	no	free

*Explanations: Land use: U - urbanized catchment area, R - rural catchment area (natural catchment area); Purpose: P - planning model, O - operational model, D - design model; Elements: P - pipe, SR - surface runoff, OCH - open channel, NF - natural flow, RE - retention element, R –hydraulics of river bed, Time step: 1 - single event, C - continuous simulation.

In addition to hydrological and hydraulic models, a third group of simulation models, quality models, also plays an important role in assessing the environmental impact of urban areas. When evaluating sewer networks and their behaviour during torrential rains, it is necessary to pay attention to the quality control of the recipient. Many world-renowned complex simulation models also have a pollution propagation prediction module implemented. This module allows monitoring and simulation of the transport of pollutants, changing the quality of wastewater in the network, sediment transport, modelling the content of organic pollution and nutrients, transport of heavy metals, and simulation of eutrophication in the receiving water. Table 1. shows a comparison of world-used hydrological models, which further categorizes according to purpose, modelled elements, and time step. Some hydrological models also have a quality module implemented. The same comparison is presented in Table 2, which focuses on the evaluation of hydraulic models.

Table 2. Comparison of hydraulic models [17], [18], [19] *.

Model	Land use	Purpose	Elements	Time scale	Water quality	Availability
HEC-RAS	R, RB	P, O	OCH	1/C	yes	free
WSPRO	RB	P, O	OCH	-	no	free
FlowMaster	U	P, O	OCH	1	no	commercial

*Explanations: Land use: U - urbanized catchment area, R - rural catchment area (natural catchment area); Purpose: P - planning model, O - operational model, D - design model; Elements: P - pipe, SR - surface runoff, OCH - open channel, NF - natural flow, RE - retention element, RB – hydraulics of river bed, Time step: 1 - single event, C - continuous simulation.

Zoppou (2001) also focused on examining qualitative modules of simulation models. He summarized the sources of surface runoff pollution and described the individual empirical relationships that are used in different modelling methods [5]. Rainwater quality models have also been discussed by McAlister et al. (2003) and emphasized the importance of time-step selection in modelling [9]. The application of pollution propagation modelling using measures to reduce rain inflow into the network has been studied by Wong et al. (2006). The study describes the development of a model that can predict the cleaning performance of water retention measures such as, e.g., wetlands, rain lakes, sedimentation reservoirs, bio filters, and grassy vegetation ditches [20].

Table 3. Comparison of hydrological – hydraulic (H-H) models [21], [22], [4; 5] *.

Model	Land use	Purpose	Elements	Time scale	Water quality	Availability	LID
PondPack	-	P, D	SR	1	no	commercial	yes
EPA SWMM	U	P,D,O	SR, P, RE	1/C	yes	free	yes
PCSWMM	U	P,D,O	SR, P, RE	1/C	yes	commercial	yes
InfoSWMM	U	P,D,O	SR, P, RE	1/C	yes	commercial	yes
xpswmm	U	P,D,O	SR, P, RE	1/C	yes	commercial	yes
MIKE URBAN (SWMM, MOUSE)	U,R	P,D,O	SR, P, OCH, RE	1/C	yes	commercial	yes
InfoWorks ICM	U	P,D,O	SR, P, OCH	1/C	yes	commercial	yes
MIKE 11	R	P,D,O	SR,OCH, RE	-	yes	commercial	-
CivilStorm	U	P, O	SR, P, OCH, RE	1	no	commercial	yes
WinHSPF	-	-	-	1/C	yes	free	yes
SWAT	U, R	-	-	1/C	yes	free	yes

QQS	U	P, D	P, OK, RE, SR	-	-	?	?
WQRRS	R	P, D	OK, R, NF, SR	1/C	yes	free	no
STORM	U, R	P	-	1	yes	free	?

*Explanations: Land use: U - urbanized catchment area, R - rural catchment area (natural catchment area); Purpose: P - planning model, O - operational model, D - design model; Elements: P - pipe, SR - surface runoff, OCH - open channel, NF - natural flow, RE - retention element, RB – hydraulics of river bed, Time step: 1 - single event, C - continuous simulation.

The software characterized in Table 3 is used most frequently when modelling the precipitation-runoff process in relation to the sewer network. This software is more widely used to simulate conditions in urbanized areas than planning, operational, or design models. The advantage of H-H models is the possibility of simulating a continuous rain event with different time steps, or a single rain event. Many of them have implemented modules to simulate LID objects.

The categorization and comparison of individual models is also included in the studies mentioned by Zoppou (2001) and Elliott and Trowsdale (2007) [5; 9]. Zoppou (2001) categorizes models according to the functionality, availability, and characteristics of the models. Elliott and Trowsdale (2007) assessed the suitability of 10 different models for simulating "Low Impact Development" (LID) objects in interaction with a sewer network. They compare models in terms of potential use, time distribution, representation, and spatial distribution of the river basin, runoff generation, and hydraulic course, and in terms of modelled pollutants [9].

5. Conclusion

The aim of this paper was to summarize knowledge about mathematical modelling of flows in the sewer network and the runoff transformation, the basic division of simulation models, and to characterize individual modelling software.

Sewerage models, including surface runoff and water quality models, are very complex in terms of mathematical prediction of natural processes. The complexity of the models is diverse. There are very simple models that predict only one process, or, conversely, complex models capable of predicting a set of interconnected processes. Therefore, before selecting a simulation tool, it is necessary to first determine the modelling objectives and the nature of the available input data. The model selection also considers whether the model complies with current regulations and corresponds to the unit system (or whether there is a choice or setting), the complexity of the hydrological process we want to model, land use (rural or urban catchment) and the purpose of the model.

Hydrodynamic modelling has great potential in water management in Slovakia because the current state of most sewer networks is unsatisfactory due to the future effects of climate change and rapid urbanization.

6. References

[1] Butler D and Davies J W 2011 *Urban Drainage 3rd edition* (Abingdon: Spon Press) p 652 ISBN 978-0415455251.

[2] Loucks D P and Van Beek E 2005 *Water resources systems planning and management: An introduction to methods, models and applications* (Paris: UNESCO) p 680 ISBN 92-3-103998-9.

[3] Cadus D and Poetsch M *Dynamic Modelling of Urban Rainfall Runoff and Drainage Coupling DHI MIKE URBAN and MIKE FLOOD* [online] [cit. 19. 02. 2021] Available: https://www.geography.at/study/analysis/cadus_poetsch_seminar-paper.pdf.

[4] Sokáč M, Sztruhár D, Sinčák P, Bundzel M 1999 *Využitie prostriedkov hydroinformatiky v mestskej hydrológii* (Bratislava: Ústav vedecko-technických informácií pre pôdohospodárstvo Nitra) ISBN 80-85330-64-4.

[5] Zoppou C 2001 Review of urban storm water models. *Environmental Modelling & Software*, **16** (3) 195-231

[6] Urcikán P and Rusnák D 2004 *Stokovanie a čistenie odpadových vôd. Stokovanie I. Navrhovanie stokových sietí* (Bratislava: Slovenská technická univerzita v Bratislave) p 324 ISBN 80-227-2136-0.

[7] Sztruhár D and Sokáč M 2005 *Hydrológia urbanizovaných území* (Bratislava: Slovenská technická univerzita v Bratislave) ISBN:80-277-2220-0.

[8] Zácha P *Metoda konečných objemů.* energetika.cvut.cz [online]. [cit. 12. 12. 2018]. Available: http://energetika.cvut.cz/wp-content/uploads/2018/06/FLUENT-prednaska-1a-Zacha-LS-2014.pdf.

[9] Elliott A H and Trowsdale S A 2007 A review of models for low impact urban stormwater drainage. *Environmental Modelling & Software* **22** (3) 394-405

[10] Trudeau M P and Richardson M 2016 Empirical assessment of effects of urbanization on event flow hydrology in watersheds of Canada's Great Lakes-St Lawrence basin. *Journal of Hydrology* **541** Part B 1456-1474

[11] Zoppou C 1999 *Review of Storm Water Models* (Canberra: CSIRO Land and Water) ISBN: 0-643-06075-8.

[12] Pina R D, Ochoa-Rodriguez S, Simoes N E, Mijic A, Marques A S, Maksimović Č 2016 Semi- vs. Fully-Distributed Urban Stormwater Models: Model Set Up and Comparison with Two Real Case Studies. *Water* **8** (2) Article No. 58. ISSN 2073-4441.

[13] EPA- United States Environmental Protection Agency. Environmental Modeling Community of Practice. Hydrological Simulation Program - FORTRAN (HSPF) [online]. [cit. 15. 02. 2021]. Available: https://www.epa.gov/ceam/hydrological-simulation-program-fortran-hspf.

[14] WILLEY, R.G. HEC-5Q: System Water Quality Modeling. 1986.

[15] HEC-RAS - Hydrologic Center - Army.mil. HEC-HMS [online]. [cit. 15. 02. 2021]. Available: https://www.hec.usace.army.mil/software/hec-hms/.

[16] EPA- United States Environmental Protection Agency. National Stormwater Calculator. Tool to help control runoff and promote the natural movement of water [online]. [cit. 15. 02. 2021]. Available: https://www.epa.gov/water-research/national-stormwater-calculator.

[17] HEC-RAS - Hydrologic Center - Army.mil. HEC-RAS [online]. [cit. 15. 02. 2021]. Available: https://www.hec.usace.army.mil/software/hec-ras/.

[18] USGS – science for a changing word. Water Resources Applications Software [online]. [cit. 15. 02. 2021]. Available: https://water.usgs.gov/cgi-bin/man_wrdapp?wspro88.

[19] Bentley. Hydraulic Calculator Software [online]. © 2021 BENTLEY SYSTEMS [cit. 15. 02. 2021] Available: https://www.bentley.com/en/products/product-line/hydraulics-and-hydrology-software/flowmaster.

[20] Wong T H F, Fletcher T D, Duncan H P, Jenkins G A 2006 Modelling urban stormwater treatment—A unified approach. *Ecological Engineering* **27** (1) 58-70. ISSN 0925-8574.

[21] Rossman A L 2015 Storm Water Management Model User's Manual Version 5.1 [online]. 353 p. [cit. 10. 04. 2021]. Available: https://www.epa.gov/sites/production/files/2019-02/documents/epaswmm5_1_manual_master_8-2-15.pdf

[22] WATER ONLINE. Wallingford Software - Software Tools For The Water Industry [online]. 28. 07. 2003 [cit. 15. 02. 2021] Available: https://www.wateronline.com/doc/wallingford-software-software-tools-for-the-w-0001.

Acknowledgments
This work is supported by the Scientific Gant Agency of the Ministry of Education, Youth and Sports of the Slovak Republic, and the Slovak Academy of Sciences within the projects VEGA 1/0574/19 and 1/0727/20, co-funded by the European Regional Development Fund and by the Slovak Research and Development Agency under the contract No. APVV-18-0203 and by the grant "Young researchers Grant by STU" under the KLIMO acronym, which is financed from the central sources of the Technical University of Bratislava.

IOP Conf. Series: Materials Science and Engineering 1252 (2022) 012069 doi:10.1088/1757-899X/1252/1/012069

Internal insulation systems and their assessment for historic buildings by hygrothermal simulation

K Hutkai[1], D Katunský[1] and M Zozulák[1]

[1] Institute of Architectural Engineering, Faculty of Civil Engineering, Technical university of Kosice, Vysokoškolská 4, 042 00 Kosice, Slovakia

E-mail: kristian.hutkai@tuke.sk

Abstract. The paper examines possible solutions for the use of internal insulation in listed buildings. Massive thick walls of historic buildings represent large heat losses and the use of thermal insulation materials from the inside is a possible solution to increase the energy efficiency of historic buildings. However, the use of thermal insulation from the interior is accompanied by a number of risks associated with condensation, mould growth and damage to the perimeter structure. It is necessary to perform dynamic simulations to verify the thermal-humidity behaviour of the structure. The article describes materials suitable for the purpose of internal insulation, their advantages and disadvantages and a description of dynamic simulations by which their effects on the thermal-humidity behaviour of the perimeter wall of a historic building are assessed.

1. Introduction

The use of thermal insulation is one of the most effective solutions to increase the level of energy efficiency of buildings [1-3]. In historic buildings, the use of insulation material from the exterior side is inadmissible, due to the high cultural value of the facade and therefore the only solution to increase the energy efficiency of these buildings is to apply insulation material from the inside. However, this solution carries a number of risks and therefore a thorough analysis of the facts and subsequent selection of the insulation material is important in the pre-project phase. The use of thermal insulation from the inside can disrupt the thermal-humidity behaviour of the structure, condensation can occur between the original structure and thermal insulation, insulation can prevent water from evaporating inwards from the structure, which can cause increased moisture in the structure and mould [4-6]. The problem can also occur with wooden beam ceilings, when due to the internal insulation, the ends of the wooden beams are no longer in the warm part of the wall, which can lead to rot of wooden structures [7]. Partition dividing structures in conjunction with the perimeter structure create thermal bridges, which is again a space for the formation of mould [8]. For the purposes of internal insulation, capillary active and hydrophilic insulating materials such as wood fibre boards, calcium silicate boards, autoclaved aerated concrete boards are most suitable, but hydrophobic insulation materials with a vapor barrier layer are also used [9-11]. The aim is to enable the functional use of historic buildings in accordance with current standard requirements using modern methods of structural and physical analysis of structural evaluation. Based on the mentioned negative impacts on the hygrothermal behaviour of internally insulated historic masonry, the aim of this publication is to define and make an overview of current building materials suitable for the purposes of internal insulation of historic buildings.

2. Thermal insulation suitable for the purpose of interior insulation of historic buildings

The decisive properties of thermal insulation suitable for the purposes of internal insulation of historic buildings are thermal conductivity λ (W / (mK)), the capillary moisture content w_{cap} (kg/m^3) and water absorption coefficient A (kg/(m^{2}s$^{0.5}$)) which are important for water transport analysis and capillary activity and also the diffusion resistance factor μ (-) [12,13].

Based on WTA 6-4 insulation systems we can divide into three systems:

- Condensate-preventing insulation systems
- Condensate-limiting insulation systems
- Condensate-tolerating insulation systems [13].

2.1. Condensate-preventing insulation systems

In the condensate preventing insulation systems, a vapor barrier prevents water vapor penetrating from the interior into the wall structure. Vapor barriers are sealing layers with a vapor diffusion equivalent air layer thickness sd (product of μ and layer thickness) of min. 1500 m. The vapor barrier layer can be a separate layer on the interior side, or it can be part of the thermal insulation, then we call it an insulation system. The efficiency of this system is independent of the indoor environment and, thanks to the restriction of the transfer of steam from the inside to the outside, it prevents the accumulation of moisture in the structure. However, this system is very sensitive to increased external moisture stress such as it is rain driven by the wind because the drying of the structure inwards is limited. Condensate-preventing insulation materials include foam glass, extruded polystyrene, polyurethane and systems that include a vapor barrier [13].

2.2. Condensate-limiting insulation systems

Condensate-limiting insulation systems contain a vapor barrier with an sd- value of min. 0.5 m and max. 1500 m. The vapor barrier in this case serves to reduce the input of water vapor from the interior to the structure, but it is necessary to sufficiently protect the outer surface of the structure against wind-driven rain. In the structure composition with the risk of condensation must not accumulate moisture and sufficient drying-out of interstitial condensate during the drying period (summer). In order for condensate-limiting insulation systems to meet the requirements, the diffusion resistance factor of the vapor barrier during the heating season should be high enough to prevent interstitial condensation and low during the summer months to allow the structure to dry out. Condensate-limiting insulation materials include vapor open insulation material with vapor barrier mineral wool-based systems (mineral wool with PE-foil or SVR), EPS or PUR based systems [13].

2.3. Condensate-tolerating insulation systems

Condensate-tolerating insulation systems form a capillary-active insulating material with adhesive mortar. Since the diffusion of water vapor in this system occurs freely and the capillary active materials are highly porous, which means that they are able to absorb and redistribute water vapor. When using capillary-active insulating materials (capillary open) from the inside, water vapor from the interior penetrates through the thermal insulation into the perimeter structure. The most commonly used capillary active insulating materials are wood fibre board, calcium silicate or aerated concrete [13].

2.3.1. Calcium silicate board

Calcium silicate is made from a mixture of fine silica from hydrated lime and fly ash reinforced with cellulose fibre. These components are mixed with water where the precursor of calcium silicate hydrate is formed during a chemical reaction. The curing method is similar to autoclaving, where at high temperature and pressure, calcium silicate crystals grow and form an open structure with fine pores. Thanks to this structure, we call the capillary active material which is able to absorb water. In addition, it has a high resistance to pressure, is non-flammable and frost-resistant [13,14]. In a case study [5] was investigated the thermal-humidity behaviour of seven insulation systems on a historic brick wall thickness of 840 mm, and the silicate calcium board improved the thermal-technical properties of the wall by 34%. In addition, the surface temperature of the interior walls increased by 2.2 ° C, which means an improvement in thermal comfort in the indoor environment.

2.3.2. Autoclaved aerated concrete board
Autoclaved aerated concrete (AAC), otherwise called cellular concrete, is a vapor-cured material made from quartz sand, mineral lime and cement. Aluminium paste is used as the expansion agent. The ingredients are mixed with the addition of water. During pre-curing, the Aluminium paste reacts with the alkaline elements of the cement to form hydrogen gas and the raw mixture itself forms blocks (blocks) as its volume increases. After evaporation of the hydrogen gas, at high temperature and pressure, the body undergoes an autoclave process. After curing, the body dries and is ready for use [14].

2.3.3 Wood fibre board
Fibreboard is produced from waste from the wood processing industry such as wood chips and sawdust. The matrix is freed of impurities, dried and mechanically crushed according to the requirements for coarse or fine particles. Fibreboard is produced by either wet or cold pressing. Wet pressing is a more energy intensive process but uses less binder. When wet-pressed from wood fibres, lignin is released, which serves as a natural binder. During dry pressing, synthetic resins are added to the mass and it is possible to produce a 40 mm thick wood fibre board. The pressing is performed at a pressure of 3-5 MPa and at a temperature of 300 ° C [15]. In [23] were examined 2 internal insulation systems and the result was a reduction of annual energy transported through the walls by 67 to 79%.

3. Dynamic simulations of hygrothermal behaviour of structures
To determine and assess the thermal properties of buildings and structures as well as to predict the risk of mould or other adverse effects associated with the disruption of thermal-humidity behaviour of structures, an accurate dynamic simulation model is needed, in which heat and moisture transfer over time can be monitored. The transfer of heat and moisture in building materials occurs simultaneously and is nonlinear [16]. In addition to the energy efficiency and durability of structures, moisture transfer also affects the quality of the indoor environment. After several decades of research and development, heat-humidity simulation models have been implemented in various construction sectors, such as energy consumption in buildings [17,18] and predicting the risk of mould growth [19,20]. WUFI is a group of software used to calculate one-dimensional and two-dimensional heat, water and energy transfer in building structures and materials. This software uses the latest knowledge regarding vapor diffusion and water transport in building materials. The simulations in this paper were performed using the WUFI Pro program, which simulates one-dimensional heat and moisture transfer in the cross-section of the structure, taking into account built-in humidity, wind-driven rain, solar radiation, capillary transport and summer condensation. [21,22].

4. Conclusion
The walls of historic buildings have low thermal resistance due to the content of materials with high thermal conductivity such as stones. Based on the facts found so far about the internal insulation of historic buildings, it can be stated that it is a suitable approach to improve the energy efficiency of these buildings. However, historic buildings are in various technical and thermal-humidity states, so it is not possible to say unambiguously what type of insulation system is most suitable. It is necessary to survey the existing state of the building and on the basis of initial conditions and using dynamic simulations to design and adapt the insulation system and its thickness for a particular building.

5. References
[1] Bjarløv S P, Finken G R and Odgaard T 2015 Retrofit with interior insulation on solid masonry walls in cool temperate climates–An evaluation of the influence of interior insulation materials on moisture condition in the building envelope *Energy Procedia* 78 1461-66
[2] Brandt E, Bunch-Nielsen T, Christensen G, Gudum C, Hansen M H and Møller E B 2009 SBi-anvisning 224: *Fugt i bygninger. Statens Byggeforskningsinstitut, Hørsholm.*
[3] Toman J, Vimmrova A and Černý R 2009 Long-term on-site assessment of hygrothermal performance of interior thermal insulation system without water vapour barrier *Energy and Buildings* 41(1) 51-55

[4] Zhao J, Grunewald J, Ruisinger U and Feng S 2017 Evaluation of capillary-active mineral insulation systems for interior retrofit solution *Building and Environment* 115 215-27
[5] Walker R and Pavía S 2015 Thermal performance of a selection of insulation materials suitable for historic buildings *Building and environment* 94 155-65
[6] Vereecken E, Van Gelder L, Janssen H and Roels S 2015 Interior insulation for wall retrofitting–A probabilistic analysis of energy savings and hygrothermal risks *Energy and Buildings* 89 231-44
[7] Kopecký P, Staněk K and Tywoniak J 2020 Hygro-thermal performance of wooden beam ends: experimental investigations of uninsulated and internally insulated brick wall *E3S Web of Conferences* Vol 172 p 01009 EDP Sciences
[8] Orlik-Kożdoń B 2019 *Interior insulation of masonry walls—selected problems in the design. Energies 12*(20) 3895
[9] Hansen T K et al. 2018 Long term in situ measurements of hygrothermal conditions at critical points in four cases of internally insulated historic solid masonry walls *Energy and Buildings* 172 235-48
[10] Finken, G R, Bjarløv S P and Peuhkuri R H 2016 Effect of façade impregnation on feasibility of capillary active thermal internal insulation for a historic dormitory–A hygrothermal simulation study *Construction and Building Materials* 113 202-14
[11] Pavlík Z and Černý R 2009 Hygrothermal performance study of an innovative interior thermal insulation system *Applied Thermal Engineering 29*(10) 1941-46
[12] RiBuild.eu 2021 [Online] Available at: https://www.ribuild.eu
[13] Blumberga A, Kašs K, Kamendere E, Žogla G, Kamenders A, Blumberga D and de Place Hansen E J 2019 Robust Internal Thermal Insulation of Historic Buildings-State of the art on historic building insulation materials and retrofit strategies
[14] Zhao J, Grunewald J, Ruisinger U and Feng S 2017 Evaluation of capillary-active mineral insulation systems for interior retrofit solution *Building and Environment 115* 215-27
[15] My house 2021 [Online] Available at: https://myhouse-sk.desigusxpro.com/strojmaterialy-i-tehnologii/dvp#i-2
[16] Dong W, Chen Y, Bao Y and Fang A 2020 A validation of dynamic hygrothermal model with coupled heat and moisture transfer in porous building materials and envelopes *Journal of Building Engineering 32* 101484
[17] Barbosa R M and Mendes N 2008 Combined simulation of central HVAC systems with a whole-building hygrothermal model *Energy Build.* 40 (3) 276–88
[18] Moon H J, Ryu S H and Kim J T 2014 The effect of moisture transportation on energy efficiency and IAQ in residential buildings *Energy Build.* 75 439–46
[19] Jensen N F, Bjarløv S P, Rode C and Odgaard T R 2018 Hygrothermal assessment of internally insulated solid masonry walls fitted with exterior hydrophobization and deliberate thermal bridge *ce/papers* 2 (4) 79–87
[20] Otiv P 2016 Hygrothermal Modelling of Internal Insulation to Solid Masonry Walls *Master's Thesis* Kgs. Lyngby Technical University of Denmark Denmark
[21] Wufi 2021 [Online] Available at:https://wufi.de/en/software/wufi-pro/
[22] Møller E B 2018 Report on material properties *RIBuild Deliverable* D2. 1
[23] Kočí V, Jerman M, Pavlík Z, Maděra J, Žák J and Černý R 2020 Interior thermal insulation systems based on wood fiberboards: Experimental analysis and computational assessment of hygrothermal and energy performance in the Central European climate *Energy and Buildings 222* 110093

Acknowledgments

The paper presents partial research results of project VEGA 1/0674/18 "Theoretical and experimental analysis of architectural-structural shapes and fragments of building envelope structures designed for harsh weather conditions".

IOP Conf. Series: Materials Science and Engineering 1252 (2022) 012070 doi:10.1088/1757-899X/1252/1/012070

Analysis of vibrations in a dynamic damper

A Jaskot

Department of Civil Engineering, Faculty of Civil Engineering, Czestochowa University of Technology, ul. Akademicka 3, 42-201 Częstochowa

E-mail: anna.jaskot@pcz.pl

Abstract. The problem of damping mechanical vibrations has been described in the paper. The subject of the research is a dynamic damper consisting of two steel plates and cuboid rubber blocks to dampen vibrations. The upper plate is subjected to forcing, causing the disturbance to propagate. This disturbance will propagate through the top plate and will pass through the damping material to the bottom plate. The damping elements have been modelled from Mooney-Rivlin material. The aim of the work is to investigate the propagation of vibrations and damping of vibrations and to assess the effects resulting from the selection of material constants with the assumed geometry of the tested system. The results included in the work have been obtained based on a comparative analysis, taking into account the linear material and the nonlinear material. Numerical tests have been carried out with the use of ANSYS software.

1. Introduction

Wave phenomena are used in many fields of science. Many wave properties are modified during propagation, which depend on the conductivity and properties of the medium. Energy propagates in free space, wave energy can be absorbed by objects in the medium. When transmitting a wave, it is important to calculate the losses that the wave can cause. In building construction, wave transmission should be limited to prevent damage. Mechanical waves propagate in a material medium under the influence of forces related to the deformation of the volume and the form of the medium's elements. Waves occurring in nature, together with an indication of possible applications, especially of weak and strong discontinuity waves, was included in [1].

The research conducted in the direction of wave analysis results from the necessity to ensure the operational reliability of machines and devices, their parts and objects exposed to dynamic loads. Forced vibrations arise as a result of displacement from the equilibrium position of particles of a material medium. The direction of wave propagation in relation to the direction of particle vibrations is determined by the type of wave (longitudinal or transverse). Due to the easy access to computer programs that improve the research process, more and more often such models are made with the use of programs based on the finite element method. Real objects are implemented and subjected to given working conditions, loads, etc. By using numerical methods, the determining not only the value, but also the place of the disturbance under the influence of a given load is possible. The work [2] proposes a method of identifying the damage - without knowing the initial state, using the finite element method. The presented in the cited work method of identifying failures in beam systems is based on changes in modal parameters with an additional mass changing the position. The solution to the problems related to the propagation of elasto-plastic stress waves is based on the description of linear differential equations, assuming that the strain amplitude is sufficiently small. The issue related to the propagation of plastic waves with finite deformations in nonlinear media requires the use of numerical methods. In the work [3], an analysis of the solutions for the propagation of longitudinal nonlinear plastic waves in the material description using the Lagrange coefficient and in the spatial manner with the use of Euler coefficients was carried out. The analysis was performed on the basis of thin rod tests. Another example illustrating wave propagation on the example of a rod was included in [4]. The description with the use of models of hyperelastic materials with the results concerning the propagation of the acceleration wave was made on the example of a bar with a variable cross-section. The constitutive equations describing the relationship between deformations and stresses in hyperelastic materials are determined on the basis of the equations of the principle of conservation of mechanical energy [5]. In the case of homogeneous isotropic elastic bodies, the constitutive equations are defined as a function of the strain energy. In the paper [6] authors present a comparison of models of hyperelastic materials and indicates the differences in stress distributions in relation to wave intensity. The authors conducted the analysis on the example of a thin segmented bar with a slowly changing cross-section. On the basis of long elements with small sizes of cross-sectional dimensions, dynamic effects and mechanical vibrations can be observed more clearly. In studies of the dynamics of systems composed of more than one element, the greatest difficulty, apart from the proper implementation of non-linear material parameters, is the appropriate modeling of the close contact of the elements in contact. In this study, the analysis covered a system consisting of two steel plates, which are separated by a modelled layer of a hyperelastic material as a vibration damping layer. Assemblies of interconnected parts made of different materials cause complex phenomena of reflection and wave transmission. For this reason, materials with different characteristics are used to obtain the most effective static and dynamic parameters. Since, for such structures, simple analytical descriptions may not be sufficient to properly model multilayer materials, the authors of [7] have attempted to develop techniques to generally understand the effects of wave propagation. In addition, the usefulness of experimental data for the validation of finite element models was investigated.

This paper proposes a solution for vibration damping through the use of rubber-based materials, which are referred to as incompressible materials. In the case of modeling linearly elastic materials according to the FEM theory, it is enough to define two material constants (Young's modulus and Poisson's number). Hyperelastic materials are able to undergo large elastic deformations under the influence of forces, but show nonlinear behavior and their deformation is not directly proportional to the applied load [8]. For hyperelastic material models, define additional material parameters. Based on this dependence and taking into account the deformation range in the simulation, a model of the deformation energy of hyperelastic incompressible materials was selected. In describing models of hyperelastic materials, it is necessary to analyse the constitutive relationships of the theory of nonlinear elasticity. The constitutive relations are postulated in accordance with the principle of conservation of mechanical energy and the existence of a unit function of elastic energy is assumed. The difficulty is the appropriate selection of the material model. The Murnaghan model is the basic model of a hyperelastic material in dynamic nonlinear elasticity. The elastic energy potential of the Murnaghan model does not meet the mathematical theory of nonlinear elasticity, so it is used in the analysis of small deformations [9]. This model can be understood as a generalized model of an elastoplastic material, assuming an alternating transition from plastic to elastic and elastic to plastic. The deformation gradient is replaced by a non-singular tensor, and the constitutive equations are written in a finite form [10]. In case of large deformations there is a need to use other material models. For hyperelastic materials in this work a Mooney-Rivlin model was adopted. In general the reduction of vibrations are performed in a way that results both from the determination of the source of vibrations, their nature and occurrence, as well as due to the constraints of the structure. The methods of reducing vibrations include propagation, energy dissipation or for active reduction methods. The solution presented in this paper concerns passive methods, which in its form use the model of a hyperelastic material in order to limit the impact of structure vibrations.

The analysis were performed in ANSYS Mechanical environment. The aim of this work was to know the difference in both qualitative and quantitative sense the influence of using hyperelastic materials in order to reduce vibrations, by observing propagation of mechanical waves in proposed system.

2. Mooney-Rivlin material model

In order to properly define rubber-like materials in numerical analysis, the material model have to be implemented. In this work a Mooney-Rivlin model has been adopted. The material characteristics of typical hyperelastic materials are shown in Figure 1.

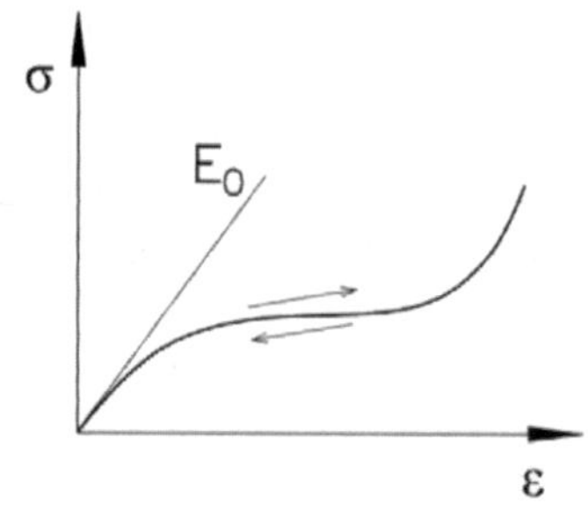

Figure 1. Characteristics of stress - strain of typical hyperelastic materials [11].

In the range from 10% to 300% deformation of the material under the influence of the applied load, the best approximation is provided by the Mooney-Rivlin model, the form of which is presented in the general formula:

$$W = C_{10}(I_1 - 3) + C_{01}(I_2 - 3) \quad (1)$$

where: - I_1 and I_2 – invariants describing the strain energy density,
- C_{10} and C_{01} – material constants.

The Mooney-Rivlin model was implemented in ANSYS in material, which represents the damping layer of proposed system.

3. Analysed model description

Studies performed in scope of this work consider numerical analysis of the vibration damping system. The construction of high degree of complexity it is necessary to perform a basic studies to obtain character and quality of the most exposed elements. The numerical analysis performed on the simple structures is needed to know the character and values of changes in mechanical systems under known operating conditions. An attempt of damping vibrations with use of the rubber-based materials was described in nis work. The model of proposed system was presented in figure 2.

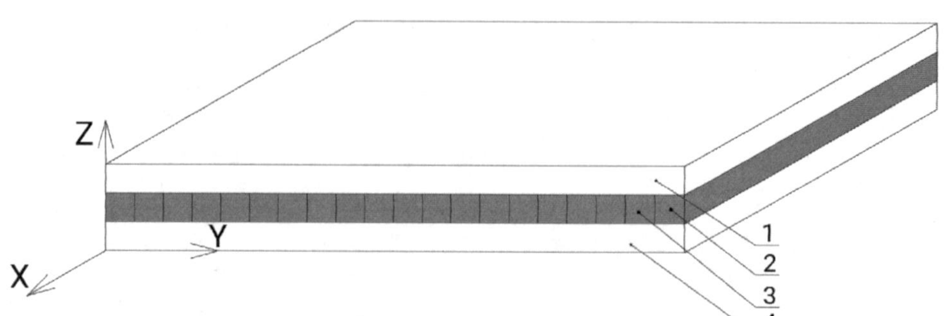

Figure 2. Model of system with hyperelastic material.

On the basis of simple structure propagation of waves according to the forcings was studied. The main purpose is to know both the quality and quantity of dynamical parameters in proposed system. The aim is to eliminate vibration. First and fourth layers in the figure 2 were modelled from the steel, with properties corresponding to the S275 steel. In steel medium it is possible to observe a wave propagation of acceleration as well as of forces and stresses. The sizes of plates are equal 1m x 1m. In proposed model the propagation of those waves was damped, through the hyperelastic material, modelled on the basis of Mooney-Rivlin material theory. The 2nd and 3rd layers represent the hyperelastic material. The geometry of second and third layer were based on a cuboids model with square base with a side length of 5cm.

The two parallel edges of the bottom plate were attached along the Y axis to the non-deformable base. To the top base the surface load with value of 1 kN.

The results of the analysis were compared with reference model results. Layers 2 and 3 of the proposed model in reference model were made of hard foam. The idea of the damper was schematically presented in figure 3.

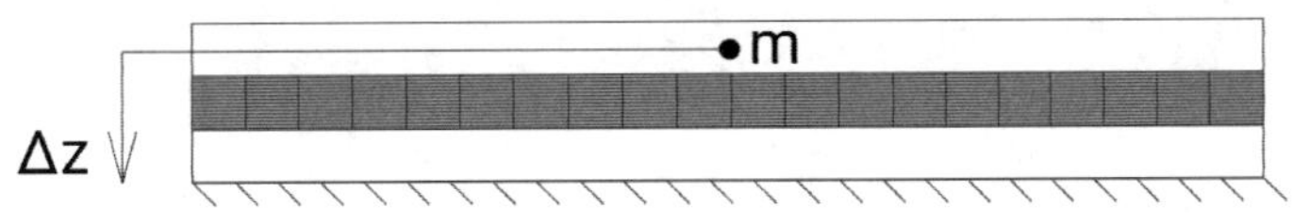

Figure 3. Dynamical scheme of the analyzed system.

Hyperelastic materials are highly deformed even under low load in very short time period. Hence, when the material properties are introduced of the undertaken material model it is necessary to properly select iteration step in numerical computation.

Strongly nonlinear behaviour and at the same time a rapid increase in load can lead to numerical instability. The step must be selected in an optimal way so as not to lead to divergence in iteration of the equilibrium. On the basis of the model described in this work the character of the dynamical interactions can be predicted, but, an estimation of quantitative benefits in the context of structural performance analysis is necessary.

4. Sample results of numerical tests

The two analyses have been performed in ANSYS Mechanical environment. First, the reference model was implemented in order to perform modal analysis. As a result the natural frequencies in the system have been found. The values of the first six modes frequencies were gathered in the table 1.

Table 1. Natural frequencies of a reference model.

Mode	1	2	3	4	5	6
Frequency [Hz]	142.46	154.06	170.49	171.14	190.05	257.25

Afterwards, the system was subjected to response analysis for the range of the obtained frequencies. In the dynamic analysis the propagation of the disturbance of the acceleration wave and the stresses under the influence of dynamic loads was examined in comparison to the proposed model. The contours of stresses (a) and strains (b) in the comparative model were presented in figure 4.

Values of analysis for dynamic load were much greater than in the case of a static load. The locations of the greatest stresses can be predicted with good approximation. Hence, numerical research is necessary, through which it is possible to determine not only the nature but also the values of the parameters.

The proposed model with rubber-like materials was implemented with the same boundary conditions, forcings and initial parameters. The stresses and strains for the proposed model were presented in figure 5.

a

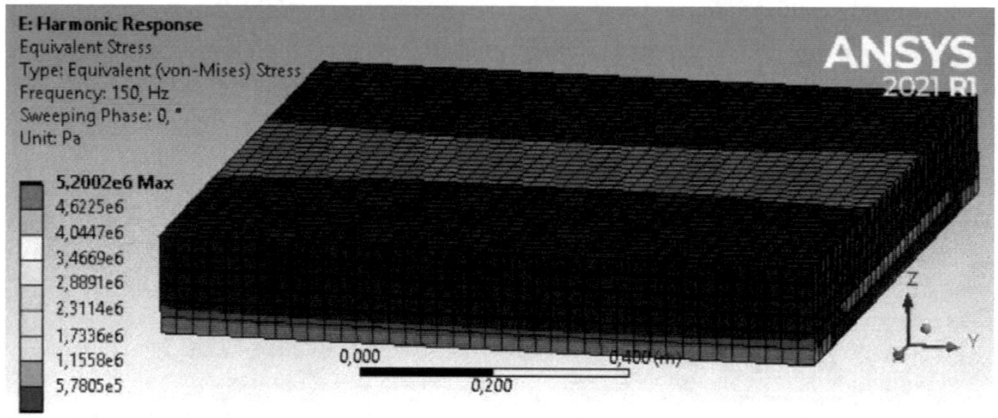

b

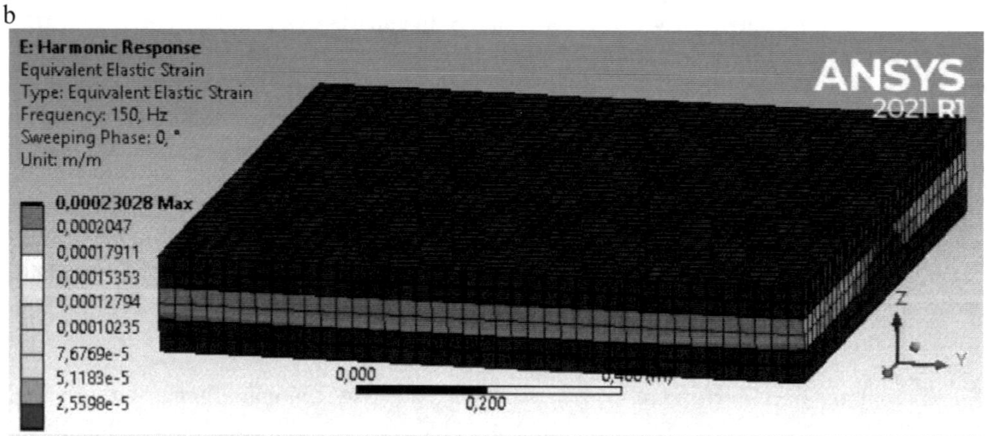

Figure 4. Stress and strain contour lines under dynamic forcings of the reference model.

a

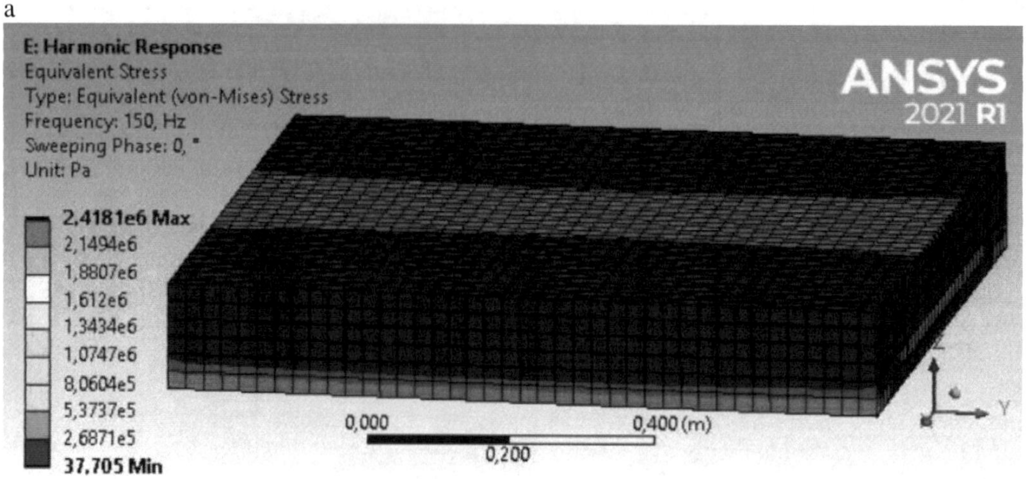

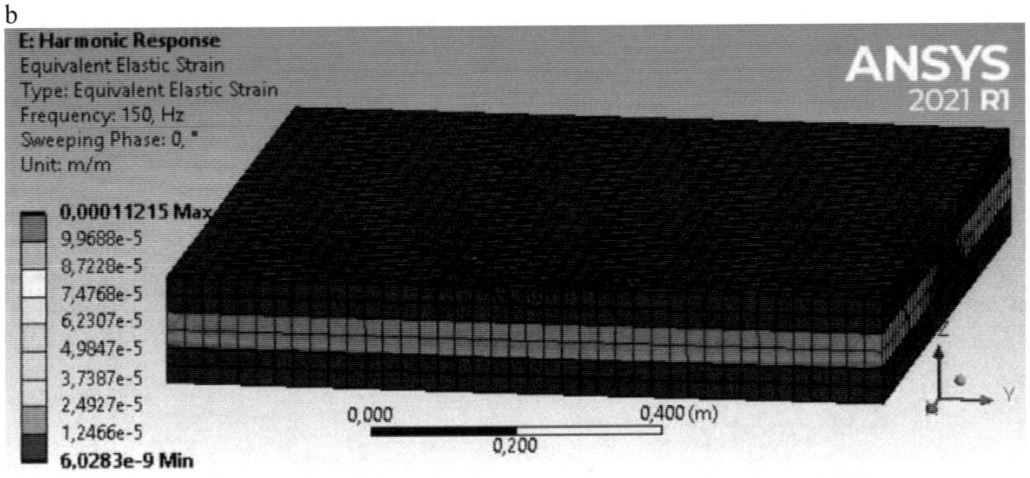

Figure 5. Stress and strain contour lines under dynamic forcings of the proposed model with hyperelastic layer.

The values of stresses and strains under the influence of dynamic loads are presented in table 2.

Table 2. Results of analyses for both systems.

	Reference model	Proposed model
Strain [m/m]	0.00023	0.000112
Stress [MPa]	5.2002	2.418

The value of the force transmitted by the system was limited due to the assumption of damping, and the values of this comparison are presented in diagrams a and b figure 6. By using an incompressible material the obtained values are more advantageous. The most visible differences can be observed in the propagation of the wave disturbance by comparing the vibration frequency of the acceleration wave, which were presented in figure 7.

In the quality sense, the character of propagation waves of acceleration and the reaction force are in both cases alike. In the case of using damping in the structure (figure 6 b and 7 b), the maximum value is much lower than the reference medium, and it occurs for a lower value of the frequency, than in reference model (figure 6 a and 7 a).

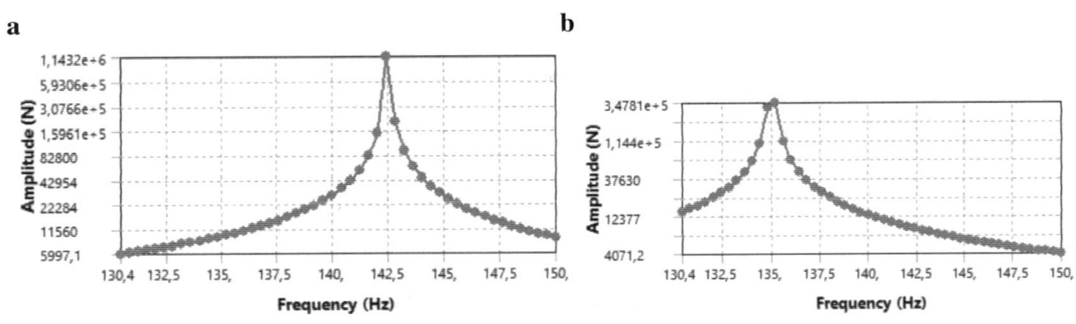

Figure 6. Character of force propagation wave in both models.

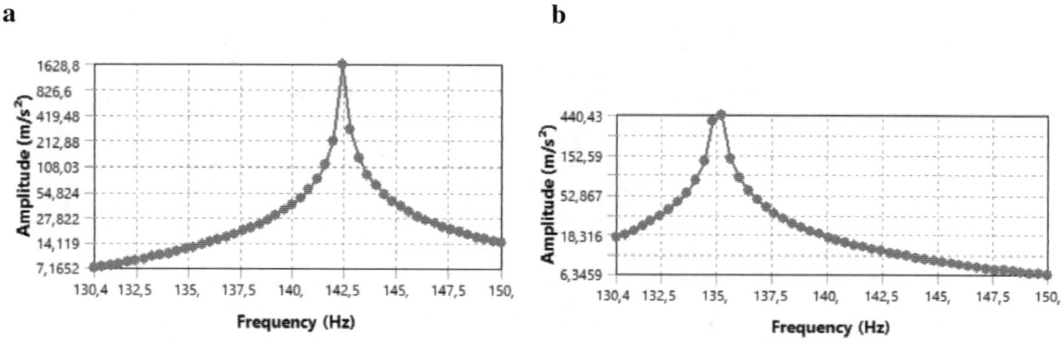

Figure 7. Character of acceleration propagation wave in both models.

The value of the force transmitted by the vibration damping system is lower than that of the system in which the damping was not used.

5. Conclusions

The research included in the work concerned the vibration damping system. A system of two steel plates connected by layers made of hyperelastic material in the Mooney-Rivlin model was modeled. The vibration analysis was performed in both systems under the influence of dynamic loads. The propagation of the acceleration wave in the gum-like medium is limited. The response of the system in the form of the substrate reaction to dynamic loading is also more favorable in the case of the proposed model compared to the reference model. Both in the qualitative and quantitative sense, the effectiveness of vibration damping has been demonstrated. The model may be useful in studies of vibration damping systems forced in other load cases. Further research will concern other load cases and tests of other material centers.

6. References

[1] Kuliński K 2015 Elastic waves - selected issues *Zeszyty naukowe Politechniki Częstochowskiej Seria Budownictwo* 21(1) pp 161-167

[2] Borowiec A and Ziemiański L 2008 Ocena Stanu Konstrukcji Belkowych Na Podstawie Zmiany Parametrów Modalnych Wywołanych Dodatkową Masą *DYNKON 8 XIII SYMPOZJUM DYNAMIKI KONSTRUKCJI Zeszyty Naukowe Politechniki Rzeszowskiej 58 Mechanika* 74 pp 9-16

[3] Sarzyński M and Włodarczyk E 2017 Propagacja nieliniowych plastycznych fal naprężenia w półnieskończonym pręcie *Biuletyn Wojskowej Akademii Technicznej* VXVI pp 13-25

[4] Major I 2002 Propagacja fali przyspieszenia w pręcie o zmiennym przekroju *Zeszyty Naukowe Politechniki Śląskiej* pp 379-386

[5] Major M and Major I 2017 Analysis of the Mechanical Wave in the Composite Made of Sandstone and Rubber *Procedia Engineering* 190 pp 223-230

[6] Selejdak J and Major M 2016 Hyperelastic Zahorski material – numerical analysis and simulation in ADINA software *Technical Transactions* 113 (3) pp 131-136

[7] Tasdemirci A and Hall I and Haque B and Guden M 2004 Stress Wave Propagation Effects in Two- and Three-Layered Composite Materials *Journal of Composite Materials* 38 pp 995-1009

[8] Jaszak P 2016 Modelowanie gumy za pomocą metody elementów skończonych *Elastomery* 20 (3) pp 31-39

[9] Murnaghan F D 1937 Finite Deformations of an Elastic Solid American *Journal of Mathematics*

59 (2) pp 235–60
[10] Shved O L 2019 Murnaghan's Elastoplastic Material Model *Mechanics of Solids* 54 pp 819–831
[11] Jaskot A and Major M 2021 Analysis of forced vibration damping with the use of hyperelastic materials, *Transactions of the VSB - Technical University of Ostrava, Civil Engineering Series 2021* 21(2) pp 21-24.

Practical deflection analysis of tubular telecommunication towers subjected to a temperature gradient

P A Król[1]

[1] Warsaw University of Technology, Faculty of Civil Engineering, Al. Armii Ludowej 16, 00-637 Warsaw, Poland

E-mail: pawel.krol@pw.edu.pl

Abstract. The subject of the paper are tubular telecommunication towers, commonly used by mobile telephony operators to build relay stations placed at relatively short distances from each other. The advantage of this type of facilities is their simple structure and small surface area they occupy. It has been observed that during operation, as a result of solar radiation, the towers experience excessive deflection, resulting in disturbances in the operation of transmitting and receiving systems mounted on the towers. The issue of a reliable and relatively precise assessment of the phenomenon of uneven heating of the body surface due to sunlight is not a trivial issue. From a practical view-point, the greatest difficulty is to estimate the temperature to which the surface of a structure exposed to sunlight heats up. Even the attempts at numerical modeling of the phenomenon do not accurately reflect its essence. The paper presents a proposal of a simplified analytical computational model that allows for a practical and uncomplicated estimation of the thermal loads to which such structures are subjected as a result of insolation. The considerations were supplemented with examples from engineering practice, showing the differences in deflection of structures protected by zinc and paint coatings.

1. Introduction

Telecommunication towers, to which this paper relates, were designed as tubular structures composed of segments made of circular hollow sections with a constant cross-section along the length of the segment, and the following dimensions, respectively: ϕ711.0x14.2, ϕ660.4x10, ϕ558.2x10, ϕ457.2x9.5, ϕ355.6x9.5, ϕ273.0x8.8 i ϕ114.3x7.1, depending on the tower type. The segments were inter-connected by the so-called bolted flanged connections. The total heights of the structures analyzed in the paper are respectively 22.43 m, 28.43 m and 34.43 m above ground level, together with the lightning rod situated on the top. They are typical commercial solutions used by one of the mobile telephony operators, known as MONOPOL NORMAL or MONOPOL STRONG, provided with an accompanying numerical symbol describing the height of the structural part. Details of dimensions of the analyzed structures are shown in figure 1, while examples of existing towers can be seen in figure 2. As a standard, the outer surface of the tower shaft was zinc coated. It has been observed that during operation, as a result of solar radiation, the towers experience excessive deflection, resulting in disturbances in the operation of antenna transmitting and receiving systems mounted on the towers. It should be mentioned that the intensification of this type of irregularities in the operation of devices occurred most often in the summer season.

IOP Conf. Series: Materials Science and Engineering 1252 (2022) 012071 doi:10.1088/1757-899X/1252/1/012071

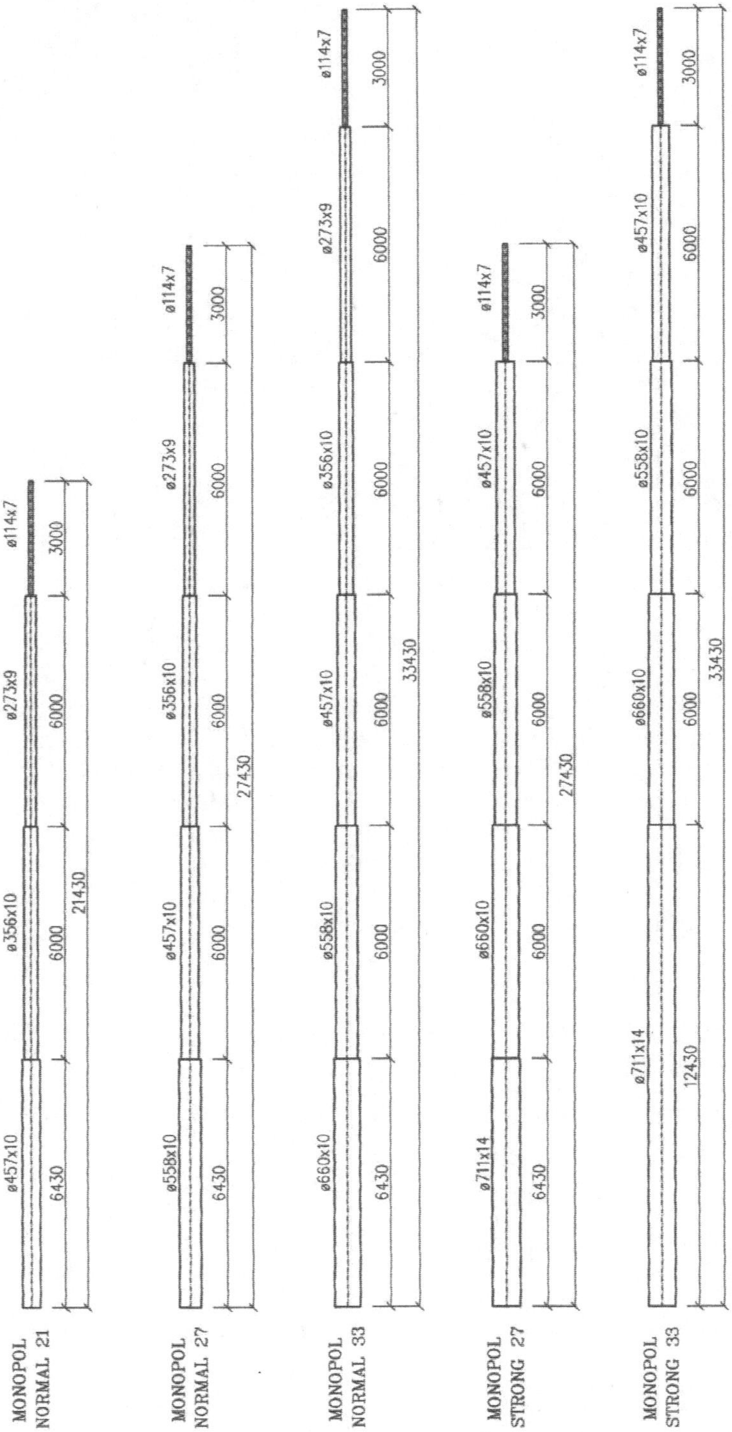

Figure 1. Dimensional details of the analyzed towers.

Figure 2. An example of a structure seen from two directions (author's photos).

Solar radiation exerting effect on the surface of any body is absorbed or reflected by it, and in the case of transparent bodies, we may also observe the phenomenon of radiation transmission inside that body. Absorbed radiation provides this body, on a continuous basis, with a certain amount of energy that becomes converted into heat; this can be noticed in the form of an increase in the body temperature. In parallel, the same body emits energy to the environment, by radiating it out. Some of the heat energy is also transferred by conduction. In the overall energy balance, the body will be heating up until it reaches the temperature at which it absorbs the same amount of energy as it simultaneously loses. This is called the equilibrium temperature. It can be presented in the form of equation (1), which, having made some simplifying assumptions, can be used to estimate the temperature of the body heated as a result of insolation [1].

$$q_{cond} - q_{conv} - q_{rad} = 0 \tag{1}$$

where: q_{cond} - heat flux exchanged by conduction [W/m^2], q_{conv} - heat flux exchanged by convection [W/m^2], q_{rad} - heat flux exchanged by radiation [W/m^2]

Although it might seem otherwise, the issue of a reliable and possibly precise assessment of the phenomenon of uneven heating of the body surface as a result of sun operation is not a trivial issue. It depends on many factors, such as: the size of the absorption coefficient and the surface emissivity coefficient (depending on the method of the surface finishing), the intensity of solar radiation, the thermal conductivity of the material from which a given body is made, the degree of contamination of the medium in which the body is immersed (in this case – air), the presence of other cooling factors (e.g. wind, contact with a body with a different thermal inertia), orientation of the surface in relation to the directions of the world, tilt angle, etc. Even some attempts made at creating a numerical model of this phenomenon do not reflect its essence with sufficient accuracy. Analytical models applied usually contain certain acceptable simplifications, thanks to which the estimation can be made in a "safe" way. However, it should be borne in mind that results of analyzes carried out on the basis of such models are usually quite conservative and slightly exaggerated. In particular, in the case of structures made of steel components with relatively small dimensions, it should be kept in mind that the thermal conductivity of the material is so high that it is difficult to expect the structure from the non-sunlit side to maintain the ambient temperature when at the same time, the sunlit side is heated to a several dozen degrees higher temperature. Simplified analytical computational models disregard the phenomenon of heat conduction, considering such simplification as having, in general, a positive effect on the structural safety. A similar approach was also proposed for the calculations made for the purposes of this study. The aim of the paper is to present a computational model that can be directly used to assess the scale of the analyzed phenomenon in engineering practice or expert activities.

2. Determination of the temperature of the sunlit surface

The following assumptions were made in the calculations:

- After a certain state of thermal equilibrium is established, the heat exchange resulting from conduction between the outer and inner surface of the mast shaft tube can be disregarded locally, so that the general equation of the energy balance of the tower shaft surface (1) in relation to the reference unit surface can be simplified to form (2) [2]:

$$q_{conv} + q_{rad} = 0 \tag{2}$$

- Radiation heat exchange can be described by equation (3), [3]:

$$q_{rad} = \varepsilon\sigma T^4 - \alpha_s I - \varepsilon\varepsilon_{sky}\sigma T_e^4 \tag{3}$$

where: ε - surface emissivity coefficient, α_s - surface absorption coefficient, I - solar radiation intensity [W/m^2], ε_{sky} - sky emissivity coefficient = 0.77, σ - Stefan-Boltzmann constant = $5.67032 \cdot 10^{-8}$ [W/(m^2·K^4)], T - temperature of the body heated by solar radiation [K], T_e - temperature of the outside medium/air [K],

- Loss of heat as a result of convection can be brought down to the form (4) [3]:

$$q_{conv} = h_c\left(T - T_e\right) = 1.24\left(T - T_e\right)^{1/3}\left(T - T_e\right) = 1.24\left(T - T_e\right)^{4/3} \tag{4}$$

where: $h_c = 1.24\left(T - T_e\right)^{1/3} W/\left(m^2K\right)$ - convective heat transfer coefficient.

As a result of substituting equations (3) and (4) to equation (2), equation (5) was obtained, which was used as the basis for further calculations:

$$1.24\left(T - T_e\right)^{4/3} + \varepsilon\sigma T^4 - \alpha_s I - \varepsilon \cdot 0.77\sigma T_e^4 = 0 \tag{5}$$

The calculations were made for weather conditions of the summer and winter seasons, using the values of the outside air temperature and the characteristic values of solar radiation for surfaces with vertical orientation on the basis of the Polish national design standard PN-B-02015:1986 [4]. These are respectively:

$$T_{e(summer)} = 27^{\circ}C = 300K$$

$$T_{e(winter)} = -24^{\circ}C = 249K$$

$$I = 874\, W/m^2$$

The choice of the national standards as a basis for the assessment of structural the deformation resulted, among others, from the fact that the facilities were designed precisely according to these standards. The national standard PN-B-02015:1986 [4], in contrast to the European standard EN 1991-1-5: 2005 [5], provides constant reference values for outside air temperature for the entire country and does not make them dependent on the precise location of the object. Considering that the objects covered by the analyses were located in various locations throughout the country, such an approach, providing an acceptable global solution, was considered the most appropriate. In addition, the standard EN 1991-1-5:2003 [5] does not address quantitatively the issue of loads resulting from sunlight, treating this issue in a casual, purely qualitative manner. In this respect, it replaces the previous national standard PN-B-02015:1986 [4] only partially, which made the UE standard useless for the purposes of the analyses undertaken. Similarly, it is difficult to find any particulars that could constitute a certain reference benchmark, also in the content of EN 1990:2004 [6].

In addition, the emissivity and absorption coefficients were adopted for the surface made of galvanized steel and that covered with white polyurethane paint [4, 7, 8] based on the literature sources. These are respectively:

- for the hot-dip galvanized surface:

$$\varepsilon = 0.28$$

$$\alpha_s = 0.60$$

- for the surface painted with white polyurethane paint:

$$\varepsilon = 0.84$$

$$\alpha_s = 0.27$$

As a result, the issue of determining the value of temperature on the sunlit surface brings down to solving four independent equilibrium equations (6) - (9):

- for the summer season and the galvanized surface:

$$1.24(T_1 - 300)^{4/3} + 0.28 \cdot 5.67 \cdot 10^{-8} T_1^4 - 0.60 \cdot 874 - 0.28 \cdot 0.77 \cdot 5.67 \cdot 10^{-8} \cdot 300^4 = 0 \quad (6)$$

- for the summer season and the surface painted with white polyurethane paint:

$$1.24(T_2 - 300)^{4/3} + 0.84 \cdot 5.67 \cdot 10^{-8} T_2^4 - 0.27 \cdot 874 - 0.84 \cdot 0.77 \cdot 5.67 \cdot 10^{-8} \cdot 300^4 = 0 \quad (7)$$

- for the winter season and the galvanized surface:

$$1.24(T_3 - 249)^{4/3} + 0.28 \cdot 5.67 \cdot 10^{-8} T_3^4 - 0.60 \cdot 874 - 0.28 \cdot 0.77 \cdot 5.67 \cdot 10^{-8} \cdot 249^4 = 0 \quad (8)$$

- for the winter season and the surface painted with white polyurethane paint:

$$1.24(T_4 - 249)^{4/3} + 0.84 \cdot 5.67 \cdot 10^{-8} T_4^4 - 0.27 \cdot 874 - 0.84 \cdot 0.77 \cdot 5.67 \cdot 10^{-8} \cdot 249^4 = 0 \quad (9)$$

As a result of the calculations made, the following temperature values of the surfaces exposed to solar radiation were obtained:

$$T_1 = 367K = 94^\circ C \qquad T_2 = 317K = 44^\circ C \qquad T_3 = 324.5K = 51.5^\circ C \qquad T_4 = 276K = 3^\circ C$$

These values were used to perform static and strength calculations aimed at determining the displacements of the tower top loaded solely with its own weight and the temperature gradient. Structural models were developed and calculations were carried out in the environment of the popular FEM engineering software RM-WIN ver. 11 [9], intended for the static and strength analysis of 2D bar structures. The structure models were loaded in such a way that a thermal load was applied to the fibers on one side of the element, corresponding to the adopted values of the outside air temperature, $T_{e\,(summer)}$ or $T_{e\,(winter)}$, respectively - depending on the design situation, and to the fibers on the other side - with the calculated appropriate temperature values T_1, T_2, T_3 or T_4. As a result, it came down to the analysis of four load variants for each of the towers, corresponding to the effect of the temperature gradient, (10) - (13):

- for the summer season and the galvanized surface:

$$\Delta T_1 = T_1 - T_{e(summer)} = 94°C - 27°C = 67°C \quad (10)$$

- for the summer season and the surface painted with white polyurethane paint:

$$\Delta T_2 = T_2 - T_{e(summer)} = 44°C - 27°C = 17°C \quad (11)$$

- for the winter season and the galvanized surface:

$$\Delta T_3 = T_3 - T_{e(winter)} = 51.5°C - (-24°C) = 75.5°C \quad (12)$$

- for the winter season and the surface painted with white polyurethane paint:

$$\Delta T_4 = T_4 - T_{e(winter)} = 3°C - (-24°C) = 27°C \quad (13)$$

The higher values of the temperature gradients obtained for winter conditions as compared to the corresponding calculation situations for summer conditions are noteworthy. This paradox is the result of the adopted simplifications that ignore the heat flow along the width of the element and will be reflected in the obtained results. The calculation results are presented in a graphical form in figure 3 and figure 4 as well as summarized in table 1.

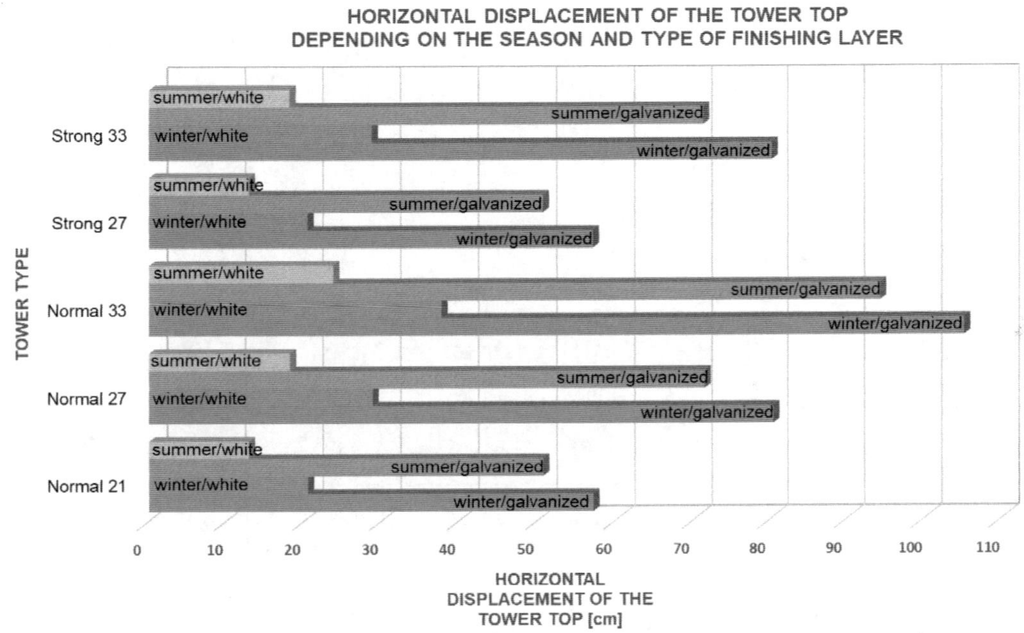

Figure 3. Horizontal displacement of the tower top, depending on the season and finishing layer.

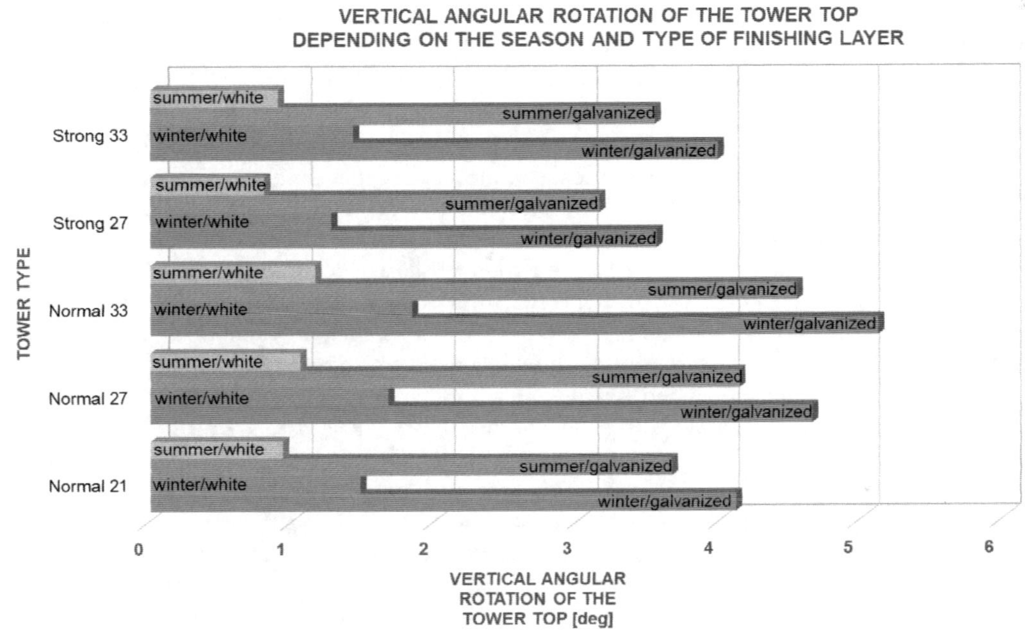

Figure 4. Vertical angular rotation of the tower top, depending on the season and finishing layer.

Table 1. Summary of calculation results of the tower top displacement.

Operating conditions	Displacement of the tower top	Tower type				
		Monopol Normal 21	Monopol Normal 27	Monopol Normal 33	Monopol Strong 27	Monopol Strong 33
Summer, galvanized surface	Horizontal [cm]	50.64	71.47	94.43	50.62	71.27
	Vertical [cm]	1.55	1.98	2.41	1.98	2.41
	Vertical angular rotation [deg]	3.647	4.134	4.547	3.144	3.533
Summer, surface covered with white paint	Horizontal [cm]	12.85	18.13	23.71	12.84	18.08
	Vertical [cm]	0.91	1.16	1.41	1.16	1.41
	Vertical angular rotation [deg]	0.925	1.049	1.154	0.798	0.896
Winter, galvanized surface	Horizontal [cm]	57.07	80.53	105.29	57.04	80.31
	Vertical [cm]	0.35	0.44	0.54	0.44	0.54
	Vertical angular rotation [deg]	4.109	4.658	5.124	3.543	3.981
Winter, surface covered with white paint	Horizontal [cm]	20.41	28.80	37.65	20.40	28.72
	Vertical [cm]	0.28	0.36	0.44	0.36	0.44
	Vertical angular rotation [deg]	1.470	1.666	1.832	1.267	1.424

The effects of wind on the tower shaft, linear equipment elements as well as transmitting and receiving systems were disregarded in the displacement calculations, although they are obviously not indifferent to both the value of horizontal displacements and the rotation angle of the tower top. Taking into account the cooling aspect of the wind, an assumption was made that such simplification of the computational model, consisting in the analysis of the structural deformations in windless conditions, is acceptable.

3. Discussion of the results and conclusions

Basing on the case study carried out and discussing the results obtained to a broader extent, the following remarks and conclusions can be formulated:

- As regards the hot-dip galvanized structure, horizontal displacements and vertical angular rotations of the tower top in each case exceed the permissible limit values specified in Polish design standard PN-B-03204:2002 [10], determined respectively at the level of: 1/100 of the tower height and 1 degree. In fact, for most broadcasting systems, such restrictions are sufficient, although there are devices with a slightly sharper requirements. European standard EN 1993-3-1:2006 [11] does not give specific limit values, limiting only to the statement that in the case of telecommunication towers the displacements cannot hinder the proper functioning of antennas or devices or disappearance of transmitted signals.
- The displacement values obtained for the winter operating conditions of the structure exceed the corresponding values obtained for the summer season. This paradox results from a greater difference in the value of the temperature gradient, determined according to the PN-B-02015:1986 design standard [4] and omission in the analysis phase of the heat flow phenomenon along the width of the element. The results obtained for the winter operating conditions should be treated with a proper degree of caution, since they are burdened with an error resulting from the model simplifications that ignore the effect of thermal conductivity of the tower material, treating them as correct only in the quality and not quantitative dimension. In fact, it is difficult to imagine a situation where one side of the tower (not sunlit) maintains a

temperature of -24.0°C, while the other (sunlit) at the same time reaches a temperature of 51.5°C. In view of the foregoing, it should be noted that the results obtained in the conditions of the same season are comparable in each of the cases, since they are burdened with the same error.

- The displacements of the top of the hot-dip galvanized tubular tower shaft can be effectively limited by painting it white with a properly selected paint having documented low absorption and, at the same time, relatively high emissivity. This has been confirmed in the form of applications on specific, real facilities. One of them is shown in figure 2.
- Painting of the galvanized structure white in the summer conditions leads to an almost four-time reduction in the tower top displacements, as a result of which the obtained values meet or almost meet the limit criteria specified in the PN-B-03204:2002 design standard [10]. In the case of normal conditions of the winter season, the "yield" ratio resulting from painting the structure white is slightly lower and amounts to 2.8. However, it should be borne in mind that this value is burdened with the calculation model error.
- The values provided in table 1 should at all times be treated as indicative. The actual values of the structure surface temperature depend on many factors mentioned earlier. From a practical, engineering point of view, they are, however, sufficient for a qualitative description of the scale of the phenomenon.
- The method proposed allows, in the event of operational problems, estimating with sufficient precision from the point of view of engineering practice, the deformation of shafts of tubular towers resulting from the effect of insulation. To obtain more precise results, it would be necessary to perform a series of land surveying field tests/measurements in various conditions of locations of the facilities as well as accompanying surface temperature measurements, and then adjust the calculated displacements using properly calibrated scaling factors.
- At the next stage of the study, some comparative analyzes using more advanced FEM tools and shell element models are planned.

4. References

[1] Bergman T L, Lavine A S, Incropera F P and Dewitt D P 2011 *Fundamentals of Heat and Mass Transfer 7th Ed.* (John Wiley & Sons, Inc.) pp.10-28

[2] Vignola F, Michalsky J and Stoffel T 2017 *Solar and Infrared Radiation Measurements* (CRC Press Taylor and Francis Group) pp.28-30

[3] Howell J R, Mengüc M P and Siegel R 2010 *Thermal Radiation Heat Transfer, 5th Ed.* (CRC Press Taylor and Francis Group) pp. 37-38 and 356-359

[4] PN-B-02015:1986 – *Actions on Building Structures. Variable environmental actions, Temperature Action*, Polish design standard, (in Polish)

[5] EN 1991-1-5:2003 – *Eurocode 1: Actions on structures – Part 1-5: General actions – Thermal actions,* CEN

[6] EN 1990:2004 – *Eurocode: Basis of structural design,* CEN

[7] Bramson M A 1968 *Infrared Radiation, A Handbook for Applications* (Springer) pp.533-552

[8] Zacharski A Z, Pogorzelski J A 1999 *Opinion on the thermal properties of the paint coating called 'Emapur', for technical approvals* (Instytut Techniki Budowlanej, Warsaw, Poland), p. 5, (in Polish)

[9] RM-WIN ver.11 engineering software, *https://www.cadsis.pl/produkty/opis/rm-win,* (in Polish)

[10] PN-B-03204:2002 – *Steel Structures. Radio and Television Masts and Towers. Design Rules*, Polish design standard, (in Polish)

[11] EN 1993-3-1:2006 – *Eurocode 3: Design of steel structures – Part 3-1: Towers, masts and chimneys – Towers and masts,* CEN

INCORPORATING NEW TECHNOLOGIES IN ROAD INDUSTRY: TEST TRACK IN ANTWERP, BELGIUM

S R Omranian[1,*], W Van den bergh[1], B Golchin[2], K Couscheir[1], C Vuye[1], and J Braspenninckx[3]

[1] University of Antwerp, Faculty of Applied Engineering, EMIB Research Group, Groenenborgerlaan 171, B-2020 Antwerp, Belgium
[2] Department of Civil Engineering, University of Mohaghegh Ardabili, Ardabil, Iran
[3] Port of Antwerp, Antwerp, Belgium

[*] E-mail: SeyedReza.Omranian@uantwerpen.be

Abstract. Numerous factors may disrupt pavement construction including human errors and unforeseen changes in the environmental conditions. These incidents can shorten pavement life span. New technologies recently came to the aid of contractors and road agencies to detect and possibly mitigate undesirable impacts of unpredictable incidents during the asphalt paving process as well as monitoring pavement performance during pavements service life. In this study, several technologies including layer thickness assessment (using both MIT-scan device coupled with metal plates and conventional topographical technique), infrared thermal camera, intelligent roller compactors, and density gauge devices were utilized to develop a coherent construction system and improve the durability of asphalt pavements in Antwerp, Belgium. The data collected in situ were compared with the characterized cores in the laboratory. It was clear that several factors can influence the outcome of the employed devices. For instance, surface temperatures obtained from an infrared camera fluctuated when the camera tilted from the perpendicular axis. Such discrepancies may be associated with the changes in the distance and environmental condition that affect the refraction of the infrared spectra. It was also found that an elaborated data collection format is essential to clearly understand the correlations between the results of the employed technologies.
Keywords: Durable asphalt pavement, Systematic construction, Advanced paving technologies, Density, Smart roller compactor, Layer thickness measurement, Infrared temperature camera

1. Introduction

Pavement performance and durability directly reflect the asphalt mixture features including mix design, properties of constituents, and precision of mixtures production and construction. Assuming the produced mixtures fall in the standard range, a reliable construction process plays a substantial role in pavement durability. It is known that a poor quality construction can result in distresses such as (early) raveling, potholes, and cracking which can considerably influence ride comfort and affect road users' safety. Currently, several new technologies are being utilized to improve the quality of asphalt mixtures during pavement construction. Some of these technologies gather and analyze data concurrent with paving procedures which can even be used for the immediate adjustment and execution in situ. For instance, increasing the number of roller compactor passes can compensate for the possible reduction in the mixture temperature during transportation from plant to construction site [1]. Several studies explored the potential of using these technologies to control the quality of constructed pavements. For instance, Sebesta and Scullion (2002) reported obvious advantages when infrared (IR) imaging and Ground Penetrating Radar (GPR) were applied to evaluate the uniformity of newly placed hot-mix overlays [2]. In order to control pavements condition, Baltrušaitis et al. (2020) measured pavement density and air void content using GPR after the end of the asphalt layer compaction process [3]. The reliability of the data was checked and confirmed by comparing them with the results measured directly on cores taken from the asphalt pavement layers. Leng et al. (2018) used electromagnetic density gauges to assess the density of asphalt mixtures after compaction, and significant effects of air voids and binder contents on the results were discovered [4].

It is known that one of the main and initial critical factors that significantly affects asphalt pavement quality is the temperature of compaction. Infrared (IR) cameras have recently been employed in the pavement industry for monitoring paving temperatures. The device is a thermographic camera and measures infrared radiation as a form of electromagnetic radiation. The amount of radiation emitted by the object reflects its temperature [5]. Keeping mixing, paving, and compaction temperatures in the designated ranges can beneficially enhance road durability. Since the mixture temperature might fluctuate due to unforeseen incidents, the infrared camera outcomes can be used for checking the quality of construction. It can also be used to locate changes in the asphalt finisher movements based on the fluctuations in temperature during asphalt delivery when coupling a truck with the finisher. Moreover, places that are more susceptible to damage or failure such as rupture can be detected based on the noticeable differences in paving temperatures [5]. The IR camera was also used for other purposes. For example, Stroup-Gardiner et al. (2000) applied infrared thermography to identify and measure segregation in hot asphalt mixtures [6]. Vayas et al. (2019) used IR thermography for debonding detection in asphalt pavements [7]. In another study, Vemuri et al. (2017) applied IR thermography and GPR as rapid scanning techniques for inspecting bridge decks with an asphalt overlay [8].

Considering production and paving temperatures satisfy the requirements outlined in specification, proper compaction plays a crucial role in a reliable pavement execution. Intelligent rollers can promote the ultimate quality of the constructed roads. Using a tablet per roller along with a GPS receiver and server in intelligent roller compactors, the compaction data can instantly be displayed in their cabin. By these measurements, the roller settings can be adjusted to determine the optimal number of roller passes during construction. Some advantages of this technology are: there is both quality control and quality assurance (QC/QA) in situ, it can be mounted on any type of roller, and rollers can be linked and synchronized together to prevent unnecessary effort and achieve desired compaction with lower energy consumption. It should be informed that such a system contains a database related to optimal compaction which should normally be provided by the roller manufacturer company. Xu et al. (2016) presented an innovative material-machine-information and human-decision integrated system for adaptive QC/QA using the intelligent compaction (IC) technology to overcome traditional compaction test controls [9]. This method developed a promising approach to improve the quality of the construction through adaptive control of compaction. Tirado et al. (2020) integrated intelligent compaction technology and modulus-based testing [10]. They proposed a protocol to address some of the main concerns frequently needed for quality control. Similarly, Wang et al. (2017) designed a compaction parameters management system for intelligent vibratory rollers in highway constructions [11].

In the past, the main method to judge roads quality and mixtures segregation was by means of visualization and core sampling. Core drilling can reduce pavement coherence and induce weaknesses that can lead to a shorter pavement lifespan. In addition, the results of cores density are generally reported after pavement construction which is too late for in situ adjustments during the compaction process. Recently, a PQI (Pavement Quality Indicator) is introduced to indirectly determine the density and air voids of asphalt pavements in situ. The device can also be employed to quantitatively evaluate the degree of segregation during the quality inspection of pavement construction [12]. PQI is a non-nuclear, electrical, non-destructive density measuring device. In this system, an electromagnetic wave is sent through the material. A difference in density results in different electrical impedances. In order to have reliable results, it is important to have a clean, dry surface and good interaction with the surface. PQI was utilized in the study conducted by Chen et al. (2006) to monitor the segregation of hot-mix asphalt pavement [12]. Although PQI were not recommended as a quality control tool by Henault et al. (2001) [13], Leng et al. (2018) stated that the measured values were comparable to analysis conducted based on conventional standard methods. It was also found that measurement direction, asphalt binder content, mixture air void content, and thickness of the testing slabs do not affect the accuracy of the density gauges, while presence of moisture, gradation of asphalt mixture and gauge calibration method can affect the outcomes [14].

The potential of using the abovementioned technologies in the Belgian road industry, particularly Flanders was explored in this study. This study also focused on the correlation between infrared camera and smart roller compactor as well as core sampling and data measured in situ. Results were finally analyzed through a statistical analysis approach. In this paper, after a brief overview of the study and previous research, methodologies as well as techniques and devices used to evaluate pavement construction quality are presented, then followed by the results and discussion section. Finally, the major conclusions and recommendations are provided.

2. Materials and methods

2.1. Test track

A test track constructed in Antwerp state, Belgium, were selected for this study. The test track is named A1 to avoid any potential commercialization. A1 was evaluated comprehensively using all the technologies addressed in Section 2.2. Different asphalt mixtures including AVS (Dutch: Asfaltmengsel met verhoogde stijdheid/EN13108-1), a dense asphalt mixture with increased stiffness modulus, recommended for base layer of higher construction classes, APO (Dutch: Asfaltmengsel met Prestatieomschrijving voor Onderlagen; EN13108-1), a dense asphalt mixture, recommended for general use as base layer, and SMA (stone mastic asphalt; EN13108-5) a typical surface layer in Belgium, were used to construct A1. The ranges of standard volumetric properties of these mixtures are introduced in Table 1. AVS mixture was used to pave base and middle layers for a part of this test track, while the other part was constructed using APO. SMA was finally used to construct the top layer for the entire road.

Table 1. Properties of asphalt layers in accordance with Flemish specifications SB250 V3.1.

Type	Apparent density (kg/m^3)	MTD (kg/m^3)	%Air Voids
SMA	2252-2399	2448	2-8
APO	2289-2438	2488	2-8
AVS	>2290	2489	<8

2.2. Data collection

Six different technologies were incorporated in this research including measurements of layer thicknesses using metal plates and topographical method, infrared thermal camera, intelligent rollers, and density. Core sampling was also used for quality control. These technologies data coupled with core outcomes were finally used for comparison and statistical analysis.

2.2.1. Layer thickness. The target values of the structural design were 40, 70, and 70 mm for surface, binder and base layers, respectively. The thicknesses of the different layers were determined in two ways including metal plates coupled with MIT-scan and topographical height. In the first method, aluminum plates were placed underneath each constructed layer as shown in Figure 1(A). To avoid miscalculation two sets of plates were positioned approximately 50cm away from each other as schematically illustrated in Figure 1(B). the measurements were then carried out using a MIT-scan T3. In the second method, the topographic measurements were conducted by means of a laser altimeter. The results were reported relative to a fixed reference point. The topographical measurements were executed at the same locations as the first method. It should be mentioned that the thicknesses were measured for at least two specific, neighboring points, using both techniques.

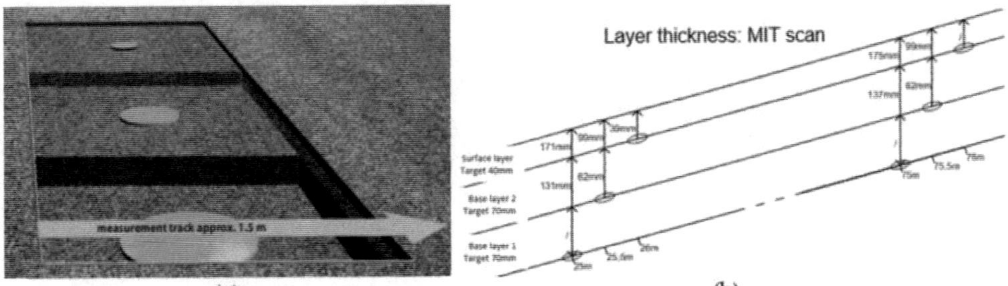

(a) (b)

Figure 1. Layer thickness results with MIT-scan and metal plates [15].

2.2.2. Infrared thermal camera. Mixture temperatures during the paving procedure were determined using an infrared thermal camera. The measurements took place over the entire width of the asphalt paver. The values were taken over a grid of 25 by 25 cm with a required accuracy of 1 °C. The collected data together with their GPS coordinates were then exported to an Excel file for further analysis. It should be informed that the outlier temperatures (below 70 °C) which could influence the precision of investigation were manually removed prior to the analysis.

2.2.3. Intelligent rollers. The number of roll passes, compaction temperature, and the EVIB value were monitored during the construction using intelligent rollers. The EVIB value is a physical measured value that correlates directly with the plate-loading test used in earthworks and road construction. With EVIB measurements, the compaction energy which is reflected during vibration is registered. The EVIB value was measured with 1 MN/m accuracy. The color code template used for the EVIB values and roller passes are illustrated in Figure 2.

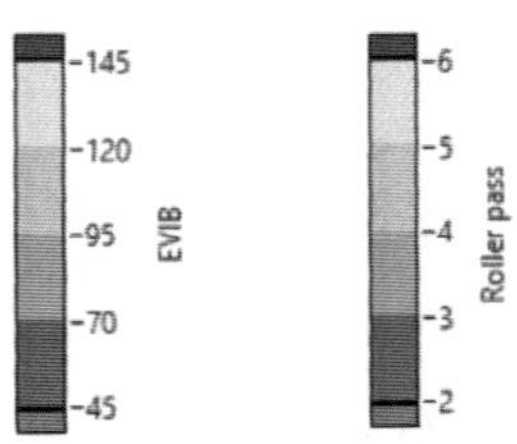

Figure 2. Color Codes for EVIB Values and Roller Passes.

2.2.4. Volumetric properties. Pavement density, air voids, and level of compaction were measured using a PQI-380 non-nuclear density gauge by TransTech at every 20m intervals. The position indication was done by indicating left/right including its distance from the starting point. The density and air voids are expressed in kg/m³ and %, respectively, with the precision of two decimals. after the title.

2.3. Core sampling

To investigate the accuracy of the measurements during construction, thirteen cores were drilled. Six of them were taken on the main road test track, while the other seven were collected from the roundabout and the T-junctions as illustrated in Figure 3. The volumetric properties of drilled cores were then determined in the laboratory.

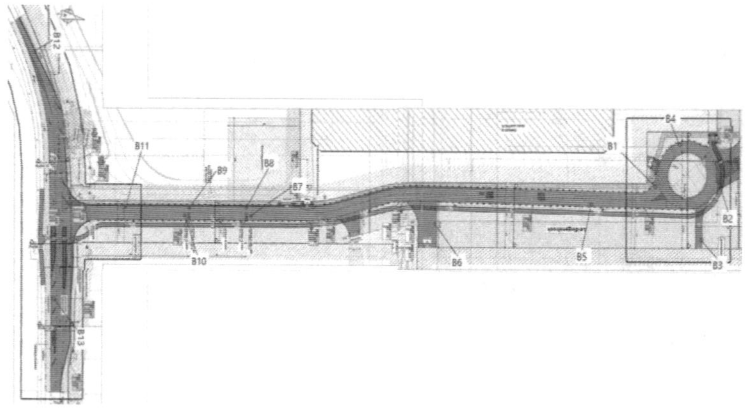

Figure 3. Location of Drilled Cores for A1.

2.4. Statistical data analysis
To find the statistical correlation between the variables, Minitab software was employed to perform Pearson correlation and the Main effect plots analysis. Minitab helps to find trends in the data, correlations between different variables as well as data visualization.

The Pearson correlation examines whether there is a linear relationship between two variables. The analysis gives the correlation value, the confidence intervals for correlation (p-value), and a matrix plot [16]. N is the sample size, while the correlation value indicates the level of dependency or relation between variables that range between -1 and +1. A positive value indicates that both variables increase together, while a negative value denotes variables behave contrarily [16].

A confidence interval needs to be determined since the analysis was performed on randomly selected data. A confidence interval is an interval of values for correlation coefficients in which the unknown correlation coefficient will be included. The 95% confidence level was chosen for this study. It should be noted that the data may not differ noticeably from a normal distribution, since it can lower the confidence intervals accuracy, regardless of the magnitude of sample size [16]. The p-value is used to determine whether the correlation value is statistically significant. A significance level of 0.05 was considered in this study. The correlation is statistically significant when the p-value is less than or equal to the significance level, and in this case, the null hypothesis can be rejected. A p-value greater than the significance level does not provide enough evidence to reject the null hypothesis [16]. The matrix plot displays the data visually and their relationship trends such as linear, monotonous, or no relationship at all. It can also be used to identify possible outliers.

The statistical analysis was extended using the main effect plot which determines the levels of effect of factors on the responses. The horizontal line means no considerable effect, while the magnitude of the main effect can be calculated form the slope of the line.

3. Results and discussions

3.1. Results obtained from incorporated technologies

3.1.1. Layer thickness. Figure 4 graphically presents the assigned thicknesses of each layer to a measured points, where TL, ML, and BL represent the top, middle, and bottom/base layers, respectively. It should be informed that several problems were encountered using the MIT-scan method in reality. After placing the plates and constructing the base (bottom) layer, measurements had to be taken. However, it was not possible to obtain values for the thickness of the specified layer since the device was out of order. The next problem was locating plates after covering with mixtures which caused failure to measure the thickness of the surface layer at some locations. Table 2 shows the percentages of deviations between the measured values and the target values. A positive percentage denotes a thicker constructed layer than its target value and vice versa. Blank cells represent the places with no measurement, the darker highlighted cells show higher deviation from the target values, while the red denotes places with the highest deviation. The MIT-scan method exhibits better approximation compared to the topographical method based on the deviations from target values between 0 and 8%. According to the Flemish standard, the surface thickness of 40 mm may vary between 34 and 46 mm, while the base layers of 70 mm should be between 62 and 78 mm. The MIT-scan results indicated that, for the surface layer, only 12% of the outcomes deviated more than 10%, whereas the corresponding value increased to approximately 28% for the base layers.

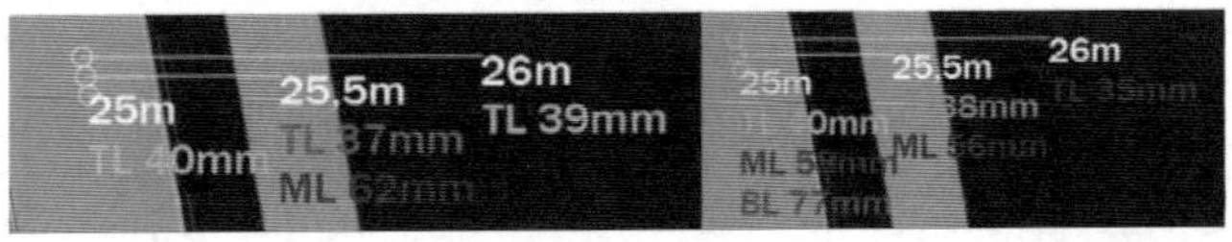

Figure 4. Layer thickness: left: metal plates, right: topographic.

On the contrary, it was found that the topographic measurements were not comparable to the target values. For the surface layer, approximately 50% of measurements deviated more than 10%, while the corresponding value for the base layers was about 43%. According to this method, a large part of the road was therefore not executed correctly. The results also show that at locations 225 m and 225.5 m, the surface layer is over-fitted, the second base layer (ML) however is too thin which denoted that this location was not correctly constructed. Hence, this location did not pass the quality control test based on a deviation of more than 10% from the target values. The results of MIT-scan and topographic methods were also compared. The differences between corresponding thicknesses were mainly between 0 and 6 mm except on locations 225 and 225.5 m where the variation increased to 8-13 mm.

Table 2. Layer thicknesses: deviation percentages to the target value.

Location		25	25.5	26	75	75.5	76	125	125.5	126	170	170.5	171	225	225.5	226	275	275.5	276	325	325.5	326
MIT-scan error [%]	TL	0	-8	-3	-5	-8		0	-8		20	18		5	3		0	-5		5	0	5
	ML		-11			-11			9			1			9			-3			-1	
Topographic error [%]	TL	0	-5	-13	-15	-15	-13	-5	-3	-3	25	23	15	25	33	33	5	-3	0	0	5	-10
	ML	-17	-20		-20	-14		14	11		-1	-3		-10	-10		-6	-1		-1	-4	
	BL	10			16			-10			-39			7			17			3		

3.1.2. Infrared line-scanner. Initially, the analysis was conducted on the data obtained from A1. Figure 5 shows an example of the average paving temperatures per square meter obtained from an infrared camera. Different colors were assigned to differentiate the temperatures. As shown in Figure 5, green, yellow, orange, and red are denoted as temperature within the design range, lower, as well as extremely low, respectively.

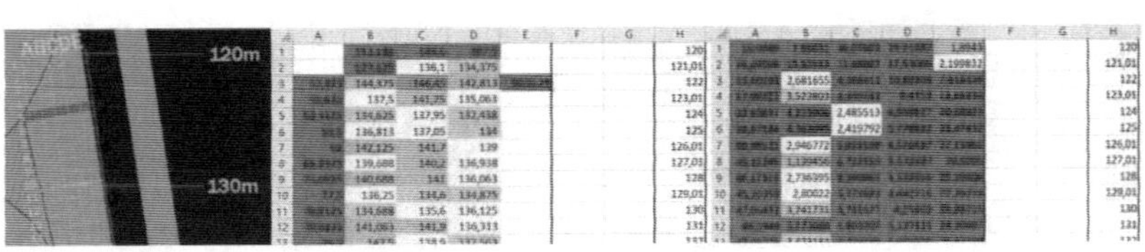

Figure 5. Example of analyzed IR thermal camera results.

Cold spots were assigned to the locations with temperatures lower than 120 °C, while the risk areas were defined as the areas with a temperature lower than 90% of the average temperature. It should be mentioned that both high risk areas and cold spots can contribute to shorter pavement service life which must be avoided for constructing more durable pavements. These values were calculated based on different width. An example of temperature deviation from average temperature in a form of horizontal profile and changes in temperature along the longitudinal profile are provided in Figure 6(A) and (B), respectively. Figure 6(A) shows an increase in the temperature variance when pavement width for analysis increases which can reflect the changes in measurement angle, while Figure 6(B) illustrates the places that trucks were coupled with paver based on temperatures decrements. Figure 7 presents the calculated cold spots and risk areas. It can be seen that the roundabout exhibits the highest risk area and cold spot followed by middle layer or second base layer as shown in the figure. The highest cold spot and risk area of the roundabout can be related to the construction difficulties due to the curves, speed of materials delivery, and manual construction. It can also be seen that the risk area and cold spot declined when road width was reduced. This variance can be associated with the changes in the accuracy of the infrared camera measurements where temperatures are reduced due to the angle deviation from the perpendicular axis to the road centre or/and lower temperature of curbs. Such difference can fluctuate based on the changes in the environmental conditions such as wind speed and humidity which influence the emitted infrared lights. To solve this problem, it is recommended to place multiple cameras next to each other on the asphalt spreader to improve the accuracy of measurements.

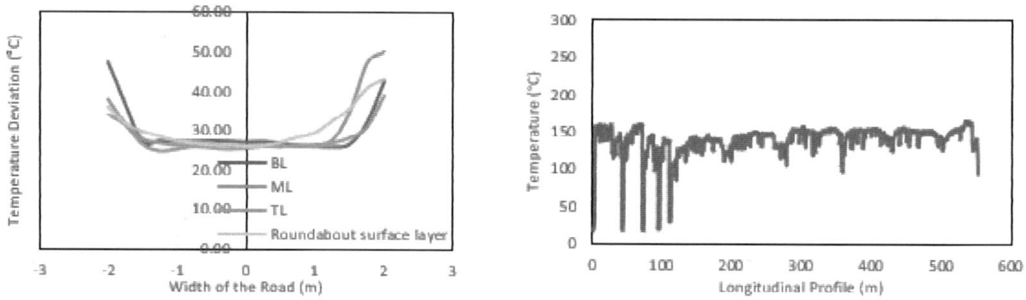

A B

Figure 6. Example of horizontal (A) and longitudinal (B) profiles.

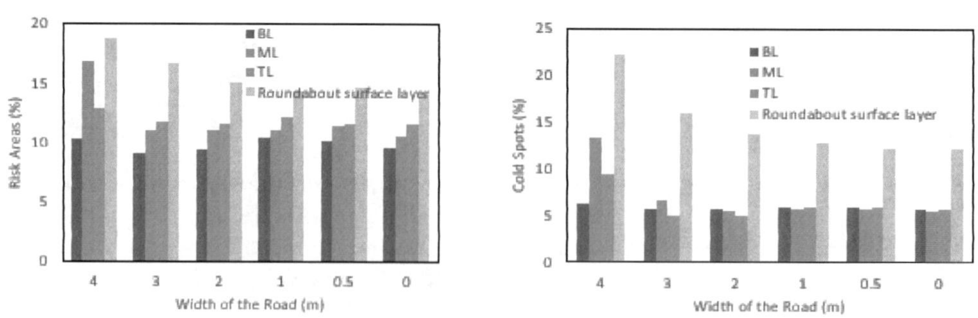

Figure 7. Risk Area (left) and Cold Spot (right).

3.1.3. Intelligent rollers. Different values including EVIB and number of roller passes were recorded by intelligent roller during A1 construction. A few problems occurred during the construction which should be avoided for the next test track. For instance, the EVIB value could not be measured at some locations. Such incidents noticeably happened during the construction of the first base layer. In addition, GPS connection was lost at some points as resulted in a gap mainly for recorded data of the second base layer. According to the intelligent roller supplier, EVIB target values are 320, 250, and 230 MN/m^2 for top, middle, and base layers, respectively. Results showed that approximately 97% of compacted layers were within these ranges. The contractor stated that some locations were compacted using a roller without intelligent system which resulted in the remaining low compacted places (3%). It is therefore recommended to equip all rollers with an intelligent system to integrate all results in the future.

3.1.4. Volumetric properties. PQI-380 was employed to measure the pavement layers density and air voids. The device is equipped with an offset option which directly converts the recorded results into the real density. The offset values of pavement layers are determined based on the procedures outlined in Belgian Road and Research Centre (BRRC) guidelines [17]. The measured PQI offset values were 190.9, 281, and 164 kg/m3 for SMA, APO, and AVS, respectively. According to the Flemish standards, the densities of SMA and APO mixtures should be between 2252 and 2399 kg/m³ and between 2289 and 2438 kg/m³, respectively, while the corresponding value for AVS mixture should be larger than 2290 kg/m³.

Table 3 presents the averages and deviations of density, compaction percentages, and air voids for paved layers and their particular mixture types. The results show that all layers are sufficiently compacted, except the AVS part of the bottom base layer which contains more than 11% air voids. Similarly, the average air voids of the top layer is relatively high. This indicates that there are possibilities that some parts of these layers were not adequately compacted. It should be mentioned that, only one measurement was taken per spot. However, such an approach is not recommended since the recorded values cannot truly reflect the real density due to a lack of precision. Hence, for better accuracy the replication should be increased to 3 per spot in the future.

Table 3. Measurements of Volumetric Properties.

Layer	Mix Type	Average density [kg/m³]	Standard deviation [kg/m³]	Average compaction [%]	Standard deviation [%]	Average Air voids [%]	Standard deviation [%]
TL	SMA	2203.54	56.73	90.18	3.16	9.82	1.15
ML	APO	2405.10	41.21	96.67	1.66	3.33	1.66
ML	AVS	2302.09	39.49	92.49	1.59	7.51	1.59
BL	APO	2357.37	51.38	93.92	1.88	6.08	1.88
BL	AVS	2206.38	55.73	88.65	2.24	11.35	2.24

3.2. Statistical analysis

Statistical analyses were performed for a better understanding of correlations between cores and collected data in situ. Table 4 provides the correlation between variables with a high correlation, i.e. values close to -1 or +1, and the variables that are statistically significant, i.e. the p-values less than or equal to 0.05, are indicated in bold format. Many variables exhibit no significant correlation which can be due to the lack of accuracy or insufficiency in the number of collected data. Therefore, the main effects analysis was only performed for those variables that exhibited significant statistical correlation. The results of cores characterizations were compared with the collected data in the field. Since some parameters such as thicknesses and densities were not measured at the same locations as the drilled cores, interpolation was used to find responses at those locations which may affect analysis precision.

Table 4. Correlation between cores and field data.

Variable 1	Variable 2	N	Correlation	P-value
Core thickness	PQI density	6	0.80	0.05
Core thickness	PQI voids	6	0.91	0.01
Core bulk density	Core voids	13	-0.88	0.00
Core bulk density	Thickness metal plate	7	-0.78	0.04
Core voids	Thickness metal plate	7	0.90	0.01
Thickness metal plate	IR temperature	3	-1.00	0.02
PQI density	PQI voids	6	-0.90	0.01
PQI density	EVIB-value	3	1.00	0.02
Roller passes	Compaction temperature	11	0.76	0.01

Figure 8 presents the relationship between densities and air voids. From Figure 8(A), a significant correlation was found between the results of 13 cores based on a p-value lower than 0.05. A linear correlation can be seen which clarifies that density increment led to air voids reduction. However, Figure 8(B) shows that the relationship between cores and PQI measurements is not statistically significant based on the p-value ($p = 0.924$). As mentioned earlier, insignificant correlation between these data can be associated with the limited number of data obtained using PQI as well as interpolation, since the corresponding values were not collected at the same coring locations.

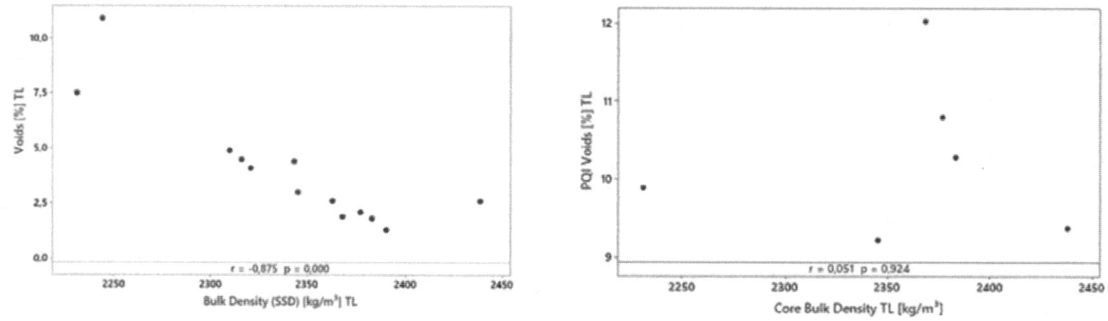

(A) (B)

Figure 8. Correlation between Densities and Air Voids for Cores (A) and in Situ (B).

Furthermore, significant relation between temperatures obtained from the thermal camera and air voids of cores was detected based on the p-value lower than 0.05 (0.011). A strong, negative correlation value (-0.912) was also observed. Figure 9 illustrates the main effects plots. From the statistical analysis and Figure 9(A), it can be concluded that when the paving temperature increases, the air voids decrease. However, Figure 9(A) shows that the percentage of air voids increases by infrared temperature increment from 78.8 to 91.4°C and from 157.4 to 158.3°C. These changes can be associated with infrared temperature bias since the core was taken from the side of the road. It was expected that at higher temperature, fewer roller passes might be required for the road compaction due to lower! bitumen viscosity or better workability. Although significant correlation between the compaction temperature and roller passes was found (p-value 0.007), no obvious trend can be observed from main effects plot as shown in Figure 9(B).

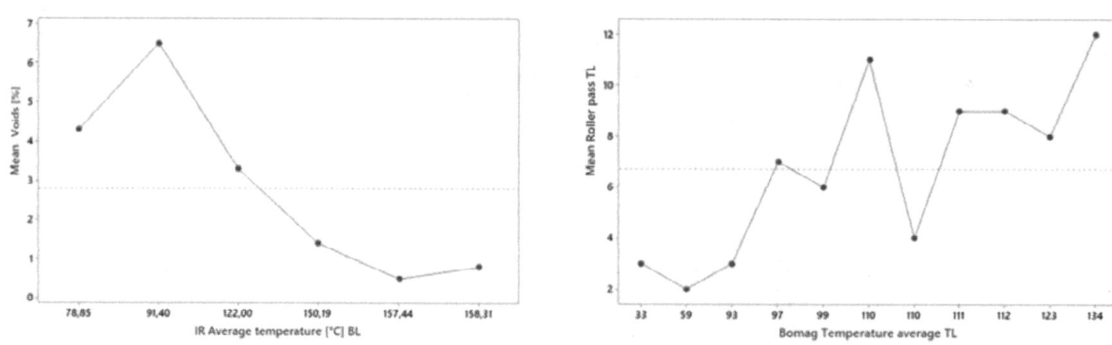

(A) (B)

Figure 9. Main Effect Plots of Temperature on Air Voids (A) and Roller Passes (B).

4. Conclusion

This paper explored the potential of applying new technologies for constructing more durable pavements in Belgium. Applications of different technologies were evaluated during the construction of two test tracks. The following conclusions and recommendations can be drawn from this study.

The layer thicknesses were determined in two different ways, using embedded metal plates and a MIT-scan device, and conventional topographic measurements. The thicknesses obtained from MIT-scan were more close to the target values. The results showed that thicknesses were out of range at the coordinates of approximately 170 m for the surface layer and at 25.5 and 75.5 m for the second base layers. The differences in thicknesses obtained from both methods were mostly between 0 to 6mm.

The proportions of cold spots and risk areas were determined from the infrared thermal camera measurements. The roundabout exhibited the highest risk area and cold spot followed by the second base layer. The calculated risk area and cold spot declined when the measured width of the road was reduced. It was also found that despite low risk area, the paving temperature and cold spot should satisfy the design to achieve sufficient compaction level and volumetric properties.

The use of intelligent rollers was tested as well. This system enables the constructor to detect the spots that may need to be further compacted during construction. Although some issues were encountered during compaction procedures, results showed that approximately 97% of compacted layers met the specification requirements.

Pavements densities and air voids were determined using the PQI device. All recorded values satisfied the specification, except the AVS part on the first base layer which was not sufficiently compacted and exhibited higher air voids.

Statistical analysis was also performed to evaluate the correlation and find links between the measurements as well as detection of variables effects such as temperature on the pavement properties. The recorded results during construction were compared with the data obtained from cores. The results showed that an increase in paving temperatures significantly reduced the cores' air voids percentage. Similar results were observed based on the PQI outcomes. However, no more particular conclusions could be made from such analysis due to the lack of accuracy or/and insufficient number of collected data. Therefore, it was concluded that more detailed instruction such as well-defined coordinates for drilling cores and a uniform format for collecting data are required to improve the evaluation of incorporating these technologies in pavement construction.

5. References

[1] Omranian SR, Hamzah MO, Pipintakos G, Vuye C, and Hasan MRM, 2020, Effects of short-term aging on the compactibility and volumetric properties of asphalt mixtures using the response surface method, *Sustainability,* **12(15)**, 6181.

[2] Sebesta S, and Scullion T, 2002, *Using infrared imaging and ground-penetrating radar to detect segregation in hot-mix overlays* (No. FHWA/TX-03/4126-1).

[3] Baltrušaitis A, Vaitkus A, and Smirnovs J, 2020, Asphalt layer density and air voids content: GPR and laboratory testing data reliance, *The Baltic Journal of Road and Bridge Engineering*, **15(3)**, 93-110.

[4] Leng Z, Zhang Z, Zhang Y, Wang Y, Yu H, and Ling T, 2018, Laboratory evaluation of electromagnetic density gauges for hot-mix asphalt mixture density measurement, *Construction and Building Materials*, **158**, 1055-1064.

[5] Bijleveld F, Hinthem PE, Oosterveld M, Dekkers RJ, and Doree AG, 2012, Brede inpassing van de PQi-methode in de praktijk, *Proc. of CROW Infradagen 2012*, Ede, Netherlands, 22-23 May 2012.

[6] Stroup-Gardiner M, Law M, and Nesmith C, 2000, Using infrared thermography to detect and measure segregation in hot mix asphalt pavements, *International journal of pavement engineering*, **1(4)**, 265-284.

[7] Vyas V, Patil VJ, Singh AP, and Srivastava A, 2019, Application of infrared thermography for debonding detection in asphalt pavements, *Journal of Civil Structural Health Monitoring,* **9(3)**, 325-337.

[8] Vemuri SH, and Atadero RA, 2017, Case study on rapid scanning techniques for concrete bridge decks with asphalt overlay: Ground-penetrating radar and infrared thermography, *Practice Periodical on Structural Design and Construction*, **22(2)**, 05016004.

[9] Xu Q, and Chang GK, 2016, Adaptive quality control and acceptance of pavement material density for intelligent road construction, *Automation in Construction*, **62**, 78-88.

[10] Tirado C, Fathi A, Rocha S, Mazari M, and Nazarian S, 2020, Integrating intelligent compaction technology and modulus-based testing for design verification, *Proceedings of the Institution of Civil Engineers-Geotechnical Engineering*, **173(4)**, 327-338.

[11] Wang GF, Hu YB, Zhu WQ, Bao ZY, Chen ZJ, Huang ZH, and Wang W, 2017, The design of a compaction parameters management system for intelligent vibratory roller, *IEEE 3rd Information Technology and Mechatronics Engineering Conference (ITOEC)*, pp. 634-638.

[12] Chen SX, and Zhang XN, 2006, Research on the Use of PQI in Evaluation of Segregation of Hot-mix Asphalt Pavement, *Journal of highway and transportation research and development*, 5.

[13] Henault JW, Lane KR, and Sime JM, 2001, *Field evaluation of a non-nuclear density pavement quality indicator, Connecticut.* Dept. of Transportation. (No. FHWA-CT-RD-2227-F-01-3).

[14] Leng Z, Zhang Z, Zhang Y, Wang Y, Yu H, and Ling T, 2018, Laboratory evaluation of electromagnetic density gauges for hot-mix asphalt mixture density measurement, *Construction and Building Materials*, **158**, 1055-1064.

[15] KSE, Testing Equipment, https://www.kesslerdcp.com/shop/asphalt-pavements/mit-scan-t2t3-reflectors/reflectors-mit-scan-t2-t3/?v=2ac843586882#.

[16] Interpret all statistics and graphs for Correlation, internet, 2020, https://support.minitab.com/en-us/minitab/19/help-and-how-to/statistics/basic-statistics/how-to/correlation/interpret-the-results/all-statistics-and-graphs/, May 30 2020.

[17] Duerinckx B, 2017, *Rapport Ringanalyse voor het gebruik van in situ dichtheidsmeters*, Ref.: BAC/3053, Belgian Road Research Centre (BRRC).

Acknowledgments

Authors would like to thank University of Antwerp, Flemish government, and port of Antwerp for supporting this study through Projects FN5531 UAntwerp-Flemish Government IWT150166 'ROAD-IT', and FN800000147 UAntwerp-Port of Antwerp 'Durable asphalt pavements for heavy duty roads'.

IOP Conf. Series: Materials Science and Engineering 1252 (2022) 012073 doi:10.1088/1757-899X/1252/1/012073

Time-space modelling of construction

A Tazikova[1], Z Strukova[1] and J Cabala[1]

[1] Institute of Construction Technology, Economics and Management, Faculty of Civil Engineering, Technical University of Kosice, Vysokoskolska 4, 040 23 Kosice, Slovakia

E-mail: alena.tazikova@tuke.sk; zuzana.strukova@tuke.sk; jozef.cabala@tuke.sk

Abstract. A successful construction project respect the construction time and the estimated costs. Such a smooth construction process ensure effective scheduling. A good time construction plan can be a daunting task for a construction contractor. The article deals with time scheduling of line construction in TILOS software. The result is an analysis of scheduling in this software. The advantages and disadvantages of working of modelling a time-distance graph are described. For better control, it is very advantageous to have a well-processed schedule, which will be to the building model by various logical links connected. Thanks to this compound, it is possible to model the building of construction in the shortest possible time and as smoothly as possible. In conclusion, it is stated, that a well-crafted time construction plan can point to possible construction collisions. An updated time-distance graph processed in a software environment can reveal construction risks and the needed rescheduling in a timely manner. Therefore, it is possible to meet the scheduled construction time and estimated costs.

1. Introduction

Time-space modelling of construction is the design of construction work in time and in a specific space. There are several display methods for construction planning. Some methods are simpler and some are more complex. For the creation of line schedules can be used e.g. MS Project or Primavera software. MS Project is a tool for project managers, which is suitable for developing a line schedule. Resources are proposed there, the progress of construction and the consumed costs are monitored [1]. Primavera is a tool that shows the time course of project implementation. The schedule shows work that is ahead, late or on time [2].

During the long construction time, it is problematic to estimate resources accurately (budget, work, machinery). The contractor solves the choice of the right type of construction schedule with the possibility of practical optimization of the Gantt chart [3].

Adherence to the contractual construction period is one of the main challenges of construction projects. The vision is to achieve almost zero delay in all construction projects. Tools already exist to improve this situation, but they are still little in practice used. These include Building Information Model (BIM), Digital twin, Integrated Project Delivery (IPD) [4].

Various experiments solve the acceptance of optimal and suboptimal schedules (Gantt charts) by project managers [5].

Line construction projects often show cost overruns and non-compliance with construction times. There is therefore a need to develop innovative approaches and tools to avoid these project risks [6].

The construction time is recorded when creating the construction model of the building itself. For better control, it is very advantageous to have a well-processed graph, which will be connected to the building model by various logical links. Thanks to this compound, we can model the construction in the shortest possible time and as smoothly as possible. With the help of computer technology and software programs, the work with modeling can be simplified. In this way, we should save resources. We should ask ourselves a few questions at the beginning of modeling. These are mostly requirements for inputs, processes and outputs of processes. After answering them, we proceed to modeling and then we should be able to determine:

- for how long the work will be completed,
- what machines and equipment will be needed,
- what costs will be required.

By using time-space modelling, it is possible to go with the trend of sustainability, to eliminate downtime of workers and machines. Appropriate timing increases the productivity of construction methods. Prolongation of construction time can be eliminated by using time-space modelling. The analysis of construction production processes and the correct logistics of resources will contribute to the correct modeling. The subject of this paper is the analysis of the space-time model for the implementation of line construction.

2. Materials and methods

For the analysis of the software processing of the time-distance graph, a section of the line construction (road communication) in Eastern Slovakia is selected. Road with a length of 3,500 km consists of several construction objects. There will be two crossroads and a river with a sluice. The first crossroads will be at 1,150 km, the second at another 0,830 km and finally at a distance of 2,470 km the bridge over the Ondava river (figure 1).

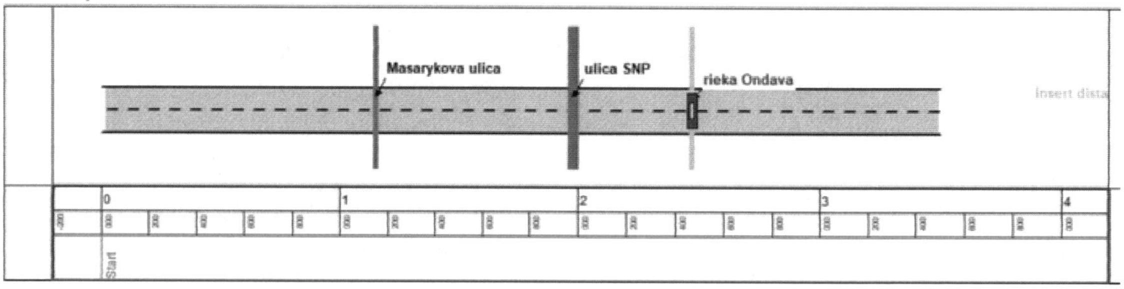

Figure 1. Solved part of the roadway.

The processing of the time-space graph is done using Tilos software. TILOS - is short for Time Location System- is the planning of processes in time and place, designed mainly for time planning of line construction projects. The time-space display is suitable for transport infrastructure for planning construction projects, such as roads, railways, tunnels, pipelines and the like (figure 2). Construction time modelling consists of several steps: creating a project, entering processes, assigning relationships and optimizing data.

TILOS time location diagrams (or otherwise called time chain diagrams, or time scale diagrams) are used to add a spatial parameter to the standard Gantt view. Thanks to this, it is possible to see where and when the work will take place. The use of time location diagrams allows you to optimize resources while preventing collisions in work activities. To speed up the work, it is possible, for example, to insert data from a line schedule into a time-space graph. The integration of Tilos software with Primavera and MS Project tools is advantageous. The result for the construction planner is reduced planning time, improved quality and saved time and money [7].

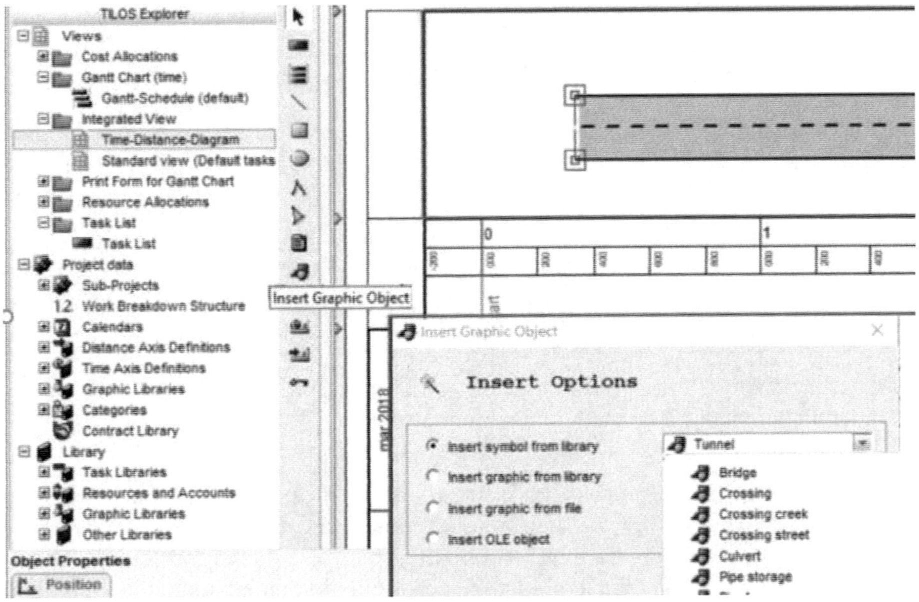

Figure 2. Roadway modelling using TILOS software.

3. Results and discussion

After importing processes into the graph, it is necessary to enter the relationships between the processes. The type of bonds used can be "start-start", "start-end", "end-start" or "end-end" (SS, SE, ES, EE). A more efficient way is to enter links by import, where the links have already been developed in one of the planning systems based on the critical path method.

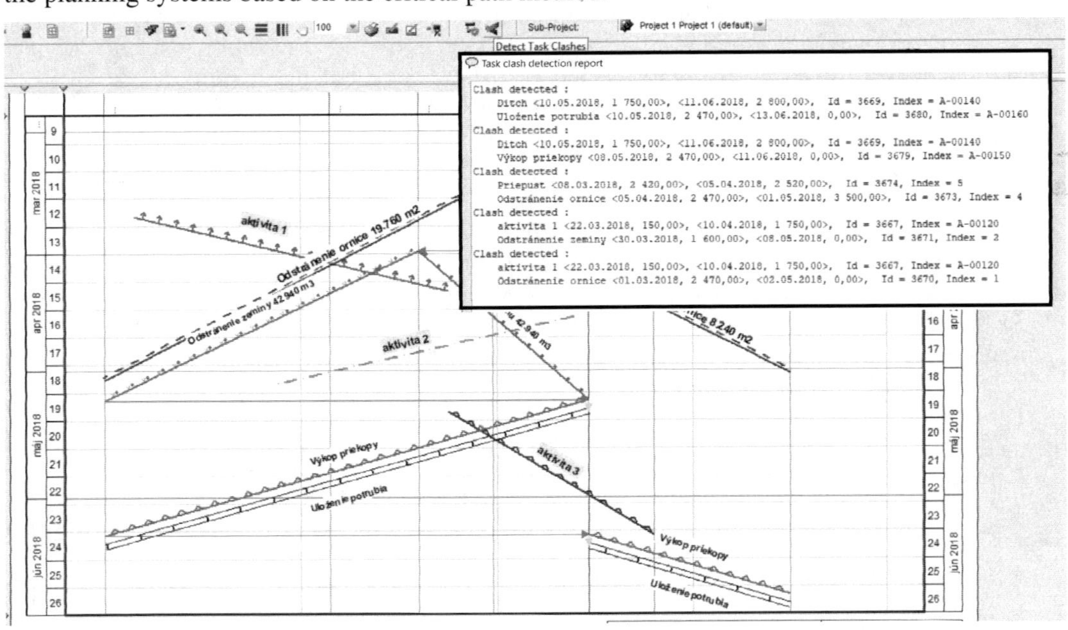

Figure 3. Time-space graph with collision detection.

In the case of trench excavation and pipe laying processes, it is no longer necessary to enter any links, as these have been imported from the library as a set of processes. They are already in the library with a binding that can be modified as needed. The resulting start and end dates for the processes are shown in the table. The graphical duration of the processes showing the location of the process performance is shown in the figure 3 in the space-time graph. Activities 1-3 are fictitious activities that show us how a collision is displayed - an inappropriate meeting of workers. The station is shown on the x-axis.

Name	Template	Start.Date	End.Date	Duration	Start.Distance	End.Distance	Quantity	Unit
Odstránenie ornice	Top soil stripping	01.03.2018	05.04.2018	24	2470	0	21600	m2
Odstránenie zeminy	Soil removal	15.03.2018	20.04.2018	24	1600	0	42940	m3
Úprava terénu	Soil installation	15.03.2018	20.04.2018	24	1600	2470	42940	m3
Odstránenie ornice	Top soil stripping	05.04.2018	17.04.2018	9	2470	3500	8240	m2
Priepust	Culvert	08.03.2018	28.03.2018	15	2420	2520	0	
Výkop priekopy	Trench opening	20.04.2018	29.05.2018	27	2470	0	2470	m
Uloženie potrubia	Pipe laying	24.04.2018	31.05.2018	27	2470	0	2470	m
Výkop priekopy	Trench opening	29.05.2018	15.06.2018	13	2470	3500	1030	m
Uloženie potrubia	Pipe laying	31.05.2018	19.06.2018	13	2470	3500	1030	m

Figure 4. Bill of quantities and process duration for the time-space diagram.

The software detects collisions and displays them in a report. In the event that some work will be performed by subcontracting and we have their time schedule, we will import this data about the project into the prepared graph. In one graph, we have the opportunity to display several projects simultaneously and identify possible collisions and critical points. If such a situation occurs, we will adjust the collisions in our project or we will agree on subcontracting, etc. Bill of quantities and process duration for the time-space diagram are shown in figure 4.

4. Conclusion

By using a time-space graph of infrastructure construction, it is possible to plan more efficient construction processes. By applying digitization, it will be possible to implement the time-space graph into the BIM model of the building. Using the TILOS software, it is possible to easily display process collisions that are not visible in the line schedule for line structures. The software allows you to display the boundary points of individual sections in a space-time graph, for better orientation, display multiple diagrams on one screen for better comparison, and create a diagram of the volume of earthworks, resource planning and costs. By incorporating this software into a project management system, it is possible to plan and manage activities much more efficiently and gain control over costs and resources.

5. References

[1] Microsoft Project Reviews, *Demo & Pricing* - 2021 (softwareadvice.com)
[2] Introducing the Primavera P6 Progress Line (tenzix.com)
[3] Bagloee SA, Sarvi M and Asadi M 2018 *Optimization for Roads' Construction: Selection, Prioritization, and Scheduling, Computer-aided civil and infra. Engineering* **33 (10)** 833-848
[4] Sepasgozar S M E, Karimi R and McCarthy D 2019 *Delay Causes and Emerging Digital Tools: A Novel Model of Delay Analysis, Including Integrated Project Delivery and PMBOK, BUILDINGS* **9 (9)**
[5] Brokman-Meltzer M, Perez D and Gelbard R 2021 *Perceived Complexity of a projects Optimal Work Plan Influences Its Likelihood of Adoption by Project Managers, Project management Journal* **52 (5)** 471-487
[6] Moret Y and Einstein HH 2016 *Construction Cost and Duration Uncertainty Model: Application to High-Speed Rail Line Project, Journal of Construction Engineering and Management* **142 (10)**
[7] Tilos-*Time Location/ Time Scales Diagram* (poprism.com)

Acknowledgments

The Slovak Research and Development Agency supported this paper under the contract no. APVV-17-0549. This work presents partial research results of project VEGA 1/0336/22 Research on the effects of Lean Production/Lean Construction methods on increasing the efficiency of on-site and off-site construction technologies.

IOP Conf. Series: Materials Science and Engineering 1252 (2022) 012074 doi:10.1088/1757-899X/1252/1/012074

Analysis of building materials used for the construction of family houses in the boundaries from Cradle to Gate with Options

E Kridlova Burdova[1] and S Vilcekova[1]

[1] Institute of Environmental Engineering, Faculty of Civil Engineering, Technical University of Kosice, Vysokoskolska 4, 042 00 Kosice, Slovak Republic

E-mail: eva.kridlova.burdova@tuke.sk; silvia.vilcekova@tuke.sk

Abstract. The purpose of the research work is to analyse environmental impacts of building materials applied to the construction of family houses in the boundaries Cradle to Gate with Options. Within Life cycle analysis (LCA), environmental impacts such as global warming potential (GWP), ozone depletion (OP), acidification potential (AP), eutrophication potential (EP), photochemical ozone creation potential (POCP), non-hazardous waste disposal (NHWD) and biogenic carbon storage (BGS) are determined by using One Click LCA tool. This approach is selected to bring information about current state of construction trend used in Slovakia, to set a starting point for further critical reviews of different strategies as well as to inform about the impact significance of current selection of building materials. Results show that the biggest contribution to GWP are triple glazed exterior wooden doors and windows with aluminium elements, thermal insulation and reinforced concrete.

1. Introduction

Quantitative environmental assessments are the key to working effectively towards sustainable production and consumption. In recent decades, life cycle assessment (LCA) has been introduced as a viable means of measuring the environmental impact of products in the supply chain [1]. The environmental impact of buildings is commonly quantified through LCA [2]. LCA is globally recognized as one of the most complete methods for environmental assessment of buildings [3]. Studies showed that up to 30% of a building's life cycle energy and emissions are associated with the embodied phase and this number could increase to 50% for energy efficient passive houses [4]. The process of building construction depends on the considerable consumption of natural resources with far-reaching impacts outside their development area. In general, a significant part of the annual consumption of resources in construction and building is the result of the application of traditional building strategies and procedures, such as design and selection of building types, building materials and structures, HVAC systems and planning of renovations and maintenance procedures. On the other hand, in addition to construction suitability, builders generally take into account the basic requirements of public owners or private building users, with costs being the main criterion for selecting building strategies and long-term environmental and socio-economic impacts being generally ignored [5]. The use of wood and products made from wood is considered as an opportunity for the mitigation of negative environmental impacts of buildings, such as greenhouse gas emissions [6]. Buildings consisting of building structures, heating, ventilation and air conditioning (HVAC) systems and indoor air are unique, so an integrated tool is needed for the environmental assessment, comparison and subsequent assessment and certification of buildings. Anthropogenic carbon dioxide, methane, ozone and chlorofluorocarbons contribute significantly to the effects of global warming [7]. This study was performed by using LCA in accordance with international standard ISO 14 067 [8] for determining the carbon footprint of four family houses. The aim of this paper is to analyse environmental impacts such as: global warming potential (GWP), ozone depletion (OP), acidification potential (AP), eutrophication potential (EP), photochemical ozone creation potential (POCP), non-hazardous waste disposal (ANHWD) and biogenic carbon storage (BGS) of building materials applied in the construction of family houses in the boundaries Cradle to Gate with Options determined by using One Click LCA tool.

2. Materials and Methods

Environmental performance of four family houses was assessed using the LCA method within the Cradle to Gate with Options boundary based on EN 15978 [9]. In the table (Table 1), the system boundary included in this study, is shown and meets the requirements according to EN 15804+A2:2019 [10]. Characteristics of investigated four family houses are described in Table 2 and Table 3. Results obtained from the One Click LCA tool were analysed using the multicriteria analysis program MCA7 [11] for the subsequent selection of the most suitable family house in terms of building materials used.

Table 1. System boundary (Cradle to Gate with Options) [10].

Product stage			**Cons-truction stage**		**Use stage**							**End of life stage**				**Beyond the system boundaries**
Raw materials	Transport	**Manufacturing**	**Transport**	Assembly	Use	Maintenance	Repair	Replacement	Refurbishment	**Operational energy use**	**Operational water use**	**Deconstruction /demolition**	**Transport**	**Waste processing**	**Disposal**	**Reuse, Recovery, Recycling**
A1	**A2**	**A3**	**A4**	A5	B1	B2	B3	B4	B5	**B6**	**B7**	**C1**	**C2**	**C3**	**C4**	**D**
x	x	x	x	MND	MND	MND	MND	MND	MND	x	x	x	x	x	x	x

Module not declared = MND

Table 2. Basic information about family houses [12].

Family house 1	Family house 2	Family house 3	Family house 4

Build-up area (290 m^2) Usable floor area (278.2 m^2) Built-in volume (1185.6 m^3) Heat pump type ground-water, fireplace, ceiling cooling	Build-up area (248.5 m^2) Usable floor area (445.1 m^2) Built-in volume (1693.7 m^3) Gas boiler, solar systems, fireplaces, forced ventilation - recuperation unit	Build-up area (145 m^2) Usable floor area (339.9) Built-in volume (441.3 m^3) Gas condensing boiler. Natural ventilation.	Build-up area (245 m^2) Usable floor area (399.5) Built-in volume (1345.2 m^3) Water heat pump, electric boiler, recuperation unit.

Table 3. Construction information about family houses [11].

	Founda-tions	Exterior walls, surface	Ceiling structu-res	Roof structures	Interior walls, wall finishes	Floor and ceiling finishes	Insulation of foundations, exterior walls and roofs	Windows
Family house 1	Rein-forced concrete base strip	ceramic blocks, reinfor-ced concrete walls 250 to 300 mm; external silicate plaster, stone cladding	Rein-forced concrete boards 200 mm	flat vegetation roof, reinforced concrete slab, substrate with vegetation mat	sandwich plaster-board constructions, thermal insulation from mineral wool 150 mm thick, ceramic tiles; plaster	wooden floor; ceramic tiles, linoleum; concrete paving; plaster-board, suspen-ded concrete	geotextiles; PVC water-proofing, expanded polystyrene 350 mm; mineral wool hr. 200 mm; expanded polystyrene hr. 200 mm; water-proofing PVC; drainage foil, filter cloth	aluminium windows with 3x glazing, plus external aluminium blinds
Family house 2	Rein-forced concrete base strip	concrete formwork blocks plus bricks 300 mm; external silicate plaster, alumi-nium compo-site wall panels	Rein-forced concrete boards 200 mm	Flat roof, reinforced concrete slabs; waterproofing strips	bricks 150 mm, plaster plasters; ceramic tiles	wooden floor; PVC; ceramic tiles; concrete floor; gypsum ceiling plasters; plaster-board	geotextiles; PVC water-proofing; extruded wall polystyrene 120 mm, mineral wool 250 mm; extruded base polystyrene 150 mm; polystyrene concrete; extruded roof polystyrene 200 mm	aluminium windows with 3x glazing; aluminium horizontal external blinds
Family house 3	Rein-forced concrete base strip	bricks 300 mm; external silicate plaster, stone exterior cladding	Rein-forced concrete boards 200 mm	flat roof; polystyrene concrete 60-200 mm; roof cladding; river gravel	bricks 250 mm; aerated concrete blocks 100 - 150 mm; ceramic tiles; lime-cement plaster	wooden floor; PVC; ceramic tiles; concrete floor; gypsum ceiling plasters; ceiling plaster-board	geotextile, PVC waterproofing, extruded polystyrene 120 mm), extruded wall polystyrene 150 mm; polystyrene concrete; mineral wool 250 mm; extruded roof polystyrene 200 mm	wooden windows 3 glazing, external blinds aluminium horizontal
Family house 4	Rein-forced concrete base strip	Reinfor-ced concrete blocks 300 mm; bricks 300 mm; external silicate plaster; wooden siding	Rein-forced concrete boards 200 mm	flat vegetation roof	bricks 140 mm; lime plaster; ceramic tiling	wooden floors; ceramic cladding, lime ceiling plaster; exposed concrete	geotextiles; PVC waterproofing; extruded polystyrene 200 mm; facade polystyrene 200 mm, foam glass 500 mm	wooden windows 3 glazing, external blinds aluminium horizontal

3. Results and Discussion

Table 4 presents results of environmental impacts categories such as: global warming potential (GWP), ozone depletion (OP), acidification potential (AP), eutrophication potential (EP), photochemical ozone creation potential (POCP), non-hazardous waste disposal (NHWD) and biogenic carbon storage (BGS) determined by using One Click LCA tool. The figure 1 shows the environmental impact categories for family houses 1 - 4 (FH1-FH4). In terms of CO_{2eq} emissions, the FH 3 achieves the worst results (393 644.1 kg CO_{2eq}) and the FH1 achieves the best results (141 175.1 kg CO_{2eq}) (Figure 3). Similar results are achieved for SO_{2eq} emissions, the worst results goes to FH 3 (1 300.09 kgSO_{2eq}) and the best goes to FH 1 (428.23 kgSO_{2eq}). Similar results are obtained for POCP, ANHWD and BCS. Family houses 1, 3 and 4 reach the same values of 0.03 kg CFC_{11eq} and family house 2 reaches the value of 0.04 kg CFC_{11eq} (Figure 5). The figure 2 shows results for global warming potential presented for analysed life-cycle stages for family houses (FH1-FH4). It can be stated that equivalent emissions of CO_2 in the production phase of building materials (A1-A3 Materials) account for the largest share, for FH 1 and FH4. These family houses are characterized by a vegetated roof and have floor and ceiling heating with a ground-water heat pump, fireplace, ceiling cooling - dry system, natural ventilation. Family houses 2 and 3 have higher equivalent emissions of CO_2 in the energy phase (B6). These family houses are built of concrete bricks, have a flat roof with waterproofing strips or gravel and have underfloor heating and radiators, source - gas boiler with other solar systems, fireplaces, forced ventilation - recuperation unit. The results show that it makes sense to pay more attention to building materials and energy sources from local and renewable sources.

Table 4. Results of evaluated categories of environmental impacts.

Family house	GWP kg CO_{2eq}	AP kg SO_{2eq}	EP kg$PO_4^{3-}{}_{eq}$	ODP kg CFC_{11eq}	POCP kg Ethenee	NHWD kg	BCS kgCO_{2e} bio
FH 1	141 175.1	428.23	77.74	0.03	28.06	107 519.5	4 569.85
FH 2	350 949.4	1 147.75	181.88	0.04	101.49	83 274.57	19 096.74
FH 3	393 644.1	1 300.09	22.55	0.03	115.41	43 172.26	23 519.85
FH 4	186 312.5	731.68	59.21	0.03	81.85	83 601.5	10 342.1

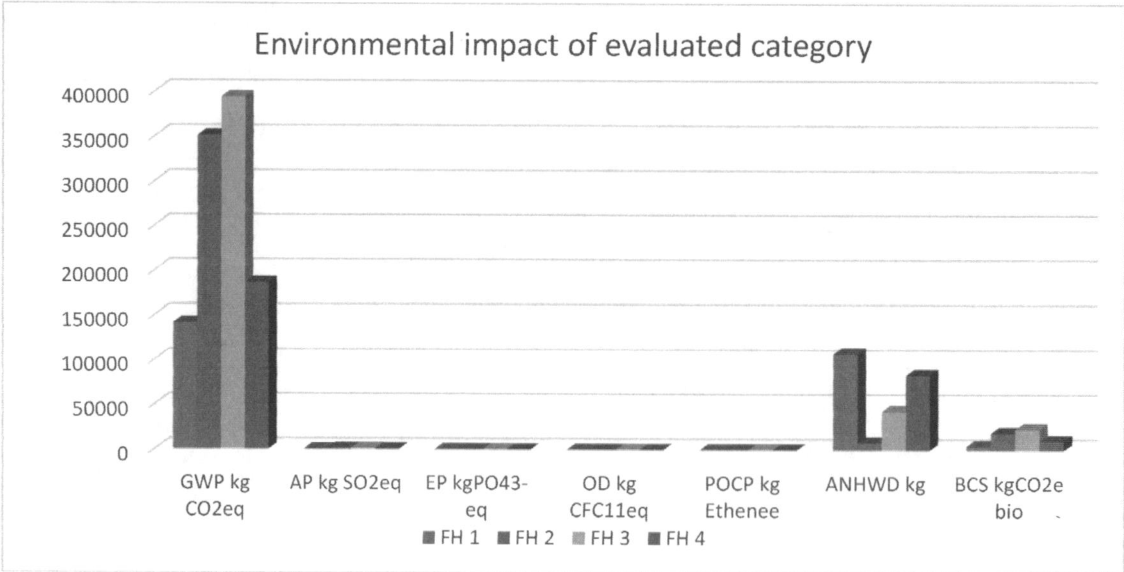

Figure 1. Environmental impact categories for family houses 1 - 4 (FH1-FH4).

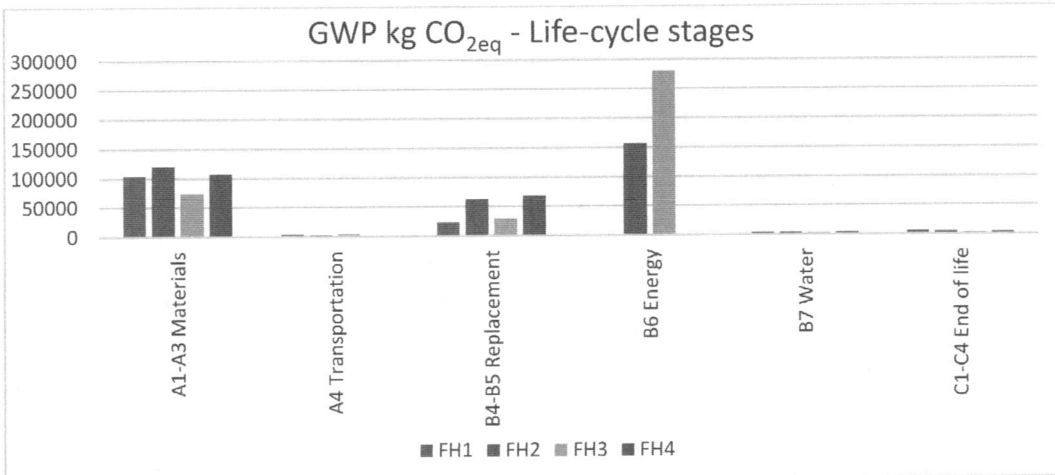

Figure 2. Equivalent emissions of CO_2 during life-cycle stages for family houses 1- 4 (FH1-FH4).

It is clear from Figure 2 that building materials and operational energy usage contribute significantly to the impacts, significantly higher compared to the transport, operational water usage and the end of the life cycle. Building materials are therefore important in terms of the need to deal with their selection while taking into account their location, requirements, thermal properties, durability and other technical or functional characteristics. Table 5 and the figure 3 show the results of individual environmental impacts stated given per square meter of floor area. The results show that family houses 1 and 4 share lower impacts in terms of GWP, AP, EP and POCP. In contrast, they contribute the most to non-hazardous waste generation. The content of biogenic carbon in the materials is reported to be lower than in the other houses (FH 2 and FH3).

Table 5. Results of environmental impacts given per meter square of total floor area.

Family house (Gross internal floor)	GWP kg CO_{2eq}/m^2	AP kg SO_{2eq}/m^2	EP kgPO_4^{3-} $_{eq}/m^2$	ODP kg CFC_{11eq}/m^2	POCP kg $Ethene_e/m^2$	NHWD kg/m^2	BCS kgCO_{2e} bio/m^2
FH 1 (278.2 m^2)	507.46	1.54	0.28	0	0.10	386.48	16.43
FH 2 (445.1 m^2)	788.54	2.58	0.41	0	0.23	187.11	42.91
FH 3 (339.9 m^2)	1 158.1	3.82	0.07	0	0.34	127.01	69.2
FH 4 (399.5 m^2)	466.36	1.83	0.15	0	0.20	209.27	25.89

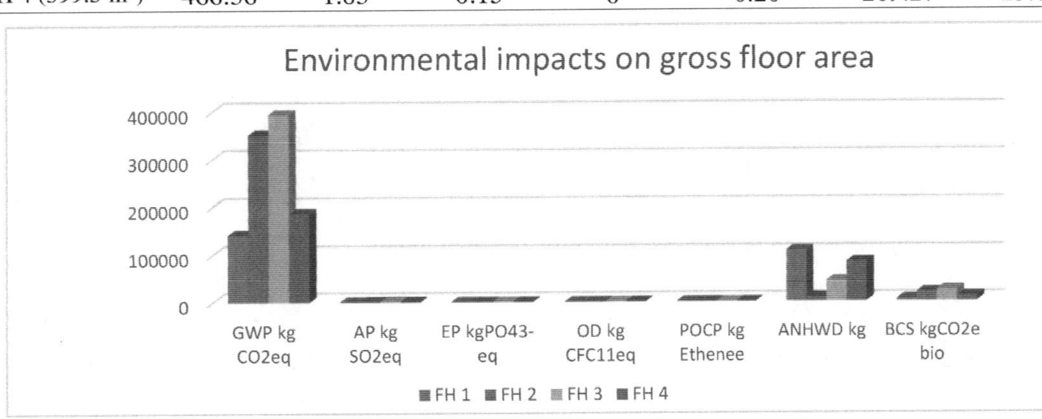

Figure 3. Environmental impact categories for family houses 1- 4 (FH1-FH4).

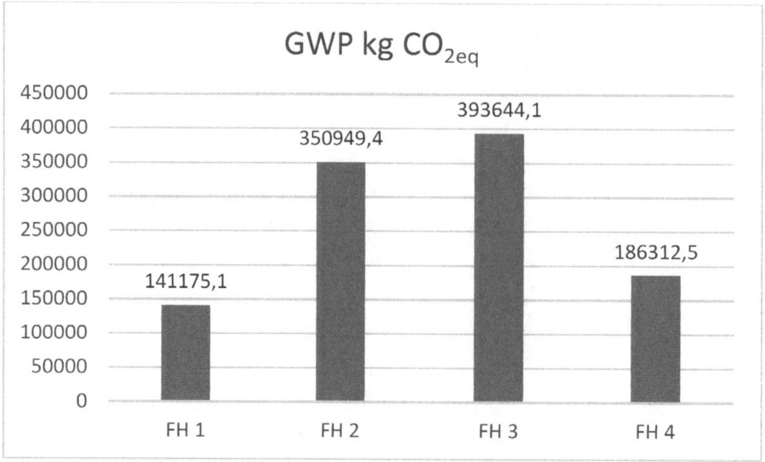

Figure 4. GWP kg CO_{2eq}.

Figures 5 – 7 depict result for AP, EP, POCP, ODP, NHWD and BSC, expressed as equivalent amounts of CO_2, PO_4^{3-}, Ethene, CFC11, kilogram non-hazardous waste and biogenic carbon.

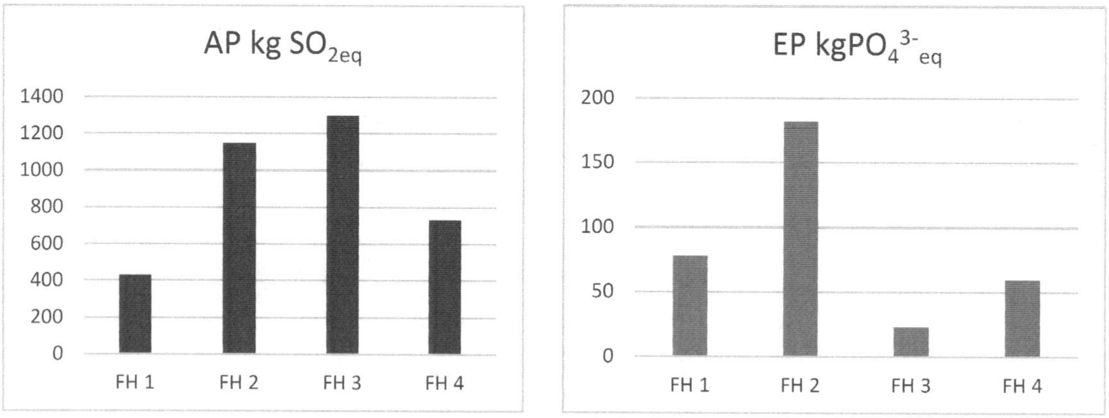

Figure 5. AP kg SO_{2eq} and EP kg$PO_4^{3-}{}_{eq}$.

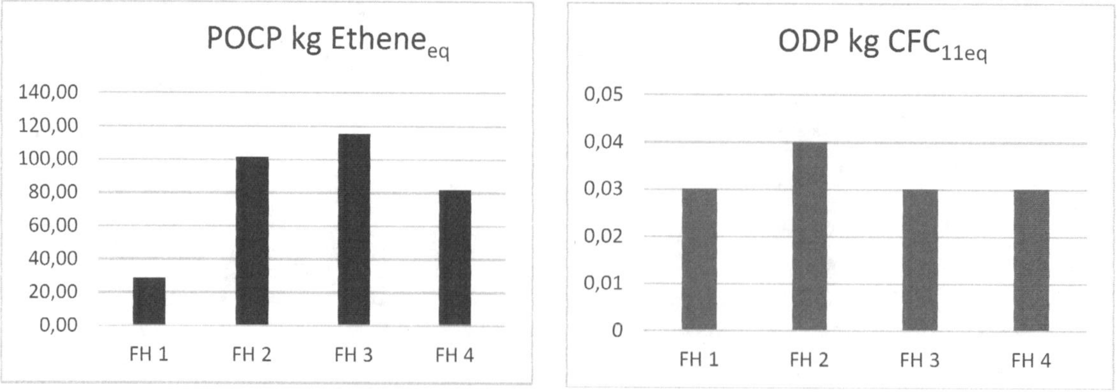

Figure 6. POCP kg $Ethene_{eq}$ and OD kg CFC_{11eq}.

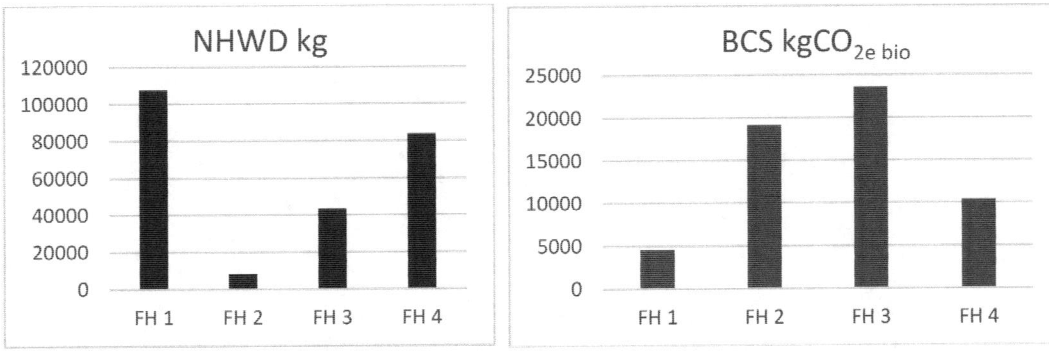

Figure 7. NHWD kg and BCS $kgCO_{2e}$ bio.

Equivalent emissions of CO_2 grouped by used materials element breakdown is shown in figure (Figure 8). Most emissions that cause global warming occur in upper floors, including horizontal structures, in terms of building materials. In terms of energy consumption, most GWP emissions are released into the environment when using fuel for gas boiler heating in family houses 2 and 3.

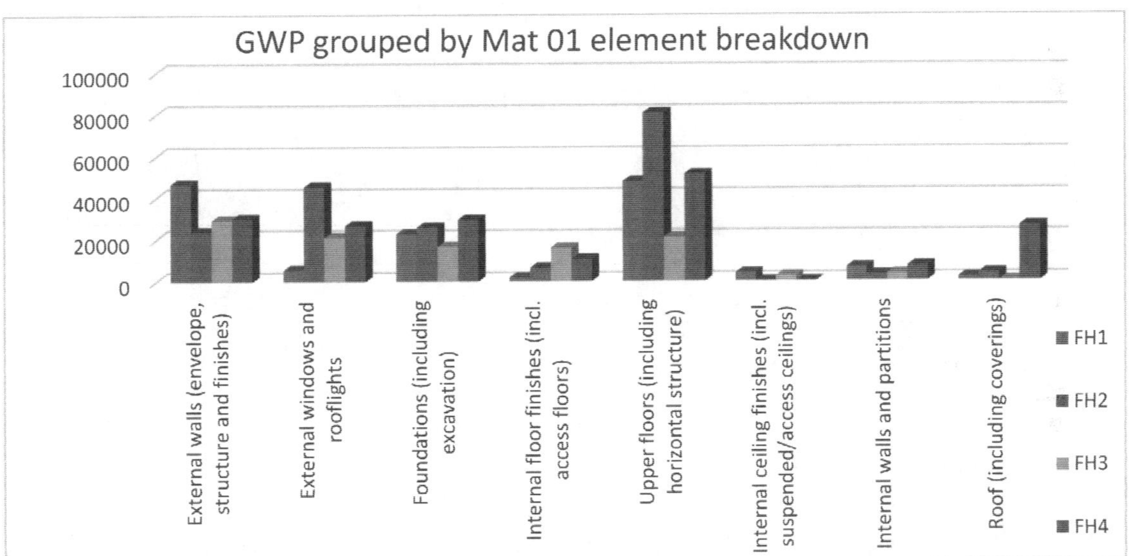

Figure 8. Global warming grouped by material element breakdown of family houses 1 – family house 4 (FH1-FH4).

The most contributing materials in terms of their impact were also analysed. In the table (Table 6) are shown five biggest material contributors to GWP in total value in kg CO_{2eq} and percentage share of building material used in family house 1- 4. The biggest contribution to the GWP in family house 1 has Rockwool insulation and reinforced concrete slab. The biggest contribution to GWP in family house 2 have triple glazed exterior wooden doors and windows with aluminium elements. The biggest contribution on GWP emissions in family house 3 have triple glazing windows with wooden frame and ready-mix concrete, C28/30. From the results, top five impacts are mostly the main structural materials of the buildings. In the case of the four selected family buildings, those materials are structural concrete with reinforcement and insulation, due to the requirement for high-energy efficiency of buildings. Also, from the point of GWP view, triple glazed exterior wooden door and windows, with aluminium elements is leading material with the largest impact (FH2).

Table 6. Five biggest material contributors to GWP. Total value in kg CO_{2eq} and percentage.

Material	FH1	FH2	FH3	FH4
Rock wool insulation panels	22 t 21.6%	- -	- -	- -
Triple glazed exterior wooden door and windows, with aluminium elements	- -	22 t 18.6%	- -	- -
Reinforced concrete slab	20.5 t 20.2%	9.7 t 8%	8.5 t 11.5%	10 t 9.6%
Ready-mix concrete, C30/37	19.4 t 18.9%	- -	8.2 t 11.2%	- -
Concrete slab	- -	19 t 16%	- -	7.4 t 6.9%
Ready-mix concrete, C25/30	- -	14 t 12.1%	- -	- -
Cork acoustic insulation	- -	12 t 10%	- -	- -
ETICS with silicate plasters and EPS insulation	6.1 t 5.9%	- -	- -	- -
Window system, with aluminium composite profile framing	5.5 t 5.3%	- -	-	- -
Clay brick	- -	- -	11 t 15.3%	- -
Triple glazing windows with wooden frame	- -	- -	10 t 14%	13 t 12.2%
Massive wooden flooring/parquet	- -	- -	7.5 t 10.1%	- -
Ready-mix concrete, normal strength, C28/30	- -	- -	- -	11 t 9.9%
Acrylic polyurethane coating				8.5 t 7.9%

Top five materials: first (green); second (green); third (green); fourth (green); fifth (green)

The results of the multicriteria analysis calculated using the MCA7 are presented in the table (Table 7). The results show that family house 1 reached the best results in all analysed statistical methods provided by the MCA7 program, namely the CDA, IPA, WSA and TOPSIS methods. Family house 2 obtained the worst results according to the IPA and WSA statistical methods, and family house 3 have the worst results according to the CDA and TOPSIS statistical methods.

Table 7. Results of multi-criteria analysis of environmental impacts of evaluated categories of four family houses.

Order	variant	CDA	variant	IPA	variant	WSA	variant	TOPSIS
1	FH1	1.2327	RD1	0.1767	RD1	0.8233	RD1	0.784
2	FH4	1.69	RD4	0.3463	RD4	0.6537	RD4	0.6834
3	FH2	3.8707	RD3	0.7351	RD3	0.2649	RD2	0.272
4	FH3	4.1903	RD2	0.7791	RD2	0.2209	RD3	0.2578

The following figures show the results for family house 1, which achieved the best results from LCA analyses of environmental impacts. There are the results of GWP during life-cycle stages (Figure 9) and

GWP structures classifications (Figure 10). In the figures (Figure 11 and Figure 12) the mass classification in kg and resource types are shown.

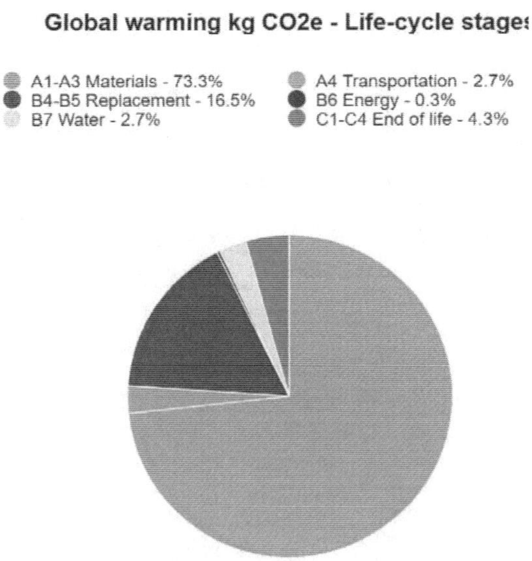

Figure 9. Global warming kg CO_{2eq} of family house 1..

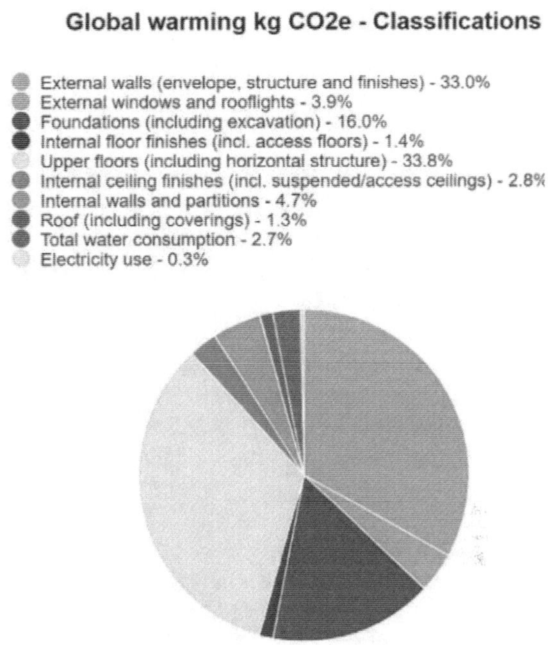

Figure 10. Life-cycle stages and Classifications of family house 1.

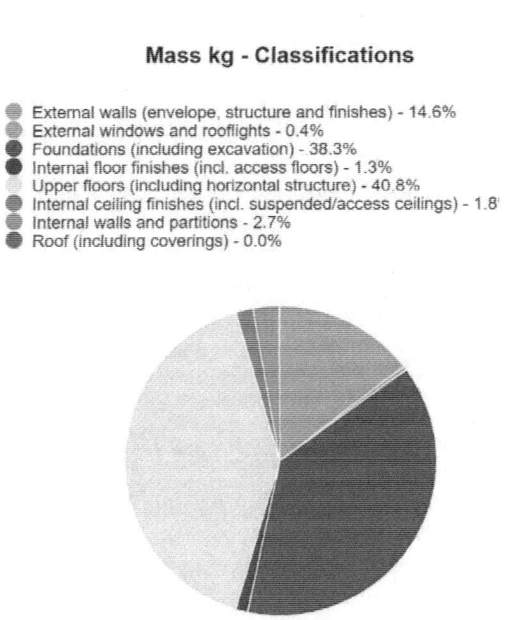

Figure 11. Mass kg – Classification of family house 1.

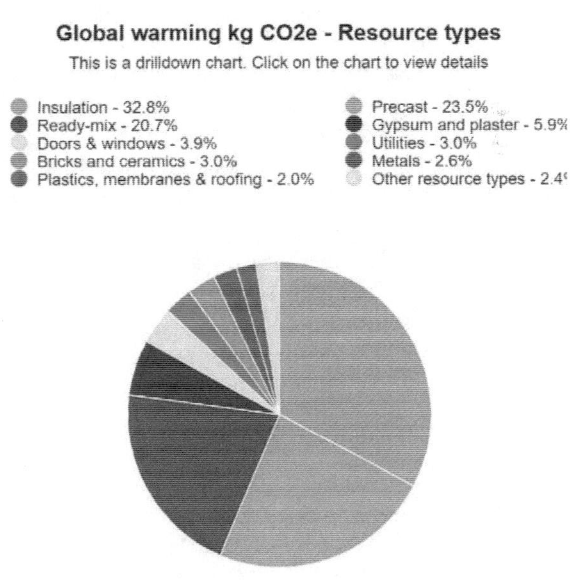

Figure 12. Global warming kg CO_{2eq} – resource types of family house 1.

The total greenhouse gas emissions for family house 1 are 141 175 kg CO_{2eq}, and for family house 3 are 393 644 kg CO_{2eq}, which means that family house 3 compared to house 1 generates 64.1 % more CO_2 emissions during its whole life cycle.

4. Conclusion

The ever-increasing requirements not only for the quality of the indoor environment in buildings, but also for the quality of the surrounding environment reflect the growing need to subject buildings to an assessment from environmental energy, social and economic dimensions. Sustainability assessment of buildings requires a multidisciplinary and multicriteria approach, the aim of which is to verify the real condition of buildings and find potential for targeted improvement of environmental performance, reliability, sustainability and safety of buildings and their surroundings. Various environmental assessment methods have been used around the world to allow comparison. LCA is an appropriate method that can be used for quantitative purposes to assess the material used, energy flows and environmental impacts of products. The construction of sustainable buildings, which are built mainly from local natural, almost energy-intensive materials, creates an opportunity to become raw material and energy self-sufficient. A path shows the way back to the nature. In this way, we can improve the quality of the indoor and outdoor environment.

5. References

[1] Suski P, Speck and M Liedtke C 2021 *Promoting sustainable consumption with LCA – A social practice based perspective, J. Cleaner Product* **283** 125234

[2] Decort Y, Steeman and M Bossche N V D 2021 Effect of a one-dimensional approach in LCA on the environmental life cycle impact of buildings: Multi-family case study in Flanders, *Build Environ* **206** 108381

[3] Soust-Verdaguer B, Llatas and C Garciá-Martínez A 2016 *Simplification in life cycle assessment of single-family houses: A review of recent developments, Build Environ* **103** 215-227

[4] Shirazi A and Ashuri B 2020 *Embodied Life Cycle Assessment (Lca) Comparison of Residential Building Retrofit Measures in Atlanta, Build Environ* **171** 106644

[5] Reza B, Sadiq R and Hewage K 2014 *Emergy-based life cycle assessment (Em-LCA) of multi-unit andsingle-family residential buildings in Canada, Int. J. Sustain. Built Environ.* **3/2** 207-224

[6] Soust-Verdaguer B, Llatas and C Moya L 2020 *Comparative BIM-based Life Cycle Assessment of Uruguayan timber and concrete-masonry single-family houses in design stage, J. Clener Product* **277** 121958

[7] Rashidi N A, Yusup S, Borhan A and Loong L H 2014 *Experimental* and modelling studies of carbon dioxide adsorption by porous biomass derived activated carbon *Clean Techn Environ Policy* **16** 1353-1361

[8] ISO 14067:2018 Greenhouse gases — Carbon footprint of products — Requirements and guidelines for quantification

[9] EN 15978:2011 *Sustainability of construction works. Assessment of environmental performance of buildings. Calculation method*

[10] EN 15804+A2:2019 *Sustainability of construction works - Environmental product declarations - Core rules for the product category of construction products*

[11] Korviny P. MCA7. Program pro multikriteriální rozhodovací analýzu. Ostrava 2003. p. 27

[12] Kridlova Burdova E 2021 *Assessment of Sustainability of Family Houses*. Habilitation work, p. 255

Acknowledgments

This study was supported by the Grant Agency of the Slovak Republic to support projects No. 1/0512/20 and APVV-18-0360 ACHIEve.

Quality of first flush of rainwater from urban area: short review

I Marko[1], R Wittmanová[2], J Hrudka[3], A Raczková[4] and I Škultétyová[5]

[1,2,3,4] Department of Sanitary and Environmental Engineering, Faculty of Civil Engineering, Slovak University of Technology in Bratislava, Radlinského 11, 810 05 Bratislava, **Slovakia**

E-mail: ivana.marko@stuba.sk[1], reka.wittmanova@stuba.sk[2], jaroslav.hrudka@stuba.sk[3], andrea.raczkova@stuba.sk[4], ivona-skultetyova@stuba.sk[5]

Abstract. As a result of changes in social, economic, and political relations, undesirable phenomena are created that have an adverse effect on the urban environment. Surface runoff represents one of several risks to environmental policy, causing the major problem in urban agglomerations. In an urbanized area, it causes an increased runoff volume and the flood peak. Also, it affects increasing the contamination of urban rainwater. The significant problem of surface runoff formation in urban river basins is the accumulation of pollutants on the surface of paved areas during the rainless period. In the case of the formation of surface runoff, these substances come into movement. It is assumed that the most polluted rainwater is in the initial phase of runoff formation. However, not all authors agree with the statement. In some literature, this phenomenon is termed the first flush. Given the above facts, the study focuses on surface runoff characteristics in terms of its quality and quantity in the initial phase of surface runoff formatting. The study aims to summarize and compare the results of several studies to evaluate the first rainwater flush in an urbanized area.

1. Introduction

Changes in social, economic, and political relations bring a number of specific problems associated with changing the city's spatial structure [1]. There is the rapid urbanization of the city, development of transport and technical infrastructure. Land use and disturbing natural conditions can affect the area of interest's hydrological, climatic, and environmental characteristics [2], [3]. These changes are reflected at the local level but can also be at the regional or global scale. Regarding the potential risk to the environment (climate change, water crisis, air pollution, etc.), the development of urban areas is a serious environmental problem [4].

The development of urbanized areas causes an increase in the percentage of impervious surfaces in the city. Today, more than 70 % of the city's areas are covered by paved surfaces [5], while of the Earth's total area, the impermeable surface represents about 29 % [5],[6]. The problem with paved areas is that they do not support the hydrological process (infiltration) due to which city's microclimate change [2],[6]. By changing the city microclimate, the air temperature increase, reducing the daily temperature range and significantly affecting the way the wind flows, limiting the dispersion of pollutants [3]. Another unwanted phenomenon of climate change is an increase in extreme precipitation events. For the last few years, dozens of flood events caused by extreme rains have been recorded. The intensive rainfall can cause the occurring of surface runoff and local floods in urban areas.

The significant problem in urban agglomerations is surface runoff, a product of land use and climate change [6]. The surface runoff, also known as urban runoff, is identified as an area source of pollution. During the dry period (period without rain), the pollutions accumulated on the surface of paved surfaces [7]. In the rainfall period, urban runoff is considered as one of the primary transport mediums of pollution draining to the sewer network or receiving water bodies [3], [7]. The problem is that water from the paved surfaces is contaminated with amount substances that can affect the physical and chemical properties of the aquatic environment [3].

Several factors condition the characteristics of pollutants in the surface runoff but mainly depend on land use pattern (building, roof, communication, etc.), type of material, geometric characteristic (slope, width), and characteristic of precipitation (rain intensity, duration, dray period) [2], [8]. Mentions factors some less, some more affect the amount of pollution in urban runoff. Generally, according to the many research the concentration of stormwater pollution depends on rainfall intensity. The rainfall intensity is a key factor in runoff formation and pollutant transport [3], [9]. In particular, many studies note that the largest amount of pollution is in the early phase of runoff that later [10]. This phase of surface runoff is called the "First Flush". The definition of "First Flush" and its characteristic of occurred is described in chapter 3.

The first flush in an urban environment represents an important phenomenon, which is disputable amongst many researchers. The problem is that not all researchers agree that the first flush occurs in the early phase of runoff (the first 30% of runoff). Due to the lack of a correct definition of the first flush, the study focuses on finding and analysing published articles, books, and studies devoted to the issues of the first flush. The study aims to analyse and compare the results of many studies evaluating the formation/occurrence of the first flush in urban areas.

2. Methodology

The paper deals with a review of the literature devoted to the issue of rain management in an urban environment with a specification of the quality of the first flush. The paper aimed to review the literature from journals, articles, books, and other studies focused on the issue. Keywords were used to search the literature as first flush, stormwater quality, stormwater, characteristic of rainfall. The searched literature was divided into two categories. The first category contained articles that described the creation of the first flush (methodology). The second category consisted of articles that assessed the creation of the first flush concerning the methodology. Based on the searched literature, a comparison of the creation of the first flush concerning the methodology was made.

3. First Flush

The "first flush" or runoff of rainwater from a paved surface is an important aspect in urban areas related to the occurrence or assumption that the first part of the rainwater runoff is most contaminated [10], [11], [12]. Most scientists share this view, while some doubt about whether the actual pollution mass of the first flush represents a significant part of the outflow. In general, the first surface runoff flush is an important aspect of surface runoff, characterized by a high concentration of pollutants in the early phase of the runoff at the beginning of the precipitation event but after the formation of the runoff [3], [10], [11], [12]. Whether the first flush exists can only be spoken of if the first outflow has a higher concentration of pollutants than the later outflow. Identifying the nature of the first flush phenomenon can have significant benefits for implementing the Sustainable Urban Drainage System measures

Understanding the quality of surface runoff and its correct definition has been the subject of many studies: Bertrand-Krajewski et al. (1998), Deletic A. (1998), Huang (2012), Seget et al., (1995), and other researchers. In general, when considering the first flush, the mass of the pollutant load and the volume of the wastewater must be taken into account. There are several ways to define this phenomenon. The most manageable way is to assess whether the first flush has occurred or not using the dimensionless curve *M(V)* [10],[13]. A dimensionless curve represents a dimensionless graph to construct which we need to know the cumulative runoff cumulative pollutant load in the time [10],[13]. For constructing a dimensionless curve, it is necessary to know the total pollution load *(M)* and the total runoff volume *(V)* for the monitored rain runoff [12].

To assess the formation of the first flush, the researchers used traditional and modified methods based on the dimensionless curve *M(V)*. The traditional method involves the use of dimensionless cumulative pollutant load curves and cumulative runoff volume [10],[13],[14]. The traditional method assumes that the most significant pollution occurs in the first 20 % or 30 % of the storm runoff volume during rainy [10] [13]. To construct a dimensionless curve, it is necessary to know the cumulative mass fraction *M(t)* and cumulative volume fraction *V(t)* values. The cumulative mass and runoff volume curves for the traditional method can be constructed from the equation (1) and equation (2)

$$M\,(t) = \frac{\sum C(t)Q(t)dt}{M} \quad (1)$$

$$V(t) = \frac{\sum Q(t)dt}{V} \quad (2)$$

where: *C(t)* represent the pollutant concentration at the time *t* [$\mu g.l^{-1}$], *Q(t)* is the instantaneous flow rate at the time *t* [$l.s^{-1}$], *M* is the total pollutant mass for the event, *V* is a total runoff volume [l], *dt* is time steps [s] [10], [13], [14].

In addition to the traditional method, there is an alternative first flush method known as Mass First Flush Ratio (MFFR). The weight fraction defines the alternative method (MFFR) by the cumulative runoff volume at a defined point [14]. The dimensionless curve MFF can be constructed using the equation:

$$MFF_n = \frac{\int_0^T C(t)Q(t)dt/M}{\int_0^T Q(t)dt/V} \quad (3)$$

where: *n* in the storm, and corresponds to the percentage of the runoff, ranging from 0 % to 100 %, *C(t)* is pollution concentration at the time t [$\mu g.l^{-1}$], *Q(t)* is runoff at the time t [$l.s^{-1}$], *V* is a total runoff volume [l], *T* is elapsed time [s] [3], [15]. According to the definition, the MFF is equal to zero at the beginning of rainfall and always equals 1.0 at the end of the rainfall event [15].

Many researchers have pointed out the shortcomings of the traditional way of assessing the formation of the first flush, given that it does not consider the impact of storm volume. According to some shortcomings of the traditional method, Bach et al. (2010) propose a new definition of the first flush. The new method quantifies the runoff volume required to reduce stormwater pollutant concentrations to background levels [14].

4. Evaluation of the First Flush

As already mentioned, the evaluation of the first flush has been the subject of many studies. These studies aimed to correctly define the methodology and understand the quality of the first flush. In order to understand the correct definition of the first flush, the study compared several studies that focused on evaluating the first flush in an urbanized area.

Bertrand-Krajewski et al. (1998) for determining the first flush has used the traditional methods. According to their research, the formation of the first flush occurs when 80 % of pollutants enter the runoff in the first 30 % of the runoff [10]. In some literature, the definition of the first flush is called 30/80. A similar statement is made by Deletic et al. (1998) [12] and Vorrieter and Hickey (1994) [16]. Deleticet al. (1998) report that the first 20 % of runoff has greater than 40 % of pollutions [13]. Vorrieter and Hickey (1994), in their study, suggest the definition of first flush 25/50 [16]. In their research, Hathaway et al. (2012) compared the different definitions of the first flush with the assumption of 20/40, 25/50, 20/80, and 30/80. According to definitions, researchers concluded that assuming the first flush 20/80 and 30/80 is very low, almost zero. Based on the definitions of 25/50 and 20/40, the occurrence of the first flush was recorded in approximately 20 rain events [12].

The assessment of the formation of the first flush according to the alternative method (MFFR) was performed by Barco et al. (2008) [16], Han et al. (2006) [17]. Barco et al., concluded that the highest mass of pollutants (40 %) is in the first 20% runoff [15]. In his research, Mamoon also evaluated the formation of the first effluent in an urbanized area, finding that the largest mass of pollutants was the first 40 % runoff [3]. Qian et al. (2021), in their research, they assessed the mass of pollutants trapped in the first 30 % runoff. They have found that in the first 30 % of runoff, is occurring 36 % to 90 % of pollutants (depending on the characteristics of pollutants, rain, and land use) [18]. In their study, Huang et al. (2012) verified that the first 30 runoff volumes carried the largest mass of pollution. Like Mamoon, he found that in that definition of the first flush (30 %), the mass of pollutants was around 35 % [19]. Not all authors agree with the statement that the first flash occurs when 80 % of pollutants enter the runoff in the first 30 % or 20 % of the runoff. In any case, it can be argued that the surface runoff generated during a rain event is the primary transport medium for pollutants.

5. Factors affecting the occurrence of the first flush

Given the different results of the assumption of the formation of the largest pollution in the surface runoff, it is important to note that the formation of the surface runoff is influenced by several factors, which may also affect the occurrence of the first runoff [3], [10], [13], [20]. Precipitation characteristics such as intensity, rain duration, length of the rainless season are important parameters in the formation of surface runoff [3], [12]. It is assumed that the longer the dry season, the more pollutants accumulate on the surface of paved surfaces. The intensity of the rain is the most important factor and determines how quickly runoff will form [3].

In terms of the characteristics of the watershed, the formation of the first flush is divided by the material, type, and geometry of the paved surface [3]. Also, land use can have an impact on the relationship between rainfall and runoff [13]. When assessing the formation of the first flush, it is important to know the relationship between the characteristics of providing and precipitation and pollutants. Mamoon et al. (2019) [3] and Qian et al. (2021) [18] report that pollutants such as dust particles are easily flushed away as heavy metals by runoff due to their weight. Detic et al. (1998) [12] also state that the first flush was not the same in all assessed river basins; a more significant first flush was recorded in river basins, where was the intensive rainfall. It can be concluded that several factors define the formation first flush as the occurrence of surface runoff.

6. Conclusion

The first rainwater flush in an urban environment is a key factor in rainwater management. Knowing the mass of pollutants at a particular runoff stage could help design measures and facilities for rainwater management. It could also be important in assessing the pollutant load of an area of interest pollutant in a rainfall event.

Despite its importance, it is a controversial phenomenon among many researchers. The problem is the difference between defining the occurrence of the reported flush and the contribution of its evaluation. Many authors agree that the mass of pollutants is found at the beginning of the runoff event (the first runoff phase). The question is when the first phase takes, how long it takes, and how many pollutants there are? Due to the above findings in the study, the occurrence of the first flush definition ranged from 20 to 40 percent. The mass of pollutants ranged from 6 % to 90 %, depending on the characteristics of pollutants, land use, rain intensity.

Due to the fact that strong factors condition the quality of the first flush, it is difficult to determine the exact definition of first flush. It is important to properly separate the first flush from the rest of the drain. At present, there are empirical methods for assessing the first flush, which provides an approximate estimate of when the first flush phase occurs. They are a traditional and alternative method, but not all authors agree with it. Many are trying to define a new method that would consider the characteristics of rain and the land used. Due to the complexity of the issue of the first flush, it is the subject of research by many studies.

7. References

[1] Gao B, Gao L, Xu D, Zhang M, Qu X and Li Y 2019 A novel method for evaluating the potential release of trace metals associated with rainfall leaching/runoff from urban soils *Science of The Total Environment* **664** 37 - 44

[2] Li Ch, Liu M, Hu Y, Shi T, Qu X and Todd Walter M 2018 Effects of urbanization on direct runoff characteristics in urban functional zones *Science of the Total Environment* **643** 301-311

[3] Mamoon A.A, Jahan S, He X, Joergensen E N and Rahman A 2019 First flush analysis using a rainfall simulator on a micro catchment in an arid climate *Science of The Total Environment* **693**

[4] Seto K C, Guneralp B and Hutyra L R 2012 Global forecasts of urban expansion to 2030 and direct impacts on biodiversity and carbon pools Proceeding of the National Academy of Sciences of the United States of America (PNAS) **109** (40) 16083-8

[5] Vítek J, Stranský D, Kabelková I, Bareš V and Vítek R 2015 Rainwater management in the Czech Republic. Praha, ČR, **127** ISBN: 978-80-260-7815-9.

[6] Angrill S, Petit-Boxi A, Morales-Pizón T, Josa A, Rieradevall J and Gabarrell X 2017 Urban rainwater runoff quantity and quality – A potential endogenous resource in cities? *Journal of Environmental Management* **189** 14 -21

[7] Wang Q, Zhang Q, Wang X C, Huang J and GE Y 2020 Impact of key factors on heavy metal accumulation in urban road-deposited sediments (RDS): Implications for RDS management. *Chemosphere* **261**

[8] Guan M, Sillanpää N and Koivusalo H 2016 Storm runoff response to rainfall pattern, magnitude and urbanization in a developing urban catchment *Hydrological Processes* **30** (4) 543-557

[9] Lee J Y, Kim H, Kim Y and Han M Y 2011 Characteristics of the eventmean concentration (EMC) from rainfall runoff on an urban highway *Environmental Pollution* **159** (4) 884-888

[10] Bertrand-Krajewski J L, Chebbo G and Saget A 1998 Distribution of pollutant mass vs volume in stormwater discharges and the first flush phenomenon. *Water Researche* **32** (8) 2341-2356

[11] Stenstrom M K and Kayhanian M 2005 First flush phenomenon characterization. Research Gate

[12] Hataway J M, Tucker R S, Spooner J M and Hunt W F 2012 A Traditional Analysis of the First Flush Effect for Nutrients in Stormwater Runoff from Two Small Urban Catchments *Water Air Soil Pollutions* **223** 5903 – 5915

[13] Deletic A 1998 The first flush load of urban surface runoff *Water Research* **32** (8) 2462-2470

[14] Bach P M, McCarthy D T and Deletic A 2010 Redefining the stormwater first flash phenomenon *Water Research* **44** 2487-2498

[15] Barco J, Papiri S and Stenstrom M K 2008 First flush in a combined sewer system *Chemosphere* **71** 827-833

[16] Vorreiter L and Hickey C 1994 Incidence of the first flush phenomenon in catchments of the Sydney region In National Conf. Publication – Institution of Engineers, Australia, **3** (94/15), 359–364

[17] Han Y H, Lau S L, Kayhanian M and Stenstrom M K 2006 Correlation analysis among highway stormwater pollutants and characteristics *Water Science and Technology* **53** 235-243

[18] Qian G, Zhang J, Li X, Yu H, Gong X and Chen J 2021 Study on pollution charakteristics of urban pavement runoff *Water Science and Technlogy* **84** (7)

[19] Huang J, Tu Z, Du P, Li Q and Lin J 2012 Analysis of rainfall runoff characteristics from a subtropical urban lawn catchment in South-east China *Front. Environ. Sci. Eng.* **6** 631-539

[20] Angrill S, Petit-Boxi A, Morales-Pizón T, Josa A, Rieradevall J and Gabarrell X 2017 Urban rainwater runoff quantity and quality – A potential endogenous resource in cities? *Journal of Environmental Management* **189** 14-21

Acknowledgments
This work is supported by the "Young researchers Grant by STU" under the KLIMO acronym, which is financed from the central sources of the Technical University of Bratislava, the Scientific Gant Agency of the Ministry of Education, Youth and Sports of the Slovak Republic, and the Slovak Academy of Sciences within the project VEGA 1/0574/19, VEGA 1/0727/20 co-funded by the European Regional Development Fund and by the Slovak Research and Development Agency under the contract No. APVV-18-0203.

IOP Conf. Series: Materials Science and Engineering 1252 (2022) 012076 doi:10.1088/1757-899X/1252/1/012076

Local loading of sandwich panels - assessment according to standards and experimental verification

M Magura[1] and J Brodniansky[1]

[1] Department of Steel and Timber Structures, Faculty of Civil Engineering STU in Bratislava, Radlinského 11, 810 05 Bratislava, Slovakia

E-mail: martin.magura@stuba.sk; jan_brodniansky@stuba.sk

Abstract. The aim of the presented paper is to show possible approaches to the design and verification of the connection of elements to sandwich panels. Rivets or rivet nuts were used. Eurocodes STN EN 1991-1 and 1993-1-8 were used for numerical design. To verify the accuracy of numerical outputs, the samples were verified by a laboratory experiment, and the results were compared.

1. Introduction – State of Art

In modern civil engineering, facade elements with several functions are often used to clad new buildings. These cladding elements have an aesthetic effect, because they create the overall design of the building, as well as waterproofing function against driving rain, load-bearing function - carrying loads from the wind - and must also fulfill the function of thermal insulation. If one of these functions fails, it is likely that the entire element will deteriorate and probably fail.

The façade element can be, for example, a sandwich panel - a composite consisting of exterior and interior steel (aluminum) sheet, between which mineral wool is inserted.

The interaction of both surface sheets during the entire functional life of the panel must also be ensured by means of wool-panel sheeting (if it is glued to the sheets).

Other elements which carry, for example, decorative panels or elements of shading technology or window fillings, are connected to the panel by connecting elements.

The connection itself can be the source of potential failures of structural safety, mainly strength of the structure, or a source of moisture penetration, and it is therefore necessary to emphasize the accuracy of the design and assessment of these details. This paper deals specifically with the issue of the element resistance. For preliminary estimations of riveted connection resistances, the standards STN EN 1999-1-1 [1] and STN EN 1999-1-4 [2] were used. Bolts, rivets, and pins are common mechanical joint elements that have been widely used in various industries due to being cost effective, flexible, highly resilient, and durable [3], [4]. Shear failure, described in [5], and fatigue by cycle loading, described in [6], can also incur structural problems.

2. Experimental samples

Two different types of façade panels were verified. **Type 1** was a 50 mm core thick composite panel, with 1 mm thick steel sheet on the outside, and 0.75 mm thick on the interior side. 50 mm thick mineral wool was glued between the sheets. Z-profiles made of sheet metal supporting the decorative panels and three steel L-brackets holding shading elements were located on the exterior side of the panel. Elements were connected by tear-off rivets with diameter of 4.8 mm. An overall view on the composite panel is shown in the Figure 1 and a detail of the connecting rivets in Figure 2.

Panel **Type 2** had a total thickness of 220 mm. 3 mm thick aluminum sheet was on the outside, and the whole panel was reinforced with aluminum L profiles 60/4. The profiles were in line of connecting rivet nuts made of stainless-steel diameter M8. The steel sheet on the inside of the panel was 1.5 mm. Mineral wool was inserted between the sheets, but wasn't glued to them - the sheets did not interact with each other as a composite cross section. The overall view of the panel is shown in the Figure 3 and the detail of the connecting rivet nut in Figures 4 and 5. Both types of connecting elements have been experimentally verified for tension loading.

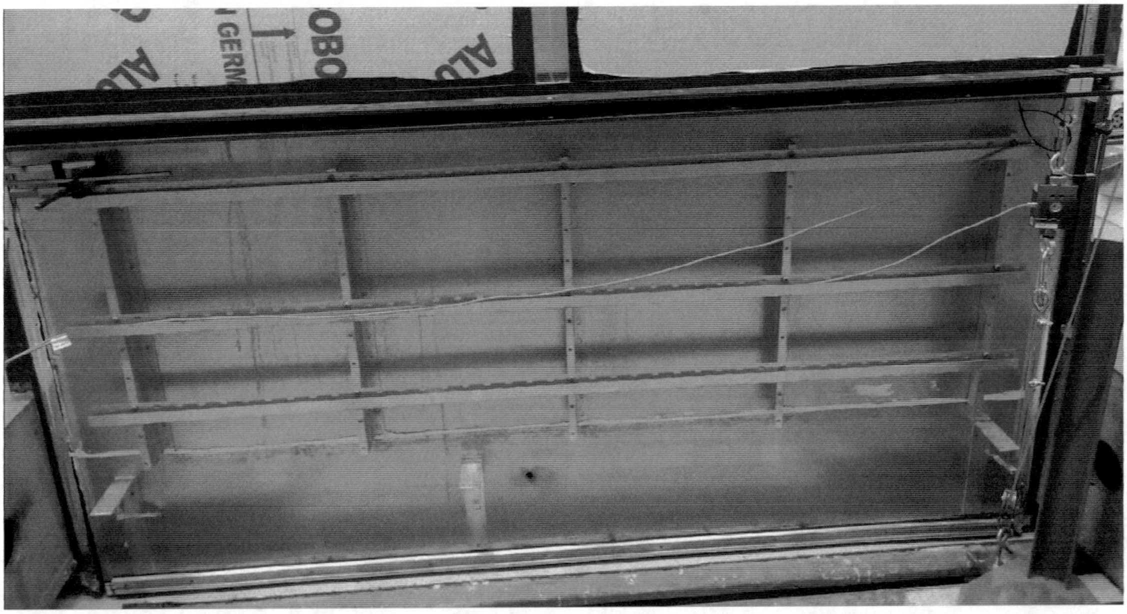

Figure 1. Panel Type 1 with riveted (4.8 mm tear-off rivets) elements for decorative panels and shade members.

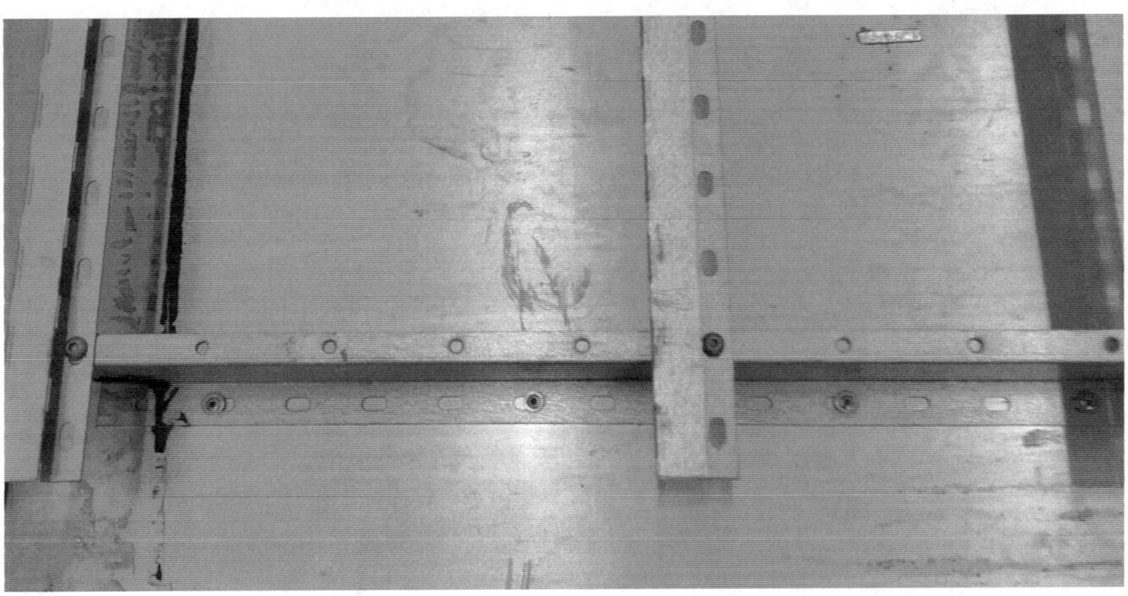

Figure 2. Panel Type 1 - detail of rivet connection of Z-profile.

Figure 3. Panel Type 2 rivet nuts M8.

Figure 4. Panel Type 2 - detail of rivet nut M8.

IOP Conf. Series: Materials Science and Engineering 1252 (2022) 012076 doi:10.1088/1757-899X/1252/1/012076

Figure 5. Panel Type 2 - detail of screws M8 in rivet nuts holding Z-profile.

3. Tear-off rivets (d = 4.8 mm) - numerical verification according to Eurocodes

The procedure of verification of tear-off rivet connections is provided by the standard [2]. The dimensional and strength characteristics of the rivet and the joined material are included in the calculation. Rivet dimensions can be taken from STN EN ISO 15977. Mechanical characteristics are taken from technical list of manufacturer AST (table 1).

Table 1. Used rivet.

Blind rivet - manufacturer AST	
Type: Standard Material: Aluminum AlMg 3,5% Material shank core: zinc coated steel Head shape: dome head	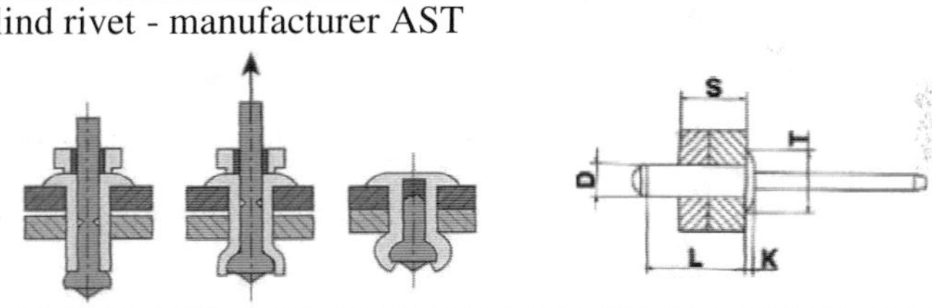
Body diameter *(D)*	4.8 mm
Body length *(L)*	6 – 19 mm
Head diameter *(T)*	9.5 mm
Head height *(K)*	1.4 mm
Tensile fracture load	2842 N
Shear-fracture load	2107 N

According to the technical sheet, the load capacity of the thread in shear is 2.1 kN and the tension is 2.8 kN. According to EN ISO 15977, the minimum load capacity of the thread obtained in the test is d = 4.8 mm Class H in a shear of 1.85 kN and a tensile of 2.6 kN. Based on [2], the thread material EN AW-5019 must achieve the following characteristics: $R_{m.min}$ = 250 MPa.

Rivet shear resistance using standard procedures in [2] clause 8.2.2.3:

$$F_{v,Rd} = 38d^2/\gamma_{M3} \tag{1}$$

$F_{v,Rd}$ – value in N

d – diameter of the rivet (hole diameter) in mm
γ_{M3} – ultimate limit state partial safety factor ($\gamma_{M3} = 1.25$)
Rivet tension resistance using standard procedures in [2] clause 8.2.3.3:

$$F_{t,Rd} = 47d^2/\gamma_{M3} \tag{2}$$

$F_{t,Rd}$ – value in N
Pull-out resistance for riveted joints loaded in tension (supporting member of steel), [2] clause 8.2.3.2:

$$F_{o,Rd} = 0.47t_{sup}df_y/\gamma_{M3} \tag{3}$$

t_{sup} – thickness of the supporting member in which the rivet is fixed
f_y – yield strength of supporting component of steel

Table 2. Calculation procedures used for the estimation of rivet shear resistance, rivet tension resistance, and pull-out resistance.

Input values			
$t_{sup} = 1$ mm	$d = 4.8$ mm	$f_y = 220$ MPa	$\gamma_{M3} = 1.25$
Used calculation procedures			
1.	$F_{v,Rd} = \frac{38d^2}{\gamma_{M3}} = 0.700$ kN		Formula (1)
2.	$F_{t,Rd} = \frac{47d^2}{\gamma_{M3}} = 0.866$ kN		Formula (2)
3.	$F_{o,Rd} = \frac{0.47t_{sup}df_y}{\gamma_{M3}} = 0.397$ kN		Formula (3)
4.	$0{,}47t_{sup}df_y = 0.496$ kN		Modified formula (3)

According to formula (3), the damage of the sandwich panel was expected on the outside of the 1mm thick plate, caused by the head core being pulled out of the rivet (ball of the rivet). Calculated design resistance of the tear-off rivet connected in 1mm steel sheet is 0.397 kN (or 0.496 kN) without using γ_{M3} factor) (table 2).

4. Experimental verification of 4.8 mm tear-off rivets

To verify the strength of the connection of the Z-profiles to the sandwich panel, a rigid board was attached to the entire surface simulating the decorative panel (all Z-profiles loaded at once) (Figure 6.). The distribution board was lifted using a mechanical pulley and the forces were measured. During loading, the Z-profiles were raised from the panel and the local deformation of the panel sheet was visible (Figure 7.). The destruction of the panel occurred due to a fault in the rivet connection of the Z-profiles to the panel. The rivets were pulled-out from the 1 mm outside sheeting (Figure 8.). The destruction occurred at a force of 9.5 kN. If the force is divided by the number of rivets (17 pcs) which held the decorative panel, the resulting force was 0.559 kN per one rivet. This value is close to the calculated design force according [2], clause 8.2.3.2 Pull-out resistance (0.377 kN).

Figure 6. Experimental loading system – rigid board connected to Z-profiles.

Figure 7. Z - profile rivet connection under loading.

IOP Conf. Series: Materials Science and Engineering 1252 (2022) 012076 doi:10.1088/1757-899X/1252/1/012076

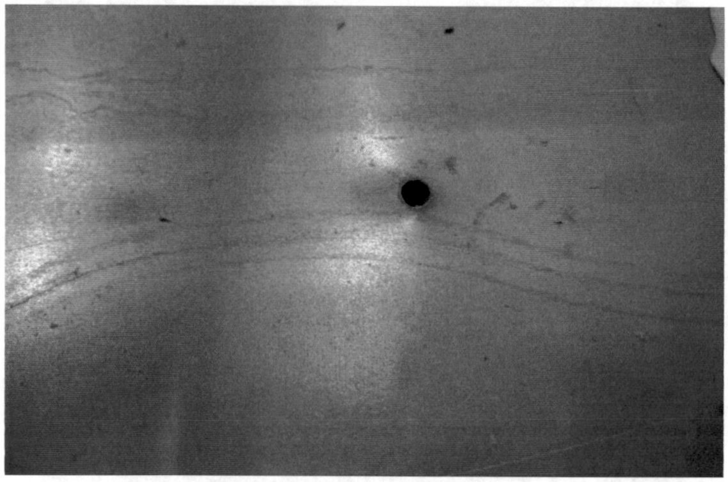

Figure 8. Detail of pulled-out rivet.

For the panel Type 1, rivets with a diameter of 4.8 mm were used as the main connecting element. Rivet connection showed low resistance and problematic performance (comparison between procedures in [2] and experimental verification).

5. Rivet nuts M8 numerical verification according to Eurocodes

In panel Type 2, rivets were replaced by M8 rivet nuts. This stainless-steel nut connects materials similarly to a classic rivet, but it is possible to screw a standard metric screw into it, which connects another element to the main structure. In our case, the rivet nut connected the 3mm aluminum cover sheet of the panel to the reinforcing aluminum L-profile 60/4, which was inside the panel. Steel Z-profiles supporting the decorative panel were connected to the rivet nut with a metric screw M8. For rivet nuts, the procedure of calculation of resistance is not specified by codes, and only characteristics of the element itself were listed in the technical sheet, without considering the joined material in the resulting resistance.

Table 3. Used rivet nut.

Rivet nut with countersunk head and knurled shank – manufacturer WÜRTH

Material: Stainless steel A2
Surface: Plain
Head type: Countersunk head
Shank detail: With knurled shank

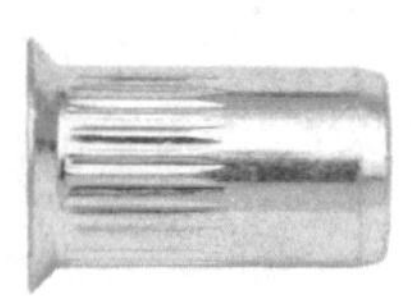

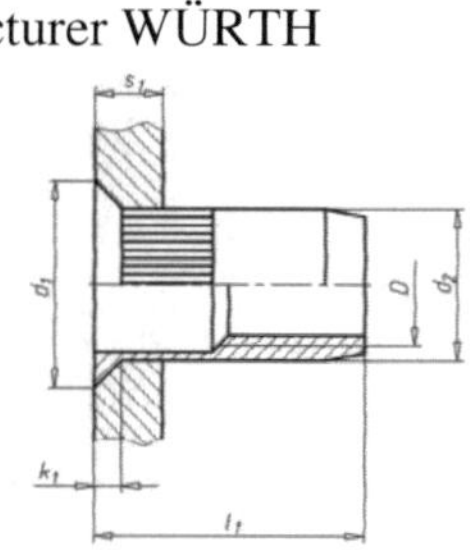

Thread type x nominal diameter *(D)*	M8
Length (l_3)	18.5 mm
Head diameter (d_1)	13.5 mm
Outer diameter (d_2)	10.9 mm
Head height (k_1)	1.5 mm
Min./max. grip range (s_1)	1.5 - 4.0 mm
Drill hole diameter	11 mm
Tensile fracture load	30840 N
Shear-fracture load	7640 N

The most suitable procedures of calculation of design resistance are in [1] and [2]. The primary verification formulas used to estimate the rivet nut resistance are described in formulas (3), (4) and (5). Resistances calculated according to Eurocodes, as well as some of the modified formulas, are shown in Table 3.

Pull-out resistance for riveted joints under tensile load (supporting member of aluminum), [2] clause 8.2.3.2:

$$F_{o,Rd} = 0.2t_{sup}df_o/\gamma_{M3} \tag{4}$$

f_o – 0,2 proof strength of supporting component of aluminum (adopting values for minimum $R_{p0,2}$ direct from the relevant product standards)

Punching shear resistance of bolts, [1] clause 8.5.5:

$$B_{p,Rd} = 0.6\pi d_m t_p f_u/\gamma_{M2} \tag{5}$$

d_m – the mean value of the across points and across flat dimensions of the bolt head or the nut or if washers are used the outer diameter of the washer, whichever is smaller

t_p – the thickness of the plate under the bolt head or the nut

f_u – ultimate strength of the member material (for aluminum member adopting values for minimum R_m direct from the relevant product standards)

γ_{M2} – ultimate limit state partial safety factor ($\gamma_{M2} = 1.25$)

EN AW 6060 T6 material characteristics according to EN 755-2.

Table 4. Calculation procedures used for the estimation of pull-out resistance and punching shear resistance of bolts.

		Input values		
$t_{sup} = 4\text{mm}$	$d = 11\text{mm}$	$f_o = 150\text{MPa}$	$f_u = 190\text{MPa}$	$\gamma_{M3} = 1.25$
$t_p = t_{sup}$	$d_m = d$	$f_y = f_o$	$\gamma_{M2} = \gamma_{M3}$	
		Used calculation procedures		
1.	$F_{o,Rd} = \frac{0.47t_{sup}df_y}{\gamma_{M3}} = 2.482\text{kN}$		Formula (3)	
2.	$F_{o,Rd} = \frac{0.2t_{sup}df_o}{\gamma_{M3}} = 1.056\text{kN}$		Formula (4)	
3.	$t_{sup}df_y = 6.600\text{kN}$		Modified formula (3) and Formula (4)	
4.	$B_{p,Rd} = \frac{0.6\pi d_m t_p f_u}{\gamma_{M2}} = 12.607\text{kN}$		Formula (5)	
5.	$\frac{0.6\pi d_m t_p f_0}{\gamma_{M2}} = 9.953\text{ kN}$		Modified formula (5)	
6.	$0,6\pi d_m t_p f_u = 15.758kN$		Modified formula (5)	
7.	$0,6\pi d_m t_p f_0 = 12.441kN$		Modified formula (5)	

The primary reason these formulas were used was to estimate possible resistance of rivet nut with M8 metric bolt. As is shown, the differences between resulting values are significant (table 4).

6. Experimental measurements of rivet nuts

Panel Type 2 was investigated by a mechanism similar to the one used in verification of the panel Type 1. Panel Type 2 performed much better and there were no faults or a failure of the rivet nut. The experimental set was the main limiting factor (maximal applied force on panel Type 2 was 12.23 kN). Therefore, one separate connector (rivet nut connection) was experimentally verified.

For the purpose of the experiment, the metric screw was replaced by an M8 threaded rod, which was screwed into the rivet nut (Figure 9.). The force was measured with a force transducer. The surroundings of the rivet nut were loaded with concrete slabs so that the panel sheet deformation was limited (Fig. 10). During the experiment, it was not possible to pull out the nut, but the panel was lifted and partially damaged - the sheet metal was separated at the edge. Results from loading process are in figures 11 and 12.

Figure 9. Axial test of rivet nut M8 – loading system.

Figure 10. Axial test of rivet nut M8 – stabilization concrete slabs.

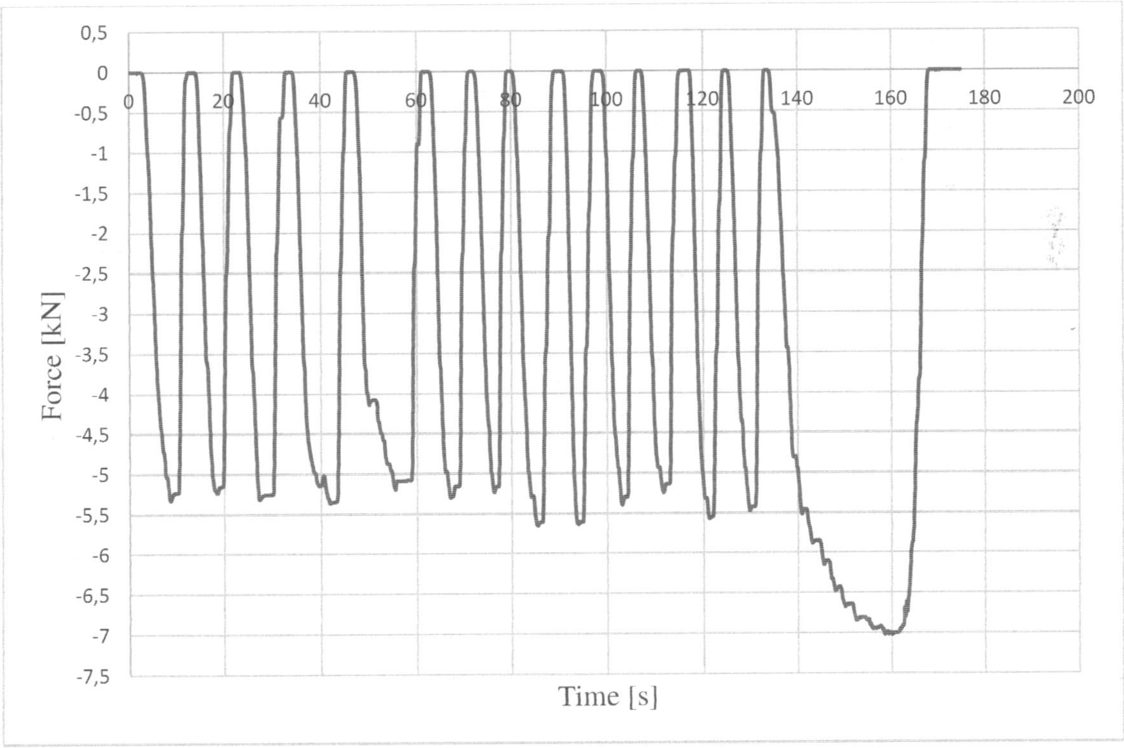

Figure 11. Results of experimental measurement no.1.

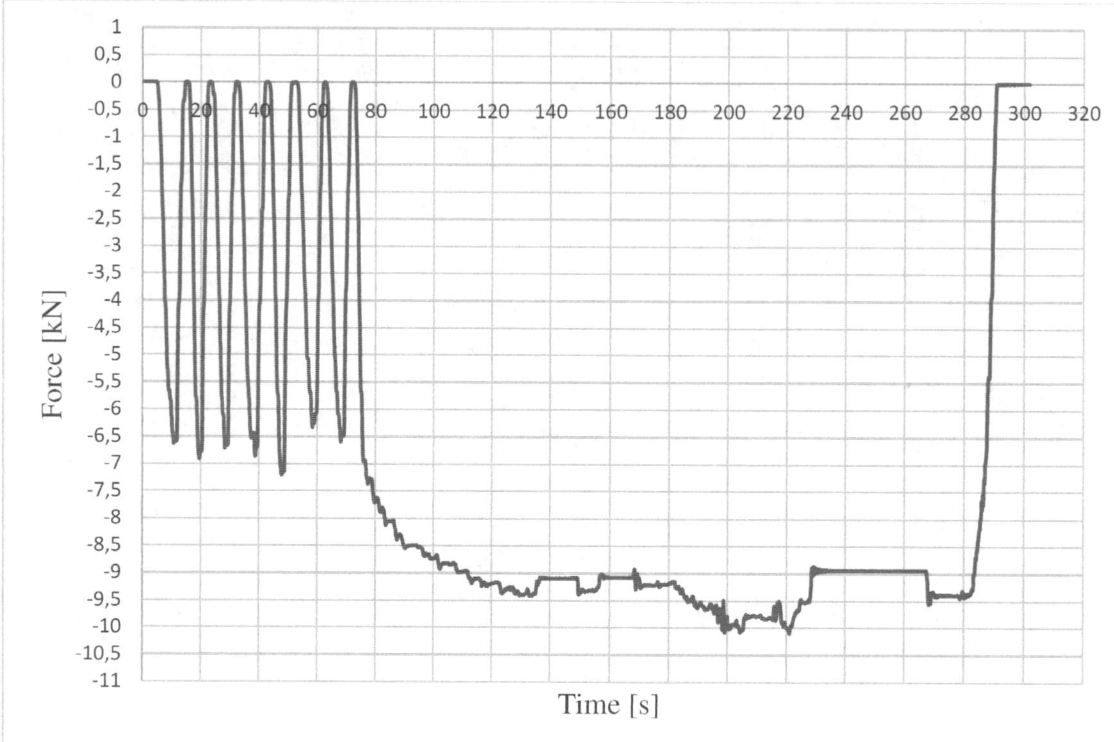

Figure 12. Results of experimental measurement no.2 – failure of the panel.

Maximal measured value: 10.098 kN – without failure of rivet nut M8. Measured force 10.098 kN exceeded calculated design results according to [2] and was close to results from [1], clause 8.5.5.

7. Conclusion

This paper presents Eurocode verification procedures for design resistance of blind rivets according to STN EN 1999-1-4. Failure of rivets with diameter 4.8 mm was experimentally verified to slightly higher value than the calculated design resistance.

By second type of panel with rivet nuts M8, which allow not only joining two layers of material, but also adding another element with a metric screw, were used. The procedure in Eurocode standards is not precisely specified for the design of these connectors. Verification procedure specified in the standard STN EN 1999-1-1 proves to be the most suitable for solving this type of connection element. However, to better determine the resistance and confront the numerical results, it is necessary to perform a larger number of mechanical tests until the connection fails.

8. References

[1] STN EN 1999-1-1 *Eurocode 9: Design of aluminum structures - Part 1-1*

[2] STN EN 1999-1-4 *Eurocode 9: Design of aluminum structures - Part 1-4: Cold formed structural sheeting*

[3] Han L, Thornton M and Shergold M 2010 *A comparison of the mechanical behavior of self-piercing riveted and resistance spot welded aluminium sheets for the automotive industry Materials and Design, 31 (3) Elsevier pp 1457-1467 10.1016/j.matdes.2009.08.031*

[4] Li D, Chrysanthou A, Patel I and Williams G 2017 *Self-piercing riveting-a review The International Journal of Advanced Manufacturing Technology) pp 1-48 10.1007/s00170-017-0156-x*

[5] Kalkowsky F, Glienke R, Blunk Chr. Dorre M and Henkel K-M *2020 Zur Bemessung von Scher-*

/Lochleibungsverbindungen mit Blindnieten in Stahlbau, Stahlbau vol 84 Issue 4 pp 304-325
[6] Hassanifard S, Adibeig M R, Mohammadpour M and Varvani-Farahani A 2019 *Fatigue life of axially loaded clamped rivet-nut joints: Experiments and analyses International Journal of Fatigue vol 129 Elsevier*

Acknowledgments
This research was supported by the Slovak Scientific Grant Agency (VEGA) in collaboration with Slovak Ministry of Education, Science, Research and Sports and Slovak Academy of Sciences (SAS), under Grant number 1/0388/19.

Exploitation of Document Management Systems in Construction Industry

T Mandicak[1], P Mesaros[1], M Spisakova[1] and A Kanalikova[2]

[1] Department of Economics, Management, and Information Systems in construction, Faculty of Civil Engineering, Technical University of Kosice, Vysokoskolska 4, 042 00 Kosice

[2] Department of Applied Mathematics and Descriptive Geometry, Faculty of Civil Engineering, Technical University of Kosice, Vysokoskolska 4, 042 00 Kosice

E-mail: tomas.mandicak@tuke.sk; peter.mesaros@tuke.sk; marcela.spisakova@tuke.sk; andrea.kanalikova@tuke.sk

Abstract. Document management systems are among the information systems that have several benefits. The main benefits include a reduction of time and of transmitting documents. These systems also allow better access to documents for all participants in construction projects. The use of Document Management Systems in the world is growing in every industry. There has been a dramatic increase in the last few years. However, some construction companies are conservative and are adopting new trends and innovations more slowly. This research addresses this issue of the use of information systems in Slovakia. The research involved investors, contractors, sub-contractors, designers operating in Slovakia. The research sample consisted of 55 respondents, including large construction companies.

1. Introduction

The current situation requires new solutions in every industrial area [1]. Current trends such as increasing the degree of digitization and robotics present a challenge for every industry [2], not excluding construction. The need to exchange information and share documents between participants in a construction project may put pressure on the adoption of document management and exchange systems [3]. The rate of digitization and work from various cities is growing. According to the study, 64% of construction project managers want better office technology, including faster internet and impressive video conferencing rooms, better displays and smart boards [4]. For this reason, it is important to implement a management system that can work with various information and documents in project activities [5].

Despite the idea that document management systems can provide easy access to and storage of project documents and information, there are a number of issues that require constant use of the systems [6]. Document management systems and technologies for the operation and exchange of documents represent a relatively broad group of tools that have several important roles specified in the management and work with documents. Effectively managing a construction project involves applying all available resources and automating and utilizing all progressive technologies to minimise each activity's cost, including the cost of managing and documenting processes.

The Document Management System (DMS) is a system whose primary function is to store and share documents in various forms and types [6]. Document management systems are an essential tool for businesses. They help streamline business processes and improve methods of cooperation between employees. It is often an electronic version of the system, so the literature often mentions the term electronic document management system (EDMS) [7]. The electronic document management system was developed for tracking and storing electronic documents while also enabling the storage of documents. The primary feature is real-time document sharing. Many such systems also allow for document editing. In this case, perfect integration with the enterprise system and other applications and software that create the documents and correct them must be ensured. These systems allow several users (participants in a construction project) to work with one document simultaneously. As already mentioned, an important function is the storage of documents.

In the study, Lin and Lee pointed out general methods of sharing documents between participants in a construction project. Construction projects involve several participants who have different documents. Therefore, exchanging and sharing documents is very important. The study shows that 25% of participants involved in construction projects use a DMS system for document sharing. Up to 20% of shared documents are in print. These two methods are among the most efficient ways to share documents, mainly to save time and money. Document sharing by personal contact (face-to-face meetings, consultations, and so on) represent a relatively high share of a total of up to 37% [8]. However, this form of document sharing represents an increased cost, as each meeting of all participants means transport costs and is also time-consuming. The study shows the need to implement knowledge systems to manage construction projects, which should be beneficial for all participants in the construction.

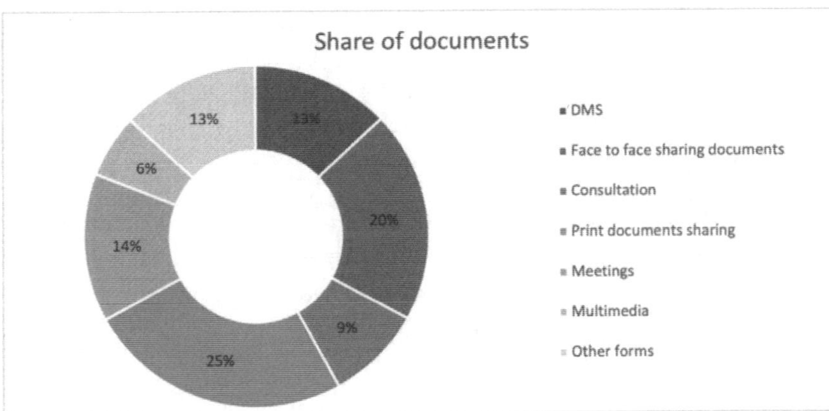

Figure 1. Documents sharing [8].

However, these study results are from the period before COVID-19 and it is therefore appropriate to compare them with newer results. The current situation has changed significantly. The pressure to digitize and transfer physical documents to the online environment with the advent of the COVID pandemic has increased significantly. The changeroom applies to every industry, architecture, engineering, and construction (AEC) as well.

Document management systems can be available from a variety of locations. Remote access is one of the main advantages of cloud-based solutions. Cloud computing in DMS is the management and storage of essential documents in one place, which is accessible from any place and at any time with the help of remote access. Within the Cloud-computing service, there is also a space (hosting) where documents are stored. Many construction companies are concerned about securing documents when using such services. Sensitive data is often involved, and unwanted access by others is a significant threat. Providers of these services assure their customers of the security of such solutions.

IOP Conf. Series: Materials Science and Engineering 1252 (2022) 012077 doi:10.1088/1757-899X/1252/1/012077

2. Methodology

2.1. Research goal and data collection

The research is focused on the use of document management systems in the management of construction projects. The aim of the research is to analyse the current state of the use of document management systems in the management of construction projects in Slovakia. The rate of use of document management systems may depend on several factors. For example, the size of the company or project and the number of participants in the construction project are possible factors that may affect the degree of use of DMS.

Data collection was carried out through a questionnaire. It included questions focused on the characteristics of the research sample. The next part focused on the use of document management systems in managing construction projects. Exploitation rate was monitored and quantified based on respondents' perceptions on a scale of 1 to 5.

2.2. Research sample and data processing

The research sample consisted of construction companies in Slovakia. As far as participants in the construction project are concerned, the research involved investors, construction contractors, subcontractors, and designers (Table 1).

Companies that use foreign capital and know-how were also involved in the research. These are affiliates of multinational companies operating worldwide.

Table 1. Characteristics of research sample.

Type of Information	Categories	Percentage
Construction Enterprise Size	**Large enterprises**	**12.73%**
	Medium-sized enterprises	**21.82%**
	Small enterprises	**30.91%**
	Microenterprises	**34.55%**
Private equity	**Slovak**	**80%**
	Foreign	**20%**
Participant of construction project	**Contractor**	**50.91%**
	Sub-contractor	**23.64%**
	Designer	**18.18%**
	Investor	**7.27%**
Age of enterprises	**0-5**	**20.00%**
	6-10	**16.36%**
	11-20	**27.27%**
	21-50	**25.45%**
	51 and more	**7.27%**
	N/A	**3.64%**
SK NACE classification	**Building Projects**	**47.27%**
	Civil Engineering Projects	**16.37%**
	Specialized construction works	**36.36%**

The data obtained through the questionnaire were evaluated based on descriptive statistics. Respondents answered questions about DMS use on a scale of 1 to 5, with 1 indicating a low exploitation rate and 5 a high exploitation rate.

These values were averaged based on research groups. The analysis of these results revealed the trend and the situation of use.

3. Results and discussion

The pre-project phase represents the initial phase of a construction project. The pre-design stage aims to analyse the opportunity and assess the feasibility of the construction project. This stage includes many activities, from the initial idea of the investor or developer to implement the construction project, through the preparation of tender documents and bids by individual participants in the construction project, evaluation of tender documents and bids, planning of resources needed for the construction project to information exchange, communication, coordination, and the exchange of documents present at each stage. These activities have been considered in research into the use of information and communication technologies.

When using DMS systems and technologies ensuring the circulation of documents, large companies use these technologies to a greater extent. However, the utilization rate of DMS systems is lower. It is 3.47 for large construction companies, 3.05 for medium-sized construction companies and 2.83 for small companies. There are more minor differences in the use of technology depending on the size of the companies when using information systems. The utilization rate for large construction companies is 3.56, and medium-sized construction companies are 3.26. This value is almost the same as for small construction companies, representing a rate of 3.27 (Figure 2).

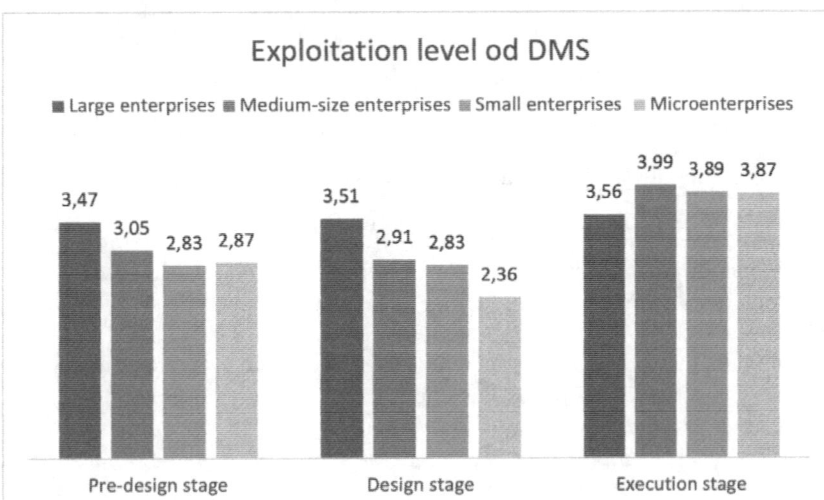

Figure 2. Exploitation of DMS according to enterprise size.

Similarly, DMS systems and document circulation technologies are very often used in large companies in the design stage. Their utilization rate is 3.51. SMEs use these systems at 2.91 and 2.83, micro-enterprises only at 2.36. In the project phase, DMS systems are of great importance for all participants in a construction project. The number of documents is growing, as is the need to share them.

The most significant difference between individual companies broken down by size is recorded when using DMS systems and technologies, ensuring the circulation of documents in the implementation phase. Large construction companies also make extensive use of DMS systems themselves. The total use of DMS systems for large companies reached 3.56. Medium-sized enterprises reach the level of 3.01. The worst are small businesses and micro-enterprises. The KS utilization rate for these companies is 2.23. Large enterprises also use cloud solutions to store and share documents.

On the contrary, smaller companies are distrustful of this technology. The relatively low level of cloud computing reflects data and document security concerns on the part of small and medium-sized enterprises. Email communication and document sharing represent the most widespread document circulation, even with a large and complex DMS system. This form is still the most widespread technology for document circulation, regardless of the size of the construction company participating in the construction company.

An overview of the use and differences in DMS use is shown in the following figure (Fig. 3). The exchange of documents most often takes place between the investor and the designers. The rate of use of DMS systems and technologies ensuring the circulation of documents in the pre-project stage is 3.49 and 3.35, respectively, for investors and developers. The main suppliers and sub-suppliers represent communication systems at the level of 2.97 and 3.02, respectively

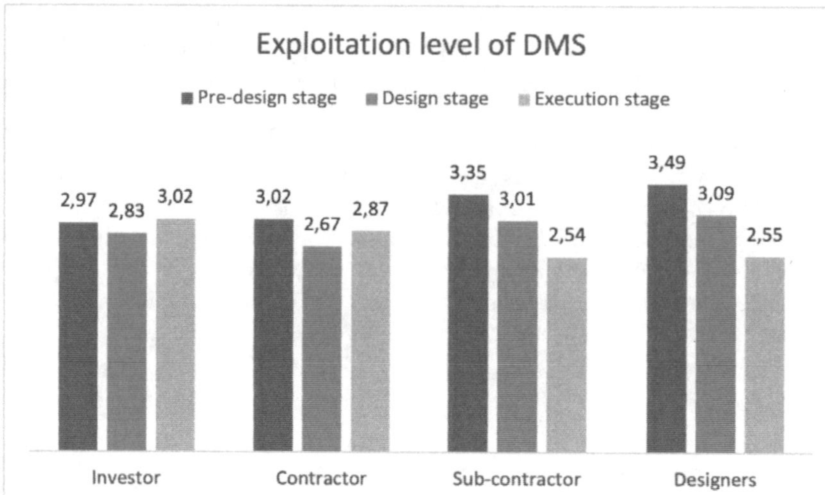

Figure 3. Exploitation of DMS according to participant of construction project.

DMS systems and technologies ensuring the circulation of documents are similarly very often used in large companies and the design stage. As part of new technologies and functionalities, sharing information and documents through these devices is also often used. Applications that support communication are used less.

When exchanging documents, servers and simple tools and applications for sharing documents such as google disk (docs), Dropbox, iCloud and the like are again used to a greater extent. Larger design companies and investors also use more costly solutions, such as DMS systems, paid cloud storage and the like. In general, however, DMS systems are used at a relatively low level by all participants in the construction project. At this stage, the most significant use is by designers and investors (3.01 and 3.09).

When using DMS systems in the implementation phase of a construction project, suppliers dominate before designers and investors. Major vendors have a utilization rate of 3.02. Medium-sized enterprises achieve a DMS utilization rate in the implementation phase of 2.87. Small micro businesses reach the 2.55 level. The rate of IS use is probably highest at the main supplier (3.65). All values are very balanced (Medium-sized enterprises 3.54; Small enterprises 3.45; Micro-enterprises 3.45). The use of DMS differs in the individual phases of the construction project. To summarize the results, it is appropriate to consider the factors that were also monitored at each stage. There was a presumption of impact or a certain degree of dependence on the use of the selected type of technology. Statistical tests of significance were performed based on established hypotheses in the use of selected groups of technologies. Large companies have more robust financial sources, and the survey shows that they invest more in DMS (Figure 4).

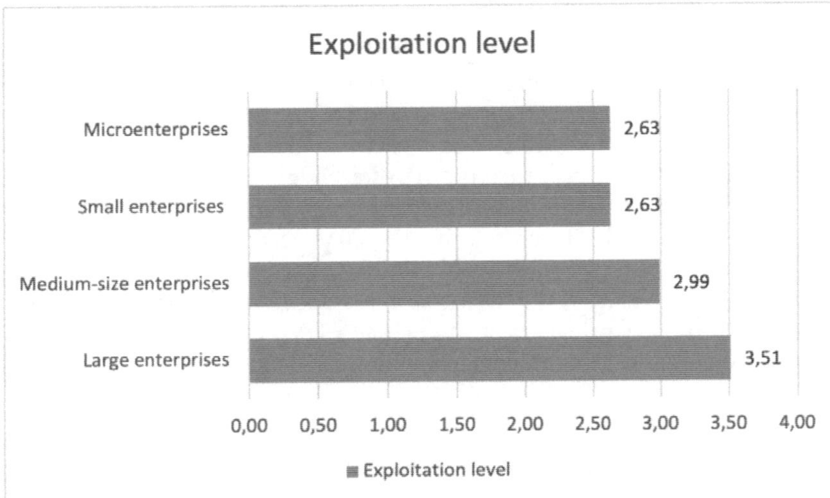

Figure 4. Total exploitation level of DMS according to enterprise size.

The use of DMS systems and applications enabling the circulation of documents is more widely used in large construction companies at the level of 3.51. Medium-sized enterprises use this tool throughout the construction project at the level of 2.99 and small and micro enterprises at the level of 2.63. Within the use of DMS systems, there were similar features of use for individual size groups of construction companies participating in the construction project (Figure 5).

Because different participants in a construction project have different activities and tasks and different interests, the pattern of their behaviour and use of information and communication technologies is likely similar.

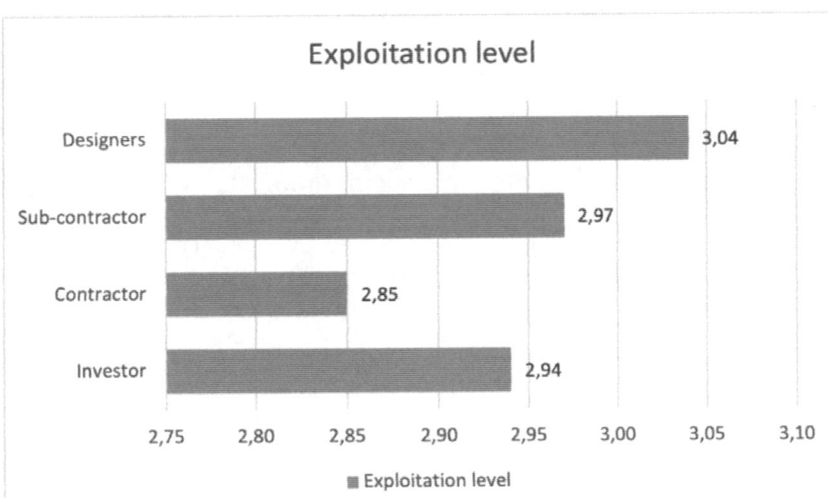

Figure 5. Total exploitation level of DMS according to participant of construction project.

The use of DMS and tools for ensuring the circulation of documents is very balanced. In many cases, no direct connection was found with the definition of a specific participant in a construction project. Nevertheless, some signs of ordinary thinking and the use of MDS systems can be observed. Designers make extensive use of cloud solutions, and on the contrary, suppliers are more conservative. All participants are conscientious and use cloud solutions to a minimum for financial documents or information.

4. Conclusion

Research on the use of document management systems has yielded exciting results and a comparison of the level of use between individual groups of participants in construction projects and the sizes of companies. No significant differences in use were found between the participants in the construction project. The more significant result and differences in use were in the different stages of the construction project. On the contrary, the differences were more significant depending on the size of the construction companies. Large construction companies have achieved a higher level of DMS use. Likewise, the size of the projects was a factor where the differences between the research groups were greater. Therefore, it is necessary to focus on other factors that could eliminate our examined factors. The results of this research need to be compared and analysed over time. Therefore, another prospect of this research is a comparison of level measures over time. Impacts such as COVID, which can significantly affect the speed and implementation of more progressive information technologies in construction and the overall rate of digitization.

5. References

[1] Juszcyk M, Výskala M and Zima K 2015 Prospects for the use of BIM in Poland and the Czech Republic – preliminary research results, *Procedia Engineerings (Creative OCnstruction Conference 2015),* vol. 123, pp. 250-259.
[2] Galić M, Završki I and Dolaček-Alduk Z 2016 Methodology and algorithm for asphalt supply chain optimization, *Tehnicki vjesnik*, vol. 23, no. 4, pp. 1193-1199.
[3] Leśniak A 2016 Bidding factors-the reduction of the data dimension with the use of PCA *I. nternational conference of numerical analysis and applied mathematics,* vol. 1863
[4] DIRTT 2021 Construction leaders look to technology, adaptability and comforts of home in the era of hybrid work, published 13.9.2021, Construction dive, Available at: https://www.constructiondive.com/spons/construction-leaders-look-to-technology-adaptability-and-comforts-of-home/606212/
[5] Fernando H, Hewavitharana T and Perera A 2019 Evaluation of electronic document management (EDM'systems for construction organizations *Moratuwa Engineering Research Conference – 5th International Multidisciplinary Engineering Research* Conference *(Proceedings, Moratuwa, Sri Lanka, 2019)*
[6] Wong C H and Sloan B 2004 Use of ICT for e-procurement in the UK construction industry: a survey of SMEs readdiness. *Khosrowsha*hi, F (Ed.), 20th Annual ARCOM Conference. 2004, vol. 1, pp. 620-8.
[7] Chassiakos A P and Sakellaropoulos, S P 2008 A web-based system for managing construction information *Advances in Eng. Software.* 2008. vol. 39. p. 865-876
[8] Lin Y and Lee H 2012 *Developingn project communities of practice.based knowledge management system in construction*

Acknowledgments

This work was supported by the Slovak Research and Development Agency under the contract no. APVV-17-0549.
The paper presents a partial research results of project VEGA 1/0336/22 "Research on the effects of Lean Production/Lean Construction methods on increasing the efficiency of on-site and off-site construction technologies".
This paper presents partial research results of project KEGA 009TUKE-4/2022 "An interactive tool for designing a safe construction site in an immersive virtual reality".

Tools of 4D modeling of construction projects

Yaroslav Nechyporchuk [1] **and Renáta Bašková** [1]

[1] Department of Construction Technology, Economy and Management, Faculty of Civil Engineering, Technical University of Košice, Vysokoškolská 4, Košice, Slovakia

E-mail: yaroslav.nechyporchuk@tuke.sk; renata.baskova@tuke.sk

Abstract. The 4D BIM model links the construction schedule and the 3D construction model. The 4D model makes possible to visualize how the construction and its structures in the construction site would look like at any point in time by simulation the construction process. The article provides basic information about BIM technology and 4D modeling. Various levels of tools for creating 4D models are presented in the core of the article. The analysis of selected software from the point of view of their conformity with requirements for creation and for usage of 4D models in construction practice is processed.

1. Introduction

Construction planning and scheduling includes the sequence of activities in space and time, also taking into account space constraints, staff, resources and other issues of the process. [1]. In today's era of innovation and constant development, the acronym BIM (Building Information Modelling) is well known in the construction industry and this technology is one of the most important in the area digitization of construction. The basis of the BIM model is usually the 3D model, the rest amount of information and its accuracy is usually dependent on the project requirements. The building information model may contain information of geometry, spatial constraints, geographic data, properties and quantity of materials, financial and time parameters, and etc. The BIM model serves as a database of information about the object and can be used throughout the whole lifecycle of a building, from conceptual design to demolition. BIM models must be configured differently and require different sets of information for different purposes and for different phases of the project.

The dimension of the BIM model (3D, 4D, 5D, 6D) depends on the information it contains and hence on the purposes for which it can be used. In the 4D model, in addition to graphical information, also introducing time information, which may include delivery time information, required for construction / demolition, time to commission, the order in which components are to be installed and dependence on other project areas. 5D adds another dimension of information to the BIM model in the form of finance data. Cost reduction and high quality of construction are always the priority, both in the construction and operation phases. The model serves as a source of information about a building and each building element can contain information about the cost value of the element and its position in space and time. 6D BIM (six-dimensional modelling of building information) helps to perform energy consumption analyzes. 7D-BIM (Seventh Dimensional Modeling of Building Information) is used by managers to operate and maintain construction equipment throughout its lifecycle. The seventh dimension of BIM allows managers to check and monitor current and relevant data on assets and components, such as maintenance / operating instructions, checklists and measures, specifications, warranty data, etc.

2. Material and methodology

In recent years, the use of 4D modeling has increased to perform condition analysis during the pre-construction phase and to monitor on-site activities during construction. The use of 4D modeling in conjunction with on-site production management methods to track actual progress and analyze the effects of delays on the overall project schedule leads to a higher level of on-site construction performance. [2].

4D BIM requires the connection of a construction plan with a 3D model, which results in a simulation of the construction process, which allows you to visualize what the building and site would look like at any time. [3]. 4D models combine the temporal and spatial aspects of a project, allowing project participants to view the entire construction process, better understand the schedule and the entire construction process, and identify collisions before implementation [4].

The construction industry most often deals with the delivery of unique products to a specific location. During this process, many different companies and firms are involved in the delivery, each of which has specific knowledge and experience on how best to deliver parts of the overall product. These companies need to coordinate their roles to enable each individual side of the project to carry out the construction work on site productively. To achieve these tasks in the project, a planning phase is organized that brings together all participants to jointly design and improve all individual design proposals, work plans and schedules. Although this approach significantly helps to reduce possible collisions on the construction site, most projects still cannot anticipate all important possible conflicts in advance, leading to problems associated with the loss of costs and time during construction work. According to many researchers, 4D models have the potential to overcome this problem [5].

The main information of BIM and 4D modeling processed in article was taken from the databases of Web of Science and electronic libraries (sciencedirect, academia.edu, etc.). For the overall and objective overview of the topic were used selected articles and books by authors from different countries, published for the last ten years.

3. Benefits and application of 4D modeling in construction

Any technology is useless if it has no application or its application does not give the user any benefits. This chapter discusses the benefits and using of 4D.

3.1. Benefits of 4D modeling

When we select a technology, tool or any product, we look at the benefits first. The benefits of 4D models are:

- Improving communication: the 4D model can be used as an effective tool for communication between different project participants [2]. Using a visual model, planners can communicate the planned construction process with all project participants. The 4D model incorporates both temporal and spatial aspects of the schedule and makes communication more efficient than the traditional Gantt chart. [1].
- Accurate bill of quantities: using 3D modeling, the scheduler can measure quantities more accurately and efficiently, allowing for a more efficient planning process. In addition, the 4D model allows the planner to determine the amount of material needed at any particular project time, leading to increased planning accuracy. [3].
- Multi-stakeholder participation: 4D models are often used in meetings to help show how a project may affect traffic, access to other buildings or other critical infrastructure issues. [1].
- Site Logistics: another advantage of 4D modeling is the positive impact on project logistics [3]. Planners can manage storage areas over time, access to construction site and movement within the construction site, location of large equipment, etc. [1].
- Trade coordination: planners can coordinate the expected time and space flow of material deliveries from subcontractors on site, as well as coordinate work in small spaces [1].
- Search for possible collisions: using the 4D model, it is possible to analyze the entire sequence of project work as well as search for possible spatial collisions in design solutions. In addition, it allows to detect spatial-temporal collisions that may occur during construction. Using 4D

models helps to analyze and prevent many problems in advance, even before construction starts [6].

- Schedule comparison and construction monitoring: project managers and planners can easily compare different schedule variants and quickly find out if a project is going according to plan or not. [1].
- Virtual reproduction: The ability to identify problems in models through the ability to simulate project processes and show how a plan would play in a "virtual world" [7].

3.2. Selected ways of application of 4D modeling in the construction

- Health and safety management: one of the focus of 4D simulation may be safety procedures. These simulations can identify methodology, various safety features and can highlight hazardous areas within the project. A number of systems based on these principles have been developed. For example, Vacharapoom and Sdhabhon discussed systems designed to automatically detect activities that have a high risk on the construction site. In addition, these systems indicate the necessary security measures. Control measures can then be incorporated into the schedule and subsequently visualized on a 4D model [7].
- Physical progress monitoring: Another goal was the application of 4D modeling to manage the physical progress of construction, which can be used as an important control tool for designers and managers who identify discrepancies between planned and implemented schedules. [3].
- Site logistics: using 4D modeling planners can more efficiently manage and plan storage areas and access to them thanks to site visualization. The implementation of 4D modeling allows productive planning of material flows within the construction site [4].
- Progress Tracking Using 4D Schedule and 3D Imaging: Yelda Turkan, Frederic Bosch, Carl T. Haas, Ralph Haas describe an automated construction progress tracking and schedule update system that can combine 3D object recognition algorithms with 4D plan data. When the recognition of the current state of the objects is completed, progress estimates are created for each activity and the plan is automatically updated based on the progress estimates. [8].
- In addition, 4D models can be used to control construction progress and create visual presentation materials.

4. Creating 4D models of construction

There are various options and tools for creating 4D models:

a) Manual CAD-Based Methods

In this method, the planners work with specialists to create rendered animations to visually present the schedule using the created animation. These animations are usually great on the visual side and are a very effective marketing tool, but they are unsuitable tools for planning and construction management. The main problem is the manual production of animations with limited possibilities to change, update and schedule scenarios in real time. When making changes to the plan, the scheduler must create new images or animations and then manually synchronize the 4D image with the plan, which is quite difficult.

b) BIM Tools with 4D Capability

Another way to create 4D models are functions that automatically filter building objects in the 3D view based on the set parameters. Nowadays there are software (Revit, Tekla) that divide objects into several phases according to preset parameters. The user can use the filters to display the desired object to display the object at a specific stage. This type of 4D functionality can be used for basic phasing and generation of 4D models according to selected requirements, but does not provide direct integration with the construction project schedule. One model object can be associated with one or more activities and one activity can be associated with one or more model objects, which allows 4D creation of a construction sequence with the creation and extinction of temporary objects. Most BIM tools do not have integrated "time" or "date" functions and require specific tools or 4D modules to connect directly to the construction plan data.

c) Export 3D/BIM to 4D tool and import schedule (Dedicated 4D BIM Tools)

Methods of creation using manual tools and 4D modeling tools based on CAD / BIM had their shortcomings, so many new specialized tools for creating a 4D model by connecting a 3D model and a construction schedule on separate platforms have emerged. These tools improve the creation of 4D models and give high-level players a lot of experience in creating and working with 4D models. These authoring tools need 3D model data imported from a 3D authoring application and schedule data imported from scheduling applications such as MS Project and Primavera, etc. The planner connects the elements from the 3D model with the construction activities from the construction schedule and we get a 4D BIM model [3].

4.1. Capabilities of specialized 4D tools.

The most popular and promising are software from the category "c". Today we have enough such kinds of software and we need to evaluate it, here are some things to by considered when evaluating specialized 4D tools:

- BIM import options: what geometry import conditions or BIM formats can users import and what types of object data can the tool import (such as geometry, connections, names, unique identifiers, and so on)? In some software, the tools only import geometry, names, and hierarchies. This may be sufficient for basic 4D modeling, but other data may be needed to allow users to find object properties, filter based on parameters, or query based on that data. [2].
- Scheduling competencies: Whether software has built-in tools for time planning and management of construction resources? Most BIM platforms don't have built-in tools for time planning and management of construction resources, and stand-alone scheduling applications are needed.
- Scheduling import options: what time schedule formats can the tool import and are the formats native or text files? Some scheduling software like Primavera work with a database. In such cases, it is good if the tool allows connection directly from the database and will not need reformatting to text files.
- Merge / update models: can users add multiple 3D files or multiple schedules to a single model and update parts or the entire model? If the project includes models created in multiple BIM software, the 4D modeling process will require the import and merging of these models. The 4D tool must provide this opportunity for the cooperation of different construction participants and for their coordination.
- Reorganization: Can the software reorganize or change the data after it is imported? The ability to easily modify and reorganize the components of the model will significantly speed up the modeling process.
- Temporary components: is it possible to add (and then remove) temporary components to the model, such as cranes, excavations, storage areas, scaffolding, etc., to the 4D model? Quite often, users have to create these components in a 3D modeling software environment and import them with the model geometry. Ideally, 4D software would need to have a library that allows users to quickly add and work with these components.
- Animation: does the software have the ability to simulate detailed site operations such as crane movement or other installation sequences? Some 4D tools allow you to animate objects for a certain period of time, which allows, for example, visualization of the movement of the device or detection of threats to the operation of surrounding objects at a certain time.
- Analysis:does the tool support specific analyzes, such as spatio-temporal conflict analysis, to identify activities that take place in the same space?
- Output: Does the software have the ability to easily output multiple frames over time or create movies with predefined views and time periods? Custom output functions make it easier to share the model with the project team.
- Automatic linking: can users automatically link spatial objects to schedule-based schedule items? This is useful for projects with standard design solutions [1].

4.2. The brief overview of tools for 4D modeling.

Selected software can be arranged into the category "b" - BIM Tools with 4D Capability:

- Revit (Autodesk): Revit is one such tool which distributes the objects into several phases according to the requirement [3] and then use Revit's view filter to show different phases and create 4D snapshots. It is not possible to play back a model, however.[1]
- Tekla Structures (Trimble): Tekla Structures, a BIM tool, provides a built-in scheduling interface enabling multiple links between the physical objects and the activities in the model. A single physical object can be linked to one or several activities and a single activity can be linked with one and several physical objects making the 4D evaluation of construction sequence possible, with emergence and disappearance of temporary facilities [3]. His interface allows definition of tasks and association of model objects to one or more tasks [1].

Further will considering selected software from the third group - Dedicated 4D BIM Tools:

- Navisworks (Autodesk): Navisworks technologies enable us to assemble all project data into a single model that enables us to make design decisions more effectively, manage the project, reduce errors. By creating a common object model in Navisworks, you can perform a collision calculation. 4D simulation technology is Navisworks, which allows us to manage a project, analyze planned and actual execution dates. Navisworks has the ability to visualize the flow of project work for efficient project management [9][10]. Navisworks does not have tools for network process analysis (Gantt chart). The software allows to create a hierarchically structured list of construction processes and assign a start and end date to the process. Displays a line schedule of processes. Processes can be associated with costs. No tools for analysis of labor and material resources (histograms). Navisworks allows to import a schedule from MS Project, Microsoft Excel and Primavera.
- Synchro uses real-time visualization and allows overview and control of the project at every stage. The CPM scheduler, which is integrated into the SW, allows you to visually view and update the project plan in real time, including logistics, temporary work and resources; resolve temporal and spatial conflicts of the project; draw attention to hazardous working conditions and optimize or test access during the project [11][12]. Synchro Pro has a Gantt Charting environment, showing a hierarchically structured list of construction processes and a line schedule of processes to which links can be added. It has tools for histograms of labor and material resources, but it does not have the tools to work with human resources (important in construction planning). Synchro Pro allows to import a schedule from Primavera, MS Project and Microsoft Excel. But to import a file MS Project you need to convert to MS Project XML.
- Vico Office Schedule Planner and 4D Manager (Trimble): Virtual Construction 5D building planning system included Constructor, Estimating, Control and 5D Presenter. In Constructor it is possible to develop a building model or import a model from another system to create a BIM. Objects are assigned to tools that define the tasks and resources needed for their production or assembly. Estimator allows you to calculate quantities and costs. Scheduled activities are defined and planned using balance sheet techniques (LOB or location-based) in the plan planner. 4D construction simulation is visualized in the Presenter application. As an alternative to planning within Vico Office, planning dates can be imported from Primavera or MS Project. Changes in the planning system are automatically reflected in the 4D visualization [1].

5. Conclusion

The article examined 4D technology and its application in construction. This technology of BIM platform has a large number of advantages and ways of applications in construction. 4D modeling makes it possible to find problems and prevent them before being implemented into the project, which makes it possible to prevent wasting of resources and time. The management process is also being improved. Also, the project becomes more understandable and real for everyone. But not all of them can be fully and correctly implemented due to the insufficient level of software development. Each software discussed in the article has its own advantages and there is no clear leader. When choosing, it is worth considering the tasks that are priorities for a particular project. Another reason, the reluctance and fear of firms to switch to new technologies. The problem for 4D modelling is lack of specialists too. But these problems are not critical. Given the fact that BIM technology has a great potential for development and various number of application possibilities, it is possible to assume these problems will be resolved soon.

6. References

[1] Sacks R, Eastman C, Lee G, Teicholz P 2018 BIM Handbook: A Guide to Building Information Modeling for Owners, Designers, Engineers, Contractors, and Facility Managers, third ed. (New Jersey: John Wiley & Sons, Inc.)

[2] Jupp J 2017 4D BIM for environmental planning and management *J. Procedia Engineering* **180** pp 190 – 201

[3] Eastman C, Teicholz P, Sacks R, Liston K 2011 BIM Handbook: A Guide to Building Information Modeling for Owners, Managers, Designers, Engineers and Contractors (New Jersey: John Wiley & Sons, Inc.)

[4] Brito D M and Ferreira E A M 2015 Strategies for representation and analyses of 4D modeling applied to construction project management J. Procedia Economics and Finance 21 pp 374-382.

[5] Trebbe M, Hartmann T, Dorée A 2015 4D CAD models to support the coordination of construction activities between contractors J. Automation in Construction 49 Part A pp 83-91

[6] Ermolov D E 4D modeling structures in Russia and abroad online on https://scienceforum.ru/2017/article/2017031031 (Available on day 08.12.2020)

[7] Swallow M and Zulu S 2019 Benefits and barriers to the adoption of 4D modeling for site health and safety management J. Frontiers in Built Environment 4 pp 1-12

[8] Turkana Y, Bosche F, Haas C T, Haas R 2012 Automated progress tracking using 4D schedule and 3D sensing technologies J. Automation in Construction 22 pp 414-421

[9] Martynets I 2010 Finding errors in 3D modeling of industrial equipment and objects online on https://sapr.ru/article/21255 (Available on day 10.12.2020)

[10] Stankevich J Disadvantages Revit in fact online on https://dmstr.ru/articles/1966/ (Available on day 10.12.2020)

[11] Chang J Synchro software review online on https://reviews.financesonline.com/p/synchro-software/ (Available on day 08.12.2020)

[12] Official page Synchro Software Ltd. online on https://www.synchroltd.com/ (Available on day 10.12.2019)

Acknowledgements

The paper presents the partial results of the project KEGA 038TUKE-4/2020 and project APVV-17-0549.

IOP Conf. Series: Materials Science and Engineering 1252 (2022) 012079 doi:10.1088/1757-899X/1252/1/012079

Benefits of BIM models and mixed reality in the implementation phase of construction projects

J Čabala[1], M Kozlovská[1], Z Struková[1] and A Tažíková[1]

[1] Institute of Technology, Economics and Management in Construction, Faculty of Civil Engineering, Technical university of Kosice, Vysokoskolska 4, 042 00 Kosice, Slovakia

E-mail: jozef.cabala@tuke.sk; maria.kozlovska@tuke.sk; zuzana.strukova@tuke.sk; alena.tazikova@tuke.sk

Abstract. The current trend based on the current needs of construction industry points to the increasing application of digital technologies in construction projects. Nowadays, digital technologies are becoming an essential part of construction in the phase of preparation and implementation of construction projects. Probably the greatest progress in the use of digital technologies is seen in the development of a virtual object data model of the building. The main advantages of this trend can be presented by the use of BIM technology, thus an information model of building, where other information attributes are assigned to the geometric model itself. The model created in this way has great information potential in the process of planning and preparation of construction, but also during its implementation. The technology of mixed reality (MR) has recently been added to the BIM technology, which significantly contributes to the expansion of workers' knowledge and at the same time has the potential to reveal the shortcomings that have arisen during the construction. The paper offers solutions of the possible use of BIM technology and mixed reality as tools to support the implementation of construction projects. These technologies were applied to the construction project Bastion Office Center in Košice - an administrative building.

1. Introduction

The main area of the construction industry is the realization of constructions. The construction has a long-term character and therefore its high quality is required. A large share of the final quality of the construction falls on the implementation phase of construction project itself. In this phase of management processes, the implementation of documents created in the preparatory phase is necessary and active. It follows from the nature and specifications of construction production that due to social and natural changes in the external environment, during the implementation of constructions, changing conditions in which the construction process takes place. Based on these changes, an update of the construction-technological documentation becomes an essential part of the implementation phase of construction project. The effectiveness of this process addicted from the speed of obtaining and processing new information during construction.

Construction technology has changed incredibly over the last hundred years. The 4th industrial revolution is currently underway, which can also be called the digital revolution. It fundamentally changes the way of life and communication in construction as well. To the construction industry receives state-of-the-art technologies, such as 3D printing, 3D construction scanning, digital construction diaries, or GPS-guided construction machinery. In the field of construction digitization, BIM and mixed reality are certainly among the most modern technologies. BIM is currently one of the topics around the world. Once with the attribute of the technology of the future, in which it is assumed that it will become a part of our everyday reality, is MR. Although the current rise in MR is mainly related to the gaming and entertainment industry, MR is gradually looking for a place in various areas of the industry. The goal of this paper is to point out the potential of possible use of BIM and MR technologies as tools to support the implementation phase of construction projects.

BIM - Building Information Modeling is a direct translation of building information modeling, more beautifully a building information model. BIM is not software, it is a process, it is a way of working and communicating over a project. It facilitates the exchange of information from project design, implementation to the use of the building throughout the life cycle of the building. It enables the design, implementation and management of building and civil engineering projects faster, more economically and with a lower impact on the environment. The main benefit of BIM technology is the saving of investment and operating costs in the field of construction and building management. The processed BIM model can be used for the first presentation in MR - mixed reality, where spatial collisions, poorly modeled structural elements and others can be revealed. MR is a combination of a virtual model (BIM model) with the surrounding real world. It's a kind of virtual reality. The modeled BIM model can be displayed simultaneously with the real world and perceive the objects actually built. [1] MR helps users efficiently interpret physical and digital information, and the spatial relations between them. MR can include augmented reality and other mixed configurations. When it comes to construction, MR provides architects, designers, engineers and service providers an unmatched capability of creating a virtual copy of what they are planning to create. They can travel around the building and find out how it is going to look once it is built. [2] This way, workers can leverage MR to compare plans against work completed. Advanced visualization extends the ability for users to view assigned tasks and capture data with onsite measurement tools. This reduces human error, making projects much more efficient. MR can transform the way construction companies consume and interact with information. This technology enables the education of on-site teams so they know exactly what task to perform and are provided feedback instantaneously. The benefits go well beyond real-time sharing of information. MR can shorten project schedules, reduce cost, eliminate rework, and support additional workflows like on-site assembly, progress tracking using 4D (time dimension) models, and even asset management. [3]

2. Analysis of BIM and MR technology

Nowadays, the use of computer technologies in construction is becoming necessary in all phases of construction. Probably the greatest progress in the use of computer systems is seen in the use of CAD systems in design, where the classic 2D modeling is increasingly using 3D modeling, or creating a virtual object data model of the construction work. The current trend, based on the current needs of construction and technological possibilities, shows the increasing application of computer technologies in the field of construction. The main advantages of this trend can be presented by the use of BIM models, ie information models of the building, where other information attributes are assigned to the geometric model of the building itself. The models created in this way have great information potential in the process of planning and preparation of construction, but also during implementation, or even during operation with the aim of facility management. [4]

The information model of the building consists of three basic information parts (see Figure 1): [5]

- Graphic parameters;
- Non-graphic parameters;
- Documentation.

Combination of graphic and non-graphic parameters, resp. information and documents belonging to the building object are stored and processed in the common data environment CDE (Common Data Environment). The difference between the traditional project information exchange in a project and when using BIM information using CDE is shown in Figure 1.

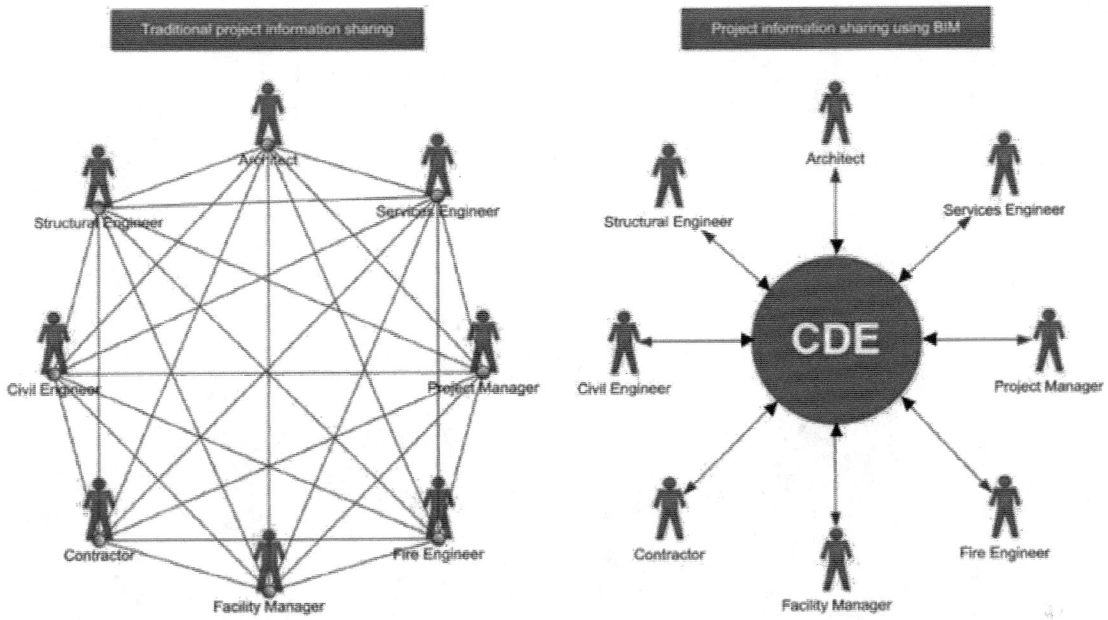

Figure 1. The difference between the tradition project information exchange and the use of BIM information using CDE [6].

To populate information about BIM models, all information from the project, each material, product, and system must be complete. The level of this information is specified by the LOD (Level of Definition). LOD describes the level of information and the level of detail of graphic and non-graphic data in the information model in individual phases of the project. [5]

Among the technologies that are looking for their place in the implementation of constructions, it is possible to include MR. MR technology in the implementation phase of construction projects offers high presentation capabilities in a realistic representation of construction (see Figure 2). Compared to conventional views, in the virtual space of the building, the participant in the construction has a better opportunity for a more objective evaluation of the dimension. MR also makes it easier for construction participants to orientate themselves in the construction space and more objectively displays the location and position of building structures.

Figure 2. Demonstration of MR on a selected part of the BIM model directly on the construction.

3. Implementation of BIM and MR technology in the construction phase

To demonstrate the potential use of MR in the construction phase, selected building structures of the construction project Bastion Office Center in Košice - an administrative building (see Figure 3) are processed in this part using BIM technology. Individual BIM models are processed as separate parts of the building. Microsoft HoloLens glasses are used to display models in MR. The digital BIM model can be holographically displayed in the field of view of the glasses at any scale and location. For example, in an office, on a green field, or directly on a construction site in a 1:1 scale.

Figure 3. Preview on the building and BIM model of the Bastion Office Center in Košice - administrative building.

In the case of BIM models of air conditioning, sewerage, water and shots of building structures, the aim was to point out selected critical details that are important in the implementation of these building structures. For example, the BIM model of air conditioning (see Figure 4) shows in MR the collision caused by piping through the ceiling structure directly on the construction site.

Figure 4. Passage of air ducts through the ceiling structure.

Microsoft HoloLens glasses contain a menu with which various properties of the model are set, e.g. setting the measure, one of the functions was measuring the distance. The measurement takes place in two steps, in the first step the first point is marked and in the second step the second point is marked, where the result was the measured distance (see Figure 5).

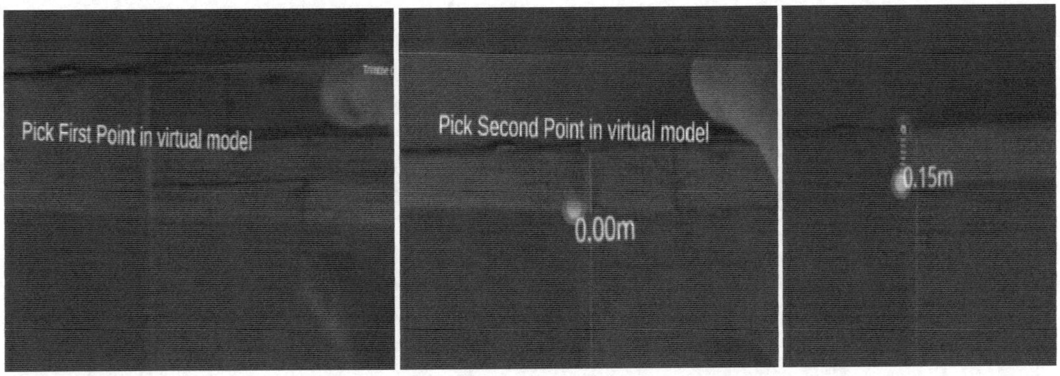

Figure 5. Example of measuring the thickness of the column head - Microsoft HoloLens glasses.

4. Result

The use of BIM and MR as tools to support the construction process can be mentioned as an innovative approach in the construction phase using the potential of digital technologies. Depending on their implementation in practice, the application of these technologies has a number of advantages, but also certain disadvantages. The benefits of the introduction of BIM technology in the implementation phase of the project from the point of view of construction contractors include:

- Cooperation and communication on the project of all construction participants;
- 3D model of the building in digital form;
- Always up-to-date documentation in digital form;
- Minimization of structural elements collisions;
- Current documents for the report - acreage;
- Easier construction planning in time and cost;
- Cooperation on the creation of the BIM model - data update;
- During reconstructions, the current state thanks to the BIM model;

Among the benefits of MR technology in the implementation phase of construction projects:

- 3D digital presentation of the building in the real world;
- Presentation of the construction outside the construction site;
- Presentation of the building in the selected scale;
- Instant information on what is poorly constructed, installed;
- Inspection of openings and transitions on the construction;
- Control of collisions on the construction between individual professions;
- Recording of digital photos and videos;

As the main negative limiting the use of BIM and MR in the construction phase is the high laboriousness of processing models of building structures in the required phase of construction. Current state of technology and price of equipment needed for BIM processing and model display in MR. However, according to the current trend that can be seen in the research and development of digital technologies, it can be argued that these restrictions will be gradually removed. The arrival of increasingly sophisticated equipment will also open up further possibilities for their use in construction practice.

5. Conclusion

The aim of this paper was to point out the potential of possible use of BIM and MR technology as tools to support the implementation phase of construction projects. These technologies were applied to the construction project Bastion Office Center in Košice - administrative building. The presented results show that the BIM technology, when extended by the MR technology, contributes to the expansion of workers' knowledge and at the same time has the potential to reveal the shortcomings that arose during construction.

IOP Conf. Series: Materials Science and Engineering 1252 (2022) 012079 doi:10.1088/1757-899X/1252/1/012079

A very interesting area of using MR, which certainly deserves attention and further research, is its use as an educational tool in teaching. MR can help students better understand and get to know building structures, work procedures, static diagrams and the like. However, MR is only a technological tool and the main condition for the success of MR in educational institutions will be the creation of quality and meaningful content.

6. References

[1] Mixed Reality In: University of Žilina. Department of Industrial Engineering, https://www.priemyselneinzinierstvo.sk/priemyselne-inzinierstvo/rozsirena-realita/
[2] Simonti P 2020 Mixed Reality in Construction. In: Geospatial Word, https://www.geospatialworld.net/blogs/mixed -reality-in-construction/
[3] Lorek S 2018 Mixed Reality In Construction Gets Real with Trimble's HoloLens, https://constructible.trimble.com/construction-industry/mixed-reality-in-construction-gets-real-with-trimble-hololens-hard-hat-video
[4] Kozlovská M and Bašková R 2017 Active use of modeling with software support in solving research tasks, http://www.svf.tuke.sk/wp-content/uploads/2015/06/B3-aktivita-1-1-modul-stavebn%C3%ADctvo-2.pdf
[5] Funtík T 2017 BIM is about data in an ordered form. In: Eurostav, č. 10, s. 70-71
[6] Burns D and kol. 2016 BIM 101 Insight, New Zealand BIM industry training group, https://bim-in-nz.squarespace.com/s/bim101-an_insight.pdf

Acknowledgments

The paper presents a partial research results of project VEGA 1/0336/22 "Research on the effects of Lean Production/Lean Construction methods on increasing the efficiency of on-site and off-site construction technologies". The article presented a partial research result of projects: KEGA No. 038TUKE-4/2020 "Supporting the platform Construction 4.0 through adoption of digital technologies in the learning process".

Green roofs on buildings – different view on its construction solution

M Kováč[1]

[1]Institute of Architectural Engineering, Faculty of Civil Engineering, Technical university of Kosice, Vysokoskolska 4, 042 00 Kosice, Slovakia

E-mail: martin.kovac@tuke.sk

Abstract. The need of green in cities or the increase of its ratio is very requested in this time of global warming. Buildings in combination with traffic cause so called urban heat islands. The high surface temperature on the facade of buildings causes the increase of surrounding air temperature in the towns. This situation creates the discomfort for their citizens. Also we have to consider with higher operating costs for cooling systems in buildings. The solution how to help is the using of green architecture for each new building but also for reconstructed existing buildings. Green roofs and walls are not new and we have different construction solutions. But this article brings different view on the construction solution of green roofs. It should be the solution that can decrease the heat load of building. This proposed solution allows using the area of flat roof for installation of air handling units and their ducts. The design of space on the flat roof overarched by greens creates better conditions for the operations of air handling units during the hot summer days when the fresh air is sucked from the shaded space that is not overheated by solar radiation.

1. Introduction

The effect of so called urban heat islands is very good known nowadays. Urban heat island (UHI) effect is a kind of heat accumulation phenomenon within urban area due to urban construction and human activities [1].The main issues are buildings, traffic and few trees, so generally little greens in the cities. Combination of these factors causes the increase of exterior air temperature in comparison with the margin areas of cities or with rural areas. The high exterior air temperature during hot summer days brings with itself the increasing the number of health complications, where some of them have also the fatal consequences for the citizens. We all good know that we need to decrease the number of cars in the streets of cities and motivate the people to using public transport. On the other side the local municipalities have to invest money into ecological public transport, effectively operate the schedules of public transport and keep the vehicles of public transport clean. We good know that we need to increase the ratio of greens in the cities, several parks is not enough, so we need green also on the buildings. This approach is very good known and it calls green building architecture. Nowadays the green roofs, green facades and green walls are used in the building architecture. Cool and green roofs are excellent mitigation technologies applied on the roof surface of buildings [2]. But the number of these types of buildings is few. Green roof is easier to design for new building than for existing building due to the load of soil on the roof construction. Especially in case where the intensive green roof is used because there the thickness of soil is higher and so heavier than in case of extensive green roof. The creation of green roof on the top of existing building can increase the investments due to need of adding the support construction elements. This article looks for another technical solution of green roof at the existing buildings. Solution that is also usable for new buildings.

Figure 1. Green roof with outdoor units of HVAC system [3].

Figure 2. Intensive green roof [4].

2. Greens on the building roof – technical solution

If we consider building the conventional green roof on the existing building, we need to consider the new load of building roof and how much money do we need for installation of new added construction elements to support the existing roof construction. Under the term conventional green roof we think the intensive and extensive green roof, where the all roof layers of green roof lie on the existing roof construction of the building. This is new added load for roof construction, especially in case of using the intensive green roof. Our proposed technical solution is the combination of green walls and partially of green roof. The difference is that this solution does not need so too much soil. The base in this case the climbing plants are, for example ivy, which covers the vertical and horizontal areas of the steel frame construction with the steel wire ropes. The steel construction, as figures 3 and 4 show, includes vertical columns with horizontal beams between them the wire ropes are installed to create the mesh, where the climbing plants can grow up. The building roof is protected from sun rays and the existing technical solution (drainage system or installed air handling units) of flat roof is preserved without the change. The plants need the soil and water for their growth up. The wooden boxes or boxes from recycled materials with soil are putted in the perimeter of the roof floor plan. These boxes are connected by water pipes in order to irrigate the soil for climbing plants. We use the rainwater harvested in the plastic water tank. Not only one big and heavy water tank, but more plastic water tanks with smaller water volume for lighter manipulation and easier maintenance. The rainwater falls on the plastic foil with outflow into water tank. The plastic foil is hanged on the steel construction of roof. In the place of runoff of rainwater from plastic foil the mechanical filter is installed to keep the mechanical particles (leafs from climbing plants) outside the water tank. In the case where the mechanical filter does not work, the rainwater fills the volume of plastic foil and then overflow the edges of plastic foil on the surface of flat roof, where the water flows into drainage system. Also the water tanks have their safety overflow. The plastic foils for harvesting rainwater will not be on their places during the winter season due to risk of damage by snow load. Also all water tanks will be without water.

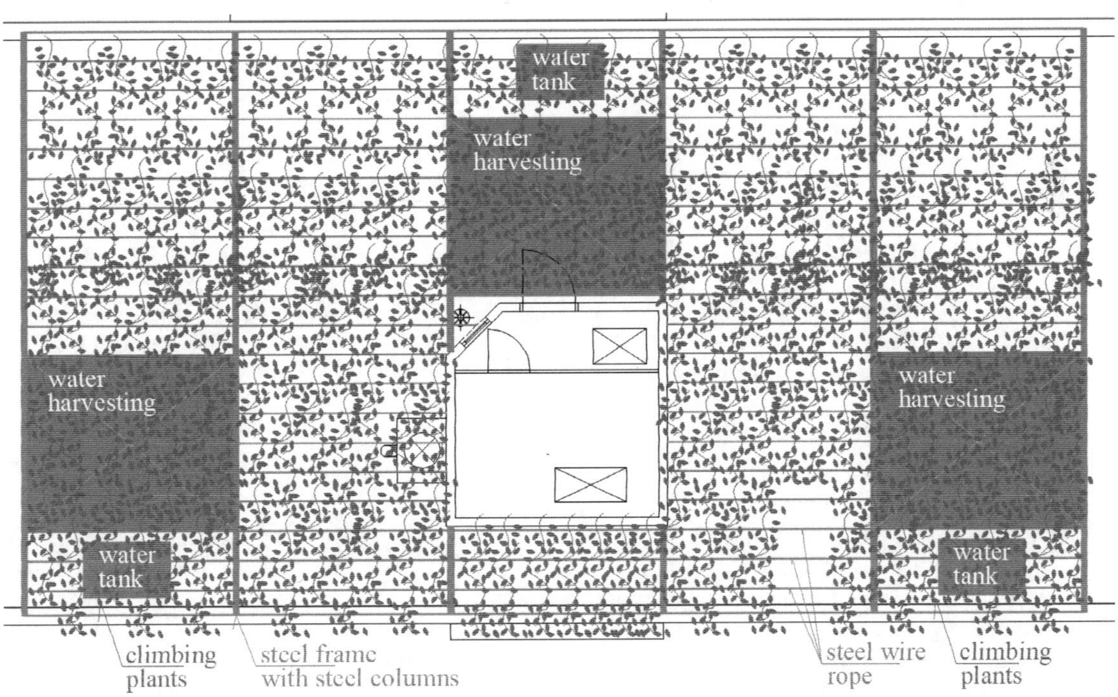

Figure 3. Top view on the steel construction with climbing plants and foils for rainwater harvesting.

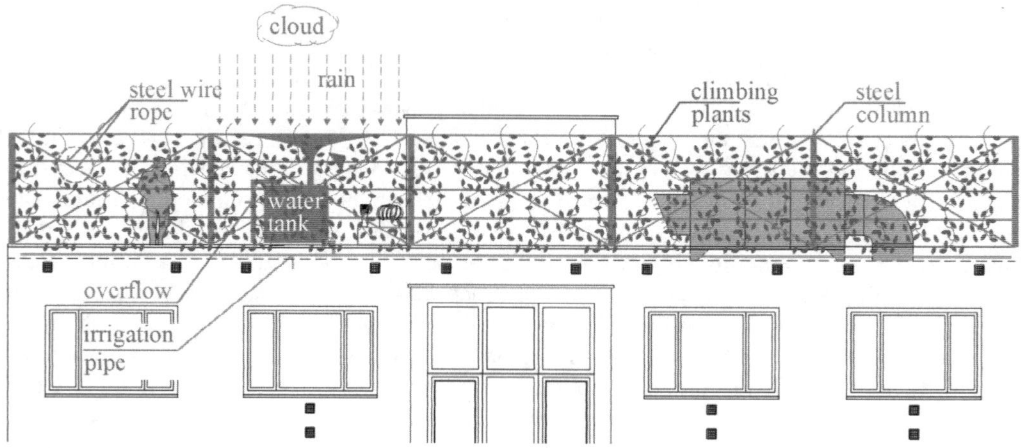

Figure 4. Front view on the steel construction with climbing plants and water tank for rainwater harvesting.

The figure 5 shows as the sun rays fall on the surface of conventional flat roofs without greens and cause the significant increase of surface temperature. The surrounding air heats up in contact with the warmer surface and so increase the air temperature above the flat roof. This fact adds to initiate of urban heat islands. If the air handling units are installed on the flat roof, they suck the overheated air during summer season and so they consume more energy for cooling the air entering the building.

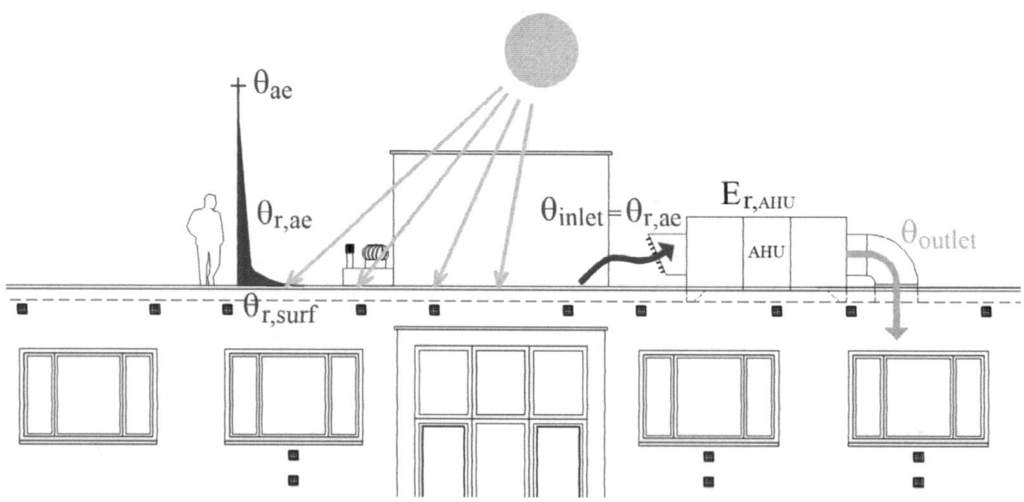

Figure 5. Graphical scheme of distribution of air temperature above the flat roof without greens.

The proposed steel construction with the mesh from steel wire ropes for climbing plants on the figure 6 offers alternative solution to conventional green roofs. The existing roof construction is not loaded by the soil layer (in case of intensive green roof is the soil layer massive). The steel columns are placed over the existing construction elements of the building (columns, beams, etc.). Different climbing plants can be used to cover vertical and horizontal areas of steel construction. This construction with climbing plants protects the building roof from sun rays. Using plastic foils for harvesting rainwater in combination with water tanks has the advantage in comparison with conventional green roof where the rainwater is not harvested. The soil layer of conventional green roof only retains the rainwater, but this layer dries up in time under the falling sun rays. The operation of air handling unit on the flat roof with this proposed steel construction with climbing plants will be more efficient due to the lower air temperature on the inlet into air handling unit. The surrounding air of air handling unit under the climbing plants will not be so overheated in comparison with flat roof without greens.

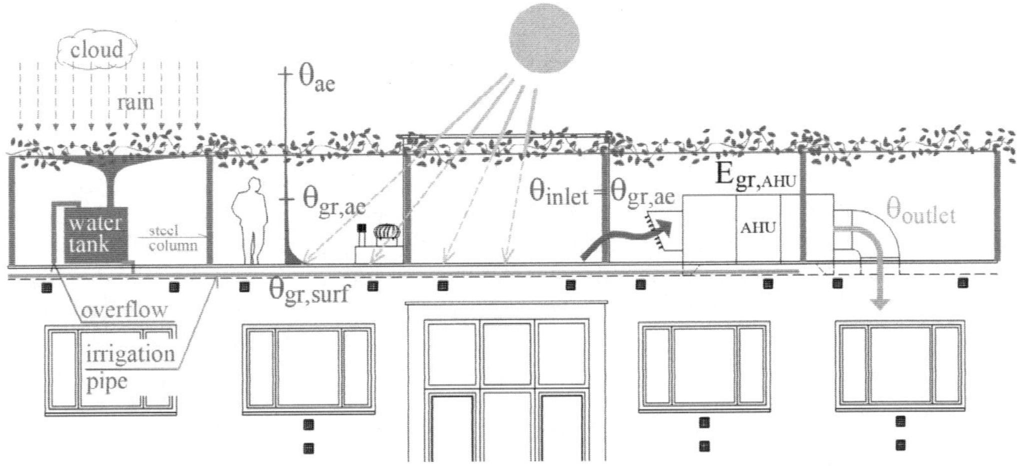

Figure 6. Graphical scheme of distribution of air temperature above the flat roof with greens.

The surface temperature difference of flat roof (without and with climbing plants) has been calculated on the base of results from computer simulation in software DesignBUILDER v7 [5]. The figure 7 shows the results for one typical hot summer day during august. We can see the trend of surface temperature of the flat roof without and with climbing plants by considering 3 different levels of shading (80 %, 90% and 100 % shading). In the case where climbing plants fully (100 % shading) covered the vertical and horizontal areas of steel construction, we can expect the surface temperature of flat roof that is almost equal to surrounding air temperature. In common situations we can predict the 80 – 90 % shading by using climbing plants, so there we can see the surface temperature difference of flat roof without and with climbing plants in the range of 2 – 12 K.

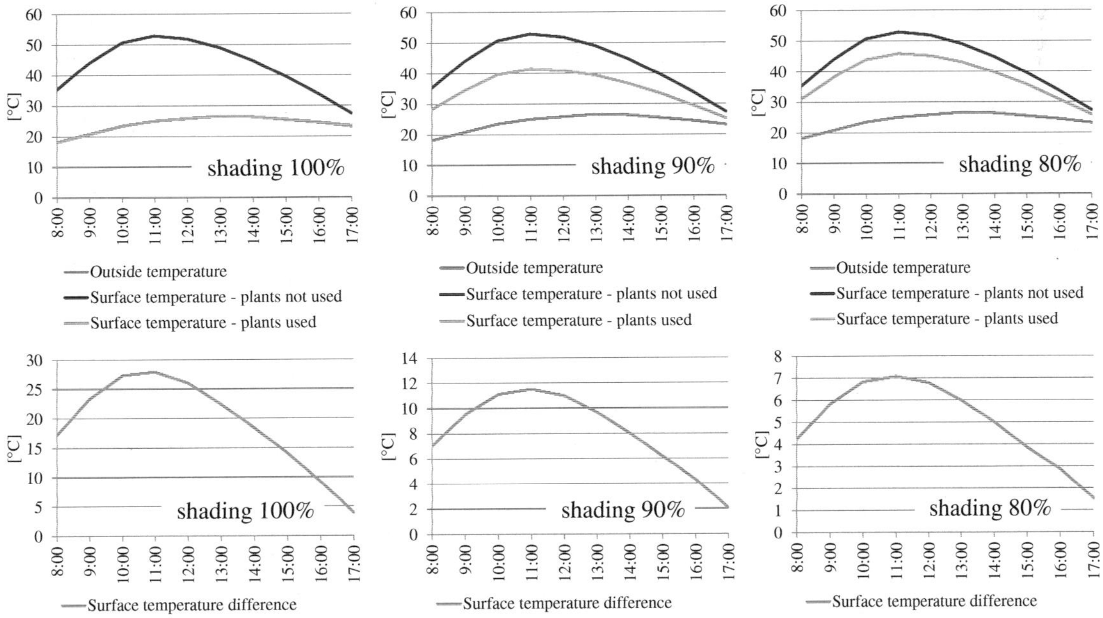

Figure 7. Trends of surface temperature of flat roof without and with climbing plants.

3. Conclusion

The proposed technical solution "how to get greens on the roof of building" offers the alternative way to the conventional green roofs (intensive or extensive) and especially in situation of existing buildings. This solution does not need to put the new layer of soil at the whole surface of flat roof, so there is no added load of roof. Combination of steel columns, beams and wire ropes creating mesh is very flexible system for different shapes of flat roofs. The steel construction have to only carry the weight of climbing plants, so the elements can by thinner and lighter. Also in winter there is no load from snow, because there are no leaves. The maintenance of existing flat roof is without change. Anytime and everywhere the waterproofing layer can be repaired. This is very important in case of existing buildings where the issues of flat roof can be oft. The variation of used climbing plants is large, so every building can have another look. This solution is also usable for new buildings. There is a large space for architects and their ideas.

References

[1] Yang Li and Qian Feng 2016 *Research on Urban Heat-Island Effect in Procedia Engineering* Volume 169 pages 11-18
[2] Kolokotsa D and Santamouris M 2013 *Green and cool roof's urban heat island mitigation potential in European climates for office buildings under free floating condition in Solar Energy* Volume 95 pages 118-130
[3] W Weissnat 2021 Green Roofs As A Modern Concept Of Green Building *The frog more* official Website [online] https://thefrogmore.com/green-roofs-as-a-modern-concept-of-green-building/
[4] Green Roofs 2019 *The Thermo House* official web site [online] https://thermohouse.ie/blog/your-guide-to-green-roof/
[5] DesignBuilder 2021 DesignBuilder Simulation Documentation for DesignBuilder v7 *The design builder* official website [online] https://designbuilder.co.uk/download/documents/468-designbuilder-printable-documentation-v7-a4-pages/file

Acknowledgments

This work was supported by the Slovak cultural and educational grant agency under the contract No. KEGA 047TUKE-4/2020.

IOP Conf. Series: Materials Science and Engineering 1252 (2022) 012081 doi:10.1088/1757-899X/1252/1/012081

Stability of travertine barrier on Milino Lake

I Štimac Grandić[1], J Drobac[1], N Bede[2], I Ružić[3]

[1] Department for Structural Engineering and Engineering Mechanics, Faculty of Civil Engineering, Univeristy of Rijeka, Radmile Matejčić 3, 51000 Rijeka, Croatia
[2] Department for Computer Modelling of Materials and Structures, Faculty of Civil Engineering, Univeristy of Rijeka, Radmile Matejčić 3, 51000 Rijeka, Croatia
[3] Department for Hydrotechnics and Geotechnics, Faculty of Civil Engineering, Univeristy of Rijeka, Radmile Matejčić 3, 51000 Rijeka, Croatia

E-mail: istimac@gradri.uniri.hr; jug.drobac@gradri.uniri.hr; natalija.bede@uniri.hr; iruzic@uniri.hr

Abstract. In March 2018, part of the travertine barrier at Milino Lake, in Plitvice Lakes National Park in Croatia, broke off, significantly altering the flow of water over the barrier. Encouraged by this event, the Faculty of Civil Engineering in Rijeka conducted a monitoring and analysis of the erosion dynamics of the travertine barriers on Plitvice Lakes. Within the project, the analyses of the stability of the travertine barriers were carried out. To the author's knowledge, physical and mechanical properties of travertine have been determined for the first time on samples excluded from travertine barriers between cascading lakes. In addition, no one has previously checked the stability of the travertine barrier located between the two lakes. The maximum compressive and tensile stresses calculated on a numerical model of the broken part of the barrier are compared with the compressive and tensile strength determined by laboratory tests on travertine samples taken from the broken barrier. By comparing the calculated stresses and determined travertine strength, it was found that the stability of the analysed part of the broken travertine barrier at Milino Lake was compromised even if only the self-weight of the collapsed part of the barrier is taken into account.

1. Introduction

Plitvice Lakes is the largest and the oldest national park in the Republic of Croatia proclaimed as the national park in 1949. The National park (NP) Plitvice Lakes have been internationally recognised in 1979 with their inscription onto the UNESCO World Heritage List [1]. Plitvice Lakes water system was created and is conditioned by the process of travertine (tufa) formation, which in turn forms travertine barriers and creates the lakes. It consists of 16 larger, named and several smaller unnamed lakes cascading one into the next. The largest lake is Kozjak (83 ha), and the smallest is Milino Lake (1.2 ha). [2, 3]

In March of 2018, part of the travertine barrier at Milino Lake broke off, after extremely snowy winter and extreme runoffs. After barrier breakdown, almost all flow was altering over the barrier (Figure 1), while the lateral travertine barrier of Milino Lake remained "dry" as it can be seen on Figure 2.

After the collapse of the part of barrier, the Protection, Maintenance, Safeguarding, Promotion and Utilisation of the National Park Department introduced extraordinary monitoring of water fluctuations (water flow and flow rates) at profiles on the lake system. Monitoring was conducted in the period up to end of November 2019.

Also, the following activities are implemented: establishment of monitoring of the current situation and possible further changes on Milino Lake, determination of hydrological conditions under which the changes occurred, determination of the characteristics of broken travertine and assessments of the stability of some barriers between cascading lakes.

The Faculty of Civil Engineering in Rijeka and department of NP for Protection, Maintenance, Safeguarding, Promotion and Utilisation conducted the project of monitoring and analysis of the dynamics of erosion of the travertine barriers [3].

Figure 1. View on the broken part of travertine barrier on Milino Lake on 9th of April 2018.

Figure 2. View of the pedestrian path with the overflow from Milino Lake: left - before the collapse of the barrier, right - after the collapse.

The expected results of the project were:

- Recording and archiving of the existing condition of Milino Lake and the surrounding area
- Quantification of the resulting morphological changes around Milino Lake based on 3D images
- Determination the hydrological and climatological conditions that led to the breaking of the travertine barrier
- Stability analysis of travertine barriers of Milino Lake based on recorded geometry and laboratory testing of travertine samples
- Preliminary assessment of stability of the downstream barriers on Prošćansko Lake, Kozjačko Lake and Galovačko Lake.

This paper presents the part of the project related to the stability of the travertine barrier of Milino Lake.

IOP Conf. Series: Materials Science and Engineering 1252 (2022) 012081 doi:10.1088/1757-899X/1252/1/012081

2. Methodology of stability evaluation

To the author's knowledge, physical and mechanical properties of travertine have been determined for the first time on samples excluded from travertine barriers between cascading lakes. In addition, no one has previously checked the stability of the travertine barrier located between the two lakes.

The evaluation of travertine barrier stability was performed by comparing the maximum compressive and tensile stresses calculated using a three-dimensional numerical model with compressive and tensile strength determined by laboratory testing of travertine samples taken from the broken part of the barrier.

The three-dimensional numerical modelling is done using the reconstruction of the geometry of broken part of the barrier and values of tested physical and mechanical properties of the material (travertine) obtained by laboratory testing.

2.1. Geometry reconstruction

The geometry reconstruction of the torn off part of the barrier was determined after the analysis of NP Lidar measurement and on-site measurements. The survey of Milino Lake and the surrounding drained areas was carried out through five on-site measurements from October 2018 to November 2019. Different methodologies were used to generate 3D point clouds: FARO Focus 3D-X130 3D scanner, SONY RX100 V digital camera and DJI Phantom 4 drone combined with SfM photogrammetry [3].

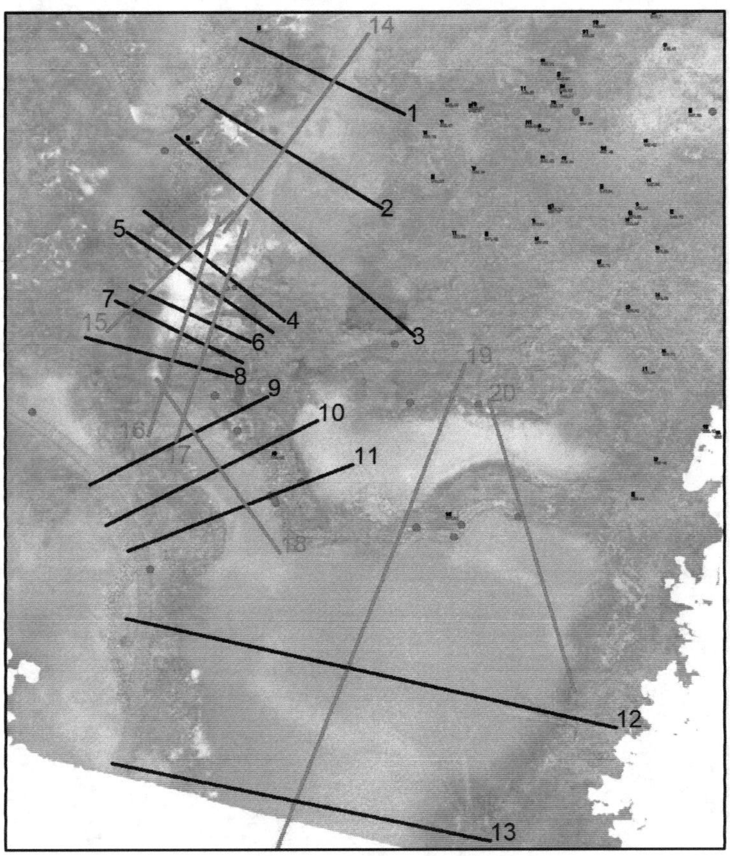

Figure 3. Digital orthophoto of Milino Lake (November 14, 2019) with the position of profiles [3].

Figure 4 shows the reconstructed geometry of torn off part of travertine barrier on Milino Lake in three profiles (15-17) shown in Figure 3, using data collected on-site measurement after collapse (red dots) and Lidar measurement data collected in 2014, before collapse (black dots).

The bottom part of cantilever protrusion of the collapsed part of the barrier before the break down have not been recorded in Lidar measurement data.

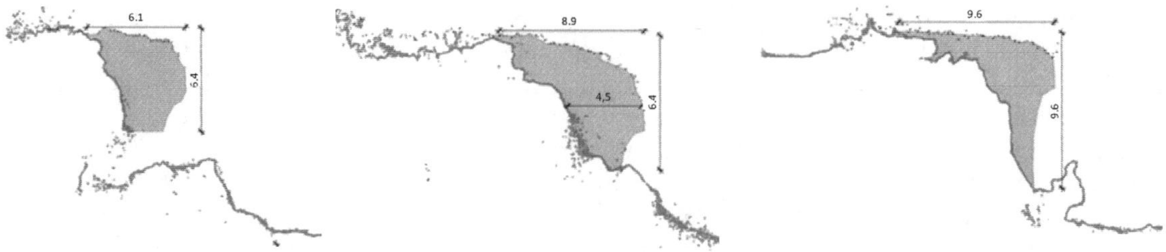

Figure 4. Changes in morphology caused by barrier collapse: profiles 15–17 [3].

2.2. Physical and mechanical properties of travertine

For numerical modelling of the collapsed part of the barrier, beside to geometric characteristics, the following physical and mechanical properties of material are required: bulk weight density (density), modulus of elasticity and Poisson's ratio.

Determination of mechanical properties of travertine is conducted in the Laboratory for Materials of the Faculty of Civil Engineering, University of Rijeka on samples taken from the collapsed part of the travertine barrier [3].

Bulk weight density is determined using the volume mass of Bulk Saturated-surface-dry sample (SSD) and specific gravity g=9.81 m/s^2. Bulk weight density of tested sample γ vary from 15.7 kN/m^3 to 18.25 kN/m^3 (mean value of 16.7 kN/m^3, standard deviation of 0.94 kN/m^3 and median of 16.4 kN/m^3).

Modulus of elasticity is calculated as the slope of the straight portion of the stress-strain curve calculated from load-displacement data recorded during compression test. Modulus of elasticity E of tested samples vary from 146 MPa to 1527 MPa (mean value of 708 MPa, standard deviation of 408 MPa and median of 629 MPa) [3].

Poisson's ratio μ was not possible to measure on samples taken from the collapsed part of the travertine barrier. Also, there is just a few paper dealing with Poisson's ratio in travertine/tufa [4-6]. According to review of Poisson's ratio for rocks [4] the typical ranges of tufa Poisson's ratio are from 0.1 to 0.27. The mean values of Poisson's ratio determined on Antalya tufa rocks are in range of 0.05-0.29 [6], while Poisson's ratios determined on specimens of the Paintbrush tufa recovered from a borehole at Yucca Mountain in Nevada, USA are in range of 0.06-0.46 [5]. Only the samples extracted from the borehole at Yucca Mountain are related to the depth of the borehole; the Poisson's ratios for samples taken from the depth between 6 and 20 m are between 0.18 and 0.21 [5] what corresponds to the depths at which the samples of the tested travertine were located.

Figure 5. A photo of varying structure of tested tufa samples [3].

The uniaxial compressive strength test on travertine samples was performed according to ASTM D7012-10 [7]. The compressive strength σ_c of the samples varies from 0.58 MPa to 3.71 MPa (mean value of 1.43 MPa, standard deviation of 0.73 MPa, and median of 1.21 MPa) [3].

To determine the tensile strength the Brazilian indirect tensile strength tests were carried out according to the instructions in the literature [8-10]. The tensile strength σ_t of the travertine samples varies from 0.19 MPa to 0.6 MPa (mean value of 0.38 MPa, standard deviation of 0.13 MPa and median of 0.35 MPa) [3].

Values of physical and mechanical properties measured on the travertine samples extracted from the collapsed barrier on Milino Lake are in very wide range due to vary heterogeneous structure (Figure 5) but comparable to the values in the literature [5,6,11].

2.3. Numerical models

The shape of the collapsed travertine barrier could not be modelled entirely as 3D model due to lack of data (the partially reconstruction is done in only 3 profiles presented in Figures 3 and 4).

The numerical model is constructed using geometry of the travertine barrier reconstruction in profile 15 (Figure 4) with the assumption of the entanglement of the broken part in the rest of the travertine barrier. The width of numerical model is 0.5 m. The geometry of the numerical model, which consists of "solid" elements, is shown in Figure 6.

Due to the very large variations of the values of physical and mechanical properties which is usual for travertine material [4-6] and the limited number of samples, the stability checking was performed by determining maximum stresses in four numerical models as follows:
Model 1 (M1): E = 146 MPa, γ = 15.7 kN/m^3
Model 2 (M2): E = 146 MPa, γ = 18.25 kN/m^3
Model 3 (M3): E = 1527 MPa, γ = 15.7 kN/m^3
Model 4 (M4): E = 1527 MPa, γ = 18.25 kN/m^3

In this way, the most unfavourable circumstances are taken into account. The value of the Poisson's ratios which is not measured on travertine samples extracted from broken part of barrier is taken as 0.2 according literature [5] and explanation in chapter 2.2.

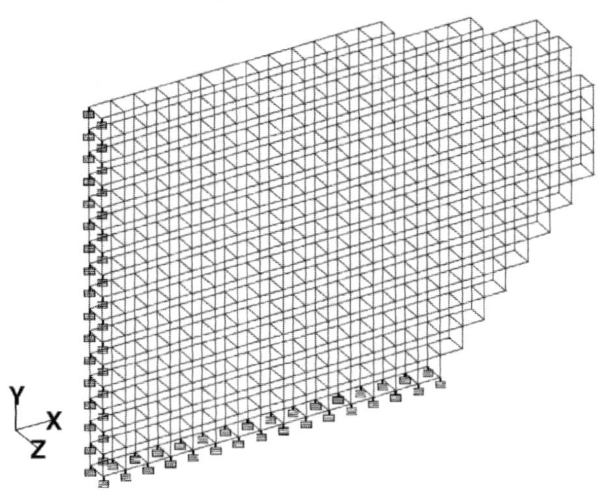

Figure 6. Numerical model geometry [3].

A linear elastic calculation, that satisfactorily describes the behaviour of the barrier under self-weight of the barrier and static water load, was performed. Although water overflow the broken part of the barrier at the time of the collapse, it is not known which column of water was overflow the barrier, thus only the self-weight of the barrier is modelled.

2.4. Numerical results

Table 1 shows the maximum calculated tensile and compressive stresses on the 4 previously described numerical models.

A graphical representation of the calculated stresses is given only for the numerical model M2 (Figure 7).

Table 1. Maximum calculated tensile and compressive stresses in MPa.

Numerical model	$\sigma_{c,xx}$	$\sigma_{c,yx}$	$\sigma_{c,zz}$	$\sigma_{t,xx}$	$\sigma_{t,yy}$	$\sigma_{t,zz}$
M1	0.136	0.433	0.104	0.190	0.037	0.045
M2	0.159	**0.503**	0.120	**0.221**	0.043	0.053
M3	0.136	0.433	0.104	0.190	0.037	0.045
M4	0.159	**0.503**	0.120	**0.221**	0.043	0.053

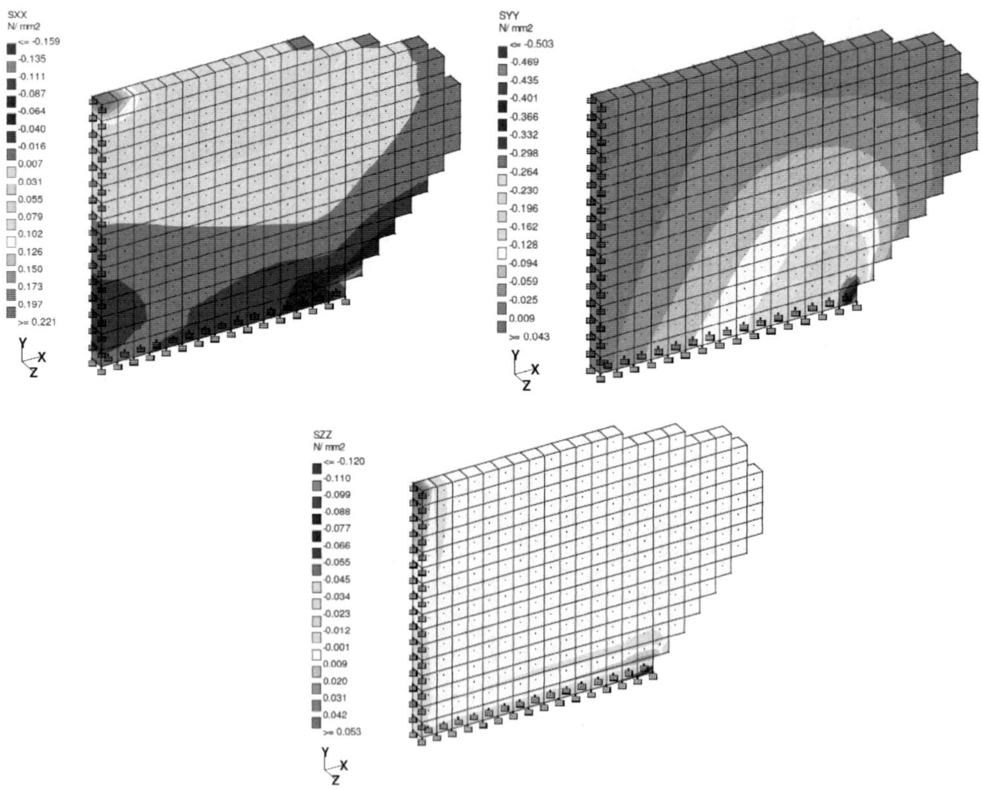

Figure 7. Calculated stresses for the numerical model M2 [3].

3. Discussion and conclusion

As written previously, no one has checked the stability of the travertine barrier located between the two lakes as well as no one has determined the physical and mechanical properties of travertine on samples excluded from travertine barriers between cascading lakes. The values of physical and mechanical properties of tested samples are comparable with the values that can be found in the literature [5,6,11].

The calculated tensile and compressive stresses on four numerical models (M1 to M4) are compared with tensile and compressive strength of tested travertine samples.

The maximum compressive stress calculated on the numerical models of the part of the collapsed barrier on Millino Lake is $\sigma_{c,yy} = 0.503$ MPa (for M2 and M4) and it is lower than the minimum measured compressive strength on the test samples, which is 0.58 MPa.

The maximum tensile stress calculated on the numerical models are $\sigma_{t,yy}$ = 0.221 MPa (for M2 and M4) and exceed the minimum measured tensile strength on the test specimens of 0.19 MPa. For the models M1 and M3 the maximum tensile stresses calculated on the numerical model is $\sigma_{t,yy}$ = 0.019 MPa which is equal to the minimum measured tensile strength on the test specimens.

From the performed numerical analyses and comparison of calculated and measured stresses it can be concluded that the stability of the analysed part of the collapsed travertine barrier on Milino Lake was endangered even if only the self-weight of the part of the barrier is taken into account in case of minimum tensile strength of travertine.

The load of water flowing over the barrier could have caused higher stresses than calculated, so it can be expected that parts of the barriers that cantilever "protrude" from the existing barriers are potentially dangerous for collapse in zones where travertine compressive or tensile strength are equal or lower than laboratory determined values.

4. References

[1] Plitvička jezera National Park official website [online] https://np-plitvicka-jezera.hr/en/about-the-park/general-information/ 12.2.2021

[2] Hrvatska enciklopedija [online] http://www.enciklopedija.hr/Natuknica.aspx?ID=48757 12.2.2011

[3] Ružić I at all 2019 *Monitoring i analiza dinamike erozije sedrene barijere na Milinom jezeru (elaborat). Građevinski fakultet u Rijeci* – in Croatian

[4] Gerceg H 2007 *Poisson's ratio values for rocks. International Journal of Rock Mechanics & Minning Sciences* **44** 1-13

[5] Martin R J, Price R H, Boyd P J and Noel J S 1995 *Bulk and mechanical properties of the paintbrush tuff recovered from borehole USW NRG-7/7A: Data report, Sandia National Laboratories, Albuquerque. New Mexico and Livermore*. California, USA

[6] Sopaci E Akgun H 2016 *Engineering geological characterization of Antalya karstic rocks, southern Turkey. Environmental Earth Scinences* **75** 366

[7] *ASTM D7012-10 (2010): Standard Test Method for Compressive Strength and Elastic Moduli of Intact Rock Core Specimens under Varying States of Stress and Temperatures*. ASTM Interanationals. US.

[8] Sopaci E Akgun H and Daemen J J K 2019 *An empirical strength criterion for the Antalya tufa rock, Souther Turkey. Environmental Earth Scinences* **78** 567

[9] Sopaci E 2012 *Geotechnical characterization and rock mass classification of the Antalya karstic rock masses. Ph.D Thesis. The Graduate School of Natural and Applied Sciences, Middle East Technical University*, Ankara, Turkey.

[10] Basu A Mishra D A and Roychowdhury K 2013 *Rock failure modes under uniaxial compression, Brazilian, and point load tests. Bull Eng Geol Environ* **72** 457-475

[11] Dipova N 2011 *The engineering properties of tufa in the Antalya area, SW Turkey. Quaterly Journal of Engineering Geology and Hydrogeology* **44** 123-134

Acknowledgments

The authors thank the University of Rijeka for supporting the publication of this paper through grant No. uniri-tehnic-18-127. Also, the authors thank the authority and employees of Plitvice Lake National Park for cooperation during the project.

IOP Conf. Series: Materials Science and Engineering 1252 (2022) 012082 doi:10.1088/1757-899X/1252/1/012082

Different ways of mechanical wave propagation damping in Ackerman ceilings using recycled rubber

K Kuliński[1]

[1] Department of Civil Engineering, Faculty of Civil Engineering, Czestochowa University of Technology, Akademicka 3, 42-201 Czestochowa, Poland

E-mail: krzysztof.kulinski@pcz.pl

Abstract. In this paper, the use of recycled rubber in the form of inserts/fillings or overlays on Ackerman ceiling blocks to partially dampen the propagation of mechanical waves from vibrating machines/devices or accidental shocks from the people movement is discussed. The numerical study is carried out with the use of ADINA software based on Finite Element Method (FEM). The energy transferred via the forced wave is observed in the form of stress plots at given time steps. Four different Ackerman blocks are chosen, which differs in height dimension and shape. In the analysis the complexity of each block supporting is simplified and each block is treated as simply supported on its external edges. The dynamic load is applied to the blocks top surface towards the bottom. For each Ackerman block the analysis is performed twice. The first sequence deals with the Ackerman block itself – without rubber filling, and the second one with insert/filling, what gives the possibility to compare the results of transferred energy. On the basis of presented in paper results, one observes that the rubber filling gives a slight to significant reduction of mechanical wave energy transfer, depending on the block configuration.

1. Introduction

Strength of materials is a field of engineering knowledge, part of mechanical engineering, dealing with the analytical and experimental study of the phenomena occurring in construction materials and structures subjected to external loads. Applied loads cause internal forces in the material, which can be described by stresses, and also cause material deformations measured as strains. When the mechanical properties of materials are examined, the most common tests are: tensile, compression, bending, shear, etc. As the history shows, i.e. Tacoma bridge collapse in 1940, not only static and/or quasi-static loads should be considered when an engineering structure is designed. Till the 1940 it seemed that if only the ultimate and serviceability limit state for the static loads are met, the construction will remain stable. Despite that, dynamic loads like: wind, vehicles/pedestrians movement, vibrations caused by eccentricities in rotatory machines etc. result in structure vibrations. The most dangerous phenomena for any kind of structures is resonance. When the vibration frequency approach and reach the frequency of the excitation force, the vibrations amplitude tends to infinity. Introduction of any engineering structure into resonant vibrations leads to its collapse.

Currently, modern material solutions are sought that allow to reduce the mechanical energy transferred through the structure being the result of single or a group of exciting forces. Moreover, due to the significant emphasis placed on environmental protection issues, researchers try to utilize as much of recycled material as possible in their solutions. One of the materials, which may be recycled and gives the possibility of reducing the effects of dynamic forces is rubber. The most common solutions in civil engineering where rubber elements are used, are bridges bearings. They not only prevent transferring significant dynamic forces to the bridge pillars but also alters (mainly damp) the vibrations. Other example of using rubber in civil engineering are special pads, placed under vibrating machines. In the case of residential construction, recycled rubber could be used as a filling or as an inserts in the beams-and-blocks floors or hollow core slabs to enhance soundproofing as well as to enhance thermal conductivity.

In the literature, one may find two different ways of enhancing dynamical properties of materials/structures using rubber. The first method concerns addition of crumbed rubber or rubber chips into the liquid material i.e. concrete mix, whereas the second one concerns the case, where rubber is glued, situated or injected on existing material. Eldin and Senouci [1] studied how rubber material change the mechanical properties of Portland cement concrete. Two different cases were considered. In the first part of specimens a coarse concrete aggregate was replaced with rubber chips, whereas in the second ones sand was replaced with crumbed rubber, respectively. As expected, the greater the volume of aggregate/sand exchanged in the specimens the lower the compressive and tensile strength was obtained. Despite that rubberized concrete exhibited ductile, plastic failure rather than brittle. Moreover, it had an ability to absorb a large amount of plastic energy under compressive and tensile loads in comparison to the normal concrete. Partial replacement of sand with recycled rubber and its influence on compressive strength, density, impact load capacity was the subject of interest by Gerges *et al.* [2]. It was proved, that the compressive strength of rubberized concrete is lowered with the rubber particles, thus it may have limited structural applications, however on the other hand it has lower density and higher impact resistance. Interesting studies concerning the temperature elevation impact on the compressive properties of recycled concrete with added recycled rubber in elevated temperatures were presented in [3]. Samples with rubber contents – 0%, 4% and 9% were taken into considerations and exposed to temperatures 20, 200, 400, and 600°C for 60 min. It is stated that rubber particles significantly reduce mass loss, increase the stain failure and improved the compressive strength. Moreover, it is shown that the specimens with 4% of rubber content are recommended for applications, however higher rubber content leads to greater material porosity when the rubber melts in high temperatures making the structure excessively loose, significantly reducing it strength. A considerable improvement of compressive, flexural, splitting-tensile strength and impact resistance may also be achieved via rubber powders what was discussed by He *et al.* [4]. The influence of rubber granulate onto different mechanical and thermal properties in concrete structures is widely discussed in the literature. [5-10]. It should be noted, that studies concerning improvement of concrete mechanical/thermal properties are not only limited to the rubber. There are many papers concerning utilization of wasted plastics, glass, polyethylene etc [11-13].

In this paper, numerical studies concerning the utilization of recycled rubber as a ceilings sound/vibrations-proofing insulation used in residential buildings is presented. These studies are based on the second method of combining recycled rubber with existing construction materials – in this case - injection/insertion of rubber material into Ackerman ceiling hollow blocks. The main advantage of numerical studies is the possibility of performing the analysis lowering the costs and reducing the time necessary for preparing the real specimens and carry out experimental tests. Numerical studies regarding to reduction of mechanical wave energy transfer in hollow wall blocks, in which rubber was injected/inserted in the hollows may be found in [14-16]. Presented in papers results show that the energy in the form of Misses-Hencky Stress (MHS) at the same point located on the opposite block surface in regard to the point of applied exciting force may be significantly reduced via rubber injects/inserts. Therefore, the question arises whether the use of rubber injects/inserts in the Ackerman ceiling blocks will have the same positive effects in reducing mechanical wave energy transferred as in the case of studied wall blocks.

2. Analysed model

For the numerical analysis purposes relating to the reduction of mechanical energy transfer via recycled rubber injects/inserts, where the mechanical wave results from the excitation force applied, frequently used in Poland ceiling solution is adopted - the Ackerman beam-and-block floor. It consists of reinforced concrete beams with special ribs driven above ceramic Ackerman hollow blocks and a layer of concrete slab formed over the top surface of blocks (sometimes with an additional steel reinforcement). Knowing that each construction material has its own damping properties, in order to obtain pure reduction of mechanical wave energy transfer, the analysis is focused only to the ceramic Ackerman block itself with and without rubber inject, neglecting other, exterior concrete layers. In Poland there are used four types of Ackerman blocks which differs with height dimension [17] i.e. h = 150 mm, 180 mm, 200 mm and 220 mm, respectively. The width of each block is equal 300 mm and depth equal 195 mm. The greater the block height, the greater the ceiling span can be achieved. It should be noted, that blocks with height h = 150 mm and 180 mm, differs slightly in shape in regard to blocks with h = 200 mm and 220 mm. In the higher blocks (h = 200 or 220 mm) there is an additional transverse reinforcement grid provided. Each block is made of red clay ceramic material. Blocks frontal shapes with their dimensions are shown in figure 1a-b, whereas in figure 1c, a typical section of raw ceiling is presented. Dimensions presented in parentheses regards to different heights for the same blocks shape. In addition, in figure 1c in some cases, where it is necessary the concrete slab located over ceramic Ackerman blocks may be reinforced with steel rebars, slightly increasing overall ceiling bending strength.

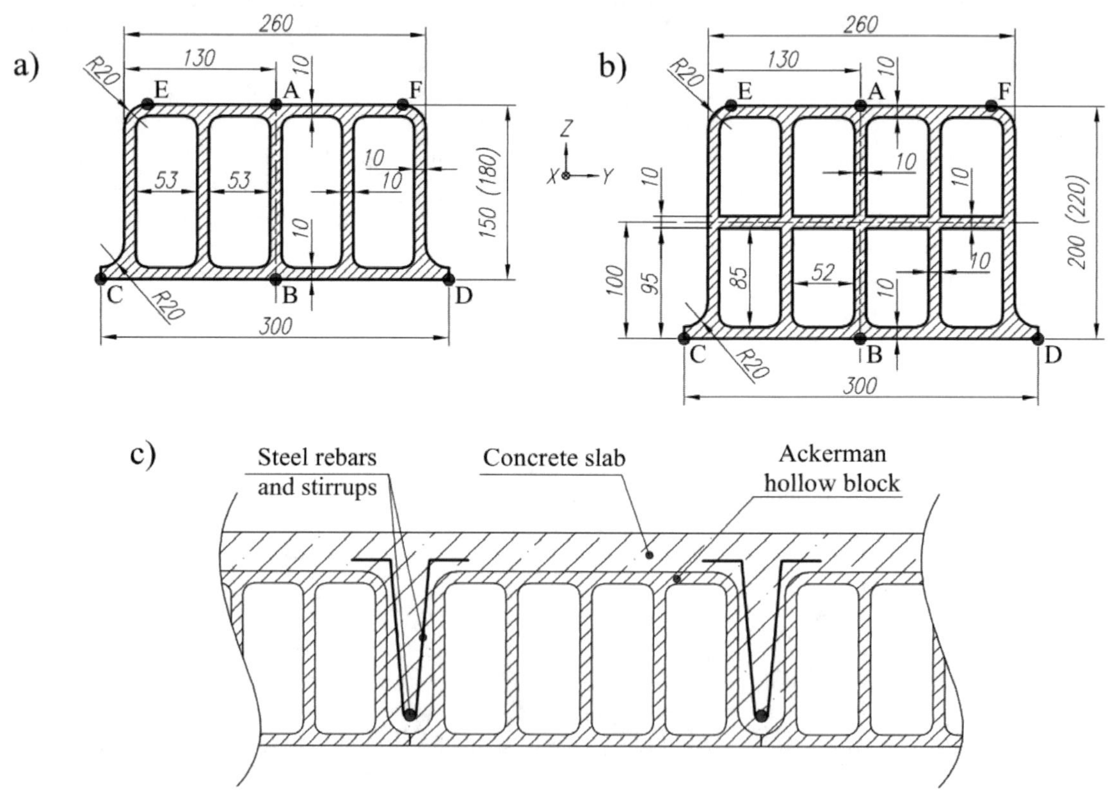

Figure 1. The Ackerman hollow blocks shapes and dimensions (a, b), section through the Ackerman ceiling (c).

The numerical study of mechanical wave propagation is performed with the use of the ADINA finite element method based software. In figure 1a and figure 1b one may notice six different points marked with filled circles and letters A-F, these are sensor points to read-out the MHS values. At the location of point A (figure 1a and figure 1b), an excitation force acting towards point B is applied with the value of 1000N. In order to observe mechanical wave propagation in the form of MHS plots, the force time of action should be relatively small [18-19]. According to that, the force value arises from the start of the analysis and reaches its maximum value of 1000N at $2\cdot10^{-6}$ s. After that, the force is removed from analysis and only introduced energy passes through the block. It should be noted, that point A is located on the Ackerman block symmetry axis, above the vertical reinforcement block wall. Hence, the analysis covers the most unfavourable force application - mechanical wave travels through the ceramic reinforcement wall to the bottom surface of the blocks – point B. The numerical model is a two-dimensional representation of real specimen laid between *YZ* axes. To achieve the highest vibration amplitudes, the hollow block is treated as simply supported. According to that in point C translation in *Y* and *Z*-axis direction is fixed, whereas at point D translation along the Z-axis is fixed and translation along *Y*-axis is free. At both points (C and D) rotation against the *X*-axis is left free. 2D-Solid 6-node tetrahedron finite elements are used to represent the ceramic Ackerman hollow block and rubber injects, with their maximum edge size equal 3 mm. The numerical study leads to some simplifications where, due to small value of impact force and neglecting of plastic/brittle behaviour, the ceramic material can be introduced as elastic/isotropic material model [20-22] defined by Young's moduli, Poisson's ratio and density. The rubber injects are introduced with Mooney-Rivlin material model. For the red clay ceramic material [23] the Young's moduli is adopted as 14 GPa, Poisson's ratio as 0.22 and density 2070 kg/m^3, respectively. The rubber material is defined through two constants C_1 and C_2 [18], where C_1 = 62780 Pa, C_2 = 8829 Pa and density is equal to 1190 kg/m^3. In order to let the mechanical wave propagate through the whole analysed model and let it slightly damp over time, 250 steps of $2\cdot10^{-6}$ s were adopted, achieving the total time of analysis equal $5\cdot10^{-4}$ s. The analysis is carried out using ADINA Dynamic-Explicit module.

3. Numerical results and discussion

In this section the numerical results are presented and discussed. The discussion is divided into two parts. The first part deals with the reduction of mechanical wave energy transfer in different configurations of Ackerman blocks. In the second part of the analysis the overall mass of blocks is investigated and their influence on lowering the ceiling bearing capacity.

In order to determine how the mechanical wave energy transfer is reduced in Ackerman blocks via rubber injects in regard to the typical blocks, the maximum values of Misses-Hencky stress (MHS) taken from all analysis time steps and obtained at sensors A-F (see figure 1a, b) are compared. Having in mind that the excitation force is applied at the crossing of block symmetry line and block top surface line, treating the rubber injects as a soundproofing layers, from the engineering point of view the results obtained at sensors B, C and D (bottom blocks surface) are of the most importance. The MHS results obtained in all sensors points for each Ackerman block configurations are shown in Table 1. Blocks are identified with their height dimension (*h*), the word *ceramic* denotes the ceramic block itself, whereas *composite* denotes the block with rubber inject, respectively. Percentage reduction of mechanical wave energy transfer in blocks with rubber filling in regard to the solid blocks is shown in Table 2. Positive values denote the percentage reduction, whereas negative ones increase of MHS. Mechanical wave propagation in the form of MHS plots for different analysis time steps in the Ackerman block (*h* = 220 mm) with rubber filling is presented in figure 2.

Table 1. Misses-Hencky stress (MHS) results obtained at sensor points A-F for different Ackerman blocks configurations.

	h = 150 mm *ceramic* [Pa]	*h* = 150 mm *composite* [Pa]	*h* = 180 mm *ceramic* [Pa]	*h* = 180 mm *composite* [Pa]	*h* = 200 mm *ceramic* [Pa]	*h* = 200 mm *composite* [Pa]	*h* = 220 mm *ceramic* [Pa]	*h* = 220 mm *composite* [Pa]
A	509152	494035	509152	494035	509149	494031	507048	491962
B	39055	39476	30569	29524	22206	21617	27317	25965
C	36431	35437	26871	25962	31935	30016	29370	27883
D	36703	35708	26843	25684	31135	29757	30121	28583
E	46981	43706	46981	43706	46979	43715	46963	43707
F	46996	43709	46997	43708	46999	43702	46949	43713

Presented results in Table 1 shows that at point A the highest MHS is obtained for each configuration and it is over ten times higher than in other sensor points, which is connected with the fact, that the sensor at point A is located directly under the applied excitation force. For each ceramic Ackerman block configurations at point A there are nearly the same values obtained. Identical situation occurs for the composite blocks. One may notice, that in each composite the MHS values are slightly lower than for their ceramic counterparts.

Table 2. Percentage reduction of mechanical wave energy transfer in Ackerman blocks with injected rubber in regard to the solid blocks.

Point	h = 150 mm [%]	h = 180 mm [%]	h = 200 mm [%]	h = 220 mm [%]
A	2.97%	2.97%	2.97%	2.98%
B	-1.08%	3.42%	2.65%	4.95%
C	2.73%	3.38%	6.01%	5.06%
D	-0.02%	4.32%	4.43%	5.11%
E	6.97%	6.97%	6.95%	6.93%
F	6.99%	7.00%	7.02%	6.89%

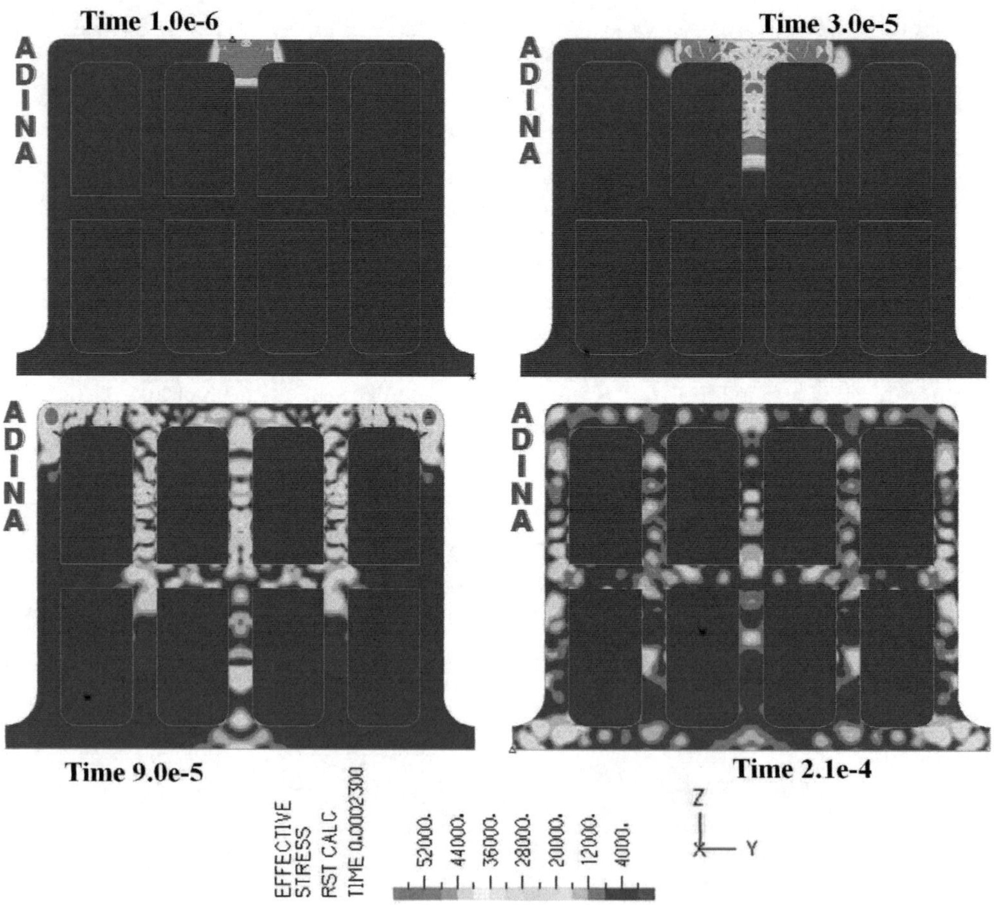

Figure 2. Mechanical wave propagation at different analysis time steps in the form of MHS plots for Ackerman block h = 220 mm with rubber injects introduced in hollows.

As shown in Table 2, at point A one obtains nearly the same value of mechanical wave energy transfer reduction, approximately 3%. At sensors E and F for any Ackerman block height, in composites the MHS values are approximately 7% lower in regard to the ceramic ones. From the engineering and soundproofing point of view, the most important are sensors B, C and D (see figure 1a and figure 1b). First of all, for any block configuration one may notice differences in results between point C and D where boundary conditions were applied. That differences results from the Y-axis translation released at point D, whereas at point C fixed, respectively. Only for configuration with h = 150 mm one obtains increase of MHS in the composites at point B and D, which is connected with propagating wave interference. At point C damping equal 2.73% is observed, however taking the entire bottom surface into account (points B, C and D), rubber injection rather hinder that helps. The positive influence of rubber injection in mechanical wave damping is observed in configurations with $h > 150$ mm. Naturally, for that configurations the lowest damping is observed at point B, by reason of wave propagating through the vertical ceramic block reinforcement wall (see figure 1a or figure 1b). At points C and D damping is slightly higher in comparison to the location of point B, however the differences vary only in small range, up to 3.36% for the block with $h = 200$ mm. Despite that, the point B would be decisive. According to that in the Ackerman block with $h = 180$ mm one obtains damping equal 3.42%, whereas for $h = 200$ mm it is only 2.65%. It should be noted that between configuration $h = 180$ mm and $h = 200$ or 220 mm there is a difference in block shape (see figure 1a and figure 1b) – additional horizontal ceramic reinforcement is provided for configurations with $h = 200$ and 220 mm. The highest percentage damping is obtained for the composite block with $h = 220$ mm, and it is equal 4.95%. Visual mechanical wave propagation shown in figure 2 proves, that the propagation takes place mainly through material with higher stiffness – in this case the ceramic. Together with increase of time, the mechanical energy transferred decreases. At time 2.1e-4 one may observe regions all over the block, where wave interference occurs (regions with approximately 28000 Pa of MHS). The wave interference is also clearly visible at 9.0e-5 s, near the top block corners. According to that, one can state, that the ceramic block hollows should have more complex or irregular shape to prevent interference and provide regions where waves could refract. Numerical studies of wall hollow blocks with rubber injects [14, 16, 24] proves, that if only rubber crosses transversely the line of acting force, the higher values of mechanical wave damping are obtained.

Discussed in this article study cases shows that injecting recycled rubber into the ceiling Ackerman blocks gives rather insignificant values of percentage damping. Moreover, with injecting rubber one increases each block weight. Mass of each Ackerman block with and without rubber injection is presented in Table 3.

Table 3. Ackerman blocks weight with and without rubber injection.

Block height [mm]	*Ceramic* [kg]	*Composite* [kg]
150	7.21	12.28
180	8.12	14.48
200	9.85	16.42
220	10.45	17.88

According to the weights presented in Table 3, one can state that injected rubber is very heavy, increasing the mass of each block by approximately 1.7 times. That significant mass increase translates directly into a reduction in ceiling carrying capacity. It would lead to nearly two times shorter spans of ceiling floors or additional steel reinforcement would be required to keep the ceiling within the assumed spans in comparison to the simple ceramic blocks. At this point it should be noted, that adding additional rebar reinforcement would increase the total cost of construction. Moreover, an injection of rubber in each block would also increase the price of blocks. Hence, it is clear that despite positive damping results obtained for wall blocks and advantages of using the recycled rubber [14, 16, 24], in ceilings achieved mechanical wave damping via rubber injects is insignificant in terms of additional ceiling costs and it has negative influence on load bearing capacity. In order to keep the standard ceiling bearing capacity, two methods may be proposed: ceramic blocks could be exchanged into the concrete blocks with expanded clay aggregate what would reduce their weight by 2-2.5 times, the second method is exchanging rubber into lighter material i.e. polyurethane foams. The ideal solution from the engineering point of view would be combining these two methods into one, which would give very light ceiling material with quite good soundproofing properties, however at present these materials are very expensive. Thus, it could turn out that the solution is not profitable.

4. Conclusion

In this paper reduction of mechanical wave energy transfer resulting from applied excitation load in the ceiling blocks with and without rubber filling is discussed. Four different configurations of ceramic Ackerman blocks produced in Poland, which differs mainly in height and slightly with shape were numerically studied. The encouraging results of the research on the achieved mechanical wave damping presented in the works [14-16, 18-19, 24] in which the recycled rubber was used to fill the wall blocks hollows, led to the question if recycled rubber injects in ceiling blocks will give approximately the same positive damping results. Through the performed dynamic analysis, obtained results in the form MHS plots exhibited only small percentage values of mechanical wave damping (range of approx. 2.5 up to 3.2%) in Ackerman blocks with height $h > 150$ mm and rubber inject in regard to the typical ceramic blocks. In the case of Ackerman composite block with $h = 150$ mm, a wave interference near the boundary points is observed resulting in slightly higher MHS (approx. 1%) in regard to the typical block. On the other hand, the rubber injects significantly increase blocks weight. In case of wall blocks it had small meaning [14-16, 18-19, 24] – the load bearing capacity is only slightly decreased, however in ceilings blocks weight have crucial importance. Through the analysis it is shown that the weight of each composite is nearly 1.7 times higher than for typical ceramic blocks. The total cost of constructing rubberized ceiling (additional rebars, rubber injection) with maintaining the same span ranges as for typical ceramic blocks would significantly increase, which makes them uneconomic. Even utilization of lighter materials for Ackerman blocks could be unprofitable due to very small values of mechanical wave damping. According to that, in the future research one should concern ceiling block elements with complex shape of hollows with injected rubber (possibility of wave refraction) or utilization of concrete blocks with some content of sand exchanged with crumbed rubber.

5. References

[1] Eldin N N and Senouci A B 1993 Rubber-tire particles as concrete aggregate *J. Mat. in Civ. Eng.* **5** pp 478-96
[2] Gerges N N, Issa C A, and Fawaz S A 2018 Rubber concrete: Mechanical and dynamical properties *Case Studies in Constr. Mat.* **9** pp 1-13
[3] Tang Y, Feng W, Feng W, Chen J, Bao D and Li L 2021 Compressive properties of rubber modified recycled aggregate concrete subjected to elevated temperatures *Constr. Build. Mat.* **268** 121181
[4] He L, Cai H, Huang Y, Ma Y, Van Den Bergh W, Gaspar L, Valentin J, Vasiliev E Y, Kowalski K J and Zhang J 2021 Research on the properties of rubber concrete containing surface-modified rubber powders *J. Build. Eng.* **35** 101991
[5] Yang F, Feng W, Liu F, Jing L, Yuan B and Chen D 2019 Experimental and numerical study of rubber concrete slabs with steel reinforcement under close-in blast loading *Constr. Build. Mat.* **198** pp 423-36
[6] Pacheco-Torres R, Cerro-Prada E, Escolano F and Varela F 2018 Fatigue performance of waste rubber concrete for rigid road pavements *Constr. Build. Mat.* **176** pp 539-548
[7] Mendis A S, Al-Deen S and Ashraf M 2017 Effect of rubber particles on the flexural behaviour of reinforced crumbed rubber concrete beams *Constr. Build. Mat.* **154** pp 644-57
[8] Mendis A S, Al-Deen S and Ashraf M 2018 Flexural shear behaviour of reinforced crumbed rubber concrete beam *Constr. Build. Mat.* **166** pp 779-791
[9] Medina N F, Medina D F, Hernández-Olivares F and Navacerrada M A 2017 Mechanical and thermal properties of concrete incorporating rubber and fibres from tyre recycling *Constr. Build. Mat.* **144** pp 563-73
[10] Bala A and Gupta S 2021 Thermal resistivity, sound absorption and vibration damping of concrete composite doped with waste tire Rubber: A review *Constr. Build. Mat.* **299** 123939
[11] Steyn Z C, Babafemi A J, Fataar H and Combrinck R 2021 Concrete containing waste recycled glass, plastic and rubber as sand replacement *Constr. Build. Mat.* **269** 121242
[12] Saikia N and De Brito J 2014 Mechanical properties and abrasion behaviour of concrete containing shredded PET bottle waste as a partial substitution of natural aggregate *Constr. Build. Mat.* **52** pp 236-44

[13] Azhdarpour A M, Nikoudel M R and Taheri M 2016 The effect of using polyethylene terephthalate particles on physical and strength-related properties of concrete; A laboratory evaluation *Constr. Build. Mat.* **109** pp 52-62
[14] Major M, Kuliński K and Major I 2017 Dynamic analysis of an impact load applied to the composite wall structure *MATEC Web Conf.* **107** pp 1-6
[15] Major M, Major I, Kuchárová D and Kuliński K 2018 Reduction of dynamic impacts in block made of concrete-rubber composites *Civ. Env. Eng.* **4** pp 61-8
[16] Major M, Kuliński K and Major I 2017 Thermal and dynamic numerical analysis of a prefabricated wall construction composite element made of concrete-polyurethane *Procedia Eng.* **190** pp 231-6
[17] http://cegielniaprzeclaw.pl/materialy-stropowe/ (date accessed: 10th of September 2021)
[18] Major I and Major M 2014 Comparative analysis of the distribution of effective stress in Mooney and Zahorski materials using ADINA software *Adv. Mat. Research* **1020** pp 165-70
[19] Major I and Major M 2009 Travelling waves in a thin layer composed of nonlinear hyperelastic Zahorski material *Theoretical Applied Mechanics* **47** pp 109-26
[20] Bakhteri J, Makhtar A M and Sambasivam S 2004 Finite element modelling of structural clay brick masonry subjected to axial compression *J. Teknologi* **41** pp 57-68
[21] Purohit S P and Sharma S 2020 Mechanical properties of red clay bricks under compression and shear: an experimental and numerical investigation *SEC18: Proc. 11th Struct. Eng. Convention* (Kolkata, India)
[22] Narayanan S P and Sirajuddin M 2013 Properties of brick masonry for FE modeling *Amer. J. Eng. Research* **1** pp 6-11
[23] Nichols J and Totoev Y Z 1997 Experimental determination of the dynamic modulus of elasticity of masonry units *Proc. 11th Int. Brick and Block Masonry Conf. At: Shanghai*
[24] Major M, Major I and Kuliński K 2019 Analysis of the mechanical wave in the composite made of concrete and rubber - numerical analysis *Trans.VSB – Tech. Univ. Ostrava. Civ. Eng. Series* **19** pp 25-28

IOP Conf. Series: Materials Science and Engineering 1252 (2022) 012083 doi:10.1088/1757-899X/1252/1/012083

Numerical simulations for time-dependent corrosion surface of unpainted structural steel plates using spatio-temporal data

F Jiang[1] and M Hirohata[2]

[1] Graduate School of Engineering, Osaka University, 2-1 Yamada-oka Suita Osaka 565-0871 Japan
[2] Graduate School of Engineering, Osaka University, 2-1 Yamada-oka Suita Osaka 565-0871 Japan

E-mail: f-jiang@civil.eng.osaka-u.ac.jp; hirohata@civil.eng.osaka-u.ac.jp

Abstract. Structural failure due to corrosion of steel members is a long-term process. A method to precisely simulate the development of corrosion on steel surfaces is of great significance. In this study, 9 specimens of unpainted steel plates, SM400A were used for the experiments. The experiments were conducted under artificial seawater corrosive environments to accelerate the corrosion process. The corrosion depth data of each unpainted steel plate for the spatial statistical analysis was measured by a laser focus measuring system when the accelerated corrosion experiments of 28, 84, and 168 days. Semivariogram was used in this spatial statistical analysis to research the spatial autocorrelation structures of the corrosion surface. By using this and the technique of ordinary kriging, the method to simulate the spatial features of corrosion surface was proposed. The results of the simulation indicated that the corrosion depth and surface topography of corrosion surfaces were highly consistent with the experiments. In addition, a prediction method for time-dependent corrosion surface was proposed by investigating the relationship between the spatial autocorrelation structures and their time-dependence data.

1. Introduction

Corrosion is one of the main causes of deterioration and damage to steel structures [1]. The main effect of the corrosion is loss of material from the steel member's surface [2], which decreases the rigidity of the structure and affects the load-carrying capacity. Nowadays, there is an increasing recognition that the monitoring and control of the corrosion status are important components in the maintenance of infrastructures. Because the corrosion is a long-term process, the judgment of the degree of corrosion according to the actual situation often takes a lot of time. As a result, the simulation and prediction of corrosion surface play a key role in corrosion damage modeling and the fast confirmation of corrosion status [3].

Due to the inadequate information on the elements that intervene in the corrosion process and its variation with time, the variation of the corrosion surface is considered as stochastic nature [4], [5], [6]. Distribution of corrosion rate probabilities associated with corrosion surface characteristics are essential for simulation and prediction models of corrosion areas [7]. In previous research, stochastic models could be used to evaluate corrosion status [8]. Fractals and spatial autocorrelation have been used to evaluate the surface properties of corroded steel plates taken from real structural members [9]. In addition, a Monte Carlo simulation method have been used to estimate the process progression of corrosion using the probability distributions of corrosion growth rate [10], [11]. And a prediction model has been established by Bayesian inference to consider the variation of corrosion rate [12]. However, the correlation between the corrosion damage and its aging is still a hard work because the target materials are limited in each method. Therefore, the change of the spatial autocorrelation structure of the corroded surface with time, which is essential for predicting the corrosion surface properties of steel structures, has not been clarified, and the prediction method of the corrosion behavior with time has not been established.

In this study, semivariogram analysis, which is one of the spatial statistical methods, was performed on the corrosion surfaces of the unpainted structural steel plates to clarify the spatial autocorrelation structure of the corrosion surface and its temporal characteristics [11]. Semivariogram is a spatial statistical method to quantitatively evaluate the spatial correlation between the distance and the variance of each data, and is a method developed from mining science to predict the spatial distribution law of ore deposits [13]. This analysis method has been applied in many fields including mining, meteorology, economics, and geography [14]. In this study, a method to simulate the spatio-temporal characteristics of the corrosion surface in the artificial seawater corrosive environment was proposed, by using the spatial autocorrelation structure of the corrosion surface uncovered by semivariogram and ordinary kriging.

2. Method

In this study, the accelerated corrosion test under artificial seawater corrosive environments was conducted to obtain the steel plates with different degrees of corrosion surface in a short time.

2.1. Specimen

The unpainted steel plates of SM400A were used in the accelerated corrosion test. The chemical composition of SM400A is shown in Table 1. Generally, the steel materials in steel structures are painted to prevent and mitigate corrosion. In this study, in order to investigate the basic corrosion behavior of the steel materials after coating degradation, the experimental specimens were unpainted. There were 9 steel plates numbered from No. 1 to No. 9 performed for this experiment. The dimension of the steel plates is shown in Figure 1. The surfaces of the steel plates were cleaned by blasting for removing the mill scale. The back and sides of the specimens were protected with anti-corrosion tape to serve as a reference surface when measuring corrosion depth.

Table 1. Chemical compositions of steel materials.

	Chemical compositions (mass%)				
	C	Si	Mn	P	S
SM400A	0.18	0.17	0.5	0.015	0.006

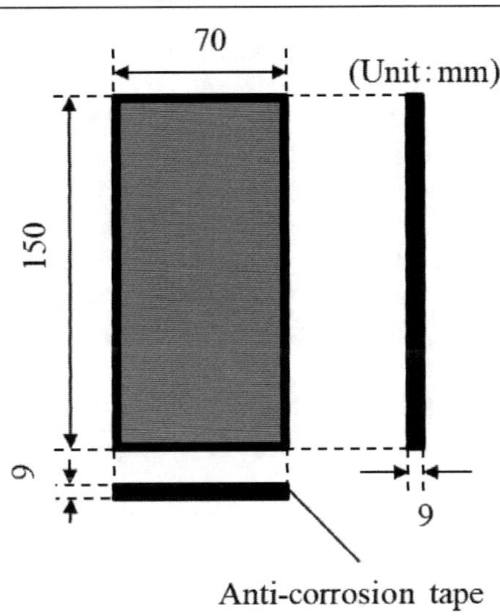

Figure 1. Dimension of specimens.

2.2. Experimental method

ISO 16539-Method B [15] was performed in this accelerated corrosion test. 9 specimens were divided equally into three groups, for the corrosion test of 28 days (No. 1 – No. 3), 84 days (No. 4 – No. 6), and 168 days (No. 7 – No. 9) respectively. After the accelerated corrosion test, the generated rust was removed from the corrosion surface by blasting. And the corrosion depth data of each specimen for the spatial statistical analysis was measured by a laser focus measuring system (resolution: 0.2 μm, measurement interval: 0.3 mm). The measurement area is shown in Figure 2.

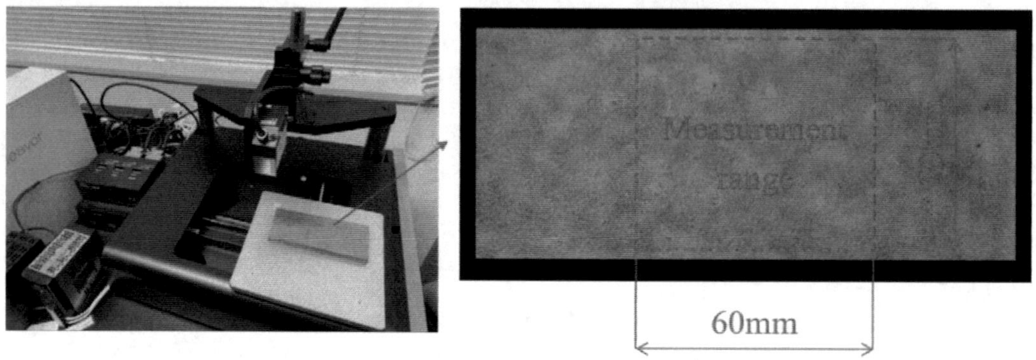

Figure 2. The measurement area of laser focus measuring system.

2.3. Semivariogram

In statistics, kriging variogram model is a method of interpolation based on Gaussian process governed by prior covariances [16]. To make a prediction with the kriging interpolation method, two tasks are necessary. The first step is to reveal the rules of dependency. The variograms and covariance functions are generated to estimate the statistical dependence values that depend on the autocorrelation model. The second step is to predict the unknown values.

Kriging weights the surrounding measured values to originate a prediction for an unmeasured location. The general formula is formed as a weighted sum of the data:

$$\hat{Z}(s_0) = \sum_{i=1}^{N} \lambda_i Z(s_i) \tag{1}$$

Where $Z(s_i)$ is the measured value at the i-th location; λ_i is an unknown weight for the measured value at the i-th location; s_0 is the prediction location; $\hat{Z}(s_0)$ is the prediction for an unmeasured location; N is the number of measured values.

Accordingly, from this formula, λ_i is the crucial problem, λ_i depends on a fitted model to the measured points, the distance to the prediction location, and the spatial relationships among the measured values around the prediction location. Semivariogram can be used in this problem. In order to clarify the spatial autocorrelation structure on the corrosion surface of the corroded specimens, semivariogram was used to extract the spatial statistics, range and sill, that represent the properties of the corrosion surface. In this study, semivariogram is an indicator of the spatial correlation between the corrosion depth at an arbitrary location and the corrosion depth at a distance from the arbitrary location on the corrosion surface. In the spatial modeling of measured points on the corrosion surface using semivariogram, all pairs of locations separated by a distance, h, are calculated with the difference squared between the corrosion depths of the paired locations. Figure 3 shows the pairing of one point with all other measured points. And this process repeats for each measured point.

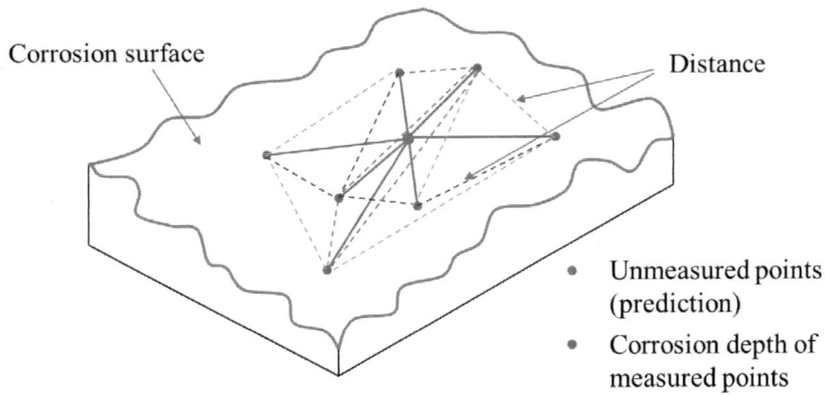

Figure 3. Spatial modeling of measured points for semivariogram analysis.

The range and sill, representing the autocorrelation structure of the corrosion surface, were calculated from the semivariogram by determining, $\gamma(h)$, shown in equation (2):

$$\gamma(h) = \frac{1}{2|N(h)|}\sum_{i=1}^{N(h)}\left(z_j - z_k\right)^2 \quad (2)$$

Where $N(h)$ is the set of all pairwise Euclidean distances j - $k = h$; $|N(h)|$ is the number of distinct pairs in $N(h)$; z_j, z_k are the corrosion depth at spatial locations j and k, respectively.

In this study, semivariogram analysis was carried out for all directions assuming isotropy because the corrosion surface properties did not have spatial variation in specific directions. Consequently, the semivariogram calculated from this analysis can be treated as a function of distance only. To fit a model to the points forming the empirical semivariogram, several types of semivariogram models are available. The spherical model is the most common model in spatial statistics [17]. It is characterized by a linear growth when the distance between 2 points is short (close to the point of origin); however, when it exceeds the range value, it will be a line parallel to the horizontal axis. In other words, this model shows a progressive decrease of spatial autocorrelation until some distance, beyond which autocorrelation is zero [18]. The formula of the spherical model is:

$$\gamma(h) = \begin{cases} 0, \ h = 0 \\ C_0 + C\left(\frac{3}{2}\frac{h}{a} - \frac{1}{2}\frac{h^3}{a^3}\right), \ 0 < h \le a \\ C_0 + C, \ h > a \end{cases} \quad (3)$$

Where C_0 is the nugget constant; $C_0 + C$ is the sill, C is the structure variance; a is the effective range.

From the above, the spatial autocorrelation structure of the whole corrosion surface in this study can be quantitatively evaluated by the spatial statistics (range and sill) calculated from the theoretical semivariogram by the spherical model.

3. Results and discussion

3.1. Experimental results

The relationship between the mean corrosion depth of the corrosion surface of each specimen, d_{mean}, and the accelerated corrosion test time, t, is shown in Figure 4. The points in the figure show the average corrosion depth of each specimen. The curve is the exponential regression. Specific spatio-temporal data was also obtained from the measurement, which will be used for spatial statistical analysis.

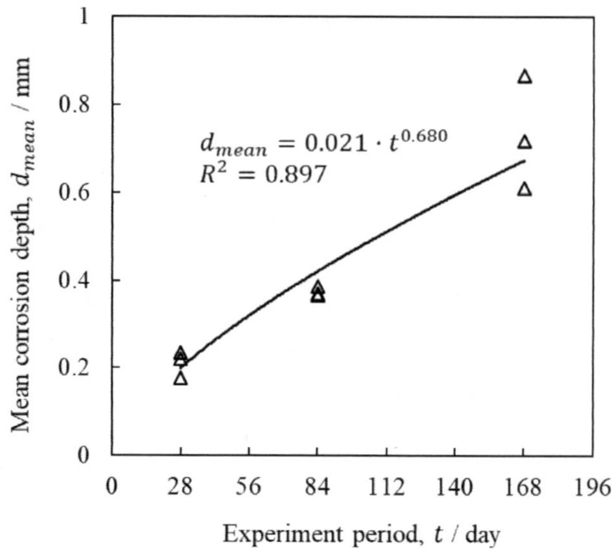

Figure 4. Mean corrosion depth d_{mean} of each tested surface as a function of experiment period t.

3.2. Results of range and sill for each specimen

Figure 5, Figure 6, and Figure 7 shows the semivariogram results of specimens for 28 days, 84 days and 168 days. Basically, the range and sill of the specimens that underwent the same period of accelerated corrosion test had the same trend. The specific values of range and sill for each specimen are shown in Table 2.

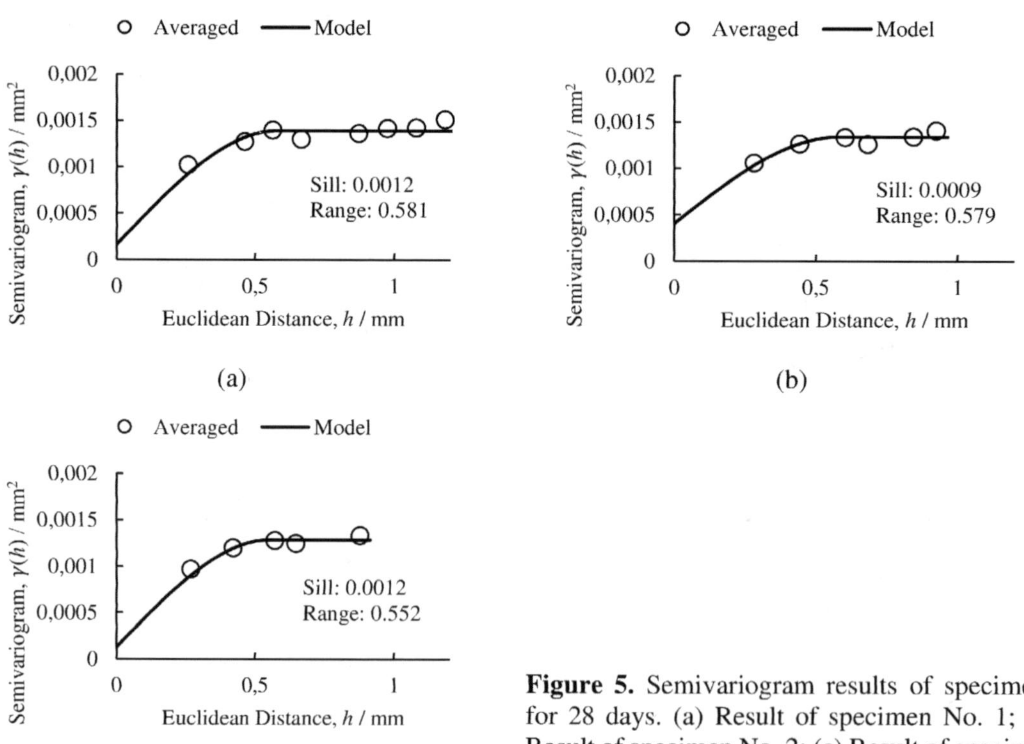

Figure 5. Semivariogram results of specimens for 28 days. (a) Result of specimen No. 1; (b) Result of specimen No. 2; (c) Result of specimen No. 3.

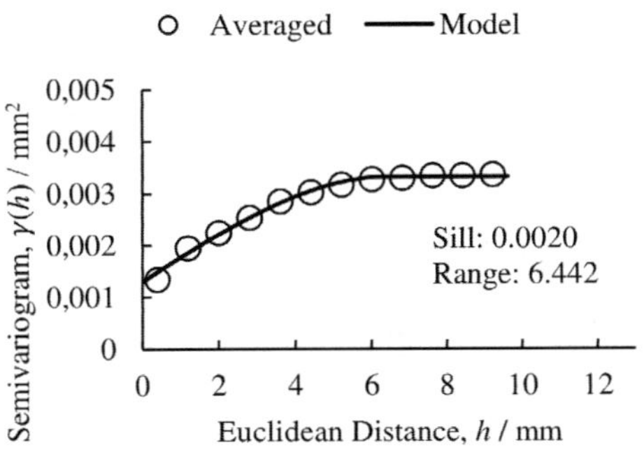

(a)

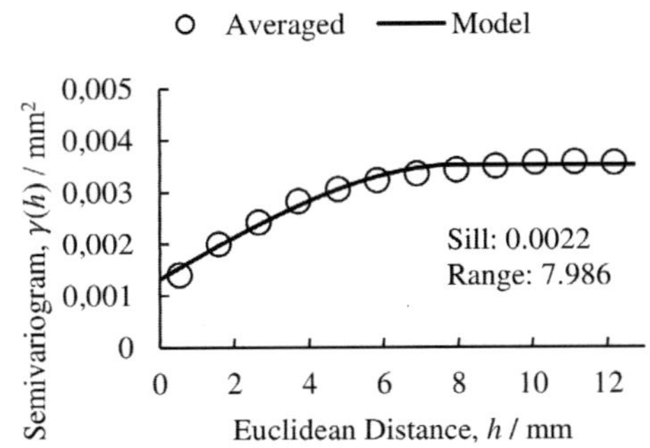

(b)

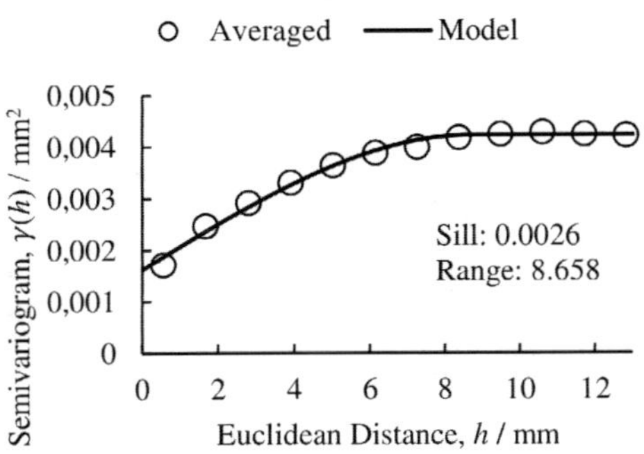

(c)

Figure 6. Semivariogram results of specimens for 84 days. (a) Result of specimen No. 4; (b) Result of specimen No. 5; (c) Result of specimen No. 6.

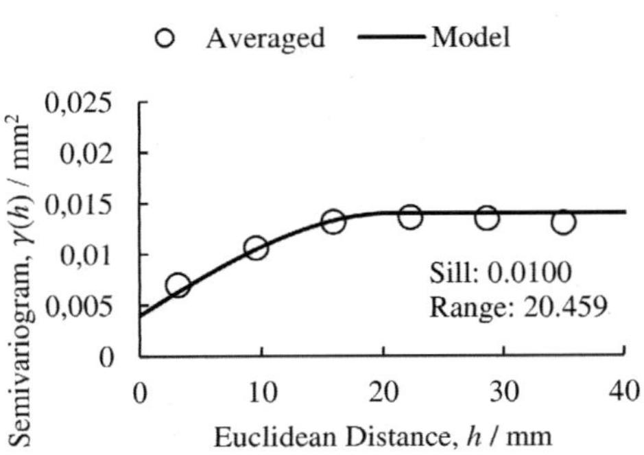

(a)

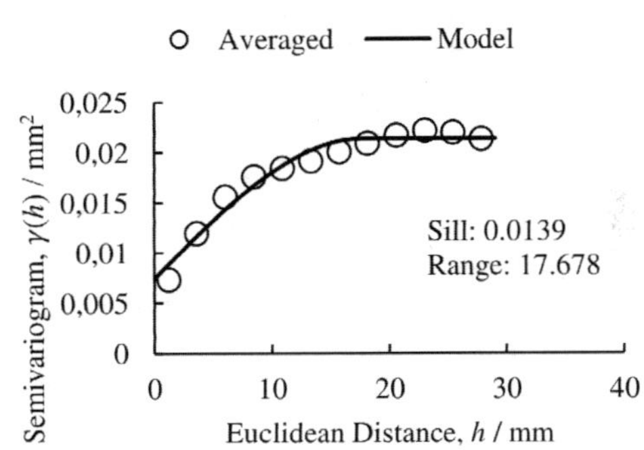

(b)

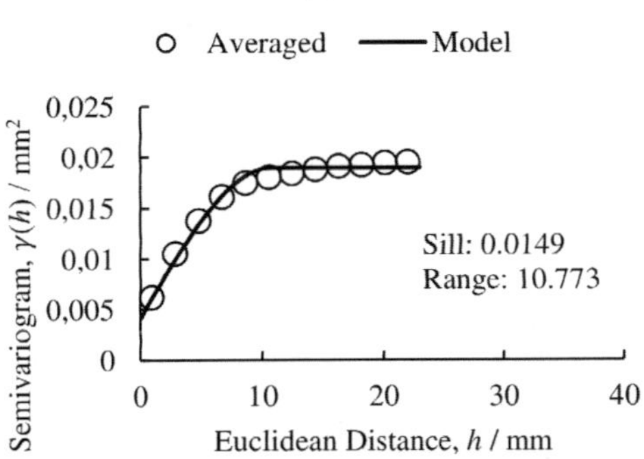

(c)

Figure 7. Semivariogram results of specimens for 168 days. (a) Result of specimen No. 7; (b) Result of specimen No. 8; (c) Result of specimen No. 9.

IOP Conf. Series: Materials Science and Engineering 1252 (2022) 012083 doi:10.1088/1757-899X/1252/1/012083

Table 2. Range and sill of all specimens.

Time	Specimen number	Sill (mm^2)	Range (mm)	Mean corrosion depth (mm)
	1	0.0012	0.581	0.220
28 days	2	0.0009	0.579	0.234
	3	0.0012	0.552	0.177
	4	0.0020	6.442	0.385
84 days	5	0.0022	7.986	0.370
	6	0.0026	8.658	0.365
	7	0.0100	20.495	0.612
168 days	8	0.0139	17.678	0.719
	9	0.0149	10.773	0.868

3.3. Simulation of corrosion surface

With the range and sill obtained in the previous section, which can represent the spatial properties of the corrosion surface, the corrosion depth of the unmeasured locations can be predicted and consequently the corrosion surface can be simulated. The result of corrosion surface simulation of No. 1 as an example is shown in Figure 8. And the accuracy of all simulations of the specimens are shown in Table 3. The RMSE results show the high simulation accuracy, which proves the effectiveness of the semivariogram method in this corrosion surface simulation.

3.4. Prediction of time-dependent corrosion surface

A prediction method of time-dependent corrosion surface is proposed in this section. As the experimental results in section 2.3, the relationship between the mean corrosion depth and the experiment period in the corrosive environment on surfaces of specimens is regressed as an exponential regression curve. On the other hand, since the mean corrosion depth and spatial statistics sill and range for the corrosion surface properties of the specimen have an obvious positive correlation. Linear regression was used to show their relationship. As a result, these spatial statistics can be evaluated by time-dependent data which makes the final prediction of the corrosion surface for future experiment period. Figure 9 shows sill γ and range h of each tested surface as a function of mean corrosion depth d_{mean}. Both have an R^2 above 0.8, which is at a high level.

With these three regressions, the spatio-temporal parameters of the prediction can be derived. These parameters can then be substituted into the semivariogram model to predict the corrosion surface. Figure 10 shows an example of the corrosion surface prediction for 168 days based on specimen No. 1 by using the corrosion depth data of specimens for 28 days and 84 days. And it also shows the comparison between the real and predicted corrosion surfaces for 168 days. From the comparison, it is confirmed that the shape of corrosion surface prediction based on specimen No. 1 is transformed from the corrosion surface of specimen No.1 for 28 days (Figure 8). Although the maximum corrosion depth and the average corrosion depth of the prediction result are close to the other specimens for 168 days, since the corroded steel plates are not the same, the shape of the corrosion is quite different between them.

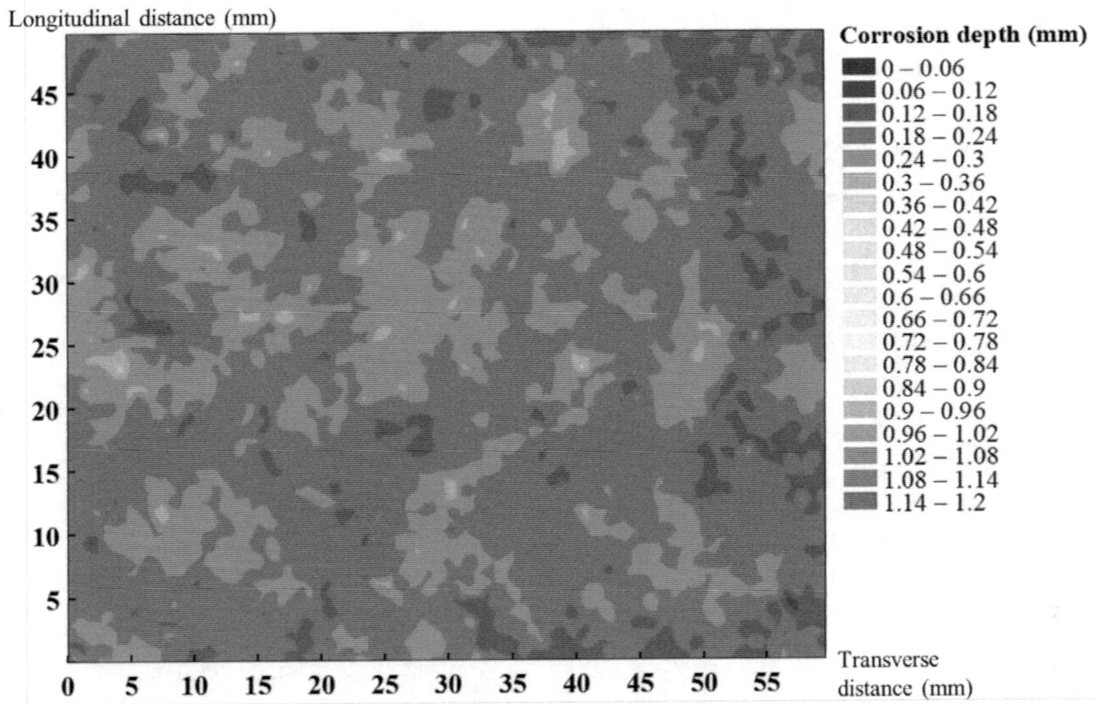

Figure 8. Corrosion surface simulation of No. 1.

Table 3. Accuracy of the corrosion surface simulation.

	No. 1	No. 2	No. 3	No. 4	No. 5	No. 6	No. 7	No. 8	No. 9
Root-Mean-Square (RMSE)	0.032	0.033	0.032	0.032	0.032	0.033	0.042	0.044	0.042

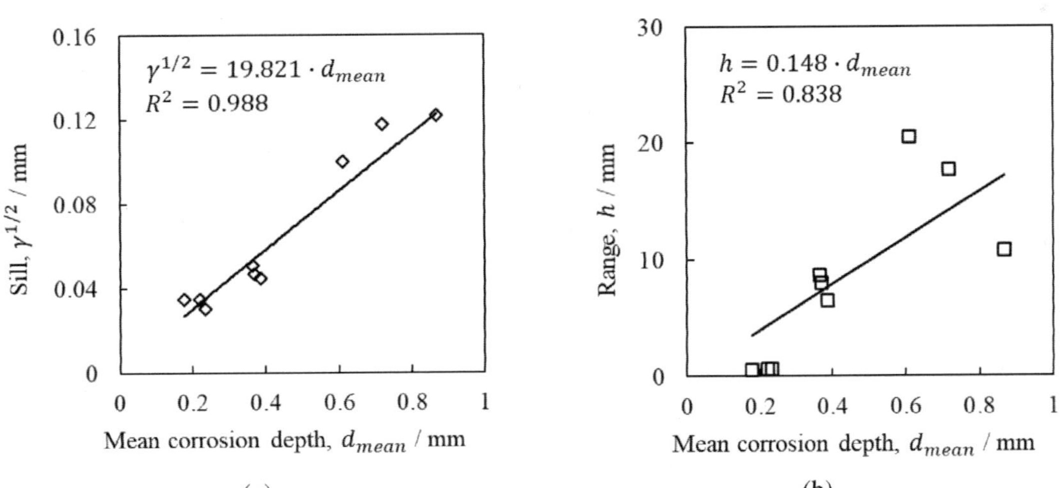

Figure 9. Sill γ and range h of each tested surface as a function of mean corrosion depth d_{mean}. (a) Sill γ as a function of d_{mean}; (b) Range h as a function of d_{mean}.

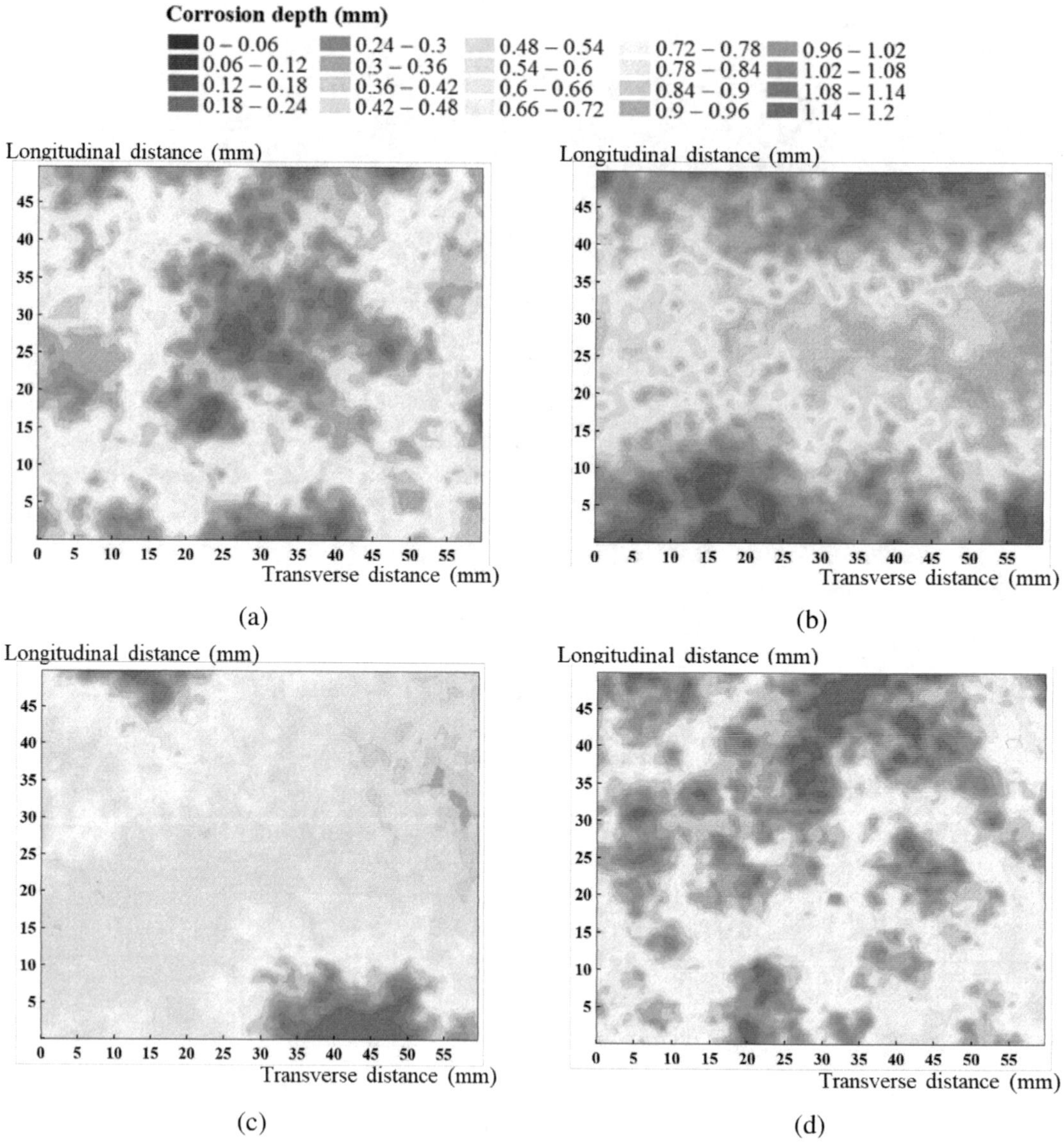

Figure 10. Corrosion surface prediction for 168 days based on specimen No. 1. (a) Prediction result; (b) Corrosion surface of No.7 (168 days); (c) Corrosion surface of No. 8 (168 days); (d) Corrosion surface of No. 9 (168 days).

4. Conclusion

In this study, a semivariogram model was used to obtain the spatial properties of the corrosion surface, and by using the technique of ordinary kriging, the method to simulate the spatial features of corrosion surface was proposed. The simulation results showed a high simulation accuracy. In addition, a prediction method for time-dependent corrosion surface was proposed by investigating the relationship between the spatial autocorrelation structures and their time-dependence data. The prediction results showed the same trend of the corrosion depth to the other specimens in the same experiment period. The same specimen could not continue the experiment after the rust was removed because the rust must be removed from the corrosion surface when measuring the corrosion depth. It means the inability to obtain spatio-temporal data on the corrosion depth of the same specimen. As consequence it was difficult to verify the prediction accuracy of this predicted corrosion surface method.

5. References

[1] Pidaparti R M, Fang L and Palakal M J 2008 *Computational simulation of multi-pit corrosion process in materials Computational Materials Science* 41(3) 255-265

[2] Secer M and Uzun E T 2017 *Corrosion damage analysis of steel frames considering lateral torsional buckling Procedia engineering* 171 1234-1241

[3] Xue L 2007 *Damage accumulation and fracture initiation in uncracked ductile solids subject to triaxial loading International journal of solids and structures* 44(16) pp 5163-5181

[4] Melchers R E 2005 *The effect of corrosion on the structural reliability of steel offshore structures Corrosion science* 47(10) pp 2391-2410

[5] Shibata T 1996 *Statistical and stochastic approaches to localized corrosion: 1996 WR Whitney Award lecture Corrosion* 52(11)

[6] Alamilla J L, Oliveros J and García-Vargas J 2009 *Probabilistic modelling of a corroded pressurized pipeline at inspection time Structure and infrastructure engineering* 5(2) pp 91-104

[7] Caleyo F, Velázquez J C, Valor A and Hallen J M 2009 *Probability distribution of pitting corrosion depth and rate in underground pipelines: A Monte Carlo study Corrosion Science* 51(9) pp 1925-1934

[8] Soares do Amaral J V, Montevechi J A B, Miranda R D C and Junior W T D S 2021 *Metamodel-based simulation optimization: A systematic literature review Simulation Modelling Practice and Theory* p 102403

[9] Fujii K, Kaita T, Hirai K and Okumura M 2002 *Applicability of spatial auto-correlation model for corroded surface modeling in corroded steel plate Journal of Structural Engineering* 48 1031-1038

[10] Kamisho T, Ishii R and Tsuda M 2021 *Estimation of hydrogen embrittlement process using statistical analysis method Corrosion Science* 192 p 109799

[11] Shigenobu K, Young S J, Kazuhiro U and Jin HA 2012 *Numerical simulations for time-dependent corrosion surfaces of unpainted carbon steel plates in atmospheric corrosive environments using spatial statistical techniques Corrosion Engineering* 61(7) pp 203-203

[12] Kim K, Lee G, Park K, Park S and Lee W B 2021 *Adaptive approach for estimation of pipeline corrosion defects via Bayesian inference Reliability Engineering & System Safety* 216 p 107998

[13] Bancroft BA and Hobbs GR 1986 *Distribution of kriging error and stationarity of the variogram in a coal property Mathematical geology* 18(7) 635-652

[14] Danyar S, Yudong S and Jumakeld M 2004 *Influence of groundwater level change on vegetation coverage and their spatial variation in arid regions Journal of Geographical Sciences* 14(3) 323-329

[15] ISO 2013 *ISO16539 Corrosion of metals and alloys-Accelerated cyclic corrosion tests with exposure to synthetic ocean water salt-deposition process-"Dry" and "wet" conditions at constant absolute humidity*

[16] Oliver M A and Webster R 1990 *Kriging: a method of interpolation for geographical information systems International Journal of Geographical Information System* 4(3) pp 313-332

[17] Haas TC 1990 *Kriging and automated variogram modeling within a moving window Atmospheric Environment Part A General Topics* 24(7) pp 1759-1769

[18] Asa E, Saafi M, Membah J and Billa A 2012 *Comparison of linear and nonlinear kriging methods for characterization and interpolation of soil data Journal of Computing in Civil Engineering* 26(1) pp 11-18

IOP Conf. Series: Materials Science and Engineering 1252 (2022) 012084 doi:10.1088/1757-899X/1252/1/012084

Numerical analysis of slope stability

S Harabinova[1] and E Panulinova[1]

[1] Institute of Structural Engineering, Faculty of Civil Engineering, Technical University of Kosice, Vysokoskolska 4, 040 23 Kosice, Slovakia

E-mail: slavka.harabinova@tuke.sk; eva.panulinova@tuke.sk

Abstract. The paper is focused on the analysis of slope stability. The problem of slope stability is processed theoretically and parametrically - parametrical study. Factors that affect slope stability include: geomorphological conditions, geological conditions, soil and rock properties, and external loads (for example, from traffic, nearby structures, possible vibrations). The shear stress parameters of soil are a keys design parameters for assessment of slope stability. The aim of the parametric study is to point out the importance of knowing the shear stress parameters of soils in assessing the slope stability. By comparing obtained result, it was founded, that change the cohesive has a significant effect on the stability factor.

1. Introduction

Slope stability that, the ability of a slope to stay on a certain gradient, is affected by several factors. These factors include conditions, that is natural conditions of the area (geological, geomorphological, hydrogeological, climatic, anthropogenic) and processes (erosion, change of slope, slope loading, relief of the heel, seismic shocks, increase in groundwater pressure, anthropogenic interventions, etc.), which change these conditions and thus cause a change in slope stability.

Slope stability is usually evaluated by the factor of safety, which is given as the ratio of passive forces (or moments) to the active forces (moments) acting on the slope. If the passive forces are greater than the active ones (i.e. the degree of stability is greater than 1), we consider the slope to be stable. If the value of the factor of safety is less than 1, we consider the slope to be unstable and a slope movement occurs, i.e. a geodynamic process during which rock masses move along the slope due to gravity [1, 2]. The result of slope movements is slope deformation (slope failure).

Analyses of slope stability are generally aimed at finding the causes of an occurred slope failure and the factors that can potentially trigger a slope movement. Landslide is a type of slope failure that occurs precisely as a result of the gravitational movement of rocks on one or more slip surfaces. Landslide is one of the most common types of slope movements, and in Slovakia, landslides make up the majority of slope deformations. Landslides account for about 90% of all registered slope deformations.

Climatic factors are a very important factor that currently impacts the stability of the slopes. In particular, rainwater, which enters the slope, causes an increase in the water pressure in the pores, especially in fine-grained soils, which also reduces the shear strength of the soils.

Based on research by Hrašna, Pavlech [3], it can be stated that the largest changes in the consistency index under the influence of precipitation and groundwater level fluctuations occur in cohesive soils with low and medium plasticity (sandy clays, clays and silty clays).

These findings were the reason for a parametric study. The aimed of parametric study to point to the effect of changes in shear strength parameters of fine-grained soils (taking into account changes in eater content and related changes in soil consistency) on the resulting factor of safety.

In general, the less water it contains, the stronger the clay soil. At the same time, its compressibility is less because water molecules "separate" clay particles into "greater" distances. Thus, soil moisture and the associated consistency and plasticity can be considered the most important physical properties of fine-grained soils.

2. Input parameters of the parametric study

Due to the sensitivity of fine-grained soils to water content change, as well as due to the relatively frequent occurrence of clays of medium or low plasticity (CI or CL, group F6) in the Kosice, this parametric study was performed on this type of soil. The height of the slope was 3 m.

The aim of the parametric study was:

- analyse the stability of the slope in fine-grained soil class F6, soft to hard consistency,
- point to the influence of shear strength parameters on the resulting factor of safety (as one of the possible ways to take into account climatic influences, especially precipitation activity related to the change of moisture of fine-grained soils),
- monitor which of the shear strength parameters has a more significant effect on the resulting factor of safety.

The slope stability has been analyzed on fine soils - group F6. The values of the geotechnical parameters of soil are given in the Table 1.

Table 1. Geotechnical parameters of soil.

Geotechnical parameters of soil	Group F6 (CL, CI)
Poisson's ratio υ [-]	0.40
Unit weight γ [kN/m^3]	21.0
Effective stress parameter – cohesion c' [kPa]	8 - 16
Effective stress parameter – angle of friction φ' [°]	17 - 21

3. Results and discussion

The stability analysis takes on the assumption that the slip surface developed will be circular. The slip circle methods of slices are often being applied to slope stability problems. However, each method is different from another and based on different assumptions on the forces acting upon the sides of the slices. In Peterson's Method, the simplest method of slices, it is assumed that the resultant of all inter-slice forces to be consistent with the direction of failure arc for the slice. Therefore, the factor of safety can be calculated without considering the forces acting on the sides of slices. Petterson Method was used for assessing of slope stability (see figure 1).

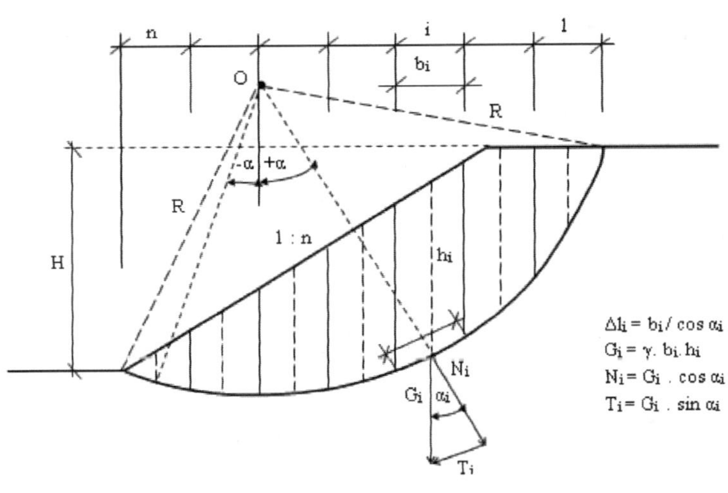

Figure 1. Petterson Method.

The factor of safety is calculated as

$$F_s = \sum_{i=1}^{n} \frac{N_i.tg\varphi + c.\Delta l_i}{Q_i} \tag{1}$$

In the parametric study, a factor of safety was calculated without the use of reduction factors, and thus the resulting degree of stability must by greater than 1.5 (*Fs*> 1.5). The calculation was made in the GEO 5 program by Fine, cz., Module "Slope stability" [4].

In the calculation of the resulting Factor of safety *Fs,* the "optimization" of the calculation was used, which consists in finding the shear surface with the smallest value of the factor of safety (see figure 2).

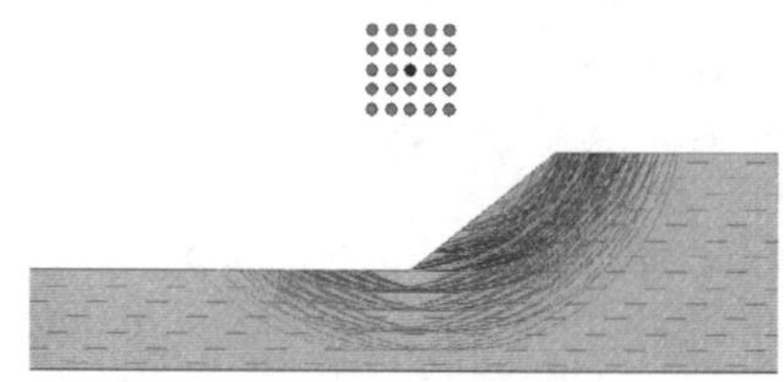

Figure 2. Calculation of the resulting degree of stability *Fs*, display of the slip surface network and "optimization" of the calculation (Petterson Method).

Figure 3 shows the change in the internal friction angle and his effect on the resulting Factor of safety calculated by the Petterson Method.

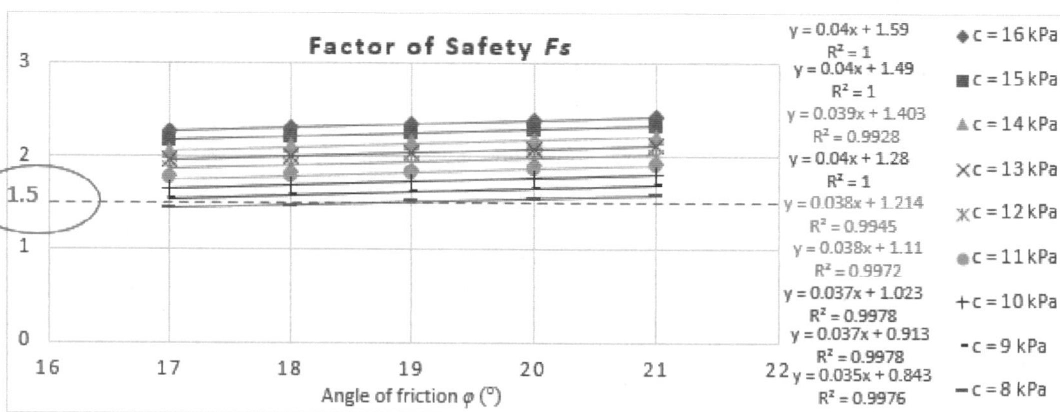

Figure 3. Factor of safety depending on the change of the friction angle.

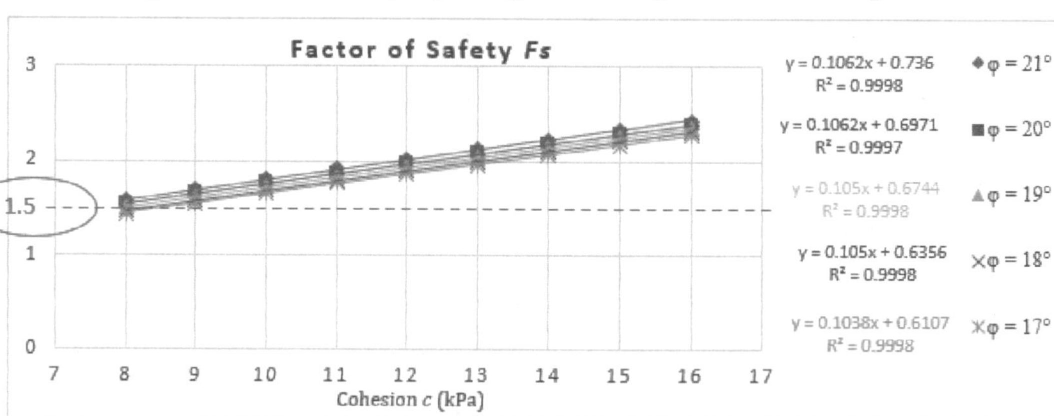

Figure 4. Factor of safety depending on the change of the cohesion.

Figure 4 shows the change in cohesion and its effect on the resulting Factor of safety calculated by the Petterson method. Each of the following linear dependences (figure 3 and 4) can be described by the equation of a line in the form:

$$Y = A_0 X + A_1, \tag{2}$$

where Y is the dependent variable which represents the calculated factor of safety (Fs), X is an independent variable, represents the change in the angle of internal friction (see figure 3), resp. change of cohesion (see figure 4). A_0 indicates how many units of measure the dependent variable Y changes, if the independent variable X changes by one unit. A_1 in the graphical representation of a line determines the point at which this line intersects the y-axis. It was found:

- When changing the angle of internal friction of the soil (φ = 17 ° to 21 °) the resulting factor of safety increases linearly.
- It is the same when changing the soil cohesion (c = 8 to 16 kPa), but the linear increase of the calculated values of the factor of safety is more pronounced when changing the soil cohesion than when changing the internal friction angle.
- -At the height of the slope H = 3m, the requirement Fs> 1.5, only the situation when the value of the internal friction angle φ = 17 ° and 18 ° and the lowest value of cohesion c = 8 kPa is considered in the calculation.

An increase in cohesion (as well as an increase in the internal friction angle) causes a linear increase in the safety factor. This finding is related to the consideration of these parameters directly in the Petterson calculation, where the degree of slope stability is calculated as the ratio of passive forces (N.tg φ + $c.L$) to active forces ($T = G.\sin\alpha$).

4. Conclusion

Water plays a very significant role in the behavior of clays, especially on the assessment of slope stability. When the water content changes, clay changes its character in a surprising manner and the shears strenght parameters. Increasing the water content in the soil reduces the shear strength parameters. A reduction in the internal friction angle (as well as a reduction in cohesion) causes a linear decrease in the factor of safety. By comparing obtained results, it was founded, that change the cohesive has a significant effect on the stability factor.

5. References

[1] Cornforth D H 2005 *Landslides in Practice, Investigation, Analysis, and Remedial Preventive Options in Soils* (John Wiley & Sons) p 323

[2] Farid S, Mojtahedi F, Tabatabaee S, Ghoroqi M, Tehrani M S, Gordan B, Ghoroqi M 2019 *A novel probabilistic simulation approach for forecasting the safety factor of slopes: a case study. Engineering with Computers,* **35** Issue 2, p 637-646

[3] Hrašna M and Pavlech. J 1981 *Changes of moisture and consistency in cohesive soils due to the influence of precipitates and subsurface water level fluctuations, Mineralia slov.* **13,** 2, p 153—162

[4] GEO 5 program © Fine Ltd.

Acknowledgments

This work was supported by the Scientific Grant Agency of the Ministry of Education of Slovak Republic and the Slovak Academy of Sciences under Project VEGA 1/0374/19.

IOP Conf. Series: Materials Science and Engineering 1252 (2022) 012085 doi:10.1088/1757-899X/1252/1/012085

The impact of the exterior green wall on the summer regime in the building – energy simulation

M Kováč[1]

[1]Institute of Architectural Engineering, Faculty of Civil Engineering, Technical university of Kosice, Vysokoskolska 4, 042 00 Kosice, Slovakia

E-mail: martin.kovac@tuke.sk

Abstract. The analysis is based on the results of computer simulation by using DesignBUILDER software. In the analysis is the multi dwelling house assessed. There are standard building constructions used like as the brick wall with polystyrene thermal insulation, Windows with plastic frame and double glazing and the flat roof with thermal insulation from polystyrene without the gravel layer. In this thermal simulation is considered with so called green walls, that are localized at eastern, southern, and western facade of the building. The light steel frame construction with the net from thin steel ropes creates the area for the climbing plants like a ivy. The function of the used green walls with climbing plants is to shade the external walls of the building in order to reduce the impact of solar radiation on the building facade during the hot summer days. We can these green walls in combination with external blinds consider like a effectively solution in order to reduce the energy consumption for cooling system in the building. By the way the used green walls help to reduce the air temperature surrounding the buildings and so they create the better town climate for their citizens.

1. Introduction

The effect of urban heat islands is good known today. It causes the increasing of air temperature in the downtown in comparison with margin areas of cities or rural areas. Buildings need more energy for their cooling systems, higher air temperature in the city decreases the life comfort of their citizens and also the number of deaths increases due to heat waves during summer days. The factors causing the effect of urban heat island are more, not only a lot of buildings but also traffic and little greens. We live in the time when we good know to design energy efficient new buildings and to design energy efficient retrofit of existing buildings. We can design buildings with minimum of heat losses in winter season, with minimum of heat load in summer season and it all in order to reduce the energy consumption and production of greenhouse gasses. We create the indoor environment in the buildings that is comfortable, clean, and healthy for their occupants to increase productivity, good thinking and keep the good and healthy life. One of all important things is also the green, plants and trees but they are few inside and outside the building. We need the ratio of green, plants and trees increase in the city, so we have to think to get greens on the buildings, not only on the roofs but also on the facades as the figure 1 shows. The plants and trees, generally greens, clean air, create oxygen, bin carbon dioxide, reduce the noise in the streets and protect indoor environment of buildings from surrounding noise (opened windows on the façade towards street). The present modern and ecological architecture can give the answer to these questions or problems [1]. We have buildings today that are called as green buildings. It is kind of buildings consuming little amount of energy for their operation, providing well good and healthy indoor environment for their occupants, using renewable energy resources and at the end this kind of building brings greens into cities by application green roofs, green facades and green walls. But these buildings are few. We have to use green architecture also for retrofit of existing buildings, not only for new projects (new buildings). We need to shift existing buildings into ecological, healthy and energy efficient buildings. This is the process we call sustainable architecture, sustainable buildings, and sustainable cities [2]. The primary target of green architecture is to improve the surrounding environment of buildings, reduce the effect of urban heat islands, improve the air quality and create positive environment for their citizens. The second target is to decrease the building energy consumption for cooling during summer season. This article analyses the second target of green architecture for selected residential building.

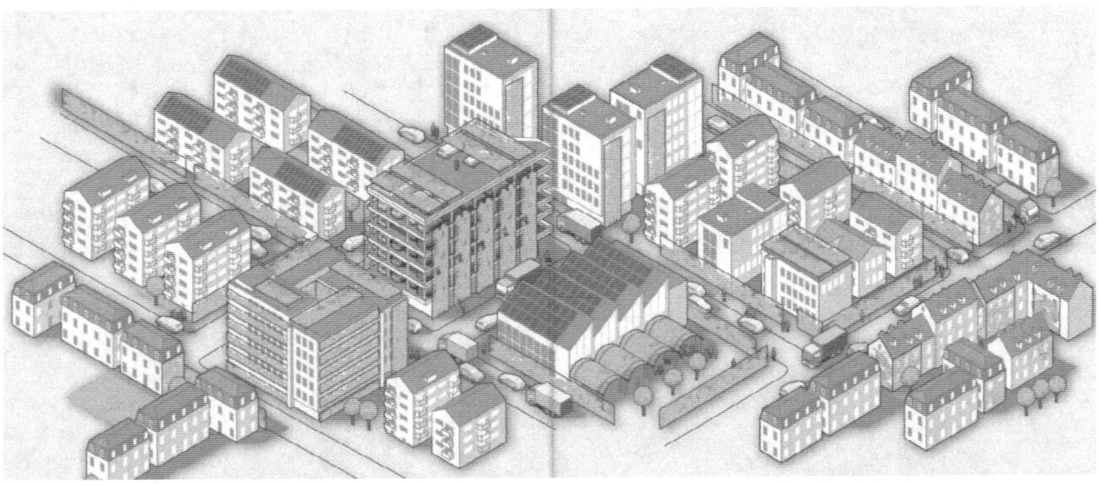

Figure 1. Buildings with elements of green architecture [3].

2. Green walls and green facades of buildings

The difference between green façade (Figure 2) and green wall (Figure 3, 4) is that green façade is included in the construction of external building wall separating the indoor and outdoor environment. The green wall has its own construction that is set away from external building wall. This separate construction is used for climbing or hanging plants. The gap between green wall and external building wall can be thin or wide usable for pavements and for pedestrian walking. The main advantage of green wall to the green façade is its flexibility due to light steel construction from steel columns and wire ropes that create a mesh for grow up of climbing plants.

Figure 2. Examples of green facades in Paris [4].

Figure 3. Example of green wall in Switzerland [5].

Figure 4. Technical solution of green walls in Switzerland [6].

3. The calculation model for energy simulation

For our analysis of impact of green wall on the energy consumption for cooling of building is the real multi dwelling house selected that was built in 2014 in Kosice. The external wall is built from ceramic brick with thermal insulation layer from polystyrene. All internal walls are built from the same material (ceramic brick). The roof construction is built from wood with thermal insulation layer from polystyrene where the final layer is plastic waterproofing layer without gravel layer. The glazing system of windows

IOP Conf. Series: Materials Science and Engineering 1252 (2022) 012085 doi:10.1088/1757-899X/1252/1/012085

is made from 2 glass panes between them air gap layer is. The next table 1 shows used thermal and optical properties for opaque and transparent constructions.

Table 1. Thermal and optical properties of building constructions.

	Coefficient of thermal transmittance *U* [W/(m^2.K)]	Thermal properties *SHGC* [-]	Optical properties τ [-]
External wall	0.17	-	-
Flat roof	0.13	-	-
Ground floor	0.21	-	-
Internal floor (floor between flats)	0.73	-	-
External floor (floor above exterior)	0.19	-	-
External window	1.40	0.70	0.744

The complete geometrical model of multi dwelling house is created in the simulation software DesignBUILDER v7 as the figure 5 shows. The boundary conditions for all building constructions are set up according values from table 1. The calculation model considers with continuous operation of natural ventilation with air exchange value of 0.5 [1/h]. This is the value that is the minimum air exchange for indoor environment of residential buildings according the national standard STN EN 73 0540-2 Z1-Z2. The cooling system of building is in operation in time from 07:00 am to 08:00 pm. In this time the cooling system provides the operative temperature at 26 °C in the indoor environment of building. We consider with second category of indoor environment for residential buildings where the indoor operative temperature is in the range of 23.0 – 26.0 °C. The solar heat gains of indoor environment are reduced by using windows shading. The interior shading blinds with slope of 30° are used in the calculation model in time from 07:00 am do 08:00 pm. Other internal heat gains such as occupants, artificial lighting and other electric appliances are not used in this calculation model. The green walls are placed on the eastern, southern and western façade of building and always between windows in front of opaque construction as the figure 6 shows. The construction of green wall is set away 200 mm from the external building wall. The solar transmission for green walls is set up on the value of 90 % in the calculation model. We consider that climbing plants almost fully cover the area.

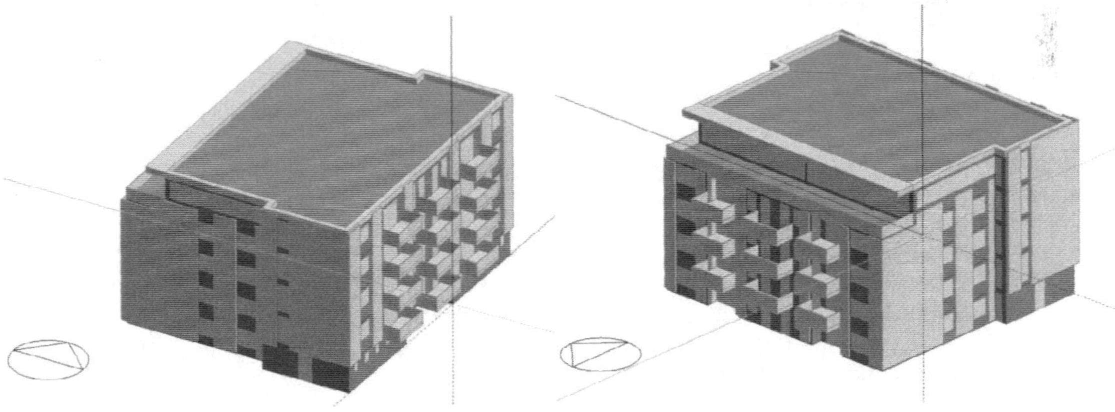

Figure 5. The geometrical model of residential building in DesignBUIDLER.

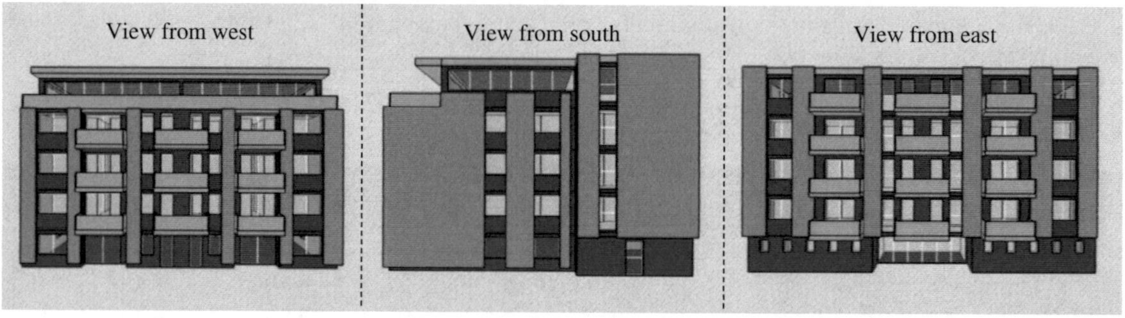

Figure 6. Visualise of building facades with green walls in DesignBUILDER.

4. Results analysis

The energy demand for cooling of building is calculated for summer season, so in condition of Slovakia it is a time frame 1. May - 30. September. The cooling system provides the operative temperature inside the building at the 26 °C in time for 07:00 am to 08:00 pm. In the rest of day could be set up the operative temperature at 30 °C or the cooling system is turn off. In the first part of analysis we compare the energy demand for cooling of building without and with green walls on eastern, southern and western façade of multi dwelling house. If the green wall is used, then the energy saving in this case is about 6.5 % globally for the whole building as the figure 7 shows. Please remember the interior windows shading with slope of blades (30°) are also used during the day from 07:00 am to 08:00 pm.

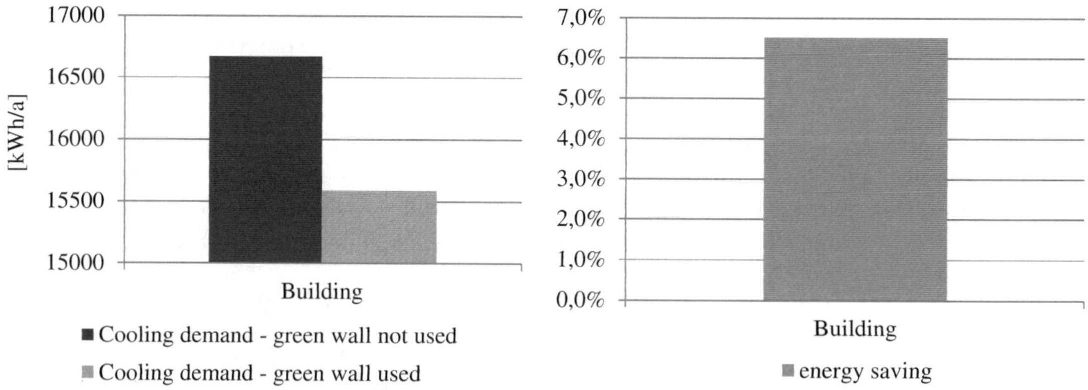

Figure 7. Energy demand for cooling of building.

In the second part of analysis we compare the energy demand for cooling for several selected rooms at the one floor and again in situation without and with green walls on eastern, southern and western façade of multi dwelling house. The next figure 8 shows selected rooms used for analysis of energy demand for cooling. If the green walls are used, then the energy saving is in the range of 5 – 20 %. In this case of analysis we mention the interior windows shading with slope of blades (30°) are also used during the day from 07:00 am to 08:00 pm.

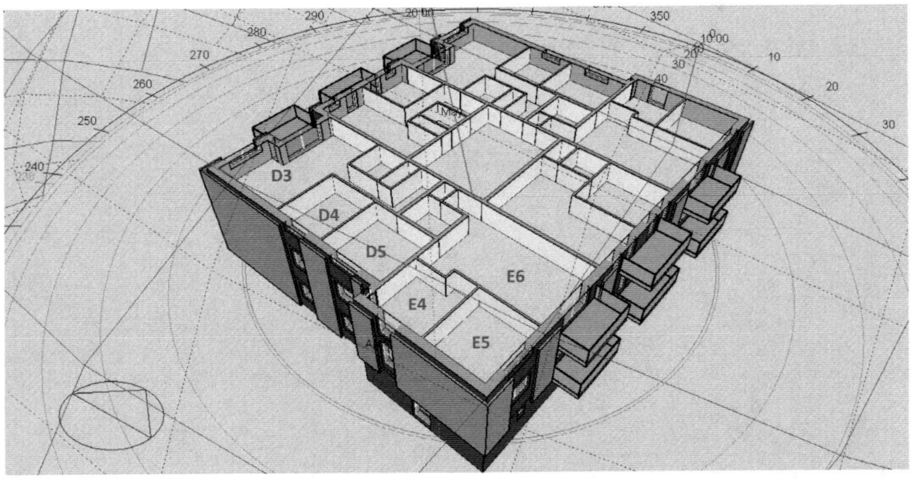

Figure 8. 3D floor plan of building with selected rooms for analysis.

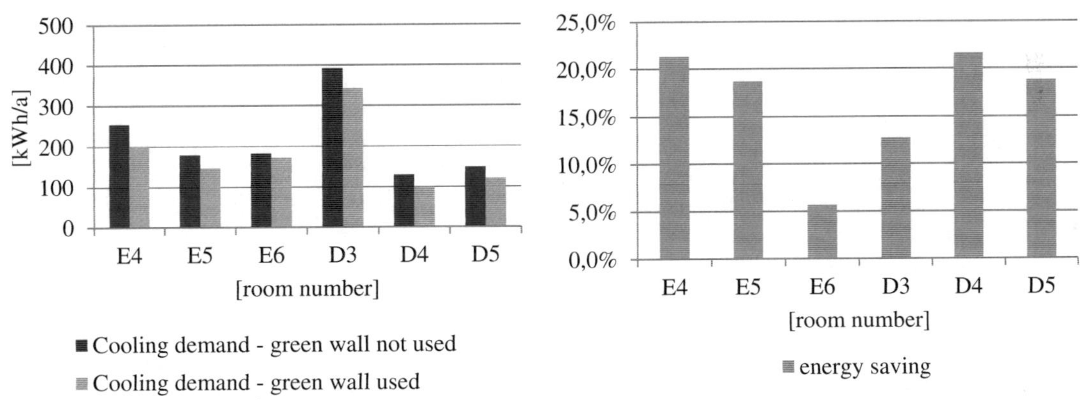

Figure 9. Energy demand for cooling of selected rooms at the one floor.

Using green walls can decreases the internal surface temperature of external wall and so decreases the indoor air temperature in the room there is no cooling system. But we cannot expect the significant temperature drops because the air temperature in the room is influenced by surface temperature of external wall, ceiling, floor, window in relation to their area. Also the amount of solar gains through the glazing system is very important. In the third part of analysis we compare air temperature drops for selected rooms in situation without and with green walls on eastern, southern and western façade of multi dwelling house. The temperature differences are in the range of 0.4 – 1.0 K as the figure 10 shows. If we compare the results of air temperature differences in the figure 10 with results in the figure 9 we can see the correlation there. Rooms with small air temperature differences until 0.5 K have the energy saving in the range of 5 -10 %. Rooms with air temperature differences between 0.5 – 1.0 K have the energy saving in the range of 15 -20 % as the figure 9 shows.

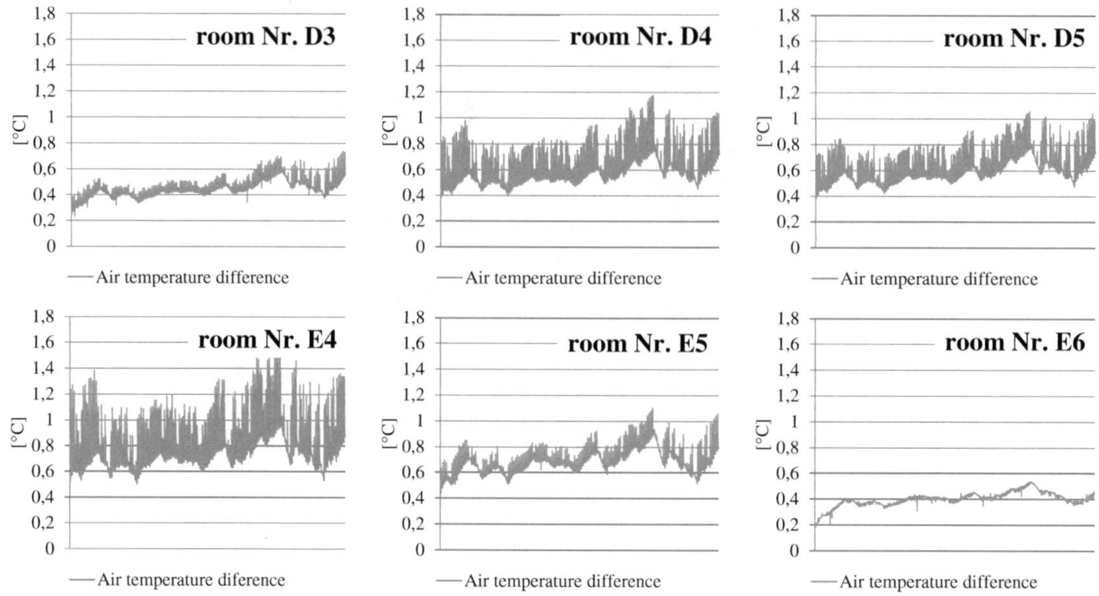

Figure 10. Trends of indoor air temperature differences for selected rooms during summer season.

5. Conclusion

The goal of article is analyse the impact of used green walls on the energy demand for cooling of building. We mention that presented results do not have the general validity because these results describe the specific type of building, orientation towards south direction, site location and building constructions with clearly defined boundary conditions. All these conditions are described in the text above. Using green walls can bring the energy saving in the cooling of building in the range of 5 – 20 % for individual rooms and about 6.5 % for the whole building. If we decrease the energy consumption so we decrease the emissions of greenhouse gasses. On the other side the greens in the cities clean the air, produce oxygen, bin carbon dioxide, reduce noise in the streets and create the positive urban environment not only for working but also for health and relax of citizens.

References

[1] Virtudes A and Manso M 2016 *Applications of green walls in urban design in Earth and Environmental Science* Volume 44 World Multidisciplinary Earth Sciences Symposium 2016
[2] Dal M 2021 *Effect of green wall systems on building heating and cooling loads in sustainable design in Architectural Sciences and Technology* pages 269-287 ISBN 978-2-38236-136-8 France
[3] W Weissnat 2021 Green Roofs As A Modern Concept Of Green Building *The frog more* official Website [online] https://thefrogmore.com/green-roofs-as-a-modern-concept-of-green-building/
[4] Green Roofs 2019 *The Thermo House* official Website [online] https://thermohouse.ie/blog/your-guide-to-green-roof/
[5] DesignBuilder 2021 DesignBuilder Simulation Documentation for DesignBuilder v7 *The designbuilder* official Website [online] https://designbuilder.co.uk/download/documents/468-designbuilder-printable-documentation-v7-a4-pages/file
[6] Green Solutions G2 *The green walls* official Website [online] https://green-walls.co.uk/wp-content/uploads/GreenSolutions-G2-References.pdf
[7] An architect´s guide to: Green walls *The architizer* official Website [online] https://architizer.com/blog/product-guides/product-guide/eantka-green-walls/
[8] M Hagenauer 2021 Hanging garden and façade greening *The Hydroplant* official Website

[online] https://www.hydroplant.ch/en/projects/novartis-campus-basel/

[9] An innovative green wall developed by the Jakob design team *The green walls* official Website [online] https://green-walls.co.uk/projects/view/innovative-green-wall/

Acknowledgment

This work was supported by the Slovak cultural and educational grant agency under the contract No. KEGA 047TUKE-4/2020.

AUTHOR INDEX

Abdallah, G Bou ... 384
Abd-Elhamid, H ... 42, 49
Almerich-Chulia, A ... 301
Assih, J ... 384
Babor, D T ... 266
Bajzecerová, V ... 1
Baka, Z ... 465
Bálint, G ... 97
Balintova, M ... 449
Barańczuk, J ... 49
Barańczuk, K ... 49
Baroš, M ... 5
Baroš, T ... 5, 211
Bašková, Renáta ... 597
Bede, N ... 615
Beglov, A ... 242
Benin, A V ... 13
Bielek, B ... 20
Biolek, V ... 399
Bisták, A ... 26
Blichova, Z ... 31
Boháček, A ... 37
Bošanská, B ... 192
Braspenninckx, J ... 548
Brodnianský, J ... 184, 184
Brodniansky, J ... 578
Bujanský, P ... 42
Burak, D ... 233
Burdova, E Kridlova ... 31, 563
Čabala, J ... 495, 603
Cabala, J ... 559
Cakyova, K ... 484
Capayova, S ... 115
Chellappa, V ... 440
Couscheir, K ... 548
Czímerová, A ... 348
Demjan, I ... 149
Diachenko, L K ... 13
Diagana, C ... 384
Dolníková, E ... 211
Dolnikova, E ... 227
Domanický, J ... 88
Dontchev, D ... 384
Drobac, J ... 615
Dubecky, D ... 257
Eštoková, A ... 64, 76, 354
Fáczányi, Zs ... 307
Flajs, R ... 364
Fošumpaur, P ... 370
Foti, D ... 81
Galla, J ... 505
Garaj, M ... 471
Gmitter, M ... 233
Gocal, J ... 217
Golchin, B ... 548
Gomboš, M ... 249
Gorás, M ... 88
Grandić, I Štimac ... 615
Hanuliak, P ... 144
Harabinova, S ... 162, 640
Harčárová, K ... 409
Hartman, P ... 144
Hirohata, M ... 318, 327, 630
Hlinková, M ... 42, 471
Hodakova, D ... 115
Hološová, M ... 354
Hraška, Jozef ... 476
Hrudka, J ... 522, 573
Hulínová, Z ... 26
Hutkai, K ... 211, 528
Ivanchev, I ... 129
Ivanova, I ... 384
Jármai, K ... 327
Jaroš, P ... 211, 289
Jaskot, A ... 532
Jiang, F ... 318, 630
Junak, J ... 58, 449
Junakova, N ... 58, 449
Kabošová, L ... 5, 211
Kaleja, P ... 440
Kanalikova, A ... 590
Kandra, B ... 249
Kanócz, J ... 1
Kapalo, P ... 64, 76, 465
Kapolka, L ... 420
Karľa, V ... 1
Karla, V ... 211
Kašpar, T ... 378
Katunská, J ... 211
Katunsky, D ... 227
Katunský, D ... 5, 31, 211, 528
Kersner, Z ... 295
Klem, J ... 20
Klosova, D ... 333
Kmet, S ... 420
Korfant, M ... 97

Kormanikova, E ... 81, 511
Kotrasova, K ... 81, 511
Kováč, M ... 465, 609, 644
Kovac, M ... 69
Kovács, R ... 348
Kozlovská, M ... 157, 175, 333, 495, 603
Krajníková, K ... 399
Król, P A ... 540
Kucharczykova, B ... 295
Kuliński, K ... 622
Kvocak, V ... 257
Labutin, N A ... 13
Laumain, X ... 301
Liu, J ... 318
Lopušniak, M ... 211
Lopusniak, M ... 227
Loukili, M ... 81
Luptáková, A ... 354
Luzay, E ... 340
Magura, M ... 578
Makro, I ... 522
Mandičák, T ... 505
Mandicak, T ... 590
Mandula, J ... 348
Mansour, S ... 81
Marko, I ... 573
Melcer, J ... 414
Merjava, V ... 233
Merta, I ... 295
Mésároš, P ... 399, 440
Mesaros, P ... 590
Mihai, P ... 266
Minárová, M ... 192, 391
Mojto, M ... 184
Mokrenko, D ... 175
Moreno-Puchalt, J ... 301
Nagy, P ... 471
Nagy, R ... 465
Nakayama, T ... 318
Nechyporchuk, Yaroslav ... 597
Odrobinak, J ... 217
Ojima, K ... 318
Omranian, S R ... 548
Pagac, S ... 140
Panulinova, E ... 162, 640
Pavelkova, D ... 249
Perc, M Nekrep ... 284
Petkanič, Š ... 340
Petrescu, T C ... 266
Pezeshki, Z ... 511
Polakovič, Ľ ... 348
Poletanovic, B ... 295
Průcha, A ... 37
Raczková, A ... 573
Ružić, I ... 615
Safaei, B ... 511
Sakamoto, T ... 318
Sanjarovsky, R ... 242
Sannikava, V H ... 201
Schlosser, P ... 97
Schlosser, T ... 97
Sedlakova, A ... 233
Shafieva, D ... 340
Sicakova, A ... 427
Simonova, H ... 295
Škultétyová, I ... 573
Šlanhof, J ... 37
Smetanková, J ... 399
Sococol, I ... 266
Špak, M ... 175, 440
Spišáková, M ... 440, 505
Spisakova, M ... 590
Stajnko, J Kramer ... 284
Stanko, Š ... 522
Šťastný, P ... 26
Stevulova, N ... 58
Strigac, J ... 58
Struková, Z ... 175, 495, 603
Strukova, Z ... 559
Stulerova, L ... 420
Sugár, V ... 307
Šutús, M ... 522
Švajlenka, J ... 157
Szabo, D ... 20
Takács, J ... 97
Tall, A ... 249
Tažiková, A ... 495
Tazikova, A ... 559
Tažíková, A ... 603
Ter-Emmanuilyan, T ... 242
Tomko, M ... 149
Tur, V V ... 201
Tvrdá, K ... 192, 391
Valašková, V ... 414
Van Den bergh, W ... 548
Vaňová, P ... 257
Vargova, A ... 484
Vavrík, I ... 26
Vertaľ, M ... 211, 289
Vertal, M ... 484
Vican, J ... 217
Vilcekova, S ... 31, 233, 563
Vilčeková, S ... 409
Vodička, R ... 167

Vojtuš, J 455
Voznyak, O 64
Vranay, F 88, 484
Vranayova, Z 484
Vuye, C 548
Weissova, M 257
Wittmanová, R 522, 573
Zeleňáková, M 340
Zelenakova, M 49
Zozulák, M 211, 528
Zuzulova, A 115